The Ultimate
BMAT Collection

ISBN 978-1-912557-25-7

Published by *RAR Medical Services Limited*
www.uniadmissions.co.uk
info@uniadmissions.co.uk
Tel: 0208 068 0438

The Ultimate
BMAT Collection

Five Books in One

Dr. Rohan Agarwal
Matthew Williams

UniAdmissions

About the Authors

Rohan is the **Director of Operations** at *UniAdmissions* and is responsible for its technical and commercial arms. He graduated from Gonville and Caius College, Cambridge and is a fully qualified doctor. Over the last five years, he has tutored hundreds of successful Oxbridge and Medical applicants. He has also authored ten books on admissions tests and interviews.

Rohan has taught physiology to undergraduates and interviewed medical school applicants for Cambridge. He has published research on bone physiology and writes education articles for the Independent and Huffington Post. In his spare time, Rohan enjoys playing the piano and table tennis.

Matthew is **Resources Editor** at *UniAdmissions* and a 5th year medical student at St Catherine's College, Oxford. As the first student from Barry Comprehensive School in South Wales to receive a place on the Oxford medicine course he embraced all aspects of university life, both social and academic. Matt Scored in the **top 5% for his UKCAT and BMAT** to secure his offer at the University of Oxford.

Matt has worked with UniAdmissions since 2014 – tutoring several applicants successfully into Oxbridge and Russell group universities. His work has been published in international scientific journals and he has presented his research at conferences across the globe. In his spare time, Matt enjoys playing rugby and golf.

Congratulations on taking the first step to your BMAT preparation! First used in 2008, the BMAT is a difficult exam and you'll need to prepare thoroughly in order to make sure you get that dream university place.

The Ultimate BMAT Collection is the most comprehensive BMAT book available – it's the culmination of five top-selling BMAT books:

➢ The Ultimate BMAT Guide
➢ BMAT Past Paper Solutions: Volume 1
➢ BMAT Past Paper Solutions: Volume 2
➢ BMAT Practice Papers: Volume 1
➢ BMAT Practice Papers: Volume 2

Whilst it might be tempting to dive straight in with mock papers, this is not a sound strategy. Instead, you should approach the BMAT in the three steps shown below. Firstly, start off by understanding the structure, syllabus and theory behind the test. Once you're satisfied with this, move onto doing the 600 practice questions found in The Ultimate BMAT Guide (not timed!).

Then, once you feel ready for a challenge, do each past paper under timed conditions. Start with the 2003 paper and work chronologically; check your solutions against the model answers given in BMAT Past Paper Worked Solutions. Finally, once you've exhausted these, go through the 8 BMAT Mock Papers found in BMAT Practice Papers – these are a final boost to your preparation.

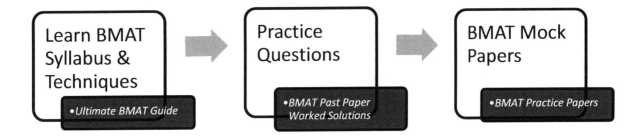

As you've probably realised by now, there are well over 3,000 questions to tackle meaning that this isn't a test that you can prepare for in a single week. From our experience, the best students will prepare anywhere between four to eight weeks (although there are some notable exceptions!).

Remember that the route to a high score is your approach and practice. Don't fall into the trap that "you can't prepare for the BMAT"– this could not be further from the truth. With knowledge of the test, some useful time-saving techniques and plenty of practice you can dramatically boost your score.

Work hard, never give up and do yourself justice. Good luck!

The Ultimate BMAT Guide

The Basics

What is the BMAT?

The BioMedical Admissions Test (BMAT) is a 2-hour written exam for medical and veterinary students who are applying for competitive universities.

What does the BMAT consist of?

Section	SKILLS TESTED	Questions	Timing
ONE	Problem-solving skills, including numerical and spatial reasoning. Critical thinking skills, including understanding argument and reasoning using everyday language.	35 MCQs	60 minutes
TWO	Ability to recall, understand and apply GCSE level principles of biology, chemistry, physics and maths. Usually the section that students find the hardest.	27 MCQs	30 minutes
THREE	Ability to organise ideas in a clear and concise manner, and communicate them effectively in writing. Questions are usually but not necessarily medical.	One essay from three	30 minutes

Why is the BMAT used?

Medical and veterinary applicants tend to be a bright bunch and therefore usually have excellent grades. For example, in 2013 over 65% of students who applied to Cambridge for Medicine had UMS greater than 90% in all of their A level subjects. This means that competition is fierce – meaning that the universities must use the BMAT to help differentiate between applicants.

When do I sit BMAT?

There are two sittings for the BMAT – the second week of September and first week of November (normally Wednesday morning). You can generally sit the BMAT on either date however, some universities will ask that you sit the test on a specific date e.g. Oxford, Lee Kong Chian and Chulalongkorn University will only accept results from the November BMAT sitting. You're highly advised to check which date you should sit the BMAT depending on your university choices.

When should I sit the BMAT?

The difficulty and content of both sittings is the same so the answer will depend on how much time you have over summer and how important it is for you to know your BMAT result before you submit your UCAS application. In general, if you're applying for two or more BMAT universities, it's a good idea to sit the BMAT in September if circumstances allow.

	SEPTEMBER	NOVEMBER
Positive	Get BMAT Results before UCAS Deadline Can use the summer to prepare thoroughly	More time to prepare overall Will have coverd more science topics in school
Negative	Less time to prepare overall May conflict with UKCAT, Personal Statement etc	Hard to balance school-work with BMAT revision Won't get BMAT results until after UCAS Deadline

Who has to sit the BMAT?

Applicants to the following universities must sit the BMAT:

University	Course
University of Cambridge	Medicine and Veterinary Medicine
University of Oxford	Medicine, Graduate Medicine, Biomedical Science
University College London	Medicine
Imperial College London	Medicine, Graduate Medicine, Biomedical Science
Brighton and Sussex	Medicine
University of Leeds	Medicine, Dentistry
Lancaster University	Medicine
Keele University	Medicine, Health Foundation Year
Royal Veterinary College	Veterinary Medicine
Lee Kong Chian (Singapore)	Medicine
Thammasat University	642901(Medicine) and 642902 (Dental Surgery)
Universidad de Navarra	Medicine
Mahidol University	Medicine

Do I have to resit the BMAT if I reapply?

You only need to resit the BMAT if you are applying to a university that requires it. You cannot use your score from any previous attempts.

How is the BMAT Scored?

Section 1 and Section 2 are marked on a scale of 0 to 9. Generally, 5 is an average score, 6 is good, and 7 is excellent. Very few people (<5%) get more than 8.

The marks for sections 1 + 2 show a normal distribution with a large range. The important thing to note is that the difference between a score of 5.0 and 6.5+ is often only 3-4 questions. Thus, you can see that even small improvements in your raw score will lead to massive improvements when they are scaled.

Section 3 is marked on 2 scales: A-E for Quality of English and 0-5 for Strength of Argument

The marks for sections 3 show a normal distribution for the strength of argument; the average mark for the strength of argument is between 3 – 3.5. The quality of English marks are negatively skewed distribution. I.e. the vast majority of students will score A or B for quality of English. The ones that don't tend to be students who are not fluent in English.

This effectively means that the letter score is used to flag students who have a comparatively weaker grasp of English- i.e. it is a test of competence rather than excellence like the rest of the BMAT. This effectively means that if you get a C or below, admissions tutors are more likely to scrutinise your essay than otherwise.

Finally, section 3 is marked by two different examiners. If there is a large discrepancy between their marks, it is marked by a third examiner.

Can I resit the BMAT?

No, you can only sit the BMAT once per admissions cycle. You can resit the BMAT if you apply for medicine again in the future.

Where do I sit the BMAT?

For the September sitting, you will need to register yourself and sit the test at one of 20 authorised centres. In November, your school will normally register you and you can usually sit the BMAT at your school or college (ask your exams officer for more information). Alternatively, if your school isn't a registered test centre or you're not attending a school or college, you can sit the BMAT at an authorised test centre.

When do I get my results?

For the September sitting, you will get your results by the end of September online. You are then responsible for informing the University of your Score. For the November sitting, The BMAT results are usually released to universities in mid-late November and then to students in late November.

How is the BMAT used?

Cambridge: Cambridge interviews more than 90% of students who apply so the BMAT score isn't vital for making the interview shortlist. However, it can play a huge role in the final decision – for example, 50% of overall marks for your application may be allocated to the BMAT. Thus, it's essential you find out as much information about the college you're applying to.

Oxford: Oxford typically receives thousands of applications each year and they use the BMAT to shortlist students for interview. Typically, 450 students are invited for interview for 150 places. Thus, if you get offered an interview-you are doing very well! Oxford centralise their short listing process and use an algorithm that uses your % A*s at GCSE along with your BMAT score to rank all their applicants of which the top are invited to interview. BMAT sections 1 + 2 count for 40% each of your BMAT score whilst section 3 counts for 20% [the strength of argument (number) contributes to 13.3% and the quality of English (letter) makes up the remaining 6.7%].

UCL: UCL make offers based on all components of the application and whilst the BMAT is important there is no magic threshold that you need to meet in order to guarantee an interview. Applicants with higher BMAT scores tend to be interviewed earlier in the year.

Imperial: Imperial employs a BMAT threshold to shortlist for interview. This exact threshold changes every year but in the past has been approximately 4.5-5.0 for sections 1 + 2 and 2.5 B for section 3.

Leeds: The BMAT contributes to 15% of your academic score at Leeds. You will be allocated marks based on your rank in the BMAT. Thus, applicants in the top 20% of the BMAT will get the full quota of marks for their application and the bottom 20% will get the lowest possible mark for their application. Thus, you can still get an interview if you perform poorly in the BMAT (it' just much harder!). Leeds will calculate your BMAT score by attributing 40% to section 1, 40% to section 2 and 20% to section 3 (lower weighting as it can come up during the interview).

Brighton: Brighton started using the BMAT in 2014 so little is known about how they use it in their decision making process. They state on their website that it "may also be used as a final discriminator if needed after interview."

Royal Veterinary College: It is unclear how the RVC use the BMAT- it has influenced applications both before and after interview and it's likely that they use it on a case-by-case basis rather than as an arbitrary cut-off.

General Advice

Start Early

It is much easier to prepare if you practice little and often. Start your preparation well in advance; ideally by mid September but at the latest by early October. This way you will have plenty of time to complete as many papers as you wish to feel comfortable and won't have to panic and cram just before the test, which is a much less effective and more stressful way to learn. In general, an early start will give you the opportunity to identify the complex issues and work at your own pace.

Prioritise

Some questions in sections 1 + 2 can be long and complex – and given the intense time pressure you need to know your limits. It is essential that you don't get stuck with very difficult questions. If a question looks particularly long or complex, mark it for review and move on. You don't want to be caught 5 questions short at the end just because you took more than 3 minutes in answering a challenging multi-step physics question. If a question is taking too long, choose a sensible answer and move on. Remember that each question carries equal weighting and therefore, you should adjust your timing in accordingly. With practice and discipline, you can get very good at this and learn to maximise your efficiency.

Positive Marking

There are no penalties for incorrect answers in the BMAT; you will gain one for each right answer and will not get one for each wrong or unanswered one. This provides you with the luxury that you can always guess should you absolutely be not able to figure out the right answer for a question or run behind time. Since each question provides you with 4 to 6 possible answers, you have a 16-25% chance of guessing correctly. Therefore, if you aren't sure (and are running short of time), then make an educated guess and move on. Before 'guessing' you should try to eliminate a couple of answers to increase your chances of getting the question correct. For example, if a question has 5 options and you manage to eliminate 2 options- your chances of getting the question increase from 20% to 33%!

Avoid losing easy marks on other questions because of poor exam technique. Similarly, if you have failed to finish the exam, take the last 10 seconds to guess the remaining questions to at least give yourself a chance of getting them right.

Practice

This is the best way of familiarising yourself with the style of questions and the timing for this section. Although the BMAT tests only GCSE level knowledge, you are unlikely to be familiar with the style of questions in all 3 sections when you first encounter them. Therefore, you want to be comfortable at using this before you sit the test.

Practising questions will put you at ease and make you more comfortable with the exam. The more comfortable you are, the less you will panic on the test day and the more likely you are to score highly. Initially, work through the questions at your own pace, and spend time carefully reading the questions and looking at any additional data. When it becomes closer to the test, **make sure you practice the questions under exam conditions**.

Past Papers

Official past papers and answers from 2003 onwards are freely available online on our website at **www.uniadmissions.co.uk/bmat-past-papers** and once you've worked your way through the questions in this book, it's a good idea to attempt as many of them as you can (ideally at least 5). Keep in mind that the specification was changed in 2009 so some things asked in earlier papers may not be representative of the content that is currently examinable in the BMAT. In general, **it is worth doing at least all the papers from 2009 onwards**. Time permitting; you can work backwards from 2009 although there is little point doing the section 3 essays pre-2009 as they are significantly different to the current style of essays.

Scoring Tables

Use these to keep a record of your scores – you can then easily see which paper you should attempt next (always the one with the lowest score).

SECTION 1	1st Attempt	2nd Attempt
2003		
2004		
2005		
2006		
2007		
2008		
2009		
2010		
2011		
2012		
2013		
2014		
2015		
2016		
2017		

SECTION 2	1st Attempt	2nd Attempt
2003		
2004		
2005		
2006		
2007		
2008		
2009		
2010		
2011		
2012		
2013		
2014		
2015		
2016		
2017		

Repeat Questions

When checking through answers, pay particular attention to questions you have got wrong. If there is a worked answer, look through that carefully until you feel confident that you understand the reasoning, and then repeat the question without help to check that you can do it. If only the answer is given, have another look at the question and try to work out why that answer is correct. This is the best way to learn from your mistakes, and means you are less likely to make similar mistakes when it comes to the test. The same applies for questions which you were unsure of and made an educated guess which was correct, even if you got it right. When working through this book, **make sure you highlight any questions you are unsure of**, this means you know to spend more time looking over them once marked.

No Calculators

You aren't permitted to use calculators in the BMAT – thus, it is essential that you have strong numerical skills. For instance, you should be able to rapidly convert between percentages, decimals and fractions. You will seldom get questions that would require calculators but you would be expected to be able to arrive at a sensible estimate. Consider for example:

Estimate 3.962 x 2.322;

3.962 is approximately 4 and 2.323 is approximately 2.33 = 7/3.

Thus, $3.962 \times 2.322 \approx 4 \times \frac{7}{3} = \frac{28}{3} = 9.33$

Since you will rarely be asked to perform difficult calculations, you can use this as a signpost of if you are tackling a question correctly. For example, when solving a physics question, you end up having to divide 8,079 by 357- this should raise alarm bells as calculations in the BMAT are rarely this difficult.

> ***Top tip!*** In general, students tend to improve the fastest in section 2 and slowest in section 1; section 3 usually falls somewhere in the middle. Thus, if you have very little time left, it's best to prioritise section 2.

A word on timing...

"If you had all day to do your BMAT, you would get 100%. But you don't."

Whilst this isn't completely true, it illustrates a very important point. Once you've practiced and know how to answer the questions, the clock is your biggest enemy. This seemingly obvious statement has one very important consequence. **The way to improve your BMAT score is to improve your speed.** There is no magic bullet. But there are a great number of techniques that, with practice, will give you significant time gains, allowing you to answer more questions and score more marks.

Timing is tight throughout the BMAT – **mastering timing is the first key to success.** Some candidates choose to work as quickly as possible to save up time at the end to check back, but this is generally not the best way to do it. BMAT questions can have a lot of information in them – each time you start answering a question it takes time to get familiar with the instructions and information. By splitting the question into two sessions (the first run-through and the return-to-check) you double the amount of time you spend on familiarising yourself with the data, as you have to do it twice instead of only once. This costs valuable time. In addition, candidates who do check back may spend 2–3 minutes doing so and yet not make any actual changes. Whilst this can be reassuring, it is a false reassurance as it is unlikely to have a significant effect on your actual score. Therefore it is usually best to pace yourself very steadily, aiming to spend the same amount of time on each question and finish the final question in a section just as time runs out. This reduces the time spent on re-familiarising with questions and maximises the time spent on the first attempt, gaining more marks.

It is essential that you don't get stuck with the hardest questions – no doubt there will be some. In the time spent answering only one of these you may miss out on answering three easier questions. If a question is taking too long, choose a sensible answer and move on. Never see this as giving up or in any way failing, rather it is the smart way to approach a test with a tight time limit. With practice and discipline, you can get very good at this and learn to maximise your efficiency. It is not about being a hero and aiming for full marks – this is almost impossible and very much unnecessary (even Oxbridge will regard any score higher than 7 as exceptional). It is about maximising your efficiency and gaining the maximum possible number of marks within the time you have.

Top tip! Ensure that you take a watch that can show you the time in seconds into the exam. This will allow you have a much more accurate idea of the time you're spending on a question. In general, if you've spent >150 seconds on a section 1 question or >90 seconds on a section 2 questions – move on regardless of how close you think you are to solving it.

Use the Options:

Some questions may try to overload you with information. When presented with large tables and data, it's essential you look at the answer options so you can focus your mind. This can allow you to reach the correct answer a lot more quickly. Consider the example below:

The table below shows the results of a study investigating antibiotic resistance in staphylococcus populations. A single staphylococcus bacterium is chosen at random from a similar population. Resistance to any one antibiotic is independent of resistance to others.

Calculate the probability that the bacterium selected will be resistant to all four drugs.

A　1 in 10^6
B　1 in 10^{12}
C　1 in 10^{20}
D　1 in 10^{25}
E　1 in 10^{30}
F　1 in 10^{35}

Antibiotic	Number of Bacteria tested	Number of Resistant Bacteria
Benzyl-penicillin	10^{11}	98
Chloramphenicol	10^9	1200
Metronidazole	10^8	256
Erythromycin	10^5	2

Looking at the options first makes it obvious that there is **no need to calculate exact values**- only in powers of 10. This makes your life a lot easier. If you hadn't noticed this, you might have spent well over 90 seconds trying to calculate the exact value when it wasn't even being asked for.

In other cases, you may actually be able to use the options to arrive at the solution quicker than if you had tried to solve the question as you normally would. Consider the example below:

A region is defined by the two inequalities: $x - y^2 > 1 \, and \, xy > 1$. Which of the following points is in the defined region?

A.　(10,3)
B.　(10,2)
C.　(-10,3)
D.　(-10,2)
E.　(-10,-3)

Whilst it's possible to solve this question both algebraically or graphically by manipulating the identities, by far **the quickest way is to actually use the options**. Note that options C, D and E violate the second inequality, narrowing down to answer to either A or B. For A: $10 - 3^2 = 1$ and thus this point is on the boundary of the defined region and not actually in the region. Thus the answer is B (as $10-4 = 6 > 1$.)

In general, it pays dividends to look at the options briefly and see if they can be help you arrive at the question more quickly. Get into this habit early – it may feel unnatural at first but it's guaranteed to save you time in the long run.

Keywords

If you're stuck on a question; pay particular attention to the options that contain key modifiers like "**always**", "**only**", "**all**" as examiners like using them to test if there are any gaps in your knowledge. E.g. the statement "arteries carry oxygenated blood" would normally be true; "All arteries carry oxygenated blood" would be false because the pulmonary artery carries deoxygenated blood.

SECTION 1

This is the first section of the BMAT and as you walk in, it is inevitable that you will feel nervous. Make sure that you have been to the toilet because once it starts you cannot simply pause and go. Take a few deep breaths and calm yourself down. Remember that panicking will not help and may negatively affect your marks- so try and avoid this as much as possible.

You have one hour to answer 35 questions in section 1. The questions fall into three categories:
-   Problem solving
-   Data handling
-   Critical thinking

Whilst this section of the BMAT is renowned for being difficult to prepare for, there are powerful shortcuts and techniques that you can use to save valuable time on these types of questions.

You have approximately 100 seconds per question; this may sound like a lot but given that you're often required to read and analyse passages or graphs- it can often not be enough. Nevertheless, this section is not as time pressured as section 2 so most students usually finish the majority of questions in time. However, some questions in this section are very tricky and can be a big drain on your limited time. **The people who fail to complete section 1 are those who get bogged down on a particular question**.

Therefore, it is vital that you start to get a feel for which questions are going to be easy and quick to do and which ones should be left till the end. The best way to do this is through practice and the questions in this book will offer extensive opportunities for you to do so.

SECTION 1: Critical Thinking

BMAT Critical thinking questions require you to understand the constituents of a good argument and be able to pick them apart. The majority of BMAT Critical thinking questions tend to fall into 3 major categories:

1.   Identifying Conclusions
2.   Identifying Assumptions + Flaws
3.   Strengthening and Weakening arguments

Having a good grasp of language and being able to filter unnecessary information quickly and efficiently is a vital skill in medical school – you simply do not have the time to sit and read vast numbers of textbooks cover to cover, you need to be able to filter the information and realise which part is important and this will contribute to your success in your studies. Similarly, when you have qualified and are on the wards, you need to be able to pick out key information from patient notes and make healthcare decisions from them, so getting to grips with verbal reasoning goes a long way and do not underestimate its importance.

Only use the Passage

Your answer must only be based on the information available in the passage. Do not try and guess the answer based on your general knowledge as this can be a trap. For example, if the passage says that spring is followed by winter, then take this as true even though you know that spring is followed by summer.

Top tip! Though it might initially sound counter-intuitive, it is often best to read the question *before* reading the passage. Then you'll have a much better idea of what you're looking for and are therefore more likely to find it quicker.

Take your time

Unlike the problem solving questions, critical thinking questions are less time pressured. Most of the passages are well below 300 words and therefore don't take long to read and process (unlike the UKCAT in which you should skim read passages). Thus, your aim should be to understand the intricacies of the passage and identify key information so that you don't miss key information and lose easy marks.

Identifying Conclusions

Students struggle with these type of questions because they confuse a premise for a conclusion. For clarities sake:

- A **Conclusion** is a summary of the arguments being made and is usually explicitly stated or heavily implied.
- A **Premise** is a statement from which another statement can be inferred or follows as a conclusion.

Hence a conclusion is shown/implied/proven by a premise. Similarly, a premise shows/indicates/establishes a conclusion. Consider for example: *My mom, being a woman, is clever as all women are clever.*

Premise 1: My mom is a woman + **Premise 2:** Women are clever = **Conclusion:** My mom is clever.

This is fairly straightforward as it's a very short passage and the conclusion is explicitly stated. Sometimes the latter may not happen. Consider: *My mom is a woman and all women are clever.*
Here, whilst the conclusion is not explicitly being stated, both premises still stand and can be used to reach the same conclusion.

You may sometimes be asked to identify if any of the options cannot be "reliably concluded". This is effectively asking you to identify why an option **cannot** be the conclusion. There are many reasons why but the most common ones are:

1. Over-generalising: *My mom is clever therefore all women are clever.*
2. Being too specific: *All kids like candy thus my son also likes candy.*
3. Confusing Correlation vs. Causation: *Lung cancer is much more likely in patients who drink water. Hence, water causes lung cancer.*
4. Confusing Cause and Effect: *Lung cancer patients tend to smoke so it follows that having lung cancer must make people want to smoke.*

Note how conjunctives like hence, thus, therefore and it follows give you a clue as to when a conclusion is being stated. More examples of these include: "it follows that, implies that, whence, entails that".
Similarly, words like "because, as indicated by, in that, given that, due to the fact that" usually identify premises.

Assumptions + Flaws:

Other types of critical thinking questions may require you to identify assumptions and flaws in a passage's reasoning. Before proceeding it is useful to define both:

- An assumption is a reasonable assertion that can be made on the basis of the available evidence.
- A flaw is an element of an argument which is inconsistent to the rest of the available evidence. It undermines the crucial components of the overall argument being made.

Consider for example: *My mom is clever because all doctors are clever.*

Premise 1: Doctors are clever. **Assumption:** My mom is a doctor. **Conclusion:** My mom is clever.

Note that the conclusion follows naturally even though there is only one premise because of the assumption. The argument relies on the assumption to work. Thus, if you are unsure if an option you have is an assumption or not, just ask yourself:

1) *Is it in the passage?* If the answer is **no** then proceed to ask:
2) *Does the conclusion rely on this piece of information in order to work?* – If the answer is **yes** – then you've identified an assumption.

You may sometimes be asked to identify flaws in an argument – it is important to be aware of the types of flaws to look out for. In general, these are broadly similar to the ones discussed earlier in the conclusion section (over-generalising, being too specific, confusing cause and effect, confusing correlation and causation). Remember that an assumption may also be a flaw.

For example consider again: *My mom is clever because all doctors are clever.*

What if the mother was not actually a doctor? The argument would then breakdown as the assumption would be incorrect or **flawed**.

> ***Top tip!*** Don't get confused between premises and assumptions. A **premise** is a statement that is explicitly stated in the passage. An **assumption** is an inference that is made from the passage.

Strengthening and Weakening Arguments:

You may be asked to identify an answer option that would most strengthen or weaken the argument being made in the passage. Normally, you'll also be told to assume that each answer option is true. Before we can discuss how to strengthen and weaken arguments, it is important to understand "what constitutes a good argument:

1. **Evidence:** Arguments which are heavily based on value judgements and subjective statements tend to be weaker than those based on facts, statistics and the available evidence.
2. **Logic:** A good argument should flow and the constituent parts should fit well into an overriding view or belief.
3. **Balance:** A good argument must concede that there are other views or beliefs (counter-argument). The key is to carefully dismantle these ideas and explain why they are wrong.

Thus, when asked to strengthen an argument, look for options that would: Increase the evidence basis for the argument, support or add a premise, address the counter-arguments.

Similarly, when asked to weaken an argument, look for options that would: decrease the evidence basis for the argument or create doubt over existing evidence, undermine a premise, strengthen the counter-arguments.

In order to be able to strengthen or weaken arguments, you must completely understand the passage's conclusion. Then you can start testing the impact of each answer option on the conclusion to see which one strengthens or weakens it the most i.e. is the conclusion stronger/weaker if I assume this information to be true and included in the passage.

Often you'll have to decide which option strengthens/weakens the passage most – and there really isn't an easy way to do this apart from lots of practice. Thankfully, you have plenty of time for these questions.

Critical Thinking Questions

Question 1-6 are based on the passage below:

People have tried to elucidate the differences between the different genders for many years. Are they societal pressures or genetic differences? In the past it has always been assumed that it was programmed into our DNA to act in a certain more masculine or feminine way but now evidence has emerged that may show it is not our genetics that determines the way we act, but that society pre-programmes us into gender identification. Whilst it is generally acknowledged that not all boys and girls are the same, why is it that most young boys like to play with trucks and diggers whilst young girls prefer dollies and pink?

The society we live in has always been an important factor in our identity, take cultural differences; the language we speak the food we eat, the clothes we wear. All of these factors influence our identity. New research finds that the people around us may prove to be the biggest influence on our gender behaviour. It shows our parents buying gendered toys may have a much bigger influence than the genes they gave us. Girls are being programmed to like the same things as their mothers and this has lasting effects on their personality. Young girls and boys are forced into their gender stereotypes through the clothes they are bought, the hairstyle they wear and the toys they play with.

The power of society to influence gender behaviour explains the cases where children have been born with different external sex organs to those that would match their sex determining chromosomes. Despite the influence of their DNA they identify to the gender they have always been told they are. Once the difference has been detected, how then are they ever to feel comfortable in their own skin? The only way to prevent society having such a large influence on gender identity is to allow children to express themselves, wear what they want and play with what they want without fear of not fitting in.

Question 1:
What is the main conclusion from the first paragraph?
A. Society controls gender behaviour.
B. People are different based on their gender.
C. DNA programmes how we act.
D. Boys do not like the same things as girls because of their genes.

Question 2:
Which of the following, if true, points out the flaw in the first paragraph's argument?
A. Not all boys like trucks.
B. Genes control the production of hormones.
C. Differences in gender may be due to an equal combination of society and genes.
D. Some girls like trucks.

Question 3:
According to the statement, how can culture affect identity?
A. Culture can influence what we wear and how we speak.
B. Our parents act the way they do because of culture.
C. Culture affects our genetics.
D. Culture usually relates to where we live.

Question 4:
Which of these is most implied by the statement?
A. Children usually identify with the gender they appear to be.
B. Children are programmed to like the things they do by their DNA.
C. Girls like dollies and pink because their mothers do.
D. It is wrong for boys to have long hair like girls.

Question 5:

What does the statement say is the best way to prevent gender stereotyping?

A. Mothers spending more time with their sons.

B. Parents buying gender-neutral clothes for their children.

C. Allowing children to act how they want.

D. Not telling children if they have different sex organs.

Question 6:

What, according to the statement is the biggest problem for children born with different external sex organs to those which match their sex chromosomes?

A. They may have other problems with their DNA.

B. Society may not accept them for who they are.

C. They may wish to be another gender.

D. They are not the gender they are treated as which can be distressing.

Questions 7-11 are based on the passage below:

New evidence has emerged that the most important factor in a child's development could be their napping routine. It has come to light that regular napping could well be the deciding factor for determining toddlers' memory and learning abilities. The new countrywide survey of 1000 toddlers, all born in the same year showed around 75% had regular 30-minute naps. Parents cited the benefits of their child having a regular routine (including meal times) such as decreased irritability, and stated the only downfall of occasional problems with sleeping at night. Research indicating that toddlers were 10% more likely to suffer regular night-time sleeping disturbances when they regularly napped supported the parent's view.

Those who regularly took 30-minute naps were more than twice as likely to remember simple words such as those of new toys than their non-napping counterparts, who also had higher incidences of memory impairment, behavioural problems and learning difficulties. Toddlers who regularly had 30 minute naps were tested on whether they were able recall the names of new objects the following day, compared to a control group who did not regularly nap. These potential links between napping and memory, behaviour and learning ability provides exciting new evidence in the field of child development.

Question 7:

If in 100 toddlers 5% who did not nap were able to remember a new teddy's name, how many who had napped would be expected to remember?

A. 8   B. 9   C. 10   D. 12

Question 8:

Assuming that the incidence of night-time sleeping disturbances is the same in for all toddlers independent of all characteristics other than napping, what is the percentage of toddlers who suffer regular night-time sleeping disturbances as a result of napping?

A. 7.5%   B. 10%   C. 14%   D. 20%   E. 50%

Question 9:

Using the information from the passage above, which of the following is the most plausible alternative reason for the link between memory and napping?

A. Children who have bad memory abilities are also likely to have trouble sleeping.

B. Children who regularly nap, are born with better memories.

C. Children who do not nap were unable to concentrate on the memory testing exercises for the study.

D. Parents who enforce a routine of napping are more likely to conduct memory exercises with their children.

Question 10:

Which of the following is most strongly indicated?

A. Families have more enjoyable meal times when their toddlers regularly nap.

B. Toddlers have better routines when they nap.

C. Parents enforce napping to improve their toddlers' memory ability.

D. Napping is important for parents' routines.

Question 11:

Which of the following, if true, would strengthen the conclusion that there is a causal link between regular napping and improved memory in toddlers?

A. Improved memory is also associated with regular mealtimes.

B. Parents who enforce regular napping are more inclined to include their children in studies.

C. Toddlers' memory development is so rapid that even a few weeks can make a difference to performance.

D. Among toddler playgroups where napping incidence is higher and more consistent memory performance is significantly improved compared to those that do not.

Question 12:

Tom's father says to him: 'You must work for your A-levels. That is the best way to do well in your A-level exams. If you work especially hard for Geography, you will definitely succeed in your Geography A-level exam'.

Which of the following is the best statement Tom could say to prove a flaw in his father's argument?

A. 'It takes me longer to study for my History exam, so I should prioritise that.'

B. 'I do not have to work hard to do well in my Geography A-level.'

C. 'Just because I work hard, does not mean I will do well in my A-levels.'

D. 'You are putting too much importance on studying for A-levels.'

E. 'You haven't accounted for the fact that Geography is harder than my other subjects.'

Question 13:

Today the NHS is increasingly struggling to be financially viable. In the future, the NHS may have to reduce the services it cannot afford. The NHS is supported by government funds, which come from those who pay tax in the UK. Recently the NHS has been criticised for allowing fertility treatments to be free, as many people believe these are not important and should not be paid for when there is not enough money to pay the doctors and nurses.

Which of the following is the most accurate conclusion of the statement above?

A. Only taxpayers should decide where the NHS spends its money.

B. Doctors and nurses should be better paid.

C. The NHS should stop free fertility treatments.

D. Fertility treatments may have to be cut if finances do not improve.

Question 14:

'We should allow people to drive as fast as they want. By allowing drivers to drive at fast speeds, through natural selection the most dangerous drivers will kill only themselves in car accidents. These people will not have children, hence only safe people will reproduce and eventually the population will only consist of safe drivers.'

Which one of the following, if true, most weakens the above argument?

A. Dangerous drivers harm others more often than themselves by driving too fast.

B. Dangerous drivers may produce children who are safe drivers.

C. The process of natural selection takes a long time.

D. Some drivers break speed limits anyway.

Question 15:

In the winter of 2014 the UK suffered record levels of rainfall, which led to catastrophic damage across the country. Thousands of homes were damaged and even destroyed, leaving many homeless in the chaos that followed. The Government faced harsh criticism that they had failed to adequately prepare the country for the extreme weather. In such cases the Government assess the likelihood of such events happening in the future and balance against the cost of advance measures to reduce the impact should they occur versus the cost of the event with no preparative defences in place. Until recently, for example, the risk of acts of terror taking was low compared with the vast cost anticipated should they occur. However, the risk of flooding is usually low, so it could be argued that the costs associated with anti-flooding measures would have been pre-emptively unreasonable. Should the Government be expected to prepare for every conceivable threat that could come to pass? Are we to put in place expensive measures against a seismic event as well as a possible extra-terrestrial invasion?

Which of the following best expresses the main conclusion of the statement above?

A. The Government has an obligation to assess risks and costs of possible future events.

B. The Government should spend money to protect against potential extra-terrestrial invasions and seismic events.

C. The Government should have spent money to protect against potential floods.

D. The Government was justified in not spending heavily to protect against flooding.

E. The Government should assist people who lost their homes in the floods.

Question 16:

Sadly the way in which children interact with each other has changed over the years. Where once children used to play sports and games together in the street, they now sit alone in their rooms on the computer playing games on the Internet. Where in the past young children learned human interaction from active games with their friends this is no longer the case. How then, when these children are grown up, will they be able to socially interact with their colleagues?

Which one of the following is the conclusion of the above statement?

A. Children who play computer games now interact less outside of them.

B. The Internet can be a tool for teaching social skills.

C. Computer games are for social development.

D. Children should be made to play outside with their friends to develop their social skills for later in life.

E. Adults will in the future play computer games as a means of interaction.

Question 17:

Between 2006 and 2013 the British government spent £473 million on Tamiflu antiviral drugs in preparation for a flu pandemic, despite there being little evidence to support the effectiveness of the drug. The antivirals were stockpiled for a flu pandemic that never fully materialised. Only 150,000 packs were used during the swine flu episode in 2009, and it is unclear if this improved outcomes. Therefore this money could have been much better spent on drugs that would actually benefit patients.

Which option best summarises the author's view in the passage?

A. Drugs should never be stockpiled, as they may not be used.

B. Spending millions of pounds on drugs should be justified by strong evidence showing positive effects.

C. We should not prepare for flu pandemics in the future.

D. The recipients of Tamiflu in the swine flu pandemic had no difference in symptoms or outcomes to patients who did not receive the antivirals.

Question 18:

High BMI and particularly central weight are risk factors associated with increased morbidity and mortality. Many believe the development of cheap, easily accessible fast-food outlets is partly responsible for the increase in rates of obesity. An unhealthy weight is commonly associated with a generally unhealthy lifestyle, such a lack of exercise. The best way to tackle the growing problem of obesity is for the government to tax unhealthy foods so they are no longer a cheap alternative.

Why is the solution given, to tax unhealthy foods, not a logical conclusion from the passage?

A. Unhealthy eating is not exclusively confined to low-income families.

B. A more general approach to unhealthy lifestyles would be optimal.

C. People do not only choose to eat unhealthy food because it is cheaper.

D. People need to take personal responsibility for their own health.

Question 19:

As people are living longer, care in old age is becoming a larger burden. Many people require carers to come into their home numerous times a day or need full residential care. It is not right that the NHS should be spending vast funds on the care of people who are sufficiently wealthy to fund their own care. Some argue that they want their savings kept to give to their children; however this is not a right, simply a luxury. It is not right that people should be saving and depriving themselves of necessary care, or worse, making the NHS pay the bill, so they have money to pass on to their offspring. People need to realise that there is a financial cost to living longer.

Which of the following statements is the main conclusion of the above passage?

A.   We need to take a personal responsibility for our care in old age.

B.   Caring for the elderly is a significant burden on the NHS.

C.   The reason people are reluctant to pay for their own care is that they want to pass money onto their offspring.

D.   The NHS should limit care to the elderly to reduce their costs.

E.   People shouldn't save their money for old age.

Question 20:

There is much interest in research surrounding production of human stem cells from non-embryo sources for potential regenerative medicine, and a huge financial and personal gain at stake. In January 2014, a team from Japan published two papers in *Nature* that claimed to have developed totipotent stem cells from adult mouse cells by exposure to an acidic environment. However, there has since been much controversy surrounding these papers. Problems included: inability by other teams to replicate the results of the experiment, an insufficient protocol described in the paper and issues with images in one of the papers. It was dishonest of the researchers to publish the papers with such problems, and a requirement of a paper is a sufficiently detailed protocol, so that another group could replicate the experiment.

Which statement is most implied?

A.   Research is fuelled mainly by financial and personal gains.

B.   The researchers should take responsibility for publishing the paper with such flaws.

C.   Rivalry between different research groups makes premature publishing more likely.

D.   The discrepancies were in only one of the papers published in January 2014.

Question 21:

The placebo effect is a well-documented medical phenomenon in which a patient's condition undergoes improvement after being given an ineffectual treatment that they believe to be a genuine treatment. It is frequently used as a control during trials of new drugs/procedures, with the effect of the drug being compared to the effect of a placebo, and if the drug does not have a greater effect than the placebo, then it is classed as ineffective. However, this analysis discounts the fact that the drug treatment still has more of a positive effect than no action, and so we are clearly missing out on the potential to improve certain patient conditions. It follows that where there is a demonstrated placebo effect, but treatments are ineffective, we should still give treatments, as there will therefore be some benefit to the patient.

Which of the following best expresses the main conclusion of this passage?

A.   In situations where drugs are no more effective than a placebo, we should still give drugs, as they will be more effective than not taking action.

B.   Our current analysis discounts the fact that even if drug treatments have no more effect than a placebo, they may still be more effective than no action.

C.   The placebo effect is a well-recognised medical phenomenon.

D.   Drug treatments may have negative side effects that outweigh their benefit to patients.

E.   Placebos are better than modern drugs.

Question 22:

The speed limit on motorways and dual carriageways has been 70mph since 1965, but this is an out-dated policy and needs to change. Since 1965, car brakes have become much more effective, and many safety features have been introduced into cars, such as seatbelts (which are now compulsory to wear), crumple zones and airbags. Therefore, it is clear that cars no longer need to be restricted to 70mph, and the speed limit can be safely increased to 80mph without causing more road fatalities.

Which of the following best illustrates an assumption in this passage?

A. The government should increase the speed limit to 80mph.

B. If the speed limit were increased to 80mph, drivers would not begin to drive at 90mph.

C. The safety systems introduced reduce the chances of fatal road accidents for cars travelling at higher speeds.

D. The roads have not become busier since the 70mph speed limit was introduced.

E. The public want the speed limit to increase.

Question 23:

Despite the overwhelming scientific proof of the theory of evolution, and even acceptance of the theory by many high-ranking religious ministers, there are still sections of many major religions that do not accept evolution as true. One of the most prominent of these in western society is the Intelligent Design movement, which promotes the religious-based (and scientifically discredited) notion of Intelligent Design as a scientific theory. Intelligent Design proponents often point to complex issues of biology as proof that god is behind the design of human beings, much as a watchmaker is inherent in the design of a watch.

One part of anatomy that has been identified as supposedly supporting Intelligent Design is fingerprints, with some proponents arguing that they are a mark of individualism created by God, with no apparent function except to identify each human being as unique. This is incorrect, as fingerprints do have a well documented function – namely channelling away of water to improve grip in wet conditions – in which hairless, smooth skinned hands otherwise struggle to grip smooth objects. The individualism of fingerprints is accounted for by the complexity of thousands of small grooves. Development is inherently affected by stochastic or random processes, meaning that the body is unable to uniformly control its development to ensure that fingerprints are the same in each human being. Clearly, the presence of individual fingerprints does nothing to support the so-called-theory of Intelligent Design.

Which of the following best illustrates the main conclusion of this passage?

A. Fingerprints have a well-established function.

B. Evolution is supported by overwhelming scientific proof.

C. Fingerprints do not offer any support to the notion of Intelligent Design.

D. The individual nature of fingerprints is explained by stochastic processes inherent in development that the body cannot uniformly control.

E. Intelligent design is a credible and scientifically rigorous theory.

Question 24:

High levels of alcohol consumption are proven to increase the risk of many non-infectious diseases, such as cancer, atherosclerosis and liver failure. James is a PhD student, and is analysing the data from a large-scale study of over 500,000 people to further investigate the link between heavy alcohol consumption and health problems. In the study, participants were asked about their alcohol consumption, and then their medical history was recorded. His analysis displays surprising results, concluding that those with high alcohol consumption have a *decreased* risk of cancer. James decides that those carrying out the study must have incorrectly recorded the data.

Which of the following is **NOT** a potential reason why the study has produced these surprising results?

A. Previous studies were incorrect, and high alcohol consumption does lower the risk of cancer.

B. The studies didn't take account of other cancer risk factors in comparing those with high and low alcohol consumption.

C. James has made some errors in his analysis, and thus his conclusions are erroneous.

D. The participants involved in the study did not truthfully report their alcohol consumption, leading to false conclusions being drawn.

E. The studies control group data was mixed up with the test group data.

Question 25:

A train is scheduled to depart from Newcastle at 3:30pm. It stops at Durham, Darlington, York, Sheffield, Peterborough and Stevenage before arriving at Kings Cross station in London, where the train completes its journey. The total length of the journey between Newcastle and Kings Cross was 230 miles, and the average speed of the train during the journey (including time spent stood still at calling stations) is 115mph. Therefore, the train will complete its journey at 5:30pm.

Which of the following is an assumption made in this passage?

A. The various stopping points did not increase the time taken to complete the journey.

B. The train left Newcastle on time.

C. The train travelled by the most direct route available.

D. The train was due to end its journey at Kings Cross.

E. There were no signalling problems encountered on the journey.

Question 26:

There have been many arguments over the last couple of decades about government expenditure on healthcare in the various devolved regions of the UK. It is often argued that, since spending on healthcare per person is higher in Scotland than in England, that therefore the people in Scotland will be healthier. However, this view fails to take account of the different needs of these 2 populations of the UK. For example, one major factor is that Scotland gets significantly colder than England, and cold weakens the immune system, leaving people in Scotland at much higher risk of infectious disease. Thus, Scotland requires higher levels of healthcare spending per person simply to maintain the health of the populace at a similar level to that of England.

Which of the following is a conclusion that can be drawn from this passage?

A. The higher healthcare spending per person in Scotland does not necessarily mean people living in Scotland are healthier.

B. Healthcare spending should be increased across the UK.

C. Wales requires more healthcare spending per person simply to maintain population health at a similar level to England.

D. It is unfair on England that there is more spending on healthcare per person in Scotland.

E. Scotland's healthcare budget is a controversial topic.

Question 27:

Vaccinations have been hugely successful in reducing the incidence of several diseases throughout the 20th century. One of the most spectacular achievements was arguably the global eradication of Smallpox, once a deadly worldwide killer, during the 1970s. Fortunately, there was a highly effective vaccine available for Smallpox, and a major factor in its eradication was an aggressive vaccination campaign. Another disease that is potentially eradicable is Polio. However, although there is a highly effective vaccine for Polio available, attempts to eradicate it have so far been unsuccessful. It follows that we should plan and execute an aggressive vaccination campaign for Polio, in order to ensure that this disease too is eradicated.

Which of the following is the main conclusion of this passage?

A. Polio is a potentially eradicable disease.

B. An aggressive vaccination campaign was a major factor in the eradication of smallpox.

C. Both Polio and smallpox have been eradicated by effective vaccination campaigns.

D. We should execute an aggressive vaccination campaign for Polio.

E. The eradication of smallpox remains one of the most spectacular achievements of medical science.

Question 28:

The Y chromosome is one of 2 sex chromosomes found in the human genome, the other being the X chromosome. As the Y chromosome is only found in males, it can only be passed from father to son. Additionally, the Y chromosome does not exchange sections with other chromosomes (as happens with most chromosomes), meaning it is passed on virtually unchanged through the generations. All of this makes the Y chromosome a fantastic tool for genetic analysis, both to identify individual lineages and to investigate historic population movements. One famous achievement of genetic research using the Y chromosome provides further evidence of its utility, namely the identification of Genghis Khan as a descendant of up to 8% of males in 16 populations across Asia.

Which of the following best illustrates the main conclusion of this passage?

A.   The Y chromosome is a fantastic tool for genetic analysis.
B.   Research using the Y chromosome has been able to identify Genghis Khan as the descendant of up to 8% of men in many Asian populations.
C.   The Y chromosome does not exchange sections with other chromosomes.
D.   The Y chromosome is a sex chromosome.
E.   Genghis Khan had a staggering number of children.

Question 29:

In order for a bacterial infection to be cleared, a patient must be treated with antibiotics. Rachel has a minor lung infection, which is thought by her doctor to be a bacterial infection. She is treated with antibiotics, but her condition does not improve. Therefore, it must not be a bacterial infection.

Which of the following best illustrates a flaw in this reasoning?

A.   It assumes that a bacterial infection would definitely improve after treatment with antibiotics.
B.   It ignores the other potential issues that could be treated by antibiotics.
C.   It assumes that antibiotics are necessary to treat bacterial infections.
D.   It ignores the actions of the immune system, which may be sufficient to clear the infection regardless of what has caused it.
E.   It assumes that antibiotics are the only option to treat a bacterial infection.

Question 30:

The link between smoking and lung cancer has been well established for many decades by overwhelming numbers of studies and conclusive research. The answer is clear and simple, that the single best measure that can be taken to avoid lung cancer is to not smoke, or to stop smoking if one has already started. However, despite the overwhelming evidence and clear answers, many smokers continue to smoke, and seek to minimise their risk of lung cancer by focusing on other, less important risk factors, such as exercise and healthy eating. This approach is obviously severely flawed, and the fact that some smokers feel this is a good way to reduce their risk of lung cancer shows that they are delusional.

Which of the following best illustrates the main conclusion of this passage?

A.   Many smokers ignore the largest risk factor instead focussing on eating healthily and exercising.
B.   Some smokers are delusional.
C.   The biggest risk factor of lung cancer is smoking.
D.   Overwhelming studies have proven the link between smoking and lung cancer.
E.   The government should ban smoking in order to reduce the incidence of lung cancer.

Question 31:

The government should invest more money into outreach schemes in order to encourage more people to go to university. These schemes allow students to meet other people who went to university, which they may not always be able to do otherwise, even on open days.

Which of the following is the best conclusion of the above argument?

A.   Outreach schemes are the best way to encourage people to go to university.
B.   People will not go to university without seeing it first.
C.   The government wants more people to go to university.
D.   Meeting people who went to a university is a more effective method than university open days.
E.   It is easier to meet people on outreach schemes than on open days.

Question 32:

The illegal drug cannabis was recently upgraded from a class C drug to class B, which means it will be taken less in the UK, because people will know it is more dangerous. It also means if people are caught, possessing the drug they will face a longer prison sentence than before, which will also discourage its use.

Which **TWO** statements if true, most weaken the above argument?

A. Class C drugs are cheaper than class B drugs.
B. Upgrading drugs in other countries has not reduced their use.
C. People who take illegal drugs do not know what class they are.
D. Cannabis was not the only class C drug before it was upgraded.
E. Even if they are caught possessing class B drugs, people do not think they will go to prison.

Question 33:

Schools with better sports programmes such as well-performing football and netball teams tend to have better academic results, less bullying and have overall happier students. Thus, if we want schools to have the best results, reduce bullying and increase student happiness, teachers should start more sports clubs.

Which one of the following best demonstrates a flaw in the above argument?

A. Teachers may be too busy to start sports clubs.
B. Better academic results may be a precondition of better sports teams.
C. Better sports programmes may prevent students from spending time with their family.
D. Some sports teams may be seen to encourage internal bullying.
E. Sport teams that do not perform well lead to increase bulling.

Question 34:

The legal age for purchasing alcohol in the UK is 18. This should be lowered to 16 because the majority of 16 year olds drink alcohol anyway without any fear of repercussions. Even if the police catch a 16-year-old buying alcohol, they are unable to enforce any consequences. If the drinking limit was lowered the police could spend less time trying to catch underage drinkers and deal with other more important crimes. There is no evidence to suggest that drinking alcohol at 16 is any more dangerous than at 18.

Which one of the following, if true, most weakens the above argument?

A. Most 16 year olds do not drink alcohol.
B. If the legal drinking age were lowered to 16, more 15 year olds would start purchasing alcohol.
C. Most 16 year olds do not have enough money to buy alcohol.
D. Most 16 year olds are able to purchase alcohol currently.

Question 35:

There has been a recent change in the way the government helps small businesses. Whilst previously small businesses were given non-repayable grants to help them grow their profits, they can now only receive government loans that must be repaid with interest when the business turns a certain amount of profit. The government wants to support small businesses but studies have shown they are less likely to prosper under the new scheme as they have been deterred from taking government money for fear of loan repayments.

Which one of the following can be concluded from the passage above?

A. Small businesses do not want government money.
B. The government cannot afford to give out grants to small businesses anymore.
C. All businesses avoid accumulating debt.
D. The action of the government is more likely to do more harm than good to small businesses.
E. Big businesses do not need government money.

Questions 36-41 are based on the passage below:

Despite the numerous safety measures in place within the practice of medicine, these can fail when the weaknesses in the layers of defence aligns to create a clear path leading to often disastrous results. This is known as the 'Swiss cheese model of accident causation'. One such occurrence occurred where the wrong kidney was removed from a patient due to a failure in the line of defences designed to prevent such an incident occurring.

When a kidney is diseased it is removed to prevent further complications, this operation, a 'nephrectomy', is regularly performed by experienced surgeons. Where normally the consultant who knew the patient would have conducted the procedure, in this case he passed the responsibility to his registrar, who was also well experienced but had not met the patient previously. The person who had copied out the patient's notes had poor handwriting had accidentally written the 'R' for 'right' in such a way that it was read as an 'L' and subsequently copied, and not noticed by anyone who further reviewed the notes.

The patient had been put asleep before the registrar had arrived and so he proceeded without checking the procedure with the patient, as he normally would have done. The nurses present noticed this error but said nothing, fearing repercussions for questioning a senior professional. A medical student was present whom, having met the patient previously in clinical, tried to alert the registrar to the mistake he was about to make. The registrar shouted at the student that she should not interrupt surgery; she did not know what she was talking about and asked her to leave. Consequently the surgery proceeded with the end result being that the patient's healthy left kidney was removed, leaving them with only their diseased right kidney, which would eventually lead to the patient's unfortunate death. Frightening as these cases appear what is perhaps scarier is the thought of how those reported may be just the 'tip of the iceberg'.

When questioned about his action to allow his registrar to perform the surgery alone, the consultant had said that it was normal to allow capable registrars to do this. 'While the public perception is that medical knowledge steadily increases over time, this is not the case with many doctors reaching their peak in the middle of their careers.' He had found that his initial increasing interest in surgery had enhanced his abilities, but with time and practice the similar surgeries had become less exciting and so his lack of interest had correlated with worsening outcomes, thus justifying his decision to devolve responsibility in this case.

Question 36:
Which of the following, if true, most weakens the argument above?

A. If incidences are severe enough to occur they will be reported.
B. Doctors undergo extensive training to reduce risks.
C. Thousand of operations happen every year with no problems.
D. Some errors are unavoidable.
E. The patient could have passed away even if the operation had been a complete success.

Question 37:
Which one of the following is the overall conclusion of the statement?

A. The error that occurred was a result of the failure of safety precautions in place.
B. Surgeries should only be performed by surgeons who know their patients well.
C. The human element to medicine means errors will always occur.
D. The safety procedures surrounding surgical procedures need to be reviewed.
E. Some doctors are overconfident.

Question 38:
Which of the following is attributed as the original cause of the error?

A. The medical student not having asserted herself.
B. The poor handwriting in the chart.
C. The hierarchical system of medicine.
D. The registrar not having met the patient.
E. The patient being asleep.
F. The lack of the surgical skill possessed by the registrar.
G. The registrar's poor attitude.

Question 39:

What does the 'tip of the iceberg' refer to in the passage?

A. Problems we face every day.

B. The probable large numbers of medical errors that go unreported.

C. The difficulties of surgery.

D. Reported medical errors.

E. Problems within the NHS.

You may use the graphs below once, more than once, or not at all.

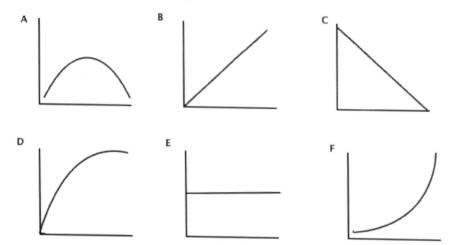

Question 40:

Which graph best describes the consultants' performance versus emotional arousal over his career?

A. A   B. B   C. C   D. D   E. E   F. F

Question 41:

Which graphs best describe the medical knowledge acquired over time?

Option	Public Perception	Consultant's Perception
A	B	B
B	B	D
C	B	F
D	D	B
E	D	D
F	D	F
G	F	B
H	F	D
I	F	F

Question 42:

Sadly, in recent times, the lack of exercise associated with sedentary lifestyles has increased in the developed world. The lack of opportunity for exercise is endemic and these countries have also seen a rise of diseases such as diabetes even in young people. In these developed countries, bodily changes such as increased blood pressure, that are usually associated with old age, are rapidly increasing. These are however still uncommon in undeveloped countries, where most people are physically active throughout the entirety of their lives.

Which one of the following can be concluded from the passage above?

A. Exercise has a greater effect on old people than young people.

B. Maintenance of good health is associated with lifelong exercise.

C. Changes in lifestyle will be necessary to cause increased life expectancies in developed countries.

D. Exercise is only beneficial when continued into old age.

E. Obesity and diabetes are the result of lack of exercise.

Questions 43 - 45 are based on the passage below:

'Midwives should now encourage women to, as often as possible, give birth at home. Not only is there evidence to suggest that normal births at home are as safe those as in hospital, but it removes the medicalisation of childbirth that emerged over the years. With the increase in availability of health resources we now, too often, use services such as a full medical team for a process that women have been completing single-handedly for thousands of years. Midwives are extensively trained to assist women during labour at home and capable enough to assess when there is a problem that requires a hospital environment. Expensive hospital births must and should move away from being standard practice, especially in an era where the NHS has far more demands on its services that it can currently afford.'

Question 43:

Which one of the following is the most appropriate conclusion from the statement?

A.   People are over dependent on healthcare.

B.   Some women prefer to have their babies in hospital.

C.   Having a baby in hospital can actually be more risky than at home.

D.   Childbirth has been over medicalised.

E.   Encouraging women to have their babies at home may relieve some of the financial pressures on the NHS.

F.   We should have more midwives than doctors.

Question 44:

Which one of the following if true most weakens the argument presented in the passage above?

A.   Some women are scared of home births.

B.   Home births are associated with poorer outcomes.

C.   Midwives do not like performing home visits.

D.   Some home births result in hospital births anyway.

Question 45:

Which one of the following describes what the statement cites as the cause for the 'medicalisation of childbirth'?

A.   Women fear giving birth without a full medical team present.

B.   Midwives are incapable of aiding childbirth without help.

C.   Giving birth at home is not as safe as it used to be.

D.   Excessive availability of health services.

E.   Women only used to give birth at home because they could not do so at hospital.

Question 46:

We need to stop focussing so much attention on the dangers of fires. In 2011 there were only 242 deaths due to exposure to smoke, fire and flames, while there were 997 deaths from hernias. We need to think more proportionally as these statistics show that campaigns such as 'fire kills' are not necessary as comparison with the risk from the death from hernias clearly shows that fires are not as dangerous as they are perceived to be.

Which of the following statements identify a weakness in the above argument?

1.   More people may die in fires if there were no campaigns about their danger and how to prevent them.

2.   The smoke of a fire is more dangerous than it flames.

3.   There may be more people with hernias than those in fires.

A.   1 only   C.   3 only   E.   1 and 3 only   G.   1, 2 and 3

B.   2 only   D.   1 and 2 only   F.   2 and 3 only

Question 47:

A survey of a school was taken to find out whether there was any correlation between the sports students played and the subjects they liked. The findings were as follows: some football players liked Maths and some of them liked History. All students liked English. None of the basketball players liked History, but all of them, as well as some rugby players liked Chemistry. All rugby players like Geography. Based on the findings, which one of the below must be true?

A.   Some of the footballers liked Maths and History.

B.   Some of the rugby players liked three subjects.

C.   Some rugby players liked History.

D.   Some of the footballers liked English but did not like Maths and History.

E.   Some basketball players like more than 3 subjects.

Question 48:

The control of illegal drug use is becoming increasingly difficult. New 'legal highs' are being manufactured which are slightly changed molecularly from illegal compounds so they are not technically illegal. These new 'legal drugs' are being brought onto the street at a rate of at least one per week, and so the authorities cannot keep up. Some health professionals therefore believe that the legality of drugs is becoming less relevant as to the potentially dangerous side effects. The fact that these new compounds are legal may however mean that the public are not aware of their equally high risks.

Which of the following are implied by the argument?
1. Some health professionals believe there is no value in making drugs illegal.
2. The major problem in controlling illegal drug use is the rapid manufacture of new drugs that are not classified as illegal.
3. The general public are not worried about the risks of legal or illegal highs.
4. There is no longer a good correlation between risk of drug taking and the legal status of the drug.

A. 1 only	C. 1 and 4	E. 2 and 3
B. 2 only	D. 2 and 4	F. 1,2,3 and 4

Question 49:

WilderTravel Inc. is a company which organises wilderness travel holidays, with activities such as trekking, mountain climbing, safari tours and wilderness survival courses. These activities carry inherent risks, so the directors of the company are drawing up a set of health regulations, with the aim of minimising the risks by ensuring that nobody participates in activities if they have medical complications meaning that doing so may endanger them. They consider the following guidelines:

'Persons with pacemakers, asthma or severe allergies are at significant risk of heart attack in low oxygen environments'. People undertaking mountain climbing activities with WilderTravel frequently encounter environments with low oxygen levels. The directors therefore decide that in order to ensure the safety of customers on WilderTravel holidays, one step that must be taken is to bar those with pacemakers, asthma or allergies from partaking in mountain climbing.

Which of the following best illustrates a flaw in this reasoning?

A. Participants should be allowed to assess the safety risks themselves, and should not be barred from activities if they decide the risk is acceptable.
B. They have assumed that all allergies carry an increased risk of heart attack, when the guidelines only say this applies to those with severe allergies.
C. The directors have failed to consider the health risks of people with these conditions taking part in other activities.
D. People with these conditions could partake in mountain climbing with other holiday organisers, and thus be exposed to danger of heart attack.

Question 50:

St John's Hospital in Northumbria is looking to recruit a new consultant cardiologist, and interviews a series of candidates. The interview panel determines that 3 candidates are clearly more qualified for the role than the others, and they invite these 3 candidates for a second interview. During this second interview, and upon further examination of their previous employment records, it becomes apparent that Candidate 3 is the most proficient at surgery of the 3, whilst Candidate 1 is the best at patient interaction and explaining the risks of procedures. Candidate 2, meanwhile, ranks between the other 2 in both these aspects.

The hospital director tells the interviewing team that the hospital already has a well-renowned team dedicated to patient interaction, but the surgical success record at the hospital is in need of improvement. The director issues instructions that therefore, it is more important that the new candidate is proficient at surgery, and patient interaction is less of a concern.

Which of the following is a conclusion that can be drawn from the Directors' comments?
A. The interviewing team should hire Candidate 2, in order to achieve a balance of good patient relations with good surgical records.
B. The interviewing team should hire Candidate 1, in order to ensure good patient interactions, as these are a vital part of a doctor's work.
C. The interviewing team should ignore the hospital director and assess the candidates further to see who would be the best fit.
D. The interviewing team should hire Candidate 3, in order to ensure that the new candidate has excellent surgical skills, to boost the hospital's success in this area.

Question 51:

Every winter in Britain, there are thousands of urgent callouts for ambulances in snowy conditions. The harsh conditions mean that ambulances cannot drive quickly, and are delayed in reaching patients. These delays cause many injuries and medical complications, which could be avoided with quicker access to treatment. Despite this, very few ambulances are equipped with winter tyres or special tyre coverings to help the ambulances deal with snow. Clearly, if more ambulances were fitted with winter tyres, then we could avoid many medical complications that occur each winter.

Which of the following is an assumption made in this passage?

A. Fitting winter tyres would allow ambulances to reach patients more quickly.
B. Ambulance trusts have sufficient funding to equip their vehicles with winter tyres.
C. Many medical complications could be avoided with quicker access to medical care.
D. There are no other alternatives to winter tyres that would allow ambulances to reach patients more quickly in snowy conditions.

Question 52:

Vaccinations have been one of the most outstanding and influential developments in medical history. Despite the huge successes, however, there is a strong anti-vaccination movement active in some countries, particularly the USA, who claim vaccines are harmful and ineffective.

There have been several high-profile events in recent years where anti-vaccine campaigners have been refused permission to enter countries for campaigns, or have had venues refuse to host them due to the nature of their campaigns. Many anti-vaccination campaigners have claimed this is an affront to free speech, and that they should be allowed to enter countries and obtain venues without hindrance. However, although free speech is desirable, an exception must be made here because the anti-vaccination campaign spreads misinformation to parents, causing vaccination to rates to drop.

When this happens, preventable infectious diseases often begin to increase, causing avoidable deaths of innocent members of the community, particularly so in children. Thus, in order to protect innocent people, we must continue to block the anti-vaccine campaigners from spreading misinformation freely by pressuring venues not to host anti-vaccination campaigners.

Which of the following best illustrates the principle that this argument follows?

A. Free speech is always desirable, and must not be compromised under any circumstances.
B. The right of innocent people to protection from infectious diseases is more important than the right of free speech.
C. The right of free speech does not apply when the party speaking is lying or spreading misinformation.
D. Public health programmes that achieve significant success in reducing the incidence of disease should be promoted.

Question 53:

In order for a tumour to grow larger than a few centimetres, it must first establish its own blood supply by promoting angiogenesis. Roger has a tumour in his abdomen, which is investigated at the Royal General Hospital. During the tests, they detect newly formed blood vessels in the tumour, showing that it has established its own blood supply. Thus, we should expect the tumour to grow significantly, and become larger than a few centimetres. Action must be taken to deal with this.

Which of the following best illustrates a flaw in this reasoning?

A. It assumes that the tumour in Roger's abdomen has established its own blood supply.
B. It assumes that a blood supply is necessary for a tumour to grow larger than a few centimetres.
C. It assumes that nothing can be done to stop the tumour once a blood supply has been established.
D. It assumes that a blood supply is sufficient for the tumour to grow larger than a few centimetres.

Question 54:

In this year's Great North Run, there are several dozen people running to raise money for the Great North Air Ambulance (GNAA), as part of a large national fundraising campaign. If the runners raise £500,000 between them, then the GNAA will be able to add a new helicopter to its fleet. However, the runners only raise a total of £420,000. Thus, the GNAA will not be able to get a new helicopter. Which of the following best illustrates a flaw in this passage?

A. It has assumed that the GNAA will not be able to acquire a new helicopter without the runners raising £500,000.
B. It has assumed that that GNAA wishes to add a new helicopter to its fleet.
C. It has assumed that the GNAA does not have better things to spend the money on.
D. It has assumed that some running in the Great North Run are raising money for the GNAA.

Question 55:

Many courses, spanning Universities, colleges, apprenticeship institutions and adult skills courses should be subsidised by the government. This is because they improve the skills of those attending them. It has been well demonstrated that the more skilled people are, the more productive they are economically. Thus, government subsidies of many courses would increase overall economic productivity, and lead to increased growth.

Which of the following would most weaken this argument?

A.   The UK already has a high level of growth, and does not need to accelerate this growth.
B.   Research has demonstrated that higher numbers of people attending adult skills courses results in increased economic growth.
C.   Research has demonstrated that the cost of many courses (to those taking them) has little effect on the number of people undertaking the courses.
D.   Employers often seek to employ those with greater skill-sets, and appoint them to higher positions.

Question 56:

Pluto was once considered the 9th planet in the solar system. However, further study of the planet led to it being reclassified as a dwarf planet in 2006. One key factor in this reclassification was the discovery of many objects in the solar system with similar characteristics to Pluto, which were also placed into this new category of 'Dwarf Planet'. Some astronomers believe that Pluto should remain classified as a planet, along with the many entities similar to Pluto that have been discovered. Considering all of this, it is clear that if we were to reclassify Pluto as a planet, and maintain consistency with classification of astronomical entities, then the number of planets would significantly increase.

Which of the following best illustrates the main conclusion of this passage?

A.   If Pluto is classified as a planet, then many other entities should also be planets, as they share similar characteristics.
B.   Some astronomers believe Pluto should be classified as a planet.
C.   Pluto should not be classified as a Planet, as this would also require many other entities to be classified as planets to ensure consistency.
D.   If Pluto is to be classified as a planet, then the number of objects classified as planets should increase significantly.

Question 57:

2 trains depart from Birmingham at 5:30 pm. One of the trains is heading to London, whilst the other is heading to Glasgow. The distance from Birmingham to Glasgow is three times larger than the distance from Birmingham to London, and the train to London arrives at 6:30 pm. Thus, the train to Glasgow will arrive at 8:30pm.

Which of the following is an assumption made in this passage?

A.   Both trains depart at the same time.
B.   Both trains depart from Birmingham.
C.   Both trains travel at the same speed.
D.   The train heading to Glasgow has to travel three times as far as the train heading to London.

Question 58:

Carcinogenesis, oncogenesis and tumorigenesis are various names given to the generation of cancer, with the term literally meaning 'creation of cancer'. In order for carcinogenesis to happen, there are several steps that must occur. Firstly, a cell (or group of cells) must achieve immortality, and escape senescence (the inherent limitation of a cell's lifespan). Then they must escape regulation by the body, and begin to proliferate in an autonomous way. They must also become immune to apoptosis and other cell death mechanisms. Finally, they must avoid detection by the immune system, or survive its responses. If a single one of these steps fails to occur, then carcinogenesis will not be able to occur.

Which of the following is a conclusion that can be reliably drawn from this passage?

A.   Several steps are essential for carcinogenesis.
B.   If all the steps mentioned occur, then carcinogenesis will definitely occur.
C.   The immune system is unable to tackle cells that have escaped regulation by the body.
D.   There are various mechanisms by which carcinogenesis can occur.
E.   The terminology for the creation of cancer is confusing.

Question 59:

P53 is one of the most crucial genes in the body, responsible for detecting DNA damage and halting cell replication until repair can occur. If repair cannot take place, P53 will signal for the cell to kill itself. These actions are crucial to prevent carcinogenesis, and a loss of functional P53 is identified in over 50% of all cancers. The huge importance of P53 towards protecting the cell from damaging mutations has led to it deservedly being known as 'the guardian of the genome'. The implications of this name are clear – any cell that has a mutation in P53 is at serious risk of developing a potentially dangerous mutation.

Which of the following **CANNOT** be reliably concluded from this passage?

A. P53 is responsible for detecting DNA damage.
B. Most cancers have lost functional P53.
C. P53 deserves its name 'guardian of the genome'.
D. A cell that has a mutation in P53 will develop damaging mutations.
E. None of the above.

Question 60:

Sam is buying a new car, and deciding whether to buy a petrol or a diesel model. He knows he will drive 9,000 miles each year. He calculates that if he drives a petrol car, he will spend £500 per 1,000 miles on fuel, but if he buys a diesel model he will only spend £300 per 1,000 miles on fuel. He calculates, therefore, that if he purchases a Diesel car, then this year he will make a saving of £1800, compared to if he bought the petrol car.

Which of the following is **NOT** an assumption that Sam has made?

A. The price of diesel will not fluctuate relative to that of petrol.
B. The cars will have the same initial purchase cost.
C. The cars will have the same costs for maintenance and garage expenses.
D. The cars will use the same amount of fuel.
E. All of the above are assumptions.

Question 61:

In the UK, cannabis is classified as a Class B drug, with a maximum penalty of up to 5 years imprisonment for possession, or up to 14 years for possession with intent to supply. The justification for drug laws in the UK is that classified drugs are harmful, addictive, and destructive to people's lives. However, available medical evidence indicates that cannabis is relatively safe, non-addictive and harmless. In particular, it is certainly shown to be less dangerous than alcohol, which is freely sold and advertised in the UK. The fact that alcohol can be freely sold and advertised, but cannabis, a less harmful drug, is banned highlights the gross inconsistencies in UK drugs policy.

Which of the following best illustrates the main conclusion of this passage?

A. Cannabis is a less dangerous drug than alcohol.
B. Alcohol should be banned, so we can ensure consistency in the UK drug policy.
C. Cannabis should not be banned, and should be sold freely, in order to ensure consistency in the UK drug policy.
D. The UK government's policy on drugs is grossly inconsistent.
E. Alcohol should not be advertised in the UK.

Question 62:

Every year in Britain, there are thousands of accidents at people's homes such as burns, broken limbs and severe cuts, which cause a large number of deaths and injuries. Despite this, very few households maintain a sufficient first aid kit equipped with bandages, burn treatments, splints and saline to clean wounds. If more households stocked sufficient first aid supplies, many of these accidents could be avoided.

Which of the following best illustrates a flaw in this argument?

A. It ignores the huge cost associated with maintaining good first aid supplies, which many households cannot afford.
B. It implies that presence of first aid equipment will lead to fewer accidents.
C. It ignores the many accidents that could not be treated even if first aid supplies were readily available.
D. It neglects to consider the need for trained first aid persons in order for first aid supplies to help in reducing the severity of injuries caused by accidents.

Question 63:

Researchers at SmithJones Inc., an international drug firm, are investigating a well-known historic compound, which is thought to reduce levels of DNA replication by inhibiting DNA polymerases. It is proposed that this may be able to be used to combat cancer by reducing the proliferation of cancer cells, allowing the immune system to combat them before they spread too far and become too damaging. Old experiments have demonstrated the effectiveness of the compound via monitoring DNA levels with a dye that stains DNA red, thus monitoring the levels of DNA present in cell clusters. They report that the compound is observed to reduce the rate at which DNA replicates. However, it is known that if researchers use the wrong solutions when carrying out these experiments, then the amount of red staining will decrease, suggesting DNA replication has been inhibited, even if it is not inhibited. As several researchers previously used this wrong solution, we can conclude that these experiments are flawed, and do not reflect what is actually happening.

Which of the following best illustrates a flaw in this argument?

A.   From the fact that the compound inhibits DNA replication, it cannot be concluded that it has potential as an anticancer drug.

B.   From the fact that the wrong solutions were used, it cannot be concluded that the experiments may produce misleading results.

C.   From the fact that the experiments are old, it cannot be concluded that the wrong solutions were used.

D.   From the fact that the compound is old, it cannot be concluded that it is safe.

Question 64:

Rotherham football club are currently top of the league, with 90 points. Their closest competitors are South Shields football club, with 84 points. Next week, the teams will play each other, and after this, they each have 2 games left before the end of the season. Each win is worth 3 points, a draw is worth 1 point, and a loss is worth 0 points. Thus, if Rotherham beat South Shields, they will win the league (as they will then be 9 points clear, and South Shields would only be able to earn 6 more points).

In the match of Rotherham vs. South Shields, Rotherham are winning until the 85th minute, when Alberto Simeone scores an equaliser for South Shields, and South Shields then go on to win the match. Thus, Rotherham will not win the league.

Which of the following best illustrates a flaw in this passage's reasoning?

A.   It has assumed that Alberto Simeone scored the winning goal for South Shields.

B.   It has assumed that beating South Shields was necessary for Rotherham to win the league, when in fact it was only sufficient.

C.   Rotherham may have scored an equaliser later in the game, and not lost the match.

D.   It has failed to consider what other teams might win the league.

Question 65:

Oakville Supermarkets is looking to build a new superstore, and a meeting of its directors has been convened to decide where the best place to build the supermarket would be. The Chairperson of the Board suggests that the best place would be Warrington, a town that does not currently have a large supermarket, and would thus give them an excellent share of the shopping market.

However, the CEO notes that the population of Warrington has been steadily declining for several years, whilst Middlesbrough has recently been experiencing high population growth. The CEO therefore argues that they should build the new supermarket in Middlesbrough, as they would then be within range of more people, and so of more potential customers.

Which of the following best illustrates a flaw in the CEO's reasoning?

A.   Middlesbrough may already have other supermarkets, so the new superstore may get a lower share of the town's shoppers.

B.   Despite the recent population changes, Warrington may still have a larger population than Middlesbrough.

C.   Middlesbrough's population is projected to continue growing, whilst Warrington's is projected to keep falling.

D.   Many people in Warrington travel to Liverpool or Manchester, 2 nearby major cities, in order to do their shopping.

Question 66:

Global warming is a key challenge facing the world today, and the changes in weather patterns caused by this phenomenon have led to the destruction of many natural habitats, causing many species to become extinct. Recent data has shown that extinctions have been occurring at a faster rate over the last 40 years than at any other point in the earth's history, exceeding the great Permian mass extinction, which wiped out 96% of life on earth. If this rate continues, over 50% of species on earth will be extinct by 2100. It is clear that in the face of this huge challenge, conservation programmes will require significantly increased levels of funding in order to prevent most of the species on earth from becoming extinct.

Which of the following are assumptions in this argument?

1. The rate of extinctions seen in the last 40 years will continue to occur without a step-up in conservation efforts.
2. Conservation programmes cannot prevent further extinctions without increased funding.
3. Global warming has caused many extinction events, directly or indirectly.

A. 1 only

B. 2 only

C. 3 only

D. 1 and 2

E. 1 and 3

F. 2 and 3

G. 1,2 and 3

Question 67:

After an election in Britain, the new government is debating what policy to adopt on the railway system, and whether it should be entirely privatised, or whether public subsidies should be used to supplement costs and ensure that sufficient services are run. Studies in Austria, which has high public funding for railways, have shown that the rail service is used by many people, and is highly thought of by the population. However, this is clearly down to the fact that Austria has many mountainous and high-altitude areas, which experience significant amounts of snow and ice. This makes many roads impassable, and travelling by road difficult. Thus, rail is often the only way to travel, explaining the high passenger numbers and approval ratings. Thus, the high public subsidies clearly have no effect.

Which of the following, if true, would weaken this argument?

1. France also has high public subsidy of railways, but does not have large areas where travel by road is difficult. The French railway also has high passenger numbers and approval ratings.
2. Italy also has high public subsidy of railways, but the local population dislike using the rail service, and it has poor passenger numbers.
3. There are many reasons affecting the passenger numbers and approval ratings of a given country's rail serviced.

A. 1 only

B. 2 only

C. 3 only

D. 1 and 2

E. 1 and 3

F. 2 and 3

Question 68:

In 2001-2002, 1,019 patients were admitted to hospital due to obesity. This figure was more than 11 times higher by 2011-12 when there were 11,736 patients admitted to hospital with the primary reason for admission being obesity. Data has shown higher percentages of both men and women were either obese or overweight in 2011 compared to 1993, with the percentage of people being overweight climbing from 58% to 65%, and female from 49% to 58%. Rates of adult obesity have increased even more steeply within this period – 13% to 24% for men and 16% to 26% for women.

Studies in 2011 found that nearly a third of children between 2 – 15 years were either overweight or obese, although this was not significantly higher than in 2008. Lifestyles are also becoming less healthy, with a decline in both children and adults eating the recommended number of fruit and vegetables each day and taking the recommended amount of exercise each week. The ease and availability of cheap fast-food outlets may be partly to blame for the rising number of obese people. Education is required to teach people the importance of a healthy lifestyle, however people must take some personal responsibility for their health.

Using only information from the passage, which of the following statements is correct?

A. In 2011, there was a higher proportion of obese men than women.

B. Obesity rates are rising steeply for both males and females of all age groups.

C. A combination of education and personal responsibility is needed to improve the population's health

D. The main reason people eat fast food is because it's cheaper

Question 69:

Tobacco companies sell cigarettes despite being fully aware that cigarettes cause significant harm to the wellbeing of those that smoke them. Diseases caused or aggravated by smoking cost billions of pounds for the NHS to treat each year. This is extremely irresponsible behaviour from the tobacco companies. Tobacco companies should be taxed, and the money raised put towards funding the NHS.

Which of the following conclusions **CANNOT** be drawn from the above?
A. There is a connection between lung cancer and smoking.
B. There is a connection between liver disease and smoking.
C. There is a connection between oral cancer and smoking.
D. All smokers drink excessively.
E. All of the above.

Question 70:

Investigations in the origins of species suggest that humans and the great apes have the same ancestors. This is suggested by the high degree of genetic similarity between humans and chimpanzees (estimated at 99%). At the same time there is an 84% homology between the human genome and that of pigs. This raises the interesting question of whether it would be possible to use pig or chimpanzee organs for the treatment of human disease.

Which conclusion can be reasonably drawn from the above article?
A. Pigs and chimpanzees have a common ancestor.
B. Pigs and humans have a common ancestor.
C. It can be assumed that chimpanzees will develop into humans if given enough time.
D. There seems to be great genetic homology across a variety of species.
E. Organs from pigs or chimpanzees present a good alternative for human organ donation.

Question 71:

Poor blood supply to a part of the body can cause damage of the affected tissue - i.e. lead to an infarction. There are a variety of known risk factors for vascular disease. Diabetes is a major risk factor. Other risk factors are more dependent on the individual as they represent individual choices such as smoking, poor dietary habits as well as little to no exercise. In some cases infarction of the limbs and in particular the feet can become very bad and extensive with patches of tissue dying. This is known as necrosis and is marked by affected area of the body turning black. Necrotic tissue is usually removed in surgery.

Which of the following statements **CANNOT** be concluded from the information in the above passage?
A. Smoking causes vascular disease.
B. Diabetes causes vascular disease.
C. Vascular disease always leads to infarctions.
D. Necrotic tissue must be removed surgically.
E. Necrotic tissue only occurs following severe infarction.
F. All of the above.

Question 72:

People who can afford to pay for private education should not have access to the state school system. This would allow more funding for students from lower income backgrounds. More funding will provide better resources for students from lower income backgrounds, and will help to bridge the gap in educational attainment between students from higher income and lower income backgrounds.

Which of the following statements, if true, would most strengthen the above argument?
A. Educational attainment is a significant factor in determining future prospects.
B. Providing better resources for students has been demonstrated to lead to an increase in educational attainment.
C. Most people who can afford to do so choose to purchase private education for their children.
D. A significant gap exists in educational attainment between students from high income and low-income backgrounds.
E. Most schools currently receive a similar amount of funding relative to the number of students in the school.

Question 73:

Increasing numbers of people are choosing to watch films on DVD in recent years. In the past few years, cinemas have lost customers, causing them to close down. Many cinemas have recently closed, removing an important focal point for many local communities and causing damage to those communities. Therefore, we should ban DVDs in order to help local communities.

Which of the following best states an assumption made in this argument?

A. The cinemas that have recently closed have done so because of reduced profits due to people choosing to watch DVDs instead.

B. Cinemas being forced to close causes damage to local communities.

C. DVDs are improving local communities by allowing people to meet up and watch films together.

D. Sales of DVDs have increased due to economic growth.

E. Local communities have called for DVDs to be banned.

Question 74:

Aeroplanes are the fastest form of transport available. An aeroplane can travel a given distance in less time than a train or a car. John needs to travel from Glasgow to Birmingham. If he wants to arrive as soon as possible, he should travel by aeroplane.

Which of the following best illustrates a flaw in this argument?

A. One day, there could be faster cars built that could travel as fast as aeroplanes.

B. Travelling by air is often more expensive.

C. It ignores the time taken to travel to an airport and check in to a flight, which may mean he will arrive later if travelling by aeroplane.

D. John may not own a car, and thus may not have any option.

E. John may not be legally allowed to make the journey.

Question 75:

During autumn, spiders frequently enter people's homes to escape the cold weather. Many people dislike spiders and seek ways to prevent them from entering properties, leading to spider populations falling as they struggle to cope with the cold weather. Studies have demonstrated that when spider populations fall, the population of flies rises. Higher numbers of flies are associated with an increase in food poisoning cases. Therefore, people must not seek to prevent spiders from entering their homes.

Which of the following best illustrates the main conclusion of this argument?

A. People should not dislike spiders being present in their homes.

B. People should seek methods to prevent flies from entering their homes.

C. People should actively encourage spiders to occupy their homes to increase biodiversity.

D. People should accept the presence of spiders in their homes to reduce the incidence of food poisoning.

E. Spiders should be cultivated and used as a biological pest control to combat flies.

Question 76:

Each year, thousands of people acquire infections during prolonged stays at hospital. Concurrently, bacteria are becoming resistant to antibiotics at an ever-increasing rate. In spite of this, progressively less pharmaceutical companies are investing in research into new antibiotics, and the number of antibiotics coming onto the market is decreasing. As a result, the number of antibiotics that can be used to treat infections is falling. If pharmaceutical companies were pressured into investing in new antibiotic research, many lives could be saved.

Which of the following best illustrates a flaw in this argument?

A. It assumes the infections acquired during stays at hospital are resulting in deaths.

B. It ignores the fact that many people never have to stay in hospital.

C. It does not take into account the fact that antibiotics do not produce much profit for pharmaceutical companies.

D. It ignores the fact that some hospital-acquired infections are caused by organisms that cannot be treated by antibiotics, such as viruses.

E. It assumes that bacterial resistance to antibiotics has not been happening for some time.

Question 77:

Katherine has shaved her armpits most of her adult life, but has now decided to stop. She explains her reasons for this to John, saying she does not like the pressures society puts on women to be shaven in this area. John listens to her reasons, but ultimately responds 'just because you explain why I should find your hairiness attractive, it does not mean I will. I find you unattractive, as I do not like girls with hair on their arm pits.'

What assumption has John made?

A. That just because he finds Katherine unattractive, he would find other girls with unshaven arm pits unattractive.

B. That Katherine is trying to make John find her armpit hair attractive.

C. That Katherine will never conceal her armpit hair.

D. Katherine must be wrong, because she is a woman.

E. That Katherine thinks women should stop shaving.

Question 78:

Medicine has improved significantly over the last century. Better medicine causes a reduction in the death rate from all causes. However, as people get older, they suffer from infectious disease more readily. Many third world countries have a high rate of deaths from infectious disease. Sunita argues that this high death rate is caused by better medicine, which has given an ageing population, thus giving a high rate of deaths from infectious disease as elderly people suffer from infectious disease more readily. Sunita believes that better medicine is thus indirectly responsible for this high death rate from infectious disease.

However, this cannot be the case. In third world countries, most people do not live to old age, often dying from infectious disease at a young age. Therefore, an ageing population cannot be the reason behind the high rate of death from infectious disease. As better medicine causes a reduction in the death rate from all causes, it is clear that better medicine will lead to a reduction in the death rate from infectious disease in third world countries.

Which of the following best states the main conclusion of this argument?

A. We can expect that improvements in medicine seen over the last century will improve.

B. Better medicine is not responsible for the increased prevalence of infectious disease in third world countries.

C. Better medicine has caused the overall death rate of third world countries to increase.

D. Better medicine will cause a decrease in the rate of death from infectious disease in third world countries.

E. As people get older, they suffer from infectious disease more readily.

Question 79:

Bristol and Cardiff are 2 cities with similar demographics, and located in a roughly similar area of the country. Bristol has higher demand for housing than Cardiff. Therefore, a house in Bristol will cost more than a similar house in Cardiff.

Which of the following best illustrates an assumption in the statement above?

A. House prices will be higher if demand for housing is higher.

B. People can commute from Cardiff to Bristol.

C. Supply of housing in Cardiff will not be lower than in Bristol.

D. Bristol is a better place to live.

E. Cardiff has sufficient housing to provide for the needs of its communities.

Question 80:

Jellicoe Motors is a small motor company in Sheffield, employing 3 people. The company is hiring a new mechanic and interviews several candidates. New research into production lines has indicated that having employees with a good ability to work as part of a team boosts a company's productivity and profits. Therefore, Jellicoe motors should hire a candidate with good team-working skills.

Which of the following best illustrates the main conclusion of this argument?

A. Jellicoe Motors should not hire a new mechanic.

B. Jellicoe motors should hire a candidate with good team-working skills in order to boost their productivity and profits.

C. Jellicoe motors should hire several new candidates in order to form a good team, and boost their productivity.

D. If Jellicoe motors does not hire a candidate with good team-working skills, they may struggle to be profitable.

E. Jellicoe motors should not listen to the new research.

Question 81:

Research into new antibiotics does not normally hold much profit for pharmaceutical firms. As a consequence many firms are not investing in antibiotic research, and very few new antibiotics are being produced. However, with bacteria becoming increasingly resistant to current antibiotics, new ones are desperately needed to avoid running the risk of thousands of deaths from bacterial infections. Therefore, the UK government must provide financial incentives for pharmaceutical companies to invest in research into new antibiotics.

Which of the following best expresses the main conclusion of this argument?

A. If bacteria continue to become resistant to antibiotics, there could be thousands of deaths from bacterial infections.

B. Pharmaceutical firms are not investing in new antibiotic research due to a lack of potential profit.

C. If the UK government invests in research into new antibiotics, thousands of lives will be saved.

D. The pharmaceutical firms should invest in areas of research that are profitable, and ignore antibiotic research.

E. The UK government must provide financial incentives for pharmaceutical firms to invest into antibiotic research if it wishes to avoid risking thousands of deaths from bacterial infections.

Question 82:

People in developing countries use far less water per person than those in developed countries. It is estimated that at present, people in the developing world use an average of 30 litres of water per person per day, whilst those in developed countries use on average 70 litres of water per person per day. It is estimated that for the current world population, an average water usage of 60 litres per person per day would be sustainable, but any higher than this would be unsustainable.

The UN has set development targets such that in 20 years, people living in developing countries will be using the same amount of water per person per day as those living in developed countries. Assuming the world population stays constant for the next 20 years, if these targets are met the world's population will be using water at an unsustainable rate.

Which of the following, if true, would most weaken the argument above?

A. The prices of water bills are dropping in developed countries like the UK.

B. The level of water usage in developed countries is falling, and may be below 60 litres per person per day in 20 years.

C. The population of all developing countries is less than the population of all developed countries.

D. Climate change is likely to decrease the amount of water available for human use over the next 20 years.

E. The UN's development targets are unlikely to be met.

Question 83:

In this Senior Management post we need someone who can keep a cool head in a crisis and react quickly to events. The applicant says he suffers from a phobia about flying, and panics especially when an aircraft is landing and that therefore he would prefer not to travel abroad on business if it could be avoided. He is obviously a very nervous type of person who would clearly go to pieces and panic in an emergency and fail to provide the leadership qualities necessary for the job. Therefore this person is not a suitable candidate for the post.

Which of the following highlights the biggest flaw in the argument above?

A. It falsely assumes that phobias are untreatable or capable of being eliminated.

B. It falsely assumes that the person appointed to the job will need to travel abroad.

C. It falsely assumes that a specific phobia indicates a general tendency to panic.

D. It falsely assumes that people who stay cool in a crisis will be good leaders.

E. It fails to take into account other qualities the person might have for the post.

Question 84:

There are significant numbers of people attending university every year, as many as 45% of 18 year olds. As a result, there are many more graduates entering the workforce with better skills and better earning potential. Going to university makes economic sense and we should encourage as many people to go there as possible.

Which of the following highlights the biggest flaw in the argument above?

A. There are no more university places left.

B. Students can succeed without going to university.

C. Not all degrees equip students with the skills needed to earn higher salaries.

D. Some universities are better than others.

Question 85:

Young people spend too much time watching television, which is bad for them. Watching excessive amounts of TV is linked to obesity, social exclusion and can cause eye damage. If young people were to spend just one evening a week playing sport or going for a walk the benefits would be manifold. They would lose weight, feel better about themselves and it would be a sociable activity. Exercise is also linked to strong performance at school and so young people would be more likely to perform well in their exams.

Which of the following highlights the biggest flaw in the argument above?

A.   Young people can watch sport on television.
B.   There are many factors that affect exam performance.
C.   Television does not necessarily have any damaging effect.
D.   Television and sport are not linked.

Question 86:

Campaigners pushing for legalisation of cannabis have many arguments for their cause. Most claim there is little evidence of any adverse affects to health caused by cannabis usage, that many otherwise law-abiding people are users of cannabis and that in any case, prohibition of drugs does not reduce their usage. Legalising cannabis would also reduce crime associated with drug trafficking and would provide an additional revenue stream for the government.

Which of the following best represents the conclusion of the passage?

A.   Regular cannabis users are unlikely to have health problems.
B.   Legalising cannabis would be good for cannabis users.
C.   There are multiple reasons to legalise cannabis.
D.   Prohibition is an effective measure to reduce drugs usage.
E.   Drug associated crime would reduce if cannabis was legal.

Question 87:

Mohan has been offered a new job in Birmingham, starting in several months with a fixed salary. In order to ensure he can afford to live in Birmingham on his new salary, Mohan compares the prices of some houses in Birmingham. He finds that a 2 bedroomed house will cost £200,000. A 3 bedroomed house will cost £250,000. A 4 bedroomed house with a garden will cost £300,000.

Mohan's bank tells him that if he is earning the salary of the job he has been offered, they will grant him a mortgage for a house costing up to £275,000. After a month of deliberation, Mohan accepts the job and decides to move to Wolverhampton. He begins searching for a house to buy. He reasons that he will not be able to purchase a 4-bedroomed house.

Which of the following is NOT an assumption that Mohan has made?

A.   A house in Wolverhampton will cost the same as a similar house in Birmingham.
B.   A different bank will not offer him a mortgage for a more expensive house on the same salary.
C.   The salary for the job could increase, allowing him to purchase a more expensive house.
D.   A 4-bedroomed house without a garden will not cost less than a 4-bedroomed house with a garden.
E.   House prices in Birmingham will not have fall in the time between now and Mohan purchasing a house.

Question 88:

We should teach the Holocaust in schools. It is important that young people see what it was like for Jewish people under Nazi rule. If we expose the harsh realities to impressionable people then this will help improve tolerance of other races. It will also prevent other such terrible events happening again.

Which is the best conclusion?

A.   We should teach about the Holocaust in schools.
B.   The Holocaust was a tragedy.
C.   The Nazis were evil.
D.   We should not let terrible events happen again.
E.   Educating people is the best solution to the world's problems.

Question 89:

The popular series 'Game of Thrones' should not be allowed on television because it shows scenes of a disturbing nature, in particular scenes of rape. Children may find themselves watching the programme on TV, and then going on to commit the terrible crime of rape, mimicking what they have watched.

Which of the following best illustrates a flaw in this argument?

A. Children may also watch the show on DVD.

B. Adults may watch the show on television.

C. Watching an action does not necessarily lead to recreating the action yourself.

D. There are lots of non-violent scenes in the show.

Question 90:

The TV series 'House of Cards' teaches us all a valuable lesson: the world is not a place that rewards kind behaviour. The protagonist of the series, Frank Underwood, uses intrigue and guile to achieve his goals, and through clever political tactics he is able to climb in rank. If he were to be kinder to people, he would not be able to be so successful. Success is predicated on his refusal to conform to conventional morality. The TV series should be shown to small children in schools, as it could teach them how to achieve their dreams.

Which of the following is an assumption made in the argument?

A. Children pay attention to school lessons.

B. The TV series is sufficiently entertaining.

C. One cannot both obey a moral code and succeed.

D. Frank Underwood is a likable character.

Question 91:

Freddy makes lewd comments on a female passer-by's body to his friend, Neil, loud enough for the woman in question to hear. Neil is uncomfortable with this, and states that it is inappropriate for Freddy to do so, and that Freddy is being sexist. Freddy refutes this, and Neil retorts that Freddy would not make these comments about a man's body. Freddy replies by saying 'it is not sexist, I am a feminist, I believe in equality for men and women.'

Which of the following describes a flaw made in Freddy's logic?

A. A self-proclaimed feminist could still say a sexist thing.

B. The female passer-by in question felt uncomfortable.

C. Neil, too, considers himself a feminist.

D. It would still not be OK to make lewd comments at male passers-by.

E. Lewd comments are always inappropriate.

Question 92:

The release of CO_2 from consumption of fossil fuels is the main reason behind global warming, which is causing significant damage to many natural environments throughout the world. One significant source of CO_2 emissions is cars, which release CO_2 as they use up petrol. In order to tackle this problem, many car companies have begun to design cars with engines that do not use as much petrol. However, engines which use less petrol are not as powerful, and less powerful cars are not attractive to the public. If a car company produces cars which are not attractive to the public, they will not be profitable.

Which of the following best illustrates the main conclusion of this argument?

A. Car companies which produce cars that use less petrol will not be profitable.

B. The public prefer more powerful cars.

C. Car companies should prioritise profits over helping the environment.

D. Car companies should seek to produce engines that use less petrol but are still just as powerful.

E. The public are not interested in helping the environment.

SECTION 1: Problem Solving Questions

Section 1 problem solving questions are arguably the hardest to prepare for. However, there are some useful techniques you can employ to solve some types of questions much more quickly:

Construct Equations

Some of the problems in Section 1 are quite complex and you'll need to be comfortable with turning prose into equations and manipulating them. For example, when you read "Mark is twice as old as Jon" – this should immediately register as M = 2J. Once you get comfortable forming equations, you can start to approach some of the harder questions in this book (and past papers) which may require you to form and solve simultaneous equations. Consider the example:

Nick has a sleigh that contains toy horses and clowns and counts 44 heads and 132 legs in his sleigh. Given that horses have one head and four legs, and clowns have one head and two legs, calculate the difference between the number of horses and clowns.

A. 0
B. 5
C. 22
D. 28
E. 132
F. More information is needed.

To start with, let C= Clowns and H= Horses.
For Heads: $C + H = 44$; For Legs: $2C + 4H = 132$
This now sets up your two equations that you can solve simultaneously.
$C = 44 - H$ so $2(44 - H) + 4H = 132$
Thus, $88 - 2H + 4H = 132$;
Therefore, $2H = 44$; $H = 22$
Substitute back in to give $C = 44 - H = 44 - 22 = 22$
Thus the difference between horses and clowns $= C - H = 22 - 22 = 0$

It's important you are able to do these types of questions quickly (and **without resorting to trial & error** as they are commonplace in section 1.

Diagrams

When a question asks about timetables, orders or sequences, draw out diagrams. By doing this, you can organise your thoughts and help make sense of the question.

"Mordor is West of Gondor but East of Rivendale. Lorien is midway between Gondor and Mordor. Erebus is West of Mordor. Eden is not East of Gondor."

*Which of the following **cannot** be concluded?*

A. Lorien is East of Erebus and Mordor.
B. Mordor is West of Gondor and East of Erebus.
C. Rivendale is west of Lorien and Gondor.
D. Gondor is East of Mordor and East of Lorien
E. Erebus is West of Mordor and West of Rivendale.

Whilst it is possible to solve this in your head, it becomes much more manageable if you draw a quick diagram and plot the positions of each town:

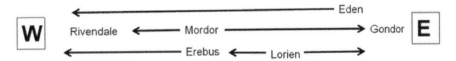

Now, it's a simple case of going through each option and seeing if it is correct according to the diagram. You can now easily see that Option E- Erebus cannot be west of Rivendale.

Don't feel that you have to restrict yourself to linear diagrams like this either – for some questions you may need to draw tables or even Venn diagrams. Consider the example:

Slifers and Osiris are not legendary. Krakens and Minotaurs are legendary. Minotaurs and Lords are both divine. Humans are neither legendary nor divine.

A.   Krakens may be only legendary or legendary and divine.
B.   Humans are not divine.
C.   Slifers are only divine.
D.   Osiris may be divine.
E.   Humans and Slifers are the same in terms of both qualities.

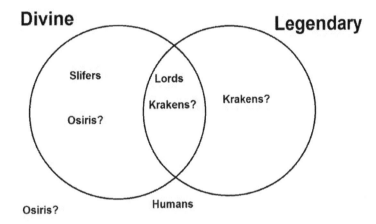

Constructing a Venn diagram allows us to quickly see that the position of Osiris and Krakens aren't certain. Thus, A and D must be true. Humans are neither so B is true. Krakens may be divine so A is true. E cannot be concluded as Slifers are divine but are humans are not. Thus, E is False.

Spatial Reasoning

There are usually 1-2 spatial reasoning questions every year. They usually give nets for a shape or a patterned cuboid and ask which options are possible rotations. Unfortunately, they are extremely difficult to prepare for because the skills necessary to solve these types of questions can take a very long time to improve. The best thing you can do to prepare is to familiarise yourself with the basics of how cube nets work and what the effect of transformations are e.g. what happens if a shape is reflected in a mirror etc.

It is also a good idea to try to learn to draw basic shapes like cubes from multiple angles if you can't do so already. Finally, remember that if the shape is straightforward like a cube, it might be easier for you to draw a net, cut it out and fold it yourself to see which of the options are possible.

Problem Solving Questions

Question 93:

Pilbury is south of Westside, which is south of Harrington. Twotown is north of Pilbury and Crewville but not further north than Westside. Crewville is:

A. South of Westside, Pilbury and Harrington but not necessarily Twotown.
B. North of Pilbury, and Westside.
C. South of Westside and Twotown, but north of Pilbury.
D. South of Westside, Harrington and Twotown but not necessarily Pilbury.
E. South of Harrington, Westside, Twotown and Pilbury.

Question 94:

The hospital coordinator is making the rota for the ward for next week; two of Drs Evans, James and Luca must be working on weekdays, none of them on Sundays and all of them on Saturdays. Dr Evans works 4 days a week including Mondays and Fridays. Dr Luca cannot work Monday or Thursday. Only Dr James can work 4 days consecutively, but he cannot do 5.

What days does Dr James work?

A. Saturday, Sunday and Monday.
B. Monday, Tuesday, Wednesday, Thursday and Saturday.
C. Monday, Thursday Friday and Saturday.
D. Tuesday, Wednesday, Friday and Saturday.
E. Monday, Tuesday, Wednesday, Thursday and Friday.

Question 95:

Michael, a taxi driver, charges a call out rate and a rate per mile for taxi rides. For a 4 mile ride he charges £11, and for a 5 mile ride, £13.

How much does he charge for a 9-mile ride?

A. £15   B. £17   C. £19   D. £20   E. £21

Question 96:

Goblins and trolls are not magical. Fairies and goblins are both mythical. Elves and fairies are magical. Gnomes are neither mythical nor magical.

Which of the following is **FALSE**?

A. Elves may be only magical or magical and mythical.
B. Gnomes are not mythical.
C. Goblins are only mythical.
D. Trolls may be mythical.
E. Gnomes and goblins are the same in terms of both qualities.

Question 97:

Jessica runs a small business making bespoke wall tiles. She has just had a rush order for 100 tiles placed that must be ready for today at 7pm. The client wants the tiles packed all together, a process which will take 15 minutes. Only 50 tiles can go in the kiln at any point and they must be put in the kiln to heat for 45 minutes. The tiles then sit in the kiln to cool before they can be packed, a process which takes 20 minutes. While tiles are in the kiln Jessica is able to decorate more tiles at a rate of 1 tile per minute.

What is the latest time Jessica can start making the tiles?

A. 2:55pm   B. 3:15pm   C. 3:30pm   D. 3:45pm

Question 98:

Pain nerve impulses are twice as fast as normal touch impulses. If Yun touches a boiling hot pan this message reaches her brain, 1 metre away, in 1 millisecond.
What is the speed of a normal touch impulse?

A. 5 m/s   B. 20 m/s   C. 50 m/s   D. 200m/s   E. 500 m/s

Question 99:

A woman has two children Melissa and Jack, yearly, their birthdays are 3 months apart, both being on the 22nd. The woman wishes to continue the trend of her children's names beginning with the same letter as the month they were born. If her next child, Alina is born on the 22nd 2 months after Jack's birthday, how many months after Alina is born will Melissa have her next birthday?

A. 2 months   B. 4 months   C. 5 months   D. 6 months   E. 7 months

Question 100:

Policemen work in pairs. PC Carter, PC Dirk, PC Adams and PC Bryan must work together but not for more than seven days in a row, which PC Adams and PC Bryan now have. PC Dirk has worked with PC Carter for 3 days in a row. PC Carter does not want to work with PC Adams if it can be avoided.

Who should work with PC Bryan?

A. PC Carter
B. PC Dirk
C. PC Adams
D. Nobody is available under the guidelines above.

Question 101:

My hair-dressers charges £30 for a haircut, £50 for a cut and blow-dry, and £60 for a full hair dye. They also do manicures, of which the first costs £15, and includes a bottle of nail polish, but are subsequently reduced by £5 if I bring my bottle of polish. The price is reduced by 10% if I book and pay for the next 5 appointments in advance and by 15% if I book at least the next 10.

I want to pay for my next 5 cut and blow-dry appointments, as well as for my next 3 manicures. How much will it cost?

A. £170   B. £255   C. £260   D. £285   E. £305

Question 102:

Alex, Bertha, David, Gemma, Charlie, Elena and Frankie are all members of the same family consisting of three children, two of whom, Frankie and Gemma are girls. No other assumption of gender based on name can be established. There are also four adults. Alex is a doctor and is David's brother. One of them is married to Elena, and they have two children. Bertha is married to David; Gemma is their child.

Who is Charlie?

A. Alex's daughter
B. Frankie's father
C. Gemma's brother
D. Elena's son
E. Gemma's sister

Question 103:

At 14:30 three medical students were asked to examine a patient's heart. Having already watched their colleague, the second two students were twice as fast as the first to examine. During the 8 minutes break after the final student had finished, they were told by their consultant that they had taken too long and so should go back and do the examinations again. The second time all the students took half as long as they had taken the first time with the exception of the first student who, instead took the same time as his two colleagues' second attempt. Assuming there was a one minute change over time between each student and they were finished by 15:15, how long did the second student take to examine the first time?

A. 3 minutes   B. 4 minutes   C. 6 minutes   D. 7 minutes   E. 8 minutes

Question 104:

I pay for 2 chocolate bars that cost £1.65 each with a £5 note. I receive 8 coins change, only 3 of which are the same.

Which **TWO** coins do I not receive in my change?

A. 1p   C. 5p   E. 20p   G. £1
B. 2p   D. 10p   F. £2

Question 105:

Two 140m long trains are running at the same speed in opposite directions. If they cross each other in 14 seconds then what is speed of each train?

A.   10 km/hr   B.   18 km/hr   C.   32 km/hr   D.   36 km/hr   E.   42 km/hr

Question 106:

Anil has to refill his home's swimming pool. He has four hoses which all run at different speeds. Alone, the first would completely fill the pool with water in 6 hours, the second in two days, the third in three days and the fourth in four days.

Using all the hoses together, how long will it take to fill the pool to the nearest quarter of an hour?

A.   4 hours 15 minutes
B.   4 hours 30 minutes
C.   4 hours 45 minutes
D.   5 hours
E.   5 hours 15 minutes

Question 107:

An ant is stuck in a 30 cm deep ditch. When the ant reaches the top of the ditch he will be able to climb out straight away. The ant is able to climb 3 cm upwards during the day, but falls back 2 cm at night.

How many days does it take for the ant to climb out of the ditch?

A.   27   B.   28   C.   29   D.   30   E.   31

Question 108:

When buying his ingredients a chef gets a discount of 10% when he buys 10 or more of each item, and 20% discount when he buys 20 or more. On one order he bought 5 sausages and 10 Oranges and paid £8.50. On another, he bought 10 sausages and 10 apples and paid £9, on a third he bought 30 oranges and paid £12.

How much would an order of 2 oranges, 13 sausages and 12 apples cost?

A.   £12.52   B.   £12.76   C.   £13.52   D.   £13.76   E.   £13.80

Question 109:

My hairdressers encourage all of its clients to become members. By paying an annual member fee, the cost of haircuts decreases. VIP membership costs £125 annually with a £10 reduction on haircuts. Executive VIP membership costs £200 for the year with a £15 reduction per haircut. At the moment I am not a member and pay £60 per haircut. I know how many haircuts I have a year, and I work out that by becoming a member on either programme it would work out cheaper, and I would save the same amount of money per year on either programme.

How much will I save this year by buying membership?

A.   £10   B.   £15   C.   £25   D.   £30   E.   £50

Question 110:

If criminals, thieves and judges are represented below:

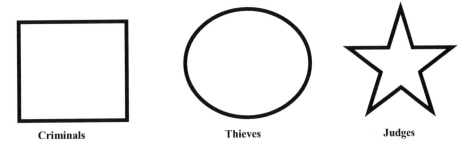

| Criminals | Thieves | Judges |

Assuming that judges must have clean record, all thieves are criminals and all those who are guilty are convicted of their crimes, which of one of the following best represents their interaction?

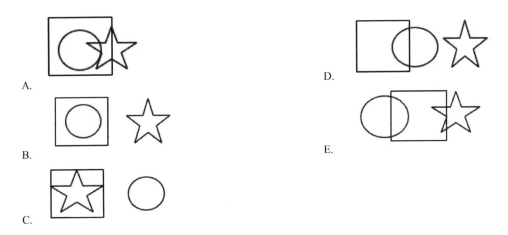

Question 111:

The months of the year have been made into number codes. The code is comprised of three factors, including two of these being related the letters that make up the name of the month. No two months would have the same first number. But some such as March, which has the code 3513, have the same last number as others, such as May, which has the code 5313. October would be coded as 10715 while February is 286.

What would be the code for April?
A. 154   B. 441   C. 451   D. 514   E. 541

Question 112:

A mother gives yearly birthday presents of money to her children based on the age and their exam results. She gives them £5 each plus £3 for every year they are older than 5, and a further £10 for every A* they achieved in their results. Josie is 16 and gained 9 A*s in her results. Although Josie's brother Carson is 2 years older he receives £44 less a year for his birthday.

How many more A*s did Josie get than Carson?
A. 2   B. 3   C. 4   D. 5   E. 10

Question 113:

Apples are more expensive than pears, which are more expensive than oranges. Peaches are more expensive than oranges. Apples are less expensive than grapes.

Which two of the following must be true?
A. Grapes are less expensive than oranges.
B. Peaches may be less expensive than pears.
C. Grapes are more expensive than pears.
D. Pears and peaches are the same price.
E. Apples and peaches are the same price.

Question 114:

What is the minimum number of straight cutting motions needed to slice a cylindrical cake into 8 equally sized pieces?

A. 2   B. 3   C. 4   D. 5   E. 6   F. 8

Question 115

Three friends, Mark, Russell and Tom had agreed to meet for lunch at 12 PM on Sunday. Daylight saving time (GMT+1) had started at 2 AM the same day, where clocks should be put forward by one hour. Mark's phone automatically changes the time but he does not realise this so when he wakes up he puts his phone forward an hour and uses his phone to time his arrival to lunch. Tom puts all of his clocks forward one hour at 7 AM. Russell forgets that the clocks should go forward, wakes at 10 AM doesn't change his clocks. All of the friends arrive on time as far as they are concerned.

Assuming that none of the friends realise any errors before arriving, which **TWO** of the following statements are **FALSE**?

A. Tom arrives at 12 PM (GMT +1).
B. All three friends arrive at the same time.
C. There is a 2 hour difference between when the first and last friend arrive.
D. Mark arrives late.
E. Mark arrives at 1 PM (GMT+3).
F. Russell arrives at 12 PM (GMT+0).

Question 116:

A class of young students has a pet spider. Deciding to play a practical joke on their teacher, one day during morning break one of the students put the spider in their teachers' desk. When first questioned by the head teacher, Mr Jones, the five students who were in the classroom during morning break all lied about what they saw. Realising that the students were all lying, Mr Jones called all 5 students back individually and, threatened with suspension, all the students told the truth. Unfortunately Mr Jones only wrote down the student's statements not whether they had been told in the truthful or lying questioning.

The students' two statements appear below:

Archie: "It wasn't Edward. "
"It was Bella."

Charlotte: "It was Edward."
"It wasn't Archie"

Darcy: "It was Charlotte"
"It was Bella"

Bella: "It wasn't Charlotte."
"It wasn't Edward."

Edward: "It was Darcy"
"It wasn't Archie"

Who put the spider in the teacher's desk?

A.   Edward
B.   Bella
C.   Darcy

D.   Charlotte
E.   More information needed.

Question 117:

Dr Massey wants to measure out 0.1 litres of solution. Unfortunately the lab assistant dropped the 200 ml measuring cylinder, and so the scientist only has a 300 ml and a half litre-measuring beaker. Assuming he cannot accurately use the beakers to measure anything less than their full capacity, what is the minimum volume he will have to use to be able to ensure he measures the right amount?

A.   100 ml
B.   200 ml

C.   300 ml
D.   400 ml

E.   500 ml
F.   600 ml

Question 118:

Francis lives on a street with houses all consecutively numbered evenly. When one adds up the value of all the house numbers it totals 870.

In order to determine Francis' house number:
1.   The relative position of Francis' house must be known.
2.   The number of houses in the street must be known.
3.   At least three of the house numbers must be known.

A.   1 only
B.   2 only
C.   3 only

D.   1 and 2
E.   2 and 3

Question 119:

There were 20 people exercising in the cardio room of a gym. Four people were about to leave when suddenly a man collapsed on one of the machines. Fortunately a doctor was on the machine beside him. Emerging from his office, one of the personal trainers called an ambulance. In the 5 minutes that followed before the two paramedics arrived, half of the people who were leaving, left upon hearing the commotion, and eight people came in from the changing rooms to hear the paramedics pronouncing the man dead.

How many living people were left in the room?

A. 25   B. 26   C. 27   D. 28   E. 29   F. 30

Question 120:

A man and woman are in an accident. They both suffer the same trauma, which causes both of them to lose blood at a rate of 0.2 Litres/minute. At normal blood volume the man has 8 litres and the woman 7 litres, and people collapse when they lose 40% of their normal blood volume.

Which **TWO** of the following are true?

A.   The man will collapse 2 minutes before the woman.
B.   The woman collapses 2 minutes before the man.
C.   The total blood loss is 5 litres.
D.   The woman has 4.2 litres of blood in her body when she collapses.
E.   The man's blood loss is 4.8 litres when he collapses.
F.   Blood loss is at a rate of 2 litres every 12 minutes.

Question 121:

Jenny, Helen and Rachel have to run a distance of 13 km. Jenny runs at a pace of 8 kmph, Helen at a pace of 10 kmph, and Rachel 11 kmph.

If Jenny sets off 15 minutes before Helen, and 25 minutes before Rachel, what order will they arrive at the destination?

A.   Jenny, Helen, Rachel.
B.   Helen, Rachel, Jenny.
C.   Helen, Jenny, Rachel.

D.   Rachel, Helen, Jenny.
E.   Jenny, Rachel, Helen.
F.   None of the above.

Question 122:

On a specific day at a GP surgery 150 people visited the surgery and common complaints were recorded as a percentage of total patients. Each patient could use their appointment to discuss up to 2 complaints. 56% flu-like symptoms, 48% pain, 20% diabetes, 40% asthma/COPD and 30% high blood pressure.

Which statement **must** be true?
A.   A minimum of 8 patients complained of pain and flu-like symptoms.
B.   No more than 45 patients complained of high blood pressure and diabetes.
C.   There were a minimum of 21 patients who did not complain about flu-like symptoms or high blood pressure.
D.   There were actually 291 patients who visited the surgery.
E.   None of the above.

Question 123:

All products in a store were marked up by 15%. They were subsequently reduced in a sale with quoted saving of 25% from the higher price. What is the true reduction from the original price?

A. 5%
B. 10%
C. 13.75%

D. 18.25%
E. 20%
F. None of the above.

Question 124:

A recipe states it makes 12 pancakes and requires the following ingredients: 2 eggs, 100g plain flour, and 300ml milk. Steve is cooking pancakes for 15 people and wants to have sufficient mixture for 3 pancakes each.

What quantities should Steve use to ensure this whilst using whole eggs?

A. 2½ eggs, 125g plain flour, 375ml milk
B. 3 eggs , 150g plain flour, 450 ml milk
C. 7½ eggs, 375g plain flour, 1125 ml milk
D. 8 eggs, 400g plain flour, 1200 ml milk
E. 12 eggs, 600g plain flour, 1800 ml milk
F. None of the above.

Question 125:

Spring Cleaning cleaners buy industrial bleach from a warehouse and dilute it twice before using it domestically. The first dilution is by 9:1 and then the second, 4:1.

If the cleaners require 6 litres of diluted bleach, how much warehouse bleach do they require?

A. 30 ml
B. 120 ml
C. 166 ml

D. 666 ml
E. 1,200 ml
F. None of the above

Question 126:

During a GP consultation in 2015, Ms Smith tells the GP about her grandchildren. Ms Smith states that Charles is the middle grandchild and was born in 2002. In 2010, Bertie was twice the age of Adam and that in 2015 there are 5 years between Bertie and Adam. Charles and Adam are separated by 3 years.

How old are the 3 grandchildren in 2015?

A. Adam = 16, Bertie = 11, Charles = 13
B. Adam = 5, Bertie = 10, Charles = 8
C. Adam = 10, Bertie = 15, Charles = 13
D. Adam = 10, Bertie = 20, Charles = 13
E. Adam = 11, Bertie = 10, Charles = 8
F. More information needed.

Question 127:

Kayak Hire charges a fixed flat rate and then an additional half-hourly rate. Peter hires the kayak for 3 hours and pays £14.50, and his friend Kevin hires 2 kayaks for 4hrs30mins each and pays £41. How much would

Tom pay to hire one kayak for 2 hours?

A. £8
B. £10.50
C. £15

D. £33.20
E. £35.70
F. None of the above.

Question 128:

A ticketing system uses a common digital display of numbers 0 – 9. The number 7 is showing. However, a number of the light elements are not currently working.

Which set of the following digits is possible?

A. 3, 4, 7
B. 0, 1, 9

C. 2, 7, 8
D. 0, 5, 9

E. 3, 8, 9
F. 3, 4, 9

Question 129:

A team of 4 builders take 12 days of 7 hours work to complete a house. The company decides to recruit 3 extra builders.

How many 8 hour days will it take the new workforce to build a house?

A. 2 days
B. 6 days
C. 7 days
D. 10 days
E. 12 days
F. More information needed

Question 130:

All astragalus are fabacaea as are all gummifer. Acacia are not astragalus. Which of the following statements is true?

A. Acacia are not fabacaea.
B. No astragalus are also gummifer.
C. All fabacae are astragalus or gummifer.
D. Some acacia may be fabacaea.
E. Gummifer are all acacia.
F. None of the above.

Question 131:

The Smiths want to reupholster both sides of their seating cushions (dimensions shown on diagram). The fabric they are using costs £10/m, can only be bought in whole metre lengths and has a standard width of 1m. Each side of a cushion must be made from a single piece of fabric. The seamstress changes a flat rate of £25 per cushion. How much will it cost them to reupholster 4 cushions?

A. £ 20
B. £ 80
C. £ 110
D. £ 130
E. £ 150
F. £ 200

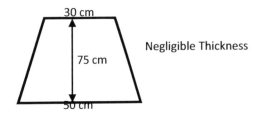

30 cm

75 cm

Negligible Thickness

50 cm

Question 132:

Lisa buys a cappuccino from either Milk or Beans Coffee shops each day. The quality of the coffee is the same but she wishes to work out the relative costs once the loyalty scheme has been taken into account. In Milk, a regular cappuccino is £2.40, and in Beans, £2.15. However, the loyalty scheme in Milk gives Lisa a free cappuccino for every 9 she buys, whereas Beans use a points system of 10 points per full pound spent (each point is worth 1p) which can be used to cover the cost of a full cappuccino.

If Lisa buys a cappuccino each day of September, which coffee shop would work out cheaper, and by how much?

A. Milk, by £4.60
B. Beans by £6.30
C. Beans, by £4.60
D. Beans, by £2.45
E. Milk, by £2.45
F. Milk, by £6.25

Question 133:

Paula needs to be at a meeting in Notting Hill at 11am. The route requires her to walk 5 minutes to the 283 bus which takes 25 minutes, and then change to the 220 bus which takes 14 minutes. Finally she walks for 3 minutes to her meeting. If the 283 bus comes every 10 minutes, and the 220 bus at 0 minutes, 20 minutes and 40 minutes past the hour, what is the latest time she can leave and still be at her meeting on time?

A. 09.45   B. 09.58   C. 10.01   D. 10.05   E. 10.10   F. 10.15

Question 134:

Two trains, a high speed train A and a slower local train B, travel from Manchester to London. Train A travels the first 20 km at 100 km/hr and then at an average speed of 150km/hr. Train B travels at a constant average speed of 90 km/hr. If train B leaves 20 minutes before train A, at what distance will train A pass train B?

A. 75 km   B. 90 km   C. 100 km   D. 120 km   E. 150 km

Question 135:

The university gym has an upfront cost of £35 with no contract fee, but classes are charged at £3 each. The local gym has no joining fee and is £15 per month. What is the minimum number of classes I need to attend in a 12 month period to make the local gym cheaper than the university gym?

A. 40   B. 48   C. 49   D. 50   E. 55   F. 60

Question 136:

"All medicines are drugs, but not all drugs are medicines", goes a well-known saying. If we accept this statement as true, and consider that all antibiotics are medicines, but no herbal drugs are medicines, then which of the following is definitely **FALSE**?

A. Some herbal drugs are not medicines.
B. All antibiotics are drugs.

C. Some herbal drugs are antibiotics.
D. Some medicines are antibiotics

Question 137:

Sonia has been studying the paths taken by various trains travelling between London and Edinburgh on the East coast. Trains can stop at the following stations: Newark, Peterborough, Doncaster, York, Northallerton, Darlington, Durham and Newcastle. She notes the following:

- All trains stop at Peterborough, York, Darlington and Newcastle.
- All trains which stop at Northallerton also stop at Durham.
- Each day, 50% of the trains stop at both Newark *and* Northallerton.
- All designated "Fast" trains make less than 5 stops. All other trains make 5 stops or more.
- On average, 16 trains run each day.

Which of the following can be reliably concluded from these observations?
A. All trains, which are not designated "fast" trains, must stop at Durham.
B. No more than 8 trains on any 1 day will stop at Northallerton.
C. No designated "Fast" trains will stop at Durham.
D. It is possible for a train to make 5 stops, including Northallerton.
E. A train which stops at Newark will also stop at Durham.

Question 138:

Rakton is 5 miles directly north of Blueville. Gallford is 8 miles directly south of Haston. Lepstone is situated 5 miles directly east of Blueville, and 5 miles directly west of Gallford.

Which of the following **CANNOT** be reliably concluded from this information?
A. Lepstone is South of Rakton
B. Haston is North of Rakton
C. Gallford is East of Rakton

D. Blueville is East of Haston
E. Haston is North of Lepstone

Question 139:

The Eastminster Parliament is undergoing a new set of elections. There are 600 seats up for election, each of which will be elected separately by the people living in that constituency. 6 parties win at least 1 seat in the election, the Blue Party, the Red party, the Orange party, the Yellow party, the Green party and the Purple party. In order to form a government, a party (or coalition) must hold *over* 50% of the seats. After the election, a political analysis committee produces the following report:

- No party has gained more than 45% of the seats, so nobody is able to form a government by themselves.
- The red and the blue party each gained over 40% of the seats.
- No other party gained more than 4% of the seats.
- The Yellow party did not win the fewest seats

The red party work out that if they collaborate with the green party and the orange party, between the 3 of them, they will have enough seats to form a coalition government.

What is the minimum number of seats that the green party could have?

A. 5   C. 13   E. 23
B. 6   D. 14   F. 24

Questions 140-144 are based on the following information:

A grandmother wants to give her 5 grandchildren £100 between them for Christmas this year. She wants to grade the money she gives to each grandchild exactly so that the older children receive more than the younger ones. She wants share the money such that she will give the 2nd youngest child as much more than the youngest, as the 3rd youngest gets than the 2nd youngest, as the 4th youngest gets from the 3rd youngest and so on. The result will be that the two youngest children together will get seven times as less money than the three oldest.

M is the amount of money the youngest child receives, and D the difference between the amount the youngest and 2nd youngest children receive.

Question 140:

What is the expression for the amount the oldest child receives?

A. M C. $2M$ E. $M + 4D$
B. $M + D$ D. $4M^2$ F. None of the above.

Question 141:

What is the correct expression for the total money received?

A. $5M = £100$ D. $5M + 10D = £100$

B. $5D + 10M = £100$ E. $M = \frac{2D}{11}$

C. $D = \frac{M}{100}$

Question 142:

"The two youngest children together will get seven times less money than the three oldest."

Which one of the following best expresses the above statement?

A. $7(3M + 9D) = 2M + D$ C. $7(2M + D) = 3M + 9D$
B. $7D = M$ D. $2(7M + D) = 3M + 9D$

Question 143

Using the statement in the previous question, what is the correct expression for M?

A. $\frac{2D}{11}$ B. $\frac{2}{11}$ C. $\frac{10D}{11}$ D. $\frac{120}{11}$

Question 144:

Express £100 in terms of D.

A. $£100 = \frac{120D}{11}$ C. $£100 = \frac{120}{11D}$

B. $£100 = \frac{120D}{10}$ D. $£100 = 21D$

                                        E. $£100 = 5M + 10D$

Question 145:

Four young girls entered a local baking competition. Though a bit burnt, Ellen's carrot cake did not come last. The girl who baked a Madeira sponge had practiced a lot, and so came first, while Jaya came third with her entry. Aleena did better than the girl who made the Tiramisu, and the girl who made the Victoria sponge did better than Veronica.

Which **TWO** of the following were **NOT** results of the competition?
A.   Veronica made a tiramisu
B.   Ellen came second
C.   Aleena made a Victoria sponge
D.   The Victoria sponge came in 3rd place
E.   The carrot cake came 3rd

Question 146:

In a young children's football league of 5 teams were; Celtic Changers, Eire Lions, Nordic Nesters, Sorten Swipers and the Whistling Winners. One of the boys playing in the league, after being asked by his parents, said that while he could remember the other teams' total points he could not remember his own, the Eire Lions, score. He said that all the teams played each other and when teams lost they were given 0 points, when they drew, 1 point, and 3 for a win. He remembered that the Celtic Changers had a total of 2 points; the Sorten Swipers had 5; the Nordic Nesters had 8, and the Whistling Winners 1.

How many did the boy's team score?
A.   1   C.   8   E.   11
B.   4   D.   10   F.   None of the above.

Question 147:

T is the son of Z, Z and J are sisters, R is the mother of J and S is the son of R.

Which one of the following statements is correct?
A.   T and J are cousins
B.   S and J are sisters
C.   J is the maternal uncle of T
D.   S is the maternal uncle of T
E.   R is the grandmother of Z.

Question 148:

John likes to shoot bottles off a shelf. In the first round he places 16 bottles on the shelf and knocks off 8 bottles. 3 of the knocked off bottles are damaged and can no longer be used, whilst 1 bottle is lost. He puts the undamaged bottles back on the shelf before continuing. In the second round he shoots six times and misses 50% of these shots. He damages two bottles with every shot which does not miss. 2 bottles also fall off the shelf at the end. He puts up 2 new bottles before continuing. In the final round, John misses all his shots and in frustration, knocks over gets angry and knocks over 50% of the remaining bottles.

How many bottles were left on the wall after the final round?
A.   2   C.   4   E.   6
B.   3   D.   5   F.   More information needed.

Questions 149 - 155 are based on the information below:

All lines are named after a station they serve, apart from the Oval and Rectangle lines, which are named for their recognisable shapes. Trains run in both directions.

➤ There are express trains that run from end to end of the St Mark's and Straightly lines in 5 and 6 minutes respectively.

➤ It takes 2 minutes to change between St Mark's and both Oval and Rectangle lines, 1 minute between Rectangle and Oval.

➤ It takes 3 minutes to change between the Straightly and all other lines, except with the St Mark's line which only takes 30 seconds

➤ The Straightly line is a fast line and takes only 2 minutes between stops apart from to and from Keyton, which only takes 1 minute, and to and from Lime St which takes 3 minutes.

➤ The Oval line is much slower and takes 4 minutes between stops, apart from between Baxton and Marven, and also Archite and West Quays, which takes 5 minutes.

➤ The Rectangle line a reliable line; never running late but as a consequence is much slower taking 6 minutes between stops.

➤ The St Mark's line is fast and takes 2 and half minutes between stations.

➤ If a passenger reaches the end of the line, it takes three minutes to change onto a train travelling back in the opposite direction.

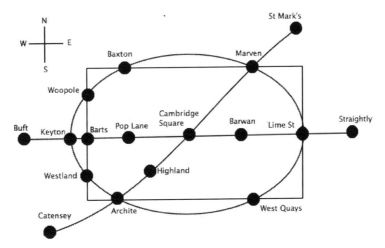

Question 149:

Assuming all lines are running on time, how long does it take to go from St Mark's to Archite on the St Mark's line?

A. 5 minutes
B. 6 minutes
C. 7.5 minutes

D. 10 minutes
E. 12.5 minutes

Question 150:

Assuming all lines are running on time, what's the shortest time it will take to go from Buft to Straightly?

A. 6 minutes
B. 10 minutes
C. 12 minutes

D. 14 minutes
E. 16 minutes

Question 151:

What is the shortest time it will take to go from Baxton to Pop Lane?

A. 11 minutes
B. 12 minutes
C. 13 minutes

D. 14 minutes
E. 15 minutes

Question 152

Which station, even at the quickest journey time, is furthest in terms of time from Cambridge Square?

A. Catensey   B. Buft   C. Woopole   D. Westland

Questions 153-155 use this additional information:

On a difficult day there are signal problems whereby all lines except the reliable line are delayed, such that train travel times between stations are doubled. These delays have caused overcrowding at the platforms which means that while changeover times between lines are still the same, passengers always have to wait an extra 5 minutes on all of the platforms before catching the next train.

Question 153

At best, how long will it now take to go from Westland to Marven?

A. 25 minutes
B. 29 minutes
C. 30 minutes

D. 33 minutes
E. 35 minutes

Question 154:

There is a bus that goes from Baxton to Archite and takes 27-31 minutes. Susan lives in Baxton and needs to get to her office in Archite as quickly as possible. With all the delays and lines out of service,

How should you advise Susan best to get to work?

A. Baxton to Archite via Barts using the Rectangle line.
B. Baxton to Woopole on the Rectangle line, then Oval to Archite via Keyton.
C. It is not possible to tell between the fastest two options.
D. Baxton to Woopole on the Rectangle line, then Oval to Archite via Keyton.
E. Baxton to Archite on the Oval line.
F. Baxton to Archite using the bus.

Question 155:

In addition to the delays the Oval line signals fail completely, so the line falls out of service. How long will it now take to go from St Mark's to West Quays as quickly as possible?

A. 35 minutes
B. 30 minutes

C. 33 minutes
D. 29 minutes

E. 30.5 minutes
F. None of the above.

Question 156:

In an unusual horserace, only 4 horses, each with different racing colours and numbers competed. Simon's horse wore number 1. Lila's horse wasn't painted yellow nor blue, and the horse that wore 3, which was wearing red, beat the horse that came in third. Only one horse wore the same number as the position it finished in. Arthur's horse beat Simon's horse, whereas Celia's horse beat the horse that wore number 1. The horse wearing green, Celia's, came second, and the horse wearing blue wore number 4. Which one of the following must be true?

A. Simon's horse was yellow and placed 3rd.
B. Celia's horse was red.
C. Celia's horse was in third place.
D. Arthur's horse was blue.
E. Lila's horse wore number 4.

Question 157:

Jessie plants a tree with a height of 40 cm. The information leaflet states that the plant should grow by 20% each year for the first 2 years, and then 10% each year thereafter.

What is the expected height at 4 years?

A. 58.08 cm
B. 64.89 cm

C. 69.696 cm
D. 89.696 cm

E. 82.944 cm
F. None of the above

Question 158

A company is required to pay each employee 10% of their wage into a pension fund if their annual total wage bill is above £200,000. However, there is a legal loophole that if the company splits over two sites, the £200,000 bill is per site. The company therefore decides to have an east site, and a west site.

Name	Annual Salary (£)
Luke	47,000
John	78,400
Emma	68,250
Nicola	88,500
Victoria	52,500
Daniel	63,000

Which employees should be grouped at the same site to minimise the cost to the company?

A. John, Nicola, Luke
B. Nicola, Victoria, Daniel
C. Nicola, Daniel, Luke

D. John, Daniel, Emma
E. Luke, Victoria, Emma

Question 159:

A bus takes 24 minutes to travel from White City to Hammersmith with no stops. Each time the bus stops to pick up and/or drop off passengers, it takes approximately 90 seconds. This morning, the bus picked up passengers from 5 stops, and dropped off passengers at 7 stops.

What is the minimum journey time from White City to Hammersmith this morning?

A. 28 minutes
B. 34 minutes

C. 34.5 minutes
D. 36 minutes

E. 37.5 minutes
F. 42 minutes

Question 160:

Sally is making a Sunday roast for her family and is planning her schedule regarding cooking times. The chicken takes 15 minutes to prepare, 75 minutes to cook, and needs to stand for exactly 5 minutes after cooking. The potatoes take 18 minutes to prepare, 5 minutes to boil, then 50 minutes to roast, and must be roasted immediately after boiling, and then served immediately. The vegetables require only 5 minutes preparation time and 8 minutes boiling time before serving, and can be kept warm to be served at any time after cooking. Given that the cooker can only be cooking two items at any given time and Sally can prepare only one item at a time, what should Sally's schedule be if she wishes to serve dinner at 4pm and wants to start cooking each item as late as possible?

A. Chicken 2.25, potatoes 2.47, vegetables 2.42
B. Chicken 2.25, potatoes 2.47, vegetables 3.47
C. Chicken 2.35, potatoes 3.47, vegetables 2.47

D. Chicken 2.35, potatoes 2.47, vegetables 3.47
E. Chicken 2.45, potatoes 3.47, vegetables 2.47
F. Chicken 2.45, potatoes 2.47, vegetables 3.47

Question 161:

The Smiths have 4 children whose total age is 80. Paul is double the age of Jeremy. Annie is exactly half way between the ages of Jeremy and Paul, and Rebecca is 2 years older than Paul. How old are each of the children?

A. Paul 23, Jeremy 12, Rebecca 26, Annie 19.
B. Paul 22, Jeremy, 11, Rebecca 24, Annie 16.
C. Paul 24, Jeremy 12, Rebecca 26, Annie 18.

D. Paul 28, Jeremy 14, Rebecca 30, Annie 21.
E. More information needed.

Question 162:

Sarah has a jar of spare buttons that are a mix of colours and sizes. The jar contains the following assortment of buttons:

	10mm	25mm	40mm
Cream	15	22	13
Red	6	15	7
Green	9	19	8
Blue	20	6	15
Yellow	4	8	26
Black	17	16	14
Total	71	86	83

Sarah wants to use a 25mm diameter button, but doesn't mind if it is cream or yellow. What is the maximum number of buttons she will have to remove in order to guarantee to pick a suitable button on the next attempt?

A. 210
B. 218
C. 219
D. 239
E.   None of the above

Question 163:

Ben wants to optimise his score with one throw of a dart. 50% of the time he hits a segment to either side of the one he is aiming at. With this in mind, which segment should he aim for?

[Ignore all double/triple modifiers]

A. 15
B. 16
C. 17
D. 18
E. 19
F. 20

Question 164:

Victoria is completing her weekly shop, and the total cost of the items is £8.65. She looks in her purse and sees that she has a £5 note, and a large amount of change, including all types of coins. She uses the £5 note, and pays the remainder using the maximum number of coins possible in order to remove some weight from the purse. However, the store has certain rules she has to follow when paying:

- No more than 20p can be paid in "bronze" change (the name given to any combination of 1p pieces and 2p pieces)
- No more than 50p can be paid using any combination of 5p pieces and 10p pieces.
- No more than £1.50 can be paid using any combination of 20p pieces and 50p pieces.

Victoria pays the exact amount, and does not receive any change. Under these rules, what is the *maximum* number of coins that Victoria can have paid with?

A. 30
B.   31
C.   36
D.   41
E.   46

Question 165:

I look at the clock on my bedside table, and I see the following digits:

However, I also see that there is a glass of water between me and the clock, which is in front of 2 adjacent figures. I know that this means these 2 figures will appear reversed. For example, 10 would appear as 01, and 20 would appear as 05 (as 5 on a digital clock is a reversed image of a 2). Some numbers, such as 3, cannot appear reversed because there are no numbers which look like the reverse of 3.

Which of the following could be the actual time?

A.   15:52
B.   21:25
C.   12:55
D.   12:22
E.   21:52

Question 166:

Slavica has invaded Worsid, whilst Nordic has invaded Lorkdon. Worsid, spotting an opportunity to bolster its amount of land and natural resources, invades Nordic. Each of these countries is either a dictatorship or a democracy. Slavica is a dictatorship, but Lorkdon is a democracy. 10 years ago, a treaty was signed which guaranteed that no democracy would invade another democracy. No dictatorship has both invaded another dictatorship *and* been invaded by another dictatorship.

Assuming the aforementioned treaty has been upheld, what style of government is practiced in Worsid?

A.   Worsid is a Dictatorship.

B.   Worsid is a Democracy.

C.   Worsid does not practice either of these forms of government.

D.   It is impossible to tell.

Question 167:

Sheila is on a shift at the local supermarket. Unfortunately, the till has developed a fault, meaning it cannot tell her how much change to give each customer. A customer is purchasing the following items, at the following costs:

-   A packet of grated cheese priced at £3.25
-   A whole cucumber, priced at 75p
-   A fish pie mix, priced at £4.00
-   3 DVDs, each priced at £3.00

Sheila knows there is an offer on DVDs in the store at present, in which 3 DVDs bought together will only cost £8.00. The customer pays with a £50 note.

How much change will Sheila need to give the customer?

A.   £4   B.   £33   C.   £34   D.   £36   E.   £38

Question 168:

Ryan is cooking breakfast for several guests at his hotel. He is frying most of the items using the same large frying pan, to get as much food prepared in as little time as possible. Ryan is cooking Bacon, Sausages, and eggs in this pan. He calculates how much room is taken up in the pan by each item. He calculates the following:

-   Each rasher of bacon takes up 7% of the available space in the pan
-   Each sausage takes up 3% of the available space in the pan.
-   Each egg takes up 12% of the available space in the pan.

Ryan is cooking 2 rashers of bacon, 4 sausages and 1 egg for each guest. He decides to cook all the food for each guest at the same time, rather than cooking all of each item at once.

How many guests can he cook for at once?

A.   1   B.   2   C.   3   D.   4   E.   5

Question 169:

SafeEat Inc. is a national food development testing agency. The Manchester-based laboratory has a system for recording all the laboratory employees' birthdays, and presenting them with cake on their birthday, in order to keep staff morale high. Certain amounts of petty cash are set aside each month in order to fund this. 40% of the staff have their birthday in March, and the secretary works out that £60 is required to fund the birthday cake scheme during this month.

If all birthdays cost £2 to provide a cake for, how many people work at the laboratory?

A.   45   B.   60   C.   75   D.   100   E.   150

Question 170:

Many diseases, such as cancer, require specialist treatment, and thus cannot be treated by a general practitioner. Instead, these diseases must be *referred* to a specialist after an initial, more generalised, medical assessment. Bob has had a biopsy on the 1st of August on a lump found in his abdomen. The results show that it is a tumour, with a slight chance of becoming metastatic, so he is referred to a waiting list for specialist radiotherapy and chemotherapy. The average waiting time in the UK for such treatment is 3 weeks, but in Bob's local district, high demand means that it takes 50% longer for each patient to receive treatment. As he is a lower risk case, with a low risk of metastasis, his waiting time is extended by another 20%.

How many weeks will it be before Bob receives specialist treatment?

A.   4.5   B.   4.6   C.   5.0   D.   5.1   E.   5.4   F.   5.6

Question 171:

In a class of 30 seventeen year old students, 40% drink alcohol at least once a month. Of those who drink alcohol at least once a month, 75% drink alcohol at least once a week. 1 in 3 of the students who drink alcohol at least once a week also smoke marijuana. 1 in 3 of the students who drink alcohol less than once a month also smoke marijuana.

How many of the students in total smoke marijuana?

A. 3   B. 4   C. 6   D. 9   E. 10   F. 15

Question 172:

Complete the following sequence of numbers: 1, 4, 10, 22, 46, ...

A. 84   B. 92   C. 94   D. 96   E. 100

Question 173:

If the mean of 5 numbers is 7, the median is 8 and the mode is 3, what must the two largest numbers in the set of numbers add up to?

A. 14   C. 24   E. 35
B. 21   D. 26   F. More information needed.

Question 174:

Ahmed buys 1kg bags of potatoes from the supermarket. 1kg bags have to weigh between 900 and 1100 grams. In the first week, there are 10 potatoes in the bag. The next week, there are only 5. Assuming that the potatoes in the bag in week 1 are all the same weight as each other, and the potatoes in the bag in week 2 are all the same weight as each other, what is the maximum possible difference between the heaviest and lightest potato in the two bags?

A. 50g   B. 70g   C. 90g   D. 110g   E. 130g

Question 175:

A football tournament involves a group stage, then a knockout stage. In the group stage, groups of four teams play in a round robin format (i.e. each team plays every other team once) and the team that wins the most matches in each group proceeds through to a knockout stage. In addition, the single best performing second place team across all the groups gains a place in the knockout stage. In the knockout stage, sets of two teams play each other and the one that wins proceeds to the next round until there are two teams left, who play the final.

If we start with 60 teams, how many matches are played altogether?

A. 75   B. 90   C. 100   D. 105   E. 165

Question 176:

The last 4 digits of my card number are 2 times my PIN number, plus 200. The last 4 digits of my husband's card number are the last four digits of my card number doubled, plus 200. My husband's PIN number is 2 times the last 4 digits of his card number, plus 200. Given that all these numbers are 4 digits long, whole numbers, and cannot begin with 0, what is the largest number my PIN number can be?

A. 1,074   C. 2,348   E. 9,999
B. 1,174   D. 4,096   F. More information needed.

Question 177:

All women between 50 and 70 in the UK are invited for breast cancer screening every 3 years. Patients at Doddinghurst Surgery are invited for screening for the first time at any point between their 50th and 53rd birthday. If they ignore an invitation, they are sent reminders every 5 months. We can assume that a woman is screened exactly 1 month after she is sent the invitation or reminder that she accepts. The next invitation for screening is sent exactly 3 years after the previous screening.

If a woman accepts the screening on the second reminder each time, what is the youngest she can be when she has her 4th screening?

A. 60   B. 61   C. 62   D. 63   E. 64   F. 65

Question 178:

Ellie gets a pay rise of k thousand pounds on every anniversary of joining the company, where k is the number of years she has been at the company. She currently earns £40,000, and she has been at the company for 5.5 years. What was her salary when she started at the company?

A. £25,000	C. £28,000	E. £31,000
B. £27,000	D. £30,000	F. £32,000

Question 179:

Northern Line trains arrive into Kings Cross station every 8 minutes, Piccadilly Line trains every 5 minutes and Victoria Line trains every 2 minutes. If trains from all 3 lines arrived into the station exactly 15 minutes ago, how long will it be before they do so again?

A. 24 minutes	C. 40 minutes	E. 65 minutes
B. 25 minutes	D. 60 minutes	F. 80 minutes

Question 180:

If you do not smoke or drink alcohol, your risk of getting Disease X is 1 in 12. If you smoke, you are half as likely to get Disease X as someone who does not smoke. If you drink alcohol, you are twice as likely to get Disease X. A new drug is released that halves anyone's total risk of getting Disease X for each tablet taken. How many tablets of the drug would someone who drinks alcohol have to take to reduce their risk to the same level as someone who smoked but did not take the drug?

A. 0	B. 1	C. 2	D. 3	E. 4	F. 5

Questions 181 – 183 refer to the following information:

There are 20 balls in a bag. 1/2 are red. 1/10 of those that are not red are yellow. The rest are green except 1, which is blue.

Question 181:

If I draw 2 balls from the bag (without replacement), what is the most likely combination to draw?

A. Red and green	C. Red and red
B. Red and yellow	D. Blue and yellow

Question 182:

If I draw 2 balls from the bag (without replacement), what is the least likely (without being impossible) combination to draw?

A. Blue and green	C. Yellow and yellow
B. Blue and yellow	D. Yellow and green

Question 183:

How many balls do you have to draw (without replacement) to guarantee getting at least one of at least three different colours?

A. 5	B. 12	C. 13	D. 17	E. 18	F. 19

Question 184:

A general election in the UK resulted in a hung parliament, with no single party gaining more than 50% of the seats. Thus, the main political parties are engaged in discussion over the formation of a coalition government. The results of this election are shown below:

Political Party	Seats won
Conservatives	260
Labour	270
Liberal Democrats	50
UKIP	35
Green Party	20
Scottish National Party	17
Plaid Cymru	13
Sinn Fein	9
Democratic Unionist Party (DUP)	11
Other	14 (14 other parties won 1 seat each)

There are a total of 699 seats, meaning that in order to form a government, any coalition must have at least 350 seats between them. Several of the party leaders have released statements about who they are and are not willing to form a coalition with, which are summarised as follows:

– The Conservative party and Labour are not willing to take part in a coalition together.
– The Liberal Democrats refuse to take part in any coalition which also involves UKIP.
– The Labour party will only form a coalition with UKIP if the Green party are also part of this coalition.
– The Conservative party are not willing to take part in any coalition with UKIP unless the Liberal Democrats are also involved.

Considering this information, what is the minimum number of parties required to form a coalition government?

A) 2   B) 3   C) 4   D) 5   E) 6

Question 185:

On Tuesday, 360 patients attend appointments at Doddinghurst Surgery. Of the appointments that are booked in, only 90% are attended. Of the appointments that are booked in, 1 in 2 are for male patients, the remaining appointments are for female patients. Male patients are three times as likely to miss their booked appointment as female patients.

How many male patients attend appointments at Doddinghurst Surgery on Tuesday?

A. 30   B. 60   C. 130   D. 150   E.   170

Question 186:

Every A Level student at Greentown Sixth Form studies Maths. Additionally, 60% study Biology, 50% study Economics and 50% study Chemistry. The other subject on offer at Greentown Sixth Form is Physics. Assuming every student studies 3 subjects and that there are 60 students altogether, how many students study Physics?

A. 15   C. 30   E. 60
B. 24   D. 40   F. More information needed

Question 187:

100,000 people are diagnosed with chlamydia each year in the UK. An average of 0.6 sexual partners are informed per diagnosis. Of these, 80% have tests for chlamydia themselves. Half of these tests come back positive.

Assuming that each of the people diagnosed has had an average of 3 sexual partners (none of them share sexual partners or have sex with each other) and that the likelihood of having chlamydia is the same for those partners who are tested and those who are not, how many of the sexual partners who were not tested (whether they were informed or not) have chlamydia?

A. 120,000	C. 136,000	E. 240,000
B. 126,000	D. 150,000	F. 252,000

Question 188:

In how many different positions can you place an additional tile to make a straight line of 3 tiles?

A. 6
B. 7
C. 8
D. 9
E. 10
F. 11
G. 12

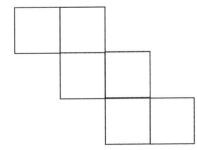

Question 189:

Harry is making orange squash for his daughter's birthday party. He wants to have a 200ml glass of squash for each of the 20 children attending and a 300ml glass of squash for him and each of 3 parents who are helping him out. He has 1,040ml of the concentrated squash.

What ratio of water: concentrated squash should he use in the dilution to ensure he has the right amount to go around?

A. 2:1	C. 4:1	E. 6:1
B. 3:1	D. 5:1	F. 5:2

Question 190:

4 children, Alex, Beth, Cathy and Daniel are each sitting on one of the 4 swings in the park. The swings are in a straight line. One possible arrangement of the children is, left to right, Alex, Beth, Cathy, Daniel.

How many other possible arrangements are there?

A. 5	C. 23	E. 64
B. 12	D. 24	F. 256

Question 191:

A delivery driver is looking to make deliveries in several towns. He is given the following map of the various towns in the area. The lines indicate roads between the towns, along with the lengths of these roads.

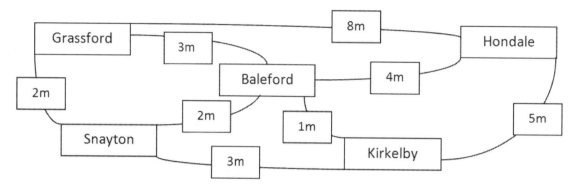

The delivery driver's vehicle has a black box which records the distance travelled and locations visited. At the end of the day, the black box recording shows that he has travelled a total of 14 miles. It also shows that he has visited one town twice, but has not visited any other town more than once. Which of the following is a possible route the driver could have taken?

A. Snayton → Baleford→ Grassford → Snayton→ Kirkelby
B. Baleford → Kirkelby→ Hondale → Grassford→ Baleford→ Snayton
C. Kirkelby → Hondale→ Baleford →Grassford→ Snayton
D. Baleford → Hondale→ Grassford → Baleford→ Hondale→ Kirkelby
E. Snayton → Baleford→ Kirkelby → Hondale→ Grassford
F. None of the above.

Question 192:

Ellie, her brother Tom, her sister Georgia, her mum and her dad line up in height order from shortest to tallest for a family photograph. Ellie is shorter than her dad but taller than her mum. Georgia is shorter than both her parents. Tom is taller than both his parents.

If 1 is shortest and 5 is tallest, what position is Ellie in the line?
A. 1   B. 2   C. 3   D. 4   E. 5

Question 193:

Miss Briggs is trying to arrange the 5 students in her class into a seating plan. Ashley must sit on the front row because she has poor eyesight. Danielle disrupts anyone she sits next to apart from Caitlin, so she must sit next to Caitlin and no-one else. Bella needs to have a teaching assistant sat next to her. The teaching assistant must be sat on the left hand side of the row, near to the teacher. Emily does not get on with Bella, so they need to be sat apart from one another. The teacher has 2 tables which each sit 3 people, which are arranged 1 behind the other.

Who is sitting in the front right seat?
A. Ashley   B. Bella   C. Caitlin   D. Danielle   E. Emily

Question 194:

My aunt runs the dishwasher twice a week, plus an extra time for each person who is living in the house that week. When her son is away at university, she buys a new pack of dishwasher tablets every 6 weeks, but when her son is home she has to buy a new one every 5 weeks. How many people are living in the house when her son is home?

A. 2   C. 4   E. 6
B. 3   D. 5   F. 7

Question 195:

Dates can be written in an 8 digit form, for example 26-12-2014. How many days after 26-12-2014 would be the next time that the 8 digits were made up of exactly 4 different integers?

A. 6

B. 8

C. 10

D. 16

E. 24

F. 30

Question 196:

Redtown is 4 miles east of Greentown. Bluetown is 5 miles north of Greentown. If every town is due North, South, East or West of at least two other towns, and the only other town is Yellowtown, how many miles away from Yellowtown is Redtown, and in what direction?

A. 4 miles east of Yellowtown.

B. 5 miles south of Yellowtown.

C. 5 miles north of Yellowtown.

D. 4 miles west of Yellowtown.

E. 5 miles west of Yellowtown.

F. None of the above.

Question 197:

Jenna pours wine from two 750ml bottles into glasses. The glasses hold 250ml, but she only fills them to 4/5 of capacity, except the last glass, where she puts whatever she has left. How full is the last glass compared to its capacity?

A. 1/5

B. 2/5

C. 3/5

D. 4/5

E. 5/5

Question 198:

There are 30 children in Miss Ellis's class. Two thirds of the girls in Miss Ellis's class have brown eyes, and two thirds of the class as a whole have brown hair. Given that the class is half boys and half girls, what is the difference between the minimum and maximum number of girls that could have brown eyes and brown hair?

A. 0

B. 2

C. 5

D. 7

E. 10

F. More information needed.

Question 199:

A biased die with the numbers 1 to 6 on it is rolled twice. The resulting numbers are multiplied together, and then their sum subtracted from this result to get the 'score' of the dice roll. If the probability of getting a negative (non-zero) score is 0.75, what is the probability of rolling a 1 on a third throw of the die?

A. 0.1

B. 0.2

C. 0.3

D. 0.4

E. 0.5

F. More information needed.

Questions 200 - 202 are based on the following information:

Fares on the number 11 bus are charged at a number of pence per stop that you travel, plus a flat rate. Emma, who is 21, travels 15 stops and pays £1.70. Charlie, who is 43, travels 8 stops and pays £1.14. Children (under 16) pay half the adult flat rate plus a quarter of the adult charge "per stop".

Question 200:

How much does 17 year old Megan pay to travel 30 stops to college?

A. £0.85

B. £2.40

C. £2.90

D. £3.40

E. More information needed.

Question 201:

How much does 14-year-old Alice pay to travel 25 stops to school?

A. £0.50

B. £0.75

C. £1.25

D. £2.50

E. More information needed.

Question 202:

James, who is 24, wants to get the bus into town. The town stop is the 25th stop along a straight road from his house, but he only has £2.

Assuming he has to walk past the stop nearest his house, how many stops will he need to walk past before he gets to the stop he can afford to catch the bus from?

A. 4
B. 6
C. 7
D. 8
E. 9
F. 10

Questions 203 -205 are based on the following information:

Emma mounts and frames paintings. Each painting needs a mount which is 2 inches bigger in each dimension than the painting, and a wooden frame which is 1 inch bigger in each dimension than the mount. Mounts are priced by multiplying 50p by the largest dimension of the mount, so a mount which is 8 inches in one direction and 6 in the other would be £4. Frames are priced by multiplying £2 by the smallest dimension of the frame, so a frame which is 8 inches in one direction and 6 in the other would be £12.

Question 203:

How much would mounting and framing a painting that is 10 x 14 inches cost?

A. £8
B. £26
C. £27
D. £34
E. £42

Question 204:

How much more would mounting and framing a 10 x 10 inch painting cost than mounting and framing an 8 x 8 inch painting?

A. £ 3.00
B. £ 4.00
C. £ 5.00
D. £ 6.00
E. £ 7.00

Question 205:

What is the largest square painting that can be framed for £40?

A. 12 inches
B. 13 inches
C. 14 inches
D. 15 inches
E. 16 inches

Question 206:

If the word 'CREATURES' is coded as 'FTEAWUTEV', which itself would be coded as 'HWEAYUWEX'. What would be the second coding of the word 'MAGICAL'?

A. QCKIGAN
B. OCIIEAN
C. PAJIFAN
D. RALIHAQ
E. RCIMGEP

Question 207:

Jane's mum has asked Jane to go to the shops to get some items that they need. She tells Jane that she will pay her per kilometre that she cycles on her bike to get to the shop, plus a flat rate payment for each place she goes to. Jane receives £6 to go to the grocers, a distance of 5 km, and £4.20 to go the supermarket, a distance of 3km.

How much would she earn if she then cycles to the library to change some books, a distance of 7 km?

A. £7.50
B. £7.70
C. £7.80
D. £8.00
E. £8.10
F. £8.20

Question 208:

In 2001-2002, 1,019 patients were admitted to hospital due to obesity. This figure was more than 11 times higher by 2011-12 when there were 11,736 patients admitted to hospital with the primary reason for admission being obesity.

If the rate of admissions due to obesity continues to increase at the same linear rate as it has from 2001/2 to 2011/12, how many admissions would you expect in 2031/32?

A. 22,453
B. 23,437
C. 33,170
D. 134,964
E. 269,928
F. 300,000

Question 209:

A shop puts its dresses on sale at 20% off the normal selling price. During the sale, the shop makes a 25% profit over the price at which they bought the dresses. What is the percentage profit when the dresses are sold at full price?

A. 36%

B. 42.5%

C. 56.25%

D. 64%

E. 77%

F. 80%

Question 210:

The 'Keys MedSoc committee' is made up of 20 students from each of the 6 years at the university. However, the president and vice-president are sabbatical roles (students take a year out from studying). There must be at least two general committee students from each year, as well as the specialist roles. Additionally, the social and welfare officers must be pre-clinical students (years 1-3) but not first years, and the treasurer must be a clinical student (years 4-6).

Which **TWO** of the following statements must be true?

1. There can be a maximum of 13 preclinical (years 1-3) students on the committee.
2. There must be a minimum of 6 2nd and 3rd years.
3. There is an unequal distribution of committee members over the different year groups.
4. There can be a maximum of 10 clinical (years 4-6) students on the committee.
5. There can be a maximum of 2 first year students on the committee.
6. General committee members are equally spread across the 6 years.

A. 1 and 4

B. 2 and 3

C. 2 and 4

D. 3 and 6

E. 4 and 5

F. 4 and 6

Question 211:

Friday the 13th is superstitiously considered an 'unlucky' day. If 13th January 2012 was a Friday, when would the next Friday the 13th be?

A. March 2012

B. April 2012

C. May 2012

D. June 2012

E. July 2012

F. August 2012

G. September 2012

H. January has the only Friday 13th in 2012.

Question 212:

A farmer has 18 sheep, 8 of which are male. Unfortunately, 9 sheep die, of which 5 were female. The farmer decides to breed his remaining sheep in order to increase the size of his herd. Assuming every female gives birth to two lambs, how many sheep does the farmer have after all the females have given birth once?

A. 10

B. 14

C. 15

D. 16

E. 19

Question 213:

Piyanga writes a coded message for Nishita. Each letter of the original message is coded as a letter a specific number of characters further on in the alphabet (the specific number is the same for all letters). Piyanga's coded message includes the word "PJVN". What could the original word say?

A. CAME

B. DAME

C. FAME

D. GAME

E. LAME

Question 214:

A number of people get on the bus at the station, which is considered the first stop. At each subsequent stop, 1/2 of the people on the bus get off and then 2 people get on. Between the 4th and 5th stop after the station, there are 5 people on the bus.

How many people got on at the station?

A. 4

B. 6

C. 20

D. 24

E. 30

Question 215:

I have recently moved into a new house, and I am looking to repaint my new living room. The price of several different colours of paint is displayed in the table below. A small can contains enough to paint 10 m² of wall. A large can contains enough to paint 25 m² of wall.

Colour	Cost for a Small Can	Cost for a Large Can
Red	£4	£12
Blue	£8	£15
Black	£3	£9
White	£2	£13
Green	£7	£15
Orange	£5	£20
Yellow	£10	£12

I decide to paint my room a mixture of blue and white, and I purchase some small cans of blue paint and white paint. The cost of blue paint accounts for 50% of the total cost. I paint a total of 100 m² of wall space.
I use up all the paint. How many m² of wall space have I painted blue?

A.   10 m²   B.   20 m²   C.   40 m²   D.   50 m²   E.   80 m²

Question 216:

Cakes usually cost 42p at the bakers. The bakers want to introduce a new offer where the amount in pence you pay for each cake is discounted by the square of the number of cakes you buy. For example, buying 3 cakes would mean each cake costs 33p. Isobel says that this is not a good offer from the baker's perspective as it would be cheaper to buy several cakes than just 1. How many cakes would you have to buy for the total cost to fall below 40p?

A.   2   B.   3   C.   4   D.   5   E.   6

Question 217:

The table below shows the percentages of students in two different universities who take various courses. There are 800 students in University A and 1200 students in University B. Biology, Chemistry and Physics are counted as "Sciences".

	University A	University B
Biology	23.50	13.25
Economics	10.25	14.5
Physics	6.25	14.75
Mathematics	11.50	17.25
Chemistry	30.25	7.00
Psychology	18.25	33.25

Assuming each student only takes one course, how many more students in University A than University B study a "Science"?

A.   10   B.   25   C.   60   D.   250   E.   600

Question 218:

Traveleasy Coaches charge passengers at a rate of 50p per mile travelled, plus an additional charge of £5.00 for each international border crossed during the journey. Europremier Coaches charge £15 for every journey, plus 10p per mile travelled, with no charge for crossing international borders. Sonia is travelling from France to Germany, crossing 1 international border. She finds that both companies will charge the same price for this journey.

How many miles is Sonia travelling?

A.   10   B.   20   C.   25   D.   35   E.   40

Question 219:

Lauren, Amy and Chloe live in different cities across England. They decide to meet up together in London and have a meal together. Lauren departs from Southampton at 2:30pm, and arrives in London at 4pm. Amy's journey lasts twice as long as Lauren's journey and she arrives in London at 4:15pm. Chloe departs from Sheffield at 1:30pm, and her journey lasts an hour longer than Lauren's journey.

Which of the following statements is definitely true?

A. Chloe's journey took the longest time.
B. Amy departed after Lauren.
C. Chloe arrived last.
D. Everybody travelled by train.
E. Amy departed before Chloe.

Question 220:

Emma is packing to go on holiday by aeroplane. On the aeroplane, she can take a case of dimension 50cm by 50cm by 20cm, which, when fully packed, can weigh up to 20kg. The empty suitcase weighs 2kg. In her suitcase, she needs to take 3 books, each of which is 0.2m by 0.1m by 0.05m in size, and weighs 1000g. She would also like to take as many items of clothing as possible. Each item of clothing has volume 1500cm^3 and weighs 400 g.

Assuming each item of clothing can be squashed so as to fill any shape gap, how many items of clothing can she take in her case?

A. 28   B. 31   C. 34   D. 37   E. 40

Question 221:

Alex is buying a new bed and mattress. There are 5 bed shops Alex can buy the bed and mattress he wants from, each of which sells the bed and mattress for a different price as follows:

➤ **Bed Shop A:** Bed £120, Mattress £70
➤ **Bed Shop B:** All beds and mattresses £90 each
➤ **Bed Shop C:** Bed £140, Mattress £60. Mattress half price when you buy a bed and mattress together.
➤ **Bed Shop D:** Bed £140, Mattress £100. Get 33% off when you buy a bed and mattress together.
➤ **Bed Shop E:** Bed £175. All beds come with a free mattress.

Which is the cheapest place for Alex to buy the bed and mattress from?

A. Bed Shop A   C. Bed Shop C   E. Bed Shop E
B. Bed Shop B   D. Bed Shop D

Question 222:

In Joseph's sock drawer, there are 21 socks. 4 are blue, 5 are red, 6 are green and the rest are black. How many socks does he need to take from the drawer in order to guarantee he has a matching pair?

A. 3   B. 4   C. 5   D. 6   E. 7

Question 223:

Printing a magazine uses 1 sheet of card and 25 sheets of paper. It also uses ink. Paper comes in packs of 500 and card comes in packs of 60 which are twice the price of a pack of paper. Each ink cartridge prints 130 sheets of either paper or card. A pack of paper costs £3. Ink cartridges cost £5 each.

How many complete magazines can be printed with a budget of £300?

A. 210   B. 220   C. 230   D. 240   E. 250

Question 224:

Rebecca went swimming yesterday. After a while she had covered one fifth of her intended distance. After swimming six more lengths of the pool, she had covered one quarter of her intended distance. How many lengths of the pool did she intend to complete?

A. 40   B. 72   C. 80   D. 100   E. 120

Question 225:

As a special treat, Sammy is allowed to eat five sweets from his very large jar which contains many sweets of each of three flavours – Lemon, Orange and Strawberry. He wants to eat his five sweets in such a way that no two consecutive sweets have the same flavour.

In how many ways can he do this?

A. 32   B. 48   C. 72   D. 108   E. 162

Question 226:

Granny and her granddaughter Gill both had their birthday yesterday. Today, Granny's age in years is an even number and 15 times that of Gill. In 4 years' time Granny's age in years will be the square of Gill's age in years.

How many years older than Gill is Granny today?

A. 42   B. 49   C. 56   D. 60   E. 64

Question 227:

Pierre said, "Just one of us is telling the truth". Qadr said, "What Pierre says is not true". Ratna said, "What Qadr says is not true". Sven said, "What Ratna says is not true". Tanya said, "What Sven says is not true".

How many of them were telling the truth?

A. 0   B. 1   C. 2   D. 3   E. 4

Question 228:

Two entrants in a school's sponsored run adopt different tactics. Angus walks for half the time and runs for the other half, whilst Bruce walks for half the distance and runs for the other half. Both competitors walk at 3 mph and run at 6 mph. Angus takes 40 minutes to complete the course.

How many minutes does Bruce take?

A. 30   B. 35   C. 40   D. 45   E. 50

Question 229:

Dr Song discovers two new alien life forms on Mars. Species 8472 have one head and two legs. Species 24601 have four legs and one head. Dr Song counts a total of 73 heads and 290 legs in the area. How many members of Species 8472 are present?

A. 0   C. 72   E. 145
B. 1   D. 73   F. More information needed.

Question 230:

A restaurant menu states that:

"All chicken dishes are creamy and all vegetable dishes are spicy. No creamy dishes contain vegetables."

Which of the following **must** be true?

A.   Some chicken dishes are spicy.
B.   All spicy dishes contain vegetables.
C.   Some creamy dishes are spicy.
D.   Some vegetable dishes contain tomatoes.
E.   None of the above

Question 231:

Simon and his sister Lucy both cycle home from school. One day, Simon is kept back in detention so Lucy sets off for home first. Lucy cycles the 8 miles home at 10 mph. Simon leaves school 20 minutes later than Lucy. How fast must he cycle in order to arrive home at the same time as Lucy?

A. 10 mph   B. 14 mph   C. 17 mph   D. 21 mph   E. 24 mph

Question 232:

Dr. Whu buys 2000 shares in a company at a rate of 50p per share. He then sells the shares for 58p per share. Subsequently he buys 1000 shares at 55p per share then sells them for 61p per share. There is a charge of £20 for each transaction of buying or selling shares. What is Dr. Whu's total profit?

A. £140   B. £160   C. £180   D. £200   E. £220

Question 233:

Jina is playing darts. A dartboard is composed of equal segments, numbered from 1 to 20. She takes three throws, and each of the darts lands in a numbered segment. None land in the centre or in double or triple sections. What is the probability that her total score with the three darts is odd?

A. $^1/_4$ B. $^1/_3$ C. $^1/_2$ D. $^3/_5$ E. $^2/_3$

Question 234:

John Morgan invests £5,000 in a savings bond paying 5% interest per annum. What is the value of the investment in 5 years' time?

A. £6,250   B. £6,315   C. £6,381   D. £6,442   E. £6,570

Question 235:

Joe is 12 years younger than Michael. In 5 years the sum of their ages will be 62. How old was Michael two years ago?

A. 20   B. 24   C. 26   D. 30   E. 32

Question 236:

A book has 500 pages. Vicky tears every page out that is a multiple of 3. She then tears out every remaining page that is a multiple of 6. Finally, she tears out half of the remaining pages. If the book measures 15 cm x 30cm and is made from paper of weight 110 gm^{-2}, how much lighter is the book now than at the start?

A. 1,648 g   B. 1,698 g   C. 1,722 g   D. 1,790 g   E. 1,848 g

Question 237:

A farmer is fertilising his crops. The more fertiliser is used, the more the crops grow. Fertiliser costs 80p per kilo. Fertilising at a rate of 0.2 kgm^{-2} increases the crop yield by £1.30 m^{-2}. For each additional 100g of fertiliser above 200g, the extra yield is 30% lower than the linear projection of the stated rate. At what rate of fertiliser application is it no longer cost effective to increase the dose

A. 0.5 kgm^{-2} B. 0.6 kgm^{-2} C. 0.7 kgm^{-2} D. 0.8 kgm^{-2} E. 0.9 kgm^{-2}

Question 238:

Pet-Star, Furry Friends and Creature Cuddles are three pet shops, which each sell food for various types of pets.

Type of pet food	Amount of food required per week	Price per Kg in:		
		Pet-star	Furry Friends	Creature Cuddles
Guinea Pig	3 Kg	£2	£1	£1.50
Cat	6 Kg	£4	£6	£5
Rabbit	4 Kg	£3	£1	£2.50
Dog	8 Kg	£5	£8	£6
Chinchilla	2 Kg	£1.50	£0.50	£1

Given the information above, which of the following statements can we state is definitely *not* true?

A. Regardless of which of these shops you use, the most expensive animal to provide food for will be a dog.
B. If I own a mixture of cats and rabbits, it will be cheaper for me to shop at Pet-star.
C. If I own 3 cats and a dog, the cheapest place for me to shop is at Pet-star
D. Furry Friends sells the cheapest food for the type of pet requiring the most food
E. If I only have one pet, Creature Cuddles will not be the cheapest place to shop regardless of which type of pet I have.

Question 239:

I record my bank balance at the start of each month for six months to help me see how much I am spending each month. My salary is paid on the 10th of each month. At the start of the year, I earn £1000 a month but from March inclusive I receive a pay rise of 10%.

Date	Bank balance
January 1st	1,200
February 1st	1,029
March 1st	1,189
April 1st	1,050
May 1st	925
June 1st	1,025

In which month did I spend the most money?

A. January
B. February
C. March
D. April
E. May

Question 240:

Amy needs to travel from Southtown station to Northtown station, which are 100 miles apart. She can travel by 3 different methods: train, aeroplane or taxi. The tables below show the different times for these 3 methods. The taxi takes 1 minute to cover a distance of 1 mile. Aeroplane passengers must be at the airport 30 minutes before their flight. Southtown airport is 10 minutes travelling time from Southtown station and Northtown airport is 30 minutes travelling time from Northtown station.

If Amy wants to arrive by 1700 and wants to set off as late as possible, what method of travel should she choose and what time will she leave Southtown station?

Train	**Departs Southtown station**	**1400**	**1500**	**1600**
	Arrives Northtown station	1615	1650	1715
Flights	Departs Southtown airport	1610		
	Arrives Northtown airport	1645		

A. Flight, 1530
B. Train, 1600
C. Taxi, 1520
D. Train, 1500
E. Flight, 1610

Question 241:

In the multiplication grid below, a, b, c and d are all integers. What does d equal?

A. 18
B. 24
C. 30
D. 40
E. 45

	c	**d**
a	168	720
b	119	510

Question 242:

A sixth form college has 1,500 students. 48% are girls. 80 of the girls are mixed race.

If an equal proportion of boys and girls are mixed race, how many mixed race boys are there in the college to the nearest 10?

A. 50 B. 60 C. 70 D. 80 E. 90

Question 243:

Christine is a control engineer at the Browdon Nuclear Power Plant. On Wednesday, she is invited to a party on the Friday, and asks her manager if she can take the Friday off. She acknowledged that this will mean she will have worked less than the required number of hours this week, and offers to make this up by working extra hours next week. Her manager suggests that instead, she works 5 hours this Sunday, and 3 extra hours next Thursday to make up the required hours. Christine accepts this proposal. Christine's amended schedule for the week is shown below:

Day	Monday	Tuesday	Wednesday	Thursday	Friday	Saturday	Sunday
Hours worked	8	7	9	6	0	0	5

How many hours was Christine supposed to have worked this week, if she had completed her usual Friday shift?

A. 34   B. 35   C. 36   D. 38   E. 40   F. 42

Question 244:

Leonidas notes that the time on a normal analogue clock is 0340. What is the smaller angle between the hands on the clock?

A. 110°   B. 120°   C. 130°   D. 140°   E. 150°

Question 245:

Sheila is on a shift at the local supermarket. Unfortunately, the till has developed a fault, meaning it cannot tell her how much change to give each customer. A customer is purchasing the following items, at the following costs:

- A packet of grated cheese priced at £3.25
- A whole cucumber, priced at 75p
- A fish pie mix, priced at £4.00
- 3 DVDs, each priced at £3.00

Sheila knows there is an offer on DVDs in the store at present, in which 3 DVDs bought together will only cost £8.00. The customer pays with a £50 note. How much change will Sheila need to give the customer?

A. £33
B. £34
C. £35
D. £36
E. £37

SECTION 1: Data Analysis

Data analysis questions show a great variation in type and difficulty. The best way to improve with these questions is to do lots of practice questions in order to familiarise yourself with the style of questions.

Options First

Despite the fact that you may have lots of data to contend with, the rule about looking at the options first still stands in this section. This will allow you to register what type of calculation you are required to make and what data you might need to look at for this. Remember, Options → Question → Data/Passage.

Working with Numbers

Percentages frequently make an appearance in this section and it's vital that you're able to work comfortably with them. For example, you should be comfortable increasing and decreasing by percentages, and working out inverse percentages too. When dealing with complex percentages, break them down into their components. For example, $17.5\% = 10\% + 5\% + 2.5\%$.

Graphs and Tables

When you're working with graphs and tables, it's important that you take a few seconds to check the following before actually extracting data from it.

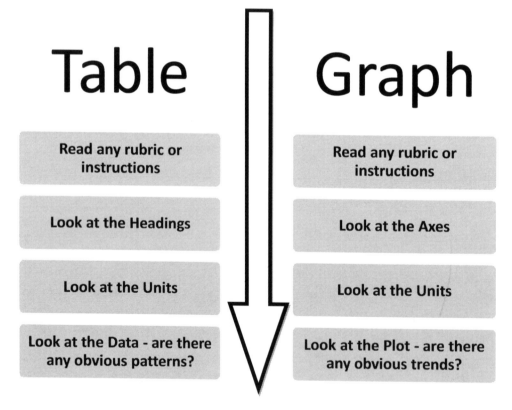

Get into the habit of doing this whenever you are faced with data and you'll find it much easier to approach these questions under time pressure.

Data Analysis Questions

Questions 246 to 248 are based on the following passage:

It has recently been questioned as to whether the recommended five fruit and vegetables a day is sufficient or if it would be more beneficial to eat 7 fruit and vegetable portions each day. A study at UCL looked at the fruit and vegetables eating habits of 65,000 people in England. Analysis of the data showed that eating more portions was beneficial and vegetables seemed to have a greater protective effect than fruit. The study however did not distinguish whether vegetables themselves have a greater protective effect, or whether these people tend to eat an overall healthier diet. A meta-analysis carried out by researchers across the world complied data from 16 studies which encompassed over 800,000 participants, of whom 56,423 had died.

They found a decline in death of around 5% from all causes for each additional portion of fruit or vegetables eaten, however they recorded no further decline for people who ate over 5 portions. Rates of cardiovascular disease, heart disease or stroke, were shown to decline 4% for each portion up to five, whereas the number of portions of fruit and vegetables eaten seemed to have little impact on cancer rates. The data from these studies points in a similar direction, that eating as much fruit and vegetables a day is preferable, but that five portions is sufficient to have a significant impact on reduction in mortality. Further studies need to look into the slight discrepancies, particularly why the English study found vegetables more protective, and if any specific cancers may be affected by fruit and vegetables even if the general cancer rates more greatly depend on other lifestyle factors.

Question 246:
Which of the following statements is correct?
A. The UCL study found no additional reduction in mortality in those who eat 7 rather than 5 portions of fruit and vegetables a day.
B. People who eat more fruit and vegetables are assumed to have an overall healthier diet which is what gives them the beneficial effect.
C. The meta analysis found fruit and vegetables are more protective against cancer than cardiovascular disease
D. The English study showed fruit had more protective effects than vegetables.
E. The meta-analysis found no additional reduction in mortality in those who eat 7 rather than 5 portions of fruit and vegetables a day.
F. The meta-analysis suggests people who eat 7 portions would have a 10% lower risk of death from any cause than those who eat 5 portions.
G. Fruit and vegetables are not protective against any specific cancers.

Question 247:
If rates of death were found to be 1% lower in the UCL study than the meta-analysis, approximately how many people died in the UCL study?
A. 3,000   B. 3,200   C. 3,900   D. 4,550   E. 5,200

Question 248
Which statement does the article **MOST** agree with?

A. Eating more fruit and vegetables does not particularly lower the risk of any specific cancers.
B. The UCL research suggests that the guideline should be 7 fruit and vegetables a day for England.
C. The results found by the UCL study and the meta-analysis were contradictory.
D. Many don't eat enough vegetables due to cost and taste.
E. Fruit and vegetables are only protective against cardiovascular disease.
F. The UCL study and meta-analysis use a similar sample of participants.
G. People should aim to eat 7 portions of fruit and vegetables a day.

Questions 249-251 relate to the following table regarding average alcohol consumption in 2010.

Country	Total	Recorded Consumption	Unrecorded consumption	Beer (%)	Wine (%)	Spirits (%)	Other (%)	2020 Consumption Projection
Belarus		14.4	3.2	17.3	5.2	46.6	30.9	17.1
Lithuania	15.4	12.9	2.5		7.8	34.1	11.6	16.2
Andorra	13.8		1.4	34.6		20.1	0	9.1
Grenada	12.5	11.9	0.7	29.3	4.3		0.2	10.4
Czech Republic	13	11.8	1.2	53.5	20.5	26	0	14.1
France	12.2	11.8		18.8	56.4	23.1	1.7	11.6
Russia		11.5	3.6	37.6	11.4	51	0	14.5
Ireland	11.9	11.4	0.5	48.1	26.1	18.7	7.7	10.9

NB: Some data is missing.

Question 249:

Which of the following countries had the highest total beer and wine consumption for 2010?

A. Belarus
B. Lithuania
C. Ireland
D. France
E. Andorra

Question 250:

Which country has the greatest difference for spirit consumption in 2010 and 2020 projection, assuming percentages stay the same?

A. Russia
B. Belarus
C. Lithuania
D. Grenada
E. Ireland

Question 251:

It was later found that some of the percentages of types of alcohol consumed had been mixed up. If the actual amount of beer consumed by each person in the Czech Republic was on average 4.9L, which country were the percentage figures mixed up with?

A. Lithuania
B. Grenada
C. Russia
D. France
E. Ireland
F. Belarus
G. Andorra

Questions 252-255 are based on the following information:

The table below shows the incidence of 6 different types of cancer in Australia:

	Prostate	Lung	Bowel	Bladder	Breast	Uterus
Men	40,000	25,000	20,000	8,000	1,000	0
Women	0	20,000	18,000	4,000	50,000	9,000

Question 252:

Supposing there are 10 million men and 10 million women in Australia, how many percentage points higher is the incidence of cancer amongst women than amongst men?

A. 0.007 %   B. 0.07 %   C. 0.093 %   D. 0.7 %   E. 0.93 %

Question 253:

Now suppose there are 11.5 million men and 10 million women in Australia. Assuming all men are equally likely to get each type of cancer and all women are equally likely to get each type of cancer, how many of the types of cancer are you more likely to develop if you are a man than if you are a woman?

A. 1   B. 2   C. 3   D. 4

Question 254:

Suppose that prostate, bladder and breast cancer patients visit hospital 1 time during the first month of 2015 and patients for all other cancers visit hospital 2 times during the first month of 2015. 10% of cancer patients in Australia are in Sydney, and patients in Sydney are not more or less likely to have certain types of cancer than other patients.

How many hospital visits are made by patients in Sydney with these 6 cancers during the first month of 2015?

A. 10,300   C. 19,500   E. 195,000
B. 18,400   D. 28,700   F. 287,000

Question 255:

Which of the graphs correctly represents the combined proportion of men versus women with bladder cancer?

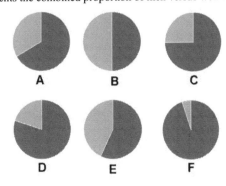

Questions 256 – 258 are based on the following information:

Units of alcohol are calculated by multiplying the alcohol percentage by the volume of liquid in litres, for example a 0.75 L bottle of wine which is 12% alcohol contains 9 units. 1 pint = 570 ml.

	Volume in bottle/barrel	Standard drinks per bottle/barrel	Percentage
Vodka	1250 ml	50	40%
Beer	10 pints	11.4	3%
Cocktail	750 ml	3	8%
Wine	750 ml	3.75	12.5%

Question 256:

Which standard drink has the most units of alcohol in?

A. Vodka   B. Beer   C. Cocktail   D. Wine

Question 257:

Some guidance suggests the recommended maximum number of units of alcohol per week for women is 14. In a week, Hannah drinks 4 standard drinks of wine, 3 standard drinks of beer, 2 standard cocktails and 5 standard vodkas. This guidance states the recommended maximum number of units per week for men is 21. In a week, Mark drinks 2 standard drinks of wine, 6 standard drinks of beer, 3 standard cocktails and 10 standard vodkas.

Who has exceeded their recommended maximum number of units by more and by how many units more have they exceeded it by than the other person?

A. Hannah, by 1 unit
B. Hannah, by 0.5 units
C. Both by the same
D. Mark, by 0.5 units
E. Mark, by 1 unit

Question 258:

How many different combinations of drinks that total 4 units are there (the same combination in a different order doesn't count).

A. 2   B. 3   C. 4   D. 5   E. 6

Questions 259-261 relate to the table below which shows information about Greentown's population:

	Female	Male	Total
Under 20	1,930		
20-39	1,960	3,760	5,720
40-59		4,130	
60 and over	2,350	2,250	4,600
Total	11,430	12,890	24,320

Question 259:

How many males under 20 are there in Greentown?

A. 2,650   B. 2,700   C. 2,730   D. 2,750   E. 2,850

Question 260:

How many females aged 40-59 are there in Greentown?

A. Between 3,000 and 4,000
B. Between 4,000 and 5,000
C. Between 5,000 and 6,000
D. Between 6,000 and 7,000

Question 261:

Which is the approximate ratio of females:males in the age group that has the highest ratio of males:females?

A. 1.4:1   B. 1.9:1   C. 1:1.9   D. 1:1.4

Questions 262-264 relate to the follow graph:

The graph below shows the average temperatures in London (top trace) and Newcastle (bottom trace).

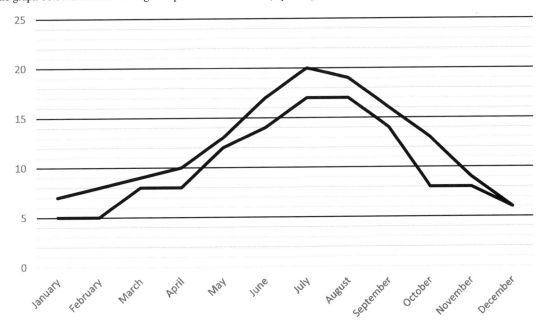

Question 262:

If the average monthly temperature is the same in every year, how many times during the period May 2007 to September 2013 inclusive is the average temperature the same in 2 consecutive months in Newcastle?

A. 20   B. 24   C. 25   D. 30

Question 263:

In how many months in the period specified in the previous question is the average temperature in London AND Newcastle lower than the previous month?

A. 19   B. 21   C. 25   D. 32

Question 264:

To the nearest 0.5 degrees Celsius, what is the average temperature difference between Newcastle and London?

A. 1.5°C   B. 2°C   C. 2.5 °C   D. 3 °C

Questions 265-267 concern the following data:

The pie chart to the right shows sales of ice cream across the four quarters of a year from January to December. Sales are lowest in the month of February. From February they increase in every subsequent month until they get to the maximum sales and from that point they decrease in every subsequent month until the end of the year.

Sales of ice cream

Question 265:

In which month are the sales highest?

A. June
B. July
C. August
D. Cannot tell

Question 266:

If total sales of ice cream were £354,720 for the year, how much of this was taken during Q1?

A. 29,480   B. £29,560   C. £29,650   D. £29,720   E. £29,800

Question 267:

Assuming total sales revenue (i.e. before costs are taken off) is £180,000, and that each tub of ice cream is sold for £2 and costs the manufacturer £1.50 in total production and transportation costs, how much profit is made during Q2?

A. £15,000   B. £30,000   C. £45,000   D. £60,000

Question 268:

Data on the amount families spend on food per month to the nearest £100 was collected for families with 1, 2 and 3 children. The percentage of families with different spending sizes is displayed below:

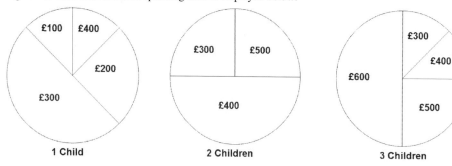

Which of the following statements is definitely true?

A. More families with 1 child than families with 2 children spent £300 a month on food.
B. The overall fraction of families spending £600 was 1/6.
C. All of the families with 2 children spent under £4000 on food per year.
D. The fraction of families with 1 child spending £400 on food per month is the same as the fraction of families with 3 children spending this amount.
E. The average amount spent on food by families with 2 children is £410 a month.

Questions 269-272 are based on the passage below:

A big secondary school recently realised that there were a large number of incidences of bullying occurring that were going unnoticed by teachers. It is possible that some believe bullying to be as much a part of student life as lessons and homework. In order to tackle the problem, the school emailed out a questionnaire to all students' parents and asked them to question their children about where they had experienced or seen bullying in school. Those children that answered yes were then asked if they had told their teachers about it, and asked why they did not if they had not. Those that had told their teacher were asked whether they had seen the teacher act upon the information and whether the bullying had stopped as a result.

Of the 2500 school students surveyed 2210 filled in the online questionnaire. The results were that, 1121 students, almost exactly half (50.7%) had seen bullying in school. Only 396 (35%) of these students told a teacher about the bullying. Of the students who told a teacher, 286 did not witness any action following sharing of the information and of those that did, 60% did not notice any direct action with the bully involved.

From those students who did not report the bullying, 146 gave the reason that they didn't think it was important. 427 cited fears of being found out. 212 students said they did not tell because they didn't think the teachers would do anything about it even if they did know. Assume that all the students who filled out the survey did so honestly.

Question 269:
To the nearest integer, what percentage of students did not respond?

A. 10%   B. 12%   C. 18%   D. 8%   E. 5%

Question 270:
If a student saw bullying occur and did not tell a teacher about it, what is the probability that the reasoning for this is that they thought it to be unimportant?

A. 0.1   B. 0.15   C. 0.2   D. 0.35   E. 0.13

Question 271:
After reporting the bullying, how many students saw the teacher act on the information directly with the bully?

A. 66
B. 44
C. 178
D. 104
E. 118

Question 272:
Which of the following does the questionnaire indicate is the best explanation for why students at the school did not report bullying?

A.   Students do not think bullying happens at their school.
B.   Students think the teachers will do nothing with the information.
C.   Students think that bullying is a part of school life.
D.   The student's were worried about others finding out.

Question 273:
The obesity epidemic is growing rapidly with reports of a three-fold rise in the period from 2007 to 2012. The rates of hospital admission have also been found to vary massively across different areas of England with the highest rates in the North-East (56 per 100,000 people), and the lowest rates in the East of England (12 per 100,000). During almost every year from 2001-12, there were around twice as many women admitted for obesity as men. The reason for this is however unclear and does not imply there are twice as many obese women as men.

What was the approximate number of admissions per 100 000 women in the North-East in 2011-12?

A. 18   B. 26   C. 37   D. 56   E. 62   F. 74

Question 274:

Health professionals are becoming increasingly worried by the decline in exercise being taken by both children and adults. Around only 40% of adults take the recommended amount of exercise which is 150 minutes per week. As well as falling rates of exercise, a shockingly low number of individuals eat five portions of fruit and vegetables a day. Figures for children aged 5-15 fell to only 16% for boys, and 20% for girls in 2011. Data for adults was only slightly better with 29% of women and 24% of men eating the recommended number of portions.

Using a figure of 8 million children between 5-15 years (equal ratio of girls to boys) in England in 2011, how many more girls than boys ate 5 portions of fruit and vegetables a day?

A. 80,000   B. 120,000   C. 160,000   D. 320,000   E. 640,000

Question 275:

The table below shows the leading causes of death in the UK.

	WOMEN		MEN	
Rank	**Cause of Death**	**Number of Deaths**	**Cause of Death**	**Number of Deaths**
1	Dementia and Alzheimer's	31,850	Coronary Heart Disease	37,797
2	Coronary Heart Disease	26,075	Lung Cancer	16,818
3	Stroke	20,706	Dementia and Alzheimer's	15,262
4	Flu and Pneumonia	15,361	Lower Respiratory Disease	15,021
5	Lower Respiratory Disease	14,927	Stroke	14,058
6	Lung Cancer	13,619	Flu and Pneumonia	11,426
7	Breast Cancer	10,144	Prostate Cancer	9,726
8	Colon Cancer	6,569	Colon Cancer	7,669
9	Urinary Infections	5,457	Lymphatic Cancer	6,311
10	Heart Failure	5,012	Liver Disease	4,661
	Total	**261,205**	**Total**	**245,585**

Using information from the table only, which of the following statements is correct?

A. More women died from cancers than men.
B. More than 30,000 women died due to respiratory causes.
C. Dementia and Alzheimer's is more common in women than men.
D. No cause of death is of the same ranking for both men and women.

Question 276 is based on the passage below:

The government has recently released a campaign leaflet saying that last year waiting times in NHS A&E departments decreased 20% compared to the year before. The opposition has criticised this statement, saying that there are several definitions which can be described as "waiting times", and the government's campaign leaflet does not make it clear what they mean by "waiting times in A&E".

The NHS watchdog has recently released the following figures describing different aspects of A&E departments, and the change from last year:

Assessment Criterion	2014	2013
Average time spent before being seen in A&E	1 hour	90 minutes
Average time between dialling 999 and receiving treatment in A&E	2 hours	3 hours
Number of people waiting for over 4 hours in A&E	3200	4000
Number of high-priority cases waiting longer than 1 hour	900	1000
Average waiting time for those seen in under 4 hours	50 minutes	40 minutes

Question 276:

Assuming these figures are correct, which criterion of assessment have the government described as "waiting times in A&E" on their campaign leaflet?

A. Number of people waiting for over 4 hours in A&E.
B. Number of people waiting for under 4 hours in A&E.
C. Number of high-priority cases waiting longer than 1 hour.
D. Average time spent before being seen in A&E.
E. Average time between dialling 999 and receiving treatment in A&E.
F. Average waiting time for those seen in less than 4 hours.

Questions 277– 279 refer to the following information:

The table below shows the final standings at the end of the season, after each team has played all the other teams twice each (once at home, once away). The teams are listed in order of how many points they got during the season. Teams get 3 points for a win, 1 point for a draw and 0 points for a loss. No team got the same number of points as another team. Some of the information in the table is missing.

Team	W	D	L
United	8	1	
Athletic	7		
City	7	2	
Town	1	4	
Rovers		0	9
Rangers		2	8

Question 277:

How many points did Rovers get?

A. 0
B. 3
C. 6
D. 9
E. More information needed.

Question 278:

How many games did Athletic lose?

A. 0
B. 1
C. 2
D. 3
E. More information needed.

Question 279:

How many more points did United get than Rangers?

A. 7
B. 15
C. 23
D. 25
E. More information needed.

Questions 280-282 use information from the graph recording A&E attendances and response times for NHS England from 2004 to 2014. Type 1 departments are major A&E units, type 2&3 are urgent care centres or minor injury units. The old target (2004 – June 2010) was 97.5%; the new target (July 2010 – 2015) is 95%.

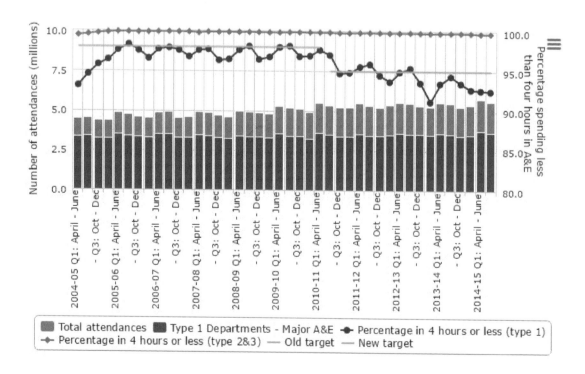

Question 280:

Which of the following statements is **FALSE**?

A.   There has been an overall increase in total A&E attendances from 2004-2014.

B.   The number of attendances in type 1 departments has been fairly constant from 2004-2014.

C.   The new target of 4 hours waiting time has only been reached in two quarters by type 1 departments.

D.   The change in attendances is largely due to an in increase people going to type 2&3 departments.

Question 281:

What percentage has the number of total attendances changed from Q1 2004-5 to Q1 2008-9?

A.   +5%

B.   −5%

C.   +10%

D.   −10%

E.   +15%

F.   −15%

Question 282:

If the new target was achieved by type 1 departments 4 times, in what percentage of the quarters was the target missed?

A.   25%

B.   60%

C.   75%

D.   90%

Questions 283-284 relate to the following data:

Ranjna is travelling from Manchester to Bali. She is required to make a stopover in Singapore for which he wants to allow at least 2 hours. It takes 14 hours to fly from Manchester to Singapore, and 2 hours from Singapore to Bali. The table below shows the departure times in local time [Manchester GMT, Singapore GMT + 8, Bali GMT + 8]:

Manchester to Singapore			Singapore to Bali			
Monday	Wednesday	Thursday	Monday	Tuesday	Wednesday	Thursday
08.00	09.30	02.30	13.00	00.00	15.30	13.00
10.45	14.00	08.30	15.30	07.30	18.00	16.00
13.30	18.00	12.30	21.00	08.30	20.30	19.00
15.00	20.00	19.00		12.00		

Question 283:

What is the latest flight Ranjna can take from Manchester to ensure she arrives at Bali Airport by Thursday 22:00?

A.   18:00 Tuesday

B.   14:00 Wednesday

C.   18:00 Wednesday

D.   20:00 Wednesday

E.   02:30 Thursday

F.   08:30 Thursday

Question 284:

Ranjna takes the 08:00 flight from Manchester to Singapore on Monday. She allows 1 hour to clear customs and collect her luggage at Bali Airport and another 45 minutes for the taxi to her hotel. At what time will she arrive at the hotel?

A. 16.45 Monday
B. 04:15 Tuesday
C. 10:30 Tuesday
D. 12:15 Tuesday
E. 12:30 Tuesday
F. 20:30 Tuesday

Question 285:

The graph below represents the percentage of adult smokers in the UK from 1974 to 2010. The top trace represents men and the bottom trace represents women. The middle trace is for both men and women.

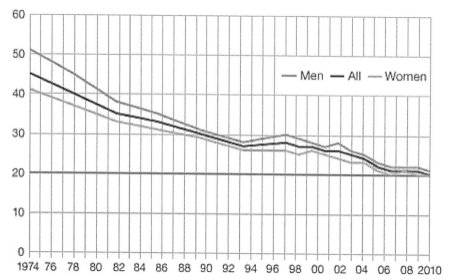

Which of the following statements can be concluded from the graph?

A. The 2007 smoking ban increased the rate in decline of smokers.
B. There has been a constant reduction in percentage of smoker since 1974.
C. The highest rate in decline in smoking for women was 2004-2006.
D. From 1974 to 2010, the smoking rate in men decreased by a half.
E. There has always been a significant difference between the smoking habits of men and women.

Question 286:

The name, age, height, weight and IQ of 11 people were recorded below in a table and a scatter plot. However, the axis labels were left out by mistake. Scale breaks are permitted.

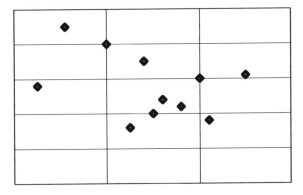

Name	Age	Height (cm)	Weight (kg)	IQ
Alice	18	180	68	110
Ben	12	160	79	120
Camilla	14	170	62	100
David	25	145	98	108
Eliza	29	165	75	96
Rohan	15	190	92	111
George	20	172	88	104
Hannah	22	168	68	115
Ian	13	182	86	98
James	17	176	90	102
Katie	27	151	66	125

Which variants are possible for the X and Y axis?

	X axis	Y axis
A	Height	Weight
B	IQ	Height
C	Age	IQ
D	Height	IQ
E	Height	Age
F	IQ	Weight

Question 287:

A group of students looked at natural variation in height and arm span within their group and got the following results:

Name	Arm span (cm)	Height (cm)
Adam	175	168
Tom	188	175
Shiv	172	184
Mary	148	142
Alice	165	156
Sarah	166	168
Emily	159	160
Matthew	165	172
Michael	185	183

They then drew a scatter plot, but forgot to include names for each point. They also forgot to plot one student.

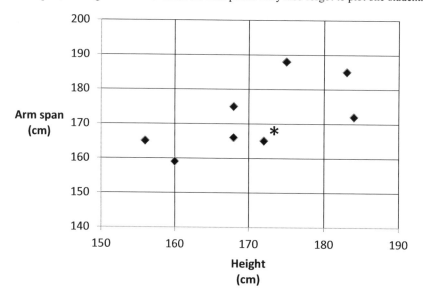

Which student is represented by the point marked with a *?

A. Alice
B. Sarah
C. Matthew
D. Adam
E. Emily
F. Michael

Questions 288 - 294 are based on the following information:

The rectangle represents women. The circle represents those that have children. The triangle represents those that work, and the square those that went to university.

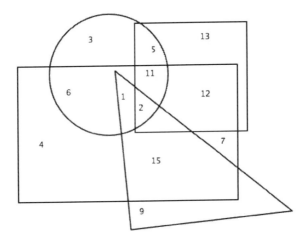

Question 288:

What is the number of non-working women who have children and who did not go to university?

A. 3   B. 5   C. 6   D. 7   E. 9

Question 289:

What is the total number of women who have children and work?

A. 1   B. 2   C. 3   D. 11   E. 14

Question 290:

How many women were surveyed in total?

A. 49   C. 58   E. 85

B. 51   D. 67   F. None of the above.

Question 291:

What is the number of people who went to university and had children?

A. 5   C. 13   E. 18

B. 11   D. 16   F. None of the above.

Question 292:

What is the total number of people who went to university, or have children but not both?

A. 18   C. 35   E. 53

B. 28   D. 41   F. None of the above.

Question 293:

The total number of men who went to university and had children was?

A. 3   B. 4   C. 5   D. 12   E. 13   F. 18

Question 294:

Which of the following people were not surveyed? Choose **TWO** options.

A. A non-working woman who went to university but did not have children.

B. A working man who went to university and has children.

C. A working woman who had children but did not go to university.

D. A non-working man who did not have children and did not go to university.

E. A working woman who went to university but did not have children.

Question 295:

Savers"R"Us is national chain of supermarkets. The price of several items in the supermarket is displayed below:

Item	Price
Beef roasting joint	£8.00
Chicken breast fillet	£6.00
Lamb shoulder	£7.00
Pork belly meat portion	£4.00
Sausages – 10 pack	£3.50

This week the supermarket has a sale on, with 50% off the normal price of all meat products. Alfred visits the supermarket during this sale and purchases a beef roasting joint, a 10 pack of sausages and a lamb shoulder, paying with a £20 note.

How much change does Alfred get?

A. £1.50

B. £5.00

C. £10.75

D. £11.75

E. £12.50

F. None of the above.

Question 296:

The local football league table is shown below, but the number of goals scored against Wilmslow is missing. Each team played the other teams in the league once at home and once away during the season.

Team Name	Points	Goals For	Goals Against
Sale	20	16	2
Wilmslow	16	11	?
Timperley	14	8	7
Altrincham	13	7	9
Mobberley	10	8	12
Hale	8	4	14

How many goals must Wilmslow have conceded?

A. 8

B. 9

C. 10

D. 11

E. 12

F. 14

Question 297:

The heights and weights of three women with BMI's 21, 22 and 23 were measured. If Julie and Lydia had different weights but the same height of 154 cm, and the weight of Emma, Lydia and Julie combined was 345 lbs, what was Emma's height?

		Weight (lbs)				
		100	105	110	115	120
	152	19	20	22	24	26
	154	18	19	21	23	25
	156	17	18	20	22	24
	158	15	17	19	21	23
Height (cm)	160	14	15	18	20	22
	162	13	14	17	19	21
	164	12	13	15	18	20
	166	11	12	14	17	19
	168	10	11	13	15	18
	170	9	10	12	14	17

A. 158 cm

B. 162 cm

C. 160 cm

D. 164 cm

E. 165 cm

Question 298:

The measurements for different types of fish appear below:

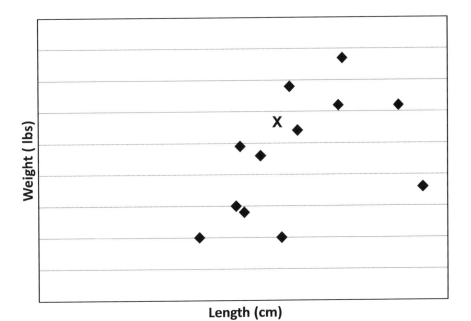

	Length (cm)	Weight (lbs)
Bluecup	78	40
Silverfinn	96	60
Starbug	98	98
Jawless	100	56
Lamprene	108	92
Scarfynne	118	40
Rayfish	122	136
Lobefin	126	108
Eringill	146	124
Whaler	148	154
Magic fish	176	124
Blondeye	188	72

Which fish is shown by the point marked **X**?

A. Silverfinn   B. Starbug   C. Lobefin   D. Blondeye   E. Eringill

The following graphs are required for questions 299-300:

The graph below shows the price of crude oil in US Dollars during 2014:

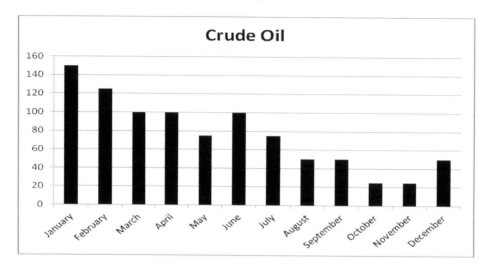

The graph below shows total oil production, in millions of barrels per day:

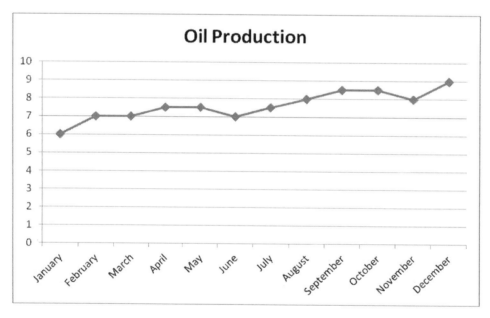

Question 299:

What was approximate total oil production in 2014?

A. 1,750 million barrels

B. 2,146 million barrels

C. 2,300 million barrels

D. 2,700 million barrels

E. 3,500 million barrels

Question 300:

How much did oil sales total in July 2014?

A. $0.56 Billion   B. $16.9 Billion   C. $17.4 Billion   D. $21.1 Billion

SECTION 2

Section 2 is undoubtedly the most time-pressured section of the BMAT. This section tests GCSE biology, chemistry, physics and maths. You have to answer 27 questions in 30 minutes. The questions can be quite difficult and it's easy to get bogged down. However, it's also the section in which you can improve the most quickly in so it's well worth spending time on it.

Although the vast majority of questions in section 2 aren't particularly difficult, the intense time pressure of having to do one question every minute makes this section the hardest in the BMAT. As with section 1, the trick is to identify and do the easy questions whilst leaving the hard ones for the end.

In general, the biology and chemistry questions in the BMAT require the least amount of time per question whilst the maths and physics are more time-draining as they usually consist of multi-step calculations.

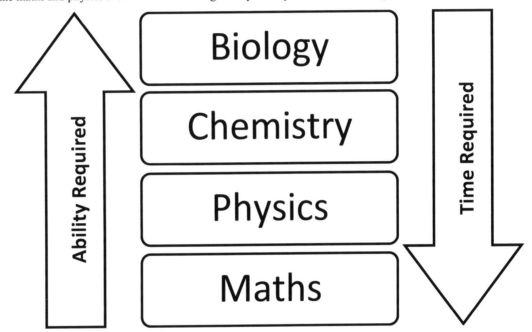

Gaps in Knowledge

The BMAT only tests GCSE level knowledge. However, there is a large variation in content between the GCSE exam boards meaning that you may not have covered some topics that are examinable. This is more likely if you didn't carry on with Biology or physics to AS level (e.g. Newtonian mechanics and parallel circuits in physics; hormones and stem cells in biology). If you fall into this category, you are highly advised to go through the BMAT Specification and ensure that you have covered all examinable topics. An electronic copy of this can be obtained from the official BMAT website at www.admissionstestingservice.org/bmat.

The questions in this book will help highlight any particular areas of weakness or gaps in your knowledge that you may have. Upon discovering these, make sure you take some time to revise these topics before carrying on – there is little to be gained by attempting section 2 questions with huge gaps in your knowledge.

Maths

Being confident with maths is extremely important for section 2. Many students find that improving their numerical and algebraic skills usually results in big improvements in their section 1 and 2 scores. Remember that maths in section 2 not only comes up in the maths question but also in physics (manipulating equations and standard form) and chemistry (mass calculations). So if you find yourself consistently running out of time in section 2, spending a few hours on brushing up your basic maths skills may do wonders for you.

SECTION 2: Biology

Thankfully, the biology questions tend to be fairly straightforward and require the least amount of time. You should be able to do the majority of these within the 60 second limit (often far less). This means that you should be aiming to make up time in these questions. In the majority of cases – you'll either know the answer or not i.e. they test advanced recall so the trick is to ensure that there are no obvious gaps in your knowledge.

Before going onto to do the practice questions in this book, ensure you are comfortable with the following commonly tested topics:

➢ Structure of animal, plant and bacterial cells

➢ Osmosis, Diffusion and Active Transport

➢ Cell Division (mitosis + meiosis)

➢ Family pedigrees and Inheritance

➢ DNA structure and replication

➢ Gene Technology & Stem Cells

➢ Enzymes – Function, mechanism and examples of digestive enzymes

➢ Aerobic and Anaerobic Respiration

➢ The central vs. peripheral nervous system

➢ The respiratory cycle including movement of ribs and diaphragm

➢ The Cardiac Cycle

➢ Hormones

➢ Basic immunology

➢ Food chains and food webs

➢ The carbon and nitrogen cycles

Top tip! If you find yourself getting less than 50% of biology questions correct in this book, make sure you revisit the syllabus before attempting more questions as this is the best way to maximise your efficiency. In general, there is no reason why you shouldn't be able to get the vast majority of biology questions correct (and in well under 60 seconds) with sufficient practice.

Biology Questions

Question 301:

In relation to the human genome, which of the following are correct?

1. The DNA genome is coded by 4 different bases.
2. The sugar backbone of the DNA strand is formed of glucose.
3. DNA is found in the nucleus of bacteria.

A. 1 only	C. 3 only	E. 1 and 3	G. 1, 2 and 3
B. 2 only	D. 1 and 2	F. 2 and 3	

Question 302:

Animal cells contain organelles that take part in vital processes. Which of the following is true?

1. The majority of energy production by animal cells occurs in the mitochondria.
2. The cell wall protects the animal cell membrane from outside pressure differences.
3. The endoplasmic reticulum plays a role in protein synthesis.

A. 1 only	C. 3 only	E. 2 and 3	G. 1, 2 and 3
B. 2 only	D. 1 and 2	F. 1 and 3	

Question 303:

With regards to animal mitochondria, which of the following is correct?

A. Mitochondria are not necessary for aerobic respiration.
B. Mitochondria are the sole cause of sperm cell movement.
C. The majority of DNA replication happens inside mitochondria.
D. Mitochondria are more abundant in fat cells than in skeletal muscle.
E. The majority of protein synthesis occurs in mitochondria.
F. Mitochondria are enveloped by a double membrane.

Question 304:

In relation to bacteria, which of the following is **FALSE**?

A. Bacteria always lead to disease.
B. Bacteria contain plasmid DNA.
C. Bacteria do not contain mitochondria.
D. Bacteria have a cell wall and a plasma membrane.
E. Some bacteria are susceptible to antibiotics.

Question 305:

In relation to bacterial replication, which of the following is correct?

A. Bacteria undergo sexual reproduction.
B. Bacteria have a nucleus.
C. Bacteria carry genetic information on circular plasmids.
D. Bacterial genomes are formed of RNA instead of DNA.
E. Bacteria require gametes to replicate.

Question 306

Which of the following are correct regarding active transport?

A. ATP is necessary and sufficient for active transport.
B. ATP is not necessary but sufficient for active transport.
C. The relative concentrations of the material being transported have little impact on the rate of active transport.
D. Transport proteins are necessary and sufficient for active transport.
E. Active transport relies on transport proteins that are powered by an electrochemical gradient.

Question 307:

Concerning mammalian reproduction, which of the following is **FALSE**?

A. Fertilisation involves the fusion of two gametes.
B. Reproduction is sexual and the offspring display genetic variation.
C. Reproduction relies upon the exchange of genetic material.
D. Mammalian gametes are diploid cells produced via meiosis.
E. Embryonic growth requires carefully controlled mitosis.

Question 308:

Which of the following apply to Mendelian inheritance?

1. It only applies to plants.
2. It treats different traits as either dominant or recessive.
3. Heterozygotes have a 25% chance of expressing a recessive trait.

A. 1 only		C. 3 only		E. 1 and 3		G. All of the above
B. 2 only		D. 1 and 2		F. 2 and 3		

Question 309:

Which of the following statements are correct?

A. Hormones are secreted into the blood stream and act over long distances at specific target organs.
B. Hormones are substances that almost always cause muscles to contract.
C. Hormones have no impact on the nervous or enteric systems.
D. Hormones are always derived from food and never synthesised.
E. Hormones act rapidly to restore homeostasis.

Question 310:

With regard to neuronal signalling in the body, which of the following are true?

1. Neuronal transmission can be caused by both electrical and chemical stimulation.
2. Synapses ultimately result in the production of an electrical current for signal transduction.
3. All synapses in humans are electrical and unidirectional.

A. 1 only		C. 3 only		E. 1 and 3		G. 1, 2 and 3
B. 2 only		D. 1 and 2		F. 2 and 3		

Question 311:

What is the **primary** reason that pH is controlled so tightly in humans?

A. To allow rapid protein synthesis.
B. To allow for effective digestion throughout the GI tract.
C. To ensure ions can function properly in neural signalling.
D. To prevent changes in electrical charge in polypeptide chains.
E. To prevent changes in core body temperature.

Question 312:
Which of the following statements are correct regarding bacterial cell walls?

1. It confers bacteria protection against external environmental stimuli.
2. It is an evolutionary remnant and now has little functional significance in most bacteria.
3. It is made up primarily of glucose in bacteria.

A. Only 1   C. Only 3   E. 2 and 3   G. 1, 2 and 3
B. Only 2   D. 1 and 2   F. 1 and 3

Question 313:
Which of the following statements are correct regarding mitosis?

1. It is important in sexual reproduction.
2. A single round of mitosis results in the formation of 2 genetically distinct daughter cells.
3. Mitosis is vital for tissue growth, as it is the basis for cell multiplication.

A. Only 1   C. Only 3   E. 2 and 3   G. 1, 2 and 3
B. Only 2   D. 1 and 2   F. 1 and 3

Question 314:
Which of the following is the best definition of a mutation?

A. A mutation is a permanent change in DNA.
B. A mutation is a permanent change in DNA that is harmful to an organism.
C. A mutation is a permanent change in the structure of intra-cellular organelles caused by changes in DNA/RNA.
D. A mutation is a permanent change in chromosomal structure caused by DNA/RNA changes.

Question 315:
In relation to mutations, which of the following are correct?

1. Mutations always lead to discernible changes in the phenotype of an organism.
2. Mutations are central to natural processes such as evolution.
3. Mutations play a role in cancer.

A. Only 1   C. Only 3   E. 2 and 3   G. 1, 2 and 3
B. Only 2   D. 1 and 2   F. 1 and 3

Question 316:
Which of the following is the most accurate definition of an antibody?

A. An antibody is a molecule that protects red blood cells from changes in pH.
B. An antibody is a molecule produced only by humans and has a pivotal role in the immune system.
C. An antibody is a toxin produced by a pathogen to damage the host organism.
D. An antibody is a molecule that is used by the immune system to identify and neutralize foreign objects and molecules.
E. Antibodies are small proteins found in red blood cells that help increase oxygen carriage.

Question 317:
Which of the following statements about the kidney are correct?

1. The kidneys filter the blood and remove waste products from the body.
2. The kidneys are involved in the digestion of food.
3. In a healthy individual, the kidneys produce urine that contains high levels of glucose.

A. Only 1   C. Only 3   E. 2 and 3   G. 1, 2 and 3
B. Only 2   D. 1 and 2   F. 1 and 3

Question 318:

Which of the following statements are correct?

1. Hormones are slower acting than nerves.
2. Hormones act for a very short time.
3. Hormones act more generally than nerves.
4. Hormones are released when you get a scare.

A. 1 only
B. 1 and 3 only

C. 2 and 4 only
D. 1, 3 and 4 only

E. 1, 2, 3 and 4

Question 319:

Which statements about homeostasis are correct?

1. Homeostasis is about ensuring the inputs within your body exceed the outputs to maintain a constant internal environment.
2. Homeostasis is about ensuring the inputs within your body are less than the outputs to maintain a constant internal environment.
3. Homeostasis is about balancing the inputs within your body with the outputs to ensure your body fluctuates with the needs of the external environment.
4. Homeostasis is about balancing the inputs within your body with the outputs to maintain a constant internal environment.

A. 1 only
B. 2 only

C. 3 only
D. 4 only

E. 1 and 3 only
F. 2 and 4 only

G. 2 and 3 only

Question 320:

Which of the following statement is true?

A. There is more energy and biomass each time you move up a trophic level.
B. There is less energy and biomass each time you move up a trophic level.
C. There is more energy but less biomass each time you move up a trophic level.
D. There is less energy but more biomass each time you move up a trophic level.
E. There is no difference in the energy or biomass when you move up a trophic level.

Question 321:

Which of the following statements are true about asexual reproduction?

1. There is no fusion of gametes.
2. There are two parents.
3. There is no mixing of chromosomes.
4. There is genetic variation.

A. 1 and 3 only
B. 1 and 4 only

C. 2 and 3 only
D. 3 and 4 only

E. 2 and 4 only
F. 1, 2, 3 and 4

Question 322:

Put the following in the order which they occur when Jonas sees a bowl of chicken and moves towards it.

1. Retina
2. Motor neuron

3. Sensory neuron
4. Brain

5. Muscle

A. 1 - 3 - 4 - 5 - 2
B. 1 - 2 - 3 - 4 - 5
C. 5 - 1 - 3 - 2 - 4

D. 1 - 3 - 2 - 4 - 5
E. 1 - 3 - 4 - 2 - 5
F. 4 - 1 - 3 - 2 - 5

Question 323:
What path does blood take from the kidney to the liver?

1. Pulmonary artery
2. Inferior vena cava

3. Hepatic artery
4. Aorta

5. Pulmonary vein
6. Renal vein

A. 2 - 1 - 4 - 3 - 5 - 6
B. 1 - 2 - 3 - 4 - 5 - 6
C. 6 - 2 - 5 - 1 - 4 - 3

D. 6 - 2 - 1 - 5 - 4 - 3
E. 3 - 2 - 1 - 4 - 6 - 5
F. 3 - 6 - 2 - 4 - 1 – 5

Question 324:
Which of the following statements are true about animal cloning?

1. Animals cloned from embryo transplants are genetically identical.
2. The genetic material is removed from an unfertilised egg during adult cell cloning.
3. Cloning can cause a reduced gene pool.
4. Cloning is only possible with mammals.

A. 1 only
B. 2 only

C. 3 only
D. 4 only

E. 1 and 2 only
F. 1, 2 and 3 only

G. 1, 2, 3 and 4

Question 325:
Which of the following statements are true with regard to evolution?

1. Individuals within a species show variation because of differences in their genes.
2. Beneficial mutations will accumulate within a population.
3. Gene differences are caused by sexual reproduction and mutations.
4. Species with similar characteristics never have similar genes.

A. 1 only
B. 1 and 4 only

C. 2 and 3 only
D. 2 and 4 only

E. 3 and 4 only
F. 1, 2 and 3 only

Question 326:
Which of the following genetic statements are correct?

1. Alleles are a similar version of different cells.
2. If you are homozygous for a trait, you have three alleles the same for that particular gene.
3. If you are heterozygous for a trait, you have two different alleles for that particular gene.
4. To show the characteristic that is caused by a recessive allele, both carried alleles for the gene have to be recessive.

A. 1 only
B. 2 only

C. 3 only
D. 4 only

E. 1 and 2 only
F. 3 and 4 only

G. 1, 2, and 3 only

Question 327:
Which of the following statements are correct about meiosis?

1. The DNA content of a gamete is half that of a human red blood cell.
2. Meiosis requires ATP.
3. Meiosis only takes place in reproductive tissue.
4. In meiosis, a diploid cell divides in such a way so as to produce two haploid cells.

A. 1 only
B. 3 only

C. 1 and 2 only
D. 2 and 3 only

E. 2 and 4 only
F. 1, 2, 3 and 4

Question 328:

Put the following statements in the correct order of events for when there is too little water in the blood.

1. Urine is more concentrated
2. Pituary gland releases ADH
3. Blood water level returns to normal
4. Hypothalamus detects too little water in blood
5. Kidney affects water level

A. 1 - 2 - 3 - 4 - 5   C. 4 - 2 - 5 - 1 - 3   E. 5 - 2 - 3 - 4 - 1
B. 5 - 4 - 3 - 2 - 1   D. 3 - 2 - 4 - 1 - 5   F. 4 - 2 – 1- 5 - 3

Question 329:

The pH of venous blood is 7.35. Which of the following is the likely pH of arterial blood?

A. 4.4   C. 6.5   E. 7.4
B. 5.2   D. 7.0   F. 7.95

Question 330:

Which of the following are true of the cytoplasm?

1. The vast majority of the cytoplasm is made up of water.
2. All contents of animal cells are contained in the cytoplasm.
3. The cytoplasm contains electrolytes and proteins.

A. 1 only   C. 3 only   E. 1 and 3 only
B. 2 only   D. 1 and 2 only   F. 1, 2 and 3

Question 331:

ATP is produced in which of the following organelles?

1. The golgi apparatus
2. The rough endoplasmic reticulum
3. The mitochondria
4. The nucleus

A. 1 only   C. 3 only   E. 1 and 2   G. 3 and 4 only
B. 2 only   D. 4 only   F. 2 and 3 only   H. 1, 2, 3 and 4

Question 332:

The cell membrane:

A. Is made up of a phospholipid bilayer which only allows active transport across it.
B. Is not found in bacteria.
C. Is a semi-permeable barrier to ions and organic molecules.
D. Consists purely of enzymes.

Question 333:

Cells of the *Polyommatus atlantica* butterfly of the Lycaenidae family have 446 chromosomes. Which of the following statements about a *P. atlantica* butterfly are correct?

1. Mitosis will produce 2 daughter cells each with 223 pairs of chromosomes
2. Meiosis will produce 4 daughter cells each with 223 chromosomes
3. Mitosis will produce 4 daughter cells each with 446 chromosomes
4. Meiosis will produce 2 daughter cells each with 223 pairs of chromosomes

A. 1 and 2 only	C. 2 and 3 only	E. 1, 2 and 3 only
B. 1 and 3 only	D. 3 and 4 only	F. 1, 2, 3 and 4

Questions 334-336 are based on the following information:

Assume that hair colour is determined by a single allele. The R allele is dominant and results in black hair. The r allele is recessive for red hair. Mary (red hair) and Bob (black hair) are having a baby girl.

Question 334:

What is the probability that she will have red hair?

A. 0% only	C. 50% only	E. 0% or 50%
B. 25% only	D. 0% or 25%	F. 25% or 50%

Question 335:

Mary and Bob have a second child, Tim, who is born with red hair. What does this confirm about Bob?

A. Bob is heterozygous for the hair allele.
B. Bob is homozygous dominant for the hair allele.
C. Bob is homozygous recessive for the hair allele.
D. Bob does not have the hair allele.

Question 336:

Mary and Bob go on to have a third child. What are the chances that this child will be born homozygous for black hair?

A. 0%	B. 25%	C. 50%	D. 75%	E. 100%

Question 337:

Why does air flow into the chest on inspiration?

1. Atmospheric pressure is smaller than intra-thoracic pressure during inspiration.
2. Atmospheric pressure is greater than intra-thoracic pressure during inspiration.
3. Anterior and lateral chest expansion decreases absolute intra-thoracic pressure.
4. Anterior and lateral chest expansion increases absolute intra-thoracic pressure.

A. 1 only	C. 2 and 3	E. 1 and 3
B. 2 only	D. 1 and 4	F. 2 and 4

Question 338:

Which of the following components of a food chain represent the largest biomass?

A. Producers	D. Secondary consumers
B. Decomposers	E. Tertiary consumers
C. Primary consumers	

~ 101 ~

Question 339:
Concerning the nitrogen cycle, which of the following are true?

1. The majority of the Earth's atmosphere is nitrogen.
2. Most of the nitrogen in the Earth's atmosphere is inert.
3. Bacteria are essential for nitrogen fixation.
4. Nitrogen fixation occurs during lightning strikes.

A. 1 and 2   C. 2 and 3   E. 3 and 4
B. 1 and 3   D. 2 and 4   F. 1, 2, 3 and 4

Question 340:
Which of the following statement are correct regarding mutations?

1. Mutations always cause proteins to lose their function.
2. Mutations always change the structure of the protein encoded by the affected gene.
3. Mutations always result in cancer.

A. Only 1   D. 1 and 2   G. 1, 2 and 3
B. Only 2   E. 2 and 3   H. None of the above
C. Only 3   F. 1 and 3

Question 341:
Which of the following is not a function of the central nervous system?

A. Coordination of movement   D. Cognition
B. Decision making and executive functions   E. Memory
C. Control of heart rate

Question 342:
Which of the following control mechanisms are involved in modulating cardiac output?

1. Voluntary control.
2. Sympathetic control to decrease heart rate.
3. Parasympathetic control to increase heart rate.

A. Only 1   D. 1 and 2   G. 1, 2 and 3
B. Only 2   E. 2 and 3   H. None of the above
C. Only 3   F. 1 and 3

Question 343:
Vijay goes to see his GP with fatty, smelly stools that float on water. Which of the following enzymes is most likely to be malfunctioning?

A. Amylase   D. Sucrase
B. Lipase   E. Lactase
C. Protease

Question 344:
Which of the following statements concerning the cardiovascular system is correct?

A. Oxygenated blood from the lungs flows to the heart via the pulmonary artery.
B. All arteries carry oxygenated blood.
C. All animals have a double circulatory system.
D. The superior vena cava contains oxygenated blood
E. All veins have valves.
F. None of the above.

Question 345:
Which part of the GI tract has the least amount of enzymatic digestion occurring?

A. Mouth
B. Stomach
C. Small intestine

D. Large intestine
E. Rectum

Question 346:
Oge touches a hot stove and immediately moves her hand away. Which of the following components are **NOT** involved in this reaction?

1. Thermo-receptor
2. Brain

3. Spinal Cord
4. Sensory nerve

5. Motor nerve
6. Muscle

A. 1 only
B. 2 only

C. 3 only
D. 1 and 2 only

E. 1, 2 and 3 only
F. 3, 4, 5 and 6

Question 347:
Which of the following represents a scenario with an appropriate description of the mode of transport?

1. Water moving from a hypotonic solution outside of a potato cell, across the cell wall and cell membrane and into the hypertonic cytoplasm of the potato cell→ Osmosis.
2. Carbon dioxide moving across a respiring cell's membrane and dissolving in blood plasma →Active transport.
3. Reabsorption of amino acids against a concentration gradient in the glomeruluar apparatus → Diffusion.

A. 1 only
B. 2 only

C. 3 only
D. 1 and 2 only

E. 2 and 3 only
F. 1 and 3 only

G. 1, 2 and 3

Question 348:
Which of the following equations represents anaerobic respiration?

1. Carbohydrate + Oxygen → Energy + Carbon Dioxide + Water
2. Carbohydrate → Energy + Lactic Acid + Carbon dioxide
3. Carbohydrate → Energy + Lactic Acid
4. Carbohydrate → Energy + Ethanol + Carbon dioxide

A. 1 only
B. 2 only
C. 3 only

D. 4 only
E. 1 and 2
F. 1 and 3

G. 1 and 4
H. 2 and 4 only
I. 3 and 4 only

Question 349:
Which of the following statements regarding respiration are correct?

1. The mitochondria are the centres for both aerobic and anaerobic respiration.
2. The cytoplasm is the main site of anaerobic respiration.
3. For every two moles of glucose that is respired aerobically, 12 moles of CO_2 are liberated.
4. Anaerobic respiration is more efficient than aerobic respiration.

A. 1 and 2
B. 1 and 4
C. 2 and 3

D. 2 and 4
E. 3 and 4

Question 350:
Which of the following statements are true?

1. The nucleus contains the cell's chromosomes.
2. The cytoplasm consists purely of water.
3. The plasma membrane is a single phospholipid layer.
4. The cell wall prevents plants cells from lysing due to osmotic pressure.

A. 1 and 2
B. 1 and 4
C. 1, 3 and 4
D. 1, 2 and 3
E. 1, 2 and 4
F. 2, 3 and 4

Question 351:
Which of the following statements are true about osmosis?

1. If a medium is hypertonic relative to the cell cytoplasm, the cell will gain water through osmosis.
2. If a medium is hypotonic relative to the cell cytoplasm, the cell will gain water through osmosis.
3. If a medium is hypotonic relative to the cell cytoplasm, the cell will lose water through osmosis.
4. If a medium is hypertonic relative to the cell cytoplasm, the cell will lose water through osmosis.
5. The medium's tonicity has no impact on the movement of water.

A. 1 only
B. 2 only
C. 1 and 3
D. 2 and 4
E. 5 only

Question 352:
Which of the following statements are true about stem cells?

1. Stem cells have the ability to differentiate into other mature types of cells.
2. Stem cells are unable to maintain their undifferentiated state.
3. Stem cells can be classified as embryonic stem cells or adult stem cells.
4. Stem cells are only found in embryos.

A. 1 and 3
B. 3 and 4
C. 2 and 3
D. 1 and 2
E. 2 and 4

Question 353:
Which of the following are **NOT** examples of natural selection?

1. Giraffes growing longer necks to eat taller plants.
2. Antibiotic resistance developed by certain strains of bacteria.
3. Pesticide resistance among locusts in farms.
4. Breeding of horses to make them run faster.

A. 1 only
B. 4 only
C. 1 and 3
D. 1 and 4
E. 2 and 4

Question 354:
Which of the following statements are true?

1. Enzymes stabilise the transition state and therefore lower the activation energy.
2. Enzymes distort substrates in order to lower activation energy.
3. Enzymes decrease temperature to slow down reactions and lower the activation energy.
4. Enzymes provide alternative pathways for reactions to occur.

A. 1 only
B. 1 and 2
C. 1 and 4
D. 2 and 4
E. 3 and 4

Question 355:
Which of the following are examples of negative feedback?

1. Salivating whilst waiting for a meal.
2. Throwing a dart.
3. The regulation of blood pH.
4. The regulation of blood pressure.

A. 1 only
B. 1 and 2
C. 3 and 4
D. 2, 3, and 4
E. 1, 2, 3 and

Question 356:
Which of the following statements about the immune system are true?

1. White blood cells defend against bacterial and fungal infections.
2. White blood cells can temporarily disable but not kill pathogens.
3. White blood cells use antibodies to fight pathogens.
4. Antibodies are produced by bone marrow stem cells.

A. 1 and 3
B. 1 and 4
C. 2 and 3
D. 2 and 4
E. 1, 2, and 3
F. 1, 3, and 4

Question 357:
The cardiovascular system does **NOT**:

A. Deliver vital nutrients to peripheral cells.
B. Oxygenate blood and transports it to peripheral cells.
C. Act as a mode of transportation for hormones to reach their target organ.
D. Facilitate thermoregulation.
E. Respond to exercise by increasing cardiac output to exercising muscles.

Question 358:
Which of the following statements is correct?

A. Adrenaline can sometimes decrease heart rate.
B. Adrenaline is rarely released during flight or fight responses.
C. Adrenaline causes peripheral vasoconstriction.
D. Adrenaline only affects the cardiovascular system.
E. Adrenaline travels primarily in lymphatic vessels.
F. None of the above.

Question 359:
Which of the following statements is true?

A. Protein synthesis occurs solely in the nucleus.
B. Each amino acid is coded for by three DNA bases.
C. Each protein is coded for by three amino acids.
D. Red blood cells can create new proteins to prolong their lifespan.
E. Protein synthesis isn't necessary for mitosis to take place.
F. None of the above.

Question 360:
A solution of amylase and carbohydrate is present in a beaker, where the pH of the contents is 6.3. Assuming amylase is saturated, which of the following will increase the rate of production of the product?

1. Add sodium bicarbonate
2. Add carbohydrate
3. Add amylase
4. Increase the temperature to 100° C

A. 1 only
B. 2 only
C. 3 only
D. 4 only
E. 1 and 2
F. 1 and 3
G. 1, 2 and 3
H. 1, 3 and 4

Question 361:
Celestial Necrosis is a newly discovered autosomal recessive disorder. A female carrier and a male with the disease produce two boys. What is the probability that neither boy's genotype contains the celestial necrosis allele?

A. 100%   B. 75%   C. 50%   D. 25%   E. 0%

Question 362:
Which among the following has no endocrine function?

A. The thyroid   C. The pancreas   E. The testes
B. The ovary   D. The adrenal gland   F. None of the above.

Question 363:
Which of the following statements are true?

1. Increasing levels of insulin cause a decrease in blood glucose levels.
2. Increasing levels of glycogen cause an increase in blood glucose levels.
3. Increasing levels of adrenaline decrease the heart rate.

A. 1 only   C. 3 only   E. 2 and 3   G. 1, 2 and 3
B. 2 only   D. 1 and 2   F. 1 and 3

Question 364:
Which of the following rows is correct?

	Oxygenated Blood		Deoxygenated Blood	
A.	Left atrium	Left ventricle	Right atrium	Right ventricle
B.	Left atrium	Right atrium	Left ventricle	Right ventricle
C.	Left atrium	Right ventricle	Right atrium	Right ventricle
D.	Right atrium	Right ventricle	Left atrium	Left ventricle
E.	Left ventricle	Right atrium	Left atrium	Right ventricle

Questions 365-367 are based on the following information:
The pedigree below shows the inheritance of a newly discovered disease that affects connective tissue called Nafram syndrome. Individual 1 is a normal homozygote.

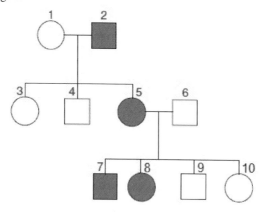

Question 365:

What is the inheritance of Nafram syndrome?

A.   Autosomal dominant
B.   Autosomal recessive
C.   X-linked dominant

D.   X-linked recessive
E.   Co-dominant

Question 366:

Which individuals must be heterozygous for Nafram syndrome?

A.   1 and 2
B.   8 and 9

C.   2 and 5
D.   5 and 6

E.   6 and 8
F.   6 and 10

Question 367:

Taking N to denote a diseased allele and n to denote a normal allele, which of the following are **NOT** possible genotypes for 6's parents?

1.   NN x NN
2.   NN x Nn
3.   Nn x nn
4.   Nn x Nn
5.   nn x nn

A.   1 and 2
B.   1 and 3

C.   2 and 3
D.   2 and 5

E.   3 and 4
F.   4 and 5

Question 368:

Which of the following correctly describes the passage of urine through the body?

	1st	2nd	3rd	4th
A	Kidney	Ureter	Bladder	Urethra
B	Kidney	Urethra	Bladder	Ureter
C	Urethra	Bladder	Ureter	Kidney
D	Ureter	Kidney	Bladder	Urethra

Question 369:

Which of the following best describes the passage of blood from the body, through the heart, back to the body?

A.   Aorta → Left Ventricle → Left Atrium → Inferior Vena Cava → Right Atrium → Right Ventricle → Lungs → Aorta
B.   Inferior vena cava → Left Atrium → Left Ventricle → Lungs → Right Atrium → Right Ventricle → Aorta
C.   Inferior vena cava → Right Ventricle → Right Atrium → Lungs → Left Atrium → Left Ventricle → Aorta
D.   Aorta → Left Atrium → Left Ventricle → Lungs → Right Atrium → Right Ventricle → Inferior Vena Cava
E.   Right Atrium → Left Atrium → Inferior vena cava → Lungs → Left Atrium → Right Ventricle → Aorta
F.   None of the above.

Question 370:

Which of the following best describes the events during inspiration?

	Intrathoracic Pressure	Intercostal Muscles	Diaphragm
A	Increases	Contract	Contracts
B	Increases	Relax	Contracts
C	Increases	Contract	Relaxes
D	Increases	Relax	Relaxes
E	Decreases	Contract	Contracts
F	Decreases	Relax	Contracts
G	Decreases	Contract	Relaxes
H	Decreases	Relax	Relaxes

Questions 371-372 are based on the following information:

DNA is made up of the four nucleotide bases: adenine, cytosine, guanine and thymine. A triplet repeat or codon is a sequence of three nucleotides which code for an amino acid. While there are only 20 amino acids there are 64 different combinations of the four DNA nucleotide bases. This means that more than one combination of 3 DNA nucleotides sequences code for the same amino acid.

Question 371:

Which property of the DNA code is described above?

A. The code is unambiguous.

B. The code is universal.

C. The code is non-overlapping.

D. The code is degenerate.

E. The code is preserved.

F. The code has no punctuation.

Question 372:

Which type of mutation does the described property protect against the most?

A. An insertion - where a single nucleotide is inserted.

B. A point mutation - where a single nucleotide is replaced for another.

C. A deletion - where a single nucleotide is deleted.

D. A repeat expansion - where a repeated trinucleotide sequence is added.

E. A duplication - where a piece of DNA is abnormally copied.

Question 373:

Which row of the table below describes what happens when external temperature decreases?

	Temperature Change Detected by	Sweat Gland Secretion	Cutaneous Blood Flow
A	Hypothalamus	Increases	Increases
B	Hypothalamus	Increases	Decreases
C	Hypothalamus	Decreases	Increases
D	Hypothalamus	Decreases	Decreases
E	Cerebral Cortex	Increases	Increases
F	Cerebral Cortex	Increases	Decreases
G	Cerebral Cortex	Decreases	Increases
H	Cerebral Cortex	Decreases	Decreases

Question 374:

Which of the following processes involve active transport?

1. Reabsorption of glucose in the kidney.
2. Movement of carbon dioxide into the alveoli in the lungs.
3. Movement of chemicals in a neural synapse.

A. 1 only

B. 2 only

C. 3 only

D. 1 and 2

E. 1 and 3

F. 2 and 3

G. 1, 2 and 3

Question 375:

Which of the following statements is correct about enzymes?

A. All enzymes are made up of amino acids only.

B. Enzymes can sometimes slow the rate of reactions.

C. Enzymes have no impact on reaction temperatures.

D. Enzymes are heat sensitive but resistant to changes in pH.

E. Enzymes are unspecific in their substrate use.

F. None of the above.

SECTION 2: Chemistry

Most students don't struggle with BMAT chemistry as they'll be studying it at A2. However, there are certain questions that even good students tend to struggle with under time pressure e.g. balancing equations and mass calculations. It is essential that you're able to do these quickly as they take up by far the most time in the chemistry questions.

Balancing Equations

For some reason, most students are rarely shown how to formally balance equations – including those studying it at A-level. Balancing equations intuitively or via trial and error will only get you so far in the BMAT as the equations you'll have to work with will be fairly complex. To avoid wasting valuable time, it is essential you learn a method that will allow you to solve these in less than 60 seconds on a consistent basis. The method shown below is the simplest way and requires you to be able to do quick mental arithmetic (which is something you should be aiming for anyway). The easiest way to do learn it is through an example:

The following equation shows the reaction between Iodic acid, hydrochloric acid and copper Iodide:

$$\textbf{a}\ HIO_3 + \textbf{b}\ CuI_2 + \textbf{c}\ HCl \rightarrow \textbf{d}\ CuCl_3 + \textbf{e}\ ICl + \textbf{f}\ H_2O$$

What values of **a**, **b**, **c**, **d**, **e** and **f** are needed in order to balance the equation?

Step 1: Pick an element and see how many atoms there are on the left and right sides.

Step 2: Form an equation to represent this. For Cu: b = d

Step 3: See if any of the answer options don't satisfy b=d. In this case, for option E, b is 8 and d is 10. This allows us to eliminate option E.

	a	b	c	d	e	f
A	5	4	25	4	13	15
B	5	4	20	4	8	15
C	5	6	20	6	8	15
D	2	8	10	8	8	15
E	6	8	24	10	16	15
F	6	10	22	10	16	15

Once you've eliminated as many options as possible, go back to step 1 and pick another element.
For Hydrogen (H): a + c = 2. Then see if any of the answer options don't satisfy a + c = 2f.

➤   Option A: 5 + 25 is equal to 2 x 15

➤   Option B: 5 + 20 is not equal to 2 x 15

➤   Option C: 5 + 20 is not equal to 2 x 15

➤   Option D: 2 + 10 is not equal to 2 x 15

This allows us to eliminate option B, C and D. E has already been eliminated. Thus, the only solution possible is A. This method works best when you get given a table above as this allows you to quickly eliminate options. However, it is still a viable method even if you don't get this information.

Chemistry Calculations

Equations you **MUST** know:

−   Atomic Mass = Mass/Moles
−   Amount (mol) = Concentration (mol/dm^3) x Volume (dm^3)

Avogadro's Constant:
One mole of anything contains 6×10^{23} of it e.g. 5 Moles of water contain $5 \times 6 \times 10^{23}$ number of water molecules.

Abundances:
The average atomic mass takes the abundances of all isotopes into account. Thus:
A_r = (Abundance of Isotope 1) x (Mass of Isotope 1) + (Abundance of Isotope 2) x (Mass of Isotope 2) +…

Chemistry Questions

Question 376:

Which of the following most accurately defines an isotope?

A. An isotope is an atom of an element that has the same number of protons in the nucleus but a different number of neutrons orbiting the nucleus.

B. An isotope is an atom of an element that has the same number of neutrons in the nucleus but a different number of protons orbiting the nucleus.

C. An isotope is any atom of an element that can be split to produce nuclear energy.

D. An isotope is an atom of an element that has the same number of protons in the nucleus but a different number of neutrons in the nucleus.

E. An isotope is an atom of an element that has the same number of protons in the nucleus but a different number of electrons orbiting it.

Question 377:

Which of the following is an example of a displacement reaction?

1. $Fe + SnSO4 \rightarrow FeSO_4 + Sn$
2. $Cl_2 + 2KBr \rightarrow Br_2 + 2KCl$
3. $H_2SO_4 + Mg \rightarrow MgSO_4 + H_2$
4. $NaHCO_3 + HCl \rightarrow NaCl + CO_2 + H_2O$

A. 1 only   C. 2 and 3 only   E. 1, 2 and 3 only

B. 1 and 2 only   D. 3 and 4 only   F. 1,2, 3 and 4

Question 378:

What values of **a**, **b** and **c** are needed to balance the equation below?

$$aCa(OH)_2 + bH_3PO_4 \rightarrow Ca_3(PO_4)_2 + cH_2O$$

A. a = 3 b = 2 c = 6
B. a = 2 b = 2 c = 4
C. a = 3 b = 2 c = 1
D. a = 1 b = 2 c = 3
E. a = 4 b = 2 c = 6
F. a = 3 b = 2 c = 4

Question 379:

What values of **s**, **t** and **u** are needed to balance the equation below?

$$sAgNO_3 + tK_3PO_4 \rightarrow 3Ag_3PO_4 + uKNO_3$$

A. s = 9 t = 3 u = 9   D. s = 9 t = 6 u = 9

B. s = 6 t = 3 u = 9   E. s = 3 t = 3 u = 9

C. s = 9 t = 3 u = 6   F. s = 9 t = 3 u = 3

Question 380:

Which of the following statements are true with regard to displacement?

1. A less reactive halogen can displace a more reactive halogen.
2. Chlorine cannot displace bromine or iodine from an aqueous solution of its salts.
3. Bromine can displace iodine because of the trend of reactivity.
4. Fluorine can displace chlorine as it is higher up the group.
5. Lithium can displace francium as it is higher up the group.

A. 3 only   D. 3 and 4 only

B. 5 only   E. 2 , 3 and 5 only

C. 1 and 2 only   F. 3, 4 and 5 only

Question 381:
What mass of magnesium oxide is produced when 75g of magnesium is burned in excess oxygen?
Relative Atomic Masses: Mg = 24, O = 16

A. 80g   B. 100g   C. 125g   D. 145g   E. 175g   F. 225g

Question 382:
Hydrogen can combine with hydroxide ions to produce water. Which process is involved in this?

A. Hydration   C. Reduction   E. Evaporation
B. Oxidation   D. Dehydration   F. Precipitation

Question 383:
Which of the following statements about Ammonia are correct?

1. It has a formula of NH_3.
2. Nitrogen contributes 82% to its mass.
3. It can be broken down again into nitrogen and hydrogen.
4. It is covalently bonded.
5. It is used to make fertilisers.

A. 1 and 2 only   D. 1, 2 and 5 only
B. 1 and 4 only   E. 3, 4 and 5 only
C. 1, 2 and 3 only   F. 1, 2, 3, 4 and 5

Question 384:
What colour will a universal indicator change to in a solution of milk and lipase?

A. From green to orange.   D. From purple to orange.
B. From red to green.   E. From yellow to purple.
C. From purple to green.   F. From purple to red.

Question 385:
Vitamin C [$C_6H_8O_6$] can be artificially synthesised from glucose [$C_6H_{12}O_6$]. What type of reaction is this likely to be?

A. Dehydration   C. Oxidation   E. Displacement
B. Hydration   D. Reduction   F. Evaporation

Question 386:
Which of the following statements are true?

1. Cu^{64} will undergo oxidation faster than Cu^{65}.
2. Cu^{65} will undergo reduction faster than Cu^{64}.
3. Cu^{65} and Cu^{64} have the same number of electrons.

A. 1 only   C. 3 only   E. 1 and 3 only
B. 2 only   D. 2 and 3 only   F. 1, 2 and 3

Question 387:
6g of Mg^{24} is added to a solution containing 30g of dissolved sulphuric acid (H_2SO_4). Which of the following statements are true?
Relative Atomic Masses: S = 32, Mg = 24, O = 16, H = 1

1. In this reaction, the magnesium is the limiting reagent
2. In this reaction, sulphuric acid is the limiting reagent
3. The mass of salt produced equals the original mass of sulphuric acid

A. 1 only   C. 3 only   E. 1 and 3 only
B. 2 only   D. 1 and 2 only   F. 2 and 3 only

Question 388:

In which of the following mixtures will a displacement reaction occur?

1. $Cu + 2AgNO_3$
2. $Cu + Fe(NO_3)_2$
3. $Ca + 2H_2O$
4. $Fe + Ca(OH)_2$

A. 1 only
B. 2 only

C. 3 only
D. 4 only

E. 1 and 2 only
F. 1 and 3 only

G. 1, 2 and 3
H. 1, 2, 3 and 4

Question 389:

Which of the following statements is true about the following chain of metals?

$$Na \rightarrow Ca \rightarrow Mg \rightarrow Al \rightarrow Zn$$

Moving from left to right:

1. The reactivity of the metals increases.
2. The likelihood of corrosion of the metals increases.
3. More energy is required to separate these metals from their ores.
4. The metals lose electrons more readily to form positive ions.

A. 1 and 2 only
B. 1and 3 only

C. 2 and 3 only
D. 1 and 4 only

E. 2, 3 and 4 only
F. 1, 2, 3 and 4

G. None of the above

Question 390:

In which of the following mixtures will a displacement reaction occur?

1. $I_2 + 2KBr$
2. $Cl_2 + 2NaBr$
3. $Br_2 + 2KI$

A. 1 only
B. 2 only

C. 3 only
D. 1 and 2 only

E. 1 and 3 only
F. 2 and 3 only

G. 1, 2 and 3

Question 391:

Which of the following statements about Al and Cu are true?

1. Al is used to build aircraft because it is lightweight and resists corrosion.
2. Cu is used to build electrical wires because it is a good insulator.
3. Both Al and Cu are good conductors of heat.
4. Al is commonly alloyed with other metals to make coins.
5. Al is resistant to corrosion because of a thin layer of aluminium hydroxide on its surface.

A. 1 and 3 only
B. 1 and 4 only

C. 1, 3 and 5 only
D. 1, 3, 4, 5 only

E. 2, 4 and 5 only
F. 2, 3, 4, 5 only

Question 392:

21g of Li^7 reacts completely with excess water. Given that the molar gas volume is 24 dm^3 under the conditions, what is the volume of hydrogen produced?

A. 12 dm^3
B. 24 dm^3

C. 36 dm^3
D. 48 dm^3

E. 72 dm^3
F. 120 dm^3

Question 393:

Which of the following statements regarding bonding are true?

1. NaCl has stronger ionic bonds than $MgCl_2$.
2. Transition metals are able to lose varying numbers of electrons to form multiple stable positive ions.
3. All covalently bonded structures have lower melting points than ionically bonded compounds.
4. All covalently bonded structures do not conduct electricity.

A. 1 only	C. 3 only	E. 1 and 2 only	G. 3 and 4 only
B. 2 only	D. 4 only	F. 2 and 3 only	H. 1, 2 and 4 only

Question 394:

Consider the following two equations:

A.	$C + O_2 \rightarrow CO_2$	$\Delta H = -394$ kJ per mole
B.	$CaCO_3 \rightarrow CaO + CO_2$	$\Delta H = +178$ kJ per mole

Which of the following statements are true?

1. Reaction **A** is exothermic and Reaction **B** is endothermic.
2. CO_2 has less energy than C and O_2.
3. CaO is more stable than $CaCO_3$.

A. 1 only	C. 3 only	E. 1 and 3	G. 1, 2 and 3
B. 2 only	D. 1 and 2	F. 2 and 3	

Question 395:

Which of the following are true of regarding the oxides formed by Na, Mg and Al?

1. All of the metals and their solid oxides conduct electricity.
2. MgO has stronger bonds than Na_2O.
3. Metals are extracted from their molten ores by fractional distillation.

A. 1 only	C. 3 only	E. 2 and 3 only
B. 2 only	D. 1 and 2 only	F. 1, 2 and 3

Question 396:

Which of the following pairs have the same electronic configuration?

1. Li^+ and Na^+
2. Mg^{2+} and Ne
3. Na^{2+} and Ne
4. O^{2+} and a Carbon atom

A. 1 only	C. 1 and 3 only	E. 2 and 4 only
B. 1 and 2 only	D. 2 and 3 only	F. 1, 2, 3 and 4

Question 397:

In relation to reactivity of elements in group 1 and 2, which of the following statements is correct?

1. Reactivity decreases as you go down group 1.
2. Reactivity increases as you go down group 2.
3. Group 1 metals are generally less reactive than group 2 metals.

A. Only 1	C. Only 3	E. 2 and 3
B. Only 2	D. 1 and 2	F. 1 and 3

Question 398:

What role do catalysts fulfil in an endothermic reaction?

A. They increase the temperature, causing the reaction to occur at a faster rate.
B. They decrease the temperature, causing the reaction to occur at a faster rate.
C. They reduce the energy of the reactants in order to trigger the reaction.
D. They reduce the activation energy of the reaction.
E. They increase the activation energy of the reaction.

Question 399:

Tritium H^3 is an isotope of Hydrogen. Why is tritium commonly referred to as 'heavy hydrogen'?

A. Because H^3 contains 3 protons making it heavier than H^1 that contains 1 proton.
B. Because H^3 contains 3 neutrons making it heavier than H^1 that contains 1 neutron.
C. Because H^3 contains 1 neutron and 2 protons making it heavier than H^1 that contains 1 neutron and 1 proton.
D. Because H^3 contains 1 proton and 2 neutrons making it heavier than H^1 that contains 1 proton.
E. Because H^3 contains 3 electrons making it heavier than H^1 that contains 1 electron.

Question 400:

In relation to redox reactions, which of the following statements are correct?

1. Oxidation describes the loss of electrons.
2. Reduction increases the electron density of an ion, atom or molecule.
3. Halogens are powerful reducing agents.

A. Only 1   C. Only 3   E. 2 and 3
B. Only 2   D. 1 and 2   F. 1 and 3

Question 401:

Which of the following statements is correct?

A. At higher temperatures, gas molecules move at angles that cause them to collide with each other more frequently.
B. Gas molecules have lower energy after colliding with each other.
C. At higher temperatures, gas molecules attract each other resulting in more collisions.
D. The average kinetic energy of gas molecules is the same for all gases at the same temperature.
E. The momentum of gas molecules decreases as pressure increases.

Question 402:

Which of the following are exothermic reactions?

1. Burning Magnesium in pure oxygen
2. The combustion of hydrogen
3. Aerobic respiration
4. Evaporation of water in the oceans
5. Reaction between a strong acid and a strong base

A. 1, 2 and 4   C. 1, 3 and 5   E. 1, 2, 3 and 5
B. 1, 2 and 5   D. 2, 3 and 4   F. 1, 2, 3, 4 and 5

Question 403:

Ethene reacts with oxygen to produce water and carbon dioxide. Which elements are oxidised/reduced?

A. Carbon is reduced and oxygen is oxidised.
B. Hydrogen is reduced and oxygen is oxidised.
C. Carbon is oxidised and hydrogen is reduced.
D. Hydrogen is oxidised and carbon is reduced.
E. Carbon is oxidised and oxygen is reduced.
F. None of the above.

Question 404:

In the reaction between Zinc and Copper (II) sulphate which elements act as oxidising + reducing agents?

A. Zinc is the reducing agent while sulfur is the oxidizing agent.
B. Zinc is the reducing agent while copper in $CuSO_4$ is the oxidizing agent.
C. Copper is the reducing agent while zinc is the oxidizing agent.
D. Oxygen is the reducing agent while copper in $CuSO_4$ is the oxidizing agent.
E. Sulfur is the reducing agent while oxygen is the oxidizing agent.
F. None of the above.

Question 405:

Which of the following statements is true?

A. Acids are compounds that act as proton acceptors in aqueous solution.
B. Acids only exist in a liquid state.
C. Strong acids are partially ionized in a solution.
D. Weak acids generally have a pH or 6 - 7.
E. The reaction between a weak and strong acid produces water and salt.

Question 406:

An unknown element, Z, has 3 isotopes: Z^5, Z^6 and Z^8. Given that the atomic mass of Z is 7, and the relative abundance of Z^5 is 20%, which of the following statements are correct?

1. Z^5 and Z^6 are present in the same abundance.
2. Z^8 is the most abundant of the isotopes.
3. Z^8 is more abundant than Z^5 and Z^6 combined.

A. 1 only
B. 2 only
C. 3 only
D. 1 and 2 only

E. 2 and 3 only
F. 1 and 3 only
G. 1, 2 and 3
H. None of the statements are correct.

Question 407:

Which of following best describes the products when an acid reacts with a metal that is more reactive than hydrogen?

A. Salt and hydrogen
B. Salt and ammonia
C. Salt and water

D. A weak acid and a weak base
E. A strong acid and a strong base
F. No reaction would occur.

Question 408:

Choose the option which balances the following equation:

$$\textbf{a } FeSO_4 + \textbf{b } K_2Cr_2O_7 + \textbf{c } H_2SO_4 \rightarrow \textbf{d } (Fe)_2(SO_4)_3 + \textbf{e } Cr_2(SO_4)_3 + \textbf{f } K_2SO_4 + \textbf{g } H_2O$$

	a	b	c	d	e	f	g
A	6	1	8	3	1	1	7
B	6	1	7	3	1	1	7
C	2	1	6	2	1	1	6
D	12	1	14	4	1	1	14
E	4	1	12	4	1	1	12
F	8	1	8	4	2	1	8

Question 409:

Which of the following statements is correct?

A. Matter consists of atoms that have a net electrical charge.
B. Atoms and ions of the same element have different numbers of protons and electrons but the same number of neutrons.
C. Over 80% of an atom's mass is provided by protons.
D. Atoms of the same element that have different numbers of neutrons react at significantly different rates.
E. Protons in the nucleus of atoms repel each other as they are positively charged.
F. None of the above.

Question 410:

Which of the following statements is correct?

A. The noble gasses are chemically inert and therefore useless to man.
B. All the noble gasses have a full outer electron shell.
C. The majority of noble gasses are brightly coloured.
D. The boiling point of the noble gasses decreases as you progress down the group.
E. Neon is the most abundant noble gas.

Question 411:

In relation to alkenes, which of the following statements is correct?
1. They all contain double bonds.
2. They can all be reduced to alkanes.
3. Aromatic compounds are also alkenes as they contain double bonds.

A. Only 1   D. 1 and 2   G. All of the above.
B. Only 2   E. 2 and 3   H. None of the above.
C. Only 3   F. 1 and 3

Question 412:

Chlorine is made up of two isotopes, Cl^{35} (atomic mass 34.969) and Cl^{37} (atomic mass 36.966). Given that the atomic mass of chlorine is 35.453, which of the following statements is correct?

A. Cl^{35} is about 3 times more abundant than Cl^{37}.
B. Cl^{35} is about 10 times more abundant than Cl^{37}.
C. Cl^{37} is about 3 times more abundant than Cl^{35}.
D. Cl^{37} is about 10 times more abundant than Cl^{35}.
E. Both isotopes are equally abundant.

Question 413:

Which of the following statements regarding transition metals is correct?

A. Transition metals form ions that have multiple colours.
B. Transition metals usually form covalent bonds.
C. Transition metals cannot be used as catalysts as they are too reactive.
D. Transition metals are poor conductors of electricity.
E. Transition metals are frequently referred to as f-block elements.

Question 414:

20 g of impure Na^{23} reacts completely with excess water to produce 8,000 cm^3 of hydrogen gas under standard conditions. What is the percentage purity of sodium?
[Under standard conditions 1 mole of gas occupies 24 dm^3]

A. 88.0%   B. 76.5%   C. 66.0%   D. 38.0%   E. 15.3%

Question 415:

An organic molecule contains 70.6% Carbon, 5.9% Hydrogen and 23.5% Oxygen. It has a molecular mass of 136. What is its chemical formula?

A. C_4H_4O B. C_5H_4O C. $C_8H_8O_2$ D. $C_{10}H_8O_2$ E. C_2H_2O

Question 416:

Choose the option which balances the following reaction:

$$aS + bHNO_3 \rightarrow cH_2SO_4 + dNO_2 + eH_2O$$

	a	b	c	d	e
A	3	5	3	5	1
B	1	6	1	6	2
C	6	14	6	14	2
D	2	4	2	4	4
E	2	3	2	3	2
F	4	4	4	4	2

Question 417:

Which of the following statements is true?

1. Ethane and ethene can both dissolve in organic solvents.
2. Ethane and ethene can both be hydrogenated in the presence of Nickel.
3. Breaking C=C requires double the energy needed to break C-C.

A. 1 only
B. 2 only
C. 3 only
D. 1 and 2 only
E. 2 and 3 only
F. 1 and 3 only
G. 1, 2 and 3

Question 418:

Diamond, Graphite, Methane and Ammonia all exhibit covalent bonding. Which row adequately describes the properties associated with each?

	Compound	Melting Point	Able to conduct electricity	Soluble in water
1.	Diamond	High	Yes	No
2.	Graphite	High	Yes	No
3.	$CH_{4\,(g)}$	Low	No	No
4.	$NH_{3\,(g)}$	Low	No	Yes

A. 1 and 2 only
B. 2 and 3 only
C. 1 and 3 only
D. 1 and 4 only
E. 1, 2 and 3
F. 2, 3 and 4
G. 1,2 and 4
H. 1, 2, 3 and 4

Question 419:

Which of the following statements about catalysts are true?

1. Catalysts reduce the energy required for a reaction to take place.
2. Catalysts are used up in reactions.
3. Catalysed reactions are almost always exothermic.

A. 1 only
B. 2 only
C. 1 and 2
D. 2 and 3
E. 1, 2 and

Question 420:

What is the name of the molecule below?

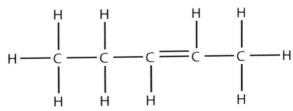

A. But-1-ene
B. But-2-ene
C. Pent-3-ene
D. Pent-1-ene
E. Pent-2-ene
F. Pentane
G. Pentanoic acid

Question 421:

Which of the following statements is correct regarding Group 1 elements? [Excluding Hydrogen]

A. The oxidation number of Group 1 elements usually decreases in most reactions.
B. Reactivity decreases as you progress down Group 1.
C. Group 1 elements do not react with water.
D. All Group 1 elements react spontaneously with oxygen.
E. All of the above.
F. None of the above.

Question 422:

Which of the following statements about electrolysis are correct?

1. The cathode attracts negatively charged ions.
2. Atoms are reduced at the anode.
3. Electrolysis can be used to separate mixtures.

A. Only 1
B. Only 2
C. Only 3
D. 1 and 2
E. 2 and 3
F. 1 and 3
G. 1, 2 and 3
H. None of the above.

Question 423:

Which of the following is **NOT** an isomer of pentane?

A. $CH_3CH_2CH_2CH_2CH_3$
B. $CH_3C(CH_3)CH_3CH_3$
C. $CH_3(CH_2)_3CH_3$
D. $CH_3C(CH_3)_2CH_3$

Question 424:

Choose the option which balances the following reaction:

$Cu + HNO_3 \rightarrow Cu(NO_3)_2 + NO + H_2O$

A. $8\ Cu + 3\ HNO_3 \rightarrow 8\ Cu(NO_3)_2 + 4\ NO + 2\ H_2O$
B. $3\ Cu + 8\ HNO_3 \rightarrow 2\ Cu(NO_3)_2 + 3\ NO + 4\ H_2O$
C. $5Cu + 7HNO_3 \rightarrow 5\ Cu(NO_3)_2 + 4\ NO + 8\ H_2O$
D. $6\ Cu + 10\ HNO_3 \rightarrow 6\ Cu(NO_3)_2 + 3\ NO + 7\ H_2O$
E. $3\ Cu + 8\ HNO_3 \rightarrow 3\ Cu(NO_3)_2 + 2\ NO + 4\ H_2O$

Question 425:

What of the following statements regarding alkenes is correct?

A. Alkenes are an inorganic homologous series.
B. Alkenes always have three times as many hydrogen atoms as they do carbon atoms.
C. Bromine water changes from clear to brown in the presence of an alkene.
D. Alkenes are more reactive than alkanes because they are unsaturated.
E. Alkenes frequently take part in subtraction reactions.
F. All of the above.

Question 426:

Which of the following statements is correct regarding Group 17?

A. All Group 17 elements are electrophilic and therefore form negatively charged ions.
B. All Group 17 elements are gasses a room temperature.
C. The reaction between Sodium and Fluorine is less vigorous than Sodium and Iodine.
D. All Group 17 elements are non-coloured.
E. Some Group 17 elements are found naturally as unbonded atoms.
F. All of the above.
G. None of the above.

Question 427:

Why does the electrolysis of NaCl solution (brine) require the strict separation of the products of anode and cathode?

A. To prevent the preferential discharge of ions.
B. In order to prevent spontaneous combustion.
C. In order to prevent production of H_2.
D. In order to prevent the formation of HCl.
E. In order to avoid CO poisoning.
F. All of the above.

Question 428:

In relation to the electrolysis of brine (NaCl), which of the following statements are correct?

1. Electrolysis results in the production of hydrogen and chlorine gas.
2. Electrolysis results in the production of sodium hydroxide.
3. Hydrogen gas is released at the anode and chlorine gas is released at the cathode.

A. Only 1
B. Only 2
C. Only 3
D. 1 and 2
E. 1 and 3
F. 2 and 3
G. All of the above

Question 429:

Which of the following statements is correct?

A. Alkanes consist of multiple C-H bonds that are very weak.
B. An alkane with 14 hydrogen atoms is called Heptane.
C. All alkanes consist purely of hydrogen and carbon atoms.
D. Alkanes burn in excess oxygen to produce carbon monoxide and water.
E. Bromine water is decolourised in the presence of an alkane.
F. None of the above.

Question 430:

Which of the following statements are correct?

1. All alcohols contain a hydroxyl functional group.
2. Alcohols are highly soluble in water.
3. Alcohols are sometimes used a biofuels.

A. Only 1
B. Only 2
C. Only 3
D. 1 and 2
E. 2 and 3
F. 1 and 3
G. 1, 2 and 3

Question 431:

Which row of the table below is correct?

	Non-Reducible Hydrocarbon			Reducible Hydrocarbon		
A	C_nH_{2n}	$Br_{2(aq)}$ remains brown	Saturated	C_nH_{2n+2}	Turns $Br_{2(aq)}$ colourless	Unsaturated
B	C_nH_{2n+2}	Turns $Br_{2(aq)}$ colourless	Unsaturated	C_nH_{2n}	$Br_{2(aq)}$ remains brown	Saturated
C	C_nH_{2n}	$Br_{2(aq)}$ remains brown	Unsaturated	C_nH_{2n+2}	Turns $Br_{2(aq)}$ colourless	Saturated
D	C_nH_{2n+2}	Turns $Br_{2(aq)}$ colourless	Saturated	C_nH_{2n}	$Br_{2(aq)}$ remains brown	Unsaturated
E	C_nH_{2n+2}	$Br_{2(aq)}$ remains brown	Saturated	C_nH_{2n}	Turns $Br_{2(aq)}$ colourless	Unsaturated

Question 432:

How many grams of magnesium chloride are formed when 10 grams of magnesium oxide are dissolved in excess hydrochloric acid? Relative atomic masses: Mg = 24, O = 16, H = 1, Cl = 35.5

A. 10.00
B. 14.95
C. 20.00
D. 23.75
E. 47.55
F. More information needed

Question 433:

Pentadecane has the molecular formula $C_{15}H_{32}$. Which of the following statements is true?

A. Pentadecane has a lower boiling point than pentane.
B. Pentadecane is more flammable than pentane.
C. Pentadecane is more volatile than pentane.
D. Pentadecane is more viscous than pentane.
E. All of the above.
F. None of the above.

Question 434:

The rate of reaction is normally dependent upon:
1. The temperature.
2. The concentration of reactants.
3. The concentration of the catalyst.
4. The surface area of the catalyst.

A. 1 and 2
B. 2 and 3
C. 2, 3 and 4
D. 1, 3 and 4
E. 1, 2 and 3
F. 1, 2, 3 and 4

Question 435:

The equation below shows the complete combustion of a sample of unknown hydrocarbon in excess oxygen.

$$C_aH_b + O_2 \rightarrow cCO_2 + dH_2O$$

The product yielded 176 grams of CO_2 and 108 grams of H_2O. What is the most likely formula of the unknown hydrocarbon? Relative atomic masses: $H = 1$, $C = 12$, $O = 16$.

A. CH_4
B. CH_3
C. C_2H_6
D. C_3H_9
E. C_2H_4
F. C_4H_{10}

Question 436:

What type of reaction must ethanol undergo in order to be converted to ethylene oxide (C_2H_4O)?

A. Oxidation
B. Reduction
C. Dehydration
D. Hydration
E. Redox
F. All of the above

Question 437:

What values of a, b and c balance the equation below?

$$a\ Ba_3N_2 + 6H_2O \rightarrow b\ Ba(OH)_2 + c\ NH_3$$

	a	b	c
A	1	2	3
B	1	3	2
C	2	1	3
D	2	3	1
E	3	1	2
F	3	2	1

Question 438:

What values of a, b and c balance the equation below?

$$a\ FeS + 7O_2 \rightarrow b\ Fe_2O_3 + c\ SO_2$$

	a	b	c
A	3	2	2
B	2	4	1
C	3	1	5
D	4	1	3
E	4	2	4

Question 439:
Magnesium consists of 3 isotopes: Mg^{23}, Mg^{25}, and Mg^{26} which are found naturally in a ratio of 80:10:10.
Calculate the relative atomic mass of magnesium.

A. 23.3   B. 23.4   C. 23.5   D. 23.6   E. 24.6   F. 25.2   G. 25.5

Question 440:
Consider the three reactions:
1. $Cl_2 + 2Br^- \rightarrow 2Cl^- + Br_2$
2. $Cu^{2+} + Mg \rightarrow Cu + Mg^{2+}$
3. $Fe_2O_3 + 3CO \rightarrow 2Fe + 3CO_2$
Which of the following statements are correct?

A. Cl_2 and Fe_2O_3 are reducing agents.
B. CO and Cu^{2+} are oxidising agents.
C. Br_2 is a stronger oxidising agent than Cl_2.
D. Mg is a stronger reducing agent than Cu.
E. All of the above.
F. None of the above.

Question 441:
Which row best describes the properties of NaCl?

	Melting Point	Solubility in Water	Conducts electricity?	
			As solid	In solution
A	High	Yes	Yes	Yes
B	High	No	Yes	No
C	High	Yes	No	Yes
D	High	No	No	No
E	Low	Yes	Yes	Yes
F	Low	No	Yes	No
G	Low	Yes	No	Yes
H	Low	No	No	No

Question 442:
80g of Sodium hydroxide reacts with excess zinc nitrate to produce zinc hydroxide. Calculate the mass of zinc hydroxide produced. Relative atomic mass: N = 14, Zn = 65, O = 16, Na = 23.

A. 49g
B. 95g
C. 99g
D. 100g
E. 198g
F. More information needed.

Question 443:
Which of the following statements is correct?

A. The reaction between all Group 1 metals and water is exothermic.
B. Sodium reacts less vigorously with water than Potassium does.
C. All Group 1 metals react with water to produce elemental hydrogen.
D. All Group 1 metals react with water to produce a metal hydroxide.
E. All of the above.
F. None of the above.

Question 444:
Which of the following statements is correct?

A. NaCl can be separated using sieves.
B. CO_2 can be separated using electrolysis.
C. Dyes in a sample of ink can be separated using chromatography.
D. Oil and water can be separated using fractional distillation.
E. Methane and diesel can be separated using a separating funnel.
F. None of the above.

Question 445:

Which of the following statements about the reaction between caesium and fluoride are correct?

1. It is an exothermic reaction and therefore requires catalysts.
2. It results in the formation of a salt.
3. The addition of water will make the reaction safer.

A. Only 1
B. Only 2
C. Only 3

D. 1 and 2
E. 2 and 3
F. 1 and 3

G. All of the above.
H. None of the above.

Question 446:

Which of the following statements is generally true about stable isotopes?
1. The nucleus contains an equal number of neutrons and protons.
2. The nuclear charge is equal and opposite to the peripheral charge due to the orbiting electrons.
3. They can all undergo radioactive decay into more stable isotopes.

A. Only 1
B. Only 2
C. Only 3

D. 1 and 2
E. 2 and 3
F. 1 and 3

G. All of the above.
H. None of the above.

Question 447:

Why do most salts have very high melting temperatures?
A. Their surface is able to radiate away a significant portion of heat to their environment.
B. The ionic bonds holding them together are very strong.
C. The covalent bonds holding them together are very strong.
D. They tend to form large macromolecules as each salt molecule bonds with multiple other molecules.
E. All of the above.

Question 448:

A bottle of water contains 306ml of pure deionised water. How many protons are in the bottle from the water? Avogadro Constant = 6×10^{23}.

A. 1×10^{22}
B. 1×10^{23}
C. 1×10^{24}
D. 1×10^{25}
E. 1×10^{26}

Question 449:

On analysis, an organic substance is found to contain 41.4% Carbon, 55.2% Oxygen and 3.45% Hydrogen by mass. Which of the following could be the empirical formula of this substance?
A. $C_3O_3H_6$
B. $C_3O_3H_{12}$

C. $C_4O_2H_4$
D. $C_4O_4H_4$

E. $C_4O_2H_8$
F. More information needed.

Question 450:

A is a Group 2 element and B is a Group 17 element. Which row best describes what happens when A reacts with B?

	B is	Formula
A	Reduced	AB
B	Reduced	A_2B
C	Reduced	AB_2
D	Oxidised	AB
E	Oxidised	A_2B
F	Oxidised	AB_2

SECTION 2: Physics

If you haven't done physics at AS then you'll have to ensure that you are confident with commonly examined topics like Newtonian mechanics, electrical circuits and radioactive decay as you may not have covered these at GCSE depending on the specification you did.

The first step to improving in this section is to memorise by rote all the equations listed on the next page.

The majority of the physics questions involve a fair bit of maths – this means you need to be comfortable with converting between units and also powers of 10. **Most questions require two step calculations**. Consider the example:

A metal ball is released from the roof a 20 metre building. Assuming air resistance equals is negligible; calculate the velocity at which the ball hits the ground. [$g = 10\text{ms}^{-2}$]

A. 5 ms^{-1}
B. 10 ms^{-1}
C. 15 ms^{-1}
D. 20 ms^{-1}
E. 25 ms^{-1}

When the ball hits the ground, all of its gravitational potential energy has been converted to kinetic energy. Thus, $E_p = E_k$:

$$mg\Delta h = \frac{mv^2}{2}$$

Thus, $v = \sqrt{2gh} = \sqrt{2 \times 10 \times 20}$

$= \sqrt{400} = 20ms^{-1}$

Here, you were required to not only recall two equations but apply and rearrange them very quickly to get the answer; all in under 60 seconds. Thus, it is easy to understand why the physics questions are generally much harder than the biology and chemistry ones.

Note that if you were comfortable with basic Newtonian mechanics, you could have also solved this using a single suvat equation: $v^2 = u^2 + 2as$

$v = \sqrt{2 \times 10 \times 20} = 20ms^{-1}$

This is why you're **strongly advised to learn the 'suvat' equations** on the next page even if they're technically not on the syllabus.

SI Units

Remember that in order to get the correct answer you must always work in SI units i.e. do your calculations in terms of metres (not centimetres) and kilograms (not grams), etc.

Top tip! Knowing SI units is extremely useful because they allow you to **'work out' equations** if you ever forget them e.g. The units for density are kg/m^3. Since Kg is the SI unit for mass, and m^3 is represented by volume –the equation for density must be = Mass/Volume.

This can also work the other way, for example we know that the unit for Pressure is Pascal (Pa). But based on the fact that Pressure = Force/Area, a Pascal must be equivalent to N/m^2. Some physics questions will test your ability to manipulate units like this so it's important you are comfortable converting between them.

Formulas you MUST know:

Equations of Motion:

- $s = ut + 0.5at^2$
- $v = u + at$
- $a = (v-u)/t$
- $v^2 = u^2 + 2as$

Equations relating to Force:

- Force = mass x acceleration
- Force = Momentum/Time
- Pressure = Force / Area
- Moment of a Force = Force x Distance
- Work done = Force x Displacement

For objects in equilibrium:

- Sum of Clockwise moments = Sum of Anti-clockwise moments
- Sum of all resultant forces = 0

Equations relating to Energy:

- Kinetic Energy = $0.5 \, mv^2$
- Δ in Gravitational Potential Energy = $mg\Delta h$
- Energy Efficiency = (Useful energy/ Total energy) x 100%

Equations relating to Power:

- Power = Work done / time
- Power = Energy transferred / time
- Power = Force x velocity

Electrical Equations:

- $Q = It$
- $V = IR$
- $P = IV = I^2R = V^2/R$
- V = Potential difference (V, Volts)

- R = Resistance (Ohms)
- P = Power (W, Watts)
- Q = Charge (C, Coulombs)
- t = Time (s, seconds)

For Transformers: $\dfrac{V_p}{V_s} = \dfrac{n_p}{n_s}$ where:

- V: Potential difference
- n: Number of turns
- p: Primary
- s: Secondary

Other:

- Weight = mass x g
- Density = Mass / Volume
- Momentum = Mass x Velocity
- $g = 9.81 \, ms^{-2}$ (unless otherwise stated)

Factor	Text	Symbol
10^{12}	Tera	T
10^9	Giga	G
10^6	Mega	M
10^3	Kilo	k
10^2	Hecto	h
10^{-1}	Deci	d
10^{-2}	Centi	c
10^{-3}	Milli	m
10^{-6}	Micro	μ
10^{-9}	Nano	n
10^{-12}	Pico	p

Physics Questions

Question 451:

Which of the following statements are **FALSE**?

A. Electromagnetic waves cause things to heat up.
B. X-rays and gamma rays can knock electrons out of their orbits.
C. Loud sounds can make objects vibrate.
D. Wave power can be used to generate electricity.
E. Since waves carry energy away, the source of a wave loses energy.
F. The amplitude of a wave determines its mass.

Question 452:

A spacecraft is analysing a newly discovered exoplanet. A rock of unknown mass falls on the planet from a height of 30 m. Given that $g = 5.4$ ms^{-2} on the planet, calculate the speed of the rock when it hits the ground and the time it took to fall.

	Speed (ms^{-1})	Time (s)
A	18	3.3
B	18	3.1
C	12	3.3
D	10	3.7
E	9	2.3
F	1	0.3

Question 453:

A canoe floating on the sea rises and falls 7 times in 49 seconds. The waves pass it at a speed of 5 ms^{-1}. How long are the waves?

A. 12 m   B. 22 m   C. 25 m   D. 35 m   E. 57 m   F. 75 m

Question 454:

Miss Orrell lifts her 37.5 kg bike for a distance of 1.3 m in 5 s. The acceleration of free fall is 10 ms^{-2}. What is the average power that she develops?

A. 9.8 W
B. 12.9 W

C. 57.9 W
D. 79.5 W

E. 97.5W
F. 98.0 W

Question 455:

A truck accelerates at 5.6 ms^{-2} from rest for 8 seconds. Calculate the final speed and the distance travelled in 8 seconds.

	Final Speed (ms^{-1})	Distance (m)
A	40.8	119.2
B	40.8	129.6
C	42.8	179.2
D	44.1	139.2
E	44.1	179.7
F	44.2	129.2
G	44.8	179.2
H	44.8	179.7

Question 456:

Which of the following statements is true when a sky diver jumps out of a plane?

A. The sky diver leaves the plane and will accelerate until the air resistance is greater than their weight.
B. The sky diver leaves the plane and will accelerate until the air resistance is less than their weight.
C. The sky diver leaves the plane and will accelerate until the air resistance equals their weight.
D. The sky diver leaves the plane and will accelerate until the air resistance equals their weight squared.
E. The sky diver will travel at a constant velocity after leaving the plane.

Question 457:

A 100 g apple falls on Isaac's head from a height of 20 m. Calculate the apple's momentum before the point of impact. Take $g = 10$ ms^{-2}

A. 0.1 kgms^{-1}
B. 0.2 kgms^{-1}

C. 1 kgms^{-1}
D. 2 kgms^{-1}

E. 10 kgms^{-1}
F. 20 kgms^{-1}

Question 458:

Which of the following do all electromagnetic waves all have in common?

1. They can travel through a vacuum.
2. They can be reflected.
3. They are the same length.
4. They have the same amount of energy.
5. They can be polarised.

A. 1, 2 and 3 only
B. 1, 2, 3 and 4 only

C. 4 and 5 only
D. 3 and 4 only

E. 1, 2 and 5 only
F. 1 and 5 only

Question 459:

A battery with an internal resistance of 0.8 Ω and e.m.f of 36 V is used to power a drill with resistance 1 Ω. What is the current in the circuit when the drill is connected to the power supply?

A. 5 A   B. 10 A   C. 15 A   D. 20 A   E. 25 A   F. 30 A

Question 460:

Officer Bailey throws a 20 g dart at a speed of 100 ms^{-1}. It strikes the dartboard and is brought to rest in 10 milliseconds. Calculate the average force exerted on the dart by the dartboard.

A. 0.2 N
B. 2 N

C. 20 N
D. 200 N

E. 2,000 N
F. 20,000 N

Question 461:

Professor Huang lifts a 50 kg bag through a distance of 0.7 m in 3 s. What average power does she develop to 3 significant figures? Take $g = 10$ms^{-2}

A. 112 W
B. 113 W

C. 114 W
D. 115 W

E. 116 W
F. 117 W

Question 462:

An electric scooter is travelling at a speed of 30 ms^{-1} and is kept going against a 50 N frictional force by a driving force of 300 N in the direction of motion. Given that the engine runs at 200 V, calculate the current in the scooter.

A. 4.5 A
B. 45 A

C. 450 A
D. 4,500 A

E. 45,000 A
F. More information needed.

Question 463:

Which of the following statements about the physical definition of work are correct?

1. $Work\ done = \frac{Force}{distance}$
2. The unit of work is equivalent to Kgms^{-2}.
3. Work is defined as a force causing displacement of the body upon which it acts.

A. Only 1
B. Only 2

C. Only 3
D. 1 and 2

E. 2 and 3
F. 1 and 3

Question 464:

Which of the following statements about kinetic energy are correct?

1. It is defined as $E_k = \frac{mv^2}{2}$
2. The unit of kinetic energy is equivalent to Pa x m^3.
3. Kinetic energy is equal to the amount of energy needed to decelerate the body in question from its current speed.

A. Only 1
B. Only 2
C. Only 3

D. 1 and 2
E. 2 and 3
F. 1 and 3

G. 1, 2 and 3

Question 465:

In relation to radiation, which of the following statements is **FALSE**?

A. Radiation is the emission of energy in the form of waves or particles.
B. Radiation can be either ionizing or non-ionizing.
C. Gamma radiation has very high energy.
D. Alpha radiation is of higher energy than beta radiation.
E. X-rays are an example of wave radiation.

Question 466:

In relation to the physical definition of half-life, which of the following statements are correct?

1. In radioactive decay, the half-life is independent of atom type and isotope.
2. Half-life is defined as the time required for exactly half of the entities to decay.
3. Half-life applies to situations of both exponential and non-exponential decay.

A. Only 1
B. Only 2

C. Only 3
D. 1 and 2

E. 2 and 3
F. 1 and 3

Question 467:

In relation to nuclear fusion, which of the following statements is **FALSE**?

A. Nuclear fusion is initiated by the absorption of neutrons.
B. Nuclear fusion describes the fusion of hydrogen atoms to form helium atoms.
C. Nuclear fusion releases great amounts of energy.
D. Nuclear fusion requires high activation temperatures.
E. All of the statements above are false.

Question 468:

In relation to nuclear fission, which of the following statements is correct?

A. Nuclear fission is the basis of many nuclear weapons.
B. Nuclear fission is triggered by the shooting of neutrons at unstable atoms.
C. Nuclear fission can trigger chain reactions.
D. Nuclear fission commonly results in the emission of ionizing radiation.
E. All of the above.

Question 469:

Two identical resistors (R_a and R_b) are connected in a series circuit. Which of the following statements are true?

1. The current through both resistors is the same.
2. The voltage through both resistors is the same.
3. The voltage across the two resistors is given by Ohm's Law.

A. Only 1
B. Only 2
C. Only 3

D. 1 and 2
E. 2 and 3
F. 1 and 3

G. 1, 2 and 3
H. None of the above.

Question 470:

The Sun is 8 light-minutes away from the Earth. Estimate the circumference of the Earth's orbit around the Sun. Assume that the Earth is in a circular orbit around the Sun. Speed of light = 3 x 10^8 ms^{-1}

A. 10^{24} m
B. 10^{21} m

C. 10^{18} m
D. 10^{15} m

E. 10^{12} m
F. 10^9 m

Question 471:

Which of the following statements about the physical definition of speed are true?

1. Speed is the same as velocity.
2. The internationally standardised unit for speed is ms^{-2}.
3. Velocity = distance/time.

A. Only 1	D. 1 and 2	G. 1, 2 and 3
B. Only 2	E. 2 and 3	H. None of the above
C. Only 3	F. 1 and 3	

Question 472:

Which of the following statements best defines Ohm's Law?

A. The current through an insulator between two points is indirectly proportional to the potential difference across the two points.
B. The current through an insulator between two points is directly proportional to the potential difference across the two points.
C. The current through a conductor between two points is inversely proportional to the potential difference across the two points.
D. The current through a conductor between two points is proportional to the square of the potential difference across the two points.
E. The current through a conductor between two points is directly proportional to the potential difference across the two points.

Question 473:

Which of the following statements regarding Newton's Second Law are correct?
1. For objects at rest, Resultant Force must be 0 Newtons
2. Force = Mass x Acceleration
3. Force = Rate of change of Momentum

A. Only 1	C. Only 3	E. 2 and 3	G. 1, 2 and 3
B. Only 2	D. 1 and 2	F. 1 and 3	

Question 474:

Which of the following equations concerning electrical circuits are correct?

1. $Charge = \dfrac{Voltage \; x \; time}{Resistance}$

2. $Charge = \dfrac{Power \; x \; time}{Voltage}$

3. $Charge = \dfrac{Current \; x \; time}{Resistance}$

A. Only 1	E. 2 and 3
B. Only 2	F. 1 and 3
C. Only 3	G. 1, 2 and 3
D. 1 and 2	H. None of the equations are correct.

Question 475:

An elevator has a mass of 1,600 kg and is carrying passengers that have a combined mass of 200 kg. A constant frictional force of 4,000 N retards its motion upward. What force must the motor provide for the elevator to move with an upward acceleration of 1 ms^{-2}? Assume: $g = 10$ ms^{-2}

A. 1,190 N	D. 22,000 N
B. 11,900 N	E. 23,800 N
C. 18,000 N	

Question 476:

A 1,000 kg car accelerates from rest at 5 ms^{-2} for 10 s. Then, a braking force is applied to bring it to rest within 20 seconds. What distance has the car travelled?

A. 125 m	C. 650 m	E. 1,200 m
B. 250 m	D. 750 m	F. More information needed

Question 477:

An electric heater is connected to 120 V mains by a copper wire that has a resistance of 8 ohms. What is the power of the heater?

A. 90 W
B. 180 W
C. 900 W
D. 1800 W
E. 9,000W
F. 18,000 W
G. More information needed

Question 478:

In a particle accelerator, electrons are accelerated through a potential difference of 40 MV and emerge with an energy of 40MeV (1 MeV = 1.60 x 10^{-13} J). Each pulse contains 5,000 electrons. The current is zero between pulses. Assuming that the electrons have zero energy prior to being accelerated what is the power delivered by the electron beam?

A. 1 kW
B. 10 kW
C. 100 kW
D. 1,000 kW
E. 10,000 kW
F. More information needed

Question 479:

Which of the following statements is **true**?

A. When an object is in equilibrium with its surroundings, there is no energy transferred to or from the object and so its temperature remains constant.
B. When an object is in equilibrium with its surroundings, it radiates and absorbs energy at the same rate and so its temperature remains constant.
C. Radiation is faster than convection but slower than conduction.
D. Radiation is faster than conduction but slower than convection.
E. None of the above.

Question 480:

A 6kg block is pulled from rest along a horizontal frictionless surface by a constant horizontal force of 12 N. Calculate the speed of the block after it has moved 300 cm.

A. $2\sqrt{3} \ ms^{-1}$
B. $4\sqrt{3} \ ms^{-1}$
C. $4\sqrt{3} \ ms^{-1}$
D. $12 \ ms^{-1}$
E. $\sqrt{\frac{3}{2}} \ ms^{-1}$

Question 481:

A 100 V heater heats 1.5 litres of pure water from 10°C to 50°C in 50 minutes. Given that 1 kg of pure water requires 4,000 J to raise its temperature by 1°C, calculate the resistance of the heater.

A. 12.5 ohms
B. 25 ohms
C. 125 ohms
D. 250 ohms
E. 500 ohms
F. 850 ohms

Question 482:

Which of the following statements are **true**?

1. Nuclear fission is the basis of nuclear energy.
2. Following fission, the resulting atoms are a different element to the original one.
3. Nuclear fission often results in the production of free neutrons and photons.

A. Only 1
B. Only 2
C. Only 3
D. 1 and 2
E. 2 and 3
F. 1 and 3
G. 1, 2 and 3
H. None of the above

Question 483:

Which of the following statements are **true**? Assume g = 10 ms^{-2}.
1. Gravitational potential energy is defined as ΔE_p = m x g x Δ h.
2. Gravitational potential energy is a measure of the work done against gravity.
3. A reservoir situated 1 km above ground level with 10^6 litres of water has a potential energy of 1 Giga Joule.

A. Only 1
B. Only 2
C. Only 3
D. 1 and 2
E. 2 and 3
F. 1 and 3
G. 1, 2 and 3
H. None of the above

Question 484:

Which of the following statements are correct in relation to Newton's 3rd law?

1. For every action there is an equal and opposite reaction.
2. According to Newton's 3rd law, there are no isolated forces.
3. Rockets cannot accelerate in deep space because there is nothing to generate an equal and opposite force.

A. Only 1   C. Only 3   E. 2 and 3
B. Only 2   D. 1 and 2   F. 1 and 3

Question 485:

Which of the following statements are correct?
1. Positively charged objects have gained electrons.
2. Electrical charge in a circuit over a period of time can be calculated if the voltage and resistance are known.
3. Objects can be charged by friction.

A. Only 1   C. Only 3   E. 2 and 3   G. 1, 2 and 3
B. Only 2   D. 1 and 2   F. 1 and 3

Question 486:

Which of the following statements is true?

A. The gravitational force between two objects is independent of their mass.
B. Each planet in the solar system exerts a gravitational force on the Earth.
C. For satellites in a geostationary orbit, acceleration due to gravity is equal and opposite to the lift from engines.
D. Two objects that are dropped from the Eiffel tower will always land on the ground at the same time if they have the same mass.
E. All of the above.
F. None of the above.

Question 487:

Which of the following best defines an electrical conductor?

A. Conductors are usually made from metals and they conduct electrical charge in multiple directions.
B. Conductors are usually made from non-metals and they conduct electrical charge in multiple directions.
C. Conductors are usually made from metals and they conduct electrical charge in one fixed direction.
D. Conductors are usually made from non-metals and they conduct electrical charge in one fixed direction.
E. Conductors allow the passage of electrical charge with zero resistance because they contain freely mobile charged particles.
F. Conductors allow the passage of electrical charge with maximal resistance because they contain charged particles that are fixed and static.

Question 488:

An 800 kg compact car delivers 20% of its power output to its wheels. If the car has a mileage of 30 miles/gallon and travels at a speed of 60 miles/hour, how much power is delivered to the wheels? 1 gallon of petrol contains 9×10^8 J.

A. 10 kW   B. 20 kW   C. 40 kW   D. 50 kW   E. 100 kW

Question 489:

Which of the following statements about beta radiation are true?

1. After a beta particle is emitted, the atomic mass number is unchanged.
2. Beta radiation can penetrate paper but not aluminium foil.
3. A beta particle is emitted from the nucleus of the atom when an electron changes into a neutron.

A. 1 only   C. 1 and 3   E. 2 and 3
B. 2 only   D. 1 and 2   F. 1, 2 and 3

Question 490:

A car with a weight of 15,000 N is travelling at a speed of 15 ms^{-1} when it crashes into a wall and is brought to rest in 10 milliseconds. Calculate the average braking force exerted on the car by the wall. Take $g = 10$ ms^{-2}

A. $1.25 \times 10^4 N$
B. $1.25 \times 10^5 N$

C. $1.25 \times 10^6 N$
D. $2.25 \times 10^4 N$

E. $2.25 \times 10^5 N$
F. $2.25 \times 10^6 N$

Question 491:

Which of the following statements are correct?

1. Electrical insulators are usually metals e.g. copper.
2. The flow of charge through electrical insulators is extremely low.
3. Electrical insulators can be charged by rubbing them together.

A. Only 1
B. Only 2

C. Only 3
D. 1 and 2

E. 2 and 3
F. 1 and 3

G. 1, 2 and 3

The following information is needed for Questions 492 and 493:

The graph below represents a car's movement. At t=0 the car's displacement was 0 m.

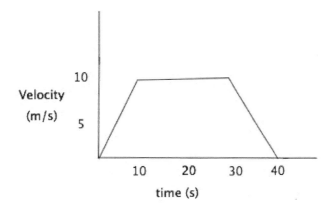

Question 492:

Which of the following statements are **NOT true**?

1. The car is reversing after t = 30.
2. The car moves with constant acceleration from t = 0 to t = 10.
3. The car moves with constant speed from t = 10 to t = 30.

A. 1 only
B. 2 only
C. 3 only

D. 1 and 3
E. 1 and 2
F. 2 and 3

G. 1, 2 and 3

Question 493:

Calculate the distance travelled by the car.

A. 200 m
B. 300 m

C. 350 m
D. 400 m

E. 500 m
F. More information needed

Question 494:

A 1,000 kg rocket is launched during a thunderstorm and reaches a constant velocity 30 seconds after launch. Suddenly, a strong gust of wind acts on it for 5 seconds with a force of 10,000 N in the direction of movement.
What is the resulting change in velocity?

A. 0.5 ms^{-1}
B. 5 ms^{-1}

C. 50 ms^{-1}
D. 500 ms^{-1}

E. 5000 ms^{-1}
F. More information needed

Question 495:

A 0.5 tonne crane lifts a 0.01 tonne wardrobe by 100 cm in 5,000 milliseconds.
Calculate the average power developed by the crane. Take $g = 10$ ms^{-2}.

A. 0.2 W   C. 5 W   E. 50 W
B. 2 W   D. 20 W   F. More information needed

Question 496:

A 20 V battery is connected to a circuit consisting of a 1 Ω and 2 Ω resistor in parallel. Calculate the overall current of the circuit.

A. 6.67 A   B. 8 A   C. 10 A   D. 12 A   E. 20 A   F. 30 A

Question 497:

Which of the following statements is correct?

A. The speed of light changes when it enters water.
B. The speed of light changes when it leaves water.
C. The direction of light changes when it enters water.
D. The direction of light changes when it leaves water.
E. All of the above.
F. None of the above.

Question 498:

In a parallel circuit, a 60 V battery is connected to two branches. Branch A contains 6 identical 5 Ω resistors and branch B contains 2 identical 10 Ω resistors.

Calculate the current in branches A and B.

	I_A (A)	I_B (A)
A	0	6
B	6	0
C	2	3
D	3	2
E	3	3
F	1	5
G	5	1

Question 499:

Calculate the voltage of an electrical circuit that has a power output of 50,000,000,000 nW and a current of 0.000000004 GA.

A. 0.0125 GV   C. 0.0125 kV   E. 0.0125 mV   G. 0.0125 nV
B. 0.0125 MV   D. 0.0125 V   F. 0.0125 μV

Question 500:

Which of the following statements about radioactive decay is correct?

A. Radioactive decay is highly predictable.
B. An unstable element will continue to decay until it reaches a stable nuclear configuration.
C. All forms of radioactive decay release gamma rays.
D. All forms of radioactive decay release X-rays.
E. An atom's nuclear charge is unchanged after it undergoes alpha decay.
F. None of the above.

Question 501:

A circuit contains three identical resistors of unknown resistance connected in series with a 15 V battery. The power output of the circuit is 60 W.
Calculate the overall resistance of the circuit when two further identical resistors are added to it.

A. 0.125 Ω C. 3.75 Ω E. 18.75 Ω
B. 1.25 Ω D. 6.25 Ω F. More information needed.

Question 502:

A 5,000 kg tractor's engine uses 1 litre of fuel to move 0.1 km. 1 ml of the fuel contains 20 kJ of energy.
Calculate the engine's efficiency. Take $g = 10$ ms^{-2}

A. 2.5 %
B. 25 %

C. 38 %
D. 50 %

E. 75 %
F. More information needed.

Question 503:

Which of the following statements are correct?

1. Electromagnetic induction occurs when a wire moves relative to a magnet.
2. Electromagnetic induction occurs when a magnetic field changes.
3. An electrical current is generated when a coil rotates in a magnetic field.

A. Only 1
B. Only 2

C. Only 3
D. 1 and 2

E. 2 and 3
F. 1 and 3

G. 1, 2 and 3

Question 504:

Which of the following statements are correct regarding parallel circuits?

1. The current flowing through a branch is dependent on the branch's resistance.
2. The total current flowing into the branches is equal to the total current flowing out of the branches.
3. An ammeter will always give the same reading regardless of its location in the circuit.

A. Only 1
B. Only 2

C. Only 3
D. 1 and 2

E. 2 and 3
F. 1 and 3

G. All of the above

Question 505:

Which of the following statements regarding series circuits are true?

1. The overall resistance of a circuit is given by the sum of all resistors in the circuit.
2. Electrical current moves from the positive terminal to the negative terminal.
3. Electrons move from the positive terminal to the negative terminal.

A. Only 1
B. Only 2

C. Only 3
D. 1 and 2

E. 2 and 3
F. 1 and 3

Question 506:

The graphs below show current vs. voltage plots for 4 different electrical components.

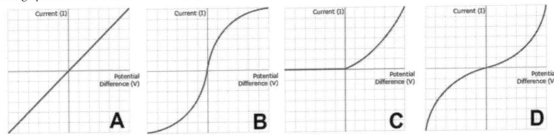

Which of the following graphs represents a resistor at constant temperature, and which a filament lamp?

	Fixed Resistor	Filament Lamp
A	A	B
B	A	C
C	A	D
D	C	A
E	C	C
F	C	D

Question 507:

Which of the following statements are true about vectors?

A. Vectors can be added or subtracted.
B. All vector quantities have a defined magnitude.
C. All vector quantities have a defined direction.
D. Displacement is an example of a vector quantity.
E. All of the above.
F. None of the above.

Question 508:

The acceleration due to gravity on the Earth is six times greater than that on the moon. Dr Tyson records the weight of a rock as 250 N on the moon.

Calculate the rock's density given that it has a volume of 250 cm³. Take $g_{Earth} = 10$ ms⁻²

A. 0.2 kg/cm³
B. 0.5 kg/cm³
C. 0.6 kg/cm³
D. 0.7 kg/cm³
E. 0.8 kg/cm³
F. More information needed.

Question 509:

A radioactive element X_{78}^{225} undergoes alpha decay. What is the atomic mass and atomic number after 5 alpha particles have been released?

	Mass Number	Atomic Number
A	200	56
B	200	58
C	205	64
D	205	68
E	215	58
F	215	73
G	225	78
H	225	83

Question 510:

A 20 A current passes through a circuit with resistance of 10 Ω. The circuit is connected to a transformer that contains a primary coil with 5 turns and a secondary coil with 10 turns. Calculate the potential difference exiting the transformer.

A. 100 V
B. 200 V
C. 400 V
D. 500 V
E. 2,000 V
F. 4,000 V
G. 5,000 V

Question 511:

A metal sphere of unknown mass is dropped from an altitude of 1 km and reaches terminal velocity 300 m before it hits the ground. Given that resistive forces do a total of 10 kJ of work for the last 100 m before the ball hits the ground, calculate the mass of the ball. Take $g = 10$ms⁻².

A. 1 kg
B. 2 kg
C. 5 kg
D. 10 kg
E. 20 kg
F. More information needed.

Question 512:

Which of the following statements is true about the electromagnetic spectrum?

A. The wavelength of ultraviolet waves is shorter than that of x-rays.
B. For waves in the electromagnetic spectrum, wavelength is directly proportional to frequency.
C. Most electromagnetic waves can be stopped with a thin layer of aluminium.
D. Waves in the electromagnetic spectrum travel at the speed of sound.
E. Humans are able to visualise the majority of the electromagnetic spectrum.
F. None of the above.

Question 513:

In relation to the Doppler Effect, which of the following statements are true?

1. If an object emitting a wave moves towards the sensor, the wavelength increases and frequency decreases.
2. An object that originally emitted a wave of a wavelength of 20 mm followed by a second reading delivering a wavelength of 15 mm is moving towards the sensor.
3. The faster the object is moving away from the sensor, the greater the increase in frequency.

A. Only 1
B. Only 2
C. Only 3
D. 1 and 2

E. 1 and 3
F. 2 and 3
G. 1, 2 and 3
H. None of the above statements are true.

Question 514:

A 5 g bullet is travels at 1 km/s and hits a brick wall. It penetrates 50 cm before being brought to rest 100 ms after impact. Calculate the average braking force exerted by the wall on the bullet.

A. 50 N
B. 500 N

C. 5,000 N
D. 50,000 N

E. 500,000 N
F. More information needed.

Question 515:

Polonium (Po) is a highly radioactive element that has no known stable isotope. Po^{210} undergoes radioactive decay to Pb^{206} and Y. Calculate the number of protons in 10 moles of Y. [Avogadro's Constant = 6×10^{23}]

A. 0
B. 1.2×10^{24}

C. 1.2×10^{25}
D. 2.4×10^{24}

E. 2.4×10^{25}
F. More information needed

Question 516:

Dr Sale measures the background radiation in a nuclear wasteland to be 1,000 Bq. He then detects a spike of 16,000 Bq from a nuclear rod made up of an unknown material. 300 days later, he visits and can no longer detect a reading higher than 1,000 Bq from the rod, even though it hasn't been disturbed.
What is the longest possible half-life of the nuclear rod?

A. 25 days
B. 50 days

C. 75 days
D. 100 days

E. 150 days
F. More information needed

Question 517:

A radioactive element Y_{89}^{200} undergoes a series of beta (β^-) and gamma decays. What are the number of protons and neutrons in the element after the emission of 5 beta particles and 2 gamma waves?

	Protons	Neutrons
A	79	101
B	84	111
C	84	116
D	89	111
E	89	106
F	94	111
G	94	106
H	109	111

Question 518:

Most symphony orchestras tune to 'standard pitch' (frequency = 440 Hz). When they are tuning, sound directly from the orchestra reaches audience members that are 500 m away in 1.5 seconds.
Estimate the wavelength of 'standard pitch'.

A. 0.05 m
B. 0.5 m

C. 0.75 m
D. 1.5 m

E. 15 m
F. More information needed

Question 519:

A 1 kg cylindrical artillery shell with a radius of 50 mm is fired at a speed of 200 ms^{-1}. It strikes an armour plated wall and is brought to rest in 500 μs.

Calculate the average pressure exerted on the entire shell by the wall at the time of impact.

A.	5 x 10^6 Pa	C.	5 x 10^8 Pa	E.	5 x 10^{10} Pa
B.	5 x 10^7 Pa	D.	5 x 10^9 Pa	F.	More information needed

Question 520:

A 1,000 W display fountain launches 120 litres of water straight up every minute. Given that the fountain is 10% efficient, calculate the maximum possible height that the stream of water could reach.
Assume that there is negligible air resistance and $g = 10$ ms^{-2}.

A.	1 m	C.	10 m	E.	50m
B.	5 m	D.	20 m	F.	More information needed

Question 521

In relation to transformers, which of the following is true?
1. Step up transformers increase the voltage leaving the transformer.
2. In step down transformers, the number of turns in the primary coil is smaller than in the secondary coil.
3. For transformers that are 100% efficient: $I_p V_p = I_s V_s$

A.	Only 1	E.	1 and 3
B.	Only 2	F.	2 and 3
C.	Only 3	G.	1, 2 and 3
D.	1 and 2	H.	None of the above.

Question 522:

The half-life of Carbon-14 is 5,730 years. A bone is found that contains 6.25% of the amount of C^{14} that would be found in a modern one. How old is the bone?

A.	11,460 years	D.	28,650 years
B.	17,190 years	E.	34,380 years
C.	22,920 years	F.	40,110 years

Question 523:

A wave has a velocity of 2,000 mm/s and a wavelength of 250 cm. What is its frequency in MHz?

A.	8 x 10^{-3} MHz	D.	8 x 10^{-6} MHz
B.	8 x 10^{-4} MHz	E.	8 x 10^{-7} MHz
C.	8 x 10^{-5} MHz	F.	8 x 10^{-8} MHz

Question 524:

A radioactive element has a half-life of 25 days. After 350 days it has a count rate of 50. What was its original count rate?

A.	102,400	E.	819,200
B.	162,240	F.	1,638,400
C.	204,800	G.	3,276,800
D.	409,600		

Question 525:

Which of the following units is **NOT** equivalent to a Volt (V)?

A.	AΩ	D.	NmC
B.	WA^{-1}	E.	JC^{-1}
C.	Nms^{-1}A^{-1}	F.	JA^{-1}s^{-1}

SECTION 2: Maths

BMAT Maths questions are designed to be time draining- if you find yourself consistently not finishing, it might be worth leaving the maths (and probably physics) questions until the very end.

Good students sometimes have a habit of making easy questions difficult; remember that the BMAT only tests GCSE level knowledge so you are not expected to know or use calculus or trigonometry in any part of the exam.

Formulas you **MUST** know:

2D Shapes		3D Shapes		
Area			**Surface Area**	**Volume**
Circle	πr^2	**Cuboid**	Sum of all 6 faces	Length x width x height
Parallelogram	Base x Vertical height	**Cylinder**	$2\pi r^2 + 2\pi rl$	πr^2 x l
Trapezium	0.5 x h x (a+b)	**Cone**	$\pi r^2 + \pi rl$	πr^2 x (h/3)
Triangle	0.5 x base x height	**Sphere**	$4\pi r^2$	$(4/3)\pi r^3$

Even good students who are studying maths at A2 can struggle with certain BMAT maths topics because they're usually glossed over at school. These include:

Quadratic Formula

The solutions for a quadratic equation in the form $ax^2 + bx + c = 0$ are given by: $x = \frac{-b \pm \sqrt{b^2 - 4ac}}{2a}$

Remember that you can also use the discriminant to quickly see if a quadratic equation has any solutions:

$$If\ b^2 - 4ac < 0: No\ solutions$$
$$If\ b^2 - 4ac = 0: One\ solution$$
$$If\ b^2 - 4ac > 2: Two\ solutions$$

Completing the Square

If a quadratic equation cannot be factorised easily and is in the format $ax^2 + bx + c = 0$ then you can rearrange it into the form $a\left(x + \frac{b}{2a}\right)^2 + \left[c - \frac{b^2}{4a}\right] = 0$

This looks more complicated than it is – remember that in the BMAT, you're extremely unlikely to get quadratic equations where $a > 1$ and the equation doesn't have any easy factors. This gives you an easier equation:

$\left(x + \frac{b}{2}\right)^2 + \left[c - \frac{b^2}{4}\right] = 0$ and is best understood with an example.

Consider: $x^2 + 6x + 10 = 0$

This equation cannot be factorised easily but note that: $x^2 + 6x - 10 = (x + 3)^2 - 19 = 0$

Therefore, $x = -3 \pm \sqrt{19}$. Completing the square is an important skill – make sure you're comfortable with it.

Difference between 2 Squares

If you are asked to simplify expressions and find that there are no common factors but it involves square numbers – you might be able to factorise by using the 'difference between two squares'.

For example, $x^2 - 25$ can also be expressed as $(x + 5)(x - 5)$.

Maths Questions

Question 526:

Robert has a box of building blocks. The box contains 8 yellow blocks and 12 red blocks. He picks three blocks from the box and stacks them up high. Calculate the probability that he stacks two red building blocks and one yellow building block, in **any** order.

A. $\frac{8}{20}$ B. $\frac{44}{95}$ C. $\frac{11}{18}$ D. $\frac{8}{19}$ E. $\frac{12}{20}$ F. $\frac{35}{60}$

Question 527:

Solve $\frac{3x+5}{5} + \frac{2x-2}{3} = 18$

A. 12.11   B. 13.49   C. 13.95   D. 14.2   E. 19   F. 265

Question 528:

Solve $3x^2 + 11x - 20 = 0$

A. 0.75 and $-\frac{4}{3}$ C. -5 and $\frac{4}{3}$ F. -12 only

B. -0.75 and $\frac{4}{3}$ D. 5 and $\frac{4}{3}$

E. 12 only

Question 529:

Express $\frac{5}{x+2} + \frac{3}{x-4}$ as a single fraction.

A. $\frac{15x-120}{(x+2)(x-4)}$ C. $\frac{8x-14}{(x+2)(x-4)}$ F. $\frac{8x-14}{x^2-8}$

B. $\frac{8x-26}{(x+2)(x-4)}$ D. $\frac{15}{8x}$

E. 24

Question 530:

The value of p is directly proportional to the cube root of q. When p = 12, q = 27. Find the value of q when p = 24.

A. 32   B. 64   C. 124   D. 128   E. 216   F. 1728

Question 531:

Write 72^2 as a product of its prime factors.

A. $2^6 \times 3^4$ C. $2^4 \times 3^4$ E. $2^6 \times 3$

B. $2^6 \times 3^5$ D. 2×3^3 F. $2^3 \times 3^2$

Question 532:

Calculate: $\dfrac{2.302 \times 10^5 + 2.302 \times 10^2}{1.151 \times 10^{10}}$

A. 0.0000202

B. 0.00020002

C. 0.00002002

D. 0.00000002

E. 0.000002002

F. 0.000002002

Question 533:

Given that $y^2 + \mathbf{a}y + \mathbf{b} = (y + 2)^2 - 5$, find the values of **a** and **b**.

	a	b
A	-1	4
B	1	9
C	-1	-9
D	-9	1
E	4	-1
F	4	1

Question 534:

Express $\dfrac{4}{5} + \dfrac{m-2n}{m+4n}$ as a single fraction in its simplest form:

A. $\dfrac{6m+6n}{5(m+4n)}$

B. $\dfrac{9m+26n}{5(m+4n)}$

C. $\dfrac{20m+6n}{5(m+4n)}$

D. $\dfrac{3m+9n}{5(m+4n)}$

E. $\dfrac{3(3m+2n)}{5(m+4n)}$

F. $\dfrac{6m+6n}{3(m+4n)}$

Question 535:

A is inversely proportional to the square root of B. When A = 4, B = 25.

Calculate the value of A when B = 16.

A. 0.8   B. 4   C. 5   D. 6   E. 10   F. 20

Question 536:

S, T, U and V are points on the circumference of a circle, and O is the centre of the circle.

Given that angle SVU = 89°, calculate the size of the smaller angle SOU.

A. 89°

B. 91°

C. 102°

D. 178°

E. 182°

F. 212°

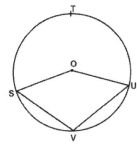

Question 537:

Open cylinder A has a surface area of 8π cm^2 and a volume of 2π cm^3. Open cylinder B is an enlargement of A and has a surface area of 32π cm^2. Calculate the volume of cylinder B.

A. 2π cm^3

B. 8π cm^3

C. 10π cm^3

D. 14π cm^3

E. 16π cm^3

F. 32π cm^3

Question 538:

Express $\frac{8}{x(3-x)} - \frac{6}{x}$ in its simplest form.

A. $\frac{3x-10}{x(3-x)}$

B. $\frac{3x+10}{x(3-x)}$

C. $\frac{6x-10}{x(3-2x)}$

D. $\frac{6x-10}{x(3+2x)}$

E. $\frac{6x-10}{x(3-x)}$

F. $\frac{6x+10}{x(3-x)}$

Question 539:

A bag contains 10 balls. 9 of those are white and 1 is black. What is the probability that the black ball is drawn in the tenth and final draw if the drawn balls are not replaced?

A. 0

B. $\frac{1}{10}$

C. $\frac{1}{100}$

D. $\frac{1}{10^{10}}$

E. $\frac{1}{362,880}$

Question 540:

Gambit has an ordinary deck of 52 cards. What is the probability of Gambit drawing 2 Kings (without replacement)?

A. 0

B. $\frac{1}{169}$

C. $\frac{1}{221}$

D. $\frac{4}{663}$

E. None of the above

Question 541:

I have two identical unfair dice, where the probability that the dice get a 6 is twice as high as the probability of any other outcome, which are all equally likely. What is the probability that when I roll both dice the total will be 12?

A. 0

B. $\frac{4}{49}$

C. $\frac{1}{9}$

D. $\frac{2}{7}$

E. None of the above

Question 542:

A roulette wheel consists of 36 numbered spots and 1 zero spot (i.e. 37 spots in total).

What is the probability that the ball will stop in a spot either divisible by 3 or 2?

A. 0

B. $\frac{25}{37}$

C. $\frac{25}{36}$

D. $\frac{18}{37}$

E. $\frac{24}{37}$

Question 543:

I have a fair coin that I flip 4 times. What is the probability I get 2 heads and 2 tails?

A. $\frac{1}{16}$

B. $\frac{3}{16}$

C. $\frac{3}{8}$

D. $\frac{9}{16}$

E. None of the above

Question 544:

Shivun rolls two fair dice. What is the probability that he gets a total of 5, 6 or 7?

A. $\frac{9}{36}$

B. $\frac{7}{12}$

C. $\frac{1}{6}$

D. $\frac{5}{12}$

E. None of the above

Question 545:

Dr Savary has a bag that contains x red balls, y blue balls and z green balls (and no others). He pulls out a ball, replaces it, and then pulls out another. What is the probability that he picks one red ball and one green ball?

A. $\frac{2(x+y)}{x+y+z}$

B. $\frac{xz}{(x+y+z)^2}$

C. $\frac{2xz}{(x+y+z)^2}$

D. $\frac{(x+z)}{(x+y+z)^2}$

E. $\frac{4xz}{(x+y+z)^4}$

F. More information necessary

Question 546:

Mr Kilbane has a bag that contains x red balls, y blue balls and z green balls (and no others). He pulls out a ball, does **NOT** replace it, and then pulls out another. What is the probability that he picks one red ball and one blue ball?

A. $\frac{2xy}{(x+y+z)^2}$

B. $\frac{2xy}{(x+y+z)(x+y+z-1)}$

C. $\frac{2xy}{(x+y+z)^2}$

D. $\frac{xy}{(x+y+z)(x+y+z-1)}$

E. $\frac{4xy}{(x+y+z-1)^2}$

F. More information needed

Question 547:

There are two tennis players. The first player wins the point with probability p, and the second player wins the point with probability 1-p. The rules of tennis say that the first player to score four points wins the game, unless the score is 4-3. At this point the first player to get two points ahead wins.

What is the probability that the first player wins in exactly 5 rounds?

A. 4p⁴(1-p)

B. p⁴(1-p)

C. 4p(1-p)

D. 4p(1-p)⁴

E. 4p⁵(1-p)

F. More information needed.

Question 548:

Solve the equation $\frac{4x+7}{2} + 9x + 10 = 7$

A. $\frac{22}{13}$

B. $-\frac{22}{13}$

C. $\frac{10}{13}$

D. $-\frac{10}{13}$

E. $\frac{13}{22}$

F. $-\frac{13}{22}$

Question 549:

The volume of a sphere is $V = \frac{4}{3}\pi r^3$, and the surface area of a sphere is $S = 4\pi r^2$. Express S in terms of V

A. $S = (4\pi)^{2/3}(3V)^{2/3}$

B. $S = (8\pi)^{1/3}(3V)^{2/3}$

C. $S = (4\pi)^{1/3}(9V)^{2/3}$

D. $S = (4\pi)^{1/3}(3V)^{2/3}$

E. $S = (16\pi)^{1/3}(9V)^{2/3}$

Question 550:

Express the volume of a cube, V, in terms of its surface area, S.

A. $V = (S/6)^{3/2}$

B. $V = S^{3/2}$

C. $V = (6/S)^{3/2}$

D. $V = (S/6)^{1/2}$

E. $V = (S/36)^{1/2}$

F. $V = (S/36)^{3/2}$

Question 551:

Solve the equations $4x + 3y = 7$ and $2x + 8y = 12$

A. $(x, y) = \left(\frac{17}{13}, \frac{10}{13}\right)$

B. $(x, y) = \left(\frac{10}{13}, \frac{17}{13}\right)$

C. $(x, y) = (1, 2)$

D. $(x, y) = (2, 1)$

E. $(x, y) = (6, 3)$

F. $(x, y) = (3, 6)$

G. No solutions possible.

Question 552:

Rearrange $\frac{(7x+10)}{(9x+5)} = 3y^2 + 2$, to make x the subject.

A. $\frac{15\,y^2}{7 - 9(3y^2 + 2)}$

B. $\frac{15\,y^2}{7 + 9(3y^2 + 2)}$

C. $-\frac{15\,y^2}{7 - 9(3y^2 + 2)}$

D. $-\frac{15\,y^2}{7 + 9(3y^2 + 2)}$

E. $-\frac{5\,y^2}{7 + 9(3y^2 + 2)}$

F. $\frac{5\,y^2}{7 + 9(3y^2 + 2)}$

Question 553:

Simplify $3x \left(\frac{3x^7}{x^{\frac{1}{3}}}\right)^3$

A. $9x^{20}$

B. $27x^{20}$

C. $87x^{20}$

D. $9x^{21}$

E. $27x^{21}$

F. $81x^{21}$

Question 554:

Simplify $2x[(2x)^7]^{\frac{1}{14}}$

A. $2x\sqrt{2}\,x^4$

B. $2x\sqrt{2x^3}$

C. $2\sqrt{2}\,x^4$

D. $2\sqrt{2x^3}$

E. $8x^3$

F. $8x$

Question 555:

What is the circumference of a circle with an area of 10π?

A. $2\pi\sqrt{10}$

B. $\pi\sqrt{10}$

C. 10π

D. 20π

E. $\sqrt{10}$

F. More information needed.

Question 556:

If $a.b = (ab) + (a + b)$, then calculate the value of $(3.4).5$

A. 19

B. 54

C. 100

D. 119

E. 132

Question 557:

If $a.b = \frac{a^b}{a}$, calculate$(2.3).2$

A. $\frac{16}{3}$

B. 1

C. 2

D. 4

E. 8

Question 558:

Solve $x^2 + 3x - 5 = 0$

A. $x = -\frac{3}{2} \pm \frac{\sqrt{11}}{2}$

B. $x = \frac{3}{2} \pm \frac{\sqrt{11}}{2}$

C. $x = -\frac{3}{2} \pm \frac{\sqrt{11}}{4}$

D. $x = \frac{3}{2} \pm \frac{\sqrt{11}}{4}$

E. $x = \frac{3}{2} \pm \frac{\sqrt{29}}{2}$

F. $x = -\frac{3}{2} \pm \frac{\sqrt{29}}{2}$

Question 559:

How many times do the curves $y = x^3$ and $y = x^2 + 4x + 14$ intersect?

A. 0   B. 1   C. 2   D. 3   E. 4

Question 560:

Which of the following graphs **do not** intersect?

1. $y = x$ 2. $y = x^2$ 3. $y = 1-x^2$ 4. $y = 2$

A. 1 and 2

B. 2 and 3

C. 3 and 4

D. 1 and 3

E. 1 and 4

F. 2 and 4

Question 561:

Calculate the product of 897,653 and 0.009764.

A. 87646.8

B. 8764.68

C. 876.468

D. 87.6468

E. 8.76468

F. 0.876468

Question 562:

Solve for x: $\frac{7x+3}{10} + \frac{3x+1}{7} = 14$

A. $\frac{929}{51}$

B. $\frac{949}{47}$

C. $\frac{949}{79}$

D. $\frac{980}{79}$

Question 563:

What is the area of an equilateral triangle with side length x.

A. $\frac{x^2\sqrt{3}}{4}$

B. $\frac{x\sqrt{3}}{4}$

C. $\frac{x^2}{2}$

D. $\frac{x}{2}$

E. x^2

F. x

Question 564:

Simplify $3 - \frac{7x(25x^2 - 1)}{49x^2(5x+1)}$

A. $3 - \frac{5x-1}{7x}$

B. $3 - \frac{5x+1}{7x}$

C. $3 + \frac{5x-1}{7x}$

D. $3 + \frac{5x+1}{7x}$

E. $3 - \frac{5x^2}{49}$

F. $3 + \frac{5x^2}{49}$

Question 565:

Solve the equation $x^2 - 10x - 100 = 0$

A. $-5 \pm 5\sqrt{5}$

B. $-5 \pm \sqrt{5}$

C. $5 \pm 5\sqrt{5}$

D. $5 \pm \sqrt{5}$

E. $5 \pm 5\sqrt{125}$

F. $-5 \pm \sqrt{125}$

Question 566:

Rearrange $x^2 - 4x + 7 = y^3 + 2$ to make x the subject.

A. $x = 2 \pm \sqrt{y^3 + 1}$

B. $x = 2 \pm \sqrt{y^3 - 1}$

C. $x = -2 \pm \sqrt{y^3 - 1}$

D. $x = -2 \pm \sqrt{y^3 + 1}$

E. x cannot be made the subject for this equation.

Question 567:

Rearrange $3x + 2 = \sqrt{7x^2 + 2x + y}$ to make y the subject.

A. $y = 4x^2 + 8x + 2$

B. $y = 4x^2 + 8x + 4$

C. $y = 2x^2 + 10x + 2$

D. $y = 2x^2 + 10x + 4$

E. $y = x^2 + 10x + 2$

F. $y = x^2 + 10x + 4$

Question 568:

Rearrange $y^4 - 4y^3 + 6y^2 - 4y + 2 = x^5 + 7$ to make y the subject.

A. $y = 1 + (x^5 + 7)^{1/4}$

B. $y = -1 + (x^5 + 7)^{1/4}$

C. $y = 1 + (x^5 + 6)^{1/4}$

D. $y = -1 + (x^5 + 6)^{1/4}$

Question 569:

The aspect ratio of my television screen is 4:3 and the diagonal is 50 inches. What is the area of my television screen?

A. 1,200 inches^2

B. 1,000 inches^2

C. 120 inches^2

D. 100 inches^2

E. More information needed.

Question 570:

Rearrange the equation $\sqrt{1 + 3x^{-2}} = y^5 + 1$ to make x the subject.

A. $x = \frac{(y^{10} + 2y^5)}{3}$

B. $x = \frac{3}{(y^{10} + 2y^5)}$

C. $x = \sqrt{\frac{3}{y^{10} + 2y^5}}$

D. $x = \sqrt{\frac{y^{10} + 2y^5}{3}}$

E. $x = \sqrt{\frac{y^{10} + 2y^5 + 2}{3}}$

F. $x = \sqrt{\frac{3}{y^{10} + 2y^5 + 2}}$

Question 571:

Solve $3x - 5y = 10$ and $2x + 2y = 13$.

A. $(x, y) = (\frac{19}{16}, \frac{85}{16})$

B. $(x, y) = (\frac{85}{16}, -\frac{19}{16})$

C. $(x, y) = (\frac{85}{16}, \frac{19}{16})$

D. $(x, y) = (-\frac{85}{16}, -\frac{19}{16})$

E. No solutions possible.

Question 572:

The two inequalities $x + y \leq 3$ and $x^3 - y^2 < 3$ define a region on a plane. Which of the following points is inside the region?

A. $(2, 1)$

B. $(2.5, 1)$

C. $(1, 2)$

D. $(3, 5)$

E. $(1, 2.5)$

F. None of the above.

Question 573:

How many times do $y = x + 4$ and $y = 4x^2 + 5x + 5$ intersect?

A. 0   B. 1   C. 2   D. 3   E. 4

Question 574:

How many times do $y = x^3$ and $y = x$ intersect?

A. 0   B. 1   C. 2   D. 3   E. 4

Question 575:

A cube has unit length sides. What is the length of a line joining a vertex to the midpoint of the opposite side?

A. $\sqrt{2}$

B. $\sqrt{\frac{3}{2}}$

C. $\sqrt{3}$

D. $\sqrt{5}$

E. $\frac{\sqrt{5}}{2}$

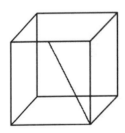

Question 576:

Solve for x, y, and z.

1. $x + y - z = -1$
2. $2x - 2y + 3z = 8$
3. $2x - y + 2z = 9$

	x	y	z
A	2	-15	-14
B	15	2	14
C	14	15	-2
D	-2	15	14
E	2	-15	14
F	No solutions possible		

Question 577:

Fully factorise: $3a^3 - 30a^2 + 75a$

A. $3a(a-3)^3$

B. $a(3a-5)^2$

C. $3a(a^2 - 10a + 25)$

D. $3a(a-5)^2$

E. $3a(a+5)^2$

Question 578:

Solve for x and y:

$$4x + 3y = 48$$
$$3x + 2y = 34$$

	x	y
A	8	6
B	6	8
C	3	4
D	4	3
E	30	12
F	12	30
G	No solutions possible	

Question 579:

Evaluate: $\frac{-(5^2 - 4 \times 7)^2}{-6^2 + 2 \times 7}$

A. $-\frac{3}{50}$ B. $\frac{11}{22}$ C. $-\frac{3}{22}$ D. $\frac{9}{50}$ E. $\frac{9}{22}$ F. 0

Question 580:

All license plates are 6 characters long. The first 3 characters consist of letters and the next 3 characters of numbers. How many unique license plates are possible?

A. 676,000

B. 6,760,000

C. 67,600,000

D. 1,757,600

E. 17,576,000

F. 175,760,000

Question 581:

How many solutions are there for: $2(2(x^2 - 3x)) = -9$

A. 0

B. 1

C. 2

D. 3

E. Infinite solutions.

Question 582:

Evaluate: $\left(x^{\frac{1}{2}} y^{-3}\right)^{\frac{1}{2}}$

A. $\frac{x^{\frac{1}{2}}}{y}$

B. $\frac{x}{y^{\frac{3}{2}}}$

C. $\frac{x^{\frac{1}{4}}}{y^{\frac{3}{2}}}$

D. $\frac{y^{\frac{1}{4}}}{x^{\frac{3}{2}}}$

Question 583:

Bryan earned a total of £ 1,240 last week from renting out three flats. From this, he had to pay 10% of the rent from the 1-bedroom flat for repairs, 20% of the rent from the 2-bedroom flat for repairs, and 30% from the 3-bedroom flat for repairs. The 3-bedroom flat costs twice as much as the 1-bedroom flat. Given that the total repair bill was £ 276 calculate the rent for each apartment.

	1 Bedroom	2 Bedrooms	3 Bedrooms
A	280	400	560
B	140	200	280
C	420	600	840
D	250	300	500
E	500	600	1,000

Question 584:

Evaluate: $5 \left[5(6^2 - 5 \times 3) + 400^{\frac{1}{2}} \right]^{1/3} + 7$

A. 0   B. 25   C. 32   D. 49   E. 56   F. 200

Question 585:

What is the area of a regular hexagon with side length 1?

A. $3\sqrt{3}$

B. $\frac{3\sqrt{3}}{2}$

C. $\sqrt{3}$

D. $\frac{\sqrt{3}}{2}$

E. 6

F. More information needed

Question 586:

Dexter moves into a new rectangular room that is 19 metres longer than it is wide, and its total area is 780 square metres. What are the room's dimensions?

A. Width = 20 m; Length = -39 m

B. Width = 20 m; Length = 39 m

C. Width = 39 m; Length = 20 m

D. Width = -39 m; Length = 20 m

E. Width = -20 m; Length = 39 m

Question 587:

Tom uses 34 meters of fencing to enclose his rectangular lot. He measured the diagonals to 13 metres long. What is the length and width of the lot?

A. 3 m by 4 m

B. 5 m by 12 m

C. 6 m by 12 m

D. 8 m by 15 m

E. 9 m by 15 m

F. 10 m by 10 m

Question 588:

Solve $\frac{3x - 5}{2} + \frac{x + 5}{4} = x + 1$

A. 1

B. 1.5

C. 3

D. 3.5

E. 4.5

F. None of the above

Question 589:

Calculate: $\frac{5.226 \times 10^6 + 5.226 \times 10^5}{1.742 \times 10^{10}}$

A. 0.033

B. 0.0033

C. 0.00033

D. 0.000033

E. 0.0000033

Question 590:

Calculate the area of the triangle shown to the right:

A. $3 + \sqrt{2}$

B. $\frac{2 + 2\sqrt{2}}{2}$

C. $2 + 5\sqrt{2}$

D. $3 - \sqrt{2}$

E. 3

F. 6

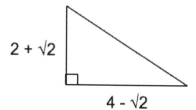

$2 + \sqrt{2}$

$4 - \sqrt{2}$

Question 591:

Rearrange $\sqrt{\frac{4}{x} + 9} = y - 2$ to make x the subject.

A. $x = \frac{11}{(y-2)^2}$

B. $x = \frac{9}{(y-2)^2}$

C. $x = \frac{4}{(y+1)(y-5)}$

D. $x = \frac{4}{(y-1)(y+5)}$

E. $x = \frac{4}{(y+1)(y+5)}$

F. $x = \frac{4}{(y-1)(y-5)}$

Question 592:

When 5 is subtracted from 5x the result is half the sum of 2 and 6x. What is the value of x?

A. 0   B. 1   C. 2   D. 3   E. 4   F. 6

Question 593:

Estimate $\frac{54.98 + 2.25^2}{\sqrt{905}}$

A. 0   B. 1   C. 2   D. 3   E. 4   F. 5

Question 594:

At a Pizza Parlour, you can order single, double or triple cheese in the crust. You also have the option to include ham, olives, pepperoni, bell pepper, meat balls, tomato slices, and pineapples. How many different types of pizza are available at the Pizza Parlour?

A. 10

B. 96

C. 192

D. 384

E. 768

F. None of the above

Question 595:

Solve the simultaneous equations $x^2 + y^2 = 1$ and $x + y = \sqrt{2}$, for x, y > 0

A. $(x, y) = (\frac{\sqrt{2}}{2}, \frac{\sqrt{2}}{2})$

B. $(x, y) = (½, \frac{\sqrt{3}}{2})$

C. $(x, y) = (\sqrt{2} - 1, 1)$

D. $(x, y) = (\sqrt{2}, ½)$

Question 596:

Which of the following statements is **FALSE**?

A.   Congruent objects always have the same dimensions and shape.

B.   Congruent objects can be mirror images of each other.

C.   Congruent objects do not always have the same angles.

D.   Congruent objects can be rotations of each other.

E.   Two triangles are congruent if they have two sides and one angle of the same magnitude.

Question 597:

Solve the inequality $x^2 \geq 6 - x$

A. $x \leq -3$ and $x \leq 2$

C. $x \geq -3$ and $x \leq 2$

E. $x \geq 2$ only

B. $x \leq -3$ and $x \geq 2$

D. $x \geq -3$ and $x \geq 2$

F. $x \geq -3$ only

Question 598:

The hypotenuse of an isosceles right-angled triangle is x cm. What is the area of the triangle in terms of x?

A. $\frac{\sqrt{x}}{2}$

B. $\frac{x^2}{4}$

C. $\frac{x}{4}$

D. $\frac{3x^2}{4}$

E. $\frac{x^2}{10}$

Question 599:

Mr Heard derives a formula: $Q = \frac{(X+Y)^2 A}{3B}$. He doubles the values of X and Y, halves the value of A and triples the value of B.

What happens to value of Q?

A.   Decreases by $\frac{1}{3}$

C.   Decreases by $\frac{2}{3}$

E.   Increases by $\frac{4}{3}$

B.   Increases by $\frac{1}{3}$

D.   Increases by $\frac{2}{3}$

F.   Decreases by $\frac{4}{3}$

Question 600:

Consider the graphs $y = x^2 - 2x + 3$, and $y = x^2 - 6x - 10$. Which of the following is true?

A.   Both equations intersect the x-axis.

B.   Neither equation intersects the x-axis.

C.   The first equation does not intersect the x-axis; the second equation intersects the x-axis.

D.   The first equation intersects the x-axis; the second equation does not intersect the x-axis.

SECTION 3

The Basics

In section 3, you have to write a one A4 page essay on one of three essay titles. Whilst different questions will inevitably demand differing levels of comprehension and knowledge, it is important to realise that one of the major skills being tested is actually your ability to construct a logical and coherent argument- and to convey it to the lay-reader.

Section 3 of the BMAT is frequently neglected by lots of students, who choose to spend their time on sections 1 & 2 instead. However, it has the highest returns per hour of work out of all three sections so is well worth putting time into.

The aim of section 3 is not to write as much as you can. Rather, the examiner is looking for you to make interesting and well supported points, and tie everything neatly together for a strong conclusion. Make sure you're writing critically and concisely; not rambling on. **Irrelevant material can actually lower your score.** You only get one side of A4 for your BMAT essay, so make it count!

Essay Structure

Most BMAT essays consist of 3 parts:

1) Explain what a quote or a statement means.
2) Argue for or against the statement.
3) Ask you "to what extent" you agree with the statement.

Number 1 should be the smallest portion of the essay (no more than 4 lines) and be used to provide a smooth segue into the rather more demanding "argue for/against" part of the question. This main part requires a firm grasp of the concept being discussed and the ability to strengthen and support the argument with a wide variety of examples from multiple fields. This section should give a balanced approach to the question, exploring **at least two distinct ideas**. Supporting evidence should be provided throughout the essay, with examples referred to when possible.

The final part effectively asks for your personal opinion and is a chance for you to shine- be brave and make an **innovative yet firmly grounded conclusion** for an exquisite mark. The conclusion should bring together all sides of the argument, in order to reach a clear and concise answer to the question. There should be an obvious logical structure to the essay, which reflects careful planning and preparation.

Paragraphs

Paragraphs are an important formatting tool which show that you have thought through your arguments and are able to structure your ideas clearly. A new paragraph should be used every time a new idea is introduced. There is no single correct way to arrange paragraphs, but it's important that each paragraph flows smoothly from the last. A slick, interconnected essay shows that you have the ability to communicate and organise your ideas effectively.

Given that you only have a limit of one A4 page to write in – **you shouldn't have more than 5 paragraphs** (use indents to show paragraphs – don't leave empty lines!). In general, 2 of these 5 will be taken up by the introduction and conclusion respectively.

Remember- the emphasis should remain on quality and not quantity. An essay with fewer paragraphs, but with well-developed ideas, is much more effective than a number of short, unsubstantial paragraphs that fail to fully grasp the question at hand.

Approaching the Essay

Section 3 can be broken down into 3 components; selecting your essay title, planning and writing it.

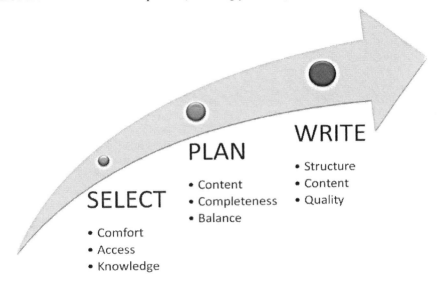

Most students think that the "writing" component is most important. This is simply not true.

The vast **majority of problems are caused by a lack of planning and essay selection**- usually because students just want to get writing as they are worried about finishing on time. Thirty minutes is long enough to be able to plan your essay well and *still* have time to write it so don't feel pressured to immediately start writing.

Step 1: Selecting

Selecting your essay is crucial- make sure you're comfortable with the topic and ensure you understand the actual question- it sounds silly but about 25% of essays that we mark score poorly because they don't actually answer the question!

Take two minutes to read all the questions. Whilst one essay might originally seem the easiest, if you haven't thought through it you might quickly find yourself running out of ideas. Likewise, a seemingly difficult essay might actually offer you a good opportunity to make interesting points.

Use this time to carefully select which question you will answer by gauging how accessible and comfortable you are with it given your background knowledge. Section 3, however, is not a test of knowledge but rather a test of how well you are able to argue.

It's surprisingly easy to change a question into something similar, but with a different meaning. Thus, you may end up answering a completely different essay title. Once you've decided which question you're going to do, read it very carefully through a few times to make sure you fully understand it. Answer all aspects of the question. Keep reading it as you answer to ensure you stay on track!

Step 2: Planning

Why should I plan my essay?
There are multiple reasons you should plan your essay for the first 5-10 minutes of section 3:

- As you don't have much space to write, make the most of it by writing a very well organised essay.
- It allows you to get all your thoughts ready before you put pen to paper.
- You'll write faster once you have a plan.
- You run the risk of missing the point of the essay or only answering part of it if you don't plan adequately.

How much time should I plan for?

There is no set period of time that should be dedicated to planning, and everyone will dedicate a different length of time to the planning process. You should spend as long planning your essay as you require, but it is essential that you leave enough time to write the essay. As a rough guide, it is **worth spending about 5-10 minutes to plan** and the remaining time on writing the essay. However, this is not a strict rule, and you are advised to tailor your time management to suit your individual style.

How should I go about the planning process?

There are a variety of methods that can be employed in order to plan essays (e.g. bullet-points, mind-maps etc). If you don't already know what works best, it's a good idea to experiment with different methods.

Generally, the first step is to gather ideas relevant to the question, which will form the basic arguments around which the essay is to be built. You can then begin to structure your essay, including the way that points will be linked. At this stage it is worth considering the balance of your argument, and confirming that you have considered arguments from both sides of the debate. Once this general structure has been established, it is useful to consider any examples or real world information that may help to support your arguments. Finally, you can begin to assess the plan as a whole, and establish what your conclusion will be based on your arguments.

Step 3: Writing

Introduction

Why are introductions important?

An introduction provides tutors with their first opportunity to examine your work. The introduction is where first impressions are formed, and these can be extremely important in producing a convincing argument. A well-constructed introduction shows that you have really thought about the question, and can indicate the logical flow of arguments that is to come.

What should an introduction do?

A good introduction should **briefly explain the statement or quote** and give any relevant background information in a concise manner. However, don't fall into the trap of just repeating the statement in a different way. The introduction is the first opportunity to suggest an answer to the question posed- the main body is effectively your justification for this answer.

Main Body

How do I go about making a convincing point?

Each idea that you propose should be supported and justified, in order to build a convincing overall argument. A point can be solidified through a basic Point → Evidence → Evaluation process. By following this process, you can be assured each sentence within a paragraph builds upon the last, and that all the ideas presented are well solidified.

How do I achieve a logical flow between ideas?

One of the most effective ways of displaying a good understanding of the question is to keep a logical flow throughout your essay. This means linking points effectively between paragraphs, and creating a congruent train of thought for the examiner as the argument develops. A good way to generate this flow of ideas is to provide ongoing comparisons of arguments, and discussing whether points support or dispute one another.

Should I use examples?

In short – yes! Examples can help boost the validity of arguments, and can help display high quality writing skills. Examples can add a lot of weight to your argument and make an essay much more relevant to the reader. When using examples, you should ensure that they are relevant to the point being made, as they will not help to support an argument if they are not.

Some questions will provide more opportunities to include examples than others so don't worry if you aren't able to use as many examples as you would have liked. There is no set rule about how many examples should be included!

> *Top tip!* Remember that there is no single correct answer to these questions and you're not expected to be able to fit everything onto one page. Instead it's better to pick a few key points to focus on.

Conclusion

The conclusion provides an opportunity to emphasise the **overall sentiment of your essay** which readers can then take away. It should summarise what has been discussed during the main body and give a definitive answer to the question.

Some students use the conclusion to **introduce a new idea that hasn't been discussed**. This can be an interesting addition to an essay, and can help make you stand out. However, it is by no means, a necessity. In fact, a well-organised, 'standard' conclusion is likely to be more effective than an adventurous but poorly executed one.

Medical Ethics

There is normally a medical ethics questions in most years so it's well worth knowing the basics. Whilst there are huge ethical textbooks available– you only need to be familiar with the basic principles for the purposes of the BMAT. **These principles can be applied to all cases** regardless what the social/ethnic background the healthcare professional or patient is from. In addition to being helpful in the BMAT, you'll need to know them for the interview stages anyway so they're well worth learning now. The principles are:

Beneficence: The wellbeing of the patient should be the doctor's first priority. In medicine this means that one must act in the patient's best interests to ensure the best outcome is achieved for them i.e. 'Do Good'.

Non-Maleficence: This is the principle of avoiding harm to the patient (i.e. Do no harm). There can be a danger that in a willingness to treat, doctors can sometimes cause more harm to the patient than good. This can especially be the case with major interventions, such as chemotherapy or surgery. Where a course of action has both potential harms and potential benefits, non-maleficence must be balanced against beneficence.

Autonomy: The patient has the right to determine their own health care. This therefore requires the doctor to be a good communicator, so that the patient is sufficiently informed to make their own decisions. 'Informed consent' is thus a vital precursor to any treatment. A doctor must respect a patient's refusal for treatment even if they think it is not the correct choice. Note that patients cannot <u>demand</u> treatment – only refuse it, e.g. an alcoholic patient can refuse rehabilitation but cannot demand a liver transplant.

There are many situations where the application of autonomy can be quite complex, for example:
> **Treating Children**: Consent is required from the parents, although the autonomy of the child is taken into account increasingly as they get older.
> **Treating adults without the capacity** to make important decisions. The first challenge with this is in assessing whether or not a patient has the capacity to make the decisions. Just because a patient has a mental illness does not necessarily mean that they lack the capacity to make decisions about their health care. Where patients do lack capacity, the power to make decisions is transferred to the next of kin (or Legal Power of Attorney, if one has been set up).

Justice: This principle deals with the fair distribution and allocation of healthcare resources for the population.

Consent: This is an extension of Autonomy- patients must agree to a procedure or intervention. For consent to be valid, it must be **voluntary informed consent.** This means that the patient must have sufficient mental capacity to make the decision and must be presented with all the relevant information (benefits, side effects and the likely complications) in a way they can understand.

Confidentiality: Patients expect that the information they reveal to doctors will be kept private- this is a key component in maintaining the trust between patients and doctors. You must ensure that patient details are kept confidential. Confidentiality can be broken if you suspect that a patient is a risk to themselves or to others e.g. Terrorism, suicides.

When answering a question on medical ethics, you need to ensure that you show an appreciation for the fact that there are often two sides to the argument. Where appropriate, you should outline both points of view and how they pertain to the main principles of medical ethics and then come to a reasoned judgement.

Common Mistakes

Ignoring the other side of the argument
Although you're normally required to support one side of the debate, it's important to **consider arguments against your judgement** in order to get the higher marks. A good way to do this is to propose an argument that might be used against you, and then to argue why it doesn't hold true. You may use the format: *"some may say that...but this doesn't seem to be important because..."* in order to dispel opposition arguments, whilst still displaying that you have considered them. For example, *"some say that fox hunting shouldn't be banned because it is a tradition. However, witch hunting was also once a tradition – we must move with the times"*.

Answering the topic/Answering only part of the question
One of the most common mistakes is to only answer a part of the question whilst ignoring the rest of it as it's inaccessible. According to the official mark scheme, **in order to get a score of 3 or more, you must write "...*an answer that addresses ALL aspects of the question*".** This should be your minimum standard- anything else that you write should then point you towards achieving 4/5.

Long Introductions
Some students can start rambling and make introductions too long and unfocussed. Although background information about the topic can be useful, it is normally not necessary. Instead, the **emphasis should be placed on responding to the question**. Some students also just **rephrase the question** rather than actually explaining it. The examiner knows what the question is, and repeating it in the introduction is simply a waste of space in an essay where you are limited to just one A4 side.

Not including a Conclusion
An essay that lacks a conclusion is incomplete and can signal that the answer has not been considered carefully or that your organisation skills are lacking. **The conclusion should be a distinct paragraph** in its own right and not just a couple of rushed lines at the end of the essay.

Sitting on the Fence
Students sometimes don't reach a clear conclusion. You need to **ensure that you give a decisive answer to the question** and clearly explain how you've reached this judgement. Essays that do not come to a clear conclusion generally have a smaller impact and score lower.

Exceeding the one page limit
The page limit is there for a reason – don't exceed it under any circumstances as any material over the limit won't be marked and it will appear that you haven't read the instructions.

Not using all the available space
Remember that you only have one A4 side to write on so ensure you make the maximum use of the space available to you. Don't leave lines to show paragraphs – instead, you should use indents. Similarly, you should also use the top-most line in the response sheet and avoid crossing entire sentences out.

Marking your Essays

Practicing section 3 can be tricky because most students don't know how to mark their essay. However, if you have a willing friend/family member, it is fairly easy to mark your own work. You can use the diagram below to get an idea of your score – keep in mind that this is just a very rough guide – examiners will look at several other factors when deciding on your overall score.

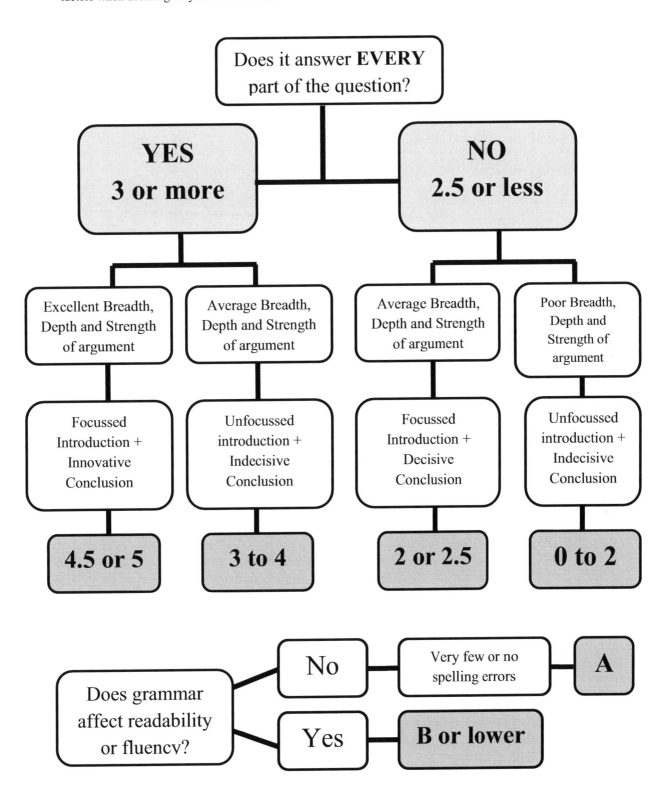

Annotated Essays

Example Essay 1:

"A doctor should never disclose medical information about his patients"
What does this statement mean? Argue to the contrary using examples to strengthen your response.
To what extent do you agree with this statement?

The statement suggests that one of a doctor's most vital qualities is maintaining confidentiality of a patient's medical record. This involves all doctors with various specialities in different work places such as the clinics.

Disclosing medical informations regarding to their patients by doctors is considered as an unacceptable act within the medical society. For Example, by informing unrelated people about the patient might result in the individual's most embarrased health situation to be exposed. For example, suffering pain from their private parts and this may disgust other. This situation would inevitably upset the patient as their health privacy has been breached by others without consent leading to a sence of distrust towards doctors.

However, disclosing such matters to certain suitable people such as family and relatives may be crucial. For example, if the patient is the head of the family or the guardian to the children. As these individuals are in charge of leading and taking are of the family, they need to be able to perform mundane task (such as providing good support to the children) at their optimum. Also, to ensure that the members realise that the patient should not over exert him or herself despite their health conditions. Also, a sudden collapse will reduce shock when the family is to rush the patient back to the hospital knowing that the illness is related to the situation.

Overall, a patient's confidentially should not be disclosed without consent or any importance by all means. This is to respect their health privacy and to avoid any inconvenience within the medical society.

Examiner's Comments:

Introduction: The student appears to have an understanding of the topic but frequently makes statements that don't add much to the argument e.g. the second sentence of the first paragraph. The introduction would be better used to set up the counter arguments that will form the bulk of the main body.

Main Body: The first paragraph actually supports the statement and is therefore not actually answering the question. The example doesn't really add much either. The key issue that needed to be discussed was a doctor's duty to the patient – not about "disgusting others". The last sentence of the first paragraph is good and starts to address the question but it comes far too late.

The second paragraph is better but misses the key points of the essay that were necessary to discuss i.e. when can patient confidentiality be broken? Examples would include suspected terrorism, notifiable diseases and criminal activity, suicides etc.

Conclusion: The conclusion doesn't really address the counter-arguments for breaking confidentiality or give a balanced answer. It also contains confusing terminology e.g. one discloses confidential information, not 'confidentiality'. The sentence concerning "inconvenience within the medical community" is also somewhat ambiguous.

Language: The grammar hinders the points that the student is trying to make throughout the essay e.g. "suffering pain from their private parts and this may disgust other". There are also frequent spelling mistakes like "embarrased" and "sence" that reduce the fluency significantly.

Score: D2

Example Essay 2

"A doctor should never disclose medical information about his patients"
What does this statement mean? Argue to the contrary using examples to strengthen your response.
To what extent do you agree with this statement?

This statement is one of the duties set out by the General Medical Council for doctors to comply with, which is to respect patient's autonomy. It means that a doctor cannot share patient's medical information to other parties unless the patients, themselves, have granted permissions to do so.
The ethical principle of respecting patients' autonomy cannot be applied in all cases, as some cases require doctors to disclose medical information about patients. First, when it involves a criminal act that has been comitted by a patient, a doctor has to report to any appropriate authority, such as the police. This is because the patient may potentially cause even more harm to others and as doctors, they have to prevent that from happening. An example of a case would be if a patient has suffered from a gunshot wound and he had told his doctor that he had gotten it when he murdered someone.

Next, another incident when a doctor has no choice but to disclose patients' medical information is when it may affect the health of society and could potentially cause an epidemic. Such patients might have an infectious disease and do not wish to let other people know about it. For instance, there has been many cases in West Africa where people who have Ebola are afraid to let their neighbours or friends find out because they do not want to be stigmatised and ostrilised from the society. However, these patients could spread the disease and so a doctor must not withold the information. Last of all, if a patient is underage, then he/ she is still not competent enough to make her own decision. Therefore, any medical information must be shared with his/ her legal guardian.

Respecting patient's autonomy by never disclosing their information is also important because patients have the right to chose who gets to know about it. It is his own body. He is the only person who knows the consequences of sharing this sensitive information. In conclusion, I believe that never disclosing patients' medical information cannot be complied in every incident. Respecting their autonomy is important but we have to treat each case separately.

Examiner's Comments:

Introduction: The introduction is well written but could be improved by making it explicitly clear that confidentiality can be breached in certain circumstances. This would then set up the main body nicely as the student would then be able to go straight into giving examples.

Main Body: The first sentence of the second paragraph is well written but should have gone in the introduction. There is good breadth of argument with the important points being covered like a doctor's duty to prevent harm to others, 'public good' and the issue of 'capacity'. However, there is unnecessary padding that doesn't add much e.g. there was no need to expand on your example of infectious diseases. The extra space from avoiding this would have allowed the student to write about the fact that although confidentiality must sometimes be breached, a doctor has certain professional duties. For example, informing the patient both before and after and explaining why they have disclosed what they have to try and mitigate any damage to the doctor patient relationship. In this way public trust in medical professionals would be maintained.

Conclusion: The conclusion concisely summarises the arguments put forth in the main body and offers a nice resolution by saying that each case is different.

Language: Whilst it is clear that the student understands the question – there is some confusion as to the difference between "autonomy" and "confidentiality" (It's important to know the basics of medical ethics as they'll be helpful for the interview stage as well).Furthermore, there are minor spelling mistakes like "comitted" and "withold". In the conclusion, they also assume that patients are only male – "it is his own body" vs. "it is their own body".

Score: B3.5

Example Essay 3

"A doctor should never disclose medical information about his patients"
What does this statement mean? Argue to the contrary using examples to strengthen your response.
To what extent do you agree with this statement?

Confidentiality is a basic patient right. The patient provides information to the doctor not to be unnecessarily shared with others without their knowledge or permission. On this basis, the statement argues that a doctor should never reveal the medical data, such as results from tests or prescriptions given.

However, it can be argued that there are many circumstances whereby it is necesary to breach patient confidentiality and disclose medical information. More specifically, if the patient poses a threat to the public health, their medical situation should be disclosed immediately so that actions can be taken to prevent the spread. For instance, under the Public Health Act 1988, if a patient is suspected of communicable diseases such as tuberculosis, the doctor is required to inform the local health authorities immediately so that they can make precautions to protect the other citizens. In addition, a doctor should also disclose medical information if the patient has broken the law. For instance, the doctor should reveal medical data to the detectives and other relevant professionals if they request for it, to enable them to come to a conclusion of the case more quickly and accurately.

However, I agree with this statement to a large extent. After all, the patient should have the right over what happens to his medical documents and information. Revealing information about the patient unnecessarily will take this basic right away, and it is extremely unfair for the patient. Furthermore, this unprofessional decision may undermine the confidence between the patient and doctor. The patient may be less willing to reveal vital personal information to the doctor in the future, in fear that he might release this information as well. This would be extremely detrimental to the diagnoses and treatment for the patient or the doctor might not be able to gain sufficient information to make a more informed decision.

In conclusion, a doctor should never disclose medical information about his patients unless there are other external circumstances that oblige the doctor to do so. Breaking this confidentiality will cause the patient-doctor relationship to collapse, compromising the trust between them. However, in some cases, the decision to disclose is not that clear-cut; if a patient had sexually-transmitted infections, should the doctor disclose this information to his spouse? Such situations have to be decided on a case-by-case basis.

Examiner's Comments:

Introduction: This is a bold introduction that catches one's attention and gets straight to the point. The student however does make a rather generalised statement - "confidentiality is a basic patient right". A pedantic examiner could easily challenge this and thus, it's important to be careful with your wording.

Main Body: There is a good level of breadth and depth of argument here. However, there are again some generalising statements that are incorrect e.g. doctors don't need to break confidentiality if a patient has broken ANY law – just serious ones e.g. committed or intend to commit murder/terrorism etc. There is however, excellent discussion of the consequences of breaking confidentiality and a good level of detail (e.g. Public health act 1988).

Conclusion: An excellent conclusion that not only summarises the main arguments from both sides but also builds upon these to offer a solution as to when to break confidentiality by treating it on a case-by-case basis.

Language: There is only one spelling mistake ("*necessry*") *and* although the somewhat general phrases stop this from being a perfect A5, it is still an excellent essay that displays good insight. Spend time making sure that you write exactly what you mean, instead of being loose with your words and conveying an incorrect message.

Score: A4.5

Example Essay 4

"Medicine is a science; not an art."

Explain what this statement means. Argue to the contrary that medicine is in fact an art using examples to illustrate your answer. To what extent, if any, is medicine a science?

I will explain the following statement "Medicine is a science, not an art" and Argue. to that medicine is in fact an art with supports and to what extent is medicine a science.

I will talk about medicine and art in my opinion why, I think medicine is in fact an art and argue with the above statement. I think medicine is an art because of the human body its like a piece of artwork with creations and it is fascinating even more than a artwork of Picccasso or Ven Gough's painting. It's like the artries and vessels ressemble the brushstroke of a painting and the heart is the meaning of the painting. To study about that and become a Doctor is like studying art to become and artist. Both Medicine and Art depend on passion if you don't have the passion you will not enjoy saving lives and will not create beautiful paintings. Medicine and Art are the most fascinating majors. They are completely different but at the sane time completely the same. Emotionally the same.

Now I will talk about the what extent is medicine a science. It depends on what course you are choosing in medicine For e.g. Biomedical it all counts as science it includes chemistry and maths which makes you a Biochemist or Phisyology, these are some interesting courses. Medicine is science because it depends on knowledge the years to study to become a doctor and save lives which is not an easy opprotunity to get.

In my essay I wrote about why I think that medicine is in face not all about sciene but about art too and argued with the above essay. I think Medicine is as fascinating and breathtaking as Art is with all the colours and bueatiful creations made my artists and saved by doctors.

Examiner's Comments:

Introduction: Although it is sometimes useful to outline what you are going to discuss when writing academic essays – you simply do not have enough space to do this in the BMAT. The introduction should be a clear and concise explanation of the statement, which isn't really done in this essay. Repeating the statement in the essay should also be avoided.

Main Body: The student doesn't really have a good grasp of what the question is asking and as a result, the argument is off topic and slightly incoherent. The question requires the student to discuss things like medicine is a science because doctors put into practice medical principles that have an empirical factual basis. Whilst the third paragraph does this to a certain extent – the message is diluted somewhat because of a lack of focus. Medicine is also an 'art' because doctors also need to be able to communicate well with patients, to interpret clinical signs etc. The student seems have interpreted the question to literally mean "medicine is art" vs. "medicine is an art".

Conclusion: When writing a conclusion it is good practice to just make your point, as opposed to telling the reader you are making it. Thus, meta-writing is again not necessary in the conclusion. The final sentence, although interesting is again, off-topic.

Language: There are frequent spelling mistakes e.g. "sane time", "ressemble" and "*bueatiful*". *The phrasing and grammatical errors are more serious and significantly affect the essay's fluency e.g. "talk about the what extent".*

Score: D1

Example Essay 5

"Medicine is a science; not an art."

Explain what this statement means. Argue to the contrary that medicine is in fact an art using examples to illustrate your answer. To what extent, if any, is medicine a science?

Medicine. Arguably one of the most advancing fields in today's society. As a result, many of us have often thought about what medicine actually encompasses. The statement, "medicine is a science; not an art." is one that is constantly the subject of debated today. It questions to what extent that medicine may be considered an art, and to what a science. I feel that the statement does not belove that medicine is an art form but instead, is well and truly a science, and is well within the definition of one.

An art form is usually something which is viewed as being expressive and emotional, as well as also being delicate. One can argue that the notion of caring for patients can be viewed as the non-scientific aspect of medicine, and therefore could be considered art as it requires one to be emotional and expressive. In order for a patient's rehabilitation process to be complete and succesfull, the care directed at the patient should be tailored for them as the recovery of the psychological side of the human body is just as important as the phsiological aspect. In order to be truly effective, the doctor should be able to be empathetic and try and understand the patient's pain when comforting them. This aspect of medicine does not involve the science of the human body or the knowledge of the intricate metabolic reactors which allow the human body to function so effectively.

However, another side of the debate could be that medicine is very much a science. This is due to the fact that medicine involves the analysis of a human body, for example, when diagnosing a patient or maybe understanding the effects a drug could bring to specific situations in the human body, something that is viewed with the utmost importance when administering a drug. The fact that in order to succesfully become a medical practitioner, one has to be aware of the physiology of the human body and also the immense and thorough knowledge of the anatomy of the human body, is the reason that medicine is associated with mainly being a science. This stems back to the days when we used to learn Biology and Chemistry back in school, and because Medicine is largely based on those two core subjects, medicine, as a result, is widely regarded as solely being a science.

Examiner's Comments:

Introduction: The opening is catchy although unnecessarily long-winded. Thus, whilst the first sentence grabs the attention, the rest of it does little to keep it the next few sentences are very wordy and don't actually say very much. This is a prime example of just rephrasing the statement rather than explaining it.

Main Body: There are a good range of examples in the second paragraph- especially those about the rehabilitation process and empathy. Whilst the pro-science arguments are also well made, they are a bit one-dimensional. It would be better to discuss how trials are done to ensure safety rather than just concentrating on the human anatomy and physiology. The last sentence also doesn't really add anything and if anything detracts from the final paragraph.

Conclusion: The essay is badly let down by a lack of a conclusion. This title required a critical analysis of the strengths of the two sides of the arguments in order to produce a well-constructed conclusion that answered the question.

Language: There are some very elegant turns of phrase that sometimes result in a loss of focus. This doesn't affect the essay's readability (and therefore the language score) but indirectly affects the strength of argument quite substantially. Nevertheless, as all parts of the question are answered to a sufficient level, it still scores a solid 3.

Score: A3

Example Essay 6

"Medicine is a science; not an art."

Explain what this statement means. Argue to the contrary that medicine is in fact an art using examples to illustrate your answer. To what extent, if any, is medicine a science?

This statement argues that medicine is more deeply rooted with the facts and set observations associated with scientific principles, and that no aspect of medicine is in itself open to interpretation; or art.

However many of the facets of Medicine could very well be regarded as art. The manual dexterity and presision required by surgeons; particularly the visually aesthetic finish reconstructive or plastic surgery aims towards has a deep basis in artistic ability. Medicine as a career has a hugely important social aspect too; healthcare proffessionals are expected to be involved with communication and, when it comes to patients, even dealing with emotions has an interpretative aspect, and as such could be viewed as art. There is not one set approach to these situations; rather the outcome is reliant on a doctor's own personal judgement and choices as to the best course of action. Theoretical medicine, too, in terms of research may find success in new, recently discovered techniques; the development of which requires thinking 'outside of the box'. The clinical aspect of diagnostic medicine too is subject to unique approached, particularly when introducing extremely complex cases.

Although to the same extent, medicine of course has a heavy involvement with biological structures, chemical processes and even principles of physics- generally all empirical; based on evidence and set in stone. In many cases there is a clear distinction between the right interpretation and thus course of action And the wrong one; for example when prescribing- for many patients (such as when allergies are involved) there are a whole host of drugs and courses of action that are unnacceptable. Therefore, the understanding of the human body and it's inner workings that is so crucial for appropriate and successful medicine could well be argued to be science. Yet it's application; it's uses by doctors and other healthcare proffessionals is less impirical, more open to interpretation, and therefore more so an art.

I feel that despite the fact medicine involves understanding and knowledge of the physiology and anatomy of the human body, it also involves the integral of caring for others, which is definitely more of an art form than a science, showing us that despite all the debates, medicine manages to combine art and science together resulting in the formation of a wondrous profession.

Examiner's Comments:

Introduction: An efficient introduction that explains the statement well. It could be improved by setting up the counter-argument.

Main Body: This is a good effort at a tricky essay – there is good breadth of argument with lots of examples like surgical precision and communication. There is also good consideration of the counter-arguments for why medicine is a science with good depth of argument e.g. drug prescriptions.

Conclusion: A strong conclusion that addresses the question well and summarises arguments from both sides concisely. It uses unnecessarily romantic language "wondrous profession" but nevertheless is an effective closing paragraph.

Language: Although there are a noticeable number of spelling mistakes (*"proffessionals" and "impirical"*), they do not detract from the essay's flow.

Score: B4.5

Example Essay 7

"The primary duty of a doctor is to prolong life as much as possible"

What does this statement mean? Argue to the contrary, that the primary duty of a doctor is not to prolong life. To what extent do you agree with this statement?

The most important responsibility of doctor is to cure diseases and extend a life of patients at his most ability acquired. The statement also states that doctors should try their best to prolong life of the sufferers as the first principle to consider. However, some might argue with this.

It is true to say that doctors are responsible for improving the conditions of diseases and alleviate the symptoms. This does not necessarily mean that is a major factor to tackle with each disease. Preventive care should be introduced at an early stage, and therefore the primary duty of medical professions, especially doctors should adopt this principle to be their main concern. For example, doctors should be aware of other health conditions of his or her patients that might be developed in the future regarding to patient's lifestyle or eating habit. To only address the present problems, relating to particular disease is not enough. Hence, prolonging life of patients is not the primary concern of doctors but to improve quality of life of the sufferers. By preventing the possible disease and acknowledging the patients are a proper most important role of a doctor.

Some people might argue with the statement as there have been a very controversial issue raised in recent years, euthanasia. Nowadays it is evidently seen that some patients in Switzerland and some other countries have their right to urge a doctor to help end their lives peacefully. Doctor may put an emphasis on methods or alternative ways to help prolong life of patient. Their prime concern is also finding the best beneficial treatment in order to fight with the disease unless there is a possible way. Therefore, putting patients at ease by ending their lives is also a primary concern to a doctor in some countries.

To some extent, I agree and support this statement as doctors have to delegate roles as healer to those who are in pain. Although it is illegal in some regions of the world to allow doctors taking life of a patient with their consent, it does mean that this method apart from prolonging life is one of the main duty for doctors to be well aware.

Examiner's Comments:

Introduction: This is a concise introduction that effectively just rephrases the statement rather than developing on it much – effectively not advancing an argument and thus wasting valuable space. A discussion of what a doctor's <u>primary</u> duty actually is would have been more appropriate here and would have set up the main body far better.

Main Body: This is a rather confusing and muddled account. The second paragraph is difficult to follow and strays from the real topic at hand. The student correctly identifies that a doctor's job is to improve quality of life and not duration. However, this isn't expressed with any degree of clarity or any examples. The third paragraph regarding euthanasia is better but again it is difficult to assess what point the student is trying to make.

Conclusion: Whilst it is perfectly fine to present a one-sided argument, you need to at least consider the counter-arguments. Thus, it was necessary here to consider that both quality of life <u>and</u> are important. In addition, the euthanasia part is somewhat misguided – not all doctors who don't put duration of life as their top priority are in favour of euthanasia.

Language: There are frequent grammatical errors e.g. "as there have been a very" which reduce the essay's fluency and. The sentence phrasing also impacts the essay's overall readability to lead to a poor language score.

Score: C2

Example Essay 8

"The primary duty of a doctor is to prolong life as much as possible"

What does this statement mean? Argue to the contrary, that the primary duty of a doctor is not to prolong life. To what extent do you agree with this statement?

A doctors' job is to cure disease through medical treatment to extend the life of the patient as much as possible. However, there is a certain limit to which doctors can go in order to prolong life as often quality of life is equally (if not more) important as quantity of life.

Doctors provide treatments to patients to help them overcome their disease so that they can live longer. This is also what the patients want. For example, for patients with kidney diseases, doctors will suggest them to have dialysis in order to remove the toxic substance in their body, which will kill them. Through dialysis, the patient's life will be extended and as this is the patient's will, a doctor's primary duty is prolong life. People take preventative method, such as endoscopy of large intestine for symptoms of cancer, to avoid late discovery of disease, which will lead to a high chance of death. Therefore as a doctor has the skills to help people to extend their, they should do as much as they can to fulfil the patient's wish, which is prolonging life.

However, doctors should also respect the patient's autonomy if a patient doesn't want a treatment, even the treatment is effective, doctors should not carry the treatment out, as everyone have the right to control their own life, even doctors want to help patient, doctors should not over-ride the will of the patients.

Although one of a doctor's duties is to prolong life – this shouldn't be at the expense of quality of life. A doctors primary duty is to offer the best possible medical advice and minimise suffering. Although the impact of this is usually to prolong life, in some cases, it may result in maximising quality of life.

Examiner's Comments:

Introduction: An excellent introduction that sets the scene very nicely for the main body and immediately conveys to the examiner that the student understands the essay of the essay.

Main Body: There are good points made throughout e.g. dialysis and the inclusion of preventative treatment shows a good insight into medicine. However, the student is not able to put together a substantial enough argument against prolonging life as they spend too much space discussing the reasons for prolonging life (which are less important). This is a perfect example of an unbalanced essay. In general, there is limited focus on the question and as a result, weak depth of argument.

Conclusion: A satisfactory conclusion that expresses the sentiments of the conclusion nicely and addresses the question well.

Language: The introduction and conclusion rescue this essay – as they convey a high degree of understanding in only a couple of sentences. The essay itself reads well and there are no obvious errors that reduce its fluency.

Score: A3.5

Example Essay 9

Animal euthanasia should be made illegal.

Explain what this statement means. Argue to the contrary that animal euthanasia should remain legal. To what extent do you agree with the statement?

The statement refers to the ethical dilemma of euthanasia. Euthanasia, synonymous of mercy killing, is the ending of someone's life because of a particular situation in which this living creature's future will be painful, sometimes short or it will ultimately be better to end it soon. The first argument for this statement is the fact that we as humans are generally kind hearted and benevolent: we see pain as a thing to be eradicated if not suppressed, and hence euthanasia could be seen as merciful, given this nature of ours. However, the natural argument against this is the undeniable fact that in the eyes of many, euthanasia is glorified murder, which goes against most people's morals since killing is seen as a negative thing in most societies nowadays. another argument in favour of this process could be the fact that killing is not only condemned but also morally wrong since it involves us "removing" an otherwise healthy living being. The counter-argument illustrated by this statement is the fact that most people would look upon this as a grateful act of mercy, where although the consequences are taken into account, it is, morally, for some, the best thing to do, particularly if an animal, a source of fondness for some, is involved. The last argument to put forth is that animals are intelligent creatures capable of feeling pain like us and should hence receive the same mercy as is sometimes shown to humans. Naturally, people would say that in some cases this would be murder rather than mercy killing, eg the killing of old horses for glue rather than because of old age. I conclude this should remain legal since animals cannot voice their own opinion and hence give more weight to a decision.

Examiner's Comments:

Introduction: This is a good albeit somewhat long introduction. The student has defined euthanasia well and then established that there is an ethical dilemma surrounding it. However, the very long sentences make it needlessly difficult to follow.

Main Body: Whilst the writing style is excellent, there is a limited amount of content here. The student presents arguments for both sides simultaneously which sometimes makes it difficult to follow. This is made more confusing by the lack of paragraphs which means that the essay doesn't flow as well.

There also some long and rambling sentences which detract from the clarity of argument. The essay would benefit from a more focussed approach in which the student gets to the point. The point about killing horses for glue is also not as relevant to euthanasia outside a slippery slope argument (which isn't expanded upon).

Conclusion: The conclusion does not really build upon any of the arguments from the main body. This gives the impression that it was rushed with little planning.

Language: There are no obvious spelling errors but colloquialisms like 'nowadays' should be avoided. Overall, the student clearly understands the topic – but the essay is let down by a limited focus on the question and poor structure due to very long sentences and a lack of paragraphs.

Score: A3

Example Essay 10

Animal euthanasia should be made illegal.

Explain what this statement means. Argue to the contrary that animal euthanasia should remain legal. To what extent do you agree with the statement?

This statement means that the purposeful act of killing animals, carried out by veterinary practitioners should be made against the law. Currently, human euthanasia is illegal and the introduction of legal euthanasia brings with it many potentially harmful implications such as the 'putting down' of healthy animals.

An ethical pillar of medicine is non-maleficence. By making animal euthanasia illegal we uphold this pillar and avoiding causing harm to potentially healthy animals. If animal euthanasia were to be legalised then some people may think it justified to slaughter animal for less than noble purposes.

However, there are many cases in which euthanasia may be the best way of progression in the medical treatment of animals. For example: if an injured or terminally ill animal has no chance of recovery and is suffering then euthanasia may be the most kind and compassionate thing to do e.g. if a horse has broken its leg and will never be able to walk again. In addition, if the quality of life of an animal is very low e.g. they have no home, are starving and there is nowhere for them to live, then euthanasia may also be the most compassionate course of action. This may be especially the case where an area is overpopulated with stray cats and dogs.

Another case where euthanasia seems the most beneficial course of action is if an animal has become infected with a disease that could spread to other animals and humans potentially causing widespread and significant harm e.g. if a dog becomes infected with rabies or a cow becomes infected with foot and mouth disease.

To conclude, I disagree with the statement as I think animal euthanasia remain legal. It should however, only be carried out by a veterinary professional and only when the animal is undergoing significant suffering.

Examiner's Comments:

Introduction: A concise and focussed introduction that answers the first part of the question well and sets up the main body nicely.

Main Body: The student presents a sophisticated argument that addresses all the aspects of the question and uses good examples to back up their points. The arguments are well thought out and naturally follow on from each other. It would be better to argue why euthanasia should remain legal **before** giving reasons for it to be made illegal. This would help improve the flow.

Conclusion: A succinct and well-supported conclusion that ties together the major arguments in the main body. It also introduces a new idea –only vets should be allowed to perform euthanasia. This is a good point but should have been developed somewhat more.

Language: The student clearly understands the essay and puts together a strong essay. There are no glaring spelling or grammatical errors.

Score: A4.5

Summary

| Intro | • Does it explain or just repeat?
• Does it set up the main body?
• Does it get to the point? |

| Main Body | • Are enough points being made? *[Breadth]*
• Are the points explained sufficiently? *[Depth]*
• Does the argument make sense? *[Strength]* |

| Conclusion | • Does it follow naturally from the main body?
• Does it consider both sides of the argument?
• Does it answer the original question? |

General Advice

✓ Always answer the question clearly – this is the key thing examiner look for in an essay.

✓ Analyse each argument made, justifying or dismissing with logical reasoning.

✓ Keep an eye on the time/space available – an incomplete essay may be taken as a sign of a candidate with poor organisational skills.

✓ Use pre-existing knowledge when possible – examples and real world data can be a great way to strengthen an argument- but don't make up statistics!

✓ Present ideas in a neat, logical fashion (easier for an examiner to absorb).

✓ Complete some practice papers in advance, in order to best establish your personal approach to the paper (particularly timings, how you plan etc.).

✗ Attempt to answer a question that you don't fully understand, or ignore part of a question.

✗ Rush or attempt to use too many arguments – it is much better to have fewer, more substantial points.

✗ Attempt to be too clever, or present false knowledge to support an argument – a tutor may call out incorrect facts etc.

✗ Panic if you don't know the answer the examiner wants – there is no right answer, the essay is not a test of knowledge but a chance to display reasoning skill.

✗ Leave an essay unfinished – if time/space is short, wrap up the essay early in order to provide a conclusive response to the question.

ANSWERS

Answer Key

Question	Answer	Question	Answer	Question	Answer	Question	Answer
1	A	51	A	101	C	151	B
2	C	52	B	102	D	152	C
3	A	53	D	103	C	153	B
4	A	54	A	104	E&F	154	C
5	C	55	C	105	D	155	D
6	D	56	D	106	C	156	D
7	D	57	C	107	B	157	C
8	A	58	A	108	A	158	C
9	A	59	D	109	C	159	C
10	B	60	D	110	B	160	A
11	D	61	D	111	C	161	C
12	C	62	B	112	D	162	A
13	D	63	C	113	B & C	163	E
14	A	64	B	114	B	164	C
15	D	65	B	115	B & D	165	B
16	A	66	D	116	D	166	B
17	B	67	E	117	F	167	C
18	B	68	C	118	D	168	B
19	A	69	E	119	D	169	C
20	B	70	D	120	B & D	170	E
21	A	71	F	121	B	171	D
22	C	72	B	122	C	172	C
23	C	73	A	123	C	173	B
24	A	74	C	124	D	174	E
25	B	75	D	125	B	175	D
26	A	76	A	126	C	176	A
27	D	77	B	127	B	177	C
28	A	78	D	128	E	178	A
29	A	79	A	129	B	179	B
30	B	80	B	130	D	180	C
31	A	81	E	131	D	181	A
32	C & E	82	B	132	C	182	B
33	B	83	C	133	C	183	F
34	B	84	C	134	B	184	C
35	D	85	D	135	C	185	E
36	A	86	C	136	C	186	B
37	A	87	C	137	C	187	B
38	B	88	A	138	D	188	C
39	D	89	C	139	C	189	C
40	A	90	C	140	E	190	C
41	B	91	A	141	D	191	F
42	B	92	A	142	C	192	C
43	E	93	D	143	A	193	A
44	B	94	B	144	A	194	C
45	D	95	E	145	C & E	195	A
46	E	96	E	146	D	196	B
47	B	97	D	147	D	197	B
48	D	98	E	148	B	198	E
49	B	99	E	149	D	199	E
50	D	100	A	150	A	200	C

Question	Answer	Question	Answer	Question	Answer	Question	Answer
201	B	251	C	301	A	351	D
202	C	252	B	302	F	352	A
203	D	253	C	303	F	353	B
204	C	254	D	304	A	354	C
205	B	255	A	305	C	355	C
206	D	256	D	306	C	356	A
207	C	257	B	307	D	357	B
208	C	258	D	308	B	358	C
209	A	259	D	309	A	359	B
210	C	260	C	310	D	360	F
211	B	261	C	311	D	361	E
212	E	262	C	312	A	362	F
213	D	263	A	313	C	363	A
214	C	264	B	314	A	364	A
215	B	265	D	315	E	365	A
216	E	266	B	316	D	366	C
217	C	267	A	317	A	367	A
218	C	268	D	318	D	368	A
219	E	269	B	319	D	369	F
220	B	270	C	320	B	370	E
221	D	271	B	321	A	371	D
222	C	272	D	322	E	372	B
223	D	273	F	323	D	373	D
224	E	274	C	324	F	374	A
225	B	275	B	325	F	375	F
226	C	276	A	326	F	376	D
227	C	277	B	327	D	377	F
228	D	278	A	328	C	378	A
229	B	279	C	329	E	379	A
230	E	280	C	330	E	380	D
231	C	281	C	331	C	381	C
232	A	282	C	332	C	382	B
233	C	283	C	333	A	383	F
234	C	284	D	334	E	384	A
235	D	285	C	335	A	385	C
236	A	286	D	336	A	386	C
237	D	287	C	337	C	387	E
238	D	288	C	338	A	388	F
239	C	289	C	339	F	389	G
240	C	290	C	340	H	390	F
241	C	291	E	341	C	391	A
242	E	292	C	342	H	392	C
243	D	293	C	343	B	393	B
244	C	294	D & E	344	F	394	D
245	B	295	C	345	E	395	B
246	E	296	C	346	B	396	E
247	C	297	C	347	A	397	B
248	B	298	C	348	I	398	D
249	E	299	D	349	C	399	D
250	D	300	C	350	B	400	D

Question	Answer	Question	Answer	Question	Answer	Question	Answer
401	D	451	F	501	D	551	D
402	E	452	A	502	F	552	A
403	E	453	D	503	G	553	F
404	B	454	E	504	D	554	D
405	B	455	G	505	A	555	A
406	G	456	C	506	E	556	D
407	A	457	D	507	C	557	D
408	B	458	E	508	D	558	F
409	E	459	D	509	D	559	D
410	B	460	D	510	C	560	C
411	D	461	F	511	D	561	B
412	A	462	B	512	F	562	C
413	A	463	C	513	B	563	A
414	B	464	G	514	A	564	A
415	C	465	D	515	C	565	C
416	B	466	E	516	C	566	B
417	A	467	A	517	G	567	D
418	F	468	E	518	C	568	C
419	A	469	G	519	B	569	A
420	E	470	E	520	B	570	C
421	D	471	H	521	E	571	C
422	H	472	E	522	C	572	C
423	B	473	G	523	E	573	B
424	E	474	D	524	E	574	D
425	D	475	E	525	D	575	E
426	A	476	D	526	B	576	D
427	D	477	G	527	C	577	D
428	D	478	F	528	C	578	B
429	C	479	B	529	C	579	E
430	G	480	A	530	E	580	E
431	E	481	C	531	A	581	B
432	D	482	G	532	C	582	C
433	D	483	D	533	E	583	A
434	F	484	D	534	E	584	C
435	C	485	E	535	C	585	B
436	A	486	B	536	E	586	B
437	B	487	A	537	E	587	B
438	E	488	E	538	E	588	C
439	C	489	D	539	B	589	C
440	D	490	F	540	C	590	A
441	C	491	E	541	B	591	C
442	C	492	A	542	B	592	D
443	E	493	B	543	C	593	C
444	C	494	C	544	D	594	D
445	B	495	D	545	C	595	A
446	B	496	F	546	B	596	C
447	B	497	E	547	A	597	B
448	E	498	C	548	F	598	B
449	D	499	C	549	D	599	A
450	C	500	B	550	A	600	C

Worked Answers

Question 1: A
Whilst **B**, **C** and **D** may be true, they are not completely stated, **A** is clearly stated and so is the correct answer.

Question 2: C
The main argument of the first paragraph is to propose the point that it is more society that controls gender behaviour not genetics. **A** and **D** do not indicate either as they only allude to the end result of gender behaviour and so are incorrect. Hormonal effects are not mentioned in the first paragraph and so **B** is incorrect. **C** would undermine the argument that society *predominately* controls gender, and so is correct.

Question 3: A
B, **C** and **D** are not stated and so are incorrect. **A** is directly stated and so is correct.

Question 4: A
B and **D** are contraindicated by the statement and so are incorrect. **C** could be true but implies children always like the same thing as their same-gendered parent irrelevant of how they are treated as a child, which is contrary to the statement and so is not correct. **A** is correct as is the overall message.

Question 5: C
D may help prevent problems with sexual identity but does not prevent stereotyping and so is incorrect. **A** is not stated, and **B** is implied but not stated and so are incorrect. **C** is the end message of how to prevent gender stereotyping and so is correct.

Question 6: D
A, **B** and **C** may be true but are not mentioned in the statement and so are incorrect. The statement implies that children born with different external organs to those that their sex chromosomes would match may find it difficult to accept this difference and be uncomfortable.

Question 7: D
The text states that 'Those who regularly took 30-minute naps were more than twice as likely to remember simple words such as those of new toys.' Which means those who napped were twice as likely to remember teddy's name than the 5% who did not, 5% x 2 = 10%, which would be twice as likely, ruling out **A** and **B**. But being 'more than twice' the only possible answer is **D**.

Question 8: A
The answer is to work out 10% (the percentage of napping toddlers more likely to suffer night disturbances) of 75% (the percentage of toddlers who regularly nap). Hence 10 % of 75% is **7.5%.**

Question 9: A
B, **C** and **D** may be true but there is nothing in the text to support them. **A** is suggested, as the passage states 'non-napping counterparts, who also had higher incidences of memory impairment, behavioural problems and learning difficulties'. If the impaired memory were the cause, as opposed to the result, of irregular sleeping then it would offer an alternative reason why those who nap less remember less.

Question 10: B
A and **C** are possible implications but not stated and so are incorrect. It is said that parents cite napping having 'the benefits of their child having a regular routine' so hence **B** is more correct than **D** as it refers to the benefit to the toddlers' rather than the parents.

Question 11: D
B, if true would counteract the conclusion, as it would imply that, the study is skewed. The same is true of **C**, which if true would imply unreliable results as the toddler sample are all the same age within a year, but not within a few weeks. **A**, if true, would not provide any additional support to the conclusion and so is incorrect. **D** if true would provide the most support for the conclusion as it proposes using groups with a higher incidence of napping in comparison to those with a lower incidence.

Question 12: C
Although it can be argued that **A**, **B**, **D** and **E** are true they are not the best answer to demonstrate a flaw in Tom's father's argument. **C** is the best because it accounts for other factors determining success for the Geography A-level exam such as aptitude for the subject.

Question 13: D

A is never stated and is incorrect. B and C are referred to being 'many people's' beliefs, and are cited as others' opinions not an argument supported by evidence in the passage, and so are not valid conclusions. It is implied that the NHS may have to reduce its services in the future, some of which could be fertility treatments hence D is the most correct answer.

Question 14: A

C does not severely affect the strength of the argument, as it is only relevant to the length of the time taken for the effects of the argument to come into place.

D is incorrect, as people breaking speed limits already would not negate the argument that speed limits should be removed, but could even be seen as supporting it. These people may count as the 'dangerous drivers' who would be ultimately weeded out of the population.

B may affect part of the argument's logic (as it undermines the idea that dangerous drivers are born to dangerous drivers), but the final conclusion that dangerous drivers will end up killing only themselves still stands, and so the ultimate population of only safe drivers may be obtained. The fact that one dead dangerous driver could have produced a safe one does not necessarily challenge the main point of this argument.

A if true would most weaken the argument as it states that fast driver is more likely to harm others and not the driver itself, which would negate the whole argument.

Question 15: D

Whilst is it stated that the Government assesses risk it is not described as an obligation, hence A is incorrect. The overall conclusion of the statement is that on balance the Government was justified in not spending money on flooding preparation, as it was unlikely to occur, so C, B and E are incorrect and D is correct.

Question 16: A

C is incorrect and D is a possible course of action rather than a conclusion. B and E are possible inferences but not the conclusion of the statement. The overall conclusion of the statement is that the way that children interact has changed to the solitary act of playing computer games.

Question 17: B

The passage does state that in this case the £473 million could have been put to better use, however, there is no mention that no drug should ever be stockpiled for a similar possible pandemic. The passage discusses the lack of evidence behind Tamiflu and therefore is stating that in a situation where there is a lack of evidence, there may not be justification for stockpiling millions of pounds worth of the drug. Stockpiling in the case of drugs with high effectiveness is not discussed so we should not assume this is a generic argument against preparation for any pandemic and stockpiling of any drug.

Question 18: B

The passage discusses the fact that unhealthy eating is associated with other aspects of an unhealthy lifestyle so the argument that tackling only the unhealthy eating aspect does not logically follow. The other statements are all possible reasons why the solution given may not be optimal, but are not directly referred to in the passage.

Question 19: A

This is a tricky question in which A, B, C and D are all true. However, the question asks for the conclusion of the passage, which is best represented by A.

B is a premise that gives justification for why the elderly should take care of themselves and C provides a justification for why they may not.

D is implied in the text but statement A is explicitly stated.

E is incorrect as the passage implies that people should spend the money that they have in old age, not stop saving altogether.

Question 20: B

The passage states stem cell research is an area where there are possible high financial and personal gains, however there is no mention of these being the main driving factors in either this area of research or others.

Although rivalry between groups may be a reason driving publishing, this is not mentioned in the passage.

The image discrepancies were in only one paper but the passage implies the protocol and replication problems were in both papers.

Question 21: A

D actually weakens the argument, and is therefore not a conclusion. C is simply a fact stated to introduce the argument, and is not a conclusion. B is a reason given in the passage to support the main conclusion. If we accept B as being true, it helps support the statement in A. E is not discussed in the passage. A is the main conclusion of this passage

Question 22: C

The passage describes improved safety features and better brakes in cars, and concludes that this means the road limit could be increased to 80mph without causing more road fatalities. However, if **C** is not true, this conclusion no longer follows on from this reasoning. At no point is it stated that **C** is true, so **C** is therefore the assumption in the passage. The statements in **B** and **D** are not *required* to be true for the argument's conclusion to lead on from its reasoning. **A** is a statement which is strengthened by this passage, and is not an assumption from the passage. **E** is not relevant to the conclusion or mentioned in the passage.

Question 23: C

Answers **A** and **D** are both reasons given to explain fingerprints under the theory of evolution, and contribute towards the notion given in **C**, that they do not offer support to intelligent design. Thus, **A** and **D** are reasons given in the passage, and **C** is the main conclusion. **B** is simply a fact stated to introduce the passage, whilst **E** actually contradicts something mentioned in the argument (namely that Intelligent Design is religious-based, and scientifically discredited). Neither of these options are conclusions.

Question 24: A

Answers **C**, **D** and **E** obviously present ways in which the conclusions drawn from the study could be wrong, without any mistakes being made by those carrying out the study, and thus are potential reasons. **B** is also a potential reason, because those with a low alcohol consumption could have many other risk factors for cancer, and end up with a higher *overall* risk. If the study does not take account of these, it could produce erroneous conclusions. **A** cannot be a valid reason because the passage *states* that it is 'proven' that alcohol increases the risk of cancer. Thus, we must accept this as true, so **A** is not a potential reason.

Question 25: B

The passage states that the average speed *including* time spent stood still at stations was 115mph. Thus, **A** is incorrect, as the stopping points have already been included in the calculations of journey time. Similarly, the passage states that the train completes its journey at Kings cross, so **D** is incorrect. **C** is not correct because we have been given the total length of the journey. Whether it took the most direct route is irrelevant. **E** is completely irrelevant and does not affect the answer. **B** is an assumption, because we have only been given the *scheduled* time of departure. If the train was delayed in leaving, it would not have left at 3:30, and so would have arrived *after* 5:30.

Question 26: A

The argument discusses healthcare spending in England and Scotland, and whether this means the population in Scotland will be healthier. It says nothing about whether this system is fair, and does not mention the expenditure in Wales. Thus, **C** and **D** are incorrect. Similarly, the argument makes no reference to whether healthcare spending should be increased, so **B** is incorrect. **E** is true but not the main message of the passage. The passage does suggest that the higher healthcare expenditure per person in Scotland does not necessarily mean that the Scottish population will be healthier, so **A** is a conclusion from this passage.

Question 27: D

C is an incorrect statement, as the passage says that Polio *hasn't* been eradicated yet. **A** and **B** are reasons given to support the conclusion, which is that given in **D**. **E**, meanwhile, is an opinion given in the passage, and is not relevant to the passage's conclusion.

Question 28: A

This passage provides various positive points of the Y chromosome, before describing how all of this means it is a fantastic tool for genetic analysis. Thus, the conclusion is clearly that given in **A**. The statement in **B** is a further point given to provide evidence of its utility, as stated in the passage. Thus **B** is not a conclusion in itself, but further evidence to support the main conclusion, given in **A**. **C** is also a reason given to support the conclusion in **A**, whilst **D** is simply a fact stated to introduce the passage. As for **E**, there is no mention of Genghis Khan's children (only his descendants).

Question 29: A

Answers **C** and **E** are not valid assumptions because the argument has *stated* that a patient *must* be treated with antibiotics for a bacterial infection to clear. **B** is not a flaw, because this does not affect whether the antibiotics would clear the infection if it were bacterial. **D** is an irrelevant statement, and also disagrees with a stated phrase in the passage (that antibiotics are required to clear a bacterial infection). **A** is a valid flaw, because the passage does not say that antibiotics are *sufficient* or *guaranteed* to clear a bacterial infection, simply that they are *necessary*. Thus, it is possible that the infection *is* bacterial but the antibiotics failed to clear it.

Question 30: B

A, **C** and **D**, if accepted as true, all contribute towards supporting the statement given in **B**, which is a valid conclusion given in this passage. Thus, **A**, **C** and **D** are all reasons given to support the main conclusion, which is the statement given in **B**. **E** is not a valid conclusion, as the passage makes no reference to action that should be taken relating to smoking, it simply discusses its position as the main risk factor for lung cancer.

Question 31: A

D is only given as a method, with no mention of its effectiveness. We do not know if **C** is true because it is not stated. **B** is not discussed in the passage. Whilst statement **E** is true, it is supporting evidence for the conclusion, not the conclusion itself.

Question 32: C & E

Whilst **A** and **B** may be true, cost is not mentioned as a deterring factor and we are only concerned with use in the UK, so they are irrelevant. Whether cannabis was the only class C drug is not important to the argument so **D** is not correct. **C** and **E** are the correct answers because the statement concerns the use of cannabis in the UK, directly stating use will decrease from people knowing it has been upgraded to a more dangerous category and from fearing longer prison sentences from higher-class drugs.

Question 33: B

Whilst **A** and **C** may be true, they are not part of the argument. **D** is a possible, but cannot be logically proposed from the information above. **E** would be a flaw if the argument were 'all levels of sports teams reduce bullying' but the passage explicitly states 'well-performing' teams. Hence **B** is correct as it undermines the whole argument, reversing the cause and effect.

Question 34: B

Options **A**, **C** and **D** do not directly weaken the argument as if any 16 year olds were buying/drinking alcohol (whether the minority or majority) – police would still be spending time catching them. The suggested benefit to reduce police time spent catching underage drinkers would be negated if **B** were true, hence it is the correct answer.

Question 35: D

A is an interpretation of the last sentence and doesn't accurately summarise the argument in the passage. **B** is untrue as there is no mention of if the government can afford to give grants or not. **C** and **E** are incorrect as the passage only talks about small businesses. **D** is correct as it best summarises the change in government policy regarding small businesses.

Question 36: A

The statement discusses a case that was reported, but aims to argue that there may be important errors occurring everyday in medicine that go unreported. Option **A** if true, would significantly weaken this argument as would negate it being a possibility. **B**, **C**, **D** and **E** may be true, but they do not negate the argument – if doctors are trained, accidents like the above may still occur. Operations that are successful do not affect those that are not, nor do unavoidable errors have any relation to avoidable ones. That the patient may have died without these errors similarly does not mean that errors, when they do occur, should not be considered errors.

Question 37: A

The main point of the statement is to highlight that although there are numerous safety precautions in place to protect patients, when the weaknesses in these precautions align big errors can occur. So **A** is correct. While **E**, **C**, **B** and **D** may well be true, they are not the overall conclusion of the statement.

Question 38: B

Though not the first to be cited, the original error is cited as being the incorrect copying of the sidedness of the kidney to be removed, hence **B** is the correct option. The other options represent errors that in the 'Swiss cheese model' would have not been allowed to occur if the original had not taken place.

Question 39: D

In this instance the 'tip of the iceberg' refers to the number of medical errors reported, implying there may be a significantly larger proportion that go unreported, hence the correct option is **D**, and not **B**.

Question 40: A

The description given about the consultant's performance versus emotional arousal, is described as initially increasing then eventually decreasing over time, which is best represented by graph **A**.

Question 41: B

The consultant says that the 'public perception is that medical knowledge increases steadily over time' which is best represented by graph **B**. The consultant says the regarding the acquisition of medical knowledge, 'many doctors [reach] their peak in the middle of their careers', which is best described by the graph **D**.

Question 42: B

Obesity is not mentioned in the passage, so **E** is incorrect. There is no mention exercise specifically as it relates to old age, so **A** and **D** are also wrong. The diseases associated with lack of exercise are not specifically stated to cause early death, only that they are associated with older people, so **C** is also incorrect. The passage does, however, argue that lack of exercise is associated with illness, and so exercise would be linked to a lack of illness, or good health, so **B** is correct.

Question 43: E

The preference of women to have their babies at hospital versus home is not commented upon so **B** is incorrect. **F** is never inferred, only that midwives are capable of assisting in normal births and assessing when women need to be transferred to be to hospital, so it is wrong. **A** and **D** are possible inferences at certain points but not conclusions of the statement. **C** is never implied, only that normal home births are no more risky than those in hospital. The overall conclusion of the statement is that the home births should be encouraged where possible as they are not more risky in the cases of normal births, and hospital births are an unsustainable cost in cash-strapped NHS.

Question 44: B

While **A**, **C** and **D** would, if true, make the practicalities of increasing home births more difficult they would not weaken the argument as **B** would. Where the statement's whole argument rests on home births being as safe as hospital **B**, if true, would negate this.

Question 45: D

The statement says 'With the increase in availability of health resources we now, too often, use services such as a full medical team for a process that women have been completing single-handedly for thousands of years.' Thus implying **D**, 'excessive availability of health resources' is the cause of 'medicalisation of childbirth'.

Question 46: E

1 and **3** identify weaknesses in the argument. If campaigns are what help keep deaths by fire low, they can be seen as 'necessary', and their necessity may be proven by the promisingly low fire-related mortalities. If there are more people with hernias than in fires, more people can possibly die from hernias, but this does not mean the fires are less dangerous to the (fewer) individuals involved in them. **2** is irrelevant, as the argument is about how dangerous fires are in their entirety, not in relation to their constituent parts. Therefore **E**, '1 and 3 only', is correct.

Question 47: B

Since 'some footballers' that like Maths are not necessarily the same 'some' who like History we can exclude **A** and **D**. Equally, while **C** may or may not be true, we are not given any information about rugby players' preference for History, so it is incorrect. We know that all basketball players like English and Chemistry, and that none of them like History, but as we do not know about a third subject they may like **E** is incorrect. We know all of the rugby players like English and Geography and some of them Chemistry, hence there must be a section of rugby players that like all three subjects so **B** is correct.

Question 48: D

The passage discusses the problems surrounding controlling drugs, and focuses on the rapid manufacture of new 'legal highs': it is therefore implied that this is the current major problem. The passage also suggests that as the authorities cannot keep up with drugs manufacture, the legality of drugs doesn't reflect their risks.
1 is incorrect as the passage says health professionals feel legality is less relevant now, but it doesn't say that it is not still important. **3** is incorrect as the last sentence says a potential problem of legal highs is that the risks are not as clear, which contradicts the statement that the public are not concerned about any risks.

Question 49: B

The passage is discussing how banning those with the mentioned medical conditions from mountain climbing are *essential* to ensuring safety. It does not claim that this is *sufficient* to ensure safety, simply that it is *necessary*. Thus **C** is irrelevant, as risks from other activities do not affect the risk from mountain climbing. **D** is also irrelevant, because the argument discusses how it is essential to ensure safety of people on WilderTravel holidays, so those using other companies are irrelevant. **A** is an irrelevant statement because the passage is discussing what should be done *to ensure safety*, not whether this is the morally correct course of action. Thus, a discussion of whether people should choose to accept the risks is not relevant. However, **B** *is* a flaw, because the guidelines only mention those with *severe* allergies, so thinking those with less severe allergies are in danger is a false assumption that has been made by the directors.

Question 50: D

The hospital director's comments make it abundantly clear that the most important aspect of the new candidate is good surgery skills, because the hospital's surgery success record requires improvement. If we accept his reasoning as being true, then it is clear that the candidate who is most proficient at surgery should be hired, and patient interaction should not be the deciding factor. Thus, Candidate 3 should be hired, as suggested by **D**.

Question 51: A

Answers **B** and **D** are irrelevant to the argument's conclusion, since the argument only talks about how medical complications could be avoided *if* winter tyres were fitted. Whether this is possible (as in **B**) or whether there are other options (as in **D**) are irrelevant to this conclusion. **C** is not an assumption because the passage states that delays cause many complications, which could be avoided with quicker treatment. However, the argument does not state that winter tyres would allow ambulances to reach patients more quickly, so **A** is an assumption.

Question 52: B

The passage discusses how anti-vaccine campaigns cause deaths by spreading misinformation and reducing vaccination rates. It claims that therefore *in order to protect* people, we should block the campaigners from spreading such misinformation freely. Thus it is made clear that this action should be taken *because the campaigners cause deaths*, not simply because they are spreading misinformation. Thus, **B** is the principle embodied in the passage, and **C** is incorrect. **A** actually demonstrates an opposite principle, whilst **D** is a somewhat irrelevant statement, as the passage makes no reference to whether we should promote successful public health programmes.

Question 53: D

The passage states that the tumour has established its own blood supply (it says this was shown during the testing), and that a blood supply is *necessary* for the tumour to grow beyond a few centimetres. Thus **A** and **B** are not assumptions. **C** is not an assumption, as it actually disagrees with something the passage has implied. The passage has actually said that action *must* be taken, implying that something *can* be done to stop the tumour. However, at no point has it been said that a blood supply is *sufficient* for a tumour to grow larger than a few centimetres. If this is not true, then the argument's conclusion that we should expect the tumour to grow larger than a few centimetres, and that action must be taken, no longer readily follows on from its reasoning. It is possible the tumour will still fail to grow larger than a few centimetres. Thus, **D** is an assumption in the passage, and a flaw in its reasoning.

Question 54: A

D is incorrect, as the passage has stated the runners are people running to raise money for the GNAA. **B** and **C**, meanwhile, are incorrect as the passage is only talking about whether the GNAA *will be able to* get a new helicopter. Thus, references to whether it wishes to, or whether this is the best use of money, are irrelevant. **A**, however, is an assumption on the part of the passage's writer. The passage says that the GNAA will be able to get a helicopter if £500,000 is raised, but this does *not* mean that it won't be able to if the £500,000 is not raised by the runners. It could well be that they secure funding from elsewhere, or that prices drop. The money being *sufficient* to get a new helicopter does not mean it is *necessary* to get one.

Question 55: C

B and **D** somewhat strengthen this argument, suggesting that more people going on courses leads to better growth, and that people who have gone on these courses are more attractive to employers. **A** does not really affect the strength of the argument, as the current rate of growth does not affect whether government subsidies would lead to increased growth. **C**, however, weakens the argument significantly by suggesting that people would not be more likely to attend the courses if the government were to subsidise them, as the cost has little effect on the numbers of people attending.

Question 56: D

B is simply a fact stated in the passage. It does not draw upon any other reasons given in the passage, so it is not a conclusion. **C** is not a conclusion because it does not follow on from the passage's reasoning. The passage discusses what should be done *if* Pluto is to be classified as a planet, it does not make any mention of whether this *should* happen. **A** and **D** are both valid conclusions from the passage. However, on closer examination we can see that if we accept **A** as being true, it gives us good reason to believe the statement in **D**. Thus, **D** is the *main* conclusion in the passage, whilst **A** is an *intermediate* conclusion, which goes on to support this main conclusion.

Question 57: C

A, **B** and **D** would all affect whether the calculation of the Glasgow train's arrival time is correct, but none are assumptions because all of these things have been stated in the passage. However, the passage has *not* stated that the trains will travel at the same speed, and if this is not true, then the conclusion that the Glasgow train will arrive at 8:30pm is no longer valid. Thus, **C** is an assumption.

Question 58: A

C can actually be seen to be probably untrue, as the passage mentions a need to escape immune responses, suggesting that the immune system *can* tackle these cells. **E** is true but not representative of the main argument made in the passage. **B** and **D** are not *definitely* true. The passage mentions several *essential* steps that *must* occur, but this does not mean that they are *sufficient* for carcinogenesis to occur, or guaranteed to allow it. Equally, the passage makes no reference to multiple mechanisms by which carcinogenesis can occur. It could be there is only one pattern in which these steps can occur. **A**, however, can be reliably concluded, because the passage does mention several steps that are *essential* for carcinogenesis to occur.

Question 59: D

Answers **A** and **C** are stated in the passage (the passage states 'deservedly known'), so these can be reliably concluded. **B** can also be concluded, as it is stated that in over 50% of cancers, a loss of functional P53 is identified. **D** however, cannot be concluded, as the passage simply states that any cell that has a mutation in P53 *is at risk* of developing dangerous mutations. Thus, it cannot be concluded that a given cell *will* develop such a mutation.

Question 60: D

D is not an assumption because Sam's calculations are based on the *cost per 1000 miles*, not on a given amount of fuel being used up. Thus, he has *not* assumed anything about whether the fuel usage is the same for each car. All of the others are assumptions, which have not been considered. Each of these will affect the total saving he will make if they are not true. For example, if the Diesel car costs £100 more than the Petrol car, the total saving will be £1700, *not* £1800 as calculated.

Question 61: D

The passage discusses how alcohol is more dangerous than cannabis, and states that this highlights the gross inconsistencies in UK drugs policy. Thus, **D** is the main conclusion of the passage, whilst **A** is a reason given to support this conclusion. The passage simply highlights that the policy is grossly inconsistent, and does not mention whether it should be changed, or how (whether alcohol should be banned or cannabis allowed).

Thus, **B** and **C** are not valid conclusions from this passage. The fact alcohol is freely advertised only mentioned briefly in the passage to add strength to the argument that alcohol is more accessible than cannabis, but no judgment is made on whether this should not be so, so **E** is also not a valid conclusion from this passage.

Question 62: B

The passage discusses how if first aid supplies were available, many accidents could be avoided. B correctly points out that this is a flaw – first aid supplies may help treat accidents and reduce the prevalence of *injuries and deaths*, but there is no reason why first aid supplies should reduce the incidence of *accidents*. Answers **C** and **D** are irrelevant, since the argument is talking about how first aid supplies could reduce *accidents*, not *injuries* or *deaths*. Thus, discussing cases in which they could not treat the injuries, or whether they need other components to do so is irrelevant. Equally A is irrelevant, as the argument is simply talking about what could happen *if* first aid supplies were stocked in homes, and makes no reference to whether this is financially viable.

Question 63: C

Answers **A** and **D** are not flaws because the passage does not conclude the things mentioned in these. No mention is made to the safety of the drug, and the argument only states that it is thought the compound *may* be of use in combating cancer. No premature conclusions are drawn, only suggestions are made. **B** is not a flaw because we can see that the experiments *may* produce misleading results if the wrong solutions are used, suggesting that DNA replication is inhibited even if it is not. **C**, however, is a valid flaw because the argument erroneously concludes that the wrong solutions must have been used when it says the experiments *do not reflect what is actually happening*. This clearly indicates a conviction that the wrong solutions were used, which does not follow on from the experiments being old.

Question 64: B

The passage has not said anything about who scored the winning goal, so **A** is not an assumption. **C** is also incorrect, because the passage states that South Shields won the game. **B** correctly identifies that whilst beating South Shields was *sufficient* to win the league, it was *not* necessary. If Rotherham wins their other 2 games, they will still win the league, so **B** demonstrates an assumption in the passage. **D** is not relevant, as it does not affect the erroneous nature of the claim that Rotherham *will not* win the league having lost the match to South Shields.

Question 65: B

C and **D** actually strengthen or reinforce the CEO's reasoning, with **C** suggesting as time progresses Middlesbrough will have more and more people compared to Warrington, whilst **D** suggests that the market share in Warrington may not be as high as suggested, adding further reasons to build in Middlesbrough. **A** somewhat weakens the CEO's argument, but it is not a flaw in the reasoning, because the CEO is simply talking about how Middlesbrough will bring them within the range of more people, so the market share comment is a counterargument, not a flaw in his reasoning. **B**, however, is a valid flaw in this argument. Just because Warrington's population is falling, and Middlesbrough's is rising, does not necessarily mean that Middlesbrough's will be higher.

Question 66: D

1 and **2** are assumptions. The information given does *not* necessarily lead on to the conclusion that these extinction events will continue without further conservation efforts. Equally, there is nothing in the passage that says conservation efforts cannot be stepped up without increased funding. However, **3** is not an assumption, because the passage *states* that global warming has caused changed weather patterns, which have caused destruction of many habitats, which have led to many extinction events. Thus, it is given that global warming has indirectly caused these extinctions, and so the answer is **D**.

Question 67: E

The argument is suggesting that in Austria, the rail service's high passenger numbers and approval ratings are accounted for by the fact that road travel is difficult in much of Austria. It then concludes that the public subsidies have no effect. We can see that **1** instantly weakens this argument by providing evidence to the contrary, (in France, difficult road travel is not prevalent and so cannot account for the high passenger numbers/approval ratings the country possesses). **3** also weakens this conclusion by suggesting multiple factors affect the situation. This makes the conclusion based on the evidence from Austria less strong. Thus, the answer is **E**. **2** actually strengthens the argument that the public subsidies do not cause high passenger numbers/approval ratings, as Italy has high subsidies but low passenger numbers/approval ratings.

Question 68: C

A is incorrect, in 2011 24% of men and 26% of women were obese (one should not confuse this with the rates of combined obese and overweight). **B** is also incorrect, as what it states is true for adults; however, the figures for children aged 2-15 have changed little over the past year. **D** is not stated or implied by the passage.

C is implied in the last two sentences of the article, and so the correct answer.

Question 69: E

All of these statements cannot be concluded from the information based on the passage.

Question 70: D

Be careful of using your own knowledge here! Whilst **A** and **B** may be true, they are not the main message of the passage. **C** may be true but is not discussed in the passage. **E** is speculative, as the passage does not say if the transplant would be a 'good alternative'. **D** is correct as it echoes the main message of the passage.

Question 71: F

Smoking and Diabetes are risk factors for vascular disease (not a cause). Vascular disease does not always lead to infarction. The passage does not give sufficient detail about necrotic tissue to conclude **C** or **D**.

Question 72: B

A is irrelevant to the argument's conclusion. Meanwhile **E** does nothing to alter the conclusion, as the fact that schools receive similar funds does not affect the fact that more funding could provide better resources, and thus improve educational attainment. **C** actually weakens the argument; by implying that banning the richer from using the state school system would not raise many funds, as most do not use it anyway. **D** does not strengthen the conclusion as stating that a gap exists does not do anything to suggest that more funding will help close it. **B** clearly supports the conclusion that more funding, and better resources, would help close the gap in educational attainment.

Question 73: A

D and **E** are irrelevant to the argument's conclusion. **C** is actually contradicting the argument. **B** is stated in the passage, so is not an assumption of the passage. **A** describes an assumption: the increase of DVDs does not, necessarily, cause the loss of cinema customers.

Question 74: C

The question refers to aeroplanes being the fastest form of transport, and states that this means that travelling by air will allow John to arrive as soon as possible. **C** correctly points out that the argument has neglected to take into account other delays induced by travelling by aeroplane. Cost and legality are irrelevant to the question, so **B** and **E** are incorrect. Meanwhile, **D** actually reinforces the argument, and **A** refers to future possible developments that will not affect John's current journey.

Question 75: D

The argument states that people should not seek to prevent spiders from entering their homes. It does not say anything about whether people should like spiders being in their home, so **A** is incorrect. The argument also makes no allusion to the notion of people preventing flies from entering their homes, so **B** is incorrect. The argument also does not mention or implies that any efforts should be made to encourage spiders to enter homes, or that they should be cultivated, so **C** and **E** are also incorrect.

Question 76: A

A correctly identifies an assumption in the argument. At no point is it stated that bacterial infections in hospitals are resulting in deaths. **B**, **C**, **D** and **E** are all valid points but they do not affect the notion that pressure for more antibiotic research would save lives. Therefore, none of these statements affect the conclusion of the argument and as such they are not assumptions in this context.

Question 77: B

The passage does not state that John disregards arguments because of the gender of the speaker, so **D** is incorrect. **A** and **C** are also wrong, as John states he finds women with armpit hair necessarily unattractive, so a different face or the knowledge of concealed hair would not make him find the female in question more appealing to his aesthetic. John does not state Katherine wants other women to stop shaving, so **E** is incorrect.

B is the correct answer, as Katherine was simply speaking about societal norms, and at no point is it said she was trying to convince John to find her, with armpit hair, attractive.

Question 78: D

A is irrelevant to the argument, which says nothing about what will happen to Medicine in the future. The argument is describing how Sunita is incorrect, and how better medicine is not responsible for a high death rate from infectious disease in third world countries, and how better medicine will actually decrease this rate. **C** is a direct contradiction to this conclusion, so is incorrect. **E** is a fact stated in the argument to explain some of its reasoning, and is not a conclusion, therefore **E** is incorrect.

Both **B** and **D** are valid conclusions from the argument. However, **B** is not the main conclusion, because the fact that 'Better medicine is not responsible for a high death rate from infectious disease in third world countries' actually supports the statement in **D**, 'Better medicine will lead to a decrease in the death rate from infectious disease in third world countries'. Therefore, **B** is an example of an intermediate conclusion in this argument, which contributes to supporting the main conclusion, which is that given in D.

Question 79: A

The statement in A, that housing prices will be higher if demand for housing is higher, is not stated in this argument. However, it is implied to be true, and if it is not true, then the argument's conclusion is not valid from the reasoning given. Therefore A correctly identifies an assumption in the argument. The other statements do not affect how the reasons given in the argument lead to the conclusion of the argument, and are therefore not assumptions in the argument.

Question 80: B

A and E are both contradictory to the argument, which concludes that because of the new research, Jellicoe motors should hire a candidate with good team-working skills. C refers to an irrelevant scenario, as the argument is referring to only one candidate being hired, and at no point does it state or imply that several should be hired.

B correctly identifies the conclusion of the argument that Jellicoe motors should hire a new candidate with good team-working skills in order to boost their productivity and profits. D meanwhile exaggerates the consequences of not following this course of action. The argument does not make any reference to the notion that Jellicoe motors will struggle to be profitable if they do not hire a candidate with good team-working skills.

Question 81: E

D is in direct contradiction to the argument, so is not the main conclusion. Meanwhile, B is a reason stated in the argument to explain some of the situations described. It is not a conclusion, as it does not follow on from the reasons given in the argument.

A and E are both valid conclusions from the argument. However, only E is the *main* conclusion. This is because both A goes on to support the statement in E. If bacterial resistance to current antibiotics could result in thousands of deaths, this supports the notion that the UK government must provide incentives for pharmaceutical firms to research new antibiotics if it does not wish to risk thousands of deaths.

Meanwhile, C appears to be another intermediate conclusion in the argument that also supports the main conclusion. However, on close inspection this is not the case. C refers to the UK government directly investing in new antibiotic research, whilst the argument refers to the government providing incentives for pharmaceutical firms to do so. Therefore, C is not a valid conclusion from the argument.

Question 82: B

E is completely irrelevant because the question is referring to an unsustainable solution *if* the UN's development targets are met, so the likelihood of them being met is irrelevant. C is irrelevant because they do not affect the fact that the situation would be unsustainable if everybody used the amount of water used by those in developed countries, as stated in the question. A is also irrelevant, as the passage does not mention price as a factor to be considered within the argument.

Meanwhile, D would actually strengthen the argument's conclusion.
Therefore, the answer is B. B correctly identifies that if those in developed countries use less water, it may be possible for everyone to use the same amount as these people and still be in a sustainable situation.

Question 83: C

There is no mention of treatment, so A is incorrect. A need to travel abroad for the post is not stated either, so B is incorrect. The need for a cool head is stated explicitly, but not necessarily that this be a leader, so D is also wrong. Other qualities are irrelevant to the argument, so E is also incorrect. C would only be relevant if there was indeed a link between 'a specific phobia' and 'a general tendency to panic'. Thus, C highlights the flaw: if a fear of flying does not necessitate a general disposition of panic, the argument for not hiring this employee crumbles.

Question 84: C

The passage does not suggest there are no more university places, nor does it make a distinction between the qualities of different universities, so A is incorrect and D is irrelevant. The argument does not deny the fact that people can be successful without a university education, so B is also wrong. C is correct, as the passage specifically states 'many more graduates', but not all, are equipped with better skills and better earning potential. This suggests not all degrees produce these skill-sets in their graduates, and so not all university places will create high-earning employees.

Question 85: D

B is unrelated to the argument, as other contributing factors would not negate the damaging potential of TV. Watching sport on television would not be akin to actually playing sport, so A is also incorrect. The possibility of eye damage is stated as caused by TV, so C is incorrect. However, if people watch television *and* partake in sport, which the passage seems to imply cannot happen, they may not suffer the negative effects of obesity and social exclusion. For example, they may play sport during the day and watch television in the evening, thus experiencing the benefits of exercise and also enjoying the sedentary activity. Therefore, various potential threats supposedly posed by watching excessive television are undermined, and D is correct.

Question 86: C

D directly counters the above argument, and so is incorrect. Though A, B and D are all suggested or stated by the passage, they each act as evidence for the main conclusion, C, describing the 'multiple reasons to legalise cannabis'.

Question 87: C

C is not an assumption as it has been explicitly stated in the question that the salary is fixed, and therefore it will not change. The rest of the statements are all assumptions that Mohan has made. At no point has it been stated that any of the other statements are true, but they are all required to be true for Mohan's reasoning to be correct. Therefore, they are all assumptions Mohan has made.

Question 88: A

The answer is not **B** because, although the Holocaust was a tragedy, this is not explicitly stated in the passage. It cannot be **C** or **E**, as these are also not directly stated above. **D** provides an intermediary conclusion that leads to the main conclusion of **A**: we should not let terrible things happen again, and through teaching we can achieve this, so therefore 'we should teach about the Holocaust in schools'.

Question 89: C

DVDs are irrelevant – though one could access disturbing material through a DVD, this does not mean the material to be seen on TV is less disturbing. The argument also is not concerned with adults, and the suggestion is that violence in any quantity may have a detrimental effect, even if a show is not entirely made up of it. **A, B** and **D** are thus not the correct answers. **C** contradicts the argument, as it suggests there is no link between witnessing and re-enacting what one has witnessed. Children may watch the scenes of rape and recognise the horror of the action, and so be sworn off ever committing that crime.

Question 90: C

A is irrelevant, as the passage states it *could* teach children, not that it necessarily would. **B** and **C** are also irrelevant, as the entertainment quality of the show or the likeability of its protagonist would not undermine the logic of the argument. **C** is the correct answer, as it shows how the question uses one model of success and projects it onto all other models, which is illogical: just because Frank succeeds without morality, does not mean all others must reject morality to succeed.

Question 91: A

B, C, D and **E** are all irrelevant to Freddy's argument that he cannot say a sexist thing because he is a feminist. The woman's discomfort, Neil's feminist stance, the appropriateness of making comments about men, or lewd comments in general do not affect his claim. The presumed link between the two (inability to say something sexist, and feminist self-description) is the flaw in Freddy's argument: someone may believe in equal rights for the genders, and still say a sexist thing.

Question 92: A

At no point is it stated or implied that car companies should prioritise profits over the environment, so C) is incorrect. Neither is it stated that the public do not care about helping the environment, so E) is incorrect.

B) is a reason given in the argument, whilst D) is impossible if we accept the argument's reasons as true, so neither of these are conclusions.

Question 93: D

The easiest thing to do is draw the relative positions. We know Harrington is north of Westside and Pilbury. We know that Twotown is between Pilbury and Westside. Crewville is south of Twotown, Westside and Harrington but we do not know but its location relative to Pilbury.

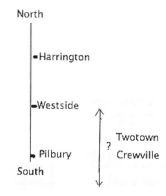

Question 94: B

By making a grid and filling in the relevant information the days Dr James works can be deduced:

	Sunday	Monday	Tuesday	Wednesday	Thursday	Friday	Saturday
Dr Evans	X	√	X	X	√	√	√
Dr James	X	√	√	√	√	X	√
Dr Luca	X	X	√	√	X	√	√

➢ No one works Sunday.
➢ All work Saturday.
➢ Dr Evans works Mondays and Fridays.
➢ Dr Luca cannot work Monday or Thursday.
➢ So, Dr James works Monday.
➢ And, Dr Evans and Dr James must work Thursday.
➢ Dr Evans cannot work 4 days consecutively so he cannot work Wednesday.
➢ Which means Dr James and Luca must work Wednesday.
➢ (mentioned earlier in the question) Dr Evans only works 4 days, so cannot work Tuesday.
➢ Which means Dr James and Luca work Tuesday.
➢ Dr James cannot work 5 days consecutively so cannot work Friday.
➢ Which means Dr Luca must work Friday.

Question 95: E

Working algebraically, using the call out rate as C, and rate per mile as M.

So, $C + 4m = 11$

$C + 5m = 13$

Hence; $(C + 5m) - (C + 4m) = £13 - £11$

$M = £2$

Substituting this back into $C + 4m = 11$

$C + (4 \times 2) = 11$

Hence, $C = £3$

Thus a ride of 9 mile will cost $£3 + (9 \times £2) = £21$.

Question 96: E

Use the information to create a Venn diagram.

We don't know the exact position of both Trolls and Elves, so **A** and **D** are true. Goblins are mythical but not magical, so **C** is true. Gnomes are neither so **B** is true. But **E** is not true.

Question 97: D

The best method may be work backwards from 7pm. The packing (15 minutes) of all 100 tiles must have started by 6:45pm, hence the cooling (20 minutes) of the last 50 tiles started by 6:25pm, and the heating (45 minutes) by 5:40pm. The first 50 heating (45 minutes) must have started by 4:35pm, and cooling (20 minutes) by 5:20pm. The decoration (50 minutes) of the second 50 can occur anytime during 4:35pm- 5:40pm as this is when the first 50 are heating and cooling in the kiln, and so does not add time. The first 50 take 50 minutes to decorate and so must be started by 3:45pm.

Question 98: E

Speed = distance/time. Hence for the faster, pain impulse the speed is 1m/ 0.001 seconds. Hence the speed of the pain impulse is 1000 metres per second. The normal touch impulse is half this speed and so is 500 metres per second.

Question 99: E

Using the months of the year, Melissa could be born in March or May, Jack in June or July and Alina in April or August. With the information that Melissa and Jack's birthdays are 3 months apart the only possible combination is March and June. Hence Alina must be born in August, which means it is another 7 months until Melissa's birthday in March.

Question 100: A

PC Bryan cannot work with PC Adams because they have already worked together for 7 days in a row, so **C** is incorrect. **B** is incorrect because if PC Dirk worked with PC Bryan that would leave PC Adams with PC Carter who does not want to work with him. PC Carter can work with PC Bryan.

Question 101: C

Paying for my next 5 appointments will cost £50 per appointment before accounting for the 10% reduction, hence the cost counting the deduction is £45 per appointment. So the total for 4 appointments = 5 x £45 = £225 for the hair. Then add £15 for the first manicure and £10 x 2 for the subsequent manicures using the same bottle of polish bringing an overall total of £260.

Question 102: D

Elena is married to Alex or David, but we are told that Bertha is married to David and so Alex must be married to Elena. Hence David, Bertha, Elena and Alex are the four adults. Bertha and David's child is Gemma. So Charlie and Frankie must be Alex and Elena's two children. Leaving only options **A** or **D** as possibilities. Only Frankie and Gemma are girls so Charlie must be a boy.

Question 103: C

Using, x (minutes) as the, unknown amount of time, the second student took to examine, we can plot the time taken with the information provided thus:

	1st student		2nd student		3rd student
1st examination:	4x	1	2x	1	2x
		Break: 8 minutes			
1st examination:	x	1	x	1	x

Hence the total time taken, 45minutes (14:30-15:15)

Is represented by, $4x + 2x + 2x + x+ x+ x+ 1+1+ 8+1+1$

$$45 \text{ minutes} = 11x \text{ (minutes)} + 12 \text{ minutes}$$
$$33 \text{ minutes} = 11x \text{ (minutes)}$$

Hence, x = 3 minutes, so the amount of time the second student took the first time, 2x, is 6 minutes.

Question 104: E & F

To work out the amount of change is the sum £5 - (2 x £1.65), which = £3.30. Logically we can then work out that the 3 coins in the change that are the same must be 1p as no other 3 coin combination can yield £1.70 when made up with 5 more coins. Thus we know that 3 of the coins are 1p, 1p & 1p. We can then deduce that there must also have been 2p and 5p coins in the change as £1.70 is divisible by ten. The only way then to make up the remaining £1.60 in 3 different coins is to have £1, 50p and 10p, Hence the change in coins is 1p, 1p, 1p, 2p, 5p, 10p, 50p and £1. So the two coins not given in change are £2 and 20p.

Question 105: D

If we express the speed of each train as W ms^{-1}. Then the relative speed of the two trains is $2W$ ms^{-1}.

Using Speed=distance/time: $2W = (140 + 140)/ 14$.

Thus, $2W = 20$, and $W = 10$. Thus, the speed of each train is 10 ms^{-1}.

To convert from metres to kilometres, divide by 1,000. To convert from seconds to hours, divide by 3,600.

Therefore, the conversion factor is to divide by $1,000/3,600 = 10/36 = 5/18$

Thus, to convert from ms^{-1} to kmph, multiply by 18/5. Therefore, the final speed of the train is 18/5 x 10 = 36km/hr.

Question 106: C

Taking the day to be 24 hours long, this means the first tap fills 1/6 of the pool in an hour, the second 1/48, the third $\frac{1}{72}$ and the fourth $\frac{1}{96}$.

Taking 288 as the lowest common denominator, this gives: $\frac{48}{288} + \frac{6}{288} + \frac{4}{288} + \frac{3}{288}$ which $= \frac{61}{288}$ full in one hour. Hence the pool will be $\frac{244}{288}$ full in 4 hours.

The pool fills by approximately $\frac{15}{288}$ every 15 minutes.

Thus, in 4 Hours 15: $\frac{244 + 15}{288} = \frac{249}{288}$

Thus, in 4 Hours 30: $\frac{244 + 30}{288} = \frac{274}{288}$

Thus, in 4 Hours 45: $\frac{244 + 45}{288} = \frac{289}{288}$

Question 107: B

Every day up until day 28 the ant gains a net distance of 1cm, so at the end of day 27 the ant is at 27cm height and therefore only 1cm below the top. On day 28 the 3cm the ant climbs in the day is enough to take it to the top of the ditch and so it is able to climb out.

Question 108: A

To solve this question three different sums are needed to use the information given to deduce the costs of the various items. With the information that 30 oranges cost £12, £12/30 = 40p per orange with the 20% discount, hence oranges must cost 50p at full price. With the information that 5 sausages and 10 oranges cost £8.50, we know that the oranges at a 10% discount account for 10 x 45p = £4.50 so 5 undiscounted sausages cost £4 so each full price sausage is £4/5 = 80p. Finally we know that 10 sausages and 10 apples cost £9, at 10% discount the sausages cost 72p each thus accounting for 10 x 72p = £7.20 of the £9, hence the 10 apples at a 10% discount must cost £1.80, so each apple costs 18p at 10% discount. So an apple is 20p full price. Now to add up the final total: 2 oranges + 13 sausages + 2 apples = (2 x 50p) + (13 x 72p) + (12 x 18p) = £12.52.

Question 109: C

If we take the number of haircuts per year to be x, the information we have can be shown:

Membership	Annual Fee	Cost per cut	Total Yearly cost
None	None	£60	60x
VIP	£125	£50	£125 + 50x
Executive VIP	£200	£45	£200 + 45x

As we know that changing to either membership option would cost the same for the year, we can express the cost for the year, y as;

VIP: y = £125 + 50x

Executive VIP: y = £200 + 45x

Therefore: £125 + 50x = £200 + 45x

Simplified   5x = £75, therefore the number of haircuts a year, x is 15.

Substituting in x, we can therefore work out:

Membership	Annual Fee	Cost per cut	Total Yearly cost
None	None	£60	£900
VIP	£125	£50	£875
Executive VIP	£200	£45	£875

Hence the amount saved by buying membership is £25.

Question 110: B

All thieves are criminals. So the circle must be fully inside the square, we are told judges cannot be criminals so the star must be completely separate from the other two.

Question 111: C

We are told that March and May have the same last number, which must be either 3 or 13. Taking the information from the question that one of the factors is related to the letters of the month names, we can interpret that 13 represents the M which starts both March and May. Therefore we know the rule is that the last number is the position of the starting letter. Knowing that there is another factor about the letters of the month that controls the code we can work out that one of the number may code for the number of letters. Which in March would be 5, which is the second letter, so we have the rule of the 2nd number. Finally through observation we may note that the first number codes for the months' relative position in the year. Hence the code of April will be 4, (for its position), 5 (for the number of letters in the name) and 1 for the position of the starting letter 'A') and so 451 is the code.

Question 112: D

If b is the number of years older than 5, and a the number of A*s, the money given to the children can be expressed:

£5 + £3b + £10a

Hence for Josie £5 + (£3 x 11) + (£10 x 9) = £128

We know that Carson receives £44 less yearly, and his b value is 13, so his amount can be expressed:

£5 + (£3 x 13) + (£10a) = £84

Simplified: £44 + £10a = £84

I.e. £10a = £40,

So Carson's 'a' value, i.e. his number of A*s is 4, so the difference between Josie and Carson is 5.

Question 113: B & C

Using the information to make a diagram:

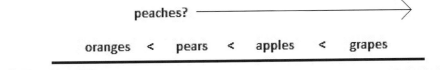

Hence **A** is incorrect. **D** and **E** may be true but we do not have enough information to say for sure. **B** is correct as we know peaches are more expensive than oranges but not about their price relative to pears. Equally we know **C** to be true as grapes are more expensive than apples so they must be more expensive than pears.

Question 114: B

It's easy to assume all the cuts should be in the vertical plane as a cake is usually sliced, however there is a way to achieve this with fewer cuts. Only three cutting motions are needed. **Start by cutting in the horizontal plane** through the centre of the cake to divide the top half from the bottom half. Then slice in the vertical plane into quarters to give 8 equally sized pieces with just three cuts.

Question 115: B & D

After the changes have been made, at 12 PM (GMT +1):

- Russell thinks it is 11 AM
- Tom thinks it is 12 PM
- Mark thinks it is 1 PM

Thus, in current GMT+1 time zone, Mark will arrive an hour early at 11 AM, Russell an hour late at 1 PM and Tom on time at 12 PM. There is therefore a two hour difference between the first and last arrival. For options E and F, be careful: the time zone listed is **NOT** GMT +1 that everyone else is working in. 1PM in GMT +3 = 11am GMT +1 (the summer time zone just entered) so that is Mark's actual arrival time; 12pm GMT +0 is the old time zone that Russell didn't change out of so that is Russell's correct arrival time.

Question 116: D

Using Bella's statements, as she must contradicted herself with her two statements, as one of them must be true, we know that it was definitely either Charlotte or Edward. Looking to the other statements, e.g. Darcy's we know that it was either Charlotte or Bella, as only one of the two statements saying it was both of them can have been a lie. Hence it must have been Charlotte.

Question 117: F

The only way to measure 0.1 litres or 100ml, is to fill the 300ml beaker, pour into the half litre/ 500 ml beaker, fill the 300ml again and pour (200ml) into the 500ml, which will make it full, leaving 100ml left in the 300ml beaker. The process requires 600ml of solution to fill the 300ml beaker twice.

Question 118: D

If you know how many houses there are on the street it is possible to work out the average, which then you can round up and down and to find the sequence of number, e.g. if you know there are 6 houses in the street 870/ 6 = 145. Which is not a house number because they are even so going up and down one even number consequentially one discovers that the numbers are 146, 144, 148, 150, 142 and 140. But it is not possible to determine Francis' house number without knowing its relative position i.e. highest, 3rd highest, lowest etc.

Question 119: D

Expressed through time:

Event	People Present
There were 20 people exercising in the cardio room	20
Four people were about to leave	20
A doctor was on the machine beside him (one of the original 20)	20
Emerging from his office one of the personal trainers called an ambulance.	21
Half of the people who were leaving, left (-2)	19
Eight people came into the room to hear the man being pronounced dead. (+8)	27
the two paramedics arrived, (+2)	29
the man was pronounced dead (-1)	28

Question 120: B & D

Blood loss can be described as 0.2 L/min.
For the man:
8 litres – 40% (3.2 L) = 4.8 L When he collapses, taking 16 minutes (3.2 / 0.2 = 16)
For the woman:
7 litres – 40% (2.8L) = 4.2: when she collapses, taking 14 minutes (2.8 / 0.2 = 14)
Hence the woman collapses 2 minutes before the man so **B** is correct, and **A** is incorrect. The total blood loss is 3.2L + 2.8L which = 6L so **C** is incorrect. The man's blood loss is 3.2L when he collapses so **E** is incorrect. The woman has a remaining blood volume of 4.2L when she collapses so **D** is correct. Blood loss is 0.2 L/min, which equates to 5 minutes per litre, which is 10 minutes per 2 litres not 12 L, so **F** is incorrect.

Question 121: B

Work out the times taken by each girl – (distance/pace) x 60 (converts to minutes) + lag time to start
Jenny: (13/8) x 60 = 97.5 minutes
Helen: (13/10) x 60 + 15 = 93 minutes
Rachel (13/11) x 60 + 25 = 95.9 minutes

Question 122: C

Work through each statement and the true figures.

A. Overlap of pain and flu-like symptoms must be at least 4% (56+48-100). 4% of 150: 0.04 x 150=6
B. 30% high blood pressure and 20% diabetes, so max percentage with both must be 20%. 20% of 150: 0.2*150 = 30
C. Total number of patients – patients with flu-like symptoms – patients with high blood pressure. Assume different populations to get min number without either. 150 – (0.56 x 150) – (0.3 x 150) = 21
D. This is an obvious trap that you might fall into if you added up the percentages and noted that the total was >100%. However, this isn't a problem as patients can discussed two problems.

Question 123: C

This is easiest to work out if you give all products an original price, I have used £100. You can then work out the higher price, and the subsequent sale price, and thus the discount from the original £100 price. As the price increases and decreases are in percentages, they will be the same for all items regardless of the price so it does not matter what the initial figure you start with is.

Marked up price: 100 x 1.15 = £115
Sale price: 115 x 0.75 = £86.25
Percentage reduction from initial price is 100 – 86.25 = 13.75%

Question 124: D

The recipe states 2 eggs makes 12 pancakes, therefore each egg makes 6 pancakes, so the number Steve must make should be a multiple of 6 to ensure he uses a whole egg.

Steve requires a minimum of 15 x 3 = 45 pancakes. To ensure use of whole eggs, this should be increased to 48 pancakes.

The original recipe is for 12 pancakes, therefore to make 48 pancakes, require 4x recipe (48/12).

Therefore quantities: 8 eggs, 400g plain flour and 1200 ml milk.

Question 125: B

Work through the question backwards.

In 6 litres of diluted bleach, there are 4.8 litres of water and 1.2 litres of partially diluted bleach.

In the 1.2 litres of partially diluted bleach, there is 9 parts water to one part original warehouse bleach. Remember that a ratio of 1:9 means 1/10 bleach and 9/10 water. Therefore working through, there is 120ml of warehouse bleach needed.

Question 126: C

We know that Charles is born in 2002, therefore in 2010 he must be 8. There are 3 years between Charles and Adam, and Charles is the middle grandchild. As Bertie is older than Adam, Adam must be younger than Charles so Adam must be 5 in 2010. In 2010, if Adam is 5, Bertie must be 10 (states he is double the age of Adam).

The question asks for ages in 2015: Adam = 10, Bertie = 15, Charles = 13

Question 127: B

Make the statements into algebraic equations and then solve them as you would simultaneous equations. Let a denote the flat fixed rate for hire, and b the price per half hour. Cost = a + b(time in mins/30)

Peter: a + 6b (6 half hours) = 14.50 (equation 1)

Kevin: 2a + 18b = 41, or this can be simplified to give cost per kayak, a + 9b = 20.5 (equation 2)

If you subtract equation 1 from equation 2: 3b = 6, therefore b = 2

Substitute b into either equation to calculate a, using equation 1, a + 12 = 14.50, therefore a = 2.50

Finally use these values to work out the cost for 2 hours:

2.50 (flat fee) + 4 x 2 (4half hours x cost/half hour) = £10.50

Question 128: E

It is most helpful to write out all the numbers from 0 – 9 in digital format to most easily see which light elements are used for each number. You can then cross out any numbers which don't use all the lights from the digit 7.

Go through the digits methodically and you can cross out: 1, 2, 4, 5, and 6. These numbers don't contain all three bars from the digit 7.

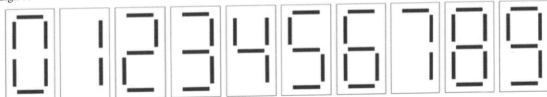

Question 129: B

In this question it is worth remembering it will take more people a shorter amount of time.

Work out how many man hours it takes to build the house. Days x hours x builders

12 x 7 x 4 = 336 hours

Work out how many hours it will take the 7man workforce: 336/7 = 48 hours. Convert to 8 hour days: 48/8 = 6 days

Question 130: D

By far the easiest way to do these type of questions is to draw a Venn diagram (use question marks if you are unsure about the exact position):

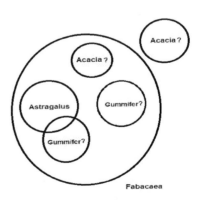

Now, it is a case of going through each statement:

A. Incorrect - Acacia may be fabacaea. Acacia are not astragalus, but does not logically follow that they therefore can't be fabacaea.
B. Incorrect – astragalus and gummifer are not necessarily separate within fabacaea.
C. Incorrect – the statement is not reversible so the fact that all astragalus and gummifer are fabacaea does not mean all facacaea are gummifer and/or astragalus. E.g. Fabacaea could be acacia.
D. Correct
E. Incorrect – Whilst some acacia could be gummifer, there is no certainty that they are.

Question 131: D

Area of a trapezium = (a+b)/2 x h

Area of cushion = (50+30)/2 x 50 = 2000cm²

Since each width of fabric is 1m wide, both sides of one cushion can fit into one width. The required length is therefore 75cm x 4 = 3m with a cost of 3 x £10 = £30.

Cost of seamstress = £25 x 4 = £100 thus total cost is £130

Question 132: C

There are 30 days in September, so Lisa will buy 30 coffees.

In Milk, every 10th coffee is free, so Lisa will pay for 27 coffees at 2.40 = £64.80

In Beans, Lisa gets 20 points each day and needs 220 points to get a free coffee, which is 11 days, with 5 points left over. Therefore, in 30 days she will get 2 free coffees. The cost for 28 coffees at 2.15 is £60.20

Beans is cheaper, and the difference is £64.80 - £60.20 = £4.60.

Question 133: C

Work backwards and take note of how often each bus comes.

Must get off 220 bus at 10.57 latest. Can get 10.40 bus therefore (arrive at 10.54).

Latest can get on 283 bus is 10.15 as to make the 220 bus connection. 283 comes every 10mins (question doesn't state at what points past the hour), so Paula should be at the bus stop at 10.06 to ensure a bus comes by 10.15 at the latest. If the bus comes every 10mins, even if a bus comes at 10.05 which Paula will miss, the next bus will come at 10.15 and therefore she will still be on time. Therefore Paula must leave at 10.01

Question 134: B

You are working out the time taken to reach the same distance (D). Make sure to take into account changing speeds of train A, and that train B leaves 20 minutes earlier.

$Speed = \frac{distance}{time}$

Make sure you keep the answers consistent in the time units you are using, the worked answer is all in minutes (hence the need to multiply by 60).

Train A: time for first $20km = \frac{20}{100} x 60 = 12\ minutes$. So the distance where it equals B is $12 + (\frac{D-20}{150}) x 60$

You need to use D-20 to account for the fact you have already calculated the time at the slower speed for the first 20km

Train B: $(\frac{D}{90}) x 60 - 20$

Make the equations equal each other as they describe the same time and distance, and solve.

Simplifies to $32 + \frac{2D}{5} - 8 = \frac{2D}{3}$ so $D = 90km$

Train B will take 60 minutes to travel 90 km and train A will take 40 minutes (but as it leaves 20 minutes later, this will be point at which it passes).

Question 135: C

Work out the annual cost of local gym: 12 x 15 = £180

Upfront cost + class costs of university gym must therefore be >£180.

Subtract upfront cost to find number of classes: 180 – 35 = £145

Divide by cost per class (£3) to find number of classes: 145/3 = 48 1/3

48 1/3 classes would make the two gyms the same price, so for the local gym to be cheaper, you would need to attend 49 classes.

Question 136: C

A is definitely true, since the question states that all herbal drugs are not medicines. **B** is also definitely true as all antibiotics are medicines which are all drugs. **C** is definitely false, because all antibiotics are medicine, yet no herbal drugs are medicines. **D** is true as all antibiotics are medicines.

Question 137: C

Answer **A** cannot be reliably concluded, because from the information given a non-"Fast" train could stop at Newark, but not at Northallerton or Durham. We have no information on whether *all* trains stopping at Newark also stop at Northallerton.

Answer **B** is not correct because 8 is the *average* number of trains that stop at Northallerton. It is possible that on some days more than 16 trains run, and more than 8 will thus stop at Northallerton. Answer **D** is incorrect because it is mentioned that *all* trains stopping at Northallerton also stop at Durham, giving a total 6 stops as a minimum for a train stopping at Northallerton (the others being the 4 stops which *all* trains stop at). Answer **E** is incorrect for a similar reason to **A**. We have no information on whether all trains stopping at Newark also stop at Northallerton, so cannot determine that they must also stop at Durham. Answer **C** is correct because "Fast" trains make less than 5 stops. Since all trains already stop at 4 stops (Peterborough, York, Darlington and Newcastle), they cannot then stop at Durham, as this would give 5 stops.

Question 138: D

From the information we are given, we can compose the following image of how these towns are located (not to scale, but shows the direction of each town with respect to the others):

From this "map", we can see that all statements apart from **D** are true. Statement **D** is definitely *not true*, since Blueville is south west of Haston it cannot be East of Haston.

Question 139: C

We are told that in order to form a government, a party (or coalition) must have *over* 50% of seats. Thus, they must have at least 50% of the total seats plus 1, which is 301 seats.

We are told that we are looking for the *minimum* number of seats the greens can have in order to form a coalition with red and orange. Thus, we are seeking for Red and Orange to have the *maximum* number of seats possible, under the criteria given.

Thus we can calculate as follows:

- No party has over 45% of seats, so the maximum that the Red party can have is 45%, which is 270 seats.
- No party except for red and blue has won more than 4% of seats. We are told that the green party won the 4th highest number of seats, so it is possible that the Orange party won the 3rd highest.
- Thus, the maximum number of seats the orange party can have won is 4% of the total, which is 24 seats.
- Thus, the maximum possible combined total of the Red and Orange party's seats won is 294.

Thus, in order to achieve a total of 301 seats in a Red-Orange-Green coalition, the Green party have to have won at least 7 seats. However, in addition to satisfy the criteria of the green party coming 4th place they must have won the majority of the remaining 36 seats giving a final breakdown of votes as: Red 270, Blue 270, Orange 24, Green 13, Yellow 12, Purple 11.

Question 140: E

Expressing the amount each child receives:

Youngest	M
2nd youngest	$M + D$
3rd youngest/ 3rd oldest	$M + 2D$
4th youngest/ 2nd oldest	$M + 3D$
Oldest	$M + 4D$

Question 141: D

The total amount of money received;
£100, $= M + M + D + M + 2D + M + 3D + M + 4D$
Simplified, thus is:
£100 $= 5M + 10D$

Question 142: C

The two youngest are expressed as M and $M + D$. Simplified as $2M + D$.
The three oldest are expressed as $M + 2D$, $M + 3D$ and $M + 4D$, Simplified as $3M + 9D$
Hence 7 times the two youngest together is expressed $7(2M + D)$, so altogether the Answer is $7(2M + D) = 3M + 9D$.

Question 143: A

To work this out, simplify the two equations:
$7(2M + D) = 3M + 9D$
$14M + 7D = 3M + 9D$
$11M = 2D$
$M = \frac{2D}{11}$

Question 144: A

Substitute M into the equation £ $100 = 5M + 10D$
$5\left(\frac{2D}{11}\right) + 10D = £100$
$\frac{10D}{11} + 10D = \frac{10D}{11} + \frac{110D}{11} = \frac{120D}{11}$

Question 145: C & E

The easiest way to work this out is using a table. With the information we know:

1st		Madeira
2nd		
3rd	Jaya	
4th		

Ellen made carrot cake and it was not last. It now cannot be 1st or 3rd as these places are taken so it must be second:

1st		Madeira
2nd	Ellen	Carrot cake
3rd	Jaya	
4th		

Aleena's was better than the tiramisu, so she can't have come last, therefore Aleena must have placed first

1st	Aleena	Madeira
2nd	Ellen	Carrot cake
3rd	Jaya	
4th		

And the girl who made the Victoria sponge was better than Veronica:

1st	Aleena	Madeira
2nd	Ellen	Carrot cake
3rd	Jaya	Victoria Sponge
4th	Veronica	Tiramisu

Question 146: D

The information given can be expressed to show the results that the teams must have had to make their points total.

The results so far total 3 wins, 6 losses and 7 draws. Since, the number of draws must be even, there must have been another draw. So we know one of the Eire Lions results is a draw.

The difference between wins (3) and losses (6) is 3. Thus, there must be another 3 wins to account for this difference. So the Eire Lions results must be 3 wins and 1 draw. Thus, they scored 3 x 3 + 1 = 10.

Team	Points	Game Results			
Celtic Changers	2	L	L	D	D
Eire Lions	?	?	?	?	?
Nordic Nesters	8	W	W	D	D
Sorten Swipers	5	W	D	D	L
Whistling Winners	1	D	L	L	L

Question 147: D

Remember to consider the gender of each person. Then draw a quick diagram to show the given information you can see that only D is correct.

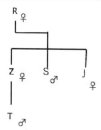

Question 148: B

After the first round; he knocks off 8 bottles to leave 8 left on the shelf. He then puts back 4 bottles. There are therefore 12 left on the shelf. After the second round, he has hit 3 bottles and damages 6 bottles in total, and an additional 2 at the end. He then puts up 2 new bottles to leave 12 – 8 + 2 = 6 bottles left on the shelf. After the final round, John knocks off 3 bottles from the shelf to leave 3 bottles standing.

Question 149: D

Based on the information we have we can plot the travel times below. Change over times are in a smaller font.

Hence on the St Mark's line, St Mark's to Archite takes 4 x 2.5 minutes = 10 minutes.

Question 150: A

Going from stop to stop on the Straightly line end Buft to Straightly would take 14 minutes, but we are told earlier on there is an express train that goes end to end and only takes 6.

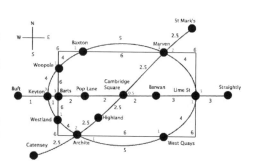

Question 151: B

The quickest route from Baxton to Pop Lane is via Marven and Cambridge Square, which takes $5 + 2 + 2.5 + 0.5 + 2 = 12$ minutes. Baxton to Pop Lane via Barts would take $4 + 1 + 6 + 3 + 2 = 16$ minutes, which is longer so **E** is incorrect. Other options include times failing to take account of, or incorrectly adding changeover times, and so are incorrect.

Question 152: C

From Cambridge Square:

➤ Catensey is ($2.5 \times 3 =$) 7.5 minutes away.
➤ Woopole, is ($4 + 3 + 1 + 2 + 2 =$) 12 minutes.
➤ Buft is ($1 + 1 + 2 + 2 =$) 6 minutes.
➤ Westland is ($4 + 2 + 2.5 + 2.5 =$) 11 minutes.

Question 153: B

With the new delay information we can plot the travel times as before, adjusted for the delays. Plus a 5 minute delay on the platforms when waiting on any platform for a train.

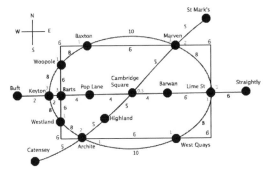

The quickest way from Westland to Marven now uses the non-delayed reliable rectangle line. Four stops on the rectangle line take 6 mins each so 24 minutes in total on the train. Add to this the additional 5 minutes platform waiting time to give a total journey time of 29 minutes.

Question 154: C

- Baxton to Archite via Barts using only the Rectangle line takes ($5 + 6 + 6 + 6 + 6 =$) 29 minutes.
- Baxton to Woopole on the Rectangle line, then Oval to Archite via Keyton takes ($5 + 6 + 1 + 5 + 8 + 8 + 8 =$) 41 minutes
- Baxton to Archite on the Oval line only takes ($5 + (8 \times 4) =$) 37 minutes
- Baxton to Woopole on the Oval line, then Rectangle to Archite via Barts takes ($5 + 8 + 1 + 5 + 6 + 6 + 6 =$) 37 minutes
- As the bus takes 27-31 minutes, it is not possible to tell from between the options which will be slower/quicker so option **C** is the right answer.

Question 155: D

Remember the 5-minute platform wait. We are not told that the St Mark's express train from end to end is no longer running so we must assume that it is, which takes 5 minutes (plus the wait at St Mark's to go to Catensey).

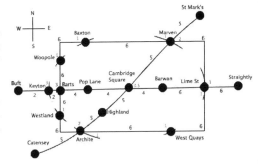

Then, there is a 5 minute wait at Catensey to Archite, and a $2 + 5$ minute changeover at Archite onto the Rectangle line which then takes 6 minutes to West Quays. $5 + 5 + 5 + 5 + 2 + 5 + 6 = 33$ minutes. Via Lime St the journey takes $5 + 5 + 5 + 2 + 5 + 6 + 6 = 29$ minutes.

Question 156: D

From the information:

- "Simon's horse wore number 1."
- "...the horse that wore 3, which was wearing red..."
- "the horse wearing blue wore number 4."

We can plot the information below:

Place	Owner	Number	Colours
	Simon	1	
		2	
		3	Red
		4	Blue

In addition: "The horse wearing green; Celia's, came second"
Which means Celia's horse must have worn number two because it cannot have worn number 1 because that is Simon's horse. Also it cannot have worn number three or four because they wore red and blue respectively. So we can plot this further deduction:

Place	Owner	Number	Colours
	Simon	1	
2nd	Celia	2	Green
		3	Red
		4	Blue

We also know that
- "Arthur's horse beat Simon's horse"
- "Celia's horse beat the horse that wore number 1." i.e. Simon's

We know Celia's horse came second, and that both Celia's and Arthur's horses beat Simon's. This means that Simon's horse must have come last. So;

Place	Owner	Number	Colours
4th	Simon	1	
2nd	Celia	2	Green
		3	Red
		4	Blue

And knowing that:
- "Only one horse wore the same number as the position it finished in."

The horses wearing numbers 3 and 4 must have placed 1st and 3rd respectively. Hence:

Place	Owner	Number	Colours
4th	Simon	1	
2nd	Celia	2	Green
1st		3	Red
3rd		4	Blue

"Lila's horse wasn't painted yellow nor blue"

So Lila's must have been red, and Simon's yellow. Leaving the only option for Arthur's to be blue. So we now know:

Place	Owner	Number	Colours
4th	Simon	1	Yellow
2nd	Celia	2	Green
1st	Lila	3	Red
3rd	Arthur	4	Blue

Question 157: C
Year 1 – 40 x 1.2 = 48
Year 2 – 48 x 1.2 = 57.6
Year 3 – 57.6 x 1.1 = 63.36
Year 4 – 63.36 x 1.1 = 69.696.

Question 158: C

To minimise the total cost to the company, they want the wage bills for each site to be less than £200,000. Working this out involves some trial and error; you can speed this up by splitting employees who earn similar amounts between the sites e.g. Nicola and John as they are the top two earners. Nicola + Daniel + Luke = £ 198,500 and John + Emma + Victoria = £ 199,150

Question 159: C

Remember that pick up and drop off stops may be the same stop, therefore the minimum number of stops the bus had to make was 7. This would take 7 x 1.5 = 10.5 minutes. Therefore the total journey time = 24 + 10.5 = 34.5 minutes.

Question 160: A

The best method here is to work backwards. We know the potatoes have to be served immediately, so they should be finished roasting at 4pm, so they should start roasting 50 minutes prior to that, at 3:10. We also know they have to be roasted immediately after boiling, so they should be prepared by 3:05, in order to boil in time. She should therefore start preparing them no later than 2:47, though she could prepare them earlier. The chicken needs to be cooked by 3:55 to give it time to stand, so it should begin roasting at 2:40, and Sally should begin to prepare it no later than 2:25. You can construct a rough timeline:

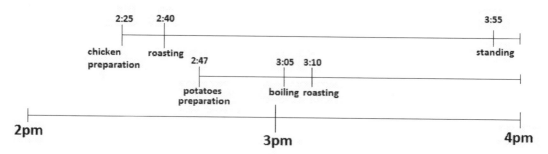

We can see from this timeline that from 2:40 onwards, there will be no long enough period of time in which there is a free space in the cooker for the vegetables to be boiled. They therefore must be finished cooking at 3:05. The latest time prior to this that Sally has time to prepare them (5 minutes) is at 2:40, between preparing the chicken and the potatoes. She should therefore begin preparing the vegetables at 2:42, then begin boiling at 2:47, so they can be finished cooking by 2:55, in time for the potatoes to boil at 3:05.

Chicken: 2:25

Potatoes: 2:47

Vegetables: 2:42

Question 161: C

The quickest way to do this is via trial and error. However, for the sake of completion: let each child's age be denoted by the letter of their name, and form an equation for their total age:

$P + J + A + R = 80$

The age of each child can be written in terms of Paul's age.

P = 2J, therefore $J = \frac{P}{2}$

$A = \frac{P+J}{2}$

Now substitute in $J = \frac{P}{2}$ to get in terms of P only: $A = \frac{P+\frac{P}{2}}{2} = \frac{P}{2} + \frac{P}{4} = \frac{3P}{4}$

$R = P + 2$

Thus: $P + \frac{P}{2} + \frac{3P}{4} + P(+2) = 80$

Simplify to give: $\frac{13P}{4} = 78$

$13P = 312$. Thus, $P = 24$

Substitute P = 24 into the equations for the other children to get: J = 12, A = 18, R = 26

Question 162: A

The total number of buttons is $71 + 86 + 83 = 240$. The total number of suitable buttons is $22 + 8 = 30$. Thus, she will have to remove a maximum of 210 buttons in order to guarantee picking a suitable button on the next attempt.

Question 163: E

This question requires you to calculate the adjusted score for Ben for each segment. If Ben has a 50% chance of hitting the segment he is aiming for, we can assume he hits each adjacent segment 25% of the time. Thus: $Adjusted\ Score =$

$\frac{Segment\ aimed\ at}{2} + \frac{First\ Adjacent\ Segment}{4} + \frac{Second\ Adjacent\ Segment}{4}$

$Adjusted\ Score = \frac{Segment\ aimed\ at}{2} + \frac{Sum\ of\ Adjacent\ Segments}{4}$

E.g. if he aims at segment 1: He will score $\frac{1}{2} + \frac{18+20}{4} = 10$

Now it is a simple case of trying the given options to see which segment gives the highest score. In this case, it is segment 19:

$\frac{19}{2} + \frac{7+3}{4} = 12$

Question 164: C

The total cost is £8.75, and Victoria uses a £5.00 note, leaving a total cost of £3.65 to be paid using change.

Up to 20p can be paid using 1p and 2p pieces, so she could use 20 1p coins to make up this amount.

Up to 50p can be paid using 5p and 10p pieces, so she could use 10 5p pieces to make up this amount. This gives a total of 30 coins, and a total payment of £0.70.

Up to £1.00 can be paid using 20p pieces and 50p pieces. Thus, she could use up to 5 20p pieces, giving a total of 35 coins used, and a total payment of £1.70.

The smallest denomination of coin that can now be used is a £1.00 coin. Hence Victoria must use 2 £1.00 coins, giving a total of 37 coins, and a total payment of £3.70. However, we know that the total cost to pay in change was £3.65, and that Victoria paid the exact amount, receiving no change. Thus, we must take away coins to the value of 5p, removing the smallest number of coins possible. This is achieved by taking away 1 5p piece, giving a grand total of 36 coins.

Question 165: B

The time could be 21:25, if first 2 digits were reversed by the glass of water (21 would be reversed to give 15). **A** cannot be the answer, because this would involve altering the last 2 digits, and we can see that 25 on a digital clock, when reversed simply gives 25 (the 2 on the left becomes a 5 on the right, and the 5 on the right becomes a 2 on the left). **C** cannot be the answer, as this involves reversing the middle 2 digits. As with the right two digits, the middle 2 digits of 2:5 would simply reverse to give itself, 2:5. **D** could be the time if the 2nd and 4th digits were reversed, as they would both become 2's. However, the question says that 2 *adjacent* digits are reversed, meaning that the 2nd and 4th digits cannot be reversed as required here. **E** is not possible as it would require all four numbers to be reversed.

Thus, the answer is **B**.

Question 166: B

We can see from the question that Lorkdon is a democracy and therefore cannot have been invaded by a democracy because of the treaty (we are assuming this treaty is upheld, as said in the question). Thus, Nordic (which has invaded Lorkdon) *must* be a dictatorship. Now, we can see that Worsid has been invaded by a dictatorship, *and* has invaded a dictatorship. The question states that no dictatorship has undergone both of these events. Thus, we know that Worsid cannot be a dictatorship. We also know from the question that each of these countries is *either* a dictatorship or a democracy. Thus, Worsid must be a democracy.

Question 167: C

The total price of all of these items would usually be £17. However, with the DVD offer, the customer saves £1, giving a total cost of £16. Thus, the customer will need to receive £34 in change.

Question 168: B

To answer this, we simply calculate how much total room in the pan will be taken up by the food for each guest:
- 2 rashers of bacon, giving a total of 14% of the available space.
- 4 sausages, taking up a total of 12% of the available space.
- 1 egg takes up 12% of the available space.

Adding these figures together, we see that each guest's food takes up a total of 38% of the available space.

Thus, Ryan can only cook for 2 guests at once, since 38% multiplied by 3 is 114%, and we cannot use up more than 100% of the available space in the pan.

Question 169: C

To calculate this, let the total number of employees be termed "Y".

We can see that £60 is the total cost for providing cakes for 40% of "Y".

We know that £2 is required for each cake. Thus, we can work out that 30 must be 40% of Y.

$0.4Y = 60/2$
$0.4Y = 30$
$Y = 75$

Thus, we can calculate that the total number of employees must be 75.

Question 170: E

The normal waiting time for treatment is 3 weeks. However, the higher demand in Bob's local district mean this waiting time is extended by 50%, giving a total of 4.5 weeks.

Then, we must consider the delay induced because Bob is a lower risk case, which extends the waiting time by another 20%. 20% of 4.5 is 0.9, so there is a delay of another 0.9 weeks for treatment.

Thus, Bob can expect to wait 5.4 weeks for specialist treatment on his tumour.

Question 171: D

In the class of 30, 40% drink alcohol at least once a month, which is 12. Of these, 75% drink alcohol once a week, which is 9. Of these, 1 in 3 smoke marijuana, which is 3.

In the class of 30, 60% drink alcohol less than once a month, which is 18. Of these, 1 in 3 smoke marijuana, which is 6.

Therefore the total number of students who smoke marijuana is 3+6, which is 9.

Question 172: C

The sequence can either be thought of as doubling the previous number then adding 2, or adding 1 then doubling. Double 46 is 92, plus 2 is 94.

Question 173: B

If the mode of 5 numbers of 3, it must feature at least two threes. If the median is 8, we know that the 3rd largest number is an 8. Hence we know that the 3 smallest numbers are 3, 3, and 8. Because the mean is 7, we know that the 5 numbers must add up to 35. The three smallest numbers add up to 14. Hence the two largest must add up to 21.

Question 174: E
The biggest difference in the weight of potatoes will be if the bag with only 5 potatoes in weighs the maximum, 1100g, and the bag with 10 potatoes weighs the minimum, 900g. If there are 5 equally heavy potatoes in a bag weighing 1100g, each weighs 220g. If there are 10 equally heavy potatoes in a 900g bag, each weighs 90g. The difference between these is 130g.

Question 175: D
There are 60 teams, and 4 teams in each group, so there are 15 groups. In each group, if each team plays each other once, there will be 6 matches in each group, making a total of 90 matches in the group stage. There are then 16 teams in the knockout stages, so 8 matches in the first round knockout, then 4, then 2, then 1 final match when only two teams are left. Hence there are 105 matches altogether (90 + 8 + 4 + 2 + 1 = 105).

Question 176: A
We know the husband's PIN number must be divisible by 8 because it has been multiplied by 2 3 times and had a multiple of 8 added to it. The largest 4 digit number which is divisible by 8 is 9992. Minus 200 is 9792. Divide by 2 is 4896. Hence the largest the husband's last 4 card digits can be is 4896. Minus 200 is 4696. Divide by 2 is 2348. Hence the largest my last 4 card digits can be is 2348. Minus 200 is 2148. Divide by 2 is 1074. Hence the largest my PIN number can be is 1074.

Question 177: C
If the first invitation is sent as early as possible, it will be sent on the 50th birthday. It will be accepted after 2 reminders and hence conducted at 50 years 11 months. The time between each screening will be 3 years 11 months. Hence, the second screening will be at 54 years 10 months. The third screening will be at 58 years 9 months. Hence, the fourth screening will be at 62 years 8 months.

Question 178: A
Ellie has worked for the company for more than five but less than six whole years. At the end of each whole year she receives a pay rise in thousands equal to the number of years of her tenure. Therefore at the end of the first year the raise is £1,000, then at the end of the second year it is £2,000 and so on to year 5. Thus the total amount of her pay comprised by the pay rises is £15,000, so the basic pay before accounting for these rises was £40,000 - £15,000 = £25,000.

Question 179: B
The trains come into the station together every 40 minutes, as the lowest common multiple of 2, 5 and 8 is 40. Hence, if the last time trains came together was 15 minutes ago, the next time will be in 25 minutes.

Question 180: C
If you smoke, your risk of getting Disease X is 1 in 24. If you drink alcohol, your risk of getting Disease X is 1 in 6. Each tablet of the drug halves your risk. Therefore a drinker taking 1 tablet means their risk is 1 in 12, and taking 2 tablets means their risk is 1 in 24, the same as someone who smokes.

Question 181: A
There are 10 red and 8 green balls. Clearly the most likely combination involves these colours only. Since there are more red balls than green, the probability of red-red is greater than green-green. However, there are **two** possible ways to draw a combination, either the red first followed by green or green first followed by red. The probability of red-red = $\left(\frac{10}{20} x \frac{9}{19}\right) = \frac{9}{38}$.

The probability of red and green = $\left(\frac{8}{20} x \frac{10}{19}\right) + \left(\frac{10}{20} x \frac{8}{19}\right) = \frac{8}{38} + \frac{8}{38} = \frac{16}{38}$. Therefore the combination of red and green is more likely.

Question 182: B
The least likely combination of balls to draw is blue and yellow. You are much more likely to draw a green ball than either a blue or yellow one because there are many more in the bag. Since the draw is taken without replacement, yellow and yellow is impossible because there is only one yellow ball.

Question 183: F
Since there is only 1 blue and 1 yellow ball, it is possible to take 18 balls which are red or green. You would need to take 19 of the 20 balls to be certain of getting either the blue ball or the yellow ball.

Question 184: C
The smallest number of parties required would theoretically be 3 – Namely Labour, the Liberal Democrats and UKIP, giving a total of 355 seats. However, the Liberal Democrats will not form a coalition with UKIP, so this will not be possible. Thus, there are 2 options:

➢ Labour can form a coalition with the Greens and UKIP, which is not contradictory to anything mentioned in the question. This would give a total of 325 seats, and would thus need the next 2 largest parties (The Scottish National Party and Plaid Cymru) in order to get more than 350 seats, meaning 5 parties would need to be involved.

> Alternatively, Labour can form a coalition with the Liberal Democrats and the Green Party. This would give a total of 340 seats. Only one more party (e.g. the Scottish National Party) would be required to exceed 350 seats, giving a grand total of 4 parties.

Thus, the smallest number of parties needed to form a coalition would be 4.

Question 185: E

360 appointments are attended and only 90% of those booked are attended, meaning there were originally 400 appointments booked in and 40 have been missed. 1 in 2 of the booked appointments were for male patients, so 200 appointments were for male patients. Male patients are three times as likely to miss booked appointments, so of the 40 that were missed, 30 were missed by men. Given that of 200 booked appointments, 30 were missed, this means 170 were attended.

Question 186: B

If every one of 60 students studies 3 subjects, this is 180 subject choices altogether. 60 of these are Maths, because everyone takes Maths. 60% of 60 is 36, so 36 are Biology. 50% of 60 is 30, so 30 are Economics and 30 are Chemistry. 60+36+30+30=156, so there are 24 subject choices left which must be Physics.

Question 187: B

If 100,000 people are diagnosed with chlamydia and 0.6 partners are informed each, this is 60,000 people, of which 80% (so 48,000) have tests. 12,000 of the partners who are informed, as well as 240,000 who are not (300,000 – 60,000) do not have tests. This makes 252,000 who are not tested. We can assume that half of these people would have tested positive for chlamydia, which is 126,000. So the answer is 126,000.

Question 188: C

Tiles can be added at either end of the 3 lines of 2 tiles horizontally or at either end of the 2 lines of 2 tiles vertically. This is a total of 10, but in two cases these positions are the same (at the bottom of the left hand vertical line and the top of the right hand vertical line). So the answer is $10 - 2 = 8$.

Question 189: C

Harry needs a total of 4000ml + 1200ml = 5200ml of squash. He has 1040ml of concentrated squash, which is a fifth of the total dilute squash he needs. So he will need 4 parts water to every 1 part concentrated squash, therefore the resulting liquid is 1/5 squash and 4/5 water.

Question 190: C

There are 24 different possible arrangements (4 x 3 x 2 x 1), which means that there are 23 other possible arrangements than Alex, Beth, Cathy, Daniel.

Question 191: F

A is incorrect because the distance travelled is only 10 miles. **B** is incorrect because the distance travelled is 19 miles. **C** is incorrect because no town is visited twice. **D** is incorrect because Hondale and Baleford are both visited twice. **E** is incorrect because no town is visited twice. Therefore **F** is the correct answer.

Question 192: C

Georgia is shorter than her Mum and Dad, and each of her siblings is at least as tall as Mum (and we know Mum is shorter than Dad because Ellie is between the two), so we know Georgia is the shortest. We know that Ellie, Tom and Dad are all taller than Mum, so Mum is second shortest. Ellie is shorter than Dad and Tom is taller than Dad, so we can work out that Ellie must be third shortest.

Question 193: A

Danielle must be sat next to Caitlin. Bella must be sat next to the teaching assistant. Hence these two pairs must sit in different rows. One pair must be sat at the front with Ashley, and the other must be sat at the back with Emily. Since the teaching assistant has to sit on the left, this must mean that Bella is sat in the middle seat and either Ashley or Emily (depending on which row they are in) is sat in the right hand seat. However, Bella cannot sit next to Emily, so this means Bella and the teaching assistant must be in the front row. So Ashley must be sat in the front right seat.

Question 194: C

The dishwasher is run 2+p times a week, where p is the number of people in the house. Let the number of people in the house when the son is not home be s, and when the son is home it is s+1. In 30 weeks when the son is home, she would buy 6 packs of dishwasher tablets. In 30 weeks when the son is not home, she would buy 5 packs of dishwasher tablets. So 1.2 times as many packs of dishwasher tablets are bought when he is home. So 2+s+1 is 1.2 time 2+s.
i.e. $2.4 + 1.2s = 2 + s + 1$. Therefore $0.2s = 0.6$. s = 3. When her son is home, there are s + 1 = 4 people in the house.

Question 195: A

No remaining days in the year obey the rule. The next date that does is 01/01/2015 (integers are 0, 1, 2, 5). This is 6 days later than the specified date.

Question 196: B

If each town is due North, South, East or West of at least 2 other towns and we know that one is east and one is north of a third, then they must be arranged in a square. So Yellowtown is 4 miles east of Bluetown to make a square, which means it must be 5 miles north of Redtown. So Redtown is 5 miles south of Yellowtown.

Question 197: B

Jenna pours 4/5 of 250 ml into each glass, which is 200 ml. Since she has 1500 ml of wine, she pours 100 ml into the last glass, which is 2/5 of the 250 ml full capacity.

Question 198: E

The maximum number of girls in Miss Ellis's class with brown eyes and brown hair is 10, because the two thirds of the girls with brown eyes could also all have brown hair. The minimum number is 0 because it could be that all the boys, and the third of the girls without brown eyes, all had brown hair, which would be 2/3 of the class.

Question 199: E

A negative "score" results from any combination of throws which includes a 1 but from no other combination. Given that a negative score has a 0.75 probability, a positive or zero score has a 0.25 probability. Therefore throwing two numbers that are not 1 twice in a row has a probability of 0.25. Hence, the probability of throwing a non-1 number on each throw is $\sqrt{0.25} = 0.5$. So the probability of throwing a 1 on an individual throw is $1 - 0.5 = 0.5$.

Question 200: C

We can work out from the information given the adult flat rate and the charge per stop. Let the charge per stop be s and the flat rate be f. Therefore: $15s + f = 1.70 \rightarrow 8s + f = 1.14$. We can hence work out that: $7s = 0.56$, so $s = 0.08$. Hence, $f = 0.50$ Megan is an adult so she pays this rate. For 30 stops, the rate will be $0.08 \times 30 + 0.50 = 2.90$.

Question 201: B

We found in the previous question that the flat rate for adults is £0.50 and the rate per stop is £0.08. We know that the child rate is half the flat rate and a quarter of the "per stop" rate, so the child flat rate is £0.25 and the rate per stop is 2p. So for 25 stops, Alice pays: $0.02 \times 25 + 0.25 = 0.75$

Question 202: C

We should first work out how many stops James can travel. For £2, he can afford to travel as many stops as £1.50 will take him once the flat rate is taken into account. The per stop rate is 8p per stop, so he can travel 18 stops, so he will need to go to the 18th stop from town. So he will need to walk past 7 stops to get to the stop he can afford to travel from.

Question 203: D

The picture will need a 12 inch by 16 inch mount, which will cost £8. It will need a 13 inch by 17 inch frame, which will cost £26. So the cost of mounting and framing the picture will be £8 + £26 = £34.

Question 204: C

Mounting and framing an 8 by 8 inch painting will cost £5 for the mount and £22 for the frame, which is £27. Mounting and framing a 10 by 10 inch painting will cost £6 for the mount and £26 for the frame, which is £32. The difference is £32 - £27 = £5.

Question 205: B

We found in the last question that mounting and framing a 10 by 10 inch painting will cost £6 for the mount and £26 for the frame, which is £32 total. We can calculate that each additional inch of mount and frame for a square painting costs £2.50; £2 for the frame and £0.50 for the mount. So an 11 inch painting will cost £34.50 to frame and mount, a 12 inch £37, a 13 inch £39.50, a 14 inch £42. The biggest painting that can be mounted and framed for £40 is a 13 inch painting.

Question 206: D

Recognise that the pattern is *"consonants move forward by two consonants; vowel stay the same"*. This allows coding of the word MAGICAL to PAJIFAN to RALIHAQ.

Forward two			Forward two
M	$\Rightarrow$	O (skips to) P	$\Rightarrow$ R
A	$\Rightarrow$	Stays the same	$\Rightarrow$ A
G	$\Rightarrow$	I (skips to) J	$\Rightarrow$ L
I	$\Rightarrow$	Stays the same	$\Rightarrow$ I
C	$\Rightarrow$	E (skips to) F	$\Rightarrow$ H
A	$\Rightarrow$	Stays the same	$\Rightarrow$ A
L	$\Rightarrow$	N	$\Rightarrow$ Q

Question 207: C

If *f* donates the flat rate, and *k* denotes the rate per km, we can form simultaneous equations:

$f + 5k = £6$ AND $f + 3k = £4.20$. Subtract equation two from equation one: $(f + 5k) - (f + 3k) = £6 - £4.20$. Thus, $2k = £1.80$ and $k = £0.90$. Therefore, $f + (5 \times 0.90) = £6$. So, $f + £4.50 = £6$. Thus, $f = £1.50$ and 7k will be $£1.50 + 7 \times £0.90 = £7.80$

Question 208: C

The increase from 2001/2 to 2011/12 was 1,019 to 11,736, which equals a linear increase of 10,717 admissions. So, in 20 years, we would expect to see an increase by $10,717 \times 2 = 21,434$. Add this to the number in 2011 to give 33,170 admissions.

Question 209: A

As the question uses percentages, it does not matter what figure you use. To make calculations easier, use an initial price of £100. When on sale, the dress is 20% off, so using a normal price of £100, the dress would be £80. When the dresses are 20% off, the shop is making a 25% profit. Therefore: £80 = 1.25 x purchase price.

Therefore, the purchase price is: $\frac{80}{1.25}$ = £64. Thus, the normal profit is £100 - £64 = £36. I.e. when a dress sells for £100, the shop makes £36 or 36% profit.

Question 210: C

1. Incorrect. There must be 6 general committee clinical students, plus the treasurer, and 2 sabbatical roles, none of whom can be preclinical, so there must be a maximum of 11 preclinical students.
2. Correct. There must two general for each year plus welfare and social officers, totalling to 6.
3. Incorrect. The committee is made up of 20 students, 2 roles are sabbatical, so there are 18 studying students, and therefore there can be 3 from each year.
4. Correct. There are 18 studying students on the committee, and there must be 6 general committee members from pre-clinical, plus welfare and social, therefore there must be a minimum of 8 pre-clinical students, so there must be 10 clinical students.
5. Incorrect. You need to count up the number of specific roles on the committee, which is 5, and there must be 2 students from each year, which is 12. This leaves 3 more positions, which the question doesn't state can't be first years. Therefore there could be up to 5 first years.
6. Incorrect. There must be at least 2 general committee members from each year. However, the worked answer to 5 shows there are 15 general committee members which are split across the 6 years, and so there must be an uneven distribution.

Question 211: B

Remember 2012 was a leap year. Work through each month, adding the correct number of days, to work out what day each 13th would be on.

If a month was 28 days, the 13th would be the same day each month, therefore to work this out quickly, you only need to count on the number of days over 28. For example, in a month with 31 days, the 13th will be 3 weekdays (31-28) later.

Thus if 13th January is a Friday, 13th February is a Monday, (February has 29 days in 2012), 13th March is a Tuesday and 13th April is a Friday.

Question 212: E

There are 18 sheep in total. The question states there are 8 male sheep, which means there are 10 female sheep before some die. 5 female sheep die, so there are 5 female sheep alive to give birth to lambs. Each delivers 2 lambs, making 10 lambs in total. There are 4 male sheep and 5 mothers so the total is $10 + 4 + 5 = 19$ sheep.

Question 213: D

We can see from the fact that all the possible answers end "AME" that the letters "AME" must be translated to the last 3 letters of the coded word, "JVN", under the code. J is the 10th letter of the alphabet so it is 9 letters on from A (V is the 21st letter of the alphabet and M is the 13th, and N is the 14th letter of the alphabet and E is the 5th, therefore these pairs are also 9 letters apart). Therefore P is the code for the letter 9 letters before it in the alphabet. P is the 16th letter of the alphabet, therefore it is the code for the 7th letter of the alphabet, G. Therefore from these solutions the only possibility for the original word is GAME.

Question 214: C

Let x be the number of people who get on the bus at the station.

It is easiest to work backwards. After the 4th stop, there are 5 people on the bus. At the 4th stop, half the people who were on the bus got off (and therefore half stayed on) and 2 people got on. Therefore, 5 is equal to 2 plus half the number of people who were on the bus after the 3rd stop. So half the number of people who were on the bus after the 3rd stop must be 3. Therefore, after the 3rd stop, there must have been 6 people on the bus.

We can then say that 6 is equal to 2 plus half the number of people who were on the bus after the 2nd stop. Therefore there were 8 people on the bus after the 2nd stop.

We can then say that 8 is equal to 2 plus half the number of people who were on the bus after the 1st stop. Therefore there were 12 people on the bus after the 1st stop.

We can then say that 12 is equal to 2 plus half the number of people who got on the bus at the station. Therefore the number of people who got on the bus at the station is 20.

Question 215: B

We know from the question that I have purchased small cans of blue and white paint, and that blue paint accounted for 50% of the total cost. Since a can of blue paint is 4 X the price of a can of white paint, we know I must have purchased 4 cans of white paint for each can of blue paint.

Each can of small paint covers a total of 10m², and I have painted a total of 100m², in doing so using up all the paint. Therefore, I must have purchased 10 cans of paint. Therefore, I must have purchased 2 cans of blue paint and 8 cans of white paint. So I must have painted 20m² of wall space blue.

Question 216: E

The cost for x cakes under this offer can be expressed as: $x(42-x^2)$

Following this formula, we can see that 2 cakes would cost 76p, 3 cakes would cost 99p, and 4 cakes would cost 104p. As the number of cakes increases beyond 4, we see that the overall price actually drops, as 5 cakes would cost 85p and 6 cakes would cost 36p. This confirms Isobel's prediction that the offer is a bad deal for the baker, as it ends up cheaper for the customer to purchase more cakes. It is clear that 6 cakes is the smallest number for which the price will be under 40p, and the price will continue to drop as more cakes are purchased.

Question 217: C

Adding up the percentages of students in University A who do "Science" subjects gives:

$23.50 + 6.25 + 30.25 = 60\%$.

60% of 800 students is 480, so 480 students in University A do "Science" subjects.

Adding up the percentages of students in University B who do "Science subjects" gives:

$13.25 + 14.75 + 7.00 = 35\%$. 35% of 1200 students is 420, so 420 students in University B do "Science" subjects. Therefore:

$480 - 420 = 60$

60 more students in University A than University B take a "Science" subject.

Question 218: C

Let the number of miles Sonia is travelling be x. Because she is crossing 1 international border, travelling by Traveleasy Coaches will cost Sonia: £(5 + 0.5x)

Travelling by Europremier coaches will cost Sonia: £(15 + 0.1x).

Because we know the cost is the same for both companies, the number of miles she is travelling can be found by setting these two expressions equal to each other: $5 + 0.5x = 15 + 0.1x$.

This equation can be rearranged to give: $0.4x = 10$

Therefore: $x = 10/0.4 = 25$

Question 219: E

To find out whether many of these statements are true it is necessary to work out the departure and arrival times, and journey time, for each girl.

Lauren departs at 2:30pm and arrives at 4pm, therefore her journey takes 1.5 hours

Chloe departs at 1:30pm and her journey takes 1 hour longer than 1.5 hours (Lauren's journey), therefore her journey takes 2.5 hours and she arrives at 4pm

Amy arrives at 4:15pm and her journey takes 2 times 1.5 hours (Lauren's journey), therefore her journey takes 3 hours and she departs at 1:15pm.

Looking at each statement, the only one which is definitely true is **E**: Amy departs at 1:15pm and Chloe departs at 1:30pm therefore Amy departed before Chloe.

D *may* be true, but nothing in the question shows it is *definitely* true, so it can be safely ignored.

Question 220: B

First consider how many items of clothing she can take by weight. The weight allowance is 20kg. Take off 2kg for the weight of the empty suitcase, then take off another 3kg (3 X 1000g) for the books she wishes to take. Therefore she can fit 15kg of clothes in her suitcase. To find out how many items of clothing this is, we can divide 15kg=15000g by 400g: 15000/400 = 150/4 = 37.5

So she can pack up to 37 items of clothing by weight.

Now consider the volume of clothes she can fit in. The total volume of the suitcase is:

50cm x 50cm x 20cm = 50000cm³

The volume of each book is: 0.2m x 0.1m x 0.05m = 1000cm³

So the volume of space available for clothes is: 50000 – (3 x 1000) = 47000cm³

To find out how many items of clothing she can fit in this space, we can divide 47000 by 1500:

47000/1500 = 470/15 = 31 1/3. So she can pack up to 31 items of clothing by volume.

Although she can fit 37 items by weight, they will not fit in the volume of the suitcase, so the maximum number of items of clothing she can pack is 31.

Question 221: D
We can work out the Answer by considering each option:
Bed Shop A: £120 + £70 = £190
Bed Shop B: £90 + £90 = £180
Bed Shop C: £140 + (1/2 x £60) = £170
Bed Shop D: (2/3) x (£140+£100) = (2/3) x (£240) = £160
Bed Shop E: £175
Therefore the cheapest is Bed Shop **D**.

Question 222: C
The numbers of socks of each colour is irrelevant, so long as there is more than one of each (which there is). There are only 4 colours of socks, so if Joseph takes 5 socks, it is guaranteed that at least 2 of them will be the same colour.

Question 223: D
Paper comes in packs of 500, and with each pack 20 magazines can be printed. Each pack costs £3.
Card comes in packs of 60, and with each pack 60 magazines can be printed. Each pack costs £3 x 2 = £6.
Each ink cartridge prints 130 sheets, which is 130/26 = 5 magazines. Each cartridge costs £5.

The lowest common multiple of 20, 60 and 5 is 60, so it is possible to work out the total cost for printing 60 magazines. Printing 60 magazines will require 3 packs of paper at £3, 1 pack of card at £6 and 12 ink cartridges at £5. So the total cost of printing 60 magazines is: (3 x 3) + 6 + (12 x 5) = £75.

The total budget is £300.
£300/£75 = 4
So we can print 4x60 magazines in this budget, which is 240 magazines.

Question 224: E
We can express the information we have as: $\frac{1}{4} - \frac{1}{5} = \frac{1}{20}$
So the six additional lengths make up 1/20 of Rebecca's intended distance. So the number of lengths she intended to complete was: 20 x 6 = 120.

Question 225: B
Sammy has a choice of 3 flavours for the first sweet that he eats. Each of the other sweets he eats cannot be the same flavour as the sweet he has just eaten. So he has a choice of 2 flavours for each of these four sweets. So the total number of ways that he can make his choices is:
$3 \times 2 \times 2 \times 2 \times 2 = 48$

Question 226: C
Suppose that today Gill is x years old. It follows that Granny is 15x years old. In 4 years' time, Gill will be (x+4) years old and Granny will be 15x+4 years old. We know that in 4 years' time, Granny's age is equal to Gill's age squared, so: $15x + 4 = (x + 4)^2$
Expanding and rearranging, we get: $x^2 - 7x + 12 = 0$
We can factorise this to get: (x - 3)(x - 4)

So x is either 3 or 4. Gill's age today is either 3 or 4 so Granny is either 45 or 60. We know Granny's age is an even number, so she must be 60 and hence Gill must be 4. So the difference in their ages is 56 years.

Question 227: C
If Pierre is telling the truth, everyone else is not telling the truth. But, also in this case, what Qadr said is not true, and hence Ratna is telling the truth. So we have a contradiction. So we deduce that Pierre is not telling the truth. Therefore, Qadr is telling the truth, and so Ratna is not telling the truth. So Sven is also telling the truth, and hence Tanya is not telling the truth. So Qadr and Sven are telling the truth and the other three are not telling the truth.

Question 228: D
Angus walks for 20 minutes at 3 mph and runs for 20 minutes at 6 mph. 20 minutes is one-third of an hour. So the number of miles that Angus covers is: $3 \times \frac{1}{3} + 6 \times \frac{1}{3} = 6$
Bruce covers the same distance. So Bruce walks $\frac{1}{2} \times 3$ miles at 3 mph which takes him 30 minutes and runs the same distance at 6 mph which takes him 15 minutes. So altogether it takes Bruce 45 minutes to finish the course.

Question 229: B
Although you could do this quickly by forming simultaneous equations, it is even quicker to note that 72 x 4 = 288. Since Species 24601 each have 4 legs; it leaves a single member of species 8472 to account for the other 2 legs.

Question 230: E

None of the options can be concluded for certain. We are not told whether any chicken dishes are spicy, only that they are all creamy. Whilst all vegetable dishes are spicy, some non-vegetable dishes could also be spicy. There is no information on whether dishes can be both creamy and spicy, nor on which, if any, dishes contain tomatoes. Remember, if you're really stuck, draw a Venn diagram for these types of questions.

Question 231: C

At 10mph, we can express the time it takes Lucy to get home as: 60 x 8/10 = 48
Since Simon sets off 20 minutes later, his time taken to get home, in order to arrive at the same time, must be:
48 – 20 = 28
Therefore his cycling speed must be: 48/28 x 10 = 17mph

Question 232: A

The total profit from the first transaction can be expressed as: 2000 x 8 = 16,000p
The total profit from the second transaction is: 1000 x 6 = 6,000p

Therefore the total profit is 22,000p or £220 before charges. There are four transactions at a cost of £20 each, therefore the overall profit is: £220 – (20 x 4) = £140

Question 233: C

For the total score to be odd, there must be either three odd or one odd and two even scores obtained. Since the solitary odd score could be either the first, second or third throw there are four possible outcomes that result in an odd total score. Additionally, there are the same number of possibilities giving an even score (either all three even or two odd and one even scores obtained), and the chance of throwing odd or even with any given dart is equal. Therefore, there is an equal probability of three darts totalling to an odd score as to an even score, and so the chance of an odd score is ½.

Question 234 C

This is a compound interest question. £5,000 must be increased by 5%, and then the answer needs to be increased by 5% for four more iterations. After one year: £5,000 x 1.05 = £5,250
Increasing sequentially gives 5512, 5788, 6077 and 6381 after five years. Therefore the answer is £6,381.

Question 235: D

If in 5 years' time the sum of their ages is 62, the sum of their ages today will be: 62 – (5 x 2) = 52
Therefore if they were the same age they would both be 26, but with a 12 year age gap they are 20 and 32 today. Michael is the older brother, so 2 years ago he would have been aged 30.

Question 236: A

Tearing out every page which is a multiple of 3 removes 166 pages. All multiples of 6 are multiples of 3, so no more pages are torn out with that instruction. Finally, half of the remaining pages are removed, which equates to an additional 167 pages. Therefore 333 pages are removed in total. The total surface area of these pages is 15 x 30 x 333 = 149,850 cm² = 14.9m². At 110 gm², 14.9 m² weighs 14.9 x 110 = 1,650g (1,648g unrounded)

Question 237: D

The cost of fertiliser is 80p/kg = 8p/100g. At 200g the incremental increase in yield is 65 pence/m. At each additional 100g it will be reduced by 30%, therefore at 300g/m it is 45.5p, at 400g/m it is 31.8p, at 500g/m it is 22.3p, at 600g/m it is 15.6p, at 700g/m it is 10.9p, and at 800g it is 7.6p. So at 800g the gain in yield is less than the cost of the fertiliser to produce the gain, and so it is no longer cost effective to fertilise more.

Question 238: D

Statements **A**, **C** and **E** are all definitely true. Meanwhile, statement **B** may be not true but is not definitely untrue, as this depends on the number of cats and rabbit owned.
Only statement **D** is definitely untrue. The type of animal requiring the most food is a dog, and as can be seen from the tables, Furry Friends actually sells the most expensive dog food, not the cheapest.

Question 239: C

The largest decrease in bank balance occurs between January 1st and February 1st, totalling £171, reflecting the amount spent during the month of January, £1171. However, because there is a pay rise beginning on March 10th, we need to consider that from April onwards, the bank balance will have increased by £1100, not £1000. This means that the same decrease in bank balance reflects £100 more spending if it occurs after March. This means that 2 months now have seen more spending than February. Between March 1st and April 1st, the bank balance has decreased by £139. With the salary increase, the salary is now £1100, so the total spending for the month of March is £1239. This is greater than the total spending during the month of January.
Similarly, the month of April has also seen more spending than January once the pay rise is considered, a total of £1225 of spending. However, this is still less than the month of March.

Question 240: C

If Amy gets a taxi, she can set off 100 minutes before 1700, which is 1520.

If Amy gets a train, she must get the 1500 train as the later train arrives after 1700, so she must set off at 1500.

Since Northtown airport is 30 minutes from Northtown station, there is no way Amy can get the flight and still arrive at Northtown station by 1700. Therefore Amy should get a taxi and should leave at 1520.

Question 241: C

We can decompose the elements of the multiplication grid into their prime factors, thus:

	c	D
a	2 x 2 x 2 x 3 x 7	2 x 2 x 2 x 2 x 3 x 3 x 5
b	7 x 17	2 x 3 x 5 x 17

$bc = 7 \times 17$, so one of b and c must be 7 and the other must be 17. b must be 17 because bd is a multiple of 17 and not of 7, and c must be 7 because ac is a multiple of 7 and not of 17. ac is 168, so a must be 168 divided by 7, which is 24. ad is 720 so d must be 720 divided by 24, which is 30. Hence the answer is 30.

Alternatively approach the question by eliminating all answers which are not factors of both 720 and 510.

Question 242: E

48% of the students are girls, which is 720 students. Hence 80 is 1/9 of the girls, so 1/9 of boys are mixed race. The remaining 780 students are boys, so 87 boys are mixed race to the nearest person. There is a shortcut to this question. Notice that 80 girls are mixed race, and the proportion is the same for boys. As there are more boys than girls we know the answer is greater than 80. Option **E** 90 is the only option for which this holds true.

Question 243: D

Don't be fooled – this is surprisingly easy. We can see that between Monday and Thursday, Christine has worked a total of 30 hours. We can also calculate how long her shift on Friday was supposed to be. She is able to make up the hours by working 3 extra hours next week, and 5 hours on Sunday. Thus, the Friday shift must have been planned to be 8 hours long. Adding this to the other 30 hours, we see that Christine was supposed to work 38 hours this week.

Question 244: C

130°. Each hour is 1/12 of a complete turn, equalling 30°. The smaller angle between 8 and 12 on the clock face is 4 gaps, therefore 120°. In addition, there is 1/3 of the distance between 3 and 4 still to turn, so an additional 10° must be added on to account for that.

Question 245: B

The total price of all of these items would usually be £17. However, with the DVD offer, the customer saves £1, giving a total cost of £16. Thus, the customer will need to receive £34 in change.

Question 246: E

A. Incorrect. UCL study found eating more portions of fruit and vegetables was beneficial.

B. Incorrect. This is a possible reason but has yet to be fully investigated.

C. Incorrect. Fruit and vegetables are more protective against cardiovascular disease, and were shown to have little effect on cancer rates.

D. Incorrect. Inconclusive – people who ate more vegetables generally had a lower mortality but unknown if this is due to eating more vegetables or other associated factors.

E. Correct. Although this has previously been the case, this study did not find so. 'they recorded no additional decline for people who ate over 5 portions'.

F. Incorrect. The 5% decline per portion was only up to 5 portions and no additional reduction in mortality for 7 than 5 portions.

G. Incorrect. Study only looks at cancers in general and states need to look into specific cancers.

Question 247: C

Deaths in meta-analysis = 56423/800000 = 0.07 or 7%

1% lower in UCL study so 6%

6% of 65,000 = 65000 x 0.06 = 3,900

Question 248: B

A. Eating more fruit and vegetables doesn't particularly lower overall risk but need research into specific cancer risk.

B. The UCL research alone found that increasing the number of fruit and vegetable portions had a beneficial effect, even though this wasn't the overall conclusion when combined with results from the meta-analysis.

C. The results were not exactly the same but showed similar overall trends.

D. Although this may be true, there is no mention of this in the passage.

E. Fruit and vegetables are protective against cardiovascular disease, but not exclusively. They also reduce the rates of death from all causes.

F. The UCL study is in England only and the meta-analysis a combination of studies from around the world.

G. Suggested by the UCL research, but not the meta-analysis, so not an overall conclusion of the article.

Question 249: E

Remember that you don't need to calculate exact values for question 249 – 251. Thus, you should round numbers frequently to make this more manageable. Work out percentage of beer and wine consumption and then the actual value using the total alcohol consumption figure:

Belarus: $17.3 + 5.2 = 22.5\%$;
$0.225 \times 17.5 = 3.94$

Lithuania: Missing figure $100 - 7.8 - 34.1 - 11.6 = 46.5$
$46.5 + 7.8 = 54.3\%$
$0.543 \times 15.4 = 8.36$

France: $18.8 + 56.4 = 75.2\%$
$0.752 \times 12.2 = 9.17$

Ireland: $48.1 + 26.1 = 74.2$
$0.742 \times 11.9 = 8.83$

Andorra: missing figure $100 - 34.6 - 20.1 = 45.3$
$34.6 + 45.3 = 79.9\%$
$0.799 \times 13.8 = 11.0$

Question 250: D

Russia:
$2010 -$ Total $= 11.5 + 3.6 = 15.1$. Spirits $= 0.51 \times 15.1 = 7.7$
$2020 -$ Total $= 14.5$. Spirits $= 0.51 \times 14.5 = 7.4$
Difference $= 0.3$ L

Belarus:
$2010 -$ Total $= 14.4 + 3.2 = 17.6$. Spirits $= 0.466 \times 17.6 = 8.2$
$2020 -$ Total $= 17.1$. Spirits $= 0.466 \times 17.1 = 8.0$
Difference $= 0.2$ L

Lithuania:
$2010 -$ Total $= 15.4$. Spirits $= 0.341 \times 15.4 = 5.3$
$2020 -$ Total $= 16.2$. Spirits $= 0.341 \times 16.2 = 5.5$
Difference $= 0.2$ L

Grenada:
$2010 -$ Total $= 12.5$. Spirits $\% = 100 - 29.3 - 4.3 - 0.2 = 66.2\%$. Spirits $= 0.662 \times 12.5 = 8.3$
$2020 -$ Total $= 10.4$. Spirits $= 0.662 \times 10.4 = 6.8$
Difference $= 1.5$ L

Ireland:
$2010 -$ Total $= 11.9$. Spirits $= 0.187 \times 11.9 = 2.2$
$2020 -$ Total $= 10.9$. Spirits $= 0.187 \times 10.9 = 2$
Difference $= 0.2$ L

Question 251: C

Work out 4.9 as a percentage of total beer consumption in Czech Republic and search other rows for similar percentage.

$4.9/13 = 0.38$, approx. 38% which is very similar to percentage consumption in Russia (37.6).

Question 252: B

We can add up the total incidence of the 6 cancers in men, which is 94,000. Then we can add up the total incidence in women, which is 101,000. As a percentage of 10 million, this is 0.94% of men and 1.01% of women. Therefore the difference is 0.07%.

Question 253: C

Given there are 1.15 times as many men as women, the incidence of each cancer amongst men needs to be greater than 1.15 times the incidence amongst women in order for a man to be more likely to develop it. The incidence is at least 1.15 higher in men for 3 cancers (prostate, lung and bladder).

Question 254: D

If 10% of cancer patients are in Sydney, there are 10,300 prostate/bladder/breast cancer patients and 9,200 lung/bowel/uterus cancer patients in Sydney. Hence the total number of hospital visits is 10,300 + 18,400, which is 28,700.

Question 255: A

The proportion of men with bladder cancer is 2/3 and women 1/3.

Question 256: D

First we work out the size of each standard drink. 50 standard drinks of vodka is equivalent to 1250ml, so one drink is 25ml or 0.025 litres. 11.4 standard drinks of beer is 10 pints of 5700ml, so one standard drink is 500ml or 0.5 litres. 3 standard drinks of cocktail is 750ml so one is 250ml or 0.25 litres. 3.75 standard drinks of wine is 750ml, so one is 200ml or 0.2 litres.

We can then work out the number of units in each drink. Vodka has 0.025 x 40 = 1 unit, Beer has 0.5 x 3 = 1.5 units, Cocktail has 0.25 x 8 = 2 units and Wine has 0.2 x 12.5 = 2.5 units. Since the drink with the most units is wine, the answer is D.

Question 257: B

We found in the last question that vodka has 1 unit, beer has 1.5, cocktail has 2 and wine has 2.5. Hence in the week, Hannah drinks 23.5 units and Mark drinks 29 units. Hence Hannah exceeds the recommended amount by 9.5 units and Mark by 9 units.

Question 258: D

We found that vodka has 1 unit, beer has 1.5, cocktail has 2 and wine has 2.5. Hence it is possible to make 5 combinations of drinks that are 4 units: 4 vodkas, 2 cocktails, 2 vodkas and a cocktail, 1 vodka and 2 beers, or a wine and a beer.

Question 259: D

The total number of males in Greentown is 12,890. Adding up the rest of the age categories, we can see that 10,140 of these are in the older age categories. Hence there are 2750 males under 20.

Question 260: C

Given that in the first question we found the number of males under 20 is 2,750, we can then add up the totals in the age categories (apart from 40-59) in order to find that 15,000 of the residents of Greentown are in other age categories. Hence 9,320 of the population are aged 40-59. We know that 4,130 of these are male, therefore 5,190 must be female.

Question 261: C

The age group with the highest ratio of males:females is 20-39, with approximately 1.9 males per females (approximately 3800:2000). As a ratio of females to males, this is 1:1.9.

Question 262: C

There are 4 instances where the line for Newcastle is flat from one month to the next per year, hence in 2008-2012 (5 years) there are 20 occasions when the average temperature is the same from month to month. During 2007, there are 2 occasions, and during 2013 there are 3.

Question 263: A

The average temperature is lower than the previous month in London for all months from August to December, which is 5 months. However, in August and November in Newcastle, the average temperature remains the same as the previous month. Hence there are only 3 months where the average temperature is lower in both cities. Hence from 2007 to 2012, there are 18 months where the average temperature is lower than the previous month. During 2013, the only included month where the temperature is lower in both cities than the previous month is September. Hence there are 19 months in total when the temperature is lower in both cities than the previous month.

Question 264: B

Firstly, work out the difference between average temperatures for each month (2, 3, 1, 2, 1, 3, 3, 2, 2, 5, 1, 0). Then sum them to give 25. Divide by the number of months (12) to give $2^1/_{12}$, which is 2°C to the nearest 0.5°C.

Question 265: D

There is not enough information to tell which month the highest sales are in. We know it increases up to a point and decreases after it, but as we don't know by how much we cannot project where the maximum sales will be.

Question 266: B

Given that by observation, Q2 and Q3 both account for 1/3 of the sales and Q4 accounts for 1/4, this leaves that Q1 accounts for 1/12 of sales. 1/12 of £354,720 is £29,560.

Question 267: A

Quarter 2 accounts for 1/3 of the sales, which is £60,000 in sales revenue. If a tub of ice cream is sold for £2 and costs the manufacturer £1.50, this means profit is 1/4 of sales revenue. Hence £15,000 profit is made during Q2. Hence the answer is A.

Question 268: D

A. and B – Incorrect. Both *could* be true but neither is *definitely* true as it is dependent on the relative number of families with each number of children, which is not given in the question. Therefore we cannot know for certain whether these statements are true.

C – Incorrect. C is definitely *untrue* as half of the families spend £400 a month on food, which totals £4800 a year.

D – Correct. This option is true as 1/6 of families with 1 child and 1/6 of families with 3 children spent £100 a month on food.

E – Incorrect. This option is definitely untrue as the average expenditure for families with 2 children is actually £400 a month.

Question 269: B

2210 out of 2500 filled in responses, meaning that 290 did not. 290 as a percentage of 2500 is roughly 12% (11.6%) of the school that did not respond.

Question 270: C

The percentage of students that saw bullying and reported it was 35%, so 65% of those who saw it did not which is equivalent to 725 students. Of this 725, 146 which roughly equals 20%, gave the reason that they did not think it was important.

Question 271: B

Of the students who told a teacher, 286 did not witness any action. Of those who did notice action, i.e. 110, only 40% noticed any direct action with the bully involved. 40% of 110 is 44, so the correct answer is B.

Question 272: D

"427 cited fears of being found out" which means about 59% out of the 725 students that did not tell about the bullying, cited that it was because they worried about others finding out.

Question 273: F

North-east: 56 per 100,000 on average. This means that there must be a higher proportion of women than this and a lower proportion of men, such that the average is 56/100,000

We must make the reasonable assumption that there are the same number of men and women in the population as the question asks us to approximate.

Therefore there are 18.6/50,000 men and 37.3/50,000 women

This scales to 74.4/100,000 women which is roughly 74/100,000.

Question 274: C

8 million children – question tells to approximate to 4 million girls and 4 million boys. Girls: 20% eat 5 portions fruit and vegetables a day. 20% of 4 million: 4 x 0.2 = 0.8 million. Boys: 16% eat 5 portions of fruit and vegetables a day. 16% of 4 million: 4 x 0.16 = 0.64 million. Number of more girls: 800,000 – 640,000 = 160,000.

Question 275: B

A.   Incorrect. Women: 13619+10144+6569 = 30332. Men: 16818 + 9726 + 7669 + 6311 = 40524

B.   Correct. Flu + pneumonia, lung cancer and chronic lower respiratory diseases = 15361 + 13619 + 14927 = 43907

C.   Incorrect. More common cause of death but no information surrounding prevalence.

D.   Incorrect. Colon cancer ranking 8 for both.

Question 276: A

The government has claimed a 20% reduction, so we are looking for an assessment criterion which has reduced 20% from 2013 to 2014. We can see that only "Number of people waiting for over 4 hours in A&E" has reduced by 20%, so this must be the criterion the government has used to describe "waiting times in A&E". Thus, the answer is A.

Question 277: B

Rovers must have played 10 games overall as they played each other's team twice. They lost 9 games scoring no points and so must have won 1 game, which scores 3 points.

Question 278: A

To have finished between City and United, Athletic must have got between 23 and 25 points. Hence they must have got 24 points because no team got the same number of points as another. Athletic won 7 games which is 21 points, so they must have also got 3 points from drawing 3 games. This accounts for all 10 games they played, so they did not lose any games.

Question 279: C

United won 8 games and drew 1, which is 25 points. Rangers drew 2 games and won none, which is 2 points. Therefore the difference in points is 23.

Question 280: C

Type 1 departments reached the new target of 95% at least three times since it was introduced. All the other statements are correct.

Question 281: C

Total attendances in Q1 08-9: 5.0 million
Total attendances in Q1 04-5: 4.5 million
The difference = 0.5 million
0.5/5 x 100 = 10% increase

Question 282: C

There are 16 quarters in total since the new target came into effect.
4/16 = 0.25, so the target has been hit 25% of the time i.e. missed 75% of the time.

Question 283: C

Ranjna must leave Singapore by 20:00 to get to Bali by 22:00. The latest flight she can therefore get is the 19:00. Thus, she must arrive in Singapore by 17:00 (accounting 2 hours for the stopover). The flight from Manchester to Singapore takes 14 hours. Manchester is 8 hours behind Singapore so she must leave Manchester 22 hours before 17:00 on Wednesday i.e. by 19:00 on Wednesday. Thus, the latest flight she can get is the 18:00 on Wednesday.

Question 284: D

The 08:00 flight will arrive at Singapore for 22:00 on Monday (GMT) or 06:00 Tuesday Singapore time. She then needs a 2 hour stopover, so earliest connecting flight she can get is 08:30 on Tuesday. The flight lands in Bali at 10:30. She then spends 1 hour and 45 minutes getting to her destination – arriving at 12:15 Tuesday.

Question 285: C

A.   Incorrect. The graph is about level, and certainly not the steepest gradient post 2007.
B.   Incorrect. Although there has been a general decline, there are some blips of increased smoking.
C.   Correct.
D.   Incorrect. The smoking rate in men decreased from 51% in 1974 to 21% in 2010. Thus, it decreased by more than a half.
E.   Incorrect. The percentage difference between men and women smokers has been minimal in the 21st century.

Question 286: D

For this type of question you will have to use trial and error after you've analysed the data pattern to find the correct answer. The quickest way to do this is to examine outliers to try and match them to data in the table e.g. the left-most point is an outlier for the X-axis but average for the y-axis. Also look for any duplicated results in the table and if they are present on the graph, e.g. Hannah and Alice weigh 68 kg but this can't be found on the graph.

Question 287: C

This is pretty straightforward; the point is at approximately 172-174 cm in height and 164 -166 cm in arm span. Matthew is the only student who fits these dimensions.

Question 288: C

This is straightforward – just label the diagram using the information in the text and it becomes obvious that C is the correct answer.

Question 289: C

Since we do not know whether they went to university or not, we must add the number of women with children who work and those who went to university, 2, to the number of women with children who work but did not go to university, 1 (2 + 1 = 3).

Question 290: C

To work this out we must add up all the numbers within the rectangle, 4 + 6 + 1 + 2 + 11 + 12 + 7 + 15 = 58

Question 291: E

Calculate the number of men + women who have children and work i.e. 11 + 5 + 2 = 18

Question 292: C

To solve this we must work out the total number of people who had children i.e. 3 + 6 + 5 + 11 + 1 + 2 =28. Then we work out the total number of people who went to university, but that do not also have children so that these are not counted twice: 13 + 12 = 25. Then we add these two numbers together, 28 + 25 = 53 and subtract the number of people who fell into both categories i.e. 53 - (5 + 11 + 2) = 35

Question 293: C

To work this out we must add up all the numbers outside the rectangle that also fall within both the circle and the square, which is 5.

Question 294: D & E

This question asks for identification of the blank space, which is the space within the triangle, the rectangle and the square i.e. indicating working women who went to university but did not have children. This also reveals non-working men who did not have children and did not go to university.

Question 295: C

The normal price of these items would be £18.50 (£8 + £7 + £3.50). However, with the 50% discount on meat products, the price in the sale for these items will be £9.25. Thus, Alfred would receive £10.75 of change from a £20 note.

Question 296: C

The number of games played and points scored is a red herring in this question. The important data is 'Goals For' and 'Goals Against'. As this is a defined league and the teams have only played each other, the 'Goals For' column must equal the 'Goals Against' column. Total Goals For = 16 + 11 + 8 + 7 + 8 + 4 = 54. Total Goals Against = 2 + Wilmslow + 7 + 9 + 12 + 14 = 44 + Wilmslow. For both columns to be equal, Wilmslow must have a total of 54 – 44 = 10 Goals Against.

Question 297: C

Working with the table it is possible to work out that the BMIs of Julie and Lydia must be 21 and 23, and hence their weights 100 and 115 lbs. Thus Emma's weight is 120 lbs, and her BMI must be 22, making her height equivalent to 160 cm.

Question 298: C

Working through the results, starting with the highest and lowest values, it is possible to plot all values and decipher which point is marked.

Question 299: D

This is a question of estimation. The average production across the year is at least 7 million barrels per day. Multiplying this by 365 gives around 2,550 million barrels per year. All other options require less than 7 million barrels daily production to be produced, and it is clear there is at least 7 million barrels per day. Therefore the answer is 2,700 million.

Alternatively we can estimate using 30 days per month, and multiplying the amount of barrels produced per day in each month by 30 (this is more accurate but more time consuming). 6+7+7+7.5+7.5+7+7.5+8+8.5+8.5+8+9 = 91.5, multiplying by 30 gives just over 2,700 million barrels.

Question 300: C

Use both graphs. For July, multiply the oil price by the amount sold in the month, and multiply by the number of days in the month. Thus, July = 7.5 million barrels x $75 per barrel x 31 days = $17,400 million = $17.4 billion

Question 301: A

DNA consists of 4 bases: adenine, guanine, thymine and cysteine. The sugar backbone consists of deoxyribose, hence the name DNA. DNA is found in the cytoplasm of prokaryotes.

Question 302: F

Mitochondria are responsible for energy production by ATP synthesis. Animal cells do not have a cell wall, only a cell membrane. The endoplasmic reticulum is important in protein synthesis, as this is where the proteins are assembled.

Question 303: F

If you aren't studying A-level biology, this question may stretch you. However, it is possible to reach an answer by process of elimination. Mitochondria are the 'powerhouse' of the cell in aerobic respiration, responsible for cell energy production rather than DNA replication or protein synthesis. As energy producers they are required in muscle cells in large numbers, and in sperm cells to drive the tail responsible for movement. They are enveloped by a double membrane, possibly because they started out as independent prokaryotes engulfed by eukaryotic cells.

Question 304: A

The majority of bacteria are commensals and don't lead to disease.

Question 305: C

Bacteria carry genetic information on plasmids and not in nuclei like animal cells. They don't need meiosis for replication, as they do not require gametes. Bacterial genomes consist of DNA, just like animal cells.

Question 306: C

Active transport requires a transport protein and ATP, as work is being done against an electrochemical gradient. Unlike diffusion, the relative concentrations of the materials being transported aren't important.

Question 307: D

Meiosis produces haploid gametes. This allows for fusion of 2 gametes to reach a full diploid set of chromosomes again in the zygote.

Question 308: B

Mendelian inheritance separates traits into dominant or recessive. It applies to all sexually reproducing organisms. Don't get confused by statement C – the offspring of 2 heterozygotes has a 25% chance of expressing a recessive trait, but it will be homozygous recessive.

Question 309: A

Hormones are released into the bloodstream and act on receptors in different organs in order to cause relatively slow changes to the body's physiology. Hormones frequently interact with the nervous system, e.g. Adrenaline and Insulin, however, they don't directly cause muscles to contract. Almost all hormones are synthesised.

Question 310: D

Neuronal signalling can happen via direct electrical stimulation of nerves or via chemical stimulation of synapses which produces a current that travels along the nerves. Electrical synapses are very rare in mammals, the majority of mammalian synapses are chemical.

Question 311: D

Remember that pH changes cause changes in electrical charge on proteins (= polypeptides) that could interfere with protein – protein interactions. Whilst the other statements are all correct to a certain extent, they are the downstream effects of what would happen if enzymes (which are also proteins) didn't work.

Question 312: A

The bacterial cell wall is made up of murein and protects the bacterium from the external environment, in particular from osmotic stresses, and is important in most bacteria.

Question 313: C

Sexual reproduction relies on formation of gametes during **meiosis**. Mitosis doesn't produce genetically distinct cells. Mitosis is, however, the basis for tissue growth.

Question 314: A

A mutation is a permanent change in the nucleotide sequence of DNA. Whilst mutations may lead to changes in organelles and chromosomes, or even be harmful, they are strictly defined as permanent changes to the DNA or RNA sequence.

Question 315: E

Mutations are fairly common, but in the vast majority of cases do not have any impact on phenotype due to the redundancy of the genome. Sometimes they can confer selective advantages and allow organisms to survive better (i.e. evolve by natural selection), or they can lead to cancers as cells start dividing uncontrollably.

Question 316: D

Antibodies represent a pivotal molecule of the immune system. They provide very pointed and selective targeting of pathogens and toxins without causing damage to the body's own cells.

Question 317: A

Kidneys are not involved in digestion, but do filter the blood of waste products. Glucose is found in high concentrations in the urine of diabetics, who cannot absorb it without working insulin.

Question 318: D

Hormones are slower acting than nerves and act for a longer time. Hormones also act in a more general way. Adrenaline is also a hormone released into the body causing the fight-or-flight response. Although it is quick acting, it still lasts for a longer time than a nervous response, as you can still feel its effects for a time after the response, e.g. shaking hands.

Question 319: D

Homeostasis is about minimising changes to the internal environment by modulating both input and output.

Question 320: B

There is less energy and biomass each time you move up a trophic level. Only 10% of consumed energy is transferred to the next trophic level, so only one tenth of the previous biomass can be sustained in the next trophic level up.

Question 321: A

In asexual reproduction, there is no fusion of gametes as the single parent cell divides. There is therefore no mixing of chromosomes and, as a result, no genetic variation.

Question 322: E

The image is first formed on the retina which conveys it to the brain via a sensory nerve. The brain then sends an impulse to the muscle via a motor neuron.

Question 323: D

Blood from the kidney returns to the heart via the renal (kidney-related) vein, which drains into the inferior vena cava. The blood then passes through the pulmonary vasculature (veins carry blood to the heart, arteries away from the heart) before going into the aorta and eventually the hepatic (liver-related) artery.

Question 324: F

Clones are genetically identical by definition, and a large number of them could conceivably reduce the gene pool of a population. In adult cell cloning, the genetic material of an egg is replaced with the genetic material of an adult cell. Cloning is possible for all DNA based life forms, including plants and other types of animals.

Question 325: F

Gene varieties cause intraspecies variation, e.g. different eye colours. If mutations confer a selective advantage, those individuals with the mutation will survive to reproduce and grow in numbers. Genetic variation is caused by mixing of parent genomes and mutations. Species with similar characteristics often do have similar genes.

Question 326: F

Alleles are different versions of the same gene. If you are a homozygous for a trait, you have two identical alleles for that particular gene, and if you are heterozygous you have two different alleles of that gene. Recessive traits only appear in the phenotype when there are no dominant alleles for that trait, i.e. two recessive alleles are carried.

Question 327: D

Remember that red blood cells don't have a nucleus and therefore have no DNA. In meiosis, a diploid cell divides in such a way so as to produce four haploid cells. Any type of cell division will require energy.

Question 328: C

The hypothalamus detects too little water in the blood, so the pituary gland releases ADH. The kidney maintains the blood water level, and allows less water to be lost in the urine until the blood water level returns to normal.

Question 329: E

Venous blood has a higher level of carbon dioxide and lower oxygen. Carbon dioxide forms carbonic acid in aqueous solution, thus making the pH of venous blood slightly more acidic than arterial blood. This leaves only E and F as possibilities, but releasing pH levels cannot fluctuate significantly gives pH 7.4.

Question 330: E

The cytoplasm is 80% water, but also contains, among other things, electrolytes and proteins. The cytoplasm doesn't contain everything, e.g. DNA is found in the nucleus.

Question 331: C

ATP is produced in mitochondria in aerobic respiration and in the cytoplasm during anaerobic respiration only.

Question 332: C

The cell membrane allows both active transport and passive transport by diffusion of certain ions and molecules, and is found in eukaryotes and prokaryotes like bacteria. It is a phospholipid bilayer.

Question 333: A

1 and 2 only: 223 PAIRS = 446 chromosomes; meiosis produces 4 daughter cells with half of the original number of chromosomes each, while mitosis produces two daughter cells with the original number of chromosomes each.

Question 334: E

If Bob is homozygous dominant (RR) the probability of having a child with red hair is 0%. However, if Bob is heterozygous (Rr), there is a 50% chance of having a child with red hair, since Mary must be homozygous recessive (rr) to have red hair. As we do not know Bob's genotype, both possibilities must be considered.

Question 335: A

If an offspring is born with red hair, it confirms Bob is heterozygous (Rr). He cannot have a red-haired child if he is homozygous dominant (RR), and would himself have red hair were he homozygous recessive (rr).

Question 336: A

Monohybrid cross rr and Rr results in 50% Rr and 50% rr offspring. 50% of offspring will have black hair, but they will be heterozygous for the hair allele.

Question 337: C

When the chest walls expand, the intra-thoracic pressure decreases. This causes the atmospheric pressure outside the chest to be greater than pressure inside the chest, resulting in a flow of air into the chest.

Question 338: A

Producers are found at the bottom of food chains and always have the largest biomass.

Question 339: F

All the statements are true; the carbon and nitrogen cycles are examinable in Section 2, so make sure you understand them! The atmosphere is 79% inert N_2 gas, which must be 'fixed' to useable forms by high-energy lightning strikes or by bacterial mediation. Humans also manually fix nitrogen for fertilisers with the Haber process.

Question 340: H

None of the above statements are correct. Mutations can be silent, cause a loss of function, or even a gain in function, depending on the exact location in the gene and the base affected. Mutations only cause a change in protein structure if the amino acids expressed by the gene affected are changed. This is normally due to a shift in reading frame. Whilst cancer arises as a result of a series of mutations, very few mutations actually lead to cancer.

Question 341: C

Remember that heart rate is controlled via the autonomic nervous system, which isn't a part of the central nervous system.

Question 342: H

None of the above are correct. There is no voluntary input to the heart in the form of a neuronal connection. Parasympathetic neurones slow the heart and sympathetic nervous input accelerates heart rate.

Question 343: B

If lipase is not working, fat from the diet will not be broken down, and will build up in the stool. Lactase, for instance, is responsible for breaking down lactose, and its malfunctioning causes lactose-intolerance.

Question 344: F

Oxygenated blood flows from the lungs to the heart via the pulmonary vein. The pulmonary artery carries deoxygenated blood from the heart to the lungs. Animals like fish have single circulatory systems. Deoxygenated blood is found in the superior vena cava, returning to the heart from the body. Veins in the arms and hands frequently don't have valves.

Question 345: E

Enzymatic digestion takes place throughout the GI tract, including in the mouth (e.g. amylase), stomach (e.g. pepsin), and small intestine (e.g. trypsin). The large intestine is primarily responsible for water absorption, whilst the rectum acts as a temporary store for faecal matter (i.e. digestion has finished by the rectum).

Question 346: B

This is an example of the monosynaptic stretch reflex; these reflexes are performed at the spinal level and therefore don't involve the brain.

Question 347: A

Statement 2 describes diffusion, as CO_2 is moving with the concentration gradient. Statement 3 describes active transport, as amino acids are moving against the concentration gradient.

Question 348: I

3 is the correct equation for animals, and 4 is correct for plants.

Question 349: C

The mitochondria are only the site for aerobic respiration, as anaerobic respiration occurs in the cytoplasm. Aerobic respiration produces more ATP per substrate than anaerobic respiration, and therefore is also more efficient. The chemical equation for glucose being respired aerobically is: $C_6H_{12}O_6 + 6O_2 \rightarrow 6CO_2 + 6H_2O$. Thus, the molar ratio is 1:6 (i.e. each mole glucose produces 6 moles of CO_2).

Question 350: B

The nucleus contains the DNA and chromosomes of the cell. The cytoplasm contains enzymes, salts and amino acids in addition to water. The plasma membrane is a bilayer. Lastly, the cell wall is indeed responsible for protecting vs. increased osmotic pressures.

Question 351: D
When a medium is hypertonic relative to the cell cytoplasm, it is more concentrated than the cytoplasm, and when it is hypotonic, it is less concentrated. So, when a medium is hypotonic relative to the cell cytoplasm, the cell will gain water through osmosis. When the medium is isotonic, there will be no net movement of water across the cell membrane. Lastly, when the medium is hypertonic relative to the cell cytoplasm, the cell will lose water by osmosis.

Question 352: A
Stem cells have the ability to differentiate and produce other kinds of cells. However, they also have the ability to generate cells of their own kind and stem cells are able to maintain their undifferentiated state. The two types of stem cells are embryonic stem cells and adult stem cells. The adult stem cells are present in both children and adults.

Question 353: B
All of the following statements are examples of natural selection, except for the breeding of horses. Breeding and animal husbandry are notable methods of artificial selection, which are brought about by humans.

Question 354: C
Enzymes create a stable environment to stabilise the transition state. Enzymes do not distort substrates. Enzymes generally have little effect on temperature directly. Lastly, they are able to provide alternative pathways for reactions to occur.

Question 355: C
A negative feedback system seeks to minimise changes in a system by modulating the response in accordance with the error that's generated. Salivating before a meal is an example of a feed-forward system (i.e. salivating is an anticipatory response). Throwing a dart does not involve any feedback (during the action). pH and blood pressure are both important homeostatic variables that are controlled via powerful negative feedback mechanisms, e.g. massive haemorrhage leads to compensatory tachycardia.

Question 356: A
One of the major functions of white blood cells is to defend the body against bacterial and fungal infections. They can kill pathogens by engulfing them and also use antibodies to help them recognise pathogens. Antibodies are produced by white blood cells.

Question 357: B
The CV system does indeed transport nutrients and hormones. It also increases blood flow to exercising muscles (via differential vasodilatation) and also helps with thermoregulation (e.g. vasoconstriction in response to cold). The respiratory system is responsible for oxygenating blood.

Question 358: C
Adrenaline always increases heart rate and is almost always released during sympathetic responses. It travels primarily in the blood and affects multiple organ systems. It is also a potent vasoconstrictor.

Question 359: B
Protein synthesis occurs in the cytoplasm. Proteins are usually coded by several amino acids. Red blood cells lack a nucleus and, therefore, the DNA to create new proteins. Protein synthesis is a key part of mitosis, as it allows the parent cell to grow prior to division.

Question 360: F
Remember that most enzymes work better in neutral environments (amylase works even better at slightly alkaline pH). Thus, adding sodium bicarbonate will increase the pH and hence increase the rate of activity. Adding carbohydrate will have no effect, as the enzyme is already saturated. Adding amylase will increase the amount of carbohydrate that can be converted per unit time. Increasing the temperature to 100° C will denature the enzyme and reduce the rate.

Question 361: E
Taking the normal allele to be C and the diseased allele to be c, one can model the scenario with the following Punnett square:

		Carrier Mother	
		C	c
Diseased	c	Cc	cc
Father	c	Cc	cc

The gender of the children is irrelevant as the inheritance is autosomal recessive, but we see that all children produced would inherit at least one diseased allele.

Question 362: F
All of the organs listed have endocrine functions. The thyroid produces thyroid hormone. The ovary produces oestrogen. The pancreas secretes glucagon and insulin. The adrenal gland secretes adrenaline. The testes produce testosterone.

Question 363: A

Insulin works to decrease blood glucose levels. Glucagon causes blood glucose levels to increase; glycogen is a carbohydrate. Adrenaline works to increase heart rate.

Question 364: A

The left side of the heart contains oxygenated blood from the lungs which will be pumped to the body. The right side of the heart contains deoxygenated blood from the body to be pumped to the lungs.

Question 365: A

Since Individual 1 is homozygous normal, and individual 5 is heterozygous and affected, the disease must be dominant. Since males only have one X-chromosome, they cannot be carriers for X-linked conditions. If Nafram syndrome was X-linked, then parents 5 and 6 would produce sons who always have no disease and daughters that always do. As this is not the case shown in individuals 7-10, the disease must be autosomal dominant.

Question 366: C

We know that the inheritance of Nafram syndrome is autosomal dominant, so using N to mean a diseased allele and n to mean a normal allele, 5, 7 and 8 must be Nn because they have an unaffected parent. 2 is also Nn, as if it was NN all its progeny would be Nn and so affected by the disease, which is not the case, as 3 and 4 are unaffected.

Question 367: A

Since 6 is disease free, his genotype must be nn. Thus, neither of 6's parents could be NN, as otherwise 6 would have at least one diseased allele.

Question 368: A

Urine passes from the kidney into the ureter and is then stored in the bladder. It is finally released through the urethra.

Question 369: F

Deoxygenated blood from the body flows through the inferior vena cava to the right atrium where it flows to the right ventricle to be pumped via the pulmonary artery to the lungs where it is oxygenated. It then returns to the heart via the pulmonary vein into the left atrium into the left ventricle where it is pumped to the body via the aorta.

Question 370: E

During inspiration, the pressure in the lungs decreases as the diaphragm contracts, increasing the volume of the lungs. The intercostal muscles contract in inspiration, lifting the rib cage.

Question 371: D

Whilst A, B, C and E are true of the DNA code, they do not represent the property described, which is that more than one combination of codons can encode the same amino acid, e.g. Serine is coded by the sequences: TCT, TCC, TCA, TCG.

Question 372: B

The degenerate nature of the code can help to reduce the deleterious effects of point mutations. The several 3-nucleotide combinations that code for each amino acid are usually similar such that a point mutation, i.e. a substitution of one nucleotide for another, can still result in the same amino acid as the one coded for by the original sequence.

The degenerate nature of the code does little to protect against deletions/insertions/duplications, which will cause the bases to be read in incorrect triplets, i.e. result in a frame shift.

Question 373: D

The hypothalamus is the site of central thermoreceptors. A decrease in environmental temperature decreases sweat secretion and causes cutaneous vasoconstriction to minimise heat loss from the blood.

Question 374: A

The movement of carbon dioxide in the lungs and neurotransmitters in a synapse are both examples of diffusion. Glucose reabsorption is an active process, as it requires work to be done against a concentration gradient.

Question 375: F

Some enzymes contain other molecules besides protein, e.g. metal ions. Enzymes can increase rates of reaction that may result in heat gain/loss, depending on if the reaction is exothermic or endothermic. They are prone to variations in pH and are highly specific to their individual substrate.

Question 376: D

Different isotopes are differentiated by the number of neutrons in the core. This gives them different molecular weights and different chemical properties with regards to stability. The number of protons defines each element, and the number of electrons its charge.

Question 377: E
A displacement reaction occurs when a more reactive element displaces a less reactive element in its compound. All 4 reactions are examples of displacement reactions as a less reactive element is being replaced by a more reactive one.

Question 378: A
There needs to be 3Ca, 12H, 14O and 2P on each side. Only option A satisfies this.

Question 379: A
To balance the equation there needs to be 9Ag, 9N, 9O$_3$, 9K, 3P on each side. Only option A satisfies this.

Question 380: D
A more reactive halogen can displace a less reactive halogen. Thus, chlorine can displace bromine and iodine from an aqueous solution of its salts, and fluorine can replace chlorine. The trend is the opposite for alkali metals, where reactivity increases down the group as electrons are further from the core and easier to lose.

Question 381: C
$2Mg + O_2 = 2MgO$. So, $2 \times 24 = 48$ and $2 \times (24 + 16) = 80$. So, 48 g of magnesium produces 80g of magnesium oxide. So 1g of magnesium produces 1g x 80g/48g = 1.666g oxide. So 75g x 1.666 = 125g

Question 382: B
$H_2 + 2OH^- \rightarrow 2H_2O + e^-$
Thus, the hydrogen loses electrons i.e. is oxidised.

Question 383: F
Ammonia is 1 nitrogen and 3 hydrogen atoms bonded covalently. N = 14g and H = 1g per mole, so percentage of N in NH$_3$ = 14g/17g = 82%. It can be produced from N$_2$ through fixation or the industrial Haber process for use in fertiliser, and may break down to its components.

Question 384: A
Milk is weakly acidic, pH 6.5-7.0, and contains fat. This is broken down by lipase to form fatty acids - turning the solution slightly more acidic.

Question 385: C
Glucose loses four hydrogen atoms; one definition of an oxidation reaction is a reaction in which there is loss of hydrogen.

Question 386: C
Isotopes have the same number of protons and electrons, but a different number of neutrons. The number of neutrons has no impact on the rate of reactions.

Question 387: E
$Mg + H_2SO_4 \rightarrow MgSO_4 + H_2$
Number of moles of Mg $= \frac{6}{24} = 0.25$ moles.
1 mole of Mg reacts with 1 mole H$_2$SO$_4$ to produce 1 mole of magnesium sulphate. Therefore, 0.25 moles H$_2$SO$_4$ will react to produce 0.25 moles of MgSO$_4$.
M$_r$ of H$_2$SO$_4$ = 2 + 32 + 64 = 98g per mole
The mass of H$_2$SO$_4$ used = 0.25 moles x 98g per mole = 24.5g.
Since 30g of H$_2$SO$_4$ is present, H$_2$SO$_4$ is in excess and the magnesium is the limiting reagent.
M$_r$ of MgSO$_4$ = 24 + 32 + 64 = 120g per mole
The mass of MgSO$_4$ produced = 0.25 moles x 120g per mole = 30g which is the same mass as that of sulphuric acid in the original reaction.

Question 388: F
Reactivity series of metals:
Cu is more reactive than Ag and will displace it.
Ca is more reactive than H and will displace it.
2 and 4 are incorrect because Fe is higher in the reactivity series than Cu and Fe is lower in the reactivity series than Ca, so no displacement will occur.

Question 389: G
Moving left to right is the equivalent of moving down the metal reactivity series (i.e. Na is most reactive and Zn is least reactive). Therefore, moving from left to right, the reactivity of the metals decreases, likelihood of corrosion decreases, less energy is required to separate metals from their ores and metals lose electrons less readily to form positive ions.

Question 390: F

Halogens become less reactive as you progress down group 17. Thus in order of increasing reactivity from left to right: I→ Br→ Cl. Therefore, I will not displace Br, Cl will displace Br and Br will displace I.

Question 391: A

Wires are made out of copper because it is a good conductor of electricity. Copper is also used in coins (not aluminium). Aluminium is resistant to corrosion but because of a layer of aluminium oxide (not hydroxide).

Question 392: C

$2Li + 2H_2O \rightarrow 2LiOH + H_2$

Therefore, 2 moles of Li react to produce 1 mole of H_2 gas (24 dm^3).

The number of moles of Li $= \frac{21}{7} = 3$ moles.

Thus, 1.5 moles of H_2 gas are produced $= 36 \text{ dm}^3$.

Question 393: B

$MgCl_2$ contains stronger bonds than NaCl because Mg ions have a 2+ charge, thus having a stronger electrostatic pull for negative chloride ions. The smaller atomic radius also means that the nucleus has less distance between it and incoming electrons. Transition metals are able to form multiple stable ions e.g. Fe^{2+} and Fe^{3+}.

Covalently bonded structures do tend to have lower MPs than ionically bonded, but the giant covalent structures (diamond and graphite for example) have very high melting points. Graphite is an example of a covalently bonded structure which conducts electricity.

Question 394: D

Energy is released from reaction **A**, as shown by a negative enthalpy. The reaction is therefore exothermic. Since energy is released, the product CO_2 has less energy than the reactants did. Therefore, CO_2 is more stable. Reaction **B** has a positive enthalpy, which means energy must be put into the reaction for it to occur i.e. it's an endothermic reaction. That means that the products (CaO and CO_2) have more energy and are less stable than the reactants ($CaCO_3$).

Question 395: B

Solid oxides are unable to conduct electricity because the ions are immobile. Metals are extracted from their molten ores by electrolysis. Fractional distillation is used to separate miscible liquids with similar boiling points. Mg^{2+} ions have a greater positive charge and a smaller ionic radius than Na^+ ions, and therefore have stronger bonds.

Question 396: E

Li^+ (2) and Na^+ (2, 8)
Mg^{2+} (2, 8) and Ne (2, 8)
Na^{2+} (2, 7) and Ne (2, 8)
O^{2+} (2, 4) and a Carbon atom (2, 4)

Question 397: B

Reactivity of both group 1 and 2 increases as you go down the groups because the valence electrons that react are further away from the positively charged nucleus (which means the electrostatic attraction between them is weaker). Group 1 metals are usually more reactive because they only need to donate one electron, whilst group 2 metals must donate two electrons.

Question 398: D

This is a straightforward question that tests basic understanding of kinetics. Catalysts help overcome energy barriers by reducing the activation energy necessary for a reaction.

Question 399: D

H^1 contains 1 proton and no neutrons. Isotopes have the same numbers of protons, but different numbers of neutrons. Thus, H^3 contains two more neutrons than H^1.

Question 400: D

Oxidation is the loss of electrons and reduction is the gain of electrons (therefore increasing electron density). Halogens tend to act as electron recipients in reactions and are therefore good oxidising agents.

Question 401: D

These statements all come from the Kinetic Theory of Gases, an idealised model of gases that allows for the derivation of the ideal gas law. The angle at which gas molecules move is not related to temperature; movement is random. Gas molecules lose no energy when they collide with each other, collisions are assumed elastic. The average kinetic energy of gas molecules is the same for all gases at the same temperature as they are assumed to be point masses. Momentum = mass x velocity. Therefore, the momentum of gas molecules increases with pressure as a greater force is exerted on each molecule.

Question 402: E
An exothermic reaction is defined as a chemical reaction that releases energy. Thus, aerobic respiration producing life energy, the burning of magnesium, and the reacting of acids/bases are almost always exothermic processes. Similarly, the combustion of most things (including hydrogen) is exothermic. Evaporation of water is a physical process in which no chemical reaction is taking place.

Question 403: E
$2 C_3H_6 + 9 O_2 \rightarrow 6 H_2O + 6 CO_2$
Assign the oxidation numbers for each element:
For C_3H_6: C = -2; H = +1
For O_2: O = 0
For H_2O: H = +1; O = -2
For CO_2: C = +4; O = -2
Look for the changes in the oxidation numbers:
H remained at +1
C changed from -2 to +4. Thus, it was oxidized
O changed from 0 to -2. Thus, it was reduced.

Question 404: B
The equation for the reaction is: $Zn + CuSO_4 \rightarrow ZnSO_4 + Cu$
Assign oxidation numbers for each element:
For Zn: Zn = 0
For $CuSO_4$: Cu = +2; S = +6; O = -2
For $ZnSO_4$: Zn = +2; S = +6; O = -2
For Cu: Cu = 0
With these oxidation numbers, we can see that Zn was oxidized and Cu in $CuSO_4$ was reduced. Thus, Zn acted as the reducing agent and Cu in $CuSO_4$ is the oxidizing agent.

Question 405: B
Acids are proton donors which only exist in aqueous solution, which is a liquid state. Strong acids are fully ionised in solution and the reaction between an acid and a base $\rightarrow$ salt + water.
The pH of weak acids is usually between 4 and 6.

Question 406: G
Let x be the relative abundance of Z^6 and y the relative abundance of Z^8.
The average atomic mass takes the abundances of all 3 isotopes into account.
Thus, (Abundance of Z^5)(Mass Z^5) + (Abundance of Z^6)(Mass Z^6) + (Abundance of Z^8)(Mass Z^8) = 7
Therefore: (5 x 0.2) + 6x + 8y = 7
So: 6x + 8y = 6
Divide by two to give: 3x + 4y = 3
The abundances of all isotopes = 100% = 1
This gives: 0.2 + x + y = 1
Solve the two equations simultaneously:
y = 0.8 − x $\rightarrow$ 3x + 4(0.8 − x) = 3 $\rightarrow$ 3x + 3.2 − 4x = 3
Therefore, x = 0.2
y = 0.8 - 0.2 = 0.6
Thus, the overall abundances are Z^5 = 20%, Z^6 = 20% and Z^8 = 60%. Therefore, all the statements are correct.

Question 407: A
If a metal is more reactive than hydrogen, a displacement reaction will occur resulting in the formation of a salt with the metal cation and hydrogen.

Question 408: B
$6 FeSO_4 + K_2Cr_2O_7 + 7 H_2SO_4 \rightarrow 3 (Fe)_2(SO_4)_3 + Cr_2(SO_4)_3 + K_2SO_4 + 7 H_2O$
In order to save time, you have to quickly eliminate options (rather than try every combination out).
The quickest way is to do this is algebraically:

For Potassium:	For Iron:	For Hydrogen:
2b = 2e = 2f	a = 2d	2c = 2g
Therefore, b = f.	Options C, D and E don't fulfil a = 2d.	Therefore, c = g.
Option F does not fulfil b = e = f.		

Option A does not fulfil c = g. This leaves option B as the answer.

Question 409: E

Atoms are electrically neutral. Ions have different numbers of electrons when compared to atoms of the same element. Protons provide just under 50% of an atom's mass, the other 50% is provided by neutrons. Isotopes don't exhibit significantly different kinetics. Protons do indeed repel each other in the nucleus (which is one reason why neutrons are needed: to reduce the electrical charge density).

Question 410: B

The noble gasses are extremely useful, e.g. helium in blimps, neon signs, argon in bulbs. They are colourless and odourless and have no valence electrons. As with the rest of the periodic table, boiling point increases as you progress down the group (because of increased Van der Waals forces). Helium is the most abundant noble gas (and indeed the 2nd most abundant element in the universe).

Question 411: D

Alkenes can be hydrogenated (i.e. reduced) to alkanes. Aromatic compounds are commonly written as cyclic alkenes, but their properties differ from those of alkenes. Therefore alkenes and aromatic compounds do not belong to the same chemical class.

Question 412: A

The average atomic mass takes the abundances of both isotopes into account:

(Abundance of Cl^{35})(Mass Cl^{35}) + (Abundance of Cl^{37})(Mass Cl^{37}) = 35.453

34.969(Abundance of Cl^{35}) + 36.966(Abundance of Cl^{37}) = 35.453

The abundances of both isotopes = 100% = 1

I.e. abundance of Cl^{35} + abundance of Cl^{37} = 1

Therefore: x + y = 1 which can be rearranged to give: y = 1-x

Therefore: x + (1 − x) = 1.

34.969x + 36.966(1-x) = 35.453

x = 0.758

1 - x = 0.242

Therefore, Cl^{35} is 3 times more abundant than Cl^{37}.

Note that you could approximate the values here to arrive at the solution even quicker, e.g. 34.969 → 35, 36.966 → 37 and 35.453 → 35.5

Question 413: A

Transition metals form multiple stable ions which may have many different colours (e.g. green Fe^{2+} and brown Fe^{3+}). They usually form ionic bonds and are commonly used as catalysts (e.g. iron in the Haber process, Nickel in alkene hydrogenation). They are excellent conductors of electricity and are known as the d-block elements.

Question 414: B

$2Na + 2H_2O$ → $2NaOH + H_2$

8000 cm^3 = 8 dm^3 = ⅓ moles of H_2

2 moles of Na react completely to form 1 mole of H_2.

Therefore, ⅔ moles of Na must have reacted to produce ⅓ moles of Hydrogen. ⅔ x 23g per mole = 15.3g.

% Purity of sample = $\frac{15.3}{20}$ x 100 = 76.5%

Question 415: C

Assume total mass of molecule is 100g. Therefore, it contains 70.6g carbon, 5.9g hydrogen and 23.5g oxygen. Now, calculate the number of moles of each element using $Moles = \frac{Mass}{Molar\ Mass}$

$Moles\ of\ Carbon = \frac{70.6}{12} \approx 6$ $Moles\ of\ Hydrogen = \frac{5.9}{1} \approx 6$ $Moles\ of\ Oxygen = \frac{23.5}{16} \approx 1.5$

Therefore, the molar ratios give an empirical formula of $C_6H_6O_{1.5}$ = C_4H_4O.

Molar mass of the empirical formula = (4 x 12) + (4 x 1) + 16 = 68.

Molar mass of chemical formula = 136. Therefore, the chemical formula = $C_8H_8O_2$.

Question 416: B

$S + 6 HNO_3$ → $H_2SO_4 + 6 NO_2 + 2 H_2O$

In order to save time, you have to quickly eliminate options (rather than try every combination out).

The quickest way to do this is algebraically:

For Hydrogen:

b = 2c + 2e

Options A, C, D, E and F don't fulfil b = 2c + 2e.

This leaves options B as the only possible answer.

Note how quickly we were able to get the correct answer here by choosing an element that appears in 3 molecules (as opposed to Sulphur or Nitrogen which only appear in 2).

Question 417: A
Alkenes undergo addition reactions, such as that with hydrogen, when catalysed by nickel, whilst alkanes do not as they are already fully saturated. The C=C bond is stronger than the C–C bond, but it is not exactly twice as strong, so will not require twice the energy to break it. Both molecules are organic and will dissolve in organic solvents.

Question 418: F
Diamond is unable to conduct electricity because all the electrons are involved in covalent bonds. Graphite is insoluble in water + organic solvents. Graphite is also able to conduct electricity because there are free electrons that are not involved in covalent bonds.
Methane and Ammonia both have low melting points. Methane is not a polar molecule, so cannot conduct electricity or dissolve in water. Ammonia is polar and will dissolve in water. It can conduct electricity in aqueous form, but not as a gas.

Question 419: A
Catalysts increase the rate of reaction by providing an alternative reaction path with a lower activation energy, which means that less energy is required and so costs are reduced. The point of equilibrium, the nature of the products, and the overall energy change are unaffected by catalysts.

Question 420: E
The 5 carbon atoms in this hydrocarbon make it a "pent" stem. The C=C bond makes it an alkene, and the location of this bond is the 2nd position, making the molecule pent-2-ene.

Question 421: D
Group 1 elements form positively charged ions in most reactions and therefore lose electrons. Thus, the oxidation number must increase. Their reactivity increases as the valence electrons are further away from the positively charged nucleus down group. All group one elements react spontaneously with oxygen – the less reactive ones form an oxide coating and the more reactive ones spontanuoesly burn.

Question 422: H
The cathode attracts positively charged ions. The cathode reduces ions and the anode oxidises ions. Electrolysis can be used to separate compounds but not mixtures (i.e. substances that are not chemically joined).

Question 423: B
Pentane, C_5H_{12}, has a total of 3 isomers. A, C and D are correctly configured. However, the 4th Carbon atom in option B has more than 4 bonds which wouldn't be possible. If you're stuck on this – draw them out!

Question 424: E
$3 Cu + 8 HNO_3 \rightarrow 3 Cu(NO_3)_2 + 2 NO + 4 H_2O$
In order to save time, you have to quickly eliminate options (rather than try every combination out). The quickest way to do this is algebraically, by first assigning coefficients to the equation:
$aCu + bHNO_3 \rightarrow cCu(NO_3)_2 + dNO + eH_2O$. For Nitrogen: $b = 2c + d$. In this case, only option E satisfies $b = 2c + d$.
Note that using copper wouldn't be as useful, as all the options satisfy $a = c$.

Question 425: D
Alkenes are an organic series and have twice as many hydrogen atoms as carbon atoms. Bromine water is decolourised in their presence and they take part in addition reactions. Alkenes are more reactive than alkanes because they contain a C=C bond.

Question 426: A
Group 17 elements are missing one valence electron, so form negative ions. Bromine is a liquid at room temperature, and is also coloured brown. Reactivity decreases as you progress down Group 17, so fluorine reacts more vigorously than iodine. All Group 17 elements are found bound to each other, e.g. F_2 and Cl_2.

Question 427: D
CO poisoning and spontaneous combustion do not occur in the electrolysis of brine. The products of cathode and anode in the electrolysis of brine are Cl_2 and H_2. If these two gases react with each other they can form HCl, which is extremely corrosive.

Question 428: D
The hydrogen produced is positively charged and therefore needs to be reduced by the addition of an electron before being released. This happens at the cathode. The chlorine produced is negatively charged and therefore needs to lose electrons. This happens at the anode. NaOH is formed in this process.

Question 429: C
Alkanes are made of chains of singly bonded carbon and hydrogen atoms. C-H bonds are very strong and confer alkanes a great deal of stability. An alkane with 14 hydrogen atoms is called Hexane, as it has 6 carbon atoms. Alkanes burn in excess oxygen to produce carbon dioxide and water. Bromine water is decolourised in the presence of alkenes.

Question 430: G

You've probably got a lot of experience of organic chemistry by now, so this should be fairly straightforward. Alcohols by definition contain an R-OH functional group and because of this polar group are highly soluble in water. Ethanol is a common biofuel.

Question 431: E

Alkanes are saturated (and therefore non-reducible), have the general formula C_nH_{2n+2} and have no effect on Bromine solution. Alkenes are unsaturated (and therefore reducible), have the general formula C_nH_{2n} and turn bromine water colourless because they can undergo an addition reaction with bromine.

Question 432: D

The balanced equation for the reaction between magnesium oxide and hydrochloric acid is:

$MgO + 2HCl \rightarrow MgCl_2 + H_2$

The relative molecular mass of MgO is $24 + 16 = 40$g per mole.

Therefore 10g of MgO represents $10/40 = 0.25$ moles.

As the ratio of MgO to $MgCl_2$ is 1:1, we know that the amount of $MgCl_2$ produced will also be 0.25 moles. One mole of $MgCl_2$ has a molecular mass of $24 + (2 \times 35.5) = 95$g per mole.

Therefore the reaction will produce $0.25 \times 95 = 23.75$g of $MgCl_2$.

Question 433: D

Moving up the alkane series, as size and mass of the molecule increases (and thus the Van der Waals forces increase), the boiling point and viscosity increase and the flammability and volatility decrease. Therefore pentadecane will be more viscous than pentane.

Question 434: F

All of the factors mentioned will affect the rate of a reaction. The temperature affects the movement rate of particles, which if moving faster in higher temperatures will collide more often, thus increasing the rate of reaction. Collision rate is also increased with a higher concentration of reactants, and with a higher concentration of a catalyst or one with larger surface area, which will provide more active sites, thus increasing the rate of reaction.

Question 435: C

The total atomic mass of the end product is $C[12 + (2 \times 16)] + D[(2 \times 1) + 16] = 44C + 18D$

We know that $176 = 44C$. Therefore $C = 4$, and that $108 = 18D$ so $D = 6$.

Thus, the equation becomes: $C_aH_b + O_2 \rightarrow 4CO_2 + 6H_2O$.

This gives a ratio of 4C to 12H, which is a ratio of 1:3 carbon to hydrogen. This means the unknown hydrocarbon must be a multiple of this ratio. By balancing the equation we can see that the unknown hydrocarbon must be ethane, C_2H_6: $2C_2H_6 + 7O_2 \rightarrow 4CO_2 + 6H_2O$.

Question 436: A

$C_2H_5OH \rightarrow C_2H_4O$. Thus, ethanol has lost two hydrogen atoms, i.e. has been oxidised. Note that although another substrate may be reduced (therefore making it a redox reaction), ethanol has only been oxidised.

Question 437: B

This is fairly straightforward but you can save time by doing it algebraically:

For Barium: $3a = b$. For Nitrogen: $2a = c$. Let $a = 1$, thus, $b = 3$ and $c = 2$

Question 438: E

There are 14 oxygen atoms on the left side. Thus: $3b + 2c = 14$.

Note also that for Sulphur: $a = c$, and for Iron: $a = 2b$.

This sets up an easy trio of simultaneous equations:

Substitute a into the first equation to give: $1.5a + 2a = 14$. Thus: $a = 14/3.5 = 4$.

Therefore, $a = c = 4$ and $b = 2$

Question 439: C

The average atomic mass takes the abundances of all isotopes into account:

Mass = (Abundance of Mg^{23})(Mass Mg^{23}) + (Abundance of Mg^{25})(Mass Mg^{25}) + (Abundance of Mg^{26})(Mass Mg^{26})

$Mass = 23 \times 0.80 + 25 \times 0.10 + 26 \times 0.10$

$= 18.4 + 2.5 + 2.6 = 23.5$

Question 440: D

Cl_2 and Fe_2O_3 are reduced in their reactions and are therefore oxidising agents. Similarly, CO and Cu^{2+} are oxidised in their reactions and are therefore reducing agents. Cl is a stronger oxidising agent than Br as it is higher up in the reactivity series, and will displace negative Br ions from its compounds to form the oxidised Br_2. Mg is a stronger reducing agent than Cu, as it is higher up in the reactivity series. Thus, Mg would displace a positive copper ion from its compound to form copper atoms. Therefore Mg reduces Cu.

Question 441: C
NaCl is an ionic compound and therefore has a high melting point. It is highly soluble in water but only conducts electricity in solution/as a liquid.

Question 442: C
The equation for the reaction is: $2NaOH + Zn(NO_3)_2 \rightarrow 2NaNO_3 + Zn(OH)_2$
Therefore, the molar ratio between NaOH and $Zn(OH)_2$ is 2:1.
Molecular Mass of NaOH = 23 + 16 + 1 = 40
Molecular Mass of $Zn(OH)_2$= 65 + 17 x 2 = 99
Thus, the number of moles of NaOH that react = 80/40 = 2 moles.
Therefore, 1 mole of $Zn(OH)_2$ is produced. Mass = 99g per mole x 1 mole = 99g

Question 443: E
Metal + Water → Hydroxide + Hydrogen gas; the reaction is always exothermic. Reactivity increases down the group, so potassium reacts more vigorously with water than sodium. Therefore all four statements are correct.

Question 444: C
Electrolysis separates NaCl into sodium and chloride ions but not CO_2 (which is a covalently bound gas). Sieves cannot separate ionically bound compounds like NaCl. Dyes are miscible liquids and can be separated by chromatography. Oil and water are immiscible liquids, so a separating funnel is necessary to separate the mixtures. Methane and diesel are separated from each other during fractional distillation, as they have different boiling points.

Question 445: B
The reaction between water and caesium can cause spontaneous combustion and therefore doesn't make the reaction safer. The reaction between caesium and fluoride is highly exothermic and does not require a catalyst. The reaction produces CsF which is a salt.

Question 446: B
The nucleus of larger elements contain more neutrons than protons to reduce the charge density, e.g. Br^{80} contains 35 protons but 45 neutrons. Stable isotopes very rarely undergo radioactive decay.

Question 447: B
The vast majority of salts contain ionic bonds that require a significant amount of heat energy to break.

Question 448: E
306ml of water is 306g, which is the equivalent of 306g/18g per mole of H_2O = 17 moles. 17 times Avogadro's constant gives the number of molecules present, which is 1.02×10^{25}. There are 10 protons and 10 electrons in each water molecule. Hence there are 1.02×10^{26} protons.

Question 449: D
The number of moles of each element = Mass/Molar Mass. Let the % represent the mass in grams: Hydrogen: 3.45g/1g per mole = 3.45 moles. Oxygen: 55.2g/16g per mole = 3.45 moles
Carbon: 41.4g/12g per mole = 3.45 moles
Thus, the molar ratio is 1:1:1. The only option that satisfies this is option D.

Question 450: C
Group 17 elements are non-metals, whilst group 2 elements are metals. Thus, the Group 17 element must gain electrons when it reacts with the Group 2 element, i.e. B is reduced. The easy way to calculate the formula is to swap the valences of both elements: A is +2 and B is -1. Thus, the compound is AB_2.

Question 451: F
That the amplitude of a wave determines its mass is false. Waves are not objects and do not have mass.

Question 452: A
We know that displacement s = 30 m, initial speed u = 0 ms^{-1}, acceleration a = 5.4 ms^{-2}, final speed v = ?, time t = ?
And that $v^2 = u^2 + 2as$
$v^2 = 0 + 2 \times 5.4 \times 30$
$v^2 = 324$ so v = 18 ms^{-1}
and $s = ut + 1/2\ at^2$ so $30 = 1/2 \times 5.4 \times t^2$
$t^2 = 30/2.7$ so t = 3.3 s

Question 453: D
The wavelength is given by: velocity $v = \lambda f$ and frequency f = 1/T so $v = \lambda/T$ giving wavelength $\lambda = vT$
The period T = 49 s/7 so λ = 5 ms^{-1} x 7 s = 35 m

Question 454: E

This is a straightforward question as you only have to put the numbers into the equation (made harder by the numbers being hard to work with).

$Power = \frac{Force \; x \; Distance}{Time} = \frac{375 \; N \; x \; 1.3 \; m}{5 \; s}$

$= 75 \; x \; 1.3 = 97.5 \; W$

Question 455: G

v = u + at

v = 0 + 5.6 x 8 = 44.8 ms^{-1}

And $s = ut + \frac{at^2}{2} = 0 + 5.6 \; x\frac{8^2}{2} = 179.2$

Question 456: C

The sky diver leaves the plane and will accelerate until the air resistance equals their weight – this is their terminal velocity. The sky diver will accelerate under the force of gravity. If the air resistance force exceeded the force of gravity the sky diver would accelerate away from the ground, and if it was less than the force of gravity they would continue to accelerate toward the ground.

Question 457: D

s = 20 m, u = 0 ms^{-1}, a = 10 ms^{-2}

and v^2 = u^2 + 2as

v^2 = 0 + 2 x 10 x 20

v^2 = 400;   v = 20 ms^{-1}

Momentum = Mass x velocity = 20 x 0.1 = 2 kgms^{-1}

Question 458: E

Electromagnetic waves have varying wavelengths and frequencies and their energy is proportional to their frequency.

Question 459: D

The total resistance = R + r = 0.8 + 1 = 1.8 Ω

and $I = \frac{e.m.f}{total \; resistance} = \frac{36}{1.8} = 20 \; A$

Question 460: D

Use Newton's second law and remember to work in SI units:

So $Force = mass \; x \; accelaration = mass \; x\frac{\Delta velocity}{time}$

$= 20 \; x \; 10^{-3} \; x \; \frac{100 - 0}{10 \; x \; 10^{-3}}$

$= 200 \; N$

Question 461: F

In this case, the work being done is moving the bag 0.7 m

i.e. $Work \; Done = Bag's \; Weight \; x \; Distance = 50 \; x \; 10 \; x \; 0.7 = 350 \; N$

$Power = \frac{Work}{Time} = \frac{350}{3} = 116.7 \; W$

= 117 W to 3 significant figures

Question 462: B

Firstly, use P = Fv to calculate the power [Ignore the frictional force as we are not concerned with the resultant force here].

So P = 300 x 30 = 9000 W

Then, use P = IV to calculate the current.

I = P/V = 9000/200 = 45 A

Question 463: C

Work is defined as W = F x s. Work can also be defined as work = force x distance moved in the direction of force. Work is measured in joules and 1 Joule = 1 Newton x 1 Metre, and 1 Newton = 1 Kg x ms^{-2} [F = ma].

Thus, 1 Joule = Kgm^{2}s^{-2}

Question 464: G

Joules are the unit of energy (and also Work = Force x Distance). Thus, 1 Joule = 1 N x 1 m.

Pa is the unit of Pressure (= Force/Area). Thus, Pa = N x m^{-2}. So J = Nm^{-2} x m^3 = Pa x m^3. Newton's third law describes that every action produces an equal and opposite reaction. For this reason, the energy required to decelerate a body is equal to the amount of energy it possess during movement, i.e. its kinetic energy, which is defined as in statement 1.

Question 465: D

Alpha radiation is of the lower energy, as it represents the movement of a fairly large particle consisting of 2 neutrons and 2 protons. Beta radiation consists of high-energy, high-speed electrons or positrons.

Question 466: E

The half-life does depend on atom type and isotope, as these parameters significantly impact on the physical properties of the atom in general, so statement 1 is false. Statement 2 is the correct definition of half-life. Statement 3 is also correct: half-life in exponential decay will always have the same duration, independent of the quantity of the matter in question; in non-exponential decay, half-life is dependent on the quantity of matter in question.

Question 467: A

In contrast to nuclear fission, where neutrons are shot at unstable atoms, nuclear fusion is based on the high speed, high-temperature collision of molecules, most commonly hydrogen, to form a new, stable atom while releasing energy.

Question 468: E

Nuclear fission releases a significant amount of energy, which is the basis of many nuclear weapons. Shooting neutrons at unstable atoms destabilises the nuclei which in turn leads to a chain reaction and fission. Nuclear fission can lead to the release of ionizing gamma radiation.

Question 469: G

The total resistance of the circuit would be twice the resistance of one resistor and proportional to the voltage, as given by Ohm's Law. Since it is a series circuit, the same current flows through each resistor and since they are identical the potential difference across each resistor will be the same.

Question 470: E

The distance between Earth and Sun = Time x Speed = 60 x 8 seconds x 3 x 10^8 ms^{-1} = 480 x 3 x 10^8 m

Approximately = 1500 x 10^8 = 1.5 x 10^{11} m.

The circumference of Earth's orbit around the sun is given by $2\pi r$ = 2 x 3 x 1.5 x 10^{11}

= 9 x 10^{11} = 10^{12} m

Question 471: H

Speed is a scalar quantity whilst velocity is a vector describing both magnitude and direction. Speed describes the distance a moving object covers over time (i.e. speed = distance/time), whereas velocity describes the rate of change of the displacement of an object (i.e. velocity = displacement/time). The internationally standardised unit for speed is meters per second (ms^{-1}), while ms^{-2} is the unit of acceleration.

Question 472: E

Ohm's Law only applies to conductors and can be mathematically expressed as $V \alpha I$. The easiest way to do this is to write down the equations for statements c, d and e. C: $I \alpha \frac{1}{V}$; D: $I \alpha V^2$; E: $I \alpha V$. Thus, statement E is correct.

Question 473: G

Any object at rest is not accelerating and therefore has no resultant force. Strictly speaking, Newton's second law is actually: Force = rate of change of momentum, which can be mathematically manipulated to give statement 2:

$$Force = \frac{momentum}{time} = \frac{mass \times velocity}{time} = mass \times accelaration$$

Question 474: D

Statement 3 is incorrect, as $Charge = Current \times time$. Statement 1 substitutes $I = \frac{V}{R}$ and statement 2 substitutes $I = \frac{P}{V}$.

Question 475: E

Weight of elevator + people = mg = 10 x (1600 + 200) = 18,000 N

Applying Newton's second law of motion on the car gives:

Thus, the resultant force is given by:

F_M = Motor Force – [Frictional Force + Weight]

F_M = M – 4,000 – 18,000

Use Newton's second law to give: F_M = M – 22,000 N = ma

Thus, M – 22,000 N = 1,800a

Since the lift must accelerate at 1ms^{-2}: M = 1,800 kg x 1 ms^{-2} + 22,000 N

M = 23,800 N

Question 476: D

Total Distance = Distance during acceleration phase + Distance during braking phase

Distance during <u>acceleration phase</u> is given by:

$$s = ut + \frac{at^2}{2} = 0 + \frac{5 \times 10^2}{2} = 250\ m$$

$$v = u + at = 0 + 5 \times 10 = 50\ ms^{-1}$$

And use $a = \frac{v-u}{t}$ to calculate the deceleration: $a = \frac{0-50}{20} = -2.5\ ms^{-2}$

Distance during the <u>deceleration phase</u> is given by:

$$s = ut + \frac{at^2}{2} = 50 \times 20 + \frac{-2.5 \times 20^2}{2} = 1000 - \frac{2.5 \times 400}{2}$$

$$s = 1000 - 500 = 500\ m$$

Thus, $Total\ Distance = 250 + 500 = 750\ m$

Question 477: G

It is not possible to calculate the power of the heater as we don't know the current that flows through it or its internal resistance. The 8 ohms refers to the external copper wire and not the heater. Whilst it's important that you know how to use equations like P = IV, it's more important that you know when you **can't** use them!

Question 478: F

This question has a lot of numbers but not any information on time, which is necessary to calculate power. You cannot calculate power by using P= IV as you don't know how many electrons are accelerated through the potential difference per unit time. Thus, more information is required to calculate the power.

Question 479: B

When an object is in equilibrium with its surroundings, it radiates and absorbs energy at the same rate and so its temperature remains constant i.e. there is no *net* energy transfer. Radiation is slower than conduction and convection.

Question 480: A

The work done by the force is given by: $Work\ Done = Force \times Distance = 12\ N \times 3\ m = 36\ J$

Since the surface is frictionless, $Work\ Done = Kinetic\ Energy$.

$$E_k = \frac{mv^2}{2} = \frac{6v^2}{2}$$

Thus, $36 = 3v^2$

$$v = \sqrt{12} = \sqrt{4}\sqrt{3} = 2\sqrt{3}\ ms^{-1}$$

Question 481: C

$Total\ energy\ supplied\ to\ water = Change\ in\ temperature \times Mass\ of\ water \times 4{,}000\ J$
$= 40 \times 1.5 \times 4{,}000 = 240{,}000\ J$

$$Power\ of\ the\ heater = \frac{Work\ Done}{time} = \frac{240{,}000}{50 \times 60} = \frac{240{,}000}{3{,}000} = 80\ W.\ \text{Using } P = IV = \frac{V^2}{R}:$$

$$R = \frac{V^2}{P} = \frac{100^2}{80} = \frac{10{,}000}{80} = 125\ ohms$$

Question 482: G

The large amount of energy released during atomic fission is the basis underlying nuclear power plants. Splitting an atom into two or more parts will by definition produce molecules of different sizes than the original atom; therefore it produces two new atoms. The free neutrons and photons produced by the splitting of atoms form the basis of the energy release.

Question 483: D

Gravitational potential energy is just an extension of the equation work done = force x distance (force is the weight of the object, *mg*, and distance is the height, *h*). The reservoir in statement 3 would have a potential energy of 10^{10} Joules i.e. 10 Giga Joules ($E_p = 10^6$ kg x 10 N x 10^3 m).

Question 484: D

Statement 1 is the common formulation of Newton's third law. Statement 2 presents a consequence of the application of Newton's third law.

Statement 3 is false: rockets can still accelerate because the products of burning fuel are ejected in the opposite direction from which the rocket needs to accelerate.

Question 485: E

Positively charged objects have lost electrons. $Charge = Current \times Time = \frac{Voltage}{Resistance} \times Time.$

Objects can become charged by friction as electrons are transferred from one object to the other.

Question 486: B

Each body of mass exerts a gravitational force on another body with mass. This is true for all planets as well.
Gravitational force is dependent on the mass of both objects. Satellites stay in orbit due to centripetal force that acts tangentially to gravity (not because of the thrust from their engines). Two objects will only land at the same time if they also have the same shape or they are in a vaccum (as otherwise air resistance would result in different terminal velocities).

Question 487: A

Metals conduct electrical charge easily and provide little resistance to the flow of electrons. Charge can also flow in several directions. However, all conductors have an internal resistance and therefore provide *some* resistance to electrical charge.

Question 488: E

First, calculate the rate of petrol consumption:

$$\frac{Speed}{Consumption} = \frac{60 \; miles/hour}{30 \; miles/gallon} = 2 \; gallons/hour$$

Therefore, the total power is:

$$2 \; gallons \; = \; 2 \times 9 \times 10^8 = 18 \times 10^8 J$$

$$1 \; hour \; = \; 60 \times 60 \; = \; 3600 \; s$$

$$Power = \frac{Energy}{Time} = \frac{18 \times 10^8}{3600} = \frac{18}{36} \times 10^6 = 5 \times 10^5 \; W$$

Since efficiency is 20%, the power delivered to the wheels $= 5 \times 10^5 \times 0.2 = 10^5 \; W = 100 \; kW$

Question 489: D

Beta radiation is stopped by a few millimetres of aluminium, but not by paper. In β^- radiation, a neutron changes into a proton plus an emitted electron. This means the atomic mass number remains unchanged.

Question 490: F

Firstly, calculate the mass of the car $= \frac{Weight}{g} = \frac{15,000}{10} = 1,500 \; kg$

Then using $v = u + at$ where v = 0 ms^{-1} and u = 15 ms^{-1} and t = 10 x 10^{-3} s

$a = \frac{0-15}{0.01} = 1500 ms^{-2}$

$F = ma = 1500 \times 1500 = 2\,250\,000 \; N$

Question 491: E

Electrical insulators offer high resistance to the flow of charge. Insulators are usually non-metals; metals conduct charge very easily. Since charge does not flow easily to even out, they can be charged with friction.

Question 492: A

The car accelerates for the first 10 seconds at a constant rate and then decelerates after t=30 seconds. It does not reverse, as the velocity is not negative.

Question 493: B

The distance travelled by the car is represented by the area under the curve (integral of velocity) which is given by the area of two triangles and a rectangle:

$$Area \; = \left(\frac{1}{2} \times 10 \times 10\right) + (20 \times 10) + \left(\frac{1}{2} \times 10 \times 10\right)$$

$$Area = 50 + 200 + 50 = 300 \; m$$

Question 494: C

Using the equation force = mass x acceleration, where the unknown acceleration = change in velocity over change in time.

Hence: $\frac{F}{m} = \frac{change \; in \; velocity}{change \; in \; time}$

We know that F = 10,000 N, mass = 1,000 kg and change in time is 5 seconds.

So, $\frac{10,000}{1,000} = \frac{change \; in \; velocity}{5}$

So change in velocity = 10 x 5 = 50 m/s

Question 495: D

This question tests both your ability to convert unusual units into SI units and to select the relevant values (e.g. the crane's mass is not important here).

0.01 tonnes = 10 kg; 100 cm = 1 m; 5,000 ms = 5 s

$Power = \frac{Work\ Done}{Time} = \frac{Force\ x\ Distance}{Time}$

In this case the force is the weight of the wardrobe = 10 x g = 10 x 10 = 100N. Thus, $Power = \frac{100\ x\ 1}{5} = 20\ W$

Question 496: F

Remember that the resistance of a parallel circuit (R_T) is given by: $\frac{1}{R_T} = \frac{1}{R_1} + \frac{1}{R_2} + \dots$

Thus, $\frac{1}{R_T} = \frac{1}{1} + \frac{1}{2} = \frac{3}{2}$ and therefore $R = \frac{2}{3}\ \Omega$

Using Ohm's Law: I $= \frac{20\ V}{\frac{2}{3}\Omega} = 20\ x\ \frac{3}{2} = 30$ A

Question 497: E

Water is denser than air. Therefore, the speed of light decreases when it enters water and increases when it leaves water. The direction of light also changes when light enters/leaves water. This phenomenon is known as refraction and is governed by Snell's Law.

Question 498: C

The voltage in a parallel circuit is the same across each branch, i.e. branch A Voltage = branch B Voltage.

The resistance of Branch A = 6 x 5 = 30 Ω; the resistance of Branch B = 10 x 2 = 20 Ω.

Using Ohm's Law: I= V/R. Thus, $I_A = \frac{60}{30} = 2\ A$; $I_B = \frac{60}{20} = 3\ A$

Question 499: C

This is a very straightforward question made harder by the awkward units you have to work with. Ensure you are able to work comfortably with prefixes of 10^9 and 10^{-9} and convert without difficulty.

50,000,000,000 nano Watts = 50 W and 0.000000004 Giga Amperes = 4 A.

Using $P = IV$: $V = \frac{P}{I} = \frac{50}{4} = 12.5\ V = 0.0125\ kV$

Question 500: B

Radioactive decay is highly random and unpredictable. Only gamma decay releases gamma rays and few types of decay release X-rays. The electrical charge of an atom's nucleus decreases after alpha decay as two protons are lost.

Question 501: D

Using $P = IV$: $I = \frac{P}{V} = \frac{60}{15} = 4\ A$

Now using Ohm's Law: $R = \frac{V}{I} = \frac{15}{4} = 3.75\ \Omega$

So, each resistor has a resistance of $\frac{3.75}{3} = 1.25\ \Omega$. If two more resistors are added, the overall resistance = 1.25 x 5 = 6.25 Ω

Question 502: F

There is not enough information to answer this question. We would be required to know the resistive forces acting against the tractor and if there is any change in height in order to calculate the useful work done and hence the efficiency.

Question 503: G

Electromagnetic induction is defined by statements 1 and 2. An electrical current is generated when a coil moves in a magnetic field.

Question 504: D

An ammeter will always give the same reading in a series circuit, not in a parallel circuit where current splits at each branch in accordance with Ohm's Law.

Question 505: D

Electrons move in the opposite direction to current (i.e. they move from negative to positive).

Question 506: A

For a fixed resistor, the current is directly proportional to the potential difference. For a filament lamp, as current increases, the metal filament becomes hotter. This causes the metal atoms to vibrate and move more, resulting in more collisions with the flow of electrons. This makes it harder for the electrons to move through the lamp and results in increased resistance. Therefore, the graph's gradient decreases as current increases.

Question 507: E

Vector quantities consist of both direction and magnitude, e.g. velocity, displacement, etc., and can be added by taking account of direction in the sum.

Question 508: C

The gravity on the moon is 6 times less than 10 ms^{-2}. Thus, $g_{moon} = \frac{10}{6} = \frac{5}{3}$ ms^{-2}.

Since weight = mass x gravity, the mass of the rock $= \frac{250}{\frac{5}{3}} = \frac{750}{5} = 150\ kg$

Therefore, the density $= \frac{mass}{volume} = \frac{150}{250} = 0.6\ kg/cm^3$

Question 509: D

An alpha particle consists of a helium nucleus. Thus, alpha decay causes the mass number to decrease by 4 and the atomic number to decrease by 2. Five iterations of this would decrease the mass number by 20 and the atomic number by 10.

Question 510: C

Using Ohm's Law: The potential difference entering the transformer (V_1) = 10 x 20 = 200 V

Now use $\frac{N1}{N2} = \frac{V1}{V2}$ to give: $\frac{5}{10} = \frac{200}{V2}$

Thus, $V_2 = \frac{2,000}{5} = 400$ V

Question 511: D

For objects in free fall that have reached terminal velocity, acceleration = 0.
Thus, the sphere's weight = resistive forces.
Using Work Done = Force x Distance: Force = 10,000 J/100 m = 100 N.
Therefore, the sphere's weight = 100 N and since $g = 10ms^{-2}$, the sphere's mass = 10 kg

Question 512: F

The wave length of ultraviolet waves is longer than that of x-rays. Wavelength is inversely proportional to frequency. Most electromagnetic waves are not stopped with aluminium (and require thick lead to stop them), and they travel at the speed of light. Humans can only see a very small part of the spectrum.

Question 513: B

If an object moves towards the sensor, the wavelength will appear to decrease and the frequency increase. The faster this happens, the faster the increase in frequency and decrease in wavelength.

Question 514: A

$Acceleration = \frac{Change\ in\ Velocity}{Time} = \frac{1,000}{0.1} = 10,000\ ms^{-2}$

Using Newton's second law: The Braking Force = Mass x Acceleration.

Thus, Braking Force = 10,000 x 0.005 = 50 N

Question 515: C

Polonium has undergone alpha decay. Thus, Y is a helium nucleus and contains 2 protons and 2 neutrons.
Therefore, 10 moles of Y contain 2 x 10 x 6 x 10^{23} protons = 120 x 10^{23} = 1.2 x 10^{25} protons.

Question 516: C

The rod's activity is less than 1,000 Bq after 300 days. In order to calculate the longest possible half-life, we must assume that the activity is just below 1,000 Bq after 300 days. Thus, the half-life has decreased activity from 16,0 00 Bq to 1,000 Bq in 300 days.
After one half-life: Activity = 8,000 Bq
After two half-lives: Activity = 4,000 Bq
After three half-lives: Activity = 2,000 Bq
After four half-lives: Activity = 1,000 Bq
Thus, the rod has halved its activity a minimum of 4 times in 300 days. 300/4 = 75 days

Question 517: G

There is no change in the atomic mass or proton numbers in gamma radiation. In β decay, a neutron is transformed into a proton (and an electron is released). This results in an increase in proton number by 1 but no overall change in atomic mass. Thus, after 5 rounds of beta decay, the proton number will be 89 + 5 = 94 and the mass number will remain at 200. Therefore, there are 94 protons and 200-94 = 106 neutrons.
NB: You are not expected to know about β$^+$ decay.

Question 518: C

Calculate the speed of the sound $= \frac{distance}{time} = \frac{500}{1.5} = 333\ ms^{-1}$

Thus, the $Wavelength = \frac{Speed}{Frequency} = \frac{333}{440}$

Approximate 333 to 330 to give: $\frac{330}{440} = \frac{3}{4} = 0.75\ m$

Question 519: B

Firstly, note the all the answer options are a magnitude of 10 apart. Thus, you don't have to worry about getting the correct numbers as long as you get the correct power of 10. You can therefore make your life easier by rounding, e.g. approximate π to 3, etc.

The area of the shell $= \pi r^2$.
$= \pi \times (50 \times 10^{-3})^2 = \pi \times (5 \times 10^{-2})^2$
$= \pi \times 25 \times 10^{-4} = 7.5 \times 10^{-3}\ m^2$

The deceleration of the shell $= \frac{u-v}{t} = \frac{200}{500 \times 10^{-6}} = 0.4 \times 10^6\ ms^{-2}$

Then, using Newton's Second Law: $Braking\ force = mass \times acceleration = 1 \times 0.4 \times 10^6 = 4 \times 10^5 N$

Finally: $Pressure = \frac{Force}{Area} = \frac{4 \times 10^5}{7.5 \times 10^{-3}} = \frac{8}{15} \times 10^8\ Pa \approx 5 \times 10^7 Pa$

Question 520: B

The fountain transfers 10% of 1,000 J of energy per second into 120 litres of water per minute. Thus, it transfers 100 J into 2 litres of water per second.

Therefore the Total Gravitational Potential Energy, $E_p = mg\Delta h$

Thus, $100\ J = 2 \times 10 \times h$

Hence, $h = \frac{100}{20} = 5\ m$

Question 521: E

In step down transformers, the number of turns of the primary coil is larger than that of the secondary coil to decrease the voltage. If a transformer is 100% efficient, the electrical power input = electrical power output (P=IV).

Question 522: C

The percentage of C^{14} in the bone halves every 5,730 years. Since it has decreased from 100% to 6.25%, it has undergone 4 half-lives. Thus, the bone is 4 × 5,730 years old = 22,920 years

Question 523: E

This is a straightforward question in principle, as it just requires you to plug the values into the equation: $Velocity = Wavelength \times Frequency$ – Just ensure you work in SI units to get the correct answer.

$Frequency = \frac{2\ m/s}{2.5\ m} = 0.8\ Hz = 0.8 \times 10^{-6} MHz = 8 \times 10^{-7}\ MHz$

Question 524: E

If an element has a half-life of 25 days, its BQ value will be halved every 25 days.

A total of 350/25 = 14 half-lives have elapsed. Thus, the count rate has halved 14 times. Therefore, to calculate the original rate, the final count rate must be doubled 14 times = 50×2^{14}.

$2^{14} = 2^5 \times 2^5 \times 2^4 = 32 \times 32 \times 16 = 16,384$.

Therefore, the original count rate = 16,384 × 50 = 819,200

Question 525: D

Remember that $V = IR = \frac{P}{I}$ and $Power = \frac{Work\ Done}{Time} = \frac{Force \times Distance}{Time} = Force \times Velocity$;

Thus, A is derived from: $V = IR$,

B is derived from: $= \frac{P}{I}$,

C is derived from: $Voltage = \frac{Power}{Current} = \frac{Force \times Velocity}{Current}$,

Since $Charge = Current \times Time$, E and F are derived from: $Voltage = \frac{Power}{Current} = \frac{Force \times Distance}{Time \times Current} = \frac{J}{As} = \frac{J}{C}$,

D is incorrect as Nm = J. Thus the correct variant would be NmC^{-1}

Question 526: B

Each three-block combination is mutually exclusive to any other combination, so the probabilities are added. Each block pick is independent of all other picks, so the probabilities can be multiplied. For this scenario there are three possible combinations:

P(2 red blocks and 1 yellow block) = P(red then red then yellow) + P(red then yellow then red) + P(yellow then red then red)

=

$(\frac{12}{20} \times \frac{11}{19} \times \frac{8}{18}) + (\frac{12}{20} \times \frac{8}{19} \times \frac{11}{18}) + (\frac{8}{20} \times \frac{12}{19} \times \frac{11}{18}) =$

$\frac{3 \times 12 \times 11 \times 8}{20 \times 19 \times 18} = \frac{44}{95}$

Question 527: C

Multiply through by 15: $3(3x + 5) + 5(2x - 2) = 18 \times 15$

Thus: $9x + 15 + 10x - 10 = 270$

$9x + 10x = 270 - 15 + 10$

$19x = 265$

$x = 13.95$

Question 528: C

This is a rare case where you need to factorise a complex polynomial:

$(3x \quad)(x \quad) = 0$, possible pairs: 2 x 10, 10 x 2, 4 x 5, 5 x 4

(3x - 4)(x + 5) = 0

3x - 4 = 0, so x = $\frac{4}{3}$

x + 5 = 0, so x = -5

Question 529: C

$\frac{5(x-4)}{(x+2)(x-4)} + \frac{3(x+2)}{(x+2)(x-4)}$

$= \frac{5x-20+3x+6}{(x+2)(x-4)} = \frac{8x-14}{(x+2)(x-4)}$

Question 530: E

p α $\sqrt[3]{q}$, so p = k $\sqrt[3]{q}$

p = 12 when q = 27 gives 12 = k $\sqrt[3]{27}$, so 12 = 3k and k = 4

so, p = 4 $\sqrt[3]{q}$

Now p = 24:

24 = 4$\sqrt[3]{q}$, so 6 = $\sqrt[3]{q}$ and q = 6^3 = 216

Question 531: A

8 x 9 = 72

8 = (4 x 2) = 2 x 2 x 2

9 = 3 x 3

$(2 \times 2 \times 2 \times 3 \times 3)^2$ = 2 x 2 x 2 x 2 x 2 x 2 x 3 x 3 x 3 x 3 = $2^6 \times 3^4$

Question 532: C

Note that 1.151 x 2 = 2.302.

Thus: $\frac{2 \times 10^5 + 2 \times 10^2}{10^{10}} = 2 \times 10^{-5} + 2 \times 10^{-8}$

$= 0.00002 + 0.00000002 = 0.00002002$

Question 533: E

$y^2 + ay + b$

$= (y +2)^2 - 5 = y^2 + 4y + 4 - 5$

$= y^2 + 4y + 4 - 5 = y^2 + 4y - 1$

So a = 4 and y = -1

Question 534: E

Take $5(m + 4n)$ as a common factor to give: $\frac{4(m+4n)}{5(m+4n)} + \frac{5(m-2n)}{5(m+4n)}$

Simplify to give: $\frac{4m+16n+5m-10n}{5(m+4n)} = \frac{9m+6n}{5(m+4n)} = \frac{3(3m+2n)}{5(m+4n)}$

Question 535: C

$A \alpha \frac{1}{\sqrt{B}}$. Thus, $= \frac{k}{\sqrt{B}}$.

Substitute the values in to give: $4 = \frac{k}{\sqrt{25}}$.

Thus, $k = 20$.

Therefore, $A = \frac{20}{\sqrt{B}}$.

When B = 16, $A = \frac{20}{\sqrt{16}} = \frac{20}{4} = 5$

Question 536: E

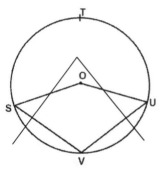

Angles SVU and STU are opposites and add up to 180°, so STU = 91°

The angle of the centre of a circle is twice the angle at the circumference so SOU = 2 x 91° = 182°

Question 537: E

The surface area of an open cylinder A = 2πrh. Cylinder B is an enlargement of A, so the increases in radius (r) and height (h) will be proportional: $\frac{r_A}{r_B} = \frac{h_A}{h_B}$. Let us call the proportion coefficient n, where $n = \frac{r_A}{r_B} = \frac{h_A}{h_B}$.

So $\frac{Area\ A}{Area\ B} = \frac{2\pi r_A h_A}{2\pi r_B h_B} = n\ x\ n = n^2$. $\frac{Area\ A}{Area\ B} = \frac{32\pi}{8\pi} = 4$, so, n = 2.

The proportion coefficient n = 2 also applies to their volumes, where the third dimension (also radius, i.e. the r^2 in V = πr^2h) is equally subject to this constant of proportionality. The cylinder's volumes are related by $n^3 = 8$.

If the smaller cylinder has volume 2π cm^3, then the larger will have volume 2π x $n^3 = 2\pi$ x 8 = 16π cm^3.

Question 538: E

$= \frac{8}{x(3-x)} - \frac{6(3-x)}{x(3-x)}$

$= \frac{8-18+6x}{x(3-x)}$

$= \frac{6x-10}{x(3-x)}$

Question 539: B

For the black ball to be drawn in the last round, white balls must be drawn every round. Thus the probability is given by $P =$

$\frac{9}{10}$ x $\frac{8}{9}$ x $\frac{7}{8}$ x $\frac{6}{7}$ x $\frac{5}{6}$ x $\frac{4}{5}$ x $\frac{3}{4}$ x $\frac{2}{3}$ x $\frac{1}{2}$

$= \frac{9\ x\ 8\ x\ 7\ x\ 6\ x\ 5\ x\ 4\ x\ 3\ x\ 2\ x\ 1}{10\ x\ 9\ x\ 8\ x\ 7\ x\ 6\ x\ 5\ x\ 4\ x\ 3\ x\ 2\ x\ 1} = \frac{1}{10}$

Question 540: C

The probability of getting a king the first time is $\frac{4}{52} = \frac{1}{13}$, and the probability of getting a king the second time is $\frac{3}{51}$. These are independent events, thus, the probability of drawing two kings is $\frac{1}{13} x \frac{3}{51} = \frac{3}{663} = \frac{1}{221}$

Question 541: B

The probabilities of all outcomes must sum to one, so if the probability of rolling a 1 is x, then: $x + x + x + x + 2x = 1$. Therefore, $x = \frac{1}{7}$.

The probability of obtaining two sixes $P_{12} = \frac{2}{7} x \frac{2}{7} = \frac{4}{49}$

Question 542: B

There are plenty of ways of counting, however the easiest is as follows: 0 is divisible by both 2 and 3. Half of the numbers from 1 to 36 are even (i.e. 18 of them). 3, 9, 15, 21, 27, 33 are the only numbers divisible by 3 that we've missed. There are 25 outcomes divisible by 2 or 3, out of 37.

Question 543: C

List the six ways of achieving this outcome: HHTT, HTHT, HTTH, and TTHH, THTH, THHT. There are 2^4 possible outcomes for 4 consecutive coin flips, so the probability of two heads and two tails is: $6 x \frac{1}{2^4} = \frac{6}{16} = \frac{5}{8}$

Question 544: D

Count the number of ways to get a 5, 6 or 7 (draw the square if helpful). The ways to get a 5 are: 1, 4; 2, 3; 3, 2; 4, 1. The ways to get a 6 are: 1, 5; 2, 4; 3, 3; 4, 2; 5, 1. The ways to get a 7 are: 1, 6; 2, 5; 3, 4; 4, 3; 5, 2; 6, 1. That is 15 out of 36 possible outcomes.

	1	2	3	4	5	6
1	2	3	4	5	6	7
2	3	4	5	6	7	8
3	4	5	6	7	8	9
4	5	6	7	8	9	10
5	6	7	8	9	10	11
6	7	8	9	10	11	12

Question 545: C

There are x+y+z balls in the bag, and the probability of picking a red ball is $\frac{x}{(x+y+z)}$ and the probability of picking a green ball is $\frac{z}{(x+y+z)}$. These are independent events, so the probability of picking red then green is $\frac{xz}{(x+y+z)^2}$ and the probability of picking green then red is the same. These outcomes are mutually exclusive, so are added.

Question 546: B

There are two ways of doing it, pulling out a red ball then a blue ball, or pulling out a blue ball and then a red ball. Let us work out the probability of the first: $\frac{x}{(x+y+z)} \times \frac{y}{x+y+z-1}$, and the probability of the second option will be the same. These are mutually exclusive options, so the probabilities may be summed.

Question 547: A

[x: Player 1 wins point, y: Player 2 wins point]

Player 1 wins in five rounds if we get: yxxxx, xyxxx, xxyxx, xxxyx.

(Note the case of xxxxy would lead to player 1 winning in 4 rounds, which the question forbids.)

Each of these have a probability of p⁴(1-p). Thus, the solution is 4p⁴(1-p).

Question 548: F

$4x + 7 + 18x + 20 = 14$

$22x + 27 = 14$

Thus, $22x = -13$

Giving $x = -\frac{13}{22}$

Question 549: D

$$r^3 = \frac{3V}{4\pi}$$

Thus, $r = \left(\frac{3V}{4\pi}\right)^{1/3}$

Therefore, $S = 4\pi\left[\left(\frac{3V}{4\pi}\right)^{\frac{1}{3}}\right]^2 = 4\pi\left(\frac{3V}{4\pi}\right)^{\frac{2}{3}}$

$$= \frac{4\pi(3V)^{\frac{2}{3}}}{(4\pi)^{\frac{2}{3}}} = (3V)^{\frac{2}{3}} \times \frac{(4\pi)^1}{(4\pi)^{\frac{2}{3}}}$$

$$= (3V)^{\frac{2}{3}}(4\pi)^{1-\frac{2}{3}} = (4\pi)^{\frac{1}{3}}(3V)^{\frac{2}{3}}$$

Question 550: A

Let each unit length be x.

Thus, $S = 6x^2$. Therefore, $x = \left(\frac{S}{6}\right)^{\frac{1}{2}}$

$V = x^3$. Thus, $V = [\left(\frac{S}{6}\right)^{\frac{1}{2}}]^3$ so $V = \left(\frac{S}{6}\right)^{\frac{3}{2}}$

Question 551: B

Multiplying the second equation by 2 we get $4x + 16y = 24$. Subtracting the first equation from this we get $13y = 17$, so $y = \frac{17}{13}$. Then solving for x we get $x = \frac{10}{13}$. You could also try substituting possible solutions one by one, although given that the equations are both linear and contain easy numbers, it is quicker to solve them algebraically.

Question 552: A

Multiply by the denominator to give: $(7x + 10) = (3y^2 + 2)(9x + 5)$

Partially expand brackets on right side: $(7x + 10) = 9x(3y^2 + 2) + 5(3y^2 + 2)$

Take x terms across to left side: $7x - 9x(3y^2 + 2) = 5(3y^2 + 2) - 10$

Take x outside the brackets: $x[7 - 9(3y^2 + 2)] = 5(3y^2 + 2) - 10$

Thus: $x = \frac{5(3y^2+2)-10}{7-9(3y^2+2)}$

Simplify to give: $x = \frac{(15y^2)}{(7-9(3y^2+2))}$

Question 553: F

$$3x\left(\frac{3x^7}{x^{\frac{1}{3}}}\right)^3 = 3x\left(\frac{3^3x^{21}}{x^{\frac{3}{3}}}\right)$$

$$= 3x\,\frac{27x^{21}}{x} = 81x^{21}$$

Question 554: D

$$2x[2^{\frac{7}{14}}x^{\frac{7}{14}}] = 2x[2^{\frac{1}{2}}x^{\frac{1}{2}}]$$

$$= 2x(\sqrt{2}\sqrt{x}) = 2\left[\sqrt{x}\sqrt{x}\right][\sqrt{2}\sqrt{x}]$$

$$= 2\sqrt{2x^3}$$

Question 555: A

$A = \pi r^2$, therefore $10\pi = \pi r^2$

Thus, $r = \sqrt{10}$

Therefore, the circumference is $2\pi\sqrt{10}$

Question 556: D

$3.4 = 12 + (3 + 4) = 19$

$19.5 = 95 + (19 + 5) = 119$

Question 557: D

$2.3 = \dfrac{2^3}{2} = 4$

$4.2 = \dfrac{4^2}{4} = 4$

Question 558: F

This is a tricky question that requires you to know how to 'complete the square':

$(x + 1.5)(x + 1.5) = x^2 + 3x + 2.25$

Thus, $(x + 1.5)^2 - 7.25 = x^2 + 3x - 5 = 0$

Therefore, $(x + 1.5)^2 = 7.25 = \dfrac{29}{4}$

Thus, $x + 1.5 = \sqrt{\dfrac{29}{4}}$

Thus $x = -\dfrac{3}{2} \pm \sqrt{\dfrac{29}{4}} = -\dfrac{3}{2} \pm \dfrac{\sqrt{29}}{2}$

Question 559: B

Whilst you definitely need to solve this graphically, it is necessary to complete the square for the first equation to allow you to draw it more easily:

$(x + 2)^2 = x^2 + 4x + 4$

Thus, $y = (x + 2)^2 + 10 = x^2 + 4x + 14$

This is now an easy curve to draw ($y = x^2$ that has moved 2 units left and 10 units up). The turning point of this quadratic is to the left and well above anything in x^3, so the only solution is the first intersection of the two curves in the upper right quadrant around (3.4, 39).

Question 560: C

By far the easiest way to solve this is to sketch them (don't waste time solving them algebraically). As soon as you've done this, it'll be very obvious that $y = 2$ and $y = 1 - x^2$ don't intersect, since the latter has its turning point at (0, 1) and zero points at $x = -1$ and 1. $y = x$ and $y = x^2$ intersect at the origin and (1, 1), and $y = 2$ runs through both.

Question 561: B

Notice that you're not required to get the actual values – just the number's magnitude. Thus, 897653 can be approximated to 900,000 and 0.009764 to 0.01. Therefore, 900,000 x 0.01 = 9,000

Question 562: C

Multiply through by 70: $7(7x + 3) + 10(3x + 1) = 14 \times 70$

Simplify: $49x + 21 + 30x + 10 = 980$

$79x + 31 = 980$

$x = \dfrac{949}{79}$

Question 563: A

Split the equilateral triangle into 2 right-angled triangles and apply Pythagoras' theorem:

$x^2 = \left(\frac{x}{2}\right)^2 + h^2$. Thus $h^2 = \frac{3}{4}x^2$

$h = \sqrt{\frac{3x^2}{4}} = \frac{\sqrt{3x^2}}{2}$

The area of a triangle $= \frac{1}{2}$ x base x height $= \frac{1}{2}x\frac{\sqrt{3x^2}}{2}$

Simplifying gives: $x\frac{\sqrt{3x^2}}{4} = x\frac{\sqrt{3}\sqrt{x^2}}{4} = \frac{x^2\sqrt{3}}{4}$

Question 564: A

This is a question testing your ability to spot 'the difference between two squares'.

Factorise to give: $3 - \frac{7x(5x-1)\,(5x+1)}{(7x)^2(5x+1)}$

Cancel out: $3 - \frac{(5x-1)}{7x}$

Question 565: C

The easiest way to do this is to 'complete the square':

$(x-5)^2 = x^2 - 10x + 25$

Thus, $(x-5)^2 - 125 = x^2 - 10x - 100 = 0$

Therefore, $(x-5)^2 = 125$

$x - 5 = \pm\sqrt{125} = \pm\sqrt{25}\sqrt{5} = \pm5\sqrt{5}$

$x = 5 \pm 5\sqrt{5}$

Question 566: B

Factorise by completing the square:

$x^2 - 4x + 7 = (x-2)^2 + 3$

Simplify: $(x-2)^2 = y^3 + 2 - 3$

$x - 2 = \pm\sqrt{y^3 - 1}$

$x = 2 \pm \sqrt{y^3 - 1}$

Question 567: D

Square both sides to give: $(3x+2)^2 = 7x^2 + 2x + y$

Thus: $y = (3x+2)^2 - 7x^2 - 2x = (9x^2 + 12x + 4) - 7x^2 - 2x$

$y = 2x^2 + 10x + 4$

Question 568: C

This is a fourth order polynomial, which you aren't expected to be able to factorise at GCSE. This is where looking at the options makes your life a lot easier. In all of them, opening the bracket on the right side involves making $(y \pm 1)^4$ on the left side, i.e. the answers are hinting that $(y \pm 1)^4$ is the solution to the fourth order polynomial.

Since there are negative terms in the equations (e.g. $-4y^3$), the solution has to be:

$(y-1)^4 = y^4 - 4y^3 + 6y^2 - 4y + 1$

Therefore, $(y-1)^4 + 1 = x^5 + 7$

Thus, $y - 1 = (x^5 + 6)^{\frac{1}{4}}$

$y = 1 + (x^5 + 6)^{1/4}$

Question 569: A

Let the width of the television be 4x and the height of the television be 3x.

Then by Pythagoras: $(4x)^2 + (3x)^2 = 50^2$

Simplify: $25x^2 = 2500$

Thus: $x = 10$. Therefore: the screen is 30 inches by 40 inches, i.e. the area is 1,200 inches^2.

Question 570: C

Square both sides to give: $1 + \frac{3}{x^2} = (y^5 + 1)^2$

Multiply out: $\frac{3}{x^2} = (y^{10} + 2y^5 + 1) - 1$

Thus: $x^2 = \frac{3}{y^{10} + 2y^5}$

Therefore: $x = \sqrt{\frac{3}{y^{10} + 2y^5}}$

Question 571: C

The easiest way is to double the first equation and triple the second to get:

$6x - 10y = 20$ *and* $6x + 6y = 39$.

Subtract the first from the second to give: $16y = 19$,

Therefore, $y = \frac{19}{16}$.

Substitute back into the first equation to give $x = \frac{85}{16}$.

Question 572: C

This is fairly straightforward; the first inequality is the easier one to work with: B and D and E violate it, so we just need to check A and C in the second inequality.

C: $1^3 - 2^2 < 3$, but A: $2^3 - 1^2 > 3$

Question 573: B

Whilst this can be done graphically, it's quicker to do algebraically (because the second equation is not as easy to sketch). Intersections occur where the curves have the same coordinates.

Thus: $x + 4 = 4x^2 + 5x + 5$

Simplify: $4x^2 + 4x + 1 = 0$

Factorise: $(2x + 1)(2x + 1) = 0$

Thus, the two graphs only intersect once at $x = -\frac{1}{2}$

Question 574: D

It's better to do this algebraically as the equations are easy to work with and you would need to sketch very accurately to get the answer. Intersections occur where the curves have the same coordinates. Thus: $x^3 = x$

$x^3 - x = 0$

Thus: $x(x^2 - 1) = 0$

Spot the 'difference between two squares': $x(x + 1)(x - 1) = 0$

Thus there are 3 intersections: at $x = 0, 1$ *and* -1

Question 575: E

Note that the line is the hypotenuse of a right angled triangle with one side unit length and one side of length ½. By Pythagoras,

$$\left(\tfrac{1}{2}\right)^2 + 1^2 = x^2$$

Thus, $x^2 = \tfrac{1}{4} + 1 = \tfrac{5}{4}$

$$x = \sqrt{\tfrac{5}{4}} = \tfrac{\sqrt{5}}{\sqrt{4}} = \tfrac{\sqrt{5}}{2}$$

Question 576: D

We can eliminate z from equation (1) and (2) by multiplying equation (1) by 3 and adding it to equation (2):

3x + 3y – 3z = -3	Equation (1) multiplied by 3
2x – 2y +3z = 8	Equation (2) then add both equations
5x + y = 5	We label this as equation (4)

Now we must eliminate the same variable z from another pair of equations by using equation (1) and (3):

2x + 2y – 2z = -2	Equation (1) multiplied by 2
2x – y + 2z = 9	Equation (3) then add both equations
4x + y = 7	We label this as equation (5)

We now use both equations (4) and (5) to obtain the value of x:

5x + y = 5	Equation (4)
- 4x - y = -7	Equation (5) multiplied by -1
x = -2	

Substitute x back in to calculate y:

4x + y = 7

4(-2) + y = 7

- 8 + y = 7

y = 15

Substitute x and y back in to calculate z:

x + y – z = -1

-2 + 15 – z = -1

13 – z = -1

-z = -14

z = 14. Thus: x = -2, y = 15, z = 14

Question 577: D

This is one of the easier maths questions. Take 3a as a factor to give:

$3a(a^2 – 10a + 25) = 3a(a – 5)(a – 5) = 3a(a – 5)^2$

Question 578: B

Note that 12 is the Lowest Common Multiple of 3 and 4. Thus:

-3 (4x + 3y) = -3 (48)	Multiply each side by -3
4 (3x + 2y) = 4 (34)	Multiply each side by 4
-12x – 9y = -144	
12x + 8y = 136	Add together
-y = -8	

$y = 8$

Substitute y back in:

$4x + 3y = 48$

$4x + 3(8) = 48$

$4x + 24 = 48$

$4x = 24$

$x = 6$

Question 579: E

Don't be fooled, this is an easy question, just obey BODMAS and don't skip steps.

$\frac{-(25-28)^2}{-36+14} = \frac{-(-3)^2}{-22}$

This gives: $\frac{-(9)}{-22} = \frac{9}{22}$

Question 580: E

Since there are 26 possible letters for each of the 3 letters in the license plate, and there are 10 possible numbers (0-9) for each of the 3 numbers in the same plate, then the number of license plates would be:

$(26) \times (26) \times (26) \times (10) \times (10) \times (10) = 17{,}576{,}000$

Question 581: B

Expand the brackets to give: $4x^2 - 12x + 9 = 0$.

Factorise: $(2x - 3)(2x - 3) = 0$.

Thus, only one solution exists, x = 1.5.

Note that you could also use the fact that the discriminant, $b^2 - 4ac = 0$ to get the answer.

Question 582: C

$= \left(x^{\frac{1}{2}}\right)^{\frac{1}{2}} (y^{-3})^{\frac{1}{2}}$

$= x^{\frac{1}{4}} y^{-\frac{3}{2}} = \frac{x^{\frac{1}{4}}}{y^{\frac{3}{2}}}$

Question 583: A

Let x, y, and z represent the rent for the 1-bedroom, 2-bedroom, and 3-bedroom flats, respectively. We can write 3 different equations: 1 for the rent, 1 for the repairs, and the last one for the statement that the 3-bedroom unit costs twice as much as the 1-bedroom unit.

(1) x + y + z = 1240

(2) 0.1x + 0.2y + 0.3z = 276

(3) z = 2x

Substitute z = 2x in both of the two other equations to eliminate z:

(4) x + y + 2x = 3x + y = 1240

(5) 0.1x + 0.2y + 0.3(2x) = 0.7x + 0.2y = 276

-2(3x + y) = -2(1240)   Multiply each side of (4) by -2

10(0.7x + 0.2y) = 10(276)   Multiply each side of (5) by 10

(6) -6x -2y = -2480   Add these 2 equations

(7) 7x + 2y = 2760

x = 280

z = 2(280) = 560   Because z = 2x

280 + y + 560 = 1240   Because x + y + z = 1240

y = 400

Thus the units rent for £ 280, £ 400, £ 560 per week respectively.

Question 584: C

Following BODMAS:

$$= 5 \left[5(6^2 - 5 \times 3) + 400^{\frac{1}{2}} \right]^{1/3} + 7$$

$$= 5 \left[5(36 - 15) + 20 \right]^{\frac{1}{3}} + 7$$

$$= 5 \left[5(21) + 20 \right]^{\frac{1}{3}} + 7$$

$$= 5 \left(105 + 20 \right)^{\frac{1}{3}} + 7$$

$$= 5 \left(125 \right)^{\frac{1}{3}} + 7$$

$$= 5 (5) + 7$$

$$= 25 + 7 = 32$$

Question 585: B

Consider a triangle formed by joining the centre to two adjacent vertices. Six similar triangles can be made around the centre — thus, the central angle is 60 degrees. Since the two lines forming the triangle are of equal length, we have 6 identical equilateral triangles in the hexagon.

Now split the triangle in half and apply Pythagoras' theorem:

$$1^2 = 0.5^2 + h^2$$

Thus, $h = \sqrt{\frac{3}{4}} = \frac{\sqrt{3}}{2}$

Thus, the area of the triangle is: $\frac{1}{2}bh = \frac{1}{2} \times 1 \times \frac{\sqrt{3}}{2} = \frac{\sqrt{3}}{4}$

Therefore, the area of the hexagon is: $\frac{\sqrt{3}}{4} \times 6 = \frac{3\sqrt{3}}{2}$

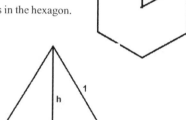

Question 586: B

Let x be the width and x+19 be the length.

Thus, the area of a rectangle is x(x + 19) = 780.

Therefore:

x² + 19x − 780 = 0

(x - 20)(x + 39) = 0

x − 20 = 0 or x + 39 = 0

x = 20 or x = -39

Since length can never be a negative number, we disregard x = -39 and use x = 20 instead.

Thus, the width is 20 metres and the length is 39 metres.

Question 587: B

The quickest way to solve is by trial and error, substituting the provided options. However, if you're keen to do this algebraically, you can do the following:

Start by setting up the equations: Perimeter = 2L + 2W = 34

Thus: L + W = 17

Using Pythagoras: $L^2 + W^2 = 13^2$

Since L + W = 17, W = 17 - L

Therefore: $L^2 + (17 - L)^2 = 169$

$L^2 + 289 - 34L + L^2 = 169$

$2L^2 - 34L + 120 = 0$

$L^2 - 17L + 60 = 0$

(L − 5) (L − 12) = 0

Thus: L = 5 and L = 12

And: W = 12 and W = 5

Question 588: C

Multiply both sides by 8: $4(3x - 5) + 2(x + 5) = 8(x + 1)$

Remove brackets: $12x - 20 + 2x + 10 = 8x + 8$

Simplify: $14x - 10 = 8x + 8$

Add 10: $14x = 8x + 18$

Subtract 8x: $6x = 18$

Therefore: $x = 3$

Question 589: C

Recognise that 1.742 x 3 is 5.226. Now, the original equation simplifies to: $= \frac{3 \times 10^6 + 3 \times 10^5}{10^{10}}$

$= 3 \times 10^{-4} + 3 \times 10^{-5} = 3.3 \times 10^{-4}$

Question 590: A

$Area = \frac{(2+\sqrt{2})(4-\sqrt{2})}{2}$

$= \frac{8 - 2\sqrt{2} + 4\sqrt{2} - 2}{2}$

$= \frac{6 + 2\sqrt{2}}{2}$

$= 3 + \sqrt{2}$

Question 591: C

Square both sides: $\frac{4}{x} + 9 = (y - 2)^2$

$\frac{4}{x} = (y - 2)^2 - 9$

Cross Multiply: $\frac{x}{4} = \frac{1}{(y-2)^2 - 9}$

$x = \frac{4}{y^2 - 4y + 4 - 9}$

Factorise: $x = \frac{4}{y^2 - 4y - 5}$

$x = \frac{4}{(y+1)(y-5)}$

Question 592: D

Set up the equation: $5x - 5 = 0.5(6x + 2)$

$10x - 10 = 6x + 2$

$4x = 12$

$x = 3$

Question 593: C

Round numbers appropriately: $\dfrac{55 + (\frac{9}{4})^2}{\sqrt{900}} = \dfrac{55 + \frac{81}{16}}{30}$

81 rounds to 80 to give: $\dfrac{55 + 5}{30} = \dfrac{60}{30} = 2$

Question 594: D

There are three outcomes from choosing the type of cheese in the crust. For each of the additional toppings to possibly add, there are 2 outcomes: 1 to include and another not to include a certain topping, for each of the 7 toppings

Thus, the number of different kinds of pizza is: $3 \times 2 \times 2 \times 2 \times 2 \times 2 \times 2 \times 2 = 3 \times 2^7$

$= 3 \times 128 = 384$

Question 595: A

Although it is possible to do this algebraically, by far the easiest way is via trial and error. The clue that you shouldn't attempt it algebraically is the fact that rearranging the first equation to make x or y the subject leaves you with a difficult equation to work with (e.g. $x = \sqrt{1 - y^2}$) when you try to substitute in the second.

An exceptionally good student might notice that the equations are symmetric in x and y, i.e. the solution is when x = y. Thus $2x^2 = 1$ and $2x = \sqrt{2}$ which gives $\dfrac{\sqrt{2}}{2}$ as the answer.

Question 596: C

If two shapes are congruent, then they are the same size and shape. Thus, congruent objects can be rotations and mirror images of each other. The two triangles in E are indeed congruent (SAS). Congruent objects must, by definition, have the same angles.

Question 597: B

Rearrange the equation: $x^2 + x - 6 \geq 0$

Factorise: $(x + 3)(x - 2) \geq 0$

Remember that this is a quadratic inequality so requires a quick sketch to ensure you don't make a silly mistake with which way the sign is.

Thus, $y = 0$ when $x = 2$ and $x = -3$. $y > 0$ when $x > 2$ or $x < -3$.

Thus, the solution is: $x \leq -3 \ and \ x \geq 2$.

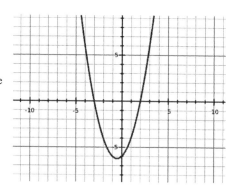

Question 598: B

Using Pythagoras: $a^2 + b^2 = x^2$

Since the triangle is isosceles: $a = b, \ so \ 2a^2 = x^2$

Area $= \dfrac{1}{2} base \ x \ height = \dfrac{1}{2}a^2$. From above, $a^2 = \dfrac{x^2}{2}$

Thus the area $= \dfrac{1}{2} x \dfrac{x^2}{2} = \dfrac{x^2}{4}$

Question 599: A

If X and Y are doubled, the value of Q increases by 4. Halving the value of A reduces this to 2. Finally, tripling the value of B reduces this to ⅔, i.e. the value decreases by ⅓.

Question 600: C

The quickest way to do this is to sketch the curves. This requires you to factorise both equations by completing the square:

$x^2 - 2x + 3 = (x-1)^2 + 2$

$x^2 - 6x - 10 = (x-3)^2 - 19$ Thus, the first equation has a turning point at $(1, 2)$ and doesn't cross the x-axis. The second equation has a turning point at $(3, -19)$ and crosses the x-axis twice.

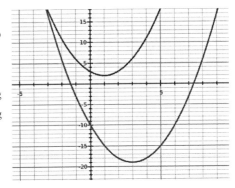

BMAT Worked Solutions

What are BMAT Past Papers?

Thousands of students take the BMAT exam each year. These exam papers are then released online to help future students prepare for the exam. Before 2013, these papers were not publically available meaning that students had to rely on the specimen papers and other resources for practice. However, since their release in 2013, BMAT past papers have become an invaluable resource in any student's preparation.

Where can I get BMAT Past Papers?

This book does not include BMAT past paper questions because it would be over 1,000 pages long if it did! However, all BMAT past papers since 2003 available for free from the official BMAT website. To save you the hassle of downloading lots of files, we've put them all into one easy-to-access folder for you at **www.uniadmissions.co.uk/bmat-past-papers**.

At the time of publication, the 2017 paper has not been released so this book only contains answers for 2003 – 2016. An updated version will be made available once the 2017 paper is released. The 2014 past paper worked solutions are also available at the link above.

How should I use BMAT Past Papers?

BMAT Past papers are one the best ways to prepare for the BMAT. Careful use of them can dramatically boost your scores in a short period of time. The way you use them will depend on your learning style and how much time you have until the exam date but here are some general pointers:

➢ 4-6 weeks of preparation is usually sufficient for most students.
➢ Students generally improve in section 2 more quickly than section 1 so if you have limited time, focus on section 2.
➢ The BMAT syllabus changed in 2009 so if you find seemingly strange questions in the earlier papers, ensure you check to see if the topic is still on the specification.
➢ Similarly, there is little point doing essays before 2009 as they are significantly different in style. We've included plans for them in this book for completeness in any case.

How should I use this book?

This book is designed to accelerate your learning from BMAT past papers. Avoid the urge to have this book open alongside a past paper you're seeing for the first time. The BMAT is difficult because of the intense time pressure it puts you under – the best way of replicating this is by doing past papers under strict exam conditions (no half measures!). Don't start out by doing past papers (see previous page) as this 'wastes' papers.

Once you've finished, take a break and then mark your answers. Then, review the questions that you got wrong followed by ones which you found tough/spent too much time on. This is the best way to learn and with practice, you should find yourself steadily improving. You should keep a track of your scores on the next page so you can track your progress.

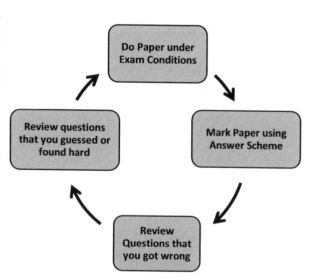

2003

Section 1

Question 1: E

There is a large increase in volume for a small increase in depth at either end, and a small increase in volume for a large increase in depth in the middle, which can only correspond to E.

Question 2: C

The argument is that ready meals should have health warnings, and the reason is because they are unhealthy. The fact that people are unaware strengthens the link between the evidence and the conclusion

Question 3: B

Sum all the rows and columns. 1 row and 1 column will be incorrect, and where these meet will be the incorrect value. Summing the year 8s gives 150, not 145 and summing the cars gives 107, not 102. Thus year 8s going by car is an incorrect value (33).

Question 4: C

The passage states that the new financial support for students must be repaid, and states that students from poorer families are more likely to be deterred by the prospect of debt. The link between these 2 can be inferred, suggesting the government's changes may deter poorer students more, said in C.

Question 5: 109

Cabbages $= n^2 + 9 = (n+1)^2 - 12$.
Rearrange, and the n^2 cancel to leave $2n = 20$
$n = 10$, cabbages $= 109$

Question 6: D

Here, the conclusion is within the middle of the argument: 'it may be undemocratic by favouring some political parties more than others.' This is paraphrased by D.

Question 7: A

A surgery session was on average 140 minutes, and each appointment is made 10 minutes apart (and the average length of a consultation is usually 10 minutes), so $140/10 = 14$ patients would be seen in an average session.

Question 8: 27

140×8.5 (surgery a week) $+ 408 = 1598$ (~ 1600)
$1600/60 = 80/3 = 26.6 = 27$ to the nearest hour

Question 9: C

10000000 people / 5000 doctors $= 2000$ patients per doctor.
155 seen a week, $155 \times 50 = 7750$ total visits a year
$7750/2000 = \sim 4$ visits per year

Question 10: D

We know the average length of a surgery session is 165 mins, and with less doctors, this will have to increase - we can thus rule out A, B and C. To work out the proportion - $165 \times 5 = 4.5 \times$ new length of surgery needed.
New length needed $= 183$ mins.

Question 11: A

We can probably get to A without having to do any calculations. We can rule out B - 1 second extra per consultation is unlikely to contribute most to the rise. The average time spent on home visits per week is less than the average length of a surgery session (of which there are several a week), hence the length of surgery sessions must contribute more. The average length of a home visit is irrelevant because we already have the time spent on home visits per week.

Question 12: A

The average numbers of patients seen per week are 135 and 155 respectively - shown by A.

Question 13: B

If x people are on the bus…

2x/3 people stay on at the first stop, 4x/9 at the second and 8x/27 at the third. 8x/27 = 8, thus x = 27.

Question 14: E

This is another causation/correlation BMAT question. The assumption is that excessive Internet use causes isolation and obesity, but no evidence of a causal link.

Question 15: C

Firstly, every team will play two matches against the other 5 teams in their pool. So team A must play B, C, D, E and F twice which is 10 matches, B must play C, D, E and F which is 8 matches (don't repeat A), etc. This makes a total of 10+8+6+4+2=30 in the pool, so a total of 60 with 2 pools. Also, each team plays each of the other teams in the other pool once, giving 6 x 6 = 36 extra matches (there are no repeats here). There is also the final. 60 + 36 + 1 = 97 matches.

Question 16: C

The line y=5 is only shown in A, B and C, and out of these y=2x only in A and C. Comparing A and C, only C shows when y is greater than 2x, whereas A shows y is less than 2x.

Question 17: B

Again, another causation/correlation question. The finding only gives a correlation, which is said by B. None of the other claims can be justified.

Question 18: 5

15 take the BMAT, 13 take Biology so 2 don't take Biology. The 2 that don't must be taking Chemistry, Physics and Maths. We also know that 3 people take Chemistry, Biology, Physics and Maths - 3+2=5. We can assume all 5 of these are boys because we are looking to maximise the number of boys.

Question 19: D

The actual time doesn't matter here. Julie's clock is one hour ahead of Clare, so she will arrive 1 hour before.

Question 20: B

The argument is that we should promote seeking advice at an earlier stage. However, if early consultation for minor symptoms incurs high costs in doctors' time then this weakens the argument.

Question 21: 7.2 seconds

The cheetah runs 50km/h faster than the Zebra and is 100m (0.1km) behind. Time = distance/speed, so 0.1/50 = 0.002 hours. To convert this into seconds, times by 3600, which gives 7.2.

Question 22: E

It had to be occupied in AD157 or there would not be any coins of this date - thus E is the only one that is definitely true.

Question 23: B

The conclusion is that the habitats of wading birds will inevitably decline, and this is due to peat. However, the assumption is that people will carry on using peat despite the alternatives. A, C, D and E are not necessarily assumed.

Question 24: C

The Fair-E coefficient is said to represent the degree of inequality in income distribution, and as this has increased, we can conclude that income inequality has not been reduced.

A is not correct as we know nothing about overall income of people.

B is not true - chart 1 suggests an increase in household income for low earners.

D is not true as we do not know about standard of living.

E is not true because we don't know anything about 'real wealth'.

Question 25: D

Chart 1 shows that for those with below average incomes, there has been a net gain. The only way this could happen with the Fair-E coefficient could have increased is if the pre-tax incomes of the rich have risen even more than the poor.

We cannot say changes are due to government fiscal policy (A), B is too general and claims too much, the use of 'better off' is too vague in C and E cannot be concluded as poverty is not mentioned.

Question 26: E

1 is correct - the income increase of the poor is less than it could have been as seen in the data.

2 could definitely be correct and we have already explored the idea that the incomes at the top end may have increased more than those at the bottom end.

3 could also increase the Fair-E coefficient while taxes and benefits change for the poor.

Question 27: C

The degree of inequality is measured by the Fair-E coefficient. Poland (0.2), Ruritania (0.35), USA (0.4) and Panama (0.6) is the order.

Question 28: B

Again, the problem is that the argument confuses cause with correlation. 2 factors may be correlated, but the causality can be either way around, or they may be linked by another factor. Thus B is the correct answer.

Question 29: E

The probability of it being yellow or blue is 70%, so 70/3% yellow and 140/3% blue. We can say there are 3 red, 7/3 yellow and 14/3 blue, but obviously there must be whole numbers. If we time them all by 3, there are 9 red, 7 yellow and 14 blue, totalling 30.

Question 30: C

1 is correct - 30% work, at least some of the time, in the commercial sector, which covers doctors working in both and the commercial sector alone. Thus, some doctors must work in the public service only.

2 is also correct - 70% work solely in the public sector, whereas 30% work in either both or just commercial.

3 cannot be deduced - we don't know how much time doctors spend at each service.

Question 31: A & C

Express each one as an inequality - $a \leq s$, $s < a$, $s \geq a$ and $a \geq s$. From this you can see A and C are equivalent. If this is not obvious to you, try considering whether Anne can be older than Susan, whether Anne and Susan can be the same age or whether Anne can be younger than Susan for each one.

Question 32: E

The conclusion is that the government must act quickly to plan for changes in holiday patterns, but the assumption here is that these changes will be more than short term.

Question 33: D

B>D and C>A to start with. D goes back 1 place so cannot start last, and B goes up 1 place so cannot start first. Thus, we can deduce that originally the order is CBDA. After the swap the order is BCAD.

Question 34: C

At the moment we have 10 parts water to 1 part concentrate, so we need to add concentrate, ruling out A and B. Test the other ones. If we add 10cm concentrate we now have 400cm water and 50cm concentrate, which is the correct 8:1 part dilution.

Question 35: C

Determine the net change for each year e.g. 800 lost for year 1. We can immediately rule out A B and D as they begin with an increase rather than a decrease. F doesn't decrease by 800 in the first year so is also wrong. We are left with C and E. E is not correct because there is an increase between years 2 and 3, so C must be correct.

Question 36: B

If 455AD is year 9, then year 72 is 455 + 63 = 518AD.

Question 37: C

We know that Gildas was born in the year of the battle of Badon so if we knew his birth date we could confirm the date of the battle.

Question 38: B

Year 93 would be 539AD in the Welsh Annals. But if this was too late by 28 years, then the battle would have been in 511AD.

Question 39: 506AD

Death of King Maelgwn was in 549AD. If Gildas was 43 at this date, the battle of Badon would be as late as possible: 549 - 43 = 506AD.

Question 40: C

The Welsh Annals give information suggesting Gildas wrote his book after the King's death, which is only shown by C.

END OF SECTION

Section 2

Question 1: B

➢   Starch is partially digested in the mouth by amylase.
➢   Proteins are digested in the stomach via proteases like pepsin.
➢   Fats are only digested in the small intestine.

Question 2: E

Mass of 8×10^6 uranium atoms $= 4 \times 10^{-25} \times 8 \times 10^6$
$32 \times 10^{-19} kg = 32 \times 10^{-13} mg$
$= 3.2 \times 10^{-12} mg$

Question 3: 2, 9, 6

The quickest way to do these type of questions is via algebraic equations. If you're unfamiliar with this approach then see the chemistry chapter in *The Ultimate BMAT Guide*.
For Carbon: 3a = 6. Hence, **a =2**
For Hydrogen: 6a = 2c; 12 = 2c. Hence, **c = 6**
For Oxygen: 2b = 12 + c; 2b = 12 + 6. Hence, **b = 9**

Question 4: B

Since the pivot is located in the middle of the bar, we can ignore the bar's weight. For the bar to balance:
$Moments\ clockwise\ =\ Moments\ anti-clockwise$
$500 \times 0.2\ =\ 0.4 \times 200\ +\ 200x$
$100\ =\ 80\ +\ 200x$
$x = \frac{20}{200} = 0.1m$

Question 5: D

Don't get confused – this is easy! At pH 5, methyl orange would be yellow, bromothymol blue would be yellow and phenolphthalein would be colourless. Thus, the solution would be yellow.

Question 6: 76.8 kJ

$Weight\ =\ mass \times g$
$Thus, the\ horse's\ mass\ = \frac{6000}{10}\ =\ 600\ kg$
$E_k\ = \frac{1}{2}\ mv^2\ = \frac{1}{2} \times 600 \times 16^2$
$=\ 300 \times 256$
$=\ 76,800\ J\ =\ 76.8\ kJ$

Question 7: B

This is no longer on the specification. Oestrogen reaches its highest level in the days prior to ovulation. Hence phase B is the correct answer.

Question 8: A, C, B and D

Don't make this more complex than it needs to be by using algebra! Resistors in parallel have a lower overall resistance than the same resistors in series. Thus A has the lowest resistance and D has the highest resistance. B only has one resistor in each branch but C has two resistors in the top branch. Thus, C has a higher resistance than B.

Question 9: B

Number of moles of $H_2 = \dfrac{9}{2} = 4.5$ moles

Number of moles of $N_2 = \dfrac{56}{28} = 2$ moles

Since 3 moles of H_2 react with 1 mole N_2, nitrogen gas is in excess.

The molar ratio between H_2 and NH_3 is 3:2 so:

4.5 moles of H_2 react with 3 moles of NH_3.

The mass of 3 moles of $NH_3 = 3 \times [14 + 3]$

$= 51g$

Question 10: B

This is just a simple exponential curve and thus is best shown by graph B. If you were unsure, you could substitute in some values and see what x is.

For example:

➤ When x = 0, y = 1 (ruling out A and C),

➤ When x = 1, y = 2 (ruling out D, as x and y increase together).

Question 11: A

Don't get confused, wavelength can change during refraction but the frequency of waves always stays the same for these phenomena.

Question 12: A

Succinic acid loses two hydrogen atoms to become fumaric acid. This is an example of an oxidation reaction.

Question 13: D

Rearrange the equation: $x^2 + 2x - 8 \geq 0$

Factorise: $(x - 2)(x + 4) \geq 0$

Remember that this is a quadratic inequality so requires a quick sketch to ensure you don't make a silly mistake with which way the sign is.

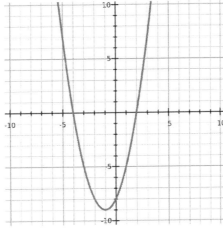

Thus, $y = 0$ when $x = 2$ and $x = -4$

$y > 0$ when $x > 2$ or $x < -4$

Thus, the solution is: $x \leq -4 \ and \ x \geq 2$

This is the same as $x^2 + 2x - 8 \geq 0$ which we can solve with an equals sign to get x = 2 and x = -4. To satisfy the original inequality we must have x ≥ 2 or x ≤ -4 (sub in numbers if unsure).

Question 14: 3, 4 and 5; E

Since 4 and 5 don't have the condition but 7 does, the disease is inherited in an autosomal recessive fashion. Thus, 4 and 5 must be heterozygotes. As 7 is affected, 3 must also be a heterozygote carrier.

Since 3 and 4 are heterozygotes, they have a 25% chance of having a child who has the disease or a 12.5% chance of having a girl with the disease.

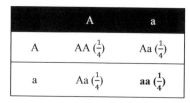

	A	a
A	AA $(\frac{1}{4})$	Aa $(\frac{1}{4})$
a	Aa $(\frac{1}{4})$	**aa** $(\frac{1}{4})$

$Probability\ (Girl\ with\ disease) = P(Girl)x\ P(Disease) = \frac{1}{2}x\frac{1}{4} = \frac{1}{8}$

Question 15: 4 minutes

The initial count rate was 120 (taking background into account) and ended with 15 counts after 12 minutes. There were 3 half-lives in this time (120 -> 60 -> 30 -> 15) so each has a time of 4 minutes.

Question 16: C

Reading off the graph, 136g of potassium chlorate dissolves in 200g of water at 70°C. Since only 80g was added, the full amount will be dissolved at 70°C.

The solution will become saturated once only 40g of potassium chlorate can be dissolved in every 100g of water. This takes place at 46°C.

The solubility at 20°C is 27g per 100g of water. Thus, as the solutions cools from 46°C to 20°C, 40 − 27 =13g of potassium chlorate will crystallise per 100g of water.

Since there is 200 g of water, a total of 13 x 2 = 26g of crystals will form.

Question 17: E

Some of the equations required for this question are now outside the BMAT Syllabus.

A. $Energy\ =\ Charge\ x\ Voltage. Thus, V\ =\ Joules/Coulombs$

B. $From\ Ohm's\ Law: V\ =\ IR$

C. $Since\ P\ =\ IV, V\ =\ \frac{P}{I}$

D. $Since\ P\ =\ IV\ and\ V\ =\ IR, P\ =\ \frac{V^2}{R}. Therefore, V\ =\ \sqrt{PR}$

E. It's not possible to manipulate equations to fit the given units.

Question 18: E, B, C, A and D

ADH travels through the vena cava to the right side of the heart, which pumps the blood through the pulmonary artery and to the lungs. The blood then returns via the pulmonary vein to left side of the heart. Blood then leaves through the aorta and travels to the kidney via the renal artery.

Question 19: C

Frequency $= \frac{5}{10^{-13}}\ =\ 5\ x\ 10^{13} Hz$

$Energy\ =\ f\ x\ h\ =\ 5\ x\ 10^{13}\ x\ 6.63\ x\ 10^{-34}$

$=\ 33.15\ x\ 10^{-21}$

$=\ 3.32\ x\ 10^{-20} J$

Question 20: C

The atom has 4 electrons in the outer shell so is in group 4. It has 2 electron shells and hence is in period 2.

Question 21: E

Simplify this by working from right to left, looking at each thing you pass and the direction it operates in. You need to know where the biceps and triceps attach to the arm, but this is GCSE level science knowledge. For the upward action, on the far right you have the fulcrum of the elbow joint, then working leftwards you have the attachment of the biceps which pulls upwards then much further on you reach the load in the hands acting downwards. This is analogous to diagram 3. Then working for the downward action, you firstly have the upwards pull of triceps, then the fulcrum and then finally the upwards effort of the load, hence diagram 5 is correct.

Question 22: E

Increasing the speed of coil rotation increases the frequency and amplitude of the A.C. as well as the output voltage.

Question 23: E

The alkali metals form ionic bonds by donating an electron. This electron is lost more easily if it is further away from the positively charged nucleus. Whilst A, B, C and D may be true, they don't explain the increase in reactivity.

Question 24: C

During expiration, the ribs move down and inwards; the diaphragm relaxes and becomes less concave (= more convex). This causes intra-thoracic volume to decrease and intra-thoracic pressure to increase.

Question 25: B

$$\left[\frac{2x^{\frac{3}{2}}y^3}{\sqrt{z}}\right]^2$$

$$=\left[\frac{2x^{\frac{3}{2}}y^3}{z^{\frac{1}{2}}}\right]^2$$

$$=\frac{4x^3y^6}{z}$$

Question 26: C

In Y, the blood oxygen concentration significantly increases after passing the organ. This means that organ Y is the lungs. This narrows the options to A or C

In organ X, the salt, glucose and urea concentration in blood has decreased. We would not expect to see these changes in the heart – but in the kidney as a result of ultra-filtration.

Question 27: A

Drag will always be a positive value (you can't get negative air resistance!)

When the parachutist falls from the aircraft, drag will increase until they reach terminal velocity. When the parachute opens, drag will increase sharply to reduce the parachutist's velocity. As their velocity decreases, drag will decrease until the new terminal velocity is reached.

The air resistance at terminal velocity will be the same. Thus A is the correct answer.

Question 28: C

Since the compounds are all electrically neutral, the charges must balance:

A. $(+2) + (-3) + (-1) = -2$
B. $(+6) + (-6) + (-1) = -1$
C. $(+10) + (-9) + (-1) = 0$
D. $(+14) + (-10) + (-1) = +3$

Compound C is the only option which is electrically balanced.

Question 29: A

This question requires the use of the sine rule and is therefore no longer on the BMAT specification.

Using the sine rule:

$$\frac{PR}{Sin\,45} = \frac{\sqrt{6}}{Sin\,60}$$

$$PR = \frac{\sqrt{6}\; x\; Sin\,45}{Sin\,60}$$

$$Sin\,45 = \frac{1}{\sqrt{2}}, Sin\,60 = \frac{\sqrt{3}}{2}$$

$$PR = \frac{\frac{\sqrt{6}}{\sqrt{2}}}{\frac{\sqrt{3}}{2}}$$

$$PR = \frac{2\sqrt{3}}{\sqrt{3}} = 2$$

END OF SECTION

Section 3

Explain what you think the author means by the term 'ethical market'.

➤ The article gives the following definition for an ethical market: "The market would be confined to a self governing geopolitical area—for example, the UK or Australia. Vendors could sell into the system, from which their family members would stand a chance of benefiting. Only citizens from that area could sell and receive organs. There would be only one purchaser, an agency like the National Health Service (NHS) or Medicare, which would buy all organs and distribute according to some fair conception of medical priority. There would be no direct sales or purchases, no exploitation of low income countries and their populations."

➤ Consider the difference of this kind of market operating in a private and public healthcare setting.

➤ If the sale of organs became commonplace, a black market may rise. We do not know how each individual will respond to the removal of the organ and sale of organs may lead to manipulation of certain people in society.

➤ The author does make a good argument for the legalisation of organs - there is a huge transplant problem and this would provide motivation for more to donate. Furthermore, there is the consideration of patient autonomy, and the money obtained may be worth more than the lost organ.

➤ Another relevant point that you may know about is that organ donation levels increase significantly with 'opt out' system rather than an 'opt in' system, which all countries could be encouraged to adopt to deal with the shortage of supply.

A little learning is a dangerous thing (Alexander Pope)

➤ Alexander Pope suggests that learning is dangerous when it is incomplete. He implies that it is safer to know nothing, than to know a little and form the misconception that you are an expert on the subject.

➤ Suggest some examples of when a little learning may be a dangerous thing. Examples include a first year medical student wanting to perform an operation after reading about it; a pilot wanting to take-off after doing a flight simulation; self-diagnosis on the internet after reading a little about an illness; etc. Knowing nothing about a particular subject means somebody is unlikely to talk/act on this subject, whereas knowing a little may give somebody the misguided confidence to believe they are an expert.

➤ However, it is unrealistic to expect everyone to either know everything or nothing about a subject and sometimes it is still useful to know a little about a subject. Examples: basic first aid or CPR knowledge can save a life; even if you do not understand the exact mechanisms of what is going on: knowing the signs of stroke (even if you don't know what it actually is); understanding the rudimentary symptoms of someone with a disease or mental illness so you can assist them and understand their behaviour; etc

➤ A conclusion somewhere in the middle of these two extreme arguments is probably wise: whether a little learning is dangerous or not depends on the particular scenario. Usually it is useful and safe when applied to straightforward day-to-day tasks. However, in order to ensure it is not dangerous in, for example, medical cases, one must acknowledge their own limitations and ensure they know that they are not experts on a subject despite some low-level learning.

It is ridiculous to treat the living body as a mechanism.

➢ The statement suggests that it is a huge over-generalisation to regard the body as purely a succession of mechanisms.

➢ The 11 physiological systems (can you name them?!) within the body can often be considered mechanistically for doctors. The systems operate in a largely similar way for everyone and thus students can learn about each system in a general manner, and thus the pathology and pharmacology associated with each. (Give a more specific example)

➢ However, people are all individual with distinct features and personalities and we must be treated as such. The human experience of the same disease by different people is always unique, and the medical treatment may also have to vary between individuals. Some phenomena such as cognition and emotion cannot yet be explained mechanistically.

➢ We should not disregard considering the body mechanistically when used in a rigorous scientific context but we should also not only look at the body in this way. Mechanisms are interlinked in networks and the word 'mechanism' suggests a purely objective approach, whereas subjectivity is also key when considering anything relating to the human body, especially for doctors working holistically.

➢ Ultimately, the body is a complex, coordinated collection of a vast number of inter-related mechanisms, some of which we do not yet understand and varying between people, instead of being one ordinary, single mechanism that can be generalised for all.

Our belief in any particular natural law cannot have a safer basis than our unsuccessful critical attempts to refute it.

➢ Popper argues that the best way to support a theory is to refute false hypotheses, and that this offers more substantiation than directly finding positive evidence supporting a theory. Indeed, if we cannot find evidence disproving a theory, this suggests that it is more likely to be true.

➢ However, scientists do often aim to prove a law by attempting to provide evidence for it, as opposed to evidence against it, and this arguably a better method. Falsification alone does not necessarily identify the truthfulness of a proposition. (If you are not sure about this, research the notion of a scientific idea being both 'necessary' AND 'sufficient' to be considered correct.)

➢ Give an example - any scientific experiment where the result was determined directly, rather than through falsification.

➢ The key point here is probably that it is vital for a hypothesis to be testable so that we can develop evidence to whether it is correct, whether this occurs through lack of falsification or otherwise.

END OF PAPER

2004

Section 1

Question 1: C

A can be made by halving diagonally and then halving each of these, as can B. D can be made by splitting the 2 triangles on the end, and making the quadrilateral in the middle with 2 equivalent triangles. E can be split into the triangle at the top and the quadrilateral at the bottom, each made with 2 triangles. If in doubt, try making what you think the right-angled triangle may be and see if that can be used to make a table. The ends of C (the 2 triangles) cannot be used to make the middle quadrilateral.

Question 2: C

The conclusion of the argument is that by taxing something, which is dangerous to health, we can improve the health of the population. An example of this working is given by C. A, D and E are unrelated to this conclusion. B may be true but would not strengthen this particular argument.

Question 3: B

We are looking for the smallest number of screws. So to minimise our final answer, we can say that all 40 slot-headed screws are 3mm diameter, making 30 cross-headed screws 3mm diameter. Of these 30 cross-headed screws, to minimise our answer, we can assume 20 of them are NOT 50mm long (i.e. 35 or 20mm long), leaving 10 that are cross-headed, 3mm diameter and 50mm long.

Question 4: A

The flaw of this argument is a BMAT favourite - the fact that a correlation between 2 factors doesn't necessarily mean causation between them. They may be linked by another factor, or the causality may be in the opposite direction to that stated. Thus the assumption made in the conclusion, that whole class teaching being more beneficial, cannot necessarily be assumed without more information. B is irrelevant, C is not necessarily assumed and D and E may not be true.

Question 5: C

Tom takes 15 minutes less than Suki as he leaves 10 minutes later and arrives 5 minutes earlier, which is 0.25 hours. We know Tom's speed is 4v, so we can set up simultaneous equations using distance = speed x time.

$2 = 4v \times (t-0.25)$

$2 = vt$

Solve these simultaneous equations to get t = 1/3 hours. They meet when t=1/3 hours and v = 6km/h, and the distance at this time is 1km.

Question 6: E

The argument suggests that 'everyone who can own a pet, should do so' and this is because these people live longer. However, this assumes that everyone should try and live longer, which is said in E.

A is not necessarily assumed - this idea is a suggestion, and saying people with allergies CAN NEVER have pets is too strong.

B is not correct as the notion of multiple pets is not mentioned.
C is not necessarily assumed - 'the emotional benefits of affectionate relationships' isn't linked to particular animals.
D is not necessarily an assumption.

Question 7: D

They share the profits equally, so £390 each for the work, so Bill owes Alf £390 here. Alf paid £240 for the materials, so Bill owes him half of this, so Bill owes Alf £120 for this. £390 + £120 is £510.

Question 8: 12.0, 12.5 or 13.0

Add up the episodes of care for each department as shown in the graph. (Roughly) 1.2 + 0.1 + 1.4 + 0.3 + 0.8 +1.4 +0.8 +1.2 + 0.8 + 5 = 13.

Question 9: A

There are around 4.34 million days in hospital for cancer, and 1.4 million episodes. 4.34/1.4 is closest to 3.1 days. If this is not immediately obvious, then try 1.4 million x 3 days which is 4.2 million days, close to the total number of days. 1.4 million x 4 days is 5.6 million days, which is not close.

Question 10: A

~1.4 million cases of cancer, 1/7 are lung cancer which is roughly 200,000 cases of lung cancer overall. Thus, the cases for men must be lower than this - A is the only feasible answer. 120,000 is indeed 50% more than 80,000.

Question 11: A & C

The average stay is half that for cancer, and that may indeed be because some cancer treatment requires a short stay (A), or because some circulatory treatment requires long stays (C). B is not relevant, because the number of episodes is weighted relative to the total number of days. D is no relevant, as death is not a factor in the average stay calculation.

Question 12: B & C

B can be correct - 6 correct answers and 6 no answers. C can also be correct - 7 correct answers, 3 incorrect answers and 2 no answers.

A cannot be correct - 3 correct answers would give 15 points, and then no more can be gained to make 18.

D cannot be correct - 8 correct answers gives 24 points, but only 4 questions are left and 6 points must be lot. Only a maximum of 1 can be lost for each go. E is incorrect for an analogous reason.

Question 13: B

The final sentence, 'without this protection the seas would freeze solid, from the bottom up; and life as we know it, which began in water, would not exist,' argues that we must have this unusual property of water to survive, thus it is a necessary condition for life. However, the passage does not suggest this structure alone allows us to survive, so the unusual property is not a sufficient condition for life.

Question 14: B

Visualisation. B cannot be correct because if the shape is folded in this way, the 2 dots for the '2' face would not be at different vertices.

Question 15: D

The conclusion is the final sentence, which his paraphrased by D (the animal Kingdom involves both animals and humans of course). The conclusion is not about the relative importance of animal and human rights (A), the argument suggests experiments are sometimes beneficial (B and C) and non-medical research is not part of the argument (E).

Question 16: A

David is given as the heaviest in all of them, so we can ignore comparisons with him. We know that Colin has lost more weight than the other 2. However, he started lighter than Barbara and lost more weight so he could not have finished heavier than her - thus A must be incorrect.

Question 17: A

The conclusion is at the start, and then the argument gives the evidence to support this. This is all effectively paraphrased by A, which is the best answer.

B is part of the evidence rather than the conclusion, C is not said in the passage, D is not as good as A in summarising the main conclusion and E is again evidence rather than the conclusion.

Question 18: C

$0.1 \times 0.1 \times 0.1$ mm cube: volume of 0.001 mm^3 and SA of 0.06 mm^2.

Big cube: volume of 1000 mm^3 and SA of 600 mm^2

Number of little cubes $= 1000/0.001 = 10^6$ cubes.

SA of all the little cubes: $10^6 \times 0.06 = 6 \times 10^4$ mm^2

SA of all the little cubes - big cube $= 60000 - 600 = 59400$ mm$^2 = 594$ cm^2

Question 19: D

D is the only one that can be drawn from the passage without being inferred - associated is not too strong a word and the passage clearly suggests than lifelong exercise and good health are linked. All the others can only be inferred with assumptions.

Question 20: D

Max recommended daily intake is 6 grams per day. We know 15% of 567 men have had 6 grams or under. $567 \times 0.15 = \sim85$.

Question 21: 7.6

$66-31 = 35\%$ have between 6 and 9 grams. Assuming linearity, 19/35 have between 6 and the median. $3 \times 19/35$ is roughly 1.6, and $6 + 1.6 = 7.6$.

Question 22: A

To work out the estimated salt intake in intervals (e.g. 9-12 and 12-15) look at the value for the larger number in the interval and take away the smaller number in the interval (e.g. 15 and under - 12 and under to work out 12-15). Thus, A is the correct graph.

Question 23: B

If this calculation were done the assumption is that the values in the survey are evenly distributed across the age range, which is not necessarily true. The other answers would not cause bias.

Question 24: 10

18 patients taking 5 mins each = 90 mins. 1 patient takes 7 minutes longer, so 97 minutes. Emergency call of 8 mins makes 105, and emergency surgery of 5 mins makes 110. 2 hours is 120 minutes, so 120-110 = 10 minutes late.

Question 25: B

The conclusion is that the habitats of wading birds will inevitably decline, and this is due to peat. However, the assumption is that people will carry on using peat despite the alternatives. A, C, D and E are not necessarily assumed.

Question 26: A

There is no straightforward mathematical method to use here, except a problem solving method. Try listing the 'off' times for the 2 lights.
Lighthouse 1: 3, 14, 25, 36, **47**, 58, 69,...
Lighthouse 2: 2, 11, 20, 29, 38, **47**, ...
Since the lights were in sync 15 seconds ago, they will next appear together in $47 - 15 = 32$ seconds

Question 27: B

Doubling the height and diameter will increase the SA fourfold and the volume eightfold. The mass is therefore 800 x 4 = 3.2kg.
The water in the first container has a mass of 15.6 - 0.8kg = 14.8kg. The second has a volume 8 times of the first - 14.8 x 8 = 118.4 kg. Add the mass of the water - 3.2 + 118.4 = 121.6kg

Question 28: C

1 is correct - 30% work, at least some of the time, in the commercial sector, which covers doctors working in both and the commercial sector alone. Thus, some doctors must work in the public service only.
2 is also correct - 70% work solely in the public sector, whereas 30% work in either both or just commercial.
3 cannot be deduced - we don't know how much time doctors spend at each service.

Question 29: E

30g/L x 0.2 L + 20g/L x 0.05 L = concentration of solution x 0.25 L
7 = concentration of solution x 0.25 L. Thus, conc = 28g/L

Question 30: B

Use distance = speed x time at different points. We can rule out A and D immediately because it is obvious from the first times on the flat that the cyclist will be at the lead at the beginning. After 1km the runner will have been travelling for 1/6 hours, which is 10 mins, and the cyclist for 1/30 hours, which is 2 mins, so the cyclist is 8 mins ahead. For the next 1.5km the runner will have been travelling for 3/8 hours, which is 22.5 mins, and the cyclist for 30 mins, so the runner makes up 7.5 mins here and is now 0.5 mins ahead at 2.5km, as seen in B only.

Question 31: A & C

Express each one as an inequality - $a \leq s$, $s < a$, $s \geq a$ and $a \geq s$. From this you can see A and C are equivalent. If this is not obvious to you, try considering whether Anne can be older than Susan, whether Anne and Susan can be the same age or whether Anne can be younger than Susan for each one.

Question 32: 14%

The last sentence tells us that collision deaths fell to 4200 from 4900 in this period. Thus the percentage decrease is 700/4900 x 100, which is 1/7. Hopefully you know this is near to 14%.

Question 33: B

The sentence 'by 1992 sprains and strains had risen to 83%, and all other injuries had fallen to 40 per cent' only makes sense if some of the claims for the former came hand in hand with claims for the latter. The others are not sufficient explanations.

Question 34: 22

Between 1980 and 1993 the claims per hundred rose by 33% to 29.3, so to find the BI claims in 1980 we must divide 29.3 by 1.33. 1.33 can be treated as 4/3 here, so we can change our sum to 29.3 x 3/4, which is 21.9, 22 to the nearest whole number.

Question 35: B and C

A doesn't offer an explanation at all, so is not correct. B identifies a possible reason for why more have made BI claims and C suggests a difference if people do not claim when there is no visible damage.

END OF SECTION

Section 2

Question 1: A

Remember that hearts are drawn as if we were looking at the patient, so the left hand side of the paper is the right hand side of the heart. Blood enters from the pulmonary vein (2) to the left aorta (3) to the left ventricle (4) and is ejected through the aorta (1). It then renters through the vena cava (7), to the right aorta (6) to the right ventricle (5) to the pulmonary artery (8). Thus, the correct sequence is given by option A.

Question 2: B

$Area = \frac{1}{2} \ x \ base \ x \ height = \frac{1}{2} x \left(2 - \sqrt{2}\right)(4 + \sqrt{2})$

$= \frac{8 + 2\sqrt{2} - 4\sqrt{2} - 2}{2}$

$= \frac{6 - 2\sqrt{2}}{2} = 3 - \sqrt{2}$

Question 3: 3, 12, 3, 6

The quickest way to do these type of questions is via algebraic equations. If you're unfamiliar with this approach then see the chemistry chapter in *The Ultimate BMAT Guide*.

For Copper: $q = s$

For Hydrogen: $r = 12$

For Oxygen: $3r = 6s + 6 + 2t$.

Since r = 12, this simplifies to: $\mathbf{15 = 3s + t}$

For Nitrogen: $r = 2s + t$.

Since $r = 12$, this simplifies to: $\mathbf{2s + t = 12}$

Solve simultaneously: $t = 15 - 3s$

Therefore $2s + 15 - 3s = 12$

Thus, $s = q = 3$

Substituting back into: $t = 15 - 3(3) = 6$

Question 4: E

Remember that $Moment = force \ x \ perpendicular \ distance \ to \ pivot$

At all points the product of force and perpendicular distance to the pivot is equal.

$Moment \ exerted = 60N \ x \ (0.16 + 0.04) = 12 \ Nm$

Therefore the moment at the piston is $12 \ Nm$

Thus the force applied to the piston $= \frac{12}{0.04} = 300N$

Question 5: 0.32A

$Charge = Current \ x \ Time$

$2 \ x \ 10^{18}$ Ions move towards the electrodes per second so:

$Current = 2 \ x \ 1.6 \ x \ 10^{-19} \ x \ 10^{18} = 0.32 \ A$

Question 6: E

This is very straightforward. During expiration, the intercostal muscles and the diaphragm relax. This results in a decrease in the volume of the thorax, which increases the Intrathoracic pressure. This causes air to move outside via the mouth.

Question 7: E, B, A, C

i)   A giant molecular structure would not conduct electricity and would have high melting and boiling points, hence E.

ii)   Metals conduct electricity when solid and molten, hence B.

iii) Ionic compounds only conduct electricity when molten, hence A.

iv) The melting point of C is -20 and the boiling point is 58 so it is a liquid in-between these temperatures.

Question 8: E

Round 79.31 → 80% to make your life easier! Thus, 20% of mass is due to oxygen.

$Moles\ of\ W = \frac{80}{184} \approx 0.4$

$Moles\ of\ O: \frac{20}{16} \approx 1.25$

Thus, the molar ratio is $0.4 \approx 1.25$

The closest ratio to this is 1:3 giving a formula of WO_3.

Question 9: 7.2m

The question heavily implies that you can assume that:

Kinetic Energy is fully transformed to Gravitational Potential Energy.

Therefore, $E_p = E_k$

This gives: $\frac{mv^2}{2} = mg\Delta h$

Therefore, $v^2 = 2g\Delta h$

$\Delta h = \frac{v^2}{2g} = \frac{144}{20} = 7.2m$

Question 10: A and B

This is no longer on the BMAT specification. It's essential that you can recall the trigonometric graphs for $\sin\theta, \cos\theta\ and\ \tan\theta$.

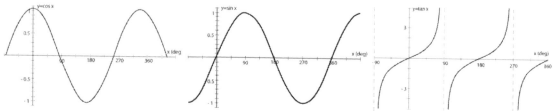

A. $\alpha + \theta = 180$. Alpha + theta = 180 so A is true.

B. Beta = 360 - theta so B is true.

C. Gamma and theta are the same as they are alternate angles, so C is false.

D. Sin 180 = 0 so D is false.

Question 11: B

Here, a ketone group (R-C=O) has been converted into an alcohol group (R-C-OH). The addition of hydrogen makes this a reduction reaction.

Question 12: D

The male carries a recessive allele on his X chromosome so this could only be passed on to his daughters - 4 and 5. If 4 had the allele, it could also be passed to either of her children - 8 or 9.

Question 13: D

Rearrange the equation: $x^2 + 2x - 8 \geq 0$

Factorise: $(x - 2)(x + 4) \geq 0$

Remember that this is a quadratic inequality so requires a quick sketch to ensure you don't make a silly mistake with which way the sign is.

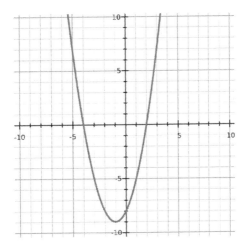

Thus, $y = 0$ when $x = 2$ and $x = -4$

$y > 0$ when $x > 2$ or $x < -4$

Thus, the solution is: $x \leq -4$ and $x \geq 2$

Question 14: D

Smaller molecules diffuse faster along the tube. The white ring is closer to the concentrated HCl so the ammonia molecule has diffused further, and thus the mass of ammonia molecules is less than the mass of HCl molecules.

Question 15: False, false, true, false, false

Closing S1 keeps the same voltage across the bulb and voltmeter which are all in parallel, and because of Ohms Law, I is therefore same too. Hence a and b are false.

Closing S2 will increase the current across A2 due to the decrease in total resistance as an alternative path is provided, but the V and hence I across A3 is the same as nothing is added in series, therefore the entire voltage is dropped across the bulb.

Closing S3 will short out the circuit therefore negligible current will flow through the ammeters and this is a huge decrease in reading to as near as makes no difference zero.

Question 16: B

Let P represent a health problem and N represent normal.

There are three permutations to choose one P and 2 x N: PNN, NPN, NNP

The overall probability $= 3 \times \frac{1}{4} \times \frac{3}{4} \times \frac{3}{4} = \frac{27}{64}$

Question 17: D

Since $g = 10ms^{-2}$, the parachutist's weight $= 90 \times 10 = 900N$

Since the parachutist is travelling at terminal velocity:

$Air\ resistance = Weight = 900N$

Question 18: D

In a hydrocarbon, a single hydrogen atom is replaced by NH_2. Thus, an amine has one more hydrogen atom and one more nitrogen atom than a hydrocarbon. Thus, D is correct.

Question 19: A, D and E

Remember that all gametes are haploid and have half the number of chromosomes of a normal body cell. Thus, C and F are abnormal.

Ova can only contain an X chromosome → B is abnormal.

Question 20: C

If you get stuck with these type of questions – write down every relevant equation e.g. $P = IV, V = IR, Q = It$ etc. This will make it easier to rearrange equations to get the correct units.

A. $P = IV$ and since $I = \frac{Q}{t}$, $P = \frac{Q}{t} \times V$

B. From Ohm's Law, $V = IR, P = IV = I \times IR = I^2 R$

C. There is no way to get these units to equal a Watt.

D. $P = IV$

E. From Ohm's Law, $I = \frac{V}{R}, P = IV = \frac{V}{R} \times V = \frac{V^2}{R}$

Question 21: D

Unlike the other vessels, the hepatic portal vein carries nutrient rich blood from one organ to another (gut → liver). Contrastingly, the others only carry blood to/from one organ.

Question 22: F

In the first decay, the atomic number increases by 1 and the mass number stays the same. This is an example of Beta decay – when a neutron transforms into an electron and proton.

In the second decay, the atomic number decreases by 2. This is an example of alpha decay- when 2 protons and 2 neutrons are emitted. Thus, X decreases by 4: $A - 4$.

Question 23: E

Oxygen displaces water and carbon monoxide displaces oxygen, so the iron-water bond is the weakest, and iron-carbon monoxide bond is the strongest. Iron-Oxygen bond is between these two.

Question 24: C

This is no longer on the BMAT Syllabus however you might be able to deduce it based on the shape of legs.

The right knee joint has extended. This is brought upon by contraction of R and relaxation of the antagonistic P.

The right ankle joint has also extended. This is brought upon by contraction of T and relaxation of the antagonistic S.

Option C is the only answer that satisfies these.

Question 25: D

$$T = 2\pi \sqrt{\frac{k^2 + h^2}{gh}}$$

$$\frac{T}{2\pi} = \sqrt{\frac{k^2 + h^2}{gh}}$$

$$\left(\frac{T}{2\pi}\right)^2 = \frac{k^2 + h^2}{gh}$$

$$\frac{T^2}{4\pi^2} = \frac{k^2 + h^2}{gh}$$

$$k^2 + h^2 = \frac{T^2 gh}{4\pi^2}$$

$$k^2 = \frac{T^2 gh}{4\pi^2} - h^2$$

$$k = \sqrt{\frac{T^2 gh}{4\pi^2} - h^2}$$

Question 26: Mitosis, both, meiosis, meiosis, both

i)   Only mitosis produces clones.

ii) Both replicate all chromosomes prior to any division.

iii) Gametes are only produced by meiosis.

iv) Haploid nuclei are only produced by meiosis.

v)   Genetic material will always appear as distinct chromosomes.

Question 27: C

Consider $(x - a)(x - b) = 0$

This can be expanded to: $x^2 - (a + b)x + ab = 0$.

$a + b = 7$ and $ab = 9$. Substitute this into the equation to give:

$x^2 - 7x + 9 = 0$

END OF SECTION

Section 3

Individual freedom and the rule of law are mutually incompatible.

➢ The statement suggests that the laws in place are against personal freedom, and that the two cannot occur synonymously.

➢ There are a number of examples of issues facing modern society whereby complete freedom is denied due to legality. One example is freedom of expression, which is considered a political right, but is restricted in content (and varies hugely between different countries). Others include euthanasia and abortion after 24 weeks - patients do not have autonomy here because the law dictates that they cannot carry this out in the UK, and patients may go abroad instead. Also, another aspect of individual freedom is the notion of ingesting any substance desired, but the personal use of many recreational drugs is illegal.

➢ However, the converse argument is that these laws do indeed allow freedom (albeit to a limited extent). The fact that freedom of expression is considered a human right, or that abortion is allowed prior to 24 weeks, allows people to do what they want within reason and suggests 'mutually incompatible' is too strong a phrase. Furthermore, the ability to express one's opinion within the boundaries of the law has been effective in making changes in society - for example the suffragettes for women's rights, Martin Luther King against racism, etc. The advent of media such as the internet and worldwide satellite broadcasting has allowed the emergence of more opportunities for freedom.

➢ Freedom is essential in any democracy in order to allow the public to be involved in decisions. However, the limits ensure that hatred and offence is not spread which would ostracise certain groups from society (arguably the exact opposite of what free speech intends to do, and may paradoxically actually reduce freedom).

There is more to healing than the application of scientific knowledge.

➢ Scientific knowledge allows the explanation of events based on objective observation. A hypothesis is developed and is tested experimentally, ensuring many factors are controlled for. Studies are tested and peer reviewed before a theory is accepted and this knowledge can then be used to develop new treatments. Along with the application of science, it is important to consider the art of managing patients. More and more, doctors are encouraged to take more of a holistic approach to care and consider the social and psychological aspects of treatment too.

➢ Treatments not wholly based upon scientific knowledge could be inferior in both efficacy and safety. Without rigorous clinical testing, patients may be put at risks and the drug may not be as effective as one designed through rational consideration of biochemical and physiological pathways. Indeed, when certain alternative treatments have been critically analysed, such as homeopathy, it can be seen that there is almost no therapeutic benefit relative to placebo.

➢ Many treatments use a 'reverse engineering' approach whereby the exact mechanism or science of the technology is not known but a clear therapeutic benefit is seen. One example is Deep Brain Stimulation for Parkinson's Disease which is currently the most efficacious treatment available but its method of action is still up for debate. Although it is desirable to understand how this technique works in order to perfect the parameters, it is not vital to know as long as it is not detrimental to the patient in other ways. Another is the increase in the clinical use of meditation/mindfulness for anxiety disorders despite not necessarily knowing this is helpful systematically. Furthermore, several studies have shown that patients who are more comfortable around their doctor are more likely to improve in health, and the placebo effect is a known example of successful healing with no obvious scientific basis.

➢ **Our genes evolved for a Stone Age life style. Therefore, we must adopt Stone Age habits if we are to be healthy.**

➢ Hundreds of millions of years of evolution led to our human ancestors, the Homo erectus, becoming the highest animal on Earth with a highly developed brain and adaptations allowing survival and hunting. Since then, our genetic makeup has changed relatively little, but we have changed significantly in a short space of time. This statement argues that because natural selection generated these very particular set of genes on Earth, it must be that the environment where our genotype was selected is where we are likely to function optimally. However, the sentences are not necessarily logically connected and we may be able to live better without Stone Age habits despite having 'Stone Age genes'. We have developed significantly, in intelligence and socially, from the Stone Age by moving away from their habits, such as the advent of systematic education, harvesting food, etc. (Also, see the concept of memes devised by Richard Dawkins, which act as cultural analogues of genes and allow changes in society that are much quicker than that which can be spread by genes.)

➢ IF we were to agree with the reasoning and implement Stone Age habits, one could argue for some benefits such as more exercise and less processed food. However, with over 7 billion humans in the world, it would be completely unrealistic to do this; for example, food demand would completely outstrip demand. Also, we would have to lose some of our greatest inventions to date and this would set back progress by thousands of years. (Admittedly this point is going quite far from the original statement.)

➢ The best conclusion would probably suggest that the statement is understandable but not viable or realistic.

END OF PAPER

2005

Section 1

Question 1: C

Count up the tiles - there are 9 black (large), 9 white (medium) and 20 grey (small) so the proportion needed is roughly 1:1:2.

Question 2: D

The conclusion is that "schools should revert to traditional sports days" and the evidence is that because "adult life [involves] competition." To link this evidence to the conclusion the assumption is that school sports should prepare children for adult life.

None of the others are assumptions that must be made, and A and B are not even relevant to the argument.

Question 3: B

Julia counted the animals - b + s = 13
Tim counted the legs (birds have 2, sheep have 4) - 2b + 4s = 36
Solve the simultaneous equations (e.g. double the first one and then take it away from the second) and you should get s = 5 and b = 8.

Question 4: A

This answer gives a reason for how the universe may be expanding overall but why galaxies can still collide.
B is not correct as it is too specific - we want a reason why galaxies in general collide.
C is not correct because it does not explain why there are collisions.
D doesn't really explain anything, and E is irrelevant.

Question 5: 30 to 49

Look at the columns which show the same years for males and females. Only for the ages 30-34, 35-39, 40-44 and 45-49 do the cancer deaths for females exceed males in all years shown. We only need to check between 30-54 because we are told that in 1991-95, every other age has more male deaths. For 50-54, there were more male deaths in 1971-75 than female.

Question 6: B

The conclusion is that "since our consumption of natural pesticides vastly outweighs that of synthetic pesticides, our health is at greater risk from natural pesticides than from synthetic ones." So to weaken this we want a reason why this may not be true. B is the only one that provides this.
A and C actually support the argument, and D is not relevant for the comparison between natural and synthetic pesticides.

Question 7: £10,500

Stamp duty will be 3% looking at the graph, and 3% of £350000 is £10500 (try multiplying by 3 and dividing by 100).

Question 8: C

There must be almost vertical parts to the graph, as for a £1 increase between e.g. £119999 and £120000 there will be a huge increase in tax, so B is incorrect.
D is also incorrect, as the tax paid will not be constant within a band - it will be relative to how much your house is within the band.
Finally, C is correct and A isn't, as the not vertical parts of the graph will become steeper as the percentage tax paid increases for each band.

Question 9: B

First £120000 is free.
Next £130000 is taxed at 1% - £1300 to pay on this.
Next £50000, to get up to the total of £300000, is taxed at 3%, so £1500 to pay on this.
So £1300 + £1500 to pay = £2800

Question 10: £3300

Price without 'cheating' - £260000 x 3% = £7800 tax
Price with cheating - £250000 x 1% (we can round up the £1 as it is basically insignificant when finding a percentage) = £2500 tax. Add the £2000 extra paid to make this £4500 (the £10000 paid is not an extra, as this would be paid in the price without cheating anyway).
£7800 - £4500 = £3300.

Question 11: A
Use the Venn Diagram - from the information we know the small circle is Spanish, the circle on the left is French and the circle on the right is German. Thus, the shaded area represents those who do French and German but not Spanish.

Question 12: E
This answer is correct, as the argument suggests more children have started using cannabis before age 15 but the levels of schizophrenia have remained stable so that cannabis cannot cause schizophrenia. However, if E is true, then we would have to wait some time to see whether cannabis can cause schizophrenia.

A supports the argument, B is irrelevant as the argument is about schizophrenia (a mental illness) and not cancer (a physical illness), C is far too general and D is not directly relevant to the conclusion.

Question 13: B
The first ray of incident light must transmit through 2 layers, each with a probability of 0.8 of transmitting, so 0.8 x 0.8 = 0.64.

However, some of the light that is transmitted through the first sheet will then be reflected by the second sheet, which will then again be reflected by the first sheet, then finally transmitted by the second sheet. The probability of this is 0.8 x 0.15 x 0.15 x 0.8 = 0.0144. The light can, in theory, be reflected infinite times between the 2 layers before being transmitted to the inside room, but the likelihood of this will be minute.
So the probability is 0.64 + 0.0144 (as any further probabilities are too small to consider if we need an answer to the nearest 1%), and 65% is the closest to this.

Question 14: E
1 is correct as the conclusion is that 'nuclear power…will have to continue to be used in 2050' and the evidence is that energy consumption will reach the target and be very high at this point, but to link these we need the assumption that the energy consumption will actually be that high in 2050.
2 is irrelevant to the conclusion - whether we can store nuclear waste or not, the argument is that we must continue using this power.
3 is correct as the evidence suggests 'these [renewable] sources will be unable to meet the shortfall in supply,' which assumes the development of these energy sources will be constant.

Question 15: D
The shopkeeper buys 4 packets (4p), and sells them for the price of 3 packets (3y), and gains 20% profit. Thus, 3y/4p = 1.2, which you can rearrange to get y=1.6p. Thus, usually, there is a 60% profit made.

Question 16: D
This is (hopefully) a quite simple, 'reading of the graph' question. When the dipstick measures 0.15m he has 400 litres of oil. If he orders 500 litres more, he has 900 litres of oil. If 900 litres are present the dipstick should read about 0.6m.

Question 17: E
The theme of the argument is that we cannot say that we are ruining the countryside's natural beauty only through wind farms, as many other factors have already caused the British countryside to change in appearance since the 12th century. Thus E is the best answer.
A is too general as the argument is specifically about the countryside.
B, C and D are irrelevant to the general theme of the argument.

Question 18: C
The argument suggests that "we should not tolerate aggressive behaviour in a civilised society" and goes on to say that players who do this should be "banned from the club's next three game." However, to link these ideas we need the assumption that the ban would help to reduce the aggressive behaviour, which is what C says.
The rest of the answers cannot be inferred/are not necessarily true and they are therefore not assumptions made.

Question 19: 3200 (3100 allowed)
The division annual rate before the installation of CCTV (the column with 'start') is 37838. We want a monthly rate, so we divide this by 12 to get just over 3150, which rounds up to 3200 (but don't worry too much about being exact as 3100 is also accepted).

Question 20: F

All of the statements are confounding factors. We have a negative correlation between CCTV installation in the target area and overall crime in the area but this may not be a causal link and the other answers give other reasons as to why this may have occurred.

Question 21: 9%

The target area % change is (131-161)/161 which is roughly -30/160, which is 18.75%. The division % change is (6442-7164)/7164 which is roughly -700/7000 which is 10%. The difference between these is roughly 9%.

Question 22: 22%

If we remove the vehicle crime figures from the overall crime figures, we have a start rate of 1526 - 279 = 1247 and an end rate of 1098 - 126 = 972. The new percentage change is (972-1247)/1247 = roughly -275/1250 = 22% change

Question 23: E

Re-read what the buffer area is - an area surrounding a target area. Providing data about this tells us if the target area crime has just relocated to an adjacent area, and it helps us to determine whether the CCTV has caused falling crime in the target area of if this is more of a general trend.

2 is not correct as it is irrelevant to the idea of a buffer area.

Question 24: A & C

There are too many students getting very high marks, suggesting that the hard questions are too easy. There are also not enough students getting low marks, suggesting that the easy questions are also too easy.

Question 25: C

The path plan shows between which letters you can travel without encountering any other letter.

➤ C to X can only be reached via I, so is wrong.
➤ A to I can only be reached via C, so is wrong.
➤ E to G can be reached directly, so is right.
➤ F to G can only be reached via E, so is wrong.
➤ J to X can only be reached via I, so is wrong.

Question 26: A

The medullary bone is only found in the female birds. Thus, the absence of it is necessary to determine whether the fossil is male - if the bone is present there is no way that it is a male. However, it is not sufficient, because the passage continues to say that the medullary bone is depleted during certain times in the female, so the absence of it alone does not mean that it is a male.

Question 27: B

1 is correct. Even if the 30% representing the 17-34 year olds were responsible for all the walking/cycling (10%) and public transport use (15%), then 5% would still use a car, and 5/30 is 1/6.

2 is correct as 30% are 16 or under, but only 15% of the total travel by public transport. Even if this was made up of all those 16 or under, at least 15/30 = 50% or under would travel by other means.

3 is not necessarily correct - 85% are under 60, and with this data alone it could be the case that all of them travel by car and public transport, and none by walking/cycling.

Question 28: B

Firstly, find the groups of 4 that add up to 14. The first that is fully available is 4 4 2 4, next is 1 5 6 2, next is 4 2 5 3 and finally we have 5 5. The next 2 numbers must add up to 4, so 4 is not viable. If it were 1 or 3, there would have been 5 consecutive odd numbers, which is not possible. Thus, 2 is the only possible answer.

Question 29: F

The conclusion is 'global travel helps to immunise the population' based on the information given.

➤ 1 is correct because if many British residents do not travel, then based on the conclusion we still need to be concerned about bird flu.
➤ 2 is correct because even if we are immune to many diseases, it does not mean that 'there is no chance that bird flu will kill thousands of people' as we may not be resistant to this one.
➤ 3 is correct as the evidence suggests the quality of nutrition has improved due to having a strong economy, but nutrition is also dependent on e.g. environmental conditions, what chemicals are used, processing, etc.

Question 30: D

Method 1: Rank the best car as 1 and the worst as 4. There are 24 possible sequences of which order they come in: 1234, 1243, 1324, 1342, 1423, 1432, 2134, 2143, 2314, 2341, 2413, 2431, 3124, 3142, 3214, 3241, 3412, 3421, 4123, 4132, 4213, 4231, 4312, 4321. 11 of these events (2134, 2143, 2314, 2341, 2413, 2431, 3124, 3142, 3412, 4123, 4132) result in him getting the best car.

Method 2: Work out that if 1 is the first, the probability of getting the best car is 0. If 2 is first, the probability of getting the best car is 1. If 3 is first, the probability of getting the best car is 1/2 (could have picked 2) and if 4 is first, the probability of getting the best car is 1/3 (could have picked 2 or 3). Add these together and divide by 4 (as each option has a ¼ chance of occurring) to get 11/24.

Question 31: A

➢ As the paragraph discusses risks and benefits of aspirin, we can conclude that 1 is correct.

➢ 2 is wrong - 'certainly' is too bold a phrase and we don't know what the risk is for each individual patient.

➢ 3 is wrong and nothing like this is included in the passage.

Question 32: D

Deduce what the sequence is based on the information given. Hopefully you should see that the symbols represent $(2 \times 5 \times 5 \times 5) + (3 \times 5) + 1 = 250 + 15 + 1 = 266$.

Question 33: B

1 in 3 had visited an alternative therapist out of a total of 247 million patients, and 247/3 is roughly 82 million visiting alternative therapists. 425 million visits were made to alternative health practitioners, so per patient this is 425/82 = 5

Question 34: C

The passage talks about how 'the time it takes is…at a premium…this is the void that alternative medicine appears to be filling.' Thus, this suggests that C is correct.

Although we know that time is at a premium, A is not necessarily true.
B is not necessarily true - the author talks about different definitions of 'Alternative Medicine' and some may claim to cure individual diseases.

D is not true - although it says 'the cornerstone of alternative medicine appears to be the belief in the body's ability to heal itself' this is not necessarily saying that doctors don't believe the same. It is too bold a statement.

E is not necessarily true for a similar reason to B - some treatments may have side effects and this does not necessarily mean it scores over conventional medicine anyway.

Question 35: D

This is true as the passage suggests that doctors having less time for patients has meant the 'void' is filled by alternative medicine.

A is not implied - even if they now have 'side-effects' this does not mean that the disease has not been cured
B is not correct - we know the 'best doctors are frustrated that combining the art of healing…is more difficult to do' but this does not mean they support alternative techniques.
C cannot be implied - there is nothing about pharm companies.

END OF SECTION

Section 2

Question 1: A

This is no longer on the BMAT syllabus.

Tidal volume is the volume of air displaced during normal ventilation in the absence of extra respiratory effort i.e. the normal volume of air that you breathe in & out.

Question 2: B

A burning splint makes:

➢ A popping sound if hydrogen is present
➢ goes out if carbon dioxide present
➢ A glowing splint relights if oxygen is present

➢ Bromine water is to test for C=C bonds
➢ Limewater is turned cloudy by carbon dioxide
➢ Litmus paper tests for pH but this is not useful for gasses.

Question 3: D

The number of protons has stayed the same (atomic number: 54+38=92) but the atomic mass has decreased by 4 (95+139=234). Thus, four neutrons must have been emitted in addition to the main products.

Question 4: C

Current will take the path of least resistance. Thus, for current to flow through the ammeter, it must follow the path:
R → Ammeter → Q or P → Ammeter → S
Since two resistors have lower resistances, current will 'prefer' to flow through them.
A. The resistance of p and q is lower so all current would travel through the top branch.
B. The resistance of p and r is lower so the current would remain split equally between the branches and carry on in the same branch.
C. The resistance of r and q is lower so the current will travel from r to q via the ammeter.
D. The resistance of r and s is lower so all current would travel through the bottom branch only.

Question 5: A

This is no longer on the BMAT Syllabus.
Smoking results in the production of carbon monoxide which irreversibly binds to haemoglobin, preventing it from binding to oxygen. This reduces the overall oxygen carrying capacity of red blood cells.

Question 6: C

$z = xy^2$
$1.2 \times 10^{13} = 3 \times 10^{-6}y^2$
$y^2 = \frac{1.2 \times 10^{13}}{3 \times 10^{-6}}$
$y^2 = 0.4 \times 10^{19} = 4 \times 10^{18}$
$y = 2 \times 10^9$

Question 7: A

Since a molecule is being broken down, this is a type of decomposition reaction (not synthesis).
Since there were 12 hydrogen atoms initially and only 8 in the products, 4 hydrogen atoms have been lost. Therefore, this is also an oxidation reaction.

Question 8: A

If the allele for long lashes was recessive, B and D would also have long lashes. Thus, the long lashes allele must be dominant. Thus, A must be heterozygous for the allele. C and E could be heterozygous but they could also be homozygous. B and D have short lashes so are homozygous recessive.

Question 9: C

$A \propto \frac{1}{B^2}$

Thus, if B is increased by 40%, then:

$A \propto \frac{1}{(1.4B)^2}$

$A \propto \frac{1}{1.96B^2}$

$A \propto \frac{100}{196B^2} = \frac{25}{49B^2}$

A is thus $\frac{25}{49}$ of what it was before, which is just over 50%. Thus, the decrease in percentage terms must be just under 50%, so a 49% decrease is the best answer.

Question 10: 39 cm

The bar is in equilibrium so: Moments clockwise = moments anti-clockwise

The bar exerts its weight at its centre of gravity which is 15 cm left of the pivot.

Therefore: $0.1 \times 800 = 200x + 10 \times 0.15$

$80 = 200x + 1.5$

$78.5 = 200x$

$x = \frac{78.5}{200} = 0.3925 \, m$

$= 39 \, cm$

Question 11: D

This question is no longer on the BMAT specification .

The formula for sodium carbonate is Na_2CO_3 and since it is a 'decahydrate', it contains 10 x H_2O molecules as well. Thus:

$Fraction = \frac{Mass\ of Water}{Mass\ of\ Sodium\ Carbonate\ Dehydrate}$

$Fraction = \frac{10\,(16+1+1)}{23x2+12+3x16+10\,x\,(16+1+1)}$

$= \frac{18\,x\,10}{46+12+48+180}$

Question 12: B

This is no longer on the BMAT specification.

When you look into the distance, the suspensory ligaments tighten to make the lens less convex so that you can cover more area, and the ciliary muscles relax.

Question 13: A

$y = \left[\frac{x^2+2ax}{b}\right]^{\frac{1}{2}}$

$y^2 = \frac{x^2+2ax}{b}$

$by^2 = x^2 + 2ax$

Complete the square to give: $by^2 = (x+a)(x+a) - a^2$

$by^2 + a^2 = (x+a)^2$

$x + a = (by^2 + a^2)^{\frac{1}{2}}$

$x = (by^2 + a^2)^{\frac{1}{2}} - a$

Question 14: B

$Distance\ travelled = Transmitter\ to\ foetus\ +\ foetus\ to\ receiver$

$= 0.1 + 0.1 = 0.2m$

$Time = \frac{distance}{speed}$

$= \frac{0.2}{500} = 0.0004s$

$= 0.4 \, ms$

Question 15: D

$The\ products\ consist\ of\ 6\ x\ N - H\ bonds\ (= 6z)$

$The\ reactants\ consist\ of\ 1\ x\ N{\equiv}N\ (= x)\ and\ 3\ x\ H - H\ bonds\ (= 3y).$

The reaction is exothermic so:

$\sum Bond\ strength\ of\ Products > \sum Bond\ strength\ of\ Reactants$

$Thus, 6z > x + 3y$

Question 16: D, C then G

This is no longer on the syllabus but you can work it out if you understand basic anatomy.

- A is bone
- B is a muscle that causes flexion at the _proximal_ hinge joint when it contracts.
- C connects two bones together like a ligament.
- D is the end of the muscle body of B and inserts into the bone, similar to a tendon.
- E is a muscle body that causes extension at the _distal_ hinge joint when it contracts.
- F is a muscle body that causes flexion at the _distal_ hinge joint when it contracts.
- G is a muscle that causes extension at the _proximal_ hinge joint when it contracts.
- E and F are antagonistic to each other; B and G are also antagonistic to each other.

Question 17: A

$$\left(\sqrt{5} - \sqrt{2}\right)^2 \left(\sqrt{5} + \sqrt{2}\right)^2$$
$$\left(5 - 2\sqrt{2}\sqrt{5} + 2\right)\left(5 + 2\sqrt{2}\sqrt{5} + 2\right)$$
$$\left(7 - 2\sqrt{10}\right)\left(7 + 2\sqrt{10}\right)$$
$$49 - 14\sqrt{10} + 14\sqrt{10} + 4(10)$$
$$49 - 40 = 9$$

Question 18: B

The maximum possible age of the rock would be if it consisted entirely of ^{235}U at the start. Since there are 8 parts in total:

Number of Half Lives	Proportion of ^{235}U	Proportion of ^{207}Pb
0	8	0
1	4	4
2	2	6
3	1	7
4	0.5	7.5

After 3 half lives, the rock would contain one part U^{235} to seven parts of Pb^{207}.

Thus, the maximum age:
$$= 3 \times 7.1 \times 10^8 = 21.3 \times 10^8 \ years$$
$$= 2.13 \times 10^9 \ years$$

Question 19: B

Using $n = cV$:

The number of moles of NaOH $= 2 \times 50 \times 10^{-3} = 100 \times 10^{-3}$
$= 0.1 \ moles$

Since the molar ratio between NaOH and H_2X is 2:1, there are 0.05 moles of H_2X.

The M_r of the acid $= \frac{mass}{moles} = \frac{4.5}{0.05} = 90$

Question 20: C

Active transport requires ATP which is produced during aerobic respiration. This requires oxygen. In options A, B and D, the substance moves from a higher concentration to a lower concentration and can therefore occur via diffusion. Contrastingly, in option C, magnesium ions move from a lower concentration to a higher concentration. This is an example of active transport and therefore requires oxygen.

Question 21: A

The triangle is equilateral so $AC = x$

Thus, the radius of the circle $= \frac{x}{2}$

Area of the Semi-circle $= \frac{\pi}{2}r^2 = \frac{\pi}{2}(\frac{x}{2})^2 = \frac{\pi x^2}{8}$

Split triangle ABC into 2 x right angled triangles:

Using Pythagoras: $AB^2 = BD^2 + AD^2$

$BD^2 = x^2 - (\frac{x}{2})^2$

$BD^2 = \frac{3x^2}{4}$

$BD = \frac{\sqrt{3x^2}}{\sqrt{4}} = \frac{x\sqrt{3}}{2}$

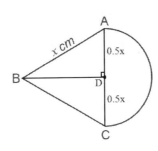

Area of Triangle ABC $= 2 \, x \, Area \, of \, AD$

$= 2 \, x\frac{1}{2} \, x\frac{x}{2} \, x \, \frac{x\sqrt{3}}{2}$

$= \frac{x^2\sqrt{3}}{4}$

Total Area $= \frac{x^2\sqrt{3}}{4} + \frac{\pi x^2}{8}$

$= \frac{2x^2\sqrt{3} + \pi x^2}{8}$

$= \frac{x^2(2\sqrt{3}+\pi)}{8}$

Question 22: D

Since g is not given, you can safely assume that $g = 10 \, ms^{-2}$

Thus, g on the new planet $= \frac{1}{4} \, x \, 10 = 2.5 \, ms^{-2}$

$E_p = mg\Delta h = 1 \, x \, 2.5 \, x \, 20 = 50J$

When the mass is dropped from the building, all of its gravitational potential energy is converted to kinetic energy, i.e. $E_p = E_k$

$E_k = \frac{mv^2}{2}$

$50 = \frac{v^2}{2}$

$100 = v^2$

$v = 10 \, m/s$

Question 23: B

Let **x** be the number of hydrogen atoms and **y** be the number of Deuterium atoms.

The mass of 4 carbon atoms $= 4 \, x \, 12 = 48$

This contributes 80% to the compound so the other 20% is made up of hydrogen and deuterium.

I.e. $H + D = 12$

Since deuterium has a mass of 2, this can be represented as $x + 2y = 12$

There are 4 carbon atoms and $x + y = 10$ (because butane has 10 hydrogens).

Now we just solve the equations simultaneously: $x = 10 - y$

Substituting into the first equation gives: $10 - y + 2y = 12$

Thus, $y = 2, \, x = 8$

Therefore, the formula is $C_4H_8D_2$.

Question 24: A

This is actually a complex question. Firstly note that:

➢ Heart rate and stroke volume would be unaffected by the opening.

➢ The opening would also have no effect on the pulmonary vein (which normally carries oxygenated blood back to the left atrium).

This leaves us with options A and E.

Blood could flow:

1) Left atrium → Right atrium:

This would result in more blood being pumped to the lungs. However, this is unlikely to cause problems to the health of the baby.

2) Right atrium → Left atrium:

This would result in some deoxygenated being pumped by the left ventricle into the aorta. This would cause the baby to become hypoxic which is a major problem.

Question 25: B

$760 \ mmHg \ = \ 100,000 \ Pa = 10^5 Pa$

Thus, $Pressure = \frac{152}{760} \ x \ 10^5 \ = \ 0.2 \ x \ 10^5 Pa$

$2 \ x \ 10^4 Pa$

$Area \ = \ 2cm^2 \ = \ 2 \ x \ 10^{-4} \ m^2$

$Force \ = \ Pressure \ x \ Area$

$Force \ = \ 2 \ x \ 10^4 \ x \ 2 \ x \ 10^{-4} \ = \ 4N$

Question 26: D

These types of questions frequently come up so it's essential you know the reactivity series off by heart. From the given elements, Na is the most reactive and therefore, is hardest to displace from its oxide. Thus reaction D would require the highest temperature.

Question 27: 0.8

When you get a question like this, start off by drawing the diagram and filling in all of the information that you know:

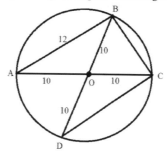

Since D is opposite B, BD is also a diameter of the circle.

Since the angle in a semicircle is always 90°, ABC = 90°

Therefore, ABC is a right angled triangle. Using Pythagoras:

$AB^2 \ + \ BC^2 \ = \ AC^2$

$BC^2 \ = \ 20^2 - 12^2$

$BC^2 = 400 - 144 = 256$

$Thus, BC \ = \ 16$

Note that BDC is also a right angled triangle. Thus:

$Sine \ BDC \ = \frac{BC}{BD} = \frac{16}{20} = 0.8$

END OF SECTION

Section 3

Animals do not feel pain as we do.

➤ This statement is probably not refuting the idea that animals can endure the same physical hurt and discomfort experienced by a noxious stimulus, which act on nociceptors common to almost all animal species (there has been a significant amount of research on this already). Instead, it is likely suggesting that they do not experience 'emotional suffering' in the same way.

➤ We cannot directly experience anyone else's pain; it is a state of consciousness. Thus, it is unlikely that we can ever answer this question unless we can get inside the head of an animal and actually be it. We cannot even know if other humans feel the same pain! However, many scientists argue that the majority of animals (with the exception of primates, dolphins and some other higher mammals) are not actively conscious and thus do not have the same subjective experience of pain.

➤ Only primates and humans have a neocortex - the 'thinking area' of the cortex, and this may be why only we are able to 'feel' pain..

➤ However, a capacity to feel emotional, conscious pain obviously enhances a species' prospects for survival, since it causes members of the species to avoid sources of injury (if they are conscious in the first place). It is surely unreasonable to suppose that nervous systems that are virtually identical physiologically, have a common origin and a common evolutionary function, and result in similar forms of behaviour in similar circumstances should actually operate in an entirely different manner on the level of subjective feelings.

➤ Behavioural changes are similar during pain; for example, many animals display behaviours associated with depression during a painful experience such as lack of motivation and lethargy.

Science should leave off making pronouncements: the river of knowledge has too often turned back on itself.

➤ A scientific pronouncement is an authoritative declaration about findings that have been made. However, this statement suggests that we should be wary to make such announcements, as the findings made can later be found to be false.

➤ Andrew Wakefield is the gentlemen who asserted the link between the MMR vaccine and the onset of autism, which resulted in the deaths of many children from measles, mumps and rubella. AW falsified his results and subjected autistic subjects to unethical procedures to prove that the MMR vaccine cause autism after being bribed by a law firm that was planning on suing the MMR vaccine company. After repeat experiments were performed and the results acquired were found to be completely different, AW's paper was withdrawn from the Lancet and he was struck off the medical register. Sadly, the myth that MMR vaccines cause autism is still believed and herd immunity has not reached pre-scare levels, highlighting the danger of making such pronouncements.

➤ Instead of making a pronouncement, scientists should accept that their result is specific to the exact experimental conditions, parameters and boundaries used, and is limited to current technology and knowledge anyway

➤ However, if pronouncements must be made to advance science then this is undoubtedly a positive. Indeed, passion, drive and intrigue are intrinsic to scientific discovery and enquiry; without these qualities, Fleming and Darwin may never have discovered antibiotics and evolution respectively. Without these pronouncements, it is unlikely that technological advances that were unplanned/unprecedented could be implemented.

➤ Furthermore, making pronouncements welcomes other scientists with different prejudices and preferences to follow what we have done to see if they get the same results through peer review, which furthers the reliability of all experiments.

With limited resources and increasing demand, doctors will not in the future be concerned about how to cure, so much as whether to cure.

➤ As the world population is growing exponentially and resources are become scarce, it is may be the case that in the future we will not be able to match the demand for medicine with supply. Then, we must then focus on how to best use the remaining resources, which will take importance over developing new treatments.

➤ Limited resources and increasing demand is likely, on a more global scale, due to the birth rate exceeding the death rate, causing overpopulation and famine. In this country, the issue is the ageing population.

➤ Thus, we may have to specifically provision and allocate specialist treatments to certain people only, if we are unable to provide them for everyone. Ethical and moral issues must be considered with each patient considered individually. Ultimately, refusing treatment to someone who will die without it, even if it is because someone 'deserves' or 'needs' it more, can be considered as passive euthanasia.

➤ The argument against this is that 'the care of the patient is your primary concern' and you 'should always act in the patients best interests,' ensuring the best quality of care possible. Withholding a treatment is certainly not fulfilling this in some instances.

➤ The whole principle of the NHS is equality of healthcare, and that is why it is often considered the 'crown jewel of the welfare state.' If we begin to tell people that they cannot have certain treatments then we are opposing the very principle of having a public healthcare system.

➤ What is the viability of the latter? The authorities can prioritise ensuring cost-effectiveness within medicine and equal consideration of all patients to try and match the supply and demand of resources and remove the concern of 'whether to cure', but achieving these is not necessarily feasible.

END OF PAPER

2006

Section 1

Question 1: C

Proportion of seabirds is 80, which decreased by 60%, so $0.4x = 80$ and $x = 200$.

Proportion of humans is 1, which increased by 25%, so $1.25y = 1$ and $y = 0.8$.

$200 : 0.8 \rightarrow 250:1$

Question 2: E

The conclusion is the last line, given away by the word 'so'. It states that 'freedom of speech...has to have limits put upon it if democracy is to be sustained.' This is clearly paraphrased in E, which is the correct answer.

Question 3: B

A 'worked solution' isn't really possible here, but it is not against the BMAT rules to bring scissors to the exam, so cut out the table and test which shapes are possible (provided you don't need the page before anymore!) 1 and 3 are possible and 2 and 4 are not. In the incorrect 2, both have inward facing lines to the middle which would not happen with these shapes. Also, as 2 and 4 are extremely similar (basically rotations of each other), you should deduce that if one is right the other probably is right, and vice versa.

Question 4: C

The conclusion is that 'political journalists are not doing their job properly' and the supporting evidence is that 'they exclude...political issues.' Thus, the unstated assumption to link this evidence to the conclusion is that it is the job of the political journalists to inform voters about the political issues. The conclusion still holds without A, B, D or E being assumed.

Question 5: 62 seconds

The second train will clearly take longer - it must travel the distance it is from the tunnel, the length of the tunnel for the front of the train to be at the end and the length of the train for the back of the train to be out of the tunnel. $d = 40 + 615 + 120 = 775$m. The speed is always 45km/h so $t = 775/45000$. Through short division of $775/45$ ($=17.2$) you can work out that this is 0.0172 hours, which you can multiply by 3600 to give 62 seconds.

Question 6: C

The phrases 'on the one hand' and 'on the other hand' and giving 2 separate examples of how the media approaches the issue of eating disorders highlights the inconsistency, so C is the correct answer.

A cannot be deduced - we do not know what they are MORE influenced by.

B cannot be inferred - we do not know if they are aware or unaware of the effects.

D is too bold - we do not know if they are ONLY interested in thin fashion models.

E is not relevant here - it is not related to how the print media approach the issue of eating disorders.

Question 7: B

If we look at the 'Total' row at the bottom, and compare GCSE performance (the initial information suggests this is done by determining how many had 5 or more GCSE passes at A* to C), Asian pupils have 58.7% and white students have 55.1%. 58.7 - 55.1 = 3.6%, so this is 4% when rounded.

Question 8: E

The column we are looking at is the 'KS 2-4 VA Measure' for students 'other than English as a first language.' The highest value in this column when looking at boys and girls combined is for the Chinese students.

Question 9: D

This is correct - if we look at % with 5A*-C for both boys and girls, and compare those who either have English as their first language or don't, we can see that only Mixed students perform better at GCSE than their equivalents.

A is not correct - Boys (977.5) have a lower value of total KS 2-4 VA Measure than Girls (997.3).

B is not correct - more Asian students (~30000/35000) have English as a second language then Chinese pupils (~1750/2250).

C is not correct - the difference between white girls and white boys is 60.2 - 50.3 = 9.9, but this difference is smaller for Chinese pupils (85.1 - 77.1 = 8).

Question 10: D

A B and C are possible reasons. This effect occurs for all ethnic groups, and the fact that there is a bigger difference in improvement for Asian girls than Asian boys is irrelevant. For example, even if we removed all the values for Asians, it would still be true that 'for all ethnic groups, girls improve more between the ages 11 and 16 than boys.'

Question 11: E

Here we have a negative correlation between cannabis use and ability to recall words. This could be due to any of the reasons given. 1 and 2 highlights that cannabis itself could be the causal factor whereas 3 highlights that it could also be because there is a difference in IQ between the groups, independent of cannabis use.

Question 12: A

At $1.50, the supplier gets 90c (60%) and the store gets 60c. With the offer, 1 box of cereal costs $1, so the supplier gets 60c and the store gets 40c. The supplier must reduce the price by 90c - 60c = 30c.

Question 13: 68p

We must go through the postal rates from cheapest to most expensive and see the first value we get that we cannot be made using only 1 or 2 stamps.

➢ 23p – 1 x 23p stamp
➢ 32p – 1 x 32p stamp
➢ 37p - 1 x 37p stamp
➢ 49p - 1 x 49p stamp
➢ 50p - 1 x 50p stamp
➢ 62p - 1 x 42p stamp and 1 x 20p stamp
➢ 68p - cannot be made with 1 or 2 stamps, so this is our answer.

Question 14: C

The conclusion is that 'experts…agree that it is a completely reliable method of identifying criminals.' However, as C says, this is an opinion and not necessarily fact, which is the weakness of the argument.

A is tempting but is not the best answer because it is too general and the argument is not about how they **apply** their method. B basically paraphrases the first line of the argument, and this is then countered within the argument, so it is not really a weakness.

D is not a weakness as it is already stated in the argument and the argument is that the technique is reliable because of the opinion of experts rather than about how long it has been used for.

Question 15: C

Remember that 'Due' means directly when thinking about directions. So if Wellbank is due east of Ruilick, and Aultviach is due north of Ruilick, Ruilick must be E (as it is the only one with something directly North and directly East of it), Wellbank must be F and Aultviach must be A. Then, if Rheindown is due south of Clashandarran then Rheindown is D and Clashandarran is C. This leaves C and G. If Beauly is south and west of Windyhill then Beauly must be G and Windyhill is C.

Question 16: B

➢ 1 is not an assumption - we do not know if less developed countries have set targets or not.
➢ 2 is not an assumption – there is no mention of whether developed countries are prioritising global warming or current poverty.
➢ 3 is not an assumption - we don't know if levels of prosperity are changing.
➢ 4 is an assumption - the argument suggests that there is a choice between either reducing poverty or decreasing greenhouse emissions.

Question 17: B

10 children overall, spending £125 - 10 x 25p (as they all get a 25p packet of bubble gum) = £122.50 spent.

2 went on the Apocalypse and the Carousel - 2 x (9 + 3.50) = £25, leaving £97.50 spent on the other 8 children.

If x went on Armageddon and dodgems and y went on Armageddon and helter skelter, $12.5x + 12y = 97.5$, and $x + y = 8$. You can either solve this, or quickly test each number given in the answer, and you should get $x = 3$.

Question 18: C

The conclusion is 'people do not have the right to complain about farmers' use of pesticides' and the evidence given is that they would 'make a loss' otherwise. The assumption made linking this evidence and conclusion is that pesticides must be used to avoid a loss, which is said in C.

A is not true, B is incorrect as you can damage the environment but still increase yield, D is irrelevant to the conclusion and E is not necessarily a flaw of the argument - we don't know if it is true or not.

Question 19: E

After a 8 day holiday she is 8x3=24 days behind, starting on the 16th. For the next 14 days she recovers her fitness, being 10 days behind before the 30th, but this is when she missed 4 further sessions, meaning she is 10 + 3 x 4 = 22 days behind from the 3rd. 22 days later on the 24th she has recovered.

Question 20: B

Roughly 31250 GPs (half way between 30000 and 32500) and 12500 women. This is slightly over 1/3, so B is the only reasonable answer.

Question 21: C

Almost 50% of 12500 women were practicing full time, so 6000 is the only reasonable answer.

Question 22: D

1994 - 27500 GPs, with roughly 12% part time. So about 3500 GPs part time and 24000 full time, and if we say full time GPs work 1 hour for ease of calculation, there are 24000 + 3500 x 0.5 = 25750 hours.

2004 - 31250 GPs, with 25% part time, so about 7800 part time and 23500 full time. 23500 + 7800 x 0.5 = 27400.

With these estimates we have a percentage increase of less than 10% but this is still the best answer relative to the alternatives.

Question 23: B

Average of the guesses is 135/6 = 22.5.

Suzie's guess is nearer than Wally's so the answer is <22 (not 22 exactly as Wally and Suzie would be equally close).

If it were 20 then Mary's guess would be closer than the average, so the answer is 21.

Question 24: C

The main idea of the passage is that H5N1 'cannot be transmitted easily between people.' Thus, in its current state, a worldwide epidemic is unlikely, but if it does mutate then this could possibly change.

A is wrong - the strain is currently not easily transmitted.

B is wrong - a mutation doesn't necessarily mean there would be a worldwide epidemic; mutations may have no/little effect, and may even have the opposite effect.

D is not correct as it is not the main idea of the passage.

Question 25: B

Let's say that 24 hours have passed.

Area covered by the minute hand = 8.4^2 x π x 12 = ~850π

Area covered by the hour hand - 6.3^2 x π = 40π

850:40 is rough ratio and only A and B are reasonable answers.

40 x 16 is much lower than 850 so 64:3 is the correct answer.

Question 26: D

The conclusion is that 'nuclear plants…are not needed.'

1 is correct because the argument suggests we do not need nuclear plants, but the demand for electricity may have increased, meaning they would be needed.

2 is correct because the 'huge rise in the amount these technologies add to the electricity supply' may also not continue, meaning there would be an electricity deficit.

3 is not correct because the argument is about whether nuclear plants are needed or not, so ways of storing waste are irrelevant.

Question 27: 5 hours

They walk 20km in total - 10km for one side of the journey.

Split the 10km as you wish - for example let's say there is 1km walked on flat land and 9km uphill, so in total they walk 2km on flat land, 9km down-hill and 9km up-hill.

Flat land - time = 2/4 = 30 minutes

Down-hill - time = 9/6 = 1 hour 30 minutes

Up-hill - time = 9/3 = 3 hours

Total - 5 hours.

Question 28: B

The first part of the graph is to be expected - as the population increases in year n the population increases in year n+1. However, the second part of the graph needs explaining - why there is a lower population in year n+1 if the population in year n is much higher. This is explained by B - population being too high in year n means that there in starvation, lowering the population in n+1.

A is not correct - the population would be constant if birth and death matched.

C is irrelevant, D is also not relevant as we can pick any year of the survey and this graph still holds true and E is not true - it can be falsified by the fact that a very low population in year n means a very low population in year n+1.

Question 29: D

➢ 1 is true, as we know that less glycidamide is produced in humans than in animals when acrylamide is broken down.

➢ 2 is not necessarily true - cooking at temperatures higher than 120C produces more acrylamide which could be harmful.

➢ 3 is correct - cooking them at lower temperatures means they absorb more fat and cooking them at higher temperatures means they produce more acrylamide, with an intermediate of both at mid-temperatures, so there will always be some health risks

Question 30: A

The total volume of the cube is $8 \times 8 \times 8 = 512cm^3$. There are 8 of the 3cm cubes, which has a volume of $3 \times 3 \times 3$ each, so these take up $8 \times 27 = 216cm^3$. The 1cm cubes on the edge overlap with the adjacent face, and there are 24 of these, taking up $24cm^3$. There are also 20 extra for each face, taking up $6 \times 20 = 120cm^3$ overall.

$512 - 216 - 24 - 120 = 152cm^3$ taken up by 2cm cubes.

Each 2cm cube takes up $8cm^3$ and $152/8 = 19$ cubes.

Question 31: B

B is the best answer - if the other companies who 'recently looked down on Google' now want to 'forge alliances', this suggests that they no longer look down on Google.

A is not necessarily true - it says Microsoft don't want this to occur, but it doesn't mean it is occurring.

C is not necessarily true - it's too bold a statement and is definitely not the only way (e.g. Microsoft could produce their own Google equivalent).

D is not necessarily true and is not a theme of the last paragraph as said in the question.

Question 32: B

You can answer this question quickly if you understand what they're asking. The revenue was boosted 80% to more than £1.2 billion, but we don't know how much more. So the revenue at the end of 2005 could have actually been any value, and we cannot put an upper limit on what this could be, so A C and D are wrong. B is therefore correct.

Question 33: C

The cartoon depicts the other companies 'ganging-up' on Google →C is correct.

A is not necessarily true - we don't know from the picture or the text that Google will 'triumph' (we don't even know what that means in this context), B is not a suitable answer, D is not true and E is not correct because the picture depicts all 3 competitors.

Question 34: C

1 is definitely true as it basically paraphrases the sentence after the conclusion given. 2 is also definitely true as it paraphrases the final sentence of this paragraph. 3 is not related to the conclusion given.

Question 35: A

Competition is a general theme of the passage, and the last sentence suggests that this 'offers opportunities for consumers.'

➢ B is too strong a statement - big companies **having** to merge -cannot be inferred.

➢ C is also too bold - the use of the word 'cannot' should put you off this answer.

➢ D- we don't know what the future holds and the author doesn't claim to know.

➢ E is not true as the passage talks about all 4 companies competing.

END OF SECTION

Section 2

Question 1: E

Glucose is stored as glycogen in skeletal muscles and the liver. Glycogen can be broken down into glucose (glycogenolysis) to increase blood glucose levels.

Insulin decreases blood glucose by stimulating glycogen synthesis. Thus, low blood insulin will result in decreased amounts of glycogen production and increased amount of glycogenolysis.

Therefore, low blood insulin results in a decrease in glycogen mass.

NB: It's easy to over think this question (especially if you're a biochemistry graduate). Keep it simple and approach it logically.

Question 2: C

Simple molecular structures with covalent bonding normally do not conduct electricity and the melting and boiling points given are reasonable. However, they also are usually insoluble in water.

Question 3: F

2 half lives will elapse in 8 years. Thus, there will now be 8×10^{20} atoms of X. Therefore, 24×10^{20} atoms of Y will have been produced. Initial $4 \times 10^{20} + 24 \times 10^{20} = 28 \times 10^{20}$.

Question 4: B

When you need to keep water in your body, the blood will become concentrated. In addition, ADH will be released from the posterior pituitary gland to cause more water to be reabsorbed in the distal convoluted tubule of the kidney. This in turn will cause the urine to become more concentrated.

Question 5: C

Non-metal elements get progressively more reactive as you go along a period because their electronegativity increases. Thus, the halogens are the most reactive non-metals within a period (as noble gasses are inert). Furthermore, the halogens get less reactive as you go down the group. Thus, Fluorine is the most reactive non-metal- which is shown by C.

Question 6: A

The sine and cosine rules are no longer on the BMAT Specification.

Nevertheless, this is a easy question if you draw a triangle with the values given, and then sub in the values to the cosine rule:

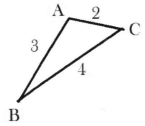

$$a^2 = b^2 + c^2 - 2bc\, CosA$$
$$4^2 = 2^2 + 3^2 - 2 \times 2 \times 3\, CosA$$
$$16 = 4 + 9 - 12CosA$$
$$3 = -12\, CosA$$
$$CosA = -\frac{1}{4}$$

Question 7: A

1. Microwaves are transverse waves, not longitudinal.
2. Microwaves do travel at the speed of light in air.
3. Ultrasound is used in pre-natal scanning
4. Infrared light is used for thermal imaging.
5. All types of electromagnetic radiation, including microwaves can travel through a vacuum.

Question 8: C

Since, 165 is the upper quartile, 25% of people will have a heart rate greater than 165. Thus, the probability of an individual having a heart rate over 165 is $= \frac{1}{4}$

There are 4 possible permutations for 3 members having a HR > 165 and 1 < 165 (the member with a HR under 165 could be in any position). Thus, the probability of this:

$$= \frac{1}{4} x \frac{1}{4} x \frac{1}{4} x \frac{3}{4} \; x \; 4 \; = \frac{12}{256}$$

The Probability of 4 members having a HR > 165:

$$= \frac{1}{4} x \frac{1}{4} x \frac{1}{4} x \frac{1}{4} \; x \; 4 \; = \frac{1}{256}$$
Total probability $= \frac{12}{256} + \frac{1}{256} = \frac{13}{256}$

Question 9: B

1. Catalysts don't affect the position of equilibrium, so will not change yield.
2. Adding more reactant will shift the equilibrium towards the products.
3. Increasing the pressure will shift the equilibrium towards the products because the reactants have two moles of gas and the products have one mole of gas.
4. Increasing the temperature will favour the reverse endothermic reaction which will decrease the salt yield.

Question 10: D

Firstly, note that only 75% of the plants survived and this is unlikely to be due to chance given the sample size. Now, let **A** be the dominant allele and let **a** be the recessive allele.

As can be seen, 2 out the 3 surviving plants are heterozygotes = 67%.

	A	a
A	$AA(\frac{1}{3})$	$Aa(\frac{1}{3})$
a	$Aa\ (\frac{1}{3})$	aa (dead)

Question 11: C

There is actually no need to do any maths here! Pressures across fluids are transmitted so the pressure at X = pressure transmitted to Y.

Question 12: E

$$r = 1 - \frac{6\sum d^2}{n(n^2-1)}$$

$$r - 1 = -\frac{6\sum d^2}{n^3-n}$$

$$1 - r = \frac{6}{n^3-n}\sum d^2$$

$$\sum d^2 = \frac{(1-r)(n^3-n)}{6}$$

Question 13: B

The 'loss in mass' = Mass of Product Gasses

The 'loss in mass' = Mass of $KHCO_3$ – Mass of K_2CO_3

M_r of the $KHCO_3$ = $39 + 1 + 12 + 3 \times 16$ = 100

Number of Moles of $KHCO_3 = \frac{50}{100}$ = $0.5\ moles$

The molar ratio between $KHCO_3$ and K_2CO_3 is 2:1. Thus, two moles of $KHCO_3$ react with one mole of K_2CO_3.

M_r of the K_2CO_3 = $2 \times 39 + 12 + 3 \times 16$ = 138

Mass of K_2CO_3 = 0.25×138 = $34.5g$

The 'loss in mass' = $50 - 34.5$ = $15.5g$

Question 14: x = 3, y = 1

$4x^2 + y^2 + 10y = 47$ and $2x - y = 5$

Thus, $y = 2x - 5$. Substitute into the first equation to give:

$4x^2 + (2x - 5)^2 + 10(2x - 5) = 47$

$4x^2 + (4x^2 - 20x + 25) + (20x - 50) = 47$

$8x^2 - 25 = 47$

$8x^2 = 72$

$x^2 = 9$

Since we want the positive solutions only; $x = 3$

Substitute back in to give: $y = 2x - 5 = 6 - 5 = 1$

Question 15: E

➤ Urea is produced when proteins are broken down in the liver, so the highest concentration will be at 3.

➤ Starch/long carbohydrates are broken down into sugar monomers in the stomach and intestines so glucose concentration will be highest at 2.

➤ Oxygen will be lowest in 3 because blood has travelled through more organs (and therefore oxygen has been used up the most).

Question 16: 2m/s²

Resultant force = Force from pulleys – Weight

Resultant force = $120 \times 2 - 20 \times 10 = 40\ N$

Now using, $F = ma$; $40 = 20a$

Thus, $a = 2\ ms^{-2}$

Question 17: E

1. The nucleus has a relative mass of 40.
2. It is not a noble gas as it has a configuration of 2,8,8,2 and thus has 2 electrons in the outer shell.
3. It has 2 electrons in the outer shell. Therefore, it would lose 2 electrons to obtain the noble gas configuration, hence it would form a positive ion.
4. The element has 2 electrons in the outer shell so it is in group 2
5. The element is in group 2 which are the alkali metals.

Question 18: C

➢ The student takes 14 x 4 = 56 breaths in 4 minutes.
➢ Each breath is 0.5 dm^3 so 56 breaths = 56 x 0.5 = 28 dm^3.
➢ The percentage of oxygen absorbed = 21 – 16 = 5%
➢ Thus, amount of oxygen absorbed = 5% of 28 dm^3 = 1.4 dm^3

Question 19: C

This is no longer on the BMAT syllabus.

The pupil dilates when the light levels decrease to allow more light in. This is accomplished by contraction of the radial muscles and relaxation of the circular muscles.

Question 20: E

You are no longer expected to be able to calculate resistances & voltages in parallel circuits for the BMAT. However, it's important that you understand the principles involved with parallel circuits as they do frequently come up.

➢ The voltage across both branches is equal.
➢ The upper branch (series) has a greater resistance than the lower branch (parallel).
➢ Since, $I = \frac{V}{R}$, the current in the upper branch is smaller than the lower branch.

Now using V = IR, we can see that there is a greater voltage running through R2 than R1.

Since the voltage splits evenly across the second branch point in the lower branch, R3 has half the voltage of R2.

Therefore, the potential difference in ascending order is: R3 → R1 → R2.

Question 21: 9

$$= \left(\frac{32^{\frac{1}{5}}+9^0}{81^{\frac{3}{4}}}\right)^{-1}$$

$$= \left(\frac{2+1}{3^3}\right)^{-1}$$

$$= \left(\frac{3}{27}\right)^{-1}$$

$$= \left(\frac{1}{9}\right)^{-1} = 9$$

Question 22: D

A. The activation energy is represented by V + W.
B. Y shows the route taken in the presence of a catalyst (not X).
C. The reaction is endothermic as the energy of the products is greater than the energy of the reactants.
D. Correct.

Question 23: C

The frequency of light doesn't change during refraction.

Blue light has a higher wavelength than red light (it's just one of those facts you need to know!). Since $c = f\lambda$, the speed of blue light must be less than the speed of red light in glass.

Question 24: E

1. Correct – aerobic respiration releases far more energy than anaerobic respiration.
2. Incorrect- Oxygen debt can only occur in the presence of anaerobic respiration.
3. Incorrect – Aerobic respiration also produces water as a waste product.
4. Correct- active transport requires ATP which would be less available if there were reduced aerobic respiration.
5. Incorrect - the muscles will still use any oxygen available for aerobic respiration.
6. Correct – lactate is the only waste product of anaerobic respiration.

Question 25: B

This is no longer on the syllabus. Don't get bogged down in the equations. 4 protons fuse to create a helium nucleus which consists of 2 protons and 2 neutrons. Therefore, 2 of the original 4 protons have to be converted into neutrons – hence B.

Question 26: A

The best way to do these is to express the relationships algebraically: $A \, \alpha \, \frac{1}{b^2}$

Therefore $A = \frac{k}{b^2}$

Substitute the values in to give: $9 = \frac{k}{4^2}$

Therefore, $k = 9 \times 16 = 144$

When A = 4; $4 = \frac{144}{b^2}$

$b^2 = 36$

$b = 6$

Question 27: 5 seconds

Charge per unit surface area = 0.25C/m^2 and the surface area = 0.04m^2

Therefore, the overall charge = $0.04 \times 0.25 = 0.01 \, C$

Recall that $Charge = current \times time$

$Time = \frac{10^{-2}}{2 \times 10^{-3}}$

$\frac{1}{2 \times 10^{-1}} = \frac{1}{0.2}$

$= 5 \, seconds$

END OF SECTION

Section 3

Our zeal to make things work better will not be our anthem: it will be our epitaph. (Appleyard)

➤ Appleyard suggests that our drive to improve things will not be our success, but our downfall.

➤ Many researchers believe that it is not long before artificial intelligence achieves the storage capacity and processing speed of the human brain. A huge number of (arguably realistic?) science fiction films portray this leading to robotic being sentient and conscious, such as Terminator and I.Robot, but even unconscious machines may have the ability to transcend humanity with the emergent property of acquiring consciousness. It is also likely that these machines would have the ability to replicate themselves more efficiently than humans.

➤ If machines had the ability to make their own decisions then one could argue that the fate of the human race would be at the mercy of these robots. Even if this was not the case, however, we may become so dependent on technology that we essentially become 'engineered' and completely lose our freedom.

➤ Talking in the present, modern technology and our insatiable desire to 'make things better' is undoubtedly causing problems such as global warming and loss of resources which will potentially have a devastating long term effect. Furthermore, rapid technological advancement has clearly led to a more stressful and fast paced lifestyle, and many epidemiologists believe that this is the cause of the exponential rise in mental health issues in the last 50 years (there is certainly, at least, a correlation).

➤ However, the only way to deal with problems such as global warming and mental health is further innovative technology, to try and find renewable, environmentally-friendly sources of energy and treatments to aid those with psychological problems.

➤ Technology is often crucial to our survival; if the Earth was to be subject to a global natural disaster, only technology would allow us to propel into space and find a new method of surviving. From a medical perspective, only new technology would be able to save us from a devastating new virus that is readily transmitted and fatal when contracted. Even on a very current basis, antibiotics need constant innovation or natural resistance would render them useless.

Higher education and great numbers - that is a contradiction in terms. (Nietzsche)

➤ Nietzsche suggests that higher education should be reserved for only a small proportion of a population, who have earned the right and privilege for further study through merit. It follows on from the idea that not all things in life cannot be common property. Higher education is commonly thought of as university, and the government has always been keen to ensure that this is not an elitist system, although the recent rise in tuition fees may suggest otherwise. However, although most 18 year old studying for their A Levels are encouraged to go to university, Nietzsche may believe that only those who have the ability to be the very best and who are keen to specialise in a precise academic area should proceed to further education - a meritocracy of sorts.

➤ Indeed, it could be argued that overcrowded institutions with overworked teachers/professors leads to a lower quality of education which is detrimental to all. Higher education should be students who are the 'cream of the crop,' taught by those who were once the 'cream of the crop,' whereas with a high number of students there is a mediocrity to the teaching and what is achieved. Also, as there are more students to teach who are not exclusively a talented and motivated academic group, bad teaching cannot be survived.

➤ Higher education is commonly thought of as university, and the government has always been keen to ensure that this is not an elitist system, although the recent rise in tuition fees may suggest otherwise. However, although most 18 year old studying for their A Levels are encouraged to go to university, Nietzsche may believe that only those who have the ability to be the very best and who are keen to specialise in a precise academic area should proceed to further education.

➤ Indeed, students often see higher education as the only way to establish a career, often meaning that the essence of 'education' itself is lost. Universities, more and more, are entering a time of commodification and marketization and we should only accept high numbers in higher education if it benefits society as a whole. How universities are run now may detract from essential 'academic' work that needs to be done.

➤ However, education as a notion is subjective and the view of Nietzsche is certainly not a general one. For many, the employability factor of higher education is huge, and for them this is a legitimate reason to pursue further study. Einstein's quote of 'don't judge a fish by its ability to climb a tree' fits in here.

➤ It can be argued that education should be a right at any level. Some may use higher education to train for a future career. Some may use it to improve society for the better and may use it to ensure that we all have a future.

➤ Indeed, over time we have had a fundamental change of university from an elite system to a mass system, and it has, on the whole, appeared a success. It allows education to go beyond your social class, race and other factors and allows equal opportunity. Students are more demanding in this age regularly criticise the teaching and learning support they get, and the education received, even though by a larger number of people, has certainly improved over the years.

The main benefit of patient consent is that it relieves doctors of blame for bad decisions.

➤ The statement suggests that, if a patient has given their permission for a treatment to go ahead, then the doctor cannot be attributable if the treatment doesn't go to plan.

➤ Although this may be true to an extent, the reason that patient consent is a main idea of Good Medical Practice issued by the GMC is that it benefits the patient and respects their autonomy.

➤ Valid, informed consent requires competence (right mind), non-coercion (right person) and information (right understanding). No consent is battery (a crime).

➤ There are many benefits of patient consent, the most obvious being the transmission of information from someone who has knowledge (the physician) to someone who does not (the patient), so that the patient is fully aware of what the treatment involves, the risks and benefits, whether there are alternative treatments and what may happen if treatment does not go ahead. It also respect patient autonomy and allows competent patients to control what happens to their body. It also protects the patient's right of self-determination and engages the patient in their own healthcare.

➤ A competent patient can decide if they want to endure chemotherapy as their cancer treatment, or whether they would like to take part in a clinical trial to try and alleviate their Parkinson's disease. However, recent Mental Health Acts protect patients that lack the capacity to make decisions for themselves, and discusses how decisions should be made in their best interests.

➤ Ultimately, in most cases, if patients have a clear appreciation and understanding of the facts, implications, and consequences of a treatment/action, they are free to choose as they wish.

END OF PAPER

2007

Section 1

Question 1: B

After 1 year the tree is 1m tall.

The max is 30m, so the difference between the height and the max is 29m. 10% of this is 2.9m, so after 2 years, the tree is 1 + 2.9 = 3.9m tall.

The difference now is 26.1m, 10% of this is 2.6m (answer is 1dp), so after 3 years, the tree is 3.9 + 2.6 = 6.5m tall.

Question 2: C

The argument discusses how WMDs are not the only means through which people are killed in warfare. C thus logically follows the idea that they are not unique in their ability to cause destruction. This is further emphasised in the last line.

'200,000 people were killed…by the atomic bombs' which suggests the WMDs are 'morally unacceptable,' 'a serious threat' and 'necessarily devastating'.

Question 3: C

Find out the increase for each representative - A 60000, B 40000, C 50000, D 100000, E 50000. You should then be able to work out that answer C is correct - for example, Darwin should have 1/3 of the pie chart because his increase is 100000, out of a total increase of 300000.

Question 4: A

The argument suggests that a higher level of mastery should be required to pass a test, which would then reduce the loss of young life. However, it does not consider that young male drivers may already be highly skilled and still be involved in accidents, as said in 1.

2 and 3 do not weaken the argument - the argument that we should increase the threshold for driving skills before allowing one to pass to prevent road deaths, still has high validity with these considerations. 2 is irrelevant to the skill of the driver and 3 actually supports the argument.

Question 5: B

From the answers given, we can see that the percentage occupancy is 92853/109505 for acute, 24291/26619 for geriatric, 27832/31667 for mental illness, 4134/4899 for learning disabilities and 5737/9095 for maternity. Hopefully you can immediately see that maternity is too low.

If we call the rest 93/110, 24.3/26.6, 27.8/31.7 and 4.1/4.9, you can hopefully see that 24.3/26.6 is the largest. 93/110 is roughly 23/27.5 (divide by 4) which is clearly smaller than 24.3/26.6.

24.3/26.6 is larger than 27.8/31.7 - 24.3 to 27.8 is a 15% increase but the denominator increase is much larger. 4.1/4.9 is around 24/30 which is also clearly larger than 24.3/26.6.

Question 6: B

1 is correct because the headline suggests there is a growing digital divide, whereas the survey only found values for social networking.

2 is correct because to show a gap is growing, you also need an earlier survey.

3 is not related to the question, because it is nothing to do with the headline.

Question 7: E

Let's say the first bulb is red - this has a 3/6 chance. With the sequence we want: the next bulb has a 3/5 chance of being yellow, the next has a 2/4 chance of being red, the next has a 2/3 chance of being yellow, the next has a ½ chance of being red and the final one has a 1/1 chance of being yellow.

So 3/6 x 3/5 x 2/4 x 2/3 x ½ x 1 = 1/20.

However, we must multiply this value by 2 because there are 2 permutations - the first bulb could be yellow, and then alternate by there. 1/10.

Question 8: C

The passage begins with 'many people think that if they cannot explain something, it must therefore be truly paranormal,' but the author goes on to criticise this by giving examples of where this thinking has been wrong, which suggests C is the correct answer.

A and B are too bold to be inferred and D doesn't have a definite basis.

Question 9: E

2kg cat - ber = 30 x 2 + 70 = 130. mer = 130 x 1.6 = 208 kcal/day

Requires 7g per 100kcal, so 14g per 200kcal, and 0.56g per 8kcal, so the total required is 14 + 0.56 = 14.56 - E is correct.

Question 10: C

BER for this cat is 70, and the MER is 70 x 1.3 = 91kcal/day. It requires 7g/100kcal of protein and thus requires 6.3g for 91kcal. CCFR has 0.06g/ml and 1kcal/ml, so dividing the former by the latter gives us 0.06g/kcal (ml cancels) which is 6g/100kcal. This means it gets 5.4g for 91kcal, which is 0.9g short.

Question 11: C

➤ A cat with renal failure requires 4g/100kcal.
➤ FCH has 1.4g/1.3kcal which is far too high.
➤ CCFR has 0.06g/1kcal which is 6g/100kcal - too high.
➤ ES has 0.04g/1kcal which is 4g/100kcal - correct.
➤ OHN has 0.05g/1.1kcal which is 5g/110kcal - too high.
➤ EMF has 0.08g/2.1kcal which is 8g/210kcal - too low.

Question 12: C

Ratio of protein to energy required is 6:100. Call the volume of ES added E.

100ml x 0.09 + 0.04E = 0.06(100+E)

9 + 0.04E = 6 + 0.06E

3 = 0.02E

300 = 2E. Therefore, E = 150ml

Question 13: C

The graph shows the percentage who had drunk on x days or more, and our percentage for all ages is about 18%.

➤ If x = 3 days or more, then we would have a percentage of 10 + 6 + 4 +3 + 11 = 34% - too high.
➤ If x = 6 days or more, then we have a percentage of 3 + 11 = 14% - too low.
➤ If x = 5 days or more, then we have a percentage of 4 + 3 + 11 = 18% - correct.

Question 14: A

The argument suggests that dogs' excellent hearing ability means they can hear sounds that occur before an earthquake, which increases their anxiety. A supports this argument because it suggests that dogs that haven't got this hearing ability did not hear the sound and thus did not show an increase in anxiety.

B is irrelevant to why dogs in particular increase in anxiety, C does not necessarily give support to the argument stated above and D weakens the argument by suggesting there are no pre-earthquake sounds.

Question 15: C

If A got 1/3, P and E got 2/3 combined, so 2x/3 = 174 if x is the total number of votes. x = 261, so x/3 = 87. Paul wins by 116 - 87 = 29 votes.

Question 16: A

The conclusion of the argument is the last line "someone who is awarded a PhD in Sweden will live longer." The only data we are given is that those aged 64 with a PhD were less likely to die within the next year than those with a BA, BSc or MA, so we cannot definitely infer that what is true for this general group is true for each individual, which is the assumption made.

B is not correct as there is nothing said about a healthy lifestyle in the argument.

C is not correct because the conclusion states those in SWEDEN will live longer, so there is no assumption about other.

D is a statement that shows an error of reasoning and the assumption is only the other way around to what is stated here.

Question 17: B

Only 2 is correct - the argument suggests that smokers are more addicted to nicotine now because they inhale, in total, a larger amount of nicotine. However, this statement suggests that the nicotine inhaled isn't necessary increased (because less cigarettes are smoked in total, despite the increase per cigarette), which weakens the argument.

1 is irrelevant to the amount of nicotine involved.

3 doesn't weaken the argument about smokers being more addicted due to increased nicotine.

Question 18: B

Paving slabs have an area of 70 x 70 = 4900 cm^2 or 0.49m^2.

Bob can buy 50 slabs with £140. 50 x 0.49 = 24.5m^2 can be covered with this.

The area covered is twice as long as it is wide.

Area = x * 2x slabs, 50 = 2x^2 so x = 10.

Length is 10m and the width is 5m.

IF we have 10 slabs along length and 5 slabs along the width, this covers 5 x 70cm + 10 x 70cm = 3.5m^2 x 7m^2 covered by the slabs.

Flowerbed width = 5m^2 - 3.5m^2 = 1.5m^2

Question 19: B

Car 1 gets 4 seconds ahead of car 2 per lap - 4 seconds after lap 1, 8 after lap 2, 12 after lap 3 etc. When car 1 laps car 2, it fulfils the criteria that 66s x number of laps = 70 x (number of laps - 1). This expands to give 70 = 4 x number of laps, so it takes 17.5 laps. 17.5 laps x 66 seconds gives 1155 seconds which is 19 minutes and 15 seconds.

Question 20: D

Out of 1000 men we know 21 are screen with a true positive and 85 are screened with a false positive. This is 106/1000 men screened positive with DRE, which is 10.6%.

Question 21: C

Draw a tree diagram here. Either men have an abnormal PSA level (0.1 chance) and have prostate cancer (0.26 chance), giving a 0.1 x 0.26 = 0.026 = 2.6% chance this way.

Or, men could have a normal PSA level (0.9 chance) and be a false negative (0.008 chance), giving a 0.9 x 0.008 = 0.0072 = 0.72% chance this way. 2.6 + 0.72 = 3.3%.

Question 22: A

1 is correct. PSA gives 0.1 x 0.74 false positives, which is 0.074 = 7.4%. DRE gives 85/1000 false positives which is 8.5% - this is higher.

2 is also correct. PSA gives 0.9 x 0.008 false negatives, which is 0.72%. DRE gives 16/1000 false positives, which is 1.6% - this is again higher.

Question 23: B

Chance of a false positive on PSA is 0.074, chance on DRE is 0.085. 0.074 x 0.085 = 0.00629 (do 74 x 85 and then work it out from there), which is 0.6% - on average 6/1000 would have false positives on both.

Question 24: D

The conclusion, 'what really drives globalisation is the availability of cheap air travel and cheap shipping,' suggests that the internet is neither necessary nor sufficient for globalisation. If it were sufficient, the argument would suggest that the internet alone, without this shipping etc., could lead to globalisation. If it were necessary, the argument would suggest that the shipping/travel isn't possible without the internet.

Question 25: B

Ethanol is currently (2007) produced on 54,000 square km of farmland, which needs to increase to 334,000 by 2017 to meet the target. This would give 334000 + 54000 = 388000 square kilometres in 2017.

This is just over 7x more than in 2007 (54 x 7 = 378) so if we can produce 35 billion gallons in 2017, we produce just over 7 times less now. 35/7 = 5, and because our denominator is slightly more than 7, we get 4.9 billion gallons.

Question 26: C

There are only 4 combinations that are possible here, and 2 of them result in Q and R sitting next to each other.

1. Window (O) (N) (M) aisle (P) ( ) ( ) which leaves Q and R to sit next to each other.
2. Window ( ) ( ) (P) aisle (M) (N) (O) which also leaves Q and R to sit next to each other.
3. Window ( ) (M) (N) aisle (P) ( ) (O) which doesn't let Q and R sit next to each other.
4. Window (O) ( ) (P) aisle (N) (M) ( ) which doesn't let Q and R sit next to each other.

Question 27: C

6/50 had rings on them when rounded up, and 50 have rings overall. So 6/50 = 50/x, x = 50 x 50/6 = ~417.

Question 28: A

The data suggests a correlation between the being light on during sleep and myopia in children, and the conclusion is that there is causality between these 2 factors. A gives another way that these 2 factors may be linked - the fact that myopic parents leave the children's light on more often, so the causal factor actually being the transfer of myopic-related genes.

B, C and D are not flaws because they do not give an alternative explanation for the correlation.

Question 29: C

The first sentence defines conflict diamonds: 'sold in order to finance the military rebellion of groups opposing recognised governments,' and the conclusion suggests that the cost of these are too high for 'war-torn countries.' However, C weakens this, suggesting that the sale of these diamonds actually helps the recognised governments of war-torn countries.

Not knowing where a diamond originates does not weaken the argument that the diamond is an unnecessary luxury and has too high a cost, so we can rule out A. B has nothing to do with diamonds, and the significance of diamonds mentioned in D has nothing to do with diamonds in the context of war/governments.

Question 30: E

Visualisation - cut the shapes out and try it yourself if you want. The triangles will fold to form squares in Q and S, but not P and R.

Question 31: B

MHR= 217 - 0.85 x 60 = 217 - 51 = 166.
Add 4 beats for 55+ elite athletes - 166 + 4 = 170.
Subtract 14 beats for swimming training - 170 - 14 = 156

Question 32: D

➢ A and B are not correct - we don't know anything about absolute wages, only proportions.
➢ C is not correct - for example the wage proportion for women in education after 21/22 (relative to those leaving education at 15) follows a general downward trend.
➢ D is correct - this is suggested by the downward trend seen for women at 21/22.
➢ E cannot be inferred - we only have proportions relative to those who left education at 15.

Question 33: C

This is yet another correlation does not mean causation question. A general positive correlation is seen between wage increase and years of education past 15, and the authors assume causality.   With averages there is no assumption that some outliers are not biasing the data, so A is wrong. B is irrelevant as the graph is about relative wages, not absolute wages. D is not an assumption made - it is actually a potential counter argument. E is clearly wrong.

Question 34: B

We have a link between being a trade union member and having lower financial returns for a degree, and we want a cause. B suggests a reason why trade union membership may mean less money.

A is irrelevant as the number of people that get trade union membership doesn't provide a reason for the link. C is wrong because we are talking about those that do have a degree. D is wrong because age is irrelevant here. E is wrong for the same reason as A.

Question 35: D

The fact that older people left education earlier but earned more actually counters the general trend in our graph, so without this, the correlation we see would be much stronger. Thus the effect is that it lowers the gradient of the graph, weakening the link.

END OF SECTION

Section 2

Question 1: B

Proteins are normally found in the blood but do not pass through the glomerulus after ultrafiltration (so A is wrong), giving an filtrate that is mainly water.

As we carry on through the kidney, more water is reabsorbed and glucose is completely reabsorbed, making the fluid more concentrated with urea. Since there is still some glucose and very little urea, this sample must be from the proximal convoluted tubule which is represented by B.

Question 2: B

A. If x was in groups 3 to 8, y would be in the third period.

B. Since x is in the second period, and y is in the second/third period, this means that both must have at least 2 electrons in their first shell.

C. Element y would only have 6 more electrons in its outer most shell if x was in group 1 or 2.

D. Since y has 6 more protons, it will likely also have a few neutrons in addition as well. Thus, the nucleon number of y will be greater than that of x by more than 6.

Question 3: $\alpha=220$, $\beta=40$

Only 60 counts pass through the sheet of paper, which is 40 when we account for the background. Only β radiation passes through a sheet of paper, so this value is 40.

The count rate is 280 at the start = 260 once adjusted for background radiation. If 40 of these counts are caused by β then 220 must be caused by α particles.

Question 4: A

The adjusted equation is: $A = \frac{(1.5x + 1.5y)^2 \times 0.8z \times Q}{2P}$

$A = \frac{1.5^2(x+y)^2 \times 0.8z \times Q}{2P}$

Since we are only interested in the percentage change, we can ignore the unknown variables:

$A = \frac{(1.5)^2 \times 0.8}{2}$

$A = 2.25 \times 0.4 = 0.9$

Thus, A has decreased by 10%.

Question 5: E

During expiration, the ribs move down and inwards (hold your ribs when exhaling if you're not sure!). The diaphragm muscles relax during expiration, not contract. The pressure in the lungs increases during expiration, so that a pressure gradient is created which causes air to move out.

Question 6: A

$t = 2\pi \sqrt{\frac{2lR^2\left(W+\frac{w}{3}\right)}{n\pi r^4 g}}$

$\left(\frac{t}{2\pi}\right)^2 = \frac{2lR^2\left(W+\frac{w}{3}\right)}{n\pi r^4 g}$

$\frac{n\pi r^4 g}{2lR^2} \times \left(\frac{t}{2\pi}\right)^2 = \left(W+\frac{w}{3}\right)$

$W = \frac{n\pi r^4 g}{2lR^2} \times \frac{t^2}{4\pi^2} - \frac{w}{3}$

$W = \frac{n r^4 g t^2}{8l\pi R^2} - \frac{w}{3}$

Question 7: 300

Using $P = IV$, the output voltage $= \frac{0.5 \times 10^3}{10} = 50 V$

Then, input these values into: $\frac{V_p}{V_s} = \frac{n_p}{n_s}$

$\frac{250}{50} = \frac{1500}{x}$

$x = \frac{1500}{5} = 300$

Question 8: B

In Fe_3O_4, the O has a redox state of -2, so O_4 has a redox state of -8. The compound is electrically neutral thus Fe_3 must consist of 2 atoms in the +3 state and 1 in the +2 state. Thus $\frac{1}{3}$ is in the +2 state.

Note that Fe_3O_4couldn't consist of 4 atoms in the +2 states because the question specifically states that the oxide contains both Fe^{2+} and Fe^{3+}.

Question 9: B

Blood is pumped from the right ventricle to the lungs and from the left ventricle to the body. Both of these processes happen at the same time. Thus, the atrio-ventricular valves are closed on both sides to prevent backflow into the ventricles and the semilunar valves are open to allow blood flow into the pulmonary artery + aorta.

Question 10: D

The greatest number of moles of reactants produces the most precipitate. Since one mole lead iodide precipitate reacts with two moles of potassium iodide, we need to double the moles of KI in the calculations.

A. $\frac{5}{1000} \times 2 + 2 \times \frac{10}{1000} \times 2 = 0.03 \ moles$

B. $\frac{2.5}{1000} \times 5 + 2 \times \frac{2.5}{1000} \times 2.5 = 0.0375 \ moles$

C. $\frac{7.5}{1000} \times 3 + 2 \times \frac{5}{1000} \times 5 = 0.0725 \ moles$

D. $\frac{5}{1000} \times 4 + 2 \times \frac{7.5}{1000} \times 5 = 0.095 \ moles$

A good student would spend little time working out the values; they would see that the value of D would be the highest just by inspection.

Question 11: E

A. A substance can lose heat energy without its temperature falling if it is changing state.

B. Thermal radiation can pass through a vacuum.

C. Steam has more heat energy at the same temperature as boiling water because the intermolecular bonds in the boiling water are still unbroken.

D. Cooling or heating a container of water will set up a convection current, regardless of the location.

Question 12: E

850 steps is the mean and 1000 steps is the upper quartile value, so ¼ of the sample have values within this range.

3 members having a value within this range has a probability:

$= \frac{1}{4} \times \frac{1}{4} \times \frac{1}{4} = \frac{1}{64}$

Question 13: C

This is no longer on the BMAT syllabus.

The fulcrum acts as a pivot. If you stand on your toes, then the calf muscles contract, and the upwards force from the calf is the effort. The downward force is the load, and this is between the pivot and effort.

Question 14: A

Elements further down in group 1 are more reactive, and elements further up in group 7 are more reactive. Thus, the element on top of group 7 (Fluorine) and on the bottom of group 1 (Caesium) will react extremely violently. This will therefore be the most exothermic reaction.

Question 15: B

This is no longer on the BMAT syllabus.

➢ The refracted rays of light will follow similar trajectories – this excludes out C & D.

➢ Since light is incident in the top half of the bubble, it will leave from the top of the bubble as well. This excludes E & F.

➢ The wavelength of red light is higher than that of violet. Thus, red light undergoes less refraction than violet light.

➢ Hence, the answer is B.

Question 16: C

The quickest way to do this is to substitute in the values to see if they satisfy the inequality:

➢ (-1, -6): $-6 \geq 1 + 3$ is incorrect

➢ (2, -1): $-1 \geq 4 + 3$ is incorrect

➢ (1, 6): $6 \geq 1 + 3$ and $1 \geq \frac{1}{6}$. This is correct for both.

➢ (2, 2): $2 \geq 4 + 3$ is incorrect

Question 17: D

We first deduce that this is a sex linked condition based on who gets the syndrome. Call the alleles X^N for the faulty gene and X^n for the normal gene.

The male in the first generation has a genotype of $X^N Y$ so he passes on his X^N gene to both his daughters, 8 and 9. The mother must provide the X^n to both otherwise they would have the condition.

This leaves answers D and E. We cannot deduce anything about 5, but we know 4 is heterozygous. 3 must have a genotype of $X^h Y$, otherwise he would have the condition. 4 must have a genotype of $X^H X^h$ - if she were homozygous dominant then she would have the condition and if she were homozygous recessive then she would not be able to have offspring with the condition.

Question 18: D

Moles of water $= \frac{mass}{Mr} = \frac{6}{18} = \frac{1}{3}\ moles$.

1 mole of gas occupies 24,000 cm^3.

Thus, $\frac{1}{3}$ moles occupy $24000 \times \frac{1}{3} = 8000\ cm^3$

Question 19: 700N

This is a difficult question as it requires a couple of conceptual leaps:

The bar is in equilibrium so: Moments clockwise = moments anti-clockwise

The bar exerts its weight in the midpoint between the pivot and the bar's end. Let y be the weight of the bar:

$1000 \times 1.5 + \frac{1.5}{2}y = 4.5 \times 100 + \frac{4.5}{2}y$

$1500 + 0.75\ y = 450 + 2.25\ y$

$1050 = 1.5\ y$

$y = 700N$

Question 20: A

Using Pythagoras' Theorem: $a^2 = c^2 - b^2$

$a^2 = \left(6 + \sqrt{5}\right)^2 - \left(3 + 2\sqrt{5}\right)^2$

$a^2 = (36 + 12\sqrt{5} + 5) - (9 + 12\sqrt{5} + 20)$

$a^2 = 36 + 12\sqrt{5} + 5 - 9 - 12\sqrt{5} - 20$

$a^2 = 12$

$a = \sqrt{12}$

$a = \sqrt{4} \times \sqrt{3}$

$a = 2\sqrt{3}$

Question 21: E

Platelets are responsible for clotting, so a decreased platelet count means less clotting. White blood cells mediate immune responses so abnormal white blood cells means reduced disease resistance. There is no information about red blood cells so we can assume oxygen transport is normal.

Question 22: D

1. Correct - NH_3 is the correct formula of ammonia.
2. Incorrect - Ammonia is a weak alkali in H_2O and thus has a pH > 7.
3. Correct - Ammonia has a molecular structure.
4. Incorrect - Ammonia is a base so turns damp litmus paper blue.
5. Incorrect - Ammonia is a gas at room temperature
6. Correct - Ammonia is covalently bonded.

Question 23: E

Let the cross sectional area be A. The section of an artery can be modelled as a cylinder with a volume Ax.

The volume of blood flowing through the artery in one second is the same as the volume of the artery divided by the time it takes for a single RBC to pass through its length:

$V = \frac{Ax}{T}$, so $A = \frac{Vx}{T}$

However, the units for this would be ml/mm, and we want the cross sectional area in mm^2. Since $1\ ml = 1\ mm^3$, we must multiply by 10^3 to get the final answer $= A = \frac{Vx}{T} \times 10^3$

Question 24: A

Volume of the hemisphere $= \frac{4}{3}\pi r^3 \times \frac{1}{2} = \frac{2}{3}\pi r^3$

Volume of the cylinder $= \pi r^2 l$

Total volume $= \frac{2}{3}\pi r^3 + \pi r^2 l$

$= \pi r^2 (\frac{2}{3}r + l)$

$= \frac{\pi r^2}{3}(2r + 3l)$

Question 25: A

The top of a fractionating column is where the hydrocarbon with the lowest boiling point fractions off, which will be at 68^0C. The temperature in the flask would be the average of both boiling points, which is 83^0C.

Question 26: C

This is no longer on the BMAT syllabus.

Oestrogen causes the lining of the uterus to thicken, progesterone maintains the lining of the uterus (e.g. during pregnancy), and a fall in progesterone causes the lining of the uterus to break down (causing bleeding in the menstrual cycle).

Question 27: D

Beta radiation involves a neutron transforming into a high energy electron and a proton. Thus, the atomic number increases by one, so 1 is correct. The tumour is attacked by gamma radiation, but the radiation attacks all cells, not just malignant cells.

END OF SECTION

Section 3

The technology of medicine has outrun its sociology.

➢ The author suggests that we have reached a stage whereby the technology of medicine has exceeded our current ability to apply it effectively as a society.

➢ Sigerist may be referring to the fact that portions of society regularly reject medical technologies due to religious, cultural or even personal beliefs, highlighting that the sociology is running behind the technology.

➢ It could also be argued that this statement refers to the notion the availability of clinical technological advancements are limited to a portion of society, independent of the people's views of the technology. Many treatments are only available privately, favouring the rich who can afford them, thus suggesting that the focus on technology has superseded the GMC's first rule of good practice: "make the care of you patient your primary concern."

➢ The technology of using embryonic stem cells can be used to, for example, make new brain cells to treat Parkinson's disease, help to replace damaged heart valves or help to rebuild cartilage and clearly has a huge range of potential applications as they can specialise into any cell. However, in order to obtain embryonic stem cells, the early embryo has to be destroyed which some regard as destroying a potential human life, and there is hence much controversy around this area and much of society does not accept this technology. Adult stem cells also exist, but these can only develop into a smaller subset of cells.

➢ To address this problem, researchers have tried to find ways to obtain embryonic stem cells without having to destroy the embryos. Methods to do this include deriving stem cells from mice embryos, and reprogramming adult stem cells to act more like embryonic stem cells - these are called induced pluripotent stem cells.

Our unprecedented survival has produced a revolution in longevity which is shaking the foundations of societies around the world and profoundly altering our attitudes to life and death

➢ Improvements in healthcare and medicine mean we have defied nature to levels that were not previously possible, which the statement suggests has affected our lifestyles and has changed our outlooks and approaches regarding life and death.

➢ Our longevity has caused overpopulation, despite a limited and finite number of resources, putting more pressure and demands on the Earth and causing increased famine and poor quality of life for some.

➢ In this country, the increased longevity has meant an ageing population who must be supported by a younger generation, putting a burden on public funds. Also, technology has led to many ethical issues about life and death, such as for how long we should keep someone on a life support machine.

➢ Furthermore, our longevity has meant people fear death less, at least in the sense of contracting a disease or illness that we consider minor now but may have been fatal many years ago. Indeed, the current obsession with risky activities such as extreme sports is probably borne from this fact, with people confident that they will recover even if something goes wrong.

➢ However, the fear of unknowns such as death still undoubtedly continues for most, and the way we treat the dead (burial, cremation etc.) has remained constant throughout history, despite longevity extending. Many societies are not experiencing this increase in longevity anyway, such as third world countries - can we definitely say their attitudes have changed?

➢ Basic human nature and lifestyles have not necessarily been 'shaken' by this increase. We are still governed by laws, we still survive through basic necessities such as food, warmth and we then must reproduce, and our attitudes and opinions towards many factors are the same.

Irrationally held truths may be more harmful than reasoned errors. (Huxley)

➢ This seemingly paradoxical statement is essentially about the concept of epistemology - how we know what we know. Huxley wants to elucidate the potential hazards of believing a concept which has not been proven correct through rigorous and logical scientific method.

➢ In stating "irrationally held truths," Huxley is referring to something that may be factually correct but the reasoning to determine this as truth was not based on evidence or rationality and cannot be justified. It can be regarded as 'blind knowledge'; for example pre-Hippocrates medicine was based only on supernatural causes of illness rather than medical observation but the likelihood is that at least some of this was factually correct.

➢ One example of something that may be considered as an irrational truth is Traditional Eastern Medicine - many of these do appear to have high efficacy despite us not knowing their mechanistic action. However, this may mean that treatment is not successful to all, and it may lead to other, untested medicines being accepted as effective. For example, homeopathy appears to 'work' on some people, although this is likely to merely be placebo and may cause danger if using it to try and cure a serious illness. For reasoned errors, the perpetrator has at least used a rational, logical method and is more likely to achieve a reasoned truth than someone who holds an irrational held truth.

➢ If someone has wrongly reasoned something then their 'logic' or thought process is open to review and criticism by those who wish to correct it, whereas irrational truths are more resistant to criticism and therefore revision and if further research is based on this 'truth', it may encounter pitfalls further along the line.

➢ However, much of medicine does take a 'reverse engineering' route whereby we test whether a treatment is useful first, and then look to see its mechanistic action. For example, Deep Brain Stimulation is currently the best treatment for Parkinson's Disease but we still do not know how it works.

➢ Reasoned errors can often be more dangerous; in a simple maths problem, a minor calculation error leads to the wrong answer whereas a completely incorrect method may lead to the correct answer. However, if this was to determine the dosage of a drug for a patient, it is clear that we would rather have the correct answer using an incorrect method.

➢ The idea of what is rational and what is irrational constantly changes over time; Darwin's ideas were dismissed as completely irrational at the time, due to the rational explanations offered by religion. Also, many people, if they are convinced they have used a logical and rational method, will never be able to see the error of their answer/explanation.

END OF PAPER

2008

Section 1

Question 1: D

The word 'some' is key here. If we know some M are Z and all Z are T, then some M (at least the subset that were Z) are T. We cannot deduce that all M are T.

Some students find it helpful to think of these in more realistic terms e.g. you could think of the question as 'Some A Level students take the BMAT. All taking the BMAT are applying to university'. And we know some A Level students are applying to uni from this, but we cannot deduce that they all are.   Another useful technique here is using a Venn Diagram.

Question 2: C

Traditional evolutionary theory suggests 'animals are never altruistic' and even though there is an example given of reciprocal altruism, the conclusion is that there is 'always clear paybacks'. Thus reciprocal altruism is not truly altruistic, which is paraphrased in C. None of the other answers are consistent with this idea.

Question 3: B

There are 250 screws overall, and he can pick a 4mm screw which is 35 or 45mm long - there are 38 in total of these. 38/250 = ~15

Question 4: A

The last line is the conclusion and the argument is that, as all wonder-drugs use animal testing at some stage, it is irrational to suggest this part of the study CAUSES the discovery of the drug. A logically fits with this idea, whereas the others do not.

Question 5: E

The grey die is conventional - if you rotate the dice downwards so that the 6 moves to the new position, you can see that the 5 remains in the correct place and there is a 4 opposite the 3.

In the other, we must rotate the 6 sideways and downwards to its new position, and this must be in the way to ensure that the 3 and 5 in the first picture are no longer there. From these rotations we can tell that the 5 is opposite the 4, the 3 is opposite the 1 so the 2 must be opposite the 6.

Question 6: D

1 is correct because the headline suggests there is a growing digital divide, whereas the survey only found values for social networking. 2 is correct because to show a gap is growing, you also need an earlier survey. 3 is not related to the question, because it is nothing to do with the headline.

Question 7: D

$20m^2$ patio.

6 lots of 60 x 60 black tiles - $3600cm^2$ x 6 = $2.16m^2$

2 lots of 40 x 40 black tiles - $1600cm^2$ x 2 = $0.32m^2$

6 lots of 40 x 60 black tiles - $2400cm^2$ x 6 = $1.44m^2$

$2.16 + 0.32 + 1.44 = 3.92m^2$

$3.92/20 = 19.6\%$

Question 8: 39%

The relative risk was 1 for non-left handed women, and 1.39 for left-handed women. The percentage increase from 1 to 1.39 is 39%.

Question 9: B

$361 + 65 = 426$ total cases.

$153422 + 19119 = 172541$ total (estimated) person-years lived.

$426/172541 \sim 4.25/170 = 2.5\%$

Question 10: A

$144 + 20 = 164$ total cases

65245 total person-years lived.

$164/65245 \sim 1.65/65 = 2.5\%$

Question 11: D

This is the only answer which gives a potential causal reason for the positive correlation seen in the figures.

A is irrelevant as this technique is correct.

B is irrelevant to being left-handed.

C doesn't give a causal link.

E is irrelevant because the 'relative risk' value takes into account the number of people in the group, so it doesn't matter how many people are involved (unless the sample size was too small).

Question 12: A

Call the number of adults x and the ticket price y.

Takings last week = xy = £1560.

40% more adults (1.4x) and 25% drop in price (0.75y)

$1.4x \cdot 0.75y = 1.05xy = £1638$

Question 13: E

We cannot be certain that people's opinions have necessarily been influenced by pressure groups - pressure groups are not mentioned in the argument.

2 is correct because if 34% think cars contribute more to climate change and 40% think planes, then at least some of the people are wrong.

3 is correct because if 47% think air travel should be limited but only 15% are willing to fly less often, there must be some who think air travel should be limited but are not willing to fly less often (say all 15% that are willing to fly less often also want to limit air travel, there are still 32% who follow this criteria).

Question 14: A

Steve Cram set his time in July 1985. The next date is August 1985 (S.A.), and is a quicker time, so this must have been the new WR. The next date with a quicker time is July 1995 (N.M.), so this must have been the new WR. The next date with a quicker time is July 1998 (H.E.G) and this remains the WR. So there are 3 in total.

Question 15: D

A is not necessarily correct - it may affect cell behaviour but we don't know that it will have a negative effect on health.

B is not correct - we have evidence that low level radiation does affect cell behaviour (not a 'negligible reaction').

C cannot be deduced from the passage.

D is correct - we are told that cells react to low level radiation from phones.

Question 16: D

For the first part of the tank, for every increase in measurement on the dipstick, there is an even greater volume of liquid, so we have an accelerating curve at the start - D or E. Then, for a while, the measurement of dipstick is constant relative to the volume of liquid, giving us a straight line - still D of E. Then, as the tank narrows, an increase in measurement on the dipstick means a lower increase in volume - so a decelerating curve at the end, and only D is right.

Question 17: A

Only 1 is correct - the passage argues that pop stars and celebrities shouldn't be promoting environmental awareness for various reasons, but if they have a broad appeal this suggests that they should.

2 is not correct - the idea that they are becoming more aware of themselves is irrelevant.

3 is not correct - the fact they did not gain financially has nothing to do with environmental awareness.

Question 18: A

We want to make sure that the resulting solution doesn't have more than 1% of chemical, so we must assume we have got as 'unlucky' as possible. The droplet could be as high as $0.025cm^3$ based on rounding, and the water could be as low as $5cm^3$ based on rounding. So to ensure we do not exceed 1% we can only have 2 drops - $0.05cm^3/5cm^3$.

Question 19: 79

The extension is on average 3.6 years with 30 minutes of running a day. This increase would be from 75 to 78.6, which we round up to 79.

Question 20: C

There are 3.6 years extra lifespan for running 30 minutes, and 1.4 for walking. Percentage increase is $(3.6 - 1.4) \times 100/1.4$, which is 157%.

Question 21: C

This observation is not stated in the passage, and cannot be inferred from the passage - there is nothing about when the people started exercising. It doesn't, however, counter the passage.

Question 22: D

The conclusion is that "people who exercise regularly really do live longer," suggesting a causal link between exercising and living longer (although there may be another factor linking these 2 variables).

A doesn't necessarily need to be assumed for the conclusion to make sense. The vigour of walking is not an idea of the passage, one can be overweight but still exercise a lot which counters C and E is stated.

Question 23: D

Hopefully you can see the pattern of flashes for each letter. If not, you can work out each flash combination (as every single one is unique), and their duration.

We know there is 1 flash for A and B, 2 flashes for C, D, E and F, 3 for G H I J K L M and N and 4 flashes for the rest. That makes $2 \times 1 + 4 \times 2 + 8 \times 3 + 12 \times 4$ flashes which are a second each, which totals to 82 seconds. We must add 25 seconds for the gaps - $82 + 25 = 107$ seconds.

Question 24: D

A is not correct because although the passage suggests the British competitors have affected their chances of winning major championships we cannot deduce that the standard of play has declined.

B cannot be inferred - although they are making a lot of money we cannot say they are more interested in this.

C cannot be assumed - although the passage suggests that the current generation are 'having lunch together' we cannot say they did not used to be on friendly terms.

D is correct - the first line of the second paragraph suggests they used to have to win many times to even make a living, whereas now they can make millions from just appearances, and we can definitely assume here that making that much money from appearances is more than you need to live.

E is not correct - we don't know whether the other commitments makes it harder to win.

Question 25: 245

Previously - r had 2.5 times more seats as b, and b had x seats, so r had 2.5x seat

Now - If r's lead was reduced by 56, then b gained 28 reps and r lost 28. B has $x + 28$ and r has $2.5x - 28$.

R now has 1.5 times more seats than b, so $2.5x - 28 = 1.5 (x+28)$.

$x = 70$.

So number of seats - $2.5 x + x = 245$.

Question 26: D

1 is correct, because the argument suggests that only those who have committed crimes before should have their DNA kept, because only these people will commit sexual and violent crimes. However, they could be first time offenders.

2 is correct because those who are not found guilty may have committed a crime and then they are not first time offenders, and should have their DNA sample kept according to the author's logic.

3 is not correct - solving sexual and violent crime without DNA evidence is irrelevant to this particular argument.

Question 27: C

Let's call the convoy's speed x and the courier's speed y. Then use distance = speed x time.

Time to get from the front to the back = 1/120 hours, and to get here we can add the speeds of the convoy and courier to get the net speed of him travelling backwards. 1km = (1/120)(x+y), so x + y = 120

Time to get from the back to the front = 1/20 hours, and to get there we have the find the difference between the speed of the courier and the convoy to get the net speed of him travelling forwards. 1km = (1/20)(y-x), so y - x = 20.

Solve simultaneous equations x+y=120 and y-x=20 and we find y = 70.

Question 28: E

All answers are correct. The 3rd sentence assumes point 1 - that increased climate change will cause increased flooding. The last sentence assumes point 2 - that there will be limited geographic mobility due to flooding, which assumes there will be no action taken to prevent this flooding. Point 3 is correct because the argument suggests people in flooded areas will find it even more difficult to sell their houses

Question 29: C

Test the combination of numbers it could be - we know there is only 1 correct combination. For example, the 7s must either be the second digit of a number or just be a 7. We also know that numbers cannot be repeated. Hopefully you should get that the combination is 34 37 4 27 33 7, so the highest number is 37.

Question 30: E

1 is correct - just because there is another more common cause of accidents it doesn't mean that these speed traps are not of use. 2 is not correct - the fact still remains that drivers under 25 cause more accidents than drivers who speed. 3 is correct - the last sentence suggests safe and responsible driving doesn't include obeying the speed limit.

Question 31: B

Visualisation - 1 way to do this is to draw a 3D grid that is 3 x 2 x 2, see which parts of the grid that piece F and G would cover, and you should see that the only piece to fit in would be B.

Question 32: D

Average of 14 cycle trips per person per year, which is 1% of their trips. Thus there are 1400 trips per person per year, which is roughly 1400/50 ~ 27 per week.

Question 33: D

A is not correct - we know 85% cycle less than once a week, but we don't know what distance. B cannot be inferred - there is not enough info.
C cannot be inferred - not enough info. D can be inferred - 15% cycle at least once a week, 8% at least once a month, and 69% less than once a year. This leaves 100 - 8 - 15 - 69 = 8% that cycle between 1 and 11 times a year.

Question 34: C

The graph suggests that having more cars means you cycle less. The text suggests that having more income means you cycle more. Thus we need a reason to explain the inverse relation of income and cars. C is the only one that does this.

Question 35: B

1 is not correct. Males between 21 and 29 cycle 78 miles a year, making 27 trips, so 78/27 ~ 2.6 miles per trip. Males between 11 and 16 cycle 74 miles a year, making 46 trips, so 74/46 ~ 1.5 miles per trip. The first result is less than double the second.

2 is correct. We can see that males between 11 and 16 cycle 46 times a year and males between 17 to 20 cycle 29 times a year, so the former cycle more often. However, they average less per trip (74/46 ~ 1.5 miles, 59/29 ~ 2 miles per trip).

END OF SECTION

Section 2

Question 1: D

D is the only answer where there is a negative concentration gradient. E, the loss of urine occurs as a result of changes in pressure. All the others have a positive concentration gradient so could occur by diffusion alone.

Question 2: C

Y^{3-} has gained 3 electrons to have a configuration of 2, 8, 8. So it usually has a configuration of 2, 8, 5, meaning it is in group 5. Since it has 3 electron shells, it is in period 3.

Question 3: B

The amount of Y must increase over time because X decays into Y so. Since there is 100 Y at the end every graph, the correct graph will show a mass of 50 at 20 seconds.

Question 4: D

$P = 2 \ (3x10^{-3})^2 x \ (2.5 \times 10^4)$
$P = 9 \ x \ 10^{-6} x \ 5 \times 10^4$
$P = 45 \ x \ 10^{-2}$
$P = 4.5 \ x \ 10^{-1}$

Question 5: C
Let **a** be the recessive allele and A be the dominant allele. Crossing two heterozygotes gives:

Proportion of the live offspring which are homozygous dominant $= \frac{1}{3}$

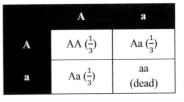

	A	a
A	AA $(\frac{1}{3})$	Aa $(\frac{1}{3})$
a	Aa $(\frac{1}{3})$	aa (dead)

Question 6: E
The quickest way to do this is by setting up algebraic equations:
If you're unsure how to do this then check out the chemistry chapter in *The Ultimate BMAT Guide*.
For Chlorine: $B = 3A + 2Y$

A. $8 \neq 3 + 6$

B. $8 \neq 3 + 8$

C. $4 \neq 3 + 6$

D. $16 \neq 6 + 12$

E. $16 = 6 + 10$

E is the answer since it is the only option that satisfies $B = 3A + 2Y$.

Question 7: B
$= (n + 1)^{th} \, term - n^{th} \, term$

$= \frac{n+1}{n+2} - \frac{n}{n+1}$

$= \frac{n+1(n+1) - n(n+2)}{(n+2)(n+1)}$

$= \frac{n^2 + 2n + 1 - n^2 - 2n}{(n+2)(n+1)}$

$= \frac{1}{(n+2)(n+1)}$

Question 8: A
Height lifted $= 0.4 \, x \, 5 = 2 \, m$
$Gravitational \, Potential \, Energy = mg\Delta h = 100 \, x \, 10 \, x \, 2 = 2,000 \, J$
The tension in the cable $= weight = mg$
$= 10 \, x \, 100 = 1,000 N$
The load does not accelerate as the crane lifts the load at a constant speed.

Question 9: D
Between 0 and 11 seconds the oxygen demand exceeds the oxygen supply. When this happens, the muscle cells use the oxygen that is available for aerobic respiration, but must gain the extra energy required from anaerobic respiration, so both occur.

Question 10: C
C_6H_{16} is not a plausible product - even if a 6 carbon atom molecule was fully saturated it would have 14 hydrogens (not 16!).

Question 11: D
A beta particle is emitted from the nucleus of an atom when a neutron changes to a proton and electron. Thus, the mass number is unchanged but the atomic number increases by 1.

Question 12: E
➢ Lipids break down to form glycerol and fatty acids - the acid would lower the pH.
➢ Proteins break down to form amino acids - the acid would lower the pH.
➢ Carbohydrates break down into sugars like glucose - this doesn't affect the pH.

Question 13: D
Triangle LMN can be rotated about the origin 90 degrees to give triangle PQR. Thus, a further 270 degree rotation will reverse this transformation. This can be accomplished by 2 and then 5.

Question 14: D
This tests your understanding of Le Chatelier's principle. Adding a catalyst does not affect the equilibrium. Adding nitrogen won't have an effect either as it is not involved in the reaction. Changing pressure will have no effect as there are the same number of moles of gas on each side of the equation. Decreasing the temperature will favour the exothermic reactions, so this would shift the equilibrium to the right- resulting in more CO being removed.

Question 15: F

Z is the car's weight because it is constant; Kinetic energy would increase until the terminal velocity was reached. Y is velocity or drag force, which increases as the car increases in speed. X can be acceleration or resultant force - this starts high but decreases to zero as terminal velocity is reached. The only possible combination that satisfies these is F.

Question 16: A

$Area = \frac{1}{2} \text{ x } base \text{ x } height$

$= \frac{1}{2} \text{ x } \left(4 - \sqrt{6}\right) \text{ x } \left(6 + \sqrt{6}\right)$

$= \frac{24 + 4\sqrt{6} - 6\sqrt{6} - 6}{2}$

$= \frac{18 - 2\sqrt{6}}{2}$

$= 9 - \sqrt{6}$

Question 17: B

Neural responses require a stimulus to be detected by a receptor. This sends an impulse via a sensory neuron to the CNS which then sends impulses via motor neuron to cause muscles to contract/relax to bring about the response. Thus, the correct order is represented by option B.

Question 18: F

Comparing an alkane and an amide from the given information, we remove 1 hydrogen atom from the alkane but add 1 Carbon, 1 Oxygen, 1 Nitrogen and 2 Hydrogen. Thus, there is a net addition of 1C 1O 1N and 1H. Add this to the alkane formula: $C_nH_{2n+2} + CONH = C_{n+1}H_{2n+3}ON$

Question 19: C

This is a tricky question that becomes much easier once you've draw a diagram. Remember that a bearing is the clockwise angle from the due North position.

Since the bearing after walking 5km is 2θ; the acute angle in the triangle must be 180 - 2θ. Therefore, the third angle is θ.

Since the triangle has two identical angles – it is an isosceles triangle:

We need to find **x**:

Splitting the triangle into two right angles:

$cos \text{ } \theta = \frac{Adjacent}{Hypotenuse} = \frac{0.5x}{5}$

$5 \text{ } cos \text{ } \theta = 0.5x$

$x = 10 \text{ } cos \text{ } \theta$

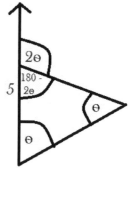

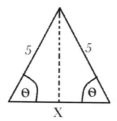

Question 20: F

With the switch closed, all current will avoid resistor Y and flow in the branch with the switch because it is the path of least resistance. Thus, the 12V will be split evenly between resistors X and Z.

When the switch was open: $V = IR; \text{ } 12 = 20mA \text{ } x \text{ } R$

Thus, $R = \frac{12}{20 x 10^{-3}} = 0.6 \text{ } x \text{ } 10^3 \Omega$

Each resistor therefore has a resistance of $0.2 \text{ } x \text{ } 10^3 \Omega$

When the switch is closed; $V = I \text{ } x \text{ } 2R; 12 = I \text{ } x \text{ } 2 \text{ } x \text{ } 0.2 \text{ } x \text{ } 10^3$

$I = \frac{12}{0.4 \text{ } x \text{ } 10^3}$

$= 30 \text{ } x \text{ } 10^{-3} = 30 \text{ } mA$

Question 21: B

1. Correct - in a double circulation, one complete circuit means that blood goes through the heart twice i.e. Lungs → Left Side of Heart → Body → Right Side of Heat→ Lungs
2. Incorrect- The lungs receive 100% of the cardiac output but since the left side of the heart pumps blood to the whole body, the kidneys + liver don't receive 100% of cardiac output.
3. Correct- all of the blood from the liver will eventually reach the lungs.
4. Incorrect- the liver has a dual blood supply.

Question 22: D

A.   False- there are strong covalent bonds between all the atoms.

B.   False- in graphite – a carbon atom is only bonded to 3 other Carbon atoms.

C.   False- graphite conducts electricity and is soft but diamond doesn't conduct electricity and is very hard.

D.   True- you might have known this or arrived at the answer via a process of elimination.

E.   False- graphite is an electrical conductor.

Question 23: C

The work done by the braking force is equal to the kinetic energy of the lorry $Work\ Done = E_k$;

$Force\ x\ Distance = \frac{mv^2}{2}$

Thus, $Distance = \frac{mv^2}{2F}$

Question 24: A

Since the solutions are a & b; this can be shown as: $(x - a)(x - b) = 0$

➤ $x^2 - ax - bx + ab = 0$

➤ $Since\ ab = 3;\ x^2 - x(a - b) + 3 = 0$

➤ $Since\ a - b = -5;\ x^2 - x(-5) + 3 = 0$

➤ $Simplify\ to\ give: x^2 + 5x + 3 = 0$

Question 25: D

The man received the Y allele from his father, and got the colour blindness X allele from his mother, so she must carry the recessive allele. The man must pass on this X allele if he has a daughter, as he only has the one recessive X allele.

Question 26: C

Using n=cV; Moles of Sulphuric acid = $12.5\ x\ 10^{-3}\ x\ 2 = 0.025\ moles$

The molar ratio between **XOH** and Sulphuric acid is 2:1 so there were 0.05 moles of **XOH** that reacted.

A_r of **XOH** $= \frac{2.8}{0.05} = 56$

A_r of **X** = A_r of **XOH** - A_r of OH

$= 56 - 16\text{-}1 = 39$

Question 27: C

Time taken for sound to travel via air $= t$

Time taken for sound to travel via steel $= t + 1.5$

$Speed\ of\ sound\ in\ air = 300\ m/s$

$Speed\ of\ sound\ in\ steel = 4800m/s$

Distance travelled by sound in air = Distance travelled by sound in steel

$300\ x\ (t + 1.5) = 4800\ x\ t$

$300t + 450 = 4800t$

$4500t = 450;\ T = 0.1$

Distance of train $= 4800\ x\ 0.1 = 480m$

END OF SECTION

Section 3

When you can measure what you are speaking about, and express it in numbers, you know something about it; but when you cannot ... your knowledge is of a meagre and unsatisfactory kind.

➤ Through this quote, Lord Kelvin argues that scientific quality depends on quantification, suggesting that the expression of unquantified knowledge is inadequate.

➤ This concept can be applied to almost anything in life; for example, how can we know and understand what makes a business successful if we cannot express their business performance numerically?

➤ Regarding science and medicine specifically, the majority of scientific method requires the quantification of the link between the independent variable and the dependent variable. We also regularly need to use statistical techniques to see if there is a statistically significant difference between our control and what we are testing, and this is vital to maintain safe practice and minimising errors. For example, in a clinical trial, the treatment in question is tested against a placebo and it is only confirmed as effective if there is a statistically significant difference between the efficacy of each treatment.

➤ However, we often cannot quantify our findings in medicine, such as for psychiatric illnesses. For example, patients with schizophrenia or bipolar disorder cannot be diagnosed using simple scientific method as patients are on a spectrum. Instead, psychoanalysis is used and there is very little quantification involved. Numerical scales that are used for diagnosis are regularly criticised and there is never an agreed consensus on how to use them.

➤ Quantification also often ignores the concept of individual differences.

Life has a natural end, and doctors and others caring for a patient need to recognise that the point may come in the progression of a patient's condition where death is drawing near.

➤ This statement suggests that awareness of the end of natural life is important, and that prolonging the dying process should not occur without regard for the patient's wishes.

➤ This guidance is important when considering whether to withhold or withdraw treatment. To ensure the best care, considerations must include the patient's desires, medical indications, benefits and burdens of treatment, opportunity cost of treatment and quality of life that may result from the treatment.

➤ One could suggest that this is support for passive euthanasia, and/or DNRs. If there is no benefit to the patient or the benefits are outweighed by the burdens, we may decide to prevent patients suffering needlessly. The risks and consequences of not recognising this is the prolonged pain and distress someone may experience if they are kept on life support.

➤ Other factors to consider: the dilemmas patients, families and doctors face when considering modern techniques to prolong life, whether there is hope of recovery, all the difficult ethical and legal issues.

Science is the great antidote to the poison of enthusiasm and superstition.

➤ This statement relates to the objectivity of scientific method, whereby we remain unbiased and develop conclusions true to our results rather than true to our beliefs. This is the opposite of what enthusiasm and superstition is based on - subjectivity.

➤ Science, by definition, is completely free of any emotion, ego or enthusiasm. Scientists should deal with everything in a logical, rational manner.

➤ Enthusiasm and superstition often cloud this rational scientific judgement, and thus science's job is to keep humanity sane. Humans are enthusiastic and emotional creatures, and superstitions flow naturally when emotion meets unrestricted intelligence. This has led to situations such as witch hunts and, notions that have later been dismissed through science.

➤ Generally, enthusiasm and superstition are based on untested and unconfirmed hypotheses. We can thence regard science as the antidote because we can use it to test whether this enthusiasm and superstition is based on rationality, or whether there is no basis. Through science, we attempt to see whether there is causality between a change in the independent variable and a change in the dependent variable, ensuring there are no confounding factors.

➤ Science is based upon accuracy and precision, whereas enthusiasm and superstition are the 'poisons' which are based on the opposite.

END OF PAPER

2009

Section 1

Question 1: 15
35 Monarchs reigned for less than 40 years, and 20 monarchs reigned for less than 20 years, so 35-20 = 15 monarchs reigned for 20-40 years.

Question 2: B
The final paragraph suggests childbirth is either done away from the labour ward **OR** surrounded by as much pain relief and medical intervention as possible, which suggests that this level of relief/intervention is not available away from a labour ward.
➢ A is not correct as the issue does not imply that the Health Secretary was either right or wrong; it instead just outlines what they believe.
➢ C cannot be implied from the information we are given.
➢ D is incorrect as the words 'strongly opposed' are far too strong relative to saying 'the delivery of babies has shifted too far in favour of medicalisation'
➢ E is not correct as the passage doesn't mention the progression of mothers/babies

Question 3: D
We must select each graph in turn, and see if each other graph has 2 bars at the wrong height. Starting with A, we can see that there are 2 bars incorrect relative to B, but 3 relative to C, so A or C cannot be right.
If we check with B, then A, B and D have 2 bars the incorrect height but E has 4 bars at a different height, so B or E cannot be correct.
This leaves D, and checking this shows that all bars have 2 incorrect bars relative to D.

Question 4: B
The conclusion, highlighted by the use 'Given that,' is 'the high cost of investing in preventative measures would be unreasonable.' This is paraphrased in B.
A   is   evidence   for   the   conclusion   above   rather   than   the   conclusion   itself.
C also relates to background information/evidence in the text, which actually suggests the opposite of this statement. D suggests the opposite of the conclusion.
E is not correct because although the argument suggests billions of pounds were lost, it doesn't say whether the government should compensate for this.

Question 5: E
There is at least 1 birthday per month - at least 12 birthdays.
There are 3 pairs of twins with birthdays in April - at least 6 in April.
There are at least 2 pairs of twins with birthdays in other months - 2 more birthdays not in April.
At least 13 birthdays not in April + 6 in April = at least 19 birthdays.
There are exactly double granddaughters as grandsons, so we cannot make 19 or 20 with this information. The minimum is thus 21 (14 granddaughters and 7 grandsons).

Question 6: G
All answers are correct.
1 is correct because the argument suggests campaigns are not needed because there are only a few deaths caused by ecstasy a year, but this may increase if the campaigns were not present.
2 is correct because the author concludes that ecstasy is not as dangerous as many people believe, but this is only mentioned relative to horse riding.
3 is correct because the author only talks about the number of deaths per year caused by each activity, rather than the rate.

Question 7: C
Gold and silver would cost the same, so if we call x the number of times equipment is used in the year, £80 + x = £100 + 0.5x so x=40 - the membership would cost £120 with this option. With bronze, it would cost 70 + 3x40 = £190, so changing saves £190 - £120 = £70.

Question 8: B
421 did not report a crime, and 133 of these thought it was insufficiently important. Thus 133/421 is the probability, and we want to find out what this is in decimals. One way of doing this is realising 133 x 3 = 399 so it is less than 0.33, but 133 x 4 = 532 so it is more than 0.25.

Question 9: G

None are correct.

1 is incorrect because we cannot infer any information about next year from this data.

2 is incorrect because we know nothing about the attitudes of those that didn't experience or witness crime.

3 is incorrect because we have no idea about the overlap of people's responses.

Question 10: A

We are given general attitudes to why people didn't report crime in the text, and a common reason is people lacking confidence in the police, which is suggested in A. Particular areas of London, drugs, crime against criminals and paperwork are not implied in the text.

Question 11: D

This can fold twice to make the shape given. Cut the shape out and try folding it if you struggle to visualise. Folding A would mean the opposite only one set of adjacent corners match. Folding B or C produces a shape similar to A, which cannot then fold to form a triangle.

Question 12: B

The conclusion is 'the government should act now to protect people…and make slimming pills subject to the same strict controls as medicines.' This statement suggests the government is already introducing legislation, weakening the argument that the government need to act now. The other answers do not link to this conclusion.

Question 13: E

Test each combination.

A is possible. For example, if the right bag has half green and half red already, then one red is added and one red is taken away.

B is possible with the same example as A.

C is possible. For example, if the right bag has two greens, one red is added and one green is removed.

D is possible with the same example as A.

E is not possible as it will not be possible to have half green and half red in the right bag if it has more than two marbles and only starts with one colour.

F is possible with the same example as C.

Question 14: A

The conclusion is that 'Colour psychology…improving the success of interrogation.' The evidence is that the colours given put people more at ease. To link the evidence and conclusion, it must be assumed that interrogation is more successful if people are more at ease. The other statements do not have to be assumed for the argument to hold.

Question 15: D

There are 7 squares which fulfil the criteria, but some cannot be used in combination with each other.

Squares - 3 between MD and OA, one 2 spaces left of AD, one 2 above JS, one 2 above NT and another to the right of this.

Question 16: s: 38, b: 36

We know USA had 36 golds and 110 medals in total, so 74 = silver + bronze medals. Based on the weighting where gold is worth 3, we know 112 = 2silver + bronze. Solving these simultaneous equation gives s = 38 and b = 36.

Question 17: 13

Germany won 16 golds, so the max Australia could have is 15. G + S + B = 46 and 3G + 2S + B = 89. Taking the first from the second gives 2G + S = 43. Since G = 15 to minimise S we get 30 + S = 43. Therefore S = 13.

Question 18: B

China have 51/100, USA have 36/110, Russia have 23/72 and GB have 19/47. China is the only above ½ and Russia and USA are under 1/3. GB have between 1/3 and ½ and are thus second.

Question 19: F

1 **could** be correct - 3 golds and 5 silvers gives a score of 19, and divided by 4.6 (population) gives a medal weighting per million population of 4.1. Note that the question doesn't mention whether Norway won any bronze medals (if it won 2 then the score is 21 → Divided by 4.6 giving = 4.56. Thus, it's **possible** Norway won at least 3 golds and 5 silvers.

2 **cannot** be correct - even if all of them were bronze, giving a score of 3, this would give a medal weighting per million population of 10.

3 **can** be correct - Slovenia have a total weighted score of 9, which could be achieved by 1 gold, 1 silver and 4 bronzes.

Question 20: C

Both are correct. GBs weighted score is 98, and 98/62 is greater than 220/142 but less than 2.5. Russia would have 139/142, which is under 1.

Question 21: 50g

The drinks are 1.5J in 200ml 2J in 400ml. With 300ml of the second drink we would have 1.5J and with 200ml of the first drink we have another 1.5J, giving 3J in total. We need 4J overall, so we need 1J extra. The supplement has 2J per 100g, so we need 50g for 1J.

Question 22: A

1 is assumed in the second to last line 'the recipient knows to be appropriate.'

2 is assumed in the last line 'only given to the extent that it is deserved.'

3 and 4 are not assumed.

Question 23: E

Month 124 is a September. From this, we can work out that month 254 is a July, 264 is a May, 274 is a March, 284 is a January and 294 is a November. November does not have a 31st day of the month so the last option is not possible.

Question 24: E

➢ None are correct.
➢ 1 We don't know anything about sexual activity of those who do and don't drink, only pregnancy rates.
➢ 2 We cannot infer this from the statistic alone.
➢ 3 We do not know how many girls have consumed alcohol by age 16, only the 40% of pregnant teenage girls below 16 were under the influence of alcohol.

Question 25: A

Trial and error to see if the data fits. Don't try and work out exact proportions, just test if the higher and lower values are in the correct place.

Question 26: E

The passage describes that ants regularly 'talk' in their nests and give an example of this aiding in the survival of blue butterflies. A is not correct as it is not suggested that ants 'talking' will actually harm their survival as a species (although it will maybe harm individuals).

B is not correct because the blue butterfly story given is just an example.

C is not correct because the slaughtering talked about is not intentional.

D is not correct because we cannot deduce that the ability to talk arose due to the threat from parasites.

Question 27: B

Tokens needed for a free meal must be over 7 otherwise Tim would not run out. If this is 8, then he will lose a token a day and he will have a while until this runs out. If this is 9, then he will lose 2 tokens a day and this would leave him with 8 tokens on Tuesday 9th. Then drawing a table is useful from here - he will have 15 tokens on Wednesday morning, and he will have 7 tokens on the Tuesday after this, which would mean no free meal.

Question 28: B

The conclusion is the last sentence: 'if the Kepler telescope finds that such planets exist, we can at last be confident that there is life on planets other than Earth', and the evidence is that the Kepler telescope can find Earth sized planets in the 'habitable zone'. The assumption to link this evidence and conclusion is that Earth sized planets in the zone = life existing.

Question 29: D

Going from left to right:
➢ 1 is unique.
➢ 2, 6 and 8 are the same.
➢ 3, 5 and 7 are the same.
➢ 4 is unique.
➢ 9 and 10 are the same.
➢ 11 is unique.
➢ 12 is unique.
➢ Hence there are 7 different patterns.

Question 30: D

The conclusion is the last sentence: 'the more systematic and organised…the more likely they will produce valid explanations,' suggesting common sense explanations are less likely to be valid than those by scientific method.

Question 31: D

The lift will begin to move to 10 (equal distance to 4, but more requests going up than down). It will then go back to 7, to 4, to 0, to 14, then to 16. This means there will be 5 stops **BEFORE** it reaches 16 - make sure you do not include the stop for 16.

Question 32: C

Cattle, wool and pigs show a decrease, potatoes go from 81.2 to 117.1 and vegetables go from 34.5 to 38.3. Hopefully you can easily see that the former percentage increase is larger than the latter.

Question 33: C

There is first an increase between 1994 to 1995, then a slight decrease to 1996, then a bigger decrease to 1997 and another decrease to 1998. The only sector that shows this out of the answers given is Livestock.

Question 34: D

This is the correct answer. In 1998, the prices of commodities being exported decreased. From 1997 to 1998 there was a decrease in farm crop £ output but quite a large increase for horticulture, suggesting that this decrease in price affected the farm crop commodities more, presumably because more were exported. The line 'if volumes of production have not changed' is key here otherwise we would not be able to infer this.

A cannot be safely inferred as, even if we know the changes in £ output, we do not know what farmers may have changed to and from. The £ output of vegetables increased more than flowers anyway.

B is not correct - from 1995 to 1996, the £ output of cattle decreased and the output of pigs increased. Anyway, we do not have any information about the price of pigs and cattle, only their £ output (it could have also been affected by the absolute amount bought)

C cannot be inferred - although the income was higher in 1998 than 1980, this may not necessarily been to do with the strength of the pound. Many other factors could have caused this e.g. a flood or a drought.

E cannot be assumed from this data alone; we do not know how other factors (such as volume of production) have influenced the £ output.

Question 35: E

The percentage change in sheep from 1997 to 1998 is $(237.4 - 245.1)/245.1 = \sim 1/30 \sim 3.3\%$.

All the sheep that are not exported are the same value as before, whereas all the sheep that were exported lost 5% value. If there was a 3.3% drop in value because of a 5% drop in value of those exported, 3.3/5 = 66% is the percentage of sheep exported.

END OF SECTION

Section 2

Question 1: C

Since is homozygous dominant and B is homozygous recessive, D must be heterozygous. If E is homozygous recessive then there is a 50:50 chance for F to be homozygous recessive and heterozygous. If E is heterozygous there is a 25:50:25 chance for F to be homozygous dominant, heterozygous and homozygous recessive.

Question 2: E

Only molecules with a double bond can take part in polymerisation. If there is one double bond, there are double the number of other atoms to carbon e.g. $C_{24}H_{48}$. CHl_3 and C_3H_7Br don't contain a double bond.

Question 3: C

$Using\ F = ma;\ a = \frac{900-600}{60}$

Thus $a = \frac{300}{60} = 5\ ms^{-2}$

The resultant force is upwards (900-600), thus the direction of acceleration is upwards.

Question 4: C

The probability of picking a red ball $= \frac{x}{x+y+z}$

The probability of picking a blue ball $= \frac{y}{x+y+z}$.

Since the first ball is replaced, the probability of picking a red ball and then a blue ball $= = \frac{x}{x+y+z}\ x\ \frac{y}{x+y+z} = \frac{xy}{(x+y+z)^2}$

Question 5: E

A. Clones don't always have the same phenotype due to environmental factors.
B. Twins aren't always members of a clone – they can be non-identical.
C. Clones can occur naturally.
D. Mutations result in variation – not clones.
E. Correct - clones contain identical DNA.

Question 6: B

This is a simple recall question- SiO_2 is the only structure that forms a giant molecular structure.

Question 7: E

$Recall\ that\ P = IV;\ V = \frac{P}{I} = Watt\ per\ Amp.$

The other units are not equivalent to a volt.

Question 8: B

Solving this requires you to use Pythagoras' theorem twice.

First, we need to find the length of the added solid line which is diagonal joining the two bottom corners. This is given by:

Since the solid line is only half the length of the diagonal, its

Thus, the length of the original line is given by: $x^2 = \left(\frac{\sqrt{2}}{2}\right)^2 +$

Thus, $x^2 = \frac{6}{4} = \frac{3}{2};\ x = \sqrt{\frac{3}{2}}$

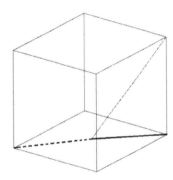

half the length of a

$\sqrt{1^2 + 1^2} = \sqrt{2}$

length $= \frac{\sqrt{2}}{2}$

$1^2 = \frac{2}{4} + 1$

Question 9: D

A. Correct because an increased number of alleles always increases the risk.
B. Correct because increased alcohol intake increases risk regardless of the allele number.
C. Correct as drinking increases the risk more than allele number.
D. Incorrect because there is an increased risk from drinking alcohol (risk value of 1.0 to 4.0 for 0 alleles) than from increased allele numbers (1.0 to 1.5 to 1.8).
E. Correct as the risk is increased more with 2 mutant alleles for heavy drinkers than for light drinkers.

Question 10: E

A_r of $CO_2 = 12 + (2\ x\ 16) = 44$

Number of moles of $CO_2 = \frac{4.77}{44} \approx 0.1\ moles$

Since the molar ratio is 1:1, there must be 0.1 moles of Carbon in the original compound.

Mass of Carbon in original compound $= 0.1\ x\ 12 = 1.2\ g$

Thus, the percentage of carbon in original compound $= \frac{1.2}{2} = 60\%$

E is the closest answer to this (note that we underestimated in the original calculations to make the maths easier so the final value is lower than the actual answer). Make sure you check the options first to see how much margin you have to round numbers.

Question 11: C

This is beta radiation because the radiation reaches a detector 30 cm away but not one 1 m away. Alpha radiation would not reach either detector.

The initial count rate is 220 and falls to 20 (= background radiation) over time. Thus, the radioactive source initially results in 200 counts/min. Thus, the half life is when the count rate decreases to $\frac{200}{2} + 20 = 120$. The time for this is at 2.40 hours.

Question 12: C

Do the calculation in brackets first, which is $\frac{2^3}{3} = \frac{8}{3}$

The next step is: $\frac{\left(\frac{8}{3}\right)^2}{2} = \frac{\frac{64}{9}}{2}$

$\frac{64}{18} = \frac{32}{9}$

Question 13: B

This is a tricky question as you can easily over interpret it depending on your level of knowledge. You need to identify which process will be severely impacted in hypoxia (low blood oxygen).

➢ A: In short, a low blood O_2 concentration won't cause a severe disruption in CO_2 diffusion. In fact, hypoxia will facilitate CO_2 to leave the blood at the alveoli via the Haldane effect (which you are not expected to know about for the BMAT!).

➢ B: Glucose absorption in the gut is an active process and would therefore require ATP derived from aerobic respiration which is dependent on oxygen. Thus, hypoxia would impair this.

➢ C: Hypoxia would increase the concentration gradient so oxygen would leave the alveoli even more quickly.

➢ D, E and F: These are just passive processes so unlikely to be impacted **immediately and severely** in hypoxia.

Question 14: B

A more reactive element can displace a less reactive element in a compound. This happens in 2, 3 and 4. In 1, 5 and 6 a less reactive element displaces a more reactive element which is not possible.

Question 15: D

The distance travelled is the area under the graph. Convert the x axis into seconds when calculating.

Distance travelled in the first minute $= 15 \times 60 + 5 \times 6 \times \frac{1}{2}$

$= 900 + 150 = 1050m$

Distance travelled between 1 and 4 minutes $= 20 \times 3 \times 60 = 3600\ m$

Assume the curve between 4 and 7 minutes is a triangle to calculate the distance: $20 \times 3 \times 60 \times \frac{1}{2} = 1800m$

Total Distance travelled $= 1050 + 3600 + 1800$

$= 6450m = 6.45$ km

This is an overestimate since the last part of the journey is < 1800 m.

Thus, the answer must be less than 6.45 km: D – 6 km.

Question 16: F

$\sqrt{\dfrac{2 \times 10^3 + 8 \times 10^2}{\frac{1}{2500} + 3 \times 10^{-4}}}$

$= \sqrt{\dfrac{2800}{\frac{4}{10000} + \frac{3}{10000}}}$

$= \sqrt{\dfrac{2800}{\frac{7}{10^4}}} = \sqrt{\dfrac{2800 \times 10^4}{7}}$

$= \sqrt{400 \times 10^4} = \sqrt{4 \times 10^6} = 2 \times 10^3 = 2000$

Question 17: F

All of the statements are correct – competition can be both between species (inter-specific) and between members of the same species (intra-specific). Natural selection can lead to both evolution and extinction.

Question 18: D

This is actually trickier than it first appears and it's easy to get bogged down in it. The best way is to form simultaneous equations and solve algebraically. If you're unfamiliar with this approach then see the chemistry chapter in *The Ultimate BMAT Guide.*

$For\ Hydrogen: B = 2C$

$For\ Nitrogen: B = 2A + 2$

$For\ Oxygen: 3B = 6A + C + 2$

$Thus, 2C = 2A + 2\ ; C = A + 1$

$3B = 3(2A + 2) = 6A + 6$

$Thus, 6A + 6 = 6A + C + 2$

$Simplify\ to\ give: C = 4$

$Since\ B = 2C = 8$

Question 19: B

Gravitational potential energy at top = Kinetic energy at bottom

$E_k = \frac{mv^2}{2} = \frac{5 \times 20^2}{2} = \frac{5 \times 400}{2} = 1,000\,J$

$E_p = mg\Delta h = 5 \times 10 \times h$

$50\,h = 1000; h = 20m$

Question 20: D

$Volume\ of\ Sphere\ = \frac{4\pi r^3}{3}$

$Volume\ of\ cylinder\ = \pi r^2 h$

$In\ this\ case, h\ =\ 2r\ so\ \pi r^2 h\ =\ 2\pi r^3$

Fraction of volume occupied by sphere $= \frac{4\pi r^3}{3} \div 2\pi r^3 = \frac{4}{6} = \frac{2}{3}$

Question 21: D

The oxygen in the alveolus is from inspired air so is high, whereas the blood in the capillary is deoxygenated. The carbon dioxide is high in the capillary after respiration whereas it is low in inspired air.

Question 22: D

No bonds are broken in 2 or 3 - bonds are only made. In reaction 1, H-H and I-I bonds are broken and in 4, the C-Br bond is broken.

Question 23: A

The weight of the system is 20,000 + 5,000 + 5,000 = 30,000 kg.

$Using\ F = ma; a = \frac{15,000}{30,000} = 0.5ms^{-2}$

To find T; apply F= ma to the individual carriage = 5000 x 0.5 = 2500 N

Question 24: A

$y = 5\left(\frac{x}{2} - 3\right)^2 - 10$

$\frac{y+10}{5} = \left(\frac{x}{2} - 3\right)^2$

$\sqrt{\frac{y+10}{5}} = \frac{x}{2} - 3$

$\frac{x}{2} = 3 + \sqrt{\frac{y+10}{5}}$

$x = 6 + 2\sqrt{\frac{y+10}{5}}$

Question 25: F

1. Insulin controls blood glucose levels, not water content (that is ADH).
2. Homeostasis depends on the hormonal and nervous systems.
3. Statements 3, 4 and 5 are correct.

Question 26: C

M_r of CH_2 = 12 + 2= 14

Thus, BrCl must have an M_r of 114 to make up 128. This is only possible if the Cl is 35 and the Br is 79.

Since Cl^{35} is three times as common as Cl^{37}; 75% of CH_2BrCl will be Cl^{35}

Since Br^{79} is equally as common as Br^{81}; 50% of CH_2BrCl will be Br^{79}

Therefore, $\frac{3}{4} x \frac{1}{2} = \frac{3}{8}$ of CH_2BrCl will have a mass of 128.

Question 27: E

➤ The first graph shows 2 cycles of the wave taking 60m.
   ○ Thus, the wavelength = 30 m
➤ The second graph shows 3 complete cycles taking 0.6s.
   ○ Thus, the frequency $= \frac{3}{0.6} = 5$

$Wave\ speed\ =\ wavelength\ x\ frequency\ =\ 30\ x\ 5\ =\ 150\ ms^{-1}$

END OF SECTION

Section 3

You must be honest and open and act with integrity.

➢ The statement argues that a good doctor is one that discloses all information to patients without exception, and builds a relationship based upon complete sincerity and integrity.

➢ These are vital attributes for a doctor to have and are highlighted in the Good Medical Practice booklet given out by the GMC. Full disclosure allows a patient to make an informed decision about their health and allows the expression of true autonomy. Furthermore, these qualities are likely to make the patient feel more at ease relative to a dishonest, closed doctor.

➢ However, of course full disclosure can often be undesirable. For example, if a doctor is running some tests for the presence of a disease, and it comes back positive, this should be run again before the patient is informed (Bayes Theorem explains that in many of these situations, a false positive can be more likely than a true positive). False positives can lead to unnecessary stress and worry. The neurologist Oliver Sacks describes in his books that he sometimes decided not to tell patients with cognitive impairments about their disorder so that they could at least attempt to have a good quality of life despite their incurable condition.

➢ This question can also be approached from the angle of the prima facie principles and patient confidentiality. Of course, if a patient presents with obvious signs of self-harm, or confides in you that they have stolen drugs in order to overdose, then you clearly have a responsibility to inform your seniors and break the trust. Other examples of breaking confidentiality include when other people's health may be at risk - HIV positive patient intending to have unprotected sex or a patient threatening to maim someone after their discharge.

➢ A doctor should obviously strive to have these attributes but what is most important is the best health of the patient and this should never be compromised.

Science is a way of trying not to fool yourself.

➢ Fooling oneself is convincing oneself that a particular belief, observation or opinion is the truth when it is actually false. The empirical nature of science means that falsehoods can be avoided and it develops concrete evidence for us to base theories on.

➢ Scientists must be objective - the ability to remain unbiased and develop conclusions true to our results rather than true to our beliefs.

➢ There are many examples of where passion and bias have blinded scientists to that which has been (seemingly) objectively proven! A famous example is Sir Fred Hoyle who rejected Hawking's presentation of the Big Bang theory despite overwhelming evidence (cosmic microwave background) until his death. Also, scientific bodies who have developed such rigorous testing methods can often build an illusion of objectivity. They sometimes think they are always right just because they came up with the objective method. This results in peer pressure which is borne out of a hierarchical structure which not only led to Darwin publishing his paper on natural selection years after writing it, but also his disgrace.

➢ Ensuring we use rigorous scientific method is the first step (no confounding variables, etc.) to guarding against this mistake. Scientists must also not fully accept a hypothesis just because it appears the logical solution at the time, without any further evidence. Experimentation must be valid and reliable, and peer review is an excellent way of testing these factors. Scientists must welcome other scientists with different prejudices and preferences to follow what they have done to see if the same results are obtained. If they are different, it may well be because of the bias of the first scientists – in this case the scientific method is flawed.

It is an obscenity that rich people can buy better medical treatment than poor people.

➢ The statement argues that it is unfair that wealthier members of society can buy better (private) healthcare than the less well off, suggesting that economic status should not determine one's health. The case is that healthcare is a human right and should thus be available with the same quality to all. In a country where the NHS is widely regarded as the 'crown jewel of the welfare state', it seems contradictory to still have a two tier healthcare system where some can still pay for better healthcare. However, simple economics does mean that in most other markets, those with more wealth can acquire 'better' goods and services.

➢ One assumption here is that private healthcare provides better treatment than public healthcare. Indeed, waiting times are often shorter, but most doctors in private clinics also work with the NHS, and all doctors are trained from the same pool of medical schools. They also use the same scientific knowledge and it is unlikely that a treatment would only be available privately.

➢ Another assumption is that only the rich use private healthcare, but health insurance is now widely available with the rise of relatively cheap insurance companies like BUPA and is included with many jobs.

➢ The NHS (arguably) already provides a good standard of care for all, so if the rich want to spend their disposable income on private healthcare they should be allowed to do so. Not allowing them to spend their money how they wish is breach of their personal rights.

➢ It can be argued that the presence of both systems is actually beneficial to all. The NHS would not be able to cope with the demands of patients using services from both sectors, and waiting times would be increased even further. Those who receive private care still pay taxes to fund the NHS, and then free up resources for those who are dependent upon it.

➢ The statement can be accepted on an ethical and moral basis, but is not entirely true due to the assumptions it makes. Upon further analysis, it appears that the existence of both systems is absolutely necessary.

END OF PAPER

2010

Section 1

Question 1: A

Jay is 150cm tall and has a BMI of 22, so his weight is 49kg. Charlie is 156cm tall and has a BMI of 24, so his weight is 58kg. The combined weight is 172kg, so Alex's BMI must be $172 - 58 - 49 = 65$kg. With a height of 162cm, his BMI is 25.

Question 2: C

This error of reasoning is in the form: x tends to have y, so to increase x we must increase y. However, y may have caused x in the first place. That is, the society being more economically robust may have made the arts flourish in the society, as answer C says.

A is irrelevant to the argument, B may be true, but wouldn't represent a flaw in the argument and D differentiates between the arts, which the argument does not.

Question 3: C

The pupil with the white diamond got the 5th highest on test 1 and the 5th highest on test 2. If you order each set of numbers you will see that the white diamond is Erin.

Question 4: C

The passage talks about the **link** between global warming and spam, and the conclusion must cover both these factors, which none of the other answers do. A is far too general to be a conclusion of the passage which has specific themes. B cannot be inferred because there is no comparison between reducing spam and easing congestion. D is tempting but does not relate to global warming which is the main theme of the passage, and long term effects are not mentioned at all in the passage so is not correct.

Question 5: D

There are 6 batches (72/12). The first batch takes $40 + 25 + 5$ minutes (70 minutes), so if they start at 1pm, then the first batch will be done by 2:10. The second batch can be started at 1:40, when the first batch is put into the oven. But after 25 minutes of preparing this second batch, the first batch must be taken out the oven and cooled which takes 5 minutes. So 15 more minutes are required to prepare after this, making preparation $25 + 5 + 15$ (45 minutes), and the next batch can be put in at 2:25. Each batch except the first will take 45 minutes to prepare due to this reason, and the oven will never be occupied when it is required so we only need to take this into account until the end.

40 minutes for the first batch + 45 minutes x 5 (for every other batch) = 265 minutes, but we must also include the 30 minutes to cook and cool the final batch, which is 30 minutes. This is 295 minutes in total, which is 5 minutes short of 5 hours, so it takes 4 hours and 55 minutes which amounts to D.

Question 6: B

1 is irrelevant as the argument is that if we did manage to stop burning fossil fuels there would be little effect. If anything, this would further that statement that 'we cannot avoid the disastrous consequences of climate change.' 2 is a fair critique of the argument, as the model they use "assumed that by then CO_2 concentrations would be double preindustrial levels" and this may not be the case if we were able to reduce carbon dioxide significantly. 3 is also irrelevant; other predictions are irrelevant to this argument.

Question 7: B

We know the first digit must be wrong (due to the score limit), and the others are not. If we inspect the similarity of the 8 to other numbers, 3 lights being permanently on could have mean the number was meant to be 2, 5, 6, 9 or 0. The number cannot be 5, 6 or 9 due to the score limit, but can be 0 or 2, so the answer is B.

Question 8: A

The first paragraph suggests that boys and girls preferring to play with different types of toys may have a biological origin rather than a social origin, which means statement A is challenged. B and C are correct according to the first paragraph, meaning they are not 'challenged'. D is irrelevant to the argument in the first paragraph.

Question 9: A

Would be a strong challenge to the study. Whether males or females were dominant, the more aggressive sex taking all the 'attractive' toys, leaving the remaining less 'attractive' toys for the other sex, would mean there is not necessarily a gender preference. B is irrelevant, as there is no talk about how long they keep their interest for. C would actually strengthen the argument, further suggesting that there was a genetic difference causing different preferences. D isn't necessarily a strong challenge, as monkey studies have more bearing on human behaviour than any other species, due to the very similar genetics. E may be true but is irrelevant to the particular topic about preferences based on sex.

Question 10: A

Is clearly an assumption made in the second paragraph, as the study talks about the interest that monkeys have for the objects, and this is measured by the amount of time spent with it. The rest do not necessarily have to be assumed for the claim to be accepted.

Question 11: D

Neither can be inferred from the photographic evidence alone. Showing the male monkey play with a car and the female play with a doll in pictures alone doesn't suggest that the respond in the same way to humans, nor have similar preferences to us. For example, for all we know the car could have been given to the male and the doll given to the female just for the purpose of the photo.

Question 12: C

Remove each tile one by one, and see what remains.

As we want the proportion of squares of each colour to be equal, and there are 36 squares and 3 colours, we want 36/3 = 12 of each colour, and we only need to test 1 colour to start with.

Removing A leaves 13 blacks, removing B leaves 14 blacks, removing C leaves 12 blacks, removing D leaves 11 blacks and removing E leaves 10 blacks so C must be the correct answer.

Question 13: E

The conclusion of the argument is "Since there will be negative stories in the press either way, we should ignore these stories and not worry about them." This suggests negative stories should be ignored, which is the flaw of the argument. None of the other statements necessarily represent a flaw in the argument.

Question 14: D

This is quite a hard question as there is not really a definite technique to use without wasting too much time. It is the kind of question to come back to at the end and go through systematically at the end if you do happen to have a few minutes spare.

You may be able to just spot the most 'different' numbers, however, which is likely to represent the maximum number of elements changing between numbers. Changing between 1 and 6 require 6 changes, and this is the maximum.

Question 15: D

The male child cyclists killed in 2006 was >400 and the female child cyclists killed was <100. This means there were >1700 male pedestrians or cyclists killed and <800 female pedestrians or cyclists. Hence, D is correct.

A is not necessarily true – for example it may be the case that males take more risks, leading to higher risk of death.

B, C and E are not necessarily true – for example, each may be negated by the fact that there may be many more males who ride bikes.

Question 16: D

Visualisation. A cannot be correct – for example if we look at the 6, if the dots were facing this particular direction with 2 rotations it would mean that the 5 should be on top or not in the picture.

B cannot be correct as the 3 can only be in the same position and orientation if it is rotated twice in the SAME direction.

C cannot be correct – for example if we look at the 6 and its orientation, it would mean the 5 should be on the right of it or not in the picture.

E cannot be correct, as at least 1 of the numbers in original picture would remain with 2 different rotations.

Question 17: C

The argument is that safety is ensured by keeping artefacts at the British Museum, citing the example of the Iraqi museum being looted as the evidence that museums in other countries are less safe.

However, an example of 1 country doesn't hold for the safety in other countries, so C is the correct answer. The other answers do not represent a flaw in the argument.

Question 18: D

You should use drawings to assist you with this question. 2 ends of the H (2cm each) can fit in the centre of the middle of another H (4cm each), so you can have 6 rows of 3, 2, 3, 2, 3, 2 where they all fit in each other in this way. This allows 15 Hs in 33cm by 24cm.

Question 19: 217

There were 228m passengers at UK airports, and 9 in 10 passengers were travelling internationally, so 10% were travelling on domestic flights. This accounts for 22.8m passengers, and if this was halved to remove the passengers counted twice we would have to take away 11.4m passengers. 228 – 11.4m = 216.6m = 217m passengers.

Question 20: C

There were 34m passenger movements to Spain out of a total of 228m, and 34m/228m represents roughly 15% (10% of 228 is 22.8 and 5% is 11.4). There were roughly 60m passengers in total in 1980 and 15% of this = ~9m.

Question 21: F

All of the answers are correct. 1 is right; the increase between 1955 to 1980 is about 55m and 1980 to 2005 is around 173m, which represents an increase of over treble (still applies if you divide each value by 25 to have the average per annum). Heathrow + Gatwick had 101m passengers, which is greater than 228 x 0.4 (as 10% is 22.8 so 40% must be below 100).

In order to reach 500m from 228m, an increase of 272m in 20 years, there must be an average increase in passengers of over 10m a year, as 272/20 > 10.

Question 22: D

The data has been correctly interpreted, using previous results to form a prediction, and this is relevant to the conclusion, so D is correct.

Question 23: E

To work out the permutations, we know that 1 can be matched with any other number except itself, 2 can be matched with any other number except itself and 1 (as this has already been counted), etc. so we have 14 + 13 + 12 + ... + 2 + 1 possibilities which amounts to 105.

We can list the combinations that follow the criteria:
➢ 2 with 10, 13, 14 or 15
➢ 3 with 10, 12, 14 or 15
➢ 4 with 10, 12, 13 or 15
➢ 5 with 10, 12, 13 or 14
➢ 6 with 10, 12, 13, 14 or 15
➢ 7 with 10, 12, 13, 14 or 15
➢ 8 and 9 as above

So there are 4 x 4 + 4 x 5 possibilities of this happening, which is 36/105, cancelling to 12/35.

Question 24: C

The argument considers 2 alternatives and suggests one in an implausible explanation, thus concludes the other explanation must be correct. However, other explanations are not considered, so the assumption is that there are no other explanations.

The other statements do not have to be assumed for the argument to hold

Question 25: A

The highest position possible is coming first (alone, not joint first). The other hands add up to 22 (11 doubled for having the same suit), 13 and 11. To beat 22, they can exchange their 2 of spades for their 7 of diamonds from the person who is first, giving them 14, which will be the highest hand (compared to player 1 who now has 6). If the 7 is not taken from player 1 then it is not possible to beat 22.

Question 26: A

The main conclusion of the argument is the first sentence, which is essential paraphrased in A. The other statements can possibly be inferred from the argument but do not address the main conclusion.

Question 27: C

➢ If Phil aims for 4, then there is the possibility of getting 1 and then no guarantee of scoring at least 30.
➢ If he aims for 7, there is the chance of getting 3 and then no guarantee of scoring at least 30.
➢ If he aims for 10, there is the chance of getting 8 and this would mean all 4 darts have not landed in different sections.

➢ If he aims for 11, there is the chance of getting 2, and then no guarantee of scoring at least 30.
➢ If he aims for 9, then he can get 4, 9 or 6. If he gets 4, then he can aim for 6 next time and guarantee over 30. If he gets 9 or 6 he can aim for 5 the next time and guarantee over 30. So C is correct.

Question 28: A

The argument here is that ONLY the name influenced the public opinion i.e. there weren't any other factors swaying public opinion. If statements 2 or 3 were true- the editors and newspaper would have influenced the public with the wording; if 1 was true, the newspaper would have actually reduced bias and the only variable would have been the name used to refer to the child.

Question 29: C

A good way to approach this is to work out the difference of each apple to 200g (what the average of the apples should roughly be), so we have -27, -18, -12, -3, +7, +19 and +24. When we combine three apples we want a total of -3 to +3. This is possible to do with -27, +19 and +7 which adds up to -1, and -18, -3 and +24 which adds up to +3. This leaves us with -12 which is 188g.

Question 30: B

It is not assumed that expensive schemes are not justified, the argument instead states that the particular idea of reducing class sizes is not justified for other reasons (and the fact that is expensive is just an additional point instead of a reason for not supporting it).

A is assumed in "candidates with lower qualifications would have to be recruited."
C is assumed in "California had risen only to 48th..."
D is assumed in the final sentence.

Question 31: D

Spatial reasoning questions in the BMAT are always hard. The key here was to analyse the relationship of the shapes e.g. the white square must be the opposite face to the cross. Then you need to look for nets that don't satisfy these relationships e.g. Net B would put the white square and cross next to each other rather than opposite.

Keep finding more relationships to exclude the options. If you're really stuck, take scissors into your exam (get it approved by your exam centre before though!).

Question 32: C

The average annual income for the bottom 20% is roughly between 16,200 and 17,300, which is 16,750. For the top 20% it is roughly between 50,000 and 160,000, which is 105,000. 6:1 appears to be the closest ratio to 105:16.75, and this can be roughly checked by seeing that 17 x 6 is 102.

Question 33: C

We can work this out by dividing the top 10%'s average income (~160k) by all the other values added together. Rounding is fine (and advised), so we have 16k + 17k + 19k + 20k + 22k + 26k + 30k + 37k + 51k + 160k which is 400k. 160/400 = 2/5 = 40%

Question 34: C

20% of the top 10% is about 160k x 0.2 = 32k, and 20% of the next top 10% is roughly 50k x 0.2 = 10k, giving 42k to distribute. If this was spread out between the other 80%, each 10% block would receive 42k/8 each, which is 5.25k. $16,200 + $5,250 is close to $21,500.

Question 35 B

1 is incorrect as the data does not concern people within a given country, only between different countries. 2 is correct as the left graph shows that countries with a lower level of inequality and hence variability have generally healthier people.

END OF SECTION

Section 2

Question 1: E

The hypothalamus detects changes in temperature. Arterioles dilate so that more heat can radiate out the body, keeping the body cool. Knowing these two factors alone will get you the correct answer.

For completeness, the hair erector muscles contract and capillaries in the skin do not move.

Question 2: E

$Moles\ of\ Iodine = \frac{63.5}{127} = 0.5\ moles$

$Moles\ of\ Oxygen = \frac{20}{16} = 1.25\ moles$

Thus the molar ratio is 2 moles of iodine to 5 moles of oxygen. I.e. I_2O_5

Question 3: A

After the end of experiment (12 minutes) there is 16g of U-234. Thus, there must have been 16g of Protactinium-234 at the start. The time at which the mass of Protactinium-234 is half its initial value (8g) is 1.2 minutes.

Question 4: C

The big container is full to start with, representing 1 unit of water. If they contain the same amount of water after pouring, each will have 0.5 units of water. The larger container thus contains 0.5 units out of a total of 1.

$$P = \frac{0.5}{Capacity} = \frac{0.5}{1} = 0.5.$$

The smaller container contains 0.5 units of water but out of a total possible volume of **less than 1** as it is a smaller container. Since Q = 0.5/Capacity, and the capacity is less than 1, Q must be > 0.5 (because dividing 0.5 by a number <1 results in a number >0.5

Question 5: C

Insulin reduces blood glucose levels by promoting the uptake of glucose into cells. Statements 2 and 3 are correct.

Question 6: D

12g of carbon = 1 mole of carbon. In the first equation, 1 mole of carbon is used. In the second equation, the initial one mole from equation 1 and an additional mole is used to produce 2CO. Thus, 2 moles of carbon are used. In the final equation, 3CO produces $3CO_2$. However, since only 2 moles of CO were produced in equation 2, only 2 moles of CO_2 are produced.

Thus, the mass of 2 moles of $CO_2 = 2 \times [12 + (2 \times 16)] = 88g$

Question 7: A

$$Amplitude = \frac{peak - trough}{2} = \frac{6}{2} = 3$$

$$Frequency = \frac{1}{period} = \frac{1}{12 \times 60 \times 60}$$

$$Frequency = \frac{1}{12 \times 3600}$$

Question 8: C

This is easier than it appears. To get back to where they started, the player must go in the opposite direction to whence they came, and the same distance. The player has a ¼ chance to go the opposite direction, and a ¼ chance to go the same distance which gives: $\frac{1}{4} x \frac{1}{4} = \frac{1}{16}$

Question 9: B

Geographic distribution is irrelevant for an individual, and a high reproductive capacity is an effect rather than a cause of natural selection. A gene pool refers to all of the alleles in a person or population. Natural selection will favour an advantageous allele within this gene pool.

Question 10: D

This is one you just have to know! The complete combustion of a fuel produces water and carbon dioxide. If you were unsure, then think about the combustion of clean hydrocarbon fuels like propane etc which result in the production of H_2O and CO_2.

Question 11: C

An alpha particle contains 2 protons and 2 neutrons. Thus, the emission of 3 alpha particles will reduce the mass number by 12 and the atomic number by 6.

In beta decay, a neutron decays into a proton and electron. Thus, two rounds of beta decay will result increase the atomic number by 2.

Overall, the atomic number decreases by 4 from 86 → 82 and the atomic mass decreases by 12 from 219 → 207.

Question 12: C

$$Mean = \frac{total\ time\ for\ all\ people}{Number\ of\ people} = 56$$

First group total time $= 20\ x\ 54 = 1080$

Second group total time $= T\ x\ P$

Substitute into original equation:

$$Mean = \frac{1080 + PT}{20 + P} = 56$$

This simplifies to: $PT - 56P = 1120 - 1080$

$P(T - 56) = 40$

$P = \frac{40}{T-56}$

Question 13: F

1. Incorrect - transmitter molecules are recognised by their receptors, but are not formed in them.
2. Incorrect - the signal is transmitted by diffusion, as osmosis only applies to water.
3. Incorrect - transmitter molecules CAUSE the signal to be transmitted across the synapse.

Statements 4 and 5 are correct.

Question 14: A

The key here is that ionic equations must be balanced for charge **and** stoichiometry. Only equations 1, 2 and 6 balance for both of these.

➢ For Equation 3: Adding 2 electrons to O^{2-} forms O^{4-}
➢ Equation 4: Taking an electron away from O^{2-} forms O^{-}
➢ For Equation 5: Taking 2 electrons away from $2I^{-}$ produces $2I$.

Question 15: B

Before the changes were made, all of the charge would bypass the 4 parallel bulbs through switch Q, so bulb Y will have the entire 12V running through it. Afterwards, all bulbs will have some charge flowing through them, so bulb X will be brighter than before and bulb Y will be dimmer than before.

Question 16: C

Notice that triangles ABC and ADE are similar. Thus: $\frac{BC}{BA} = \frac{DE}{DA}$

This gives: $\frac{x}{4} = \frac{x+3}{x}$
Cross-multiply to give: $x^2 = 4x + 12$
This sets up the quadratic equation: $x^2 - 4x - 12 = 0$
$(x - 6)(x + 2) = 0$
Since length must be a positive value, $x = 6$. Thus, $DE = x + 3$
$= 6 + 3 = 9\ cm$

Question 17: E

S, T and U do not have the condition, but could be carriers as their parents both carried the recessive allele. U must be a carrier as X has the condition and this is not possible if one parent does not have a recessive allele.

Since P and Q are carriers, the probability of both S and T being carriers is 50%.

Question 18: B

This question requires balancing the charges to equal 0:

A. A gives a charge of -2
B. B gives a charge of 0. H_2PO_4 has a charge of -1, so 2 of these groups balance the +2 on the Mg^{2+} ion.
C. C gives a charge of +2
D. D gives a charge of +2
E. E gives a charge of +2
F. F gives a charge of +3.

Question 19: B

Acceleration can be determined from velocity-time graphs but not distance-time graphs so we can immediately rule out R and S.

➤ $Acceleration = \frac{\Delta velocity}{\Delta time} = \frac{\Delta y}{\Delta x}$

➤ In P, $Acceleration = \frac{10}{24} = 0.42 \, ms^{-2}$

➤ In Q, $Acceleration = \frac{48}{20} = 2.4 \, ms^{-2}$

Question 20: A

$Total \, SA = area \, of \, 2 \, circular \, sections + longitudinal \, area$

$= 2\pi r^2 + 2\pi rh$

$Volume = \pi r^2 h$

Since Volume = Surface area: $2\pi r^2 + 2\pi rh = \pi r^2 h$

Simply to give: $2r + 2h = rh$

$2r = rh - 2h$

$h = \frac{2r}{r-2}$

Question 21: B

Statements 3 and 4 are the wrong way around; meiosis results in four nuclei and mitosis results in two nuclei. All the other statements are correct.

Question 22: C

$A_r \, of \, benzene = (6 \, x \, 12) + (6 \, x \, 1) = 78$

$Moles \, of \, benzene = \frac{3.9}{78} = 0.05 \, moles$

$A_r \, of \, nitrobenzene = (6 \, x \, 12) + 5 + 14 + (2 \, x \, 16) = 123$

$Moles \, of \, nitrobenzene = \frac{3.69}{123} = 0.03 \, moles$

$\% \, Yield = \frac{0.03}{0.05} = 60\%$

Question 23: G

$Power = \frac{work \, done}{time} = \frac{Force \, x \, Distance}{time}$

The force is the weight of the water $= mg = 5 \, x \, 10 = 50N$

Thus, $Power = \frac{50 \, x \, 5}{1} = 250 \, W$

Using $v^2 = u^2 + 2as$:

$0 = u^2 + (2)x(-10)x(5)$

$u^2 = 100$

$u = 10 \, ms^{-1}$

Question 24: C

This is a time consuming question and one you should consider leaving till the end as it could easily consume several minutes of your precious time.

Let the biggest square have a side length of 1 unit.

Second Square:

The sides of the second square are $\frac{1}{3} \times 1$ and $\frac{2}{3} \times 1$.

Using Pythagoras's theorem:

$$c^2 = \left(\frac{1}{3}\right)^2 + \left(\frac{2}{3}\right)^2 = \frac{1}{9} + \frac{4}{9}$$

$$c = \sqrt{\frac{5}{9}} = \frac{\sqrt{5}}{3}$$

Third Square:

The sides of the third square are $\frac{2}{3} \times \frac{\sqrt{5}}{3}$ and $\frac{1}{3} \times \frac{\sqrt{5}}{3}$

$$= \frac{2\sqrt{5}}{9} \text{ and } \frac{\sqrt{5}}{9}$$

Using Pythagoras's theorem:

$$c^2 = \left(\frac{2\sqrt{5}}{9}\right)^2 + \left(\frac{\sqrt{5}}{9}\right)^2 = \frac{4\times5}{81} + \frac{5}{81}$$

$$c = \sqrt{\frac{25}{81}} = \frac{5}{9}$$

Fourth Square:

The sides of the fourth square are $\frac{2}{3} \times \frac{5}{9}$ and $\frac{1}{3} \times \frac{5}{9}$

$$= \frac{10}{27} \text{ and } \frac{5}{27}$$

Using Pythagoras's theorem:

$$c^2 = \left(\frac{10}{27}\right)^2 + \left(\frac{5}{27}\right)^2 = \frac{100}{27^2} + \frac{25}{27^2}$$

$$c = \sqrt{\frac{125}{27^2}}$$

The area of the fourth square $= \sqrt{\frac{125}{27^2}} \times \sqrt{\frac{125}{27^2}} = \frac{125}{729}$

Question 25: E

Statements 1 and 4 ignore the fact that the nervous system also uses chemical transmission at the synapse. The other statements are correct.

Question 26: B

Don't be perturbed- this is a free mark. Count all the corners of the structure (don't double count if they overlap) (= 17) and the carbons on the functional groups (=3) to give a total of 20 carbon atoms.

Question 27: D

The engine is doing work against the gravity **AND** against the frictional force. Since the car moves 50m along the road, it has gained $\frac{50}{20} = 2.5m$ in height. Thus:

$Total\ work\ is\ =\ Work\ vs.gravity\ +\ Work\ vs.Frictional\ force$

$Work\ vs.gravity\ =\ Gain\ in\ Gravitational\ potential\ energy$

$E_p = 800 \times 10 \times 2.5 = 20,000\ J$

$Work\ vs.Frictional\ force\ =\ Force\ x\ Distance$

$= 500 \times 50 = 25,000 J$

$Total\ Work\ Done = 20,000 + 25,000$

$= 45,000\ J = 45\ kJ$

END OF SECTION

Section 3

Anyone who has a serious ambition to be a president or prime minister is the wrong kind of person for the job.

➤ 'Serious ambition' is considered a negative in this statement whereby somebody will scheme, conspire and exceed the boundaries of morality with the sole view of personal gain. It is previously been revealed that many world leaders have been willing to sacrifice their values, loyalty and even family in order to maximise their power at all costs.

➤ The head of a state or government is intended to be a role model and representative of the people. However, those that have dedicated their entire lives to pursue their aspiration of leading a country are more likely to lie, pander, compromise their principles, steal votes, be corrupt, manipulate the press and manipulate public opinion; traits that most of 'the people' would frown upon and consider morally unacceptable. These attributes, especially corruption, may also have longer lasting effects and lead to stunted societal and economic progress.

➤ If those that lead have the trait of over-ambition, it is also likely that they have idealistic and possibly even extreme views which are likely to conflict with those of the people. And, in a democracy, it is meant that the majority has a say, which is clearly compromised in this situation.

➤ However, some level of motivation and drive is necessary to lead a country. An ambitious attitude can lead anyone to triumph and satisfaction, which is extremely desirable for somebody that a country is meant to look up to. They will strive to perform the best they can in their role and they love the determined, passionate person they are determined to become and the feeling their work gives them, so they find no reason to quit and always aim for the very best outcomes.

➤ Ultimately need a balance of ambition without crossing any legal or moral boundaries.

People injured whilst participating in extreme sports should not be treated by a publicly funded health service.

➤ The statement argues that those participating in extreme sports knowledgeably put their health at risk and treating them on a publicly funded health service is unfair and ultimately a burden to society. The NHS has limited resources, and those who partake in these sports are more likely to require this time and money, which is borne from the taxpayers' pockets even if they themselves did not undertake such risky behaviour. This may result in reduced resources available to treat other illnesses which were not self-inflicted.

➤ In public healthcare, there should be the unstated duty to keep oneself healthy; treating patients who are not keeping themselves healthy removes the responsibility for one's own health. Indeed, in a private- or insurance-based healthcare arrangement those who encounter more hazards would be expected to pay a higher premium.

➤ However, all taxpayers pay money to the government to fund the health service, and have a right to use the facilities they have paid for. Their autonomy as contributors must be respected; we should avoid judgement and allow them to live their life as they please. It would be unfair to single participants of extreme sports as the only group not to receive treatment. The Hippocratic Oath pledges that doctors should treat all patients without exception.

➤ Everything in life is risky to some extent. One could argue, with the same logic, that drivers know the risk of having a car crash so we should not treat them if an accident were to occur.

➤ Moreover, many other illnesses and diseases that are similarly 'self-inflicted', such as those caused by smoking, excessive drinking, unsafe sex, overeating, etc. are treatable on the NHS. If we were to deny funding for those doing extreme sports, this would generate a slippery slope where other diseases were debated for exclusion from public healthcare.

➤ It is impossible to draw a definite line between an accident and self-inflicted disease, and thus whether somebody would qualify for treatment depending on the risk of the action which caused the injury.

➤ Allowing patient autonomy is key in the healthcare system, as is showing beneficence.

A pet belongs to its owner - it is their property. Thus, if a client asks for their healthy cat to be painlessly euthanized, a veterinary clinician should always agree to this request.

➢ The statement suggests that as a pet is the direct property of the owner, their wishes relating to decisions regarding the pet should be complied with, regardless of its condition.

➢ There is some support for this argument; there are many cases whereby an owner neglected a pet after euthanasia has been refused, causing more pain and suffering than a painless death at a surgery. It is true that vets are not obliged to perform euthanasia on healthy animals but if an owner is determined that euthanasia is the only option, the stress for the animal of being taken to numerous other clinics when the outcome is inevitable can be argued as unethical.

➢ However, if an owner is determined to let go of their pet, there are many options that avoid unnecessary murder. Relocation to a new, caring home or adoption to an animal shelters are both schemes that are heavily supported in this country. The definition of euthanasia is to prevent prolonged suffering and if the pet is healthy there is no need to kill it prematurely.

➢ Animals have rights and one of those is the right to live. In a similar fashion to the argument, children under 18 are considered by law the 'property' of their parents, but it would be thought ludicrous if a healthy child was pushed to be euthanised by their parents.

➢ Undoubtedly the owners are extremely important in the lives of pets, and should have influence in decisions relating to them. This does not mean necessarily mean, however, that the owners have the right to harm their pet.

Science only tells us what is possible, not what is right.

➢ Science has made astonishing progress in the pursuit answering the questions about how the universe works, and is based on empirical evidence rather than beliefs. It allows us to call a friend living on the other side of the world, has answered how the hole in the ozone layer formed and helps us determine how best to treat a patient suffering from diabetes.

➢ Science finds truth based on rational reasoning allowing objectivity, whereas opinions and natural biases would lead to subjectivity. Sam Harris, a famous neuroscientist, argues that most ethical question arise due to neurobiological factors and to answer them a thorough understanding of the brain is required.

➢ However, it is worth noting that even things we usually consider to be scientific fact, are actually termed 'theories', e.g. Germ Theory, the Theory of Relativity and even the Theory of Evolution. The nature of scientific theory means it is always open to falsification.

➢ Also, science helps us describe how the world is, and what we are able to do and achieve, but it doesn't help us make moral judgements about whether that state of affairs is right or wrong and how to use the scientific knowledge. For example, new knowledge about recombinant DNA may become available, but this doesn't mean one knows whether to use this knowledge to correct a genetic disease, develop a bruise-resistant apple, or construct a new bacterium. For almost any important scientific advance, one can imagine both positive and negative ways that knowledge could be used. Other examples include: Hiroshima atom bomb, designer babies, animal rights and euthanasia.

➢ Science can't help us make aesthetic or artistic judgements - e.g. whether a painting or a piece of music is good or bad.

➢ If it weren't for science's exploration for rational, objective answers, we wouldn't begin to consider the moral issues concerning what is right and wrong; science doesn't tell us what is right but it is necessary more indirectly.

END OF PAPER

2011

Section 1

Question 1: D

Assign each row to a bar chart based on relative values and values that are the same. Max temperature is E, wind speed is C, rain is B and cloud cover is A. D remains.

Question 2: E

None of the answers are conclusions that can be drawn from the passage.
A implies a causation between the noise of modern human life and extinction which is not a conclusion of the passage.
The passage doesn't suggest sea-based wind farms shouldn't be built so B is wrong. 'Should not' is a very bold statement and is usually involved in an incorrect answer.
Although the passage mentions that the whales are trying to adapt their communication methods, 'will be able to adapt' is too strong a phrase and thus C is wrong.
D is tempting but there is nothing to suggest that the depletion was initially caused by the growth of human noise.

Question 3: C

There is lots of irrelevant information in this question, but the maths is quite easy. Deluxe rooms are $80 a night, but there is $15 less per night being paid for no meals, so that is $65 x 6 for the room. Hiring a car is $5 + $5 x 6 (there is no taxi use). $65 x 6 + $5 + $5 x 6 = $425.

Question 4: E

The argument assumes that children either interact with each other OR play computer games, so E is correct. The other answers may be 'correct' from your previous knowledge but are unrelated to the argument. The argument is about the link between children playing computer games and social interaction, which the other answers do not address.

Question 5: F

This can be visualised to rule out 1 and 2. Also, 1 is the same as the right hand mirror, and 2 is a rotated version of the left hand mirror, which means they cannot be reflected images.

Question 6: C

The conclusion of the passage is "so all that learning...increases their memory power." This suggests that the fact their brain areas related to memory are more developed is through being a taxi driver rather than their predispositions. The other answers are not assumptions that are made in the argument.

Question 7: A

You can start by ruling out C, E and F as the range of miles per journey is too low. Out of the rest, you must work out the best fuel consumption per passenger mile. You can do this by finding out the lowest value of: fuel consumption of empty plane + fuel consumption per passenger x number of passengers. If we use 177 passengers and the values for each plane, it is easy to see that A will be the lowest per mile.

Question 8: D

Women in their 40s or 50s earned over 20% less than men on average, meaning they earned less than £0.80 for every £1 earned by a man in that range. A cannot be reliably concluded as we only have information about relative pay rather than absolute pay. B cannot be concluded, as we don't have definite information about how willing employers are to employ women based on age. C cannot be concluded, as we don't know the difference between the pay gap at the ages of 22 and 30; we only have information for the whole ranges (22-29 and 30-39).

Question 9: C

If we call male pay m, we see from the equation given that:

$22.8 = \frac{100(m-16000)}{m}$

$\frac{22.8}{100} = \frac{m-16000}{m}$

$0.228m = m - 16000$

$0.772m = 16000$

$m = \frac{16000}{0.772} \approx 20700$

Question 10: C

The assumption is that the long hours and intensity of senior positions deterred mothers in particular, but they would be happy to take these positions in any other case, which is said in C. The other answers are irrelevant.

Question 11: D

Is the only answer that could be correct, and satisfies the maths too: using estimates we know (15-9)/15 x 100 = 40%. There is no information about part-time workers over 60 so A cannot be correct. B and C assume that absolute wage values are known, whereas we only have relative proportions.

Question 12: D

There are 6 types: fully white, fully black, ¼ outlined, ¼ outlined, ¼ black, ¼ black with the other quarters outlined.

Question 13: B

Would definitely strengthen the HAR1 hypothesis. A lack of native HAR1 causes language impairment, suggesting that it is necessary for human language ability.

A, if anything, would weaken the hypothesis as the hypothesis suggests that it is the uniqueness of HAR1 that allows human language ability.

C, D and E don't strengthen the hypothesis as they are irrelevant to human language ability.

Question 14: C

Each team played 4 matches, and we can determine how many wins, draws and losses each team had. With 2 points, Central must have drawn 2 and lost 2. With 8, Northern must have won 2 and drawn 2. Southern must have won 1, drawn 2 and lost 1 whilst Western must have drawn 1 and lost 3. The discrepancies (3 wins and 6 losses) mean that Eastern must have won 3 and drawn 1.

Question 15: B

1 is incorrect as the argument is not about who, in particular, defines the human rights, and 3 is irrelevant to the argument. 2 is correct because the passage assumes something cannot be both a constitutional right and human right; it can only be one or the other.

Question 16: D

Area is length x width, and volume is area x depth. So, all we need here volume/width to determine the highest value for depth. Don't waste time working out answers for lakes that are not an answer, and we don't need to use exact numbers. For example, Caspian Sea's depth is 78,200/394,299 which we can think of as 80,000/400,000 which is roughly 0.2 km. When we do the same for the others, Baikal clearly has the highest depth which is around 0.67 km

Question 17: C

1 isn't an assumption made: the argument doesn't require the assumption that most of the population have eaten beef infected with BSE. In fact, the main conclusion is that there will be further outbreaks in the future as those who consumed the infected beef grow older, and this is irrelevant to how many have eaten the beef.

2 is clearly incorrect, as the argument discusses that inheriting the V variant in the M-V combination can lead to developing vCJD in later life. 3 is correct as the argument assumes the combination of genes is the most important factor rather than the M variant itself.

Question 18: B

Jasper earns £240 + £5 x 22 + £20 x 6 = £470 per week, so Ruby earns £510 per week. Ruby is 35, and if we call her years worked as y, £240 + £5 x 14 + £20 x y = £510, and y = 10. This means Ruby has worked for 4 more years than Jasper.

Question 19: A

The sum of the possible outcomes for drilling is -720000 x 0.1 + 400000 x 0.8 + 3800000 x 0.1 = $628000

The sum of possible outcomes for not drilling is 1000000 x 0.2 + 500000 x 0.6 = 500000. So drilling represents a more favourable option than not drilling.

Question 20: D

Is correct – the probability of a 'medium' strike is 0.8, as is the probability of selling the drilling rights at $500,000 or $1,000,000 (0.6 + 0.2). A is incorrect, as a medium strike allows a profit too. B is incorrect, as $400,000 x 0.8 is less than $500,000. C is incorrect, as there is only a 0.1 chance of making a loss, and a 0.9 chance (medium or big strike) of making a profit.

Question 21: F

All of the answers are correct. The costs would be £1,300,000 for drilling, which is higher than the returns for a medium strike (£1,200,000). The only way to make a profit is thus a big strike, which has a 10% chance. They could however make a profit if they reduced drilling costs by 25% to £600,000, so costs would be £1,100,000, which is lower than the returns for a medium strike.

Question 22: E

The sum of expected outcomes for drilling is $628,000 and is $500,000 for not drilling. Paying the insurance would make the value for drilling $428,000. Without insurance, the expected outcome for an oil spill is -$10,000,000 x 0.03 = $300,000, making the value for drilling $328,000. Not drilling has a value of $500,000. So, the order is 3, 1, 2.

Question 23: D

Determine which numbers haven't been used. On the left side the remaining numbers must add up to 13, and must be 6 and 7, but we don't know which order. On the right hand side they must add up to 10, so they are 8 and 2 but we don't know which order. This mean the number right of 9 must be 4, and the remaining 2 numbers must add up to 8 on this row, which must be 6 and 2. So between 5 and 3 is 7.

Question 24: A

The quote means Fredericks will not play if Petermass is fit, and may or may not play if Petermass is not fit. This makes 1 correct and 2 and 3 incorrect.

Question 25: C

The one with equal proportions has 90ml of oil and vinegar and the other has 120ml oil and 60ml vinegar. So if we add half of the one with equal proportions to the other, we have 45ml of each in the first and we have 165ml oil and 105ml vinegar in the other. This latter one represents 11/18 oil and 7/18 vinegar. If 90ml of this mix is taken, 55ml will be oil and 35ml will be vinegar. So there will be 110ml oil in one (165-55) and 100ml (45+55) in the other.

Question 26: C

This is the correct answer as the argument talks of the possibility that many planets could support human life.
➤ A is not correct as gravity is not the only factor that determines whether a planet can support life.
➤ B is not necessarily correct and the use of the word 'must' means it is probably too bold a statement.
➤ D is tempting but the passage uses the words 'probably in the order of 10 or 20 per cent', so saying it is 10 or 20 percent is too strong.
➤ E is again too bold a statement, as the passage only says that there 'could be' tens of billions of these systems.

Question 27: A

This is easier than it looks. The question is asking if you can form a 6 x 2 or a 4 x 3 rectangle with 3 of the shapes. However, it's not actually possible to form a rectangle of those dimensions regardless of the shapes you pick.

Question 28: C

2 is not necessary information for the argument; the distance that each bus needs to travel every week is irrelevant, so C must be the correct answer. We can check and see that the other points are reasonable, which they are.

Question 29: D

3km from the library is the distance when Claire leaves 20 minutes late. This tells you that when Claire leaves on time she is 1km from the library when Charles is 3km away (she walks 2km in 20 minutes so had she left on time she would be 2km further along). As they usually meet at the Library, Charles has to cover 3km while she covers 1km, so he cycles 3 times faster than she walks at 18km/hr.

Question 30: B

Paraphrases the last 2 sentences of the passage and is the correct answer. A and D are both too bold to be answers as the passage highlights that it is just a theory. C isn't correct as we don't know if it is the best explanation – we haven't heard any other hypotheses.

Question 31: C

There are 20 cans remaining. The view from X shows us that there are 6 cans in the shaded columns remaining. Thus, there are 14 cans in the 6 columns in the middle. Thus, a maximum of 4 cans can be missing from the middle 6 columns.

Options A, B, E and F are too full to not be possible. D is still possible if columns 3 and 4 only have one can each (therefore columns 1-6 = 14). Only C is therefore not possible as columns 1 and 2 would have to be 2 cans tall and columns 3 and 4 to be 1 can tall (6 cans missing).

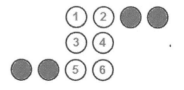

Question 32: C

Rounding is ok here because of the big difference between answers. 1500 patients per doctor, so a total of 1500 x 5 consultations a year = 7500 consultations a year per doctor. 7500/250 = 30 consultations a day, and C is the closest (it would be closer to 32 if we didn't round down earlier).

Question 33: D

The proportion in 1995 was 0.8/3.29 which is close to 0.2, and was 1.8/5.26 in 2006 which is close to 0.33. The percentage increase is (0.33-0.20)/0.20 x 100 which is roughly 65%, making D the closest.

Question 34: C

C is clearly the only viable answer, as the ratio is definitely above 2 between 15 and 35 and then starts to decrease towards 1.

Question 35: C

People may have become 11 years older, but there is still the same age group within the age demographics, so ageing is irrelevant. The others are relevant.

END OF SECTION

Section 2

Question 1: F

Carbohydrase is an enzyme – not a gland, hormone or a function so does not fit into the table. All the other words/statements fit in somewhere.

Question 2: B

X has 3 electrons in its outer most electron shell and Y needs 2 electrons to make a complete outer electron shell. I.e. Valency of X is +3 and Y is -2. The easy way to figure out formulae is to 'swap' the valencies to give: X_2Y_3.

Question 3: C

Remember that $E_k = \frac{mv^2}{2}$ and $E_p = mg\Delta h$

Since E_k is proportional to velocity^2, doubling the velocity means E_k is 4 times higher. E_p is proportional to height so doubling the height means E_p is 2 times greater. Thus, C is the correct answer.

Question 4: C

$$3x\left(3x^{-\frac{1}{3}}\right)^3$$
$$= 3x\left(3^3x^{-\frac{3}{3}}\right)$$
$$= 3x(27x^{-1})$$
$$= 3x\left(\frac{27}{x}\right)$$
$$= 81$$

Question 5: F

Mitosis produces genetically identical cells and meiosis results in variation so 1 and 2 are wrong. Statements 3,4 and 5 are correct.

Question 6: D

Raising the temperature increases the average kinetic energy of all molecules. Thus, more collisions take place per unit time and the average collision has more energy. However, temperature has no effect on the orientation of the molecules.

Question 7: E

A. Nuclear fission is the splitting of a nucleus into 2 small parts. Whilst gamma radiation may be released, this is not the definition of fission.

B. The half life of a radioactive substance is the time taken for half of it to decay (not half the time for it to decay).

C. The number of neutrons is given by the mass number – atomic number, not the other way around.

D. Nuclear power stations utilise fission, not fusion.

E. A beta particle consists of a highly charged electron. There is no change in the atomic mass.

F. When a nucleus emits an alpha particle, it loses 2 neutrons + 2 protons.

Question 8: D

At 9:45 the hour hand will be ¾ of the way to 10 from 9 and the minutes hand will be at 9. The number of degrees between the hours is $\frac{360}{12} = 30$. Thus ¾ of this is 22.5 degrees.

Question 9: C

4 is incorrect because individuals with relatively disadvantageous adaptations will still usually be able to breed, unless that adaptation actually causes death or an inability to breed. The other statements are correct.

Question 10: D

Each carbon atom can make 4 bonds and each hydrogen atom can make 1 bond. So the outer carbons are bonded to 2 hydrogens each. However, the carbon atoms between the two rings are only bound to 1 hydrogen. So if we count up, there are 10 carbons and 18 hydrogens.

Thus, $A_r = (12\ x\ 10) + (1\ x\ 18) = 138$

Question 11: B

Be careful! The current in this circuit flows clockwise (most circuits are anticlockwise). The diode is thus a break in the circuit in this direction. When the switch is open, we have 2 breaks in the circuit, so no current can flow through either of the branches. Thus, the reading on the ammeter will be 0. When the switch is closed, the current can pass through the branch without the diode in it. Since the resistance is 3 ohms and voltage is 6V: $I = \frac{6}{3} = 2\ A$

Question 12: D

The quickest way to solve this is via trial and error- assign values to each option to see if you can disprove them e.g. If W is 3 and x is 2, then option A would be incorrect (yet the inequalities in the question would still hold). The only inequality that **must** be true is x > y because $y^2 < x$.

Question 13: E

Oxygen and carbon dioxide move across membranes via diffusion, not osmosis. Oxygenated blood goes to muscles and deoxygenated blood returns from muscles. The oxygen concentration is low in the muscle cells as oxygen is required for aerobic respiration, and carbon dioxide is high because it is produced during aerobic respiration. Thus:

➢ CO_2 is high in muscles and low in plasma.

➢ O_2 is high in RBC and low in muscle.

➢ Hence, E is correct.

Question 14: C

A metal cannot form covalent bonds with a non-metal, so NaCl and Na_2O are ionic compounds. All the other compounds contain a covalent bond.

Question 15: B

This is a good example of why it's handy to know the suvat equations as it can save you a lot of time.

Using $v^2 = u^2 + 2as$

$0^2 = 300^2 + 2\ x\ a\ x\ 0.6$

Thus, $0 = 90000 + 1.2a$

$a = \frac{9\ x\ 10^4}{1.2} = 7.5\ x\ 10^4$

Now use $F = ma$ to give: $F = 0.05\ x\ 7.5\ x\ 10^4$

$F = 0.375\ x\ 10^4 = 3.75\ x\ 10^3 N$

Question 16: E

Don't try to solve these algebraically – it's much quicker to sketch them!

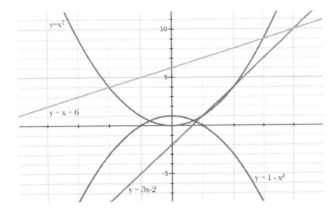

As is easily visible, $y = 1 - x^2$ and $y = x + 6$ don't intersect. Thus, the answer is E.

Question 17: D

Any are feasible. The condition could be dominant with P, Q, R and S having the recessive alleles only. The sperm from T could have carried the allele from the condition, or a mutation could have been present in an egg of S.

Question 18: E

The easiest way to solve this is to set up your own algebraic equation for Oxygen (as it appears the most number of times):

$3B = 6X + Y + 2$

Then simply see if any of the options satisfy this equation- only option E does. If you're unsure about how to setup these equations then check out the chemistry chapter in *The Ultimate BMAT Guide*.

Question 19: A

The current flowing through a resistor at a constant temperature is directly proportional to the potential difference across it. Thus, as voltage increases, current increases at a constant rate i.e. resistance does not change.

Question 20: B

This is a test of how quickly you can use Pythagoras's theorem:

Bottom Triangle:

The hypotenuse is given by $= \sqrt{1^2 + 3^2} = \sqrt{10}$

Middle Triangle:

Since the triangles are similar, the small edge must be 1/3 of $\sqrt{10} = \frac{\sqrt{10}}{3}$

The hypotenuse is given by $= \sqrt{\left(\sqrt{10}\right)^2 + \left(\frac{\sqrt{10}}{3}\right)^2} = \sqrt{10 + \frac{10}{9}}$

$= \sqrt{\frac{90+10}{9}} = \sqrt{\frac{100}{9}} = \frac{10}{3}$

Top Triangle:

Since the triangles are similar, the small edge must be 1/3 of $\frac{10}{3} = \frac{10}{9}$

The area of the triangle $= \text{½ } base \times height = \frac{1}{2} x \frac{10}{9} x \frac{10}{3}$

$= \frac{100}{54} = \frac{50}{27} \ cm^2$

Question 21: D

Cell P is haploid and can be any sex cell. Cell Q is diploid and can be almost any somatic cell. Cell R can be enucleated, or a red blood cells (because they have no nucleus). Only D satisfies these criteria.

Question 22: D

The mass of PbS in the ore $= 70\%$ of $478kg$

$= \frac{70}{100} \times 478 = 335\,kg$

The atomic mass of PbS $= 207 + 32 = 239$

The proportion of lead in PbS $= \frac{207}{239} \approx \frac{210}{240} = 0.875$

Thus, the mass of lead that can be extracted $= 0.875 \times 335 = 293\,kg$

The closest answer to this is 289.8 kg.

NB: Remember rounding numbers to make the maths easier also means you won't get the **exact** answer on the paper. Thus, always check the options to see how much room you have to round numbers. If they are close together then you should avoid rounding and vice versa.

Question 23: C

Light travels slower in glass than air, but its frequency remains the same. Since $c = f\lambda$, wavelength must decrease to accommodate for the lower speed. Option C is the only one that satisfies these requirements.

Question 24: B

12 can only be achieved by rolling 2 sixes.

The probability of rolling a 6 on the fair die $= 1/6$

$P(12) = P(6\ on\ fair\ die) \times P(6\ on\ unfair\ die)$

$\frac{1}{18} = \frac{1}{6} \times P(6\ on\ unfair\ die)$

Thus, $P(6\ on\ unfair\ die) = \frac{1}{3}$

$P(1\ to\ 5\ on\ unfair\ die) = 1 - \frac{1}{3} = \frac{2}{3}$

$P(1\ on\ unfair\ die) = \frac{\frac{2}{3}}{5} = \frac{2}{15}$

2 can only be achieved by rolling 2 ones.

$P(2) = P(1\ on\ fair\ die) \times P(1\ on\ unfair\ die) = \frac{2}{15} \times \frac{1}{6}$

Simplifies to: $\frac{1}{90} = \frac{1}{45}$

Question 25: D

The homeostatic response brings the level back to the stable, normal level. If the system is less responsive, all phases are likely to occur later (not earlier) and the deviation from the normal level will be higher.   Thus, phase 2 would be higher.

Question 26: E

The key word here is **excess** oxygen. When sulphur reacts with oxygen, the product is Sulphur dioxide. This eliminates options A-C.

Since there is excess oxygen, the other product will be carbon dioxide (not carbon monoxide). Thus, E is correct.

Question 27: B

50 beats per minute means there is a beat every $\frac{6}{5}$ seconds. Since the back soldiers put down their left foot at the same time as the front soldiers put down their right, the adjusted beat is every $\frac{3}{5}$ seconds.

Thus, the *minimum distance* $= \frac{3}{5} \times 330 = \frac{990}{5} = 198$ m

END OF SECTION

Section 3

"Democratic freedom means there should be no restriction on what may be said in public."

➢ Democracy allows power to be vested in the people; it perpetuates the idea of freedom, equality and liberty, and thus the right of expression including freedom of speech. This suggests there should be no restriction or intervention if someone wishes to express their opinion in a public setting.

➢ Freedom of speech does not necessarily remove all boundaries; a restriction on the proclamation of unwarranted, extreme and prejudiced ideas seems just.

➢ People generally believe what they hear from people that appear more 'powerful' than them. Giving misleading information to the public can be dangerous; for example, the freedom to preach that drink-driving is not a crime to a bar could endanger many.

➢ Complete freedom of speech allows those with extreme ideas to declare whatever they believe, which could be to the detriment of young, impressionable minds enticed by these opinions.

➢ There should be a limit if 'free speech' is inciting hatred. Discrimination against minorities, whether it is sexuality, race or even age, can cause friction, hatred and 'broken' societies that is against the very notion of a democratic society.

➢ A utilitarian stance appears fairest; the 'harm principle' postulates that the actions of individuals should only be limited to those that prevent harm to other individuals, and this appears a good balance between completely free speech and a restrictive stranglehold. A difference in opinion can and should be freely expressed, but it is explicable that hate speech, classified information, obscenity, etc. is regulated.

The art of medicine consists of amusing the patient while nature cures the disease.

➢ The author argues the point that medicine is an art, and suggests that physicians need qualities of empathy and compassion to help satisfy patients' worries and beliefs while nature takes its course to cure them over time.

➢ Although this statement may have been wholly true in the 18th century, nowadays medicine is an applied science and it is debatable whether the statement still has any implications for the doctors of today.

➢ Scientific knowledge now allows us to treat patients better or more quickly than nature. Drugs target specific biochemical or physiological systems in order to cure. The fact that clinical trials are double blind and tested against a placebo and require a statistically significant difference between the groups highlights that the active ingredient of a drug is having an effect greater than that of nature and the 'amusing' of a patient. Example of the swine flu pandemic in 2009; oseltamivir highly successful. Many infectious diseases through history are now effectively eradicated worldwide.

➢ However, placebo is much more effective than no treatment, suggesting that patients can seek comfort in the thought that they are being cared for by those who they believe have the knowledge and skill to do so.

➢ Example of the common cold – can only treat the symptoms and wait for the innate immune system to take control.

➢ Doctors must always use more than science; examples include the importance of teamwork, leadership and diagnosis where there is no scientific method except interacting with the patient (e.g. bipolar disorder and other psychiatric illnesses).

A scientific man ought to have no wishes, no affections - a mere heart of stone.

➢ Darwin tackles the issue of objectivity in science; the ability to remain unbiased and develop conclusions true to our empirical evidence rather than true to our beliefs. Wishes and affections greatly influence one's behaviour and one's rationality.

➢ Within science there undoubtedly exists a facet which requires absolute removal of our wishes and affections. These traits could influence experiments and therefore could and render invalid the interpretations of these studies, which would be detrimental in our aim to dispel our own ignorance and further scientific knowledge. There are many examples of where passion and bias have blinded scientists to that which has been objectively proven. A famous example is Sir Fred Hoyle who rejected Hawking's presentation of the Big Bang Theory despite overwhelming evidence (cosmic microwave background) until his death.

➢ It can be dangerous to not show complete objectivity. An example is the case of Andrew Wakefield, the gentlemen who asserted the link between the MMR vaccine and the onset of autism, which resulted in the deaths of many children from measles, mumps and rubella. (AW falsified his results and subjected autistic subjects to unethical procedures to 'prove' that the MMR vaccine causes autism after being bribed by a law firm that was planning on suing the MMR vaccine company. After repeat experiments were performed and the results acquired were found to be completely different, AW's paper was withdrawn from the Lancet and he was struck off the medical register.) Sadly, the myth that MMR vaccines cause autism is still believed and herd immunity has not reached pre-scare levels. Scientists should thus welcome other scientists with different prejudices and preferences to follow what they have done to see if they get the same results through peer reviewing.

➢ But our objectivity is not intended to quash our biases and natural passion and yearning for a scientific outcome. On the contrary, objectivity very much allows us to be passionate for science because it prevents the passion from influencing the results. So why have passion at all? Passion, drive and intrigue are intrinsic to scientific discovery and enquiry. Without such qualities Fleming or Darwin might never have discovered antibiotics and evolution. Einstein has always asserted the importance of imagination over knowledge. One reason is that imagination and creativity allows for options for technological advances that were unplanned and unprecedented (e.g. Fleming).

➢ One could argue that this is oxymoronic, as if one is "a scientific man" this suggests they have an affection and wish to proceed with scientific method. Without our wishes and affections, there would be no driving force underpinning our desire to understand how the world works.

Veterinary pet care in the UK should be free at the point of delivery, as human care is.

➢ The argument is that, as sentient beings, animals should receive free healthcare like their human owners.

➢ Animals experience disease, pain and illness in the (arguably) same manner that we do, and our experience with these should mean our empathy extends to the animal world. The mantra of public healthcare is essentially that everyone deserves the 'right' of healthcare, regardless of wealth, status and privilege. Should this not translate to animals that are often most vulnerable?

➢ For most owners, their pet is part of the family and they would do anything to protect them and ensure they stay healthy.

➢ This plan would reduce the number if abandoned pets whose owners were unable to afford veterinary care.

➢ One of the core responsibilities of pet ownership is that you should make sure that you can take care of them and can afford to pay for everything that they need for the entire duration of their lives. If one cannot afford veterinary care, then perhaps they should not have a pet. In any case, taking out pet insurance is recommended to most owners.

➢ Public healthcare through the NHS is only possible in the UK because taxpayers contribute towards its funding. In the case of veterinary care, forcing non-pet owners to pay taxes towards other people's pets - essentially, for a service that they will not use - seems very unfair. However, we do pay for the healthcare of those who smoke, heavily drink or take part in extreme sports, which is similarly 'unfair' in the sense that they are much more likely to use the publically funded money than others. Owners may agree to pay a kind of specific owner tax.

➢ It is also possible that pet owners may take less responsibility for their pet's health if they know that they will not have to pay for veterinary care.

➢ It is hard to draw a line between how well we should treat animals relative to ourselves. It is clear that in general we treat ourselves as if we are 'above' animals in society; we eat them, we test on them and we exploit them. If we have the resources to keep them healthy, then they deserve to have free healthcare, but this may be at the expense of other resources that the majority of people find more important in a democratic society.

END OF PAPER

2012

Section 1

Question 1: D

10% of the population is 8628709 x 0.1 = 862870.9, and the only islands with less than this amount are Brosnan and Dalton. 20% of the area is 26315 x 0.2 = 5263; Brosnan has less than this amount and Dalton has more, so the answer is Dalton.

HINT: As only 1 answer can be correct, for the second part of this question it can be assumed that the higher value is above 20% and the lower value is below 20% without doing any exact workings.

Question 2: C

As it is preceded by the words '*based on these findings,*' it is fairly easy to determine that the main conclusion of the passage is "*pale-skinned people should be added to the list of those for whom vitamin D supplements are recommended by the government.*"

➤ A could probably be argued as correct based on the passage with a lack of a better answer, but the conclusion above clearly mentions supplements which this answer does not. Furthermore, the use of the word 'need' immediately suggests that the answer is probably too strong and thus incorrect.

➤ B is probably true based on the information, but again is a point made to back up the argument rather than the main conclusion of the passage.

➤ C paraphrases the conclusion given above, and is clearly the best answer.

➤ D may be tempting from your previous knowledge, but is irrelevant and beyond the scope of this passage as skin cancer is not mentioned.

➤ E is not correct as the passage makes no comparison between pale and dark skinned people. Again, we can probably rule this answer out from the start for being too strong and sweeping.

Question 3: F

The best way to answer this question is probably by testing the effect of eliminating each tile. Checking whether the number of black and white tiles is equal first is probably easiest as they stand out, but if this criterion is satisfied the other patterns must be checked too.

➤ By removing A, you can see that you will be left with 6 black tiles and only 4 white tiles.

➤ Removing B leaves 5 black tiles but 6 white tiles.

➤ Removing C leaves 3 black tiles but 5 white tiles.

➤ Removing D leaves 6 black tiles and only 5 white tiles.

➤ Removing E leaves 5 black tiles and 5 white tiles. However, only 3 of the dotted pattern remain.

Removing F leaves 5 black tiles and 5 white tiles. It can also be seen that all the other patterns amount to 5 tiles.

Question 4: B

This question asks for a conclusion of the passage.

A can be inferred but is not a conclusion of the passage.

B is the best answer as it sums up one of the fundamental points of the argument, "*only fossil fuels, which produce emissions of CO_2, can provide the extra capacity.*"

The use of the word 'never' in C immediately suggests that this is the incorrect answer, and there is no mention of the future potential of wind power.

D is incorrect, as the first sentence states that electric engines are actually more economical than petrol engines.

Question 5: E

The best way to answer this question is to work out the collective area of the shrubs, veg, pond and lawn, take this away from the total area and then determine the number of slabs needed. We are given several widths and heights of these areas and must work out the other ones.

➤ The height of the pond is equal to the veg (3m) and is clearly a square, so the area of the pond is 3 x 3= $9m^2$

➤ The total width of the veg areas is 18 - 1 - 3 - 1 - 0.5 - 0.5 - 1 = 11, therefore the total area of these plots are 11 x 3 = $33m^2$

➤ The total width that the shrubs take up is 18 - 1 - 3 - 1 - 0.5 - 1 = 11.5m, thus the total area that the shrubs take up is 4 x 11.5 = $46m^2$

➤ The height that the lawn takes up is 12 - 1 - 3 - 1 - 1 = 6m, so the area of the lawn is 3 x 6 = $18cm^2$

➤ The total area of the garden is 18 x 12 = $216m^2$, so the area taken up by the patio and paths is 216 - 9 - 33 - 46 - 18 = $110m^2$

➤ $1m^2$ requires 4 ($0.5m^2$) slabs, so $110m^2$ requires 110 x 4 = 440 slabs.

Question 6: E

The main conclusion of the argument is "*if parents spend time discussing these issues with their children they will help their children read well,*" given away by the use of the word 'so' and the fact that it is at the end of the passage. (As a general rule, conclusions are more likely to be at the start or end of a passage than in the middle). Flaws are generally related to the conclusion of the passage, whereas questions asking for statements that 'weaken' the argument could potentially relate to any of the passage.

➤ A is clearly wrong as there is no mention of peer groups in the passage, and the statement has nothing to do with the conclusion above.

➤ B could be maybe be debated, but appears like a weak suggestion in the face of other flaws given, and is not directly related to the conclusion of the argument.

➤ C again could maybe be argued, but is again not linked to the conclusion of the argument, and there is clearly a more pertinent answer.

➤ D is unrelated to the conclusion and is beyond the scope of the passage.

➤ E is the best answer, as it directly addresses the conclusion above. A key principle in scientific argument, and hence the BMAT, is that correlation ≠ causation, which this statement asserts.

Question 7: A

This is a very difficult question, and you should ensure that you do not spend over 2 minutes trying to work this out and drawing all the permutations. For pattern questions like this it is best to work out what you must add to carry on repeating the pattern indefinitely (the 'repeating unit'. This repeating unit consists of 1 hexagon, 2 triangles and 3 squares. The easy way to do this was to realise that there are more squares than triangles in the diagram- this allows you to eliminate B,C and D. Intuitively, it is unlikely to be E (as that is obtained just by counting all the shapes)- leaving you with A as the only option.

Question 8: C

In questions like this it is too time-consuming to work with exact numbers and the best method is to work with close estimates. The total number of patient days we are dealing with is 11549 + 30432 = around 42000.

Method 1: The rate of infection is 9.86 per 10000 over ~30000 patient days, so over ~40000 days it will be ¾ of this value, which is slightly under 7.5, and 7.15 appears a very reasonable answer.

Method 2: There are 3 cases in 42000 days. To make this per 100000 days we have to times the number of cases by roughly. This equals 7.5 and again 7.15 appears the correct solution.

Question 9: D

The highest number of cases by far comes from organisation 3, with 26. This is out of a total of 69 (you can probably work this out using mental maths to save time). 26/69 is not intuitive, so we can use short division to work out that the first 3 digits are 0.37, suggesting that 38% is the correct answer of the ones listed.

HINT: We know that 69/3 is 23, thus 23/69 is equivalent to 1/3 or 33%. Our answer is higher than this value so we can easily rule out answers A-C.

Question 10: B

Again, don't waste time by working out exact values and use rough estimates.

➤ Organisation 2 had 16000/11 = ~1500 patient days per month up to November, giving a total of 17500 for the year.

➤ We thus have 1 case in 17500 days but we need a value per 100000 days.

➤ 100000/17500 = 100/17.5.

➤ 100/17.5 is clearly between 5 and 6 as we know 100/5 = 20 and 100/6 = 16.7. This is an easy way of ruling out C-E.

➤ We must then determine whether A or B is correct. We can work out that 17.5 x 5 = 87.5 and 17.5 x 6 = 105. 100 is closer to the latter, which means the answer is above 5.5 and thus must be 5.67.

Question 11: E

➤ E is the correct answer, which we can determine by ruling out all of the other statements.
A & C - Several of the organisations are made up of more than 1 type of hospital, and the data is not split between these so we don't know which type of hospital the cases were found in. We can thus not reliably conclude where these Cdl cases occurred.

➤ B - There is only 1 organisation with a small hospital alone, and this had only 1 case over 2009 and 2010 with a low rate of infection, so this is definitely incorrect.

➤ D - Again, using the reasoning above, there may not necessarily have been any Cdl found in either DCs or TCs. For example, only organisation 5 has a DC in the data given and here, all the cases may have arisen from the TC.

Question 12: B

Nicola wants to get the first bus of the day from the airport, which is at 09:15 on Thursday. It takes 50 minutes to get to the centre, so she will be there at 10:05. She must get the 15:20 bus from the centre to reach the airport before 17:00. Thus she is in the centre between 10:05 to 15:20, which is 15:20 - 10:05 = 5 hours and 15 minutes.

Question 13: D

This question should hopefully be relatively easy based on the BMAT tricks and tips you have learnt thus far. The correct letter can be determined just from reading the answers, as the passage is relatively passive and unforceful in its conclusions, whereas 3 of the answers are far too strong.

➢ A uses the word 'only' and C uses the word 'must' which suggests they are wrong, and the passage is much less bold than these statements.

➢ B is slightly less strong but the use of 'would not have' means the answer is not necessarily true, especially with a better answer present.

➢ D is not overly strong and paraphrases the first sentence, making it the best answer.

Question 14: D

The cube must be visualised to work out this question. A and B can be ruled out because the triangle should always point to the long end of a solid line. C and E are excluded by looking at the configuration of the X and the surrounding shapes. If you get stuck and have time, make the net yourself – it will save time in the long run.

Question 15: C

The conclusion of this argument is *"parents of children with autism are damaging their children's health by using the sprays,"* and the rest is background information.

➢ A would actually support this conclusion and does not weaken the argument

➢ B would again strengthen the argument as it reinforces the idea that parents are potentially damaging children if preliminary tests have not been done.

➢ C is the best answer, as if it were true, the claim in the conclusion becomes much less convincing.

➢ D isn't directly relevant to the question and doesn't potentially weaken the argument depending on what these cultural effects entail and is therefore not the right answer.

➢ E also strengthens the argument by suggesting that oxytocin is bad.

Question 16: C

To make this question easier, we can focus on $1/6^{th}$ of the conservatory as it represents the pattern of the whole area. There are 25 tiles in this area, of which 5 have the pattern with no black, 4 have the pattern with ¼ black, 12 have the pattern which is ½ black and 4 have the pattern with all black.

To work out the proportion of black, we need to work out the number of black tiles that these tiles add up to, and divide it by the total number of tiles. $0 \times 5 + 0.25 \times 4 + 0.5 \times 12 + 1 \times 4 = 11$, and dividing this by 25 (times by 4 and divide by 100) is 0.44 (44%).

Question 17: G

None of the statements can be drawn as a conclusion of the passage.

1 - This is a strong statement and although the passage states staffing levels are lower at weekends and there are more deaths at weekends, it does not suggest the causation that increased staffing would reduce death rates.

2 – The passage states that patients are dying in hospitals rather than at home (which implies that they would die anyway). This artificially inflates the mortality stats. Thus, enhancing weekend provision of primary care services wouldn't help mortality rates- only massage the statistics.

3 - This again assumes the causation that low staffing levels leads to patient deaths, which is not what the passage suggests.

Question 18: A

If x is the percentage of people that own both a tumble dryer and a dishwasher, we must work out the smallest and largest possible values of x.

75 to 85 - x is the number of people with only a dishwasher, 35to40 - x is the number of people with only a tumble dryer, 0to5 is the number of people with neither and x is the number of people with both. Thus 75to85 - x + 35to40 - x + x + 0to5 = 100, meaning x = 75to85 + 35to40 + 0to5 - 100.

Smallest value: x = 75 + 35 + 0 - 100 = 10%, largest value: x = 85 + 40 + 5 = 30%

Question 19: B

The number of category A calls in 2011 was 2.23 million. 74.9% (we can take this as 75%) were responded to within 8 minutes, leaving 25% not responded to within 8 minutes. 2.23 x 0.25 is slightly above 0.5, and working it out fully gives an answer of 0.5575.

Question 20: D

We are told that category A calls made up around 34%, category B calls made up roughly 40% and so category C calls made up the remainder (26%). We thus want a pie chart showing Category C taking up very slightly more than a quarter of the graph and category A taking up around a third. Only pie chart D fits this bill.

Question 21: B

➢ **B i**s the only sensible answer.

➢ A is completely irrelevant to how many calls led to treatment/transport at the scene.

➢ C suggests that something being a 'genuine emergency' completely determines whether people are treated/transported at a scene, which is not the case. B is the better answer.

➢ D is wrong as category C cases may still require treatment/transport, but with different timings. Also, 26% of calls were category C, which amounts to over 2 million.

Question 22: A

In 2010, 2.08 million incidents were category A, and ~75% were attended within 8 minutes. 2.08 x 0.75 = 1.56m
In 2011, 2.23 million incidents were category B, and ~75% were attended within 8 minutes. 2.23 x 0.75 = 1.68m
1.68 - 1.56 = 0.12 million

Question 23: B

Working horizontally, we can see that patterns 1, 3, 8 and 10 are equivalent, patterns 6, 7 and 11 are equivalent and patterns 4 and 12 are equivalent. This leaves patterns 2, 5 and 9. 2 and 9 are equivalent through rotation, but pattern 5 cannot be rotated to match either of these. There are thus 5 distinct patterns.

Question 24: E

The conclusion is clearly the first line: "*Police should be given clear permission to use water cannons against rioters and rules about when it is appropriate.*"

➢ A weakens the argument as it suggests that water cannons also affecting the innocent means it is not a good solution for targeting crime.

➢ B could maybe be debated as the answer, but the police are unlikely to be against the use of training and resources.

➢ C weakens the arguments as it suggests there are other equally useful strategies.

➢ D mentions the high expense of water cannons and thus weakens the argument.

➢ E is the best answer based on the information given, as it strengthens the idea that water cannons should be used.

Question 25: C

We can work out that the highest score in a full turn is 18 and the lowest is 2, which gives 17 possible scores. We must then work out if any scores between 2 and 18 are not possible (and we only need one example of a score being possible).

2 - 2, 2, miss.
3 - 2, miss, 2
4 - 4, 4, miss
5 - 4, 6, miss
The rest of the even numbers can be made with combinations of 2, 4 and 6
7 - 6, miss, 4
9 - 6, miss, 6
Higher odd numbers cannot be made, leaving 13 possibilities in total.

Question 26: D

This argument makes the reasoning error of x is z, y is z so z is x. There may be cases of z that are not x, or z may not link to x at all.

➢ A follows x is y and y is z so x is z, which is reasonable.

➢ B suggests x is y and z so everything z is y, which is wrong but a different error to that in the passage

➢ C suggests x is y so not-x is not-y, which is (generally) fairly reasonable and definitely different to the reasoning error in the argument.

➢ D follows x is z and y is z so x is z, which is the same reasoning argument as above.

Question 27: C

We know that furniture costs 3 months to pay, with ½ paid in the 1st month, and ¼ in the next 2 - this means that it does not matter whether sales of e.g. $2000 came from 1 sale or from several sales, as the money received per month will be equivalent in either case. We also know that there were no sales in May and June. Using these 2 pieces of information we can deduce the answer.

June must have been the final payment for 1 piece/pieces of furniture due to the closure, which means $2000 must also have come in from this April sale in May and $4000 in April. This gives a total of $8000 of sales from April The extra $1000 in April and May earned must have come from a sale in March, which would have amounted to a total of $4000 in March, and a total of $2000 earned from this/these sales in March.

The $2000 missing from March must have come from January sales of $8000, giving $4000 in January and $2000 in February. Any excess money from January and February must have come from furniture sold before January.
Thus, the total sales in this time period are $8000 (January), $4000 (March) and $8000 (April), giving an answer of $20000.

Question 28: G

All the answers identify a weakness in the argument.

1. This is a weakness as saying "*this is nonsense*" suggests that the author thinks the ski holiday industry does not damage the environment, because all travel damages the environment.
2. This is a weakness as the argument that ski holiday resorts use less energy than other resorts is conflicted by this information. Using percentages here may mask the fact that absolute levels of energy consumption may be high, which this statement addresses
3. This is definitely a weakness as the author fails to consider that damage to the environment is not only caused by energy consumption.

Question 29: C

➤ £12240 was made from y sales.
➤ 0.4y represents 40% of the ticket sales that were refunded £5 each.
➤ So 0.6y x 20 + 0.4y x 15 = 12240. Solving for x gives x=680.
➤ 40% of 680 is 272 tickets, and £5 x 272 = £1360 refunded.

Question 30: C

Only statement 3 can be inferred. This could maybe be determined without the passage as it is least bold statement.
1 - "*authors…give a one sided view*" is very strong and the passage does not mention effectiveness or safety.
2 - There is nothing to suggest that "*companies…aim to influence the content of the articles.*"
3 - This effectively paraphrases what is said in the first 2 sentences of the passage.

Question 31: C

We can deduce several things from this passage:
We know Jill must be at least 7 points ahead of 4th place so that even if she comes last and they come first the places remain the same. With the same reasoning, she must also be at least 7 points behind 2nd place.
Karen and Gemma must have the same number of points, so that even if one comes 3rd and one comes 4th, the final race determines who finishes on top.
The person in 4th place must have 7 points less than Jill if he is going to finish last (regardless of the scores in the last round). The easiest way to do this is go through each option and see if the scores sum to 90. However, remember that 4th place gets 6 points in the last round so we need to take away 6 from each answer.

➤ Option E: 23 – If 4th place has 23 in round 10, they have 17 in round 9. Thus, Jill has 24 in round 9 and the rest have 31 each. This sums to 103. [Too high]
➤ Option D: 21 – If 4th place has 21 in round 10, they have 15 in round 9. Thus, Jill has 22 in round 9 and the rest have 29 each. This sums to 95. [Too high]
➤ Option C: 19 – If 4th place has 19 in round 10, they have 13 in round 9. Thus, Jill has 20 in round 9 and the rest have 27 each. This sums to 87. [Close enough!]

Question 32: A

This question can be worked out relatively quickly without determining exact values. There were 7000 people killed out of 2.3 million vehicles in 1930, and 3180 people killed out of 27 million today. There are roughly half the people killed for 10 times the number of vehicles, which gives a fraction of 1/20, or 0.04 times as much.

Question 33: D

It is important to read exactly what the question is asking you here. It wants reasons that are *not already in the text* which strengthen the case for roads become safer.

➢ A and C don't mean that roads are safer; if anything, it could mean that accidents are being under-reported (as mentioned in the article).

➢ B and E are valid points but are mentioned in the article already.

➢ D is a reasonable answer and gives us more of a reason to trust figures given in the article.

Question 34: C

This is probably the easiest question on the paper, and requires you to work out 40% of 319928. As the answer is wanted to the nearest 1000 anyway, 40% of 320000 is a reasonable calculation, which is 128000.

Question 35: A

➢ Although a very strong statement, it is the only answer that is plausible based on elimination and accounts for the discrepancy in the results.

➢ B is irrelevant, as roads being safer is unrelated to the discrepancy.

➢ C would mean that hospital admissions decrease, but they remain unchanged.

➢ D would mean the DfT figures should be higher than hospital figures.

➢ E would again make DfT figures higher than the hospital.

END OF SECTION

Section 2

Question 1: F

Homeostasis is defined as the maintenance of constant internal conditions. Homeostatic responses occur whether a factor rises or reduces, and internal body conditions can be affected by changes in variables inside our body (e.g. blood glucose levels) or changes in variables in the environment (e.g. temperature). Thus, all the statements could result in a homeostatic response.

Question 2: D

Atomic mass of bromobutane $= (12 \times 4) + (9 \times 1) + 80 = 137 g/mol$

Atomic mass of butanol $= (12 \times 4) + (10 \times 1) + 16 = 74 g/mol$

Since the molar ratio between bromobutane and butanol is 1:1, we can form: $\frac{2.74}{137} = \frac{x}{74}$ where x is the theoretical yield.

Rearranging gives: $x = \frac{2.74 \times 74}{137} \approx \frac{3 \times 70}{140}$

$x = 1.5 \, g$.

The actual yield is 1.1g. Therefore, percentage yield $= \frac{1.1}{1.5} x \, 100 = 73.3\%$

Note that because we rounded earlier on to make the maths easier, the answer isn't exactly 75%. This, is fine because the options are far enough apart to make **D** the only obvious answer. Remember, to **look at the options first to see how freely you can round numbers**.

Question 3: B

The first step is α decay, as 2 protons are lost, giving R-2. This means 2 neutrons are also lost, meaning the atomic mass is decreased by 4.

The second step is β decay, so a neutron changes into a proton (plus an electron). This leaves the atomic mass unchanged as seen and the atomic number increases by 1. Thus, P = N – 4 and Q = R - 1

Question 4: A

There is no quick way to do this. Shaded Area $=$

$$\textit{Largest Circle } + \textit{ 2nd largest circle } - \textit{ 3rd Largest Circle } - \textit{ Smallest Circle}$$

$$= \pi \left(\frac{4d}{2}\right)^2 + \pi \left(\frac{2d}{2}\right)^2 - \pi \left(\frac{3d}{2}\right)^2 - \pi \left(\frac{d}{2}\right)^2$$

$$= \pi [4d^2 + d^2 - \frac{9d^2}{4} - \frac{d^2}{4}]$$

$$= \pi [5d^2 + - \frac{10d^2}{4}]$$

$$= \pi d^2 [5 - 2.5]$$

$$= \frac{5}{2} \pi d^2$$

Question 5: B

1. Nicotine acts at the nicotinic acetylcholine receptors in the brain, leading to addiction.
2. Bronchitis is an infection of the bronchi, which can be caused by smoking and the bronchi is what area 2 points to.
3. Emphysema is a disease causing damage to the alveoli in the lungs, making you short of breath, and can be caused by smoking.
4. Carbon monoxide can enter your blood stream via the alveoli due to the effects of smoking.

Question 6: C

Lecithin is an emulsifier which has a hydrophilic head forming bonds with water and a hydrophobic tail forming bonds with oil, to prevent separation.

Question 7: F

None of the radiation is stopped by the paper, suggesting there is no α radiation. Some of the radiation is stopped by aluminium, but not all of it, suggesting the presence of β and γ radiation.

NB: Don't confuse yourself by bringing background radiation into this – the question strongly implies that it's not relevant here e.g. detector is **close** to the radioactive source.

Question 8: E

Remember to take a step by step approach when rearranging formulae:

$$G = 5 + \sqrt{7(9 - R)^2 + 9}$$

$$G - 5 = \sqrt{7(9 - R)^2 + 9}$$

$$(G - 5)^2 = 7(9 - R)^2 + 9$$

$$\frac{(G-5)^2 - 9}{7} = (9 - R)^2$$

$$9 - R = \sqrt{\frac{(G-5)^2 - 9}{7}}$$

$$R = 9 - \sqrt{\frac{(G-5)^2 - 9}{7}}$$

Question 9: A

➢ The patient will no longer be able to sense pain if these neurons cannot detect stimuli normally causing pain.
➢ Statement 1 must be correct as it is involved in each answer. There would be no reflex response if the neuron does not sense any pain.
➢ Statement 2 is correct. If there is a visual stimulus, such as a pin prick coming towards you, then you would still move away based on vision alone.
➢ Statement 3 and 4 suggest that the patient can sense the pain but this is incorrect.

Question 10: D

It is easier to use an elimination method rather than trying to balance this equation manually. Start by balancing for Phosphorous, as it has relatively simple values in the equation. The quickest way to do this is via algebra (for more information, see the Chemistry chapter in *The Ultimate BMAT Guide*).

For Phosphorous: $A + B = 3C$;
Only Options C + D fulfil this equation. Thus, we can eliminate, A, B + E.
Next, for Hydrogen: $2A + B = 2D$; Only D fulfil this equation.

Question 11: D

For work to be done, a force must act in a parallel direction to the object.

1. The person sat on the chair hasn't moved so no work has been done.
2. The force is acting perpendicular to the direction of motion. Thus, whilst work is being done to move the barrow, $Work\ done \neq Fd$.
3. d represents the direction in which work is being done so $Work\ done = Fd$

Question 12: E

$$= \sqrt[3]{\frac{2\ x\ 10^5}{(5\ x\ 10^{-3})^2}} - \sqrt{4\ x\ 10^3 - 4\ x\ 10^2}$$

$$= \sqrt[3]{\frac{2\ x\ 10^5}{25\ x\ 10^{-6}}} - \sqrt{4000 - 400}$$

$$= \sqrt[3]{\frac{2\ x\ 10^5}{2.5\ x\ 10^{-5}}} - \sqrt{3600}$$

$$= \sqrt[3]{0.8\ x\ 10^{10}} - 60 = \sqrt[3]{8\ x\ 10^9} - 60$$

$$= 2\ x\ 10^3 - 60$$

$$= 2000 - 60$$

$$= 1940$$

Question 13: E

Answer 1 could be an explanation as the antibiotic discs Q and R have very similar effects on the bacterial colonies. Answer 2 is unrelated to the question and the fact that the antibiotic is working suggests there is little to no resistance. Answer 3 is potentially correct as this is where the distance up to which the bacteria are destroyed.

Question 14: F

As all of the possible formulae of azurite have 3 copper atoms, we know the stoichiometry of the reaction is 2 of one of the reagents and 1 of the other. We can test both of these combinations.

$2CuCO_3 + Cu(OH)_2 \rightarrow Cu_3C_2H_2O_8$ which is as a possible answer.

$CuCO_3 + 2Cu(OH)_2 \rightarrow Cu_3CH_4O_7$ which is not a possible answer.

Question 15: B

$Wave\ speed = frequency\ x\ wavelength$

$f = \frac{3\ x\ 10^8}{0.12}$, so f= 2.5 x 10^9 Hz

As frequency stays constant, wavelength $= \frac{2.0\ x\ 10^8}{2.5\ x\ 10^9} = 0.08\ m = 8cm$

Question 16: C

There are several ways of doing this question. One of the simples is to form additional triangles by adding a point on the line MB.

$Tan\ (x) = \frac{opposite}{adjacent}$ and since:

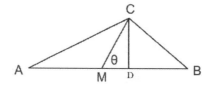

> ➤ Tan B= 2/3 → CD = 2 and DB = 3.
> ➤ Tan A = 1/6 → Since, CD is 2, $AD = 2\ x\ 6 = 12$
> ➤ $AB = AD + DB = 12 + 3 = 15$

Then, looking at right angled triangle CMD, we know the opposite length is 2 and the adjacent length can be calculated as follows:

Since, M is the midpoint, $MD = 7.5 - DB = 4.5$

Therefore $Tan\ \theta = \frac{2}{4.5} = \frac{4}{9}$

Question 17: C

It is helpful to write out the relationship described:

Alcohol → ADH decreases → Dilute Urine produced. Thus, we can conclude that ADH concentrates urine.

1 is correct as all hormones travel in the bloodstream. There is a negative correlation between dilute urine production and ADH levels, rendering 2 and 3 incorrect. 4 is correct as increased formation of dilute urine means more water is lost and dehydration may occur.

Question 18: D

This requires you to have memorised the reactivity series. Zinc, and thus vanadium, are below Chlorine, Sodium, aluminium and magnesium in the reactivity series so they can be displaced by the more reactive elements. However, Zinc and vanadium are above iron in the reactivity series. Thus, the less reactive iron will not be able to displace the more reactive vanadium from a compound.

Question 19: D

Consider each factor in turn:
- Total resistance: The circuit changes from parallel to a series which have a higher overall resistance.
- Ammeter 1 measures overall current. Since, total resistance has increased, and the voltage is the same, the total current must decrease in accordance with Ohm's Law ($V = IR$).
- Ammeter 2 was measuring the current in a branch. Since, the circuit is now in series, the current no longer 'splits' at branches. Thus, the entire current flows through ammeter 2. Therefore the current flowing through this ammeter will increase.

Question 20: B

The balls are arranged to give the smallest possible probability for the player to win. Thus, the arrangement is:
- Bag 1: 2 red and 2 yellow
- Bag 2: 2 red, 1 yellow and 1 blue

There are 3 ways to win, each with equal probability:
- 2 yellows from bag 1
- 2 reds from bag 1
- 2 reds from bag 2

The probability of any choosing the correct bag = ½

Thus, the probability of picking the above permutations is given by: $= \frac{1}{2} x \frac{2}{4} x \frac{1}{3} = \frac{1}{12}$

Since, there are 3 ways of winning, the probability of winning $= \frac{1}{12} x 3 = \frac{1}{4}$

Question 21: D

Statement 1 is wrong, as this does the opposite of explaining why there are less recessive phenotypes than expected. The other 2 reasons are correct and are fundamental ideas in scientific research (we need large sample sizes to reduce the effect of chance).

Question 22: B

This is easier than it looks. Remember that Helium is an inert gas so HeOH is not a plausible product. Tritium beta decays to a single Helium atom. Thus, Let T represent both Helium and Tritium in the equations below:

$HTO \rightarrow H_2O + O_2 + T$ or $HTO \rightarrow H_2O + H_2 + T$

The first equation can be balanced as: $4HTO \rightarrow 2H_2O + O_2 + 4T$ but the second equation cannot be balanced. Thus, the only possible products are those shown by option 2.

Question 23: D

$Loss\ in\ E_p = mg\Delta h = 100\ x\ 10\ x\ 100 = 10^5 J$

Since the cyclist descends 100m vertically, and for every 1m descended 10m is travelled along the road, the cyclist travels 1000m along the road.

Since Work is being done by the cyclist: Work Done = Force x Distance

Therefore, $Force = \frac{10^5}{10^3} = 10^2 N$

Since the cyclist is travelling at constant velocity, the resultant force must = 0. Thus: Resistive Force = Force due to weight = 100 N.

Question 24: A

This is a tough question:

➢ Let the cost of wood be y, and the cost of metal 3y
➢ Let the quantity of metal used be m_1, which is proportional to d
➢ Let the quantity of wood used be w_1, which is proportional to d^2
➢ Total Cost=3y x m_1+ y x w_1

Now we have 2d:

$m_2 = 2d = 2m_1$

$w_2 = (2d)^2 = 4d^2 = 4w_1$

3(total cost) = 3y x m_2 + y x w_2 = 6y x m_1+ 4y x w_1

Substitute for total cost: 3(3y x m_1 + y x w_1) = 6y x m_1 + 4y x w_1

$9m_1 + 3w_1 = 6m_1 + 4w_1$

$3m_1 = w_1$

Percentage of metal = $m_1/(m_1+w_1)$ = ¼ = 25%

Question 25: E

It is easiest to work backwards in each case.

Let's call the dominant allele A and the recessive allele a. Remember we are looking for the **minimum number**. When only U is recessive (aa), S and T must be heterozygous (Aa). If one was dominant, then they would not be able to produce offspring with the recessive condition.

R can be AA. Only one of P or Q must be Aa so that S is also heterozygous, giving a total of 3. When R is also recessive, P and Q must be heterozygous, along with S and T, giving a total of 4 heterozygous individuals.

Question 26: E

$C_2H_4 + H_2 \rightarrow C_2H_6$

The pressure will initially increase due to the increase in temperature. However, the final pressure will be ½ of the initial pressure because:

➢ Pressure is determined the number of moles of a gas present and
➢ The products have ½ the number of moles as the reactants.

Question 27: G

➢ Statement P is wrong as the speed of sound is constant in air (330 m/s).
➢ Q is wrong because the amplitude would be half of the distance between X and Y (2.5 mm)
➢ R is wrong because we cannot comment on wavelength with the available data.
➢ Frequency $= \frac{1}{0.2 \, x \, 10^{-3}}$ = 1,000 x 5 = 5 kHz.
➢ Thus, only statement S is correct.

END OF SECTION

Section 3

"Doubt is not a pleasant condition, but certainty is absurd." (Voltaire)

➢ Voltaire suggests that whilst questioning something can be a very daunting feeling it is better than merely accepting something as certain. This statement therefore encapsulates the concept of science, where every idea and theory can be challenged and questioned. Many theories that have previously been accepted as 'truth' in the past are later falsified and revised. Doubt is at least a logical position where one can look for more answers and evidence.

➢ It can be argued that some things are 'certain' and are not necessarily absurd. At school we are taught that it is certain that the sun will rise tomorrow, and there has never been evidence to suggest otherwise. Of course, there is the miniscule possibility of a cataclysmic event, which would have to consist of something beyond our current knowledge of physics. However, does this factor mean it is 'absurd' to declare the sun rising as a certainty? Other things that seem certain include the fact that we are all currently living, and that somebody who is deceased will not come back to life.

➢ Good statement to promote scientific, rational and logical thinking scientific, but absurd may be too strong a word.

"There is something attractive about people who don't regard their own health and longevity as the most important things in the world." (Alexander Chancellor)

➢ This statement describes admiration for those who are unselfish and consider others needs as much as theirs. The chancellor condemns living egocentrically, whereby one is concerned only for prolonging their life and well-being.

➢ It can be argued that one's own health and longevity is more important than anything else; ultimately we are biological creatures who must undergo survival of the fittest in the evolutionary process of natural selection and thus our main aim is to preserve our genetic information, meaning our fitness is more important than other's fitness. We must compete with others and this involves placing ourselves above others. If we were to ignore this self-worth and put others above us, then we may also end up suffering. We only live once, and in that sense one's health and life can be regarded as highly precious (as its virtually irreplaceable) and therefore it should have the highest worth other others or material possessions.

➢ It is very humanitarian to consider other people's health and longevity; this can be through religious influences, or through 'humanist' principles. These argue that as a human race we should value each other's survival as a whole; that is perhaps a factor why we have progressed through the ages and seen this openness gradually accepted as a social norm today. Altruistic behaviour, empathy and many other traits based on selflessness are arguably what make us 'human'.

The scientist is not someone who gives the right answers but one who asks the right questions.

➢ The statement addresses the argument that science is not just about reading textbooks, memorizing info and regurgitating it when necessary; it is instead about asking the pertinent questions that allow scientific advancement.

➢ Scientists must be comfortable with the unknown to allow progression rather than needing answers to everything.

➢ A scientist cannot formulate satisfactory answers without satisfactory questions; a fundamental principle of scientific research is that a hypothesis, derived from a scientific question, must be present before any experimentation takes place.

➢ Science is about thinking "How does that work", or "why does this happen", and then trying to answer that question. Some of the most exciting science discoveries would never have happened, without that initial spark of inspiration from an inquisitive question. If Newton hadn't asked himself how the apple came to fall to Earth, or Darwin hadn't asked why the mockingbirds on different islands were separate species, could they possibly have found the answers to these questions?

➢ Insightful questions can challenge accepted models, and turn the way we think about a concept on its head. However, you still need a curious, inquisitive mind to come up with the right answers.

➢ Questions without answers are often frustrating and undesirable, suggesting the answer is at least as important as the question. Indeed, neuroscientists studying consciousness focus on the 'easy problems' which can be viably studied and avoid the 'hard problems' for which we do not even know how to approach to determine an answer.

➢ A scientific theory is empirical and is therefore always open to falsification. Accepted beliefs and conventions are changeable in science; what we believe to be true now could easily be disproved in 100 years, 50 years, or tomorrow. Inspiration for progression can come in several ways. It can come from realising that the old questions are no longer working, highlighting the importance of scientific questions. Or, it can come from thinking of new ways to approach old questions, and this represents innovation with the answer rather than the question

➢ Einstein claimed "I have no special talents, I am just passionately curious."

"... Dolphins are very intelligent and so similar to humans that they are worthy of a special ethical status: that of 'non-human persons'."

➤ Dolphins have distinct personalities, self-awareness, forward thinking, complex social structures, empathy, and many other 'higher' functions we previously only attributed to humans. The statement suggests that this should lead to the, having an ethical status of 'non-human persons' which will give them a legally enforceable right to life.

➤ Animals are generally all bound by equal animal rights, whereas humans are bound by equal human rights, and within these groups we do not differentiate for intelligence. Why should there be an exception for dolphins?

➤ We have no concept of what being a dolphin is like, as they live in a completely different environment and have different lifestyles. We currently mistreat them and should they not be protected from exploitation?

➤ Intelligence is an extraordinary attribute, and the fact that they are self-aware and are likely to experience similar emotions in a similar way to us means it appears unethical to use them for our entertainment or kill them for food, suggesting they deserve these rights.

➤ Are there many downsides to giving dolphins rights? Most people don't aim to kill them either deliberately or inadvertently, and there isn't much to lose. If they are sentient beings whose intelligence warrants such rights then it seems like a rational idea. All life is sacred, but arguably those with higher levels of consciousness deserve more respect and more protection.

➤ If an animal is more intelligence than another should it receive greater treatment? The notion because x is smarter than y then x should not be killed brings many ethical implications along with it. Should more intelligent humans have better rights than less intelligent humans? (Or you could argue that they already do, through privilege). Again, it depends how one defines intelligence - 'don't judge a fish by its ability to climb a tree.'

➤ "Intelligent" behaviour of various kinds is found in many animals and when we measure intelligence, we do so according to a human (and culturally specific) norm. Even in testing human intelligence, there is controversy over these issues. Why should intelligence be the standard for conferring rights? Does this mean that mentally handicapped humans or those suffering from dementia should be denied rights? Then, there is a question as to whether these rights are to be extended to all cetaceans or only the most intelligent ones.

➤ One can support the protection of animals, but intelligence appears to be a rigid standard to use in order to decide which animals deserve what protection.

END OF PAPER

2013

Section 1

Question 1: A

This question can be worked out sequentially using the information given. Carla isn't working on Monday, which means Bob and Amy must be. That fills Bob quota of 3 days in the week. This means Carla and Amy must be working on Wednesday and Thursday. Amy cannot also work on Tuesday, as that would make it 4 consecutive days, so Carla must work on Tuesday too.

Question 2: C

➢ A is a very bold statement and probably too strong to even be considered. The passage doesn't say that life can't exist on Kepler-22b and in fact even suggests that its new reclassification as uninhabitable may be inaccurate.

➢ B suggests the opposite of the background we are given. Cosmologists now suggest that less planets are habitable than previously thought.

➢ C summarises a main conclusion of the passage and is thus the best answer. The passage gives the example of Earth; that it is close to being outside the habitable zone but is robustly life-friendly, doubting the accuracy of the criteria.

➢ D cannot be inferred from this information alone. The passage doesn't describe a link between clouds and Kepler-22b.

Question 3: C

There are 6 combinations, and each one can be tested by determining the number of days between birthdays. If this equals a multiple of 7 then their birthdays are always on the same day of the week. 281 - 218 = 63, so Adam and Tara have their birthdays on the same day of the week every year.

Question 4: C

The main conclusion of the argument is easy to spot in this argument, as it is at the end of the passage and begins with the word 'therefore.'

➢ A, B and D are points which are mentioned in the text, but only as background information and they do not constitute the conclusion of the passage.

➢ C paraphrases the conclusion that "*the secret to losing weight is painfully simple - do more and/or eat less*" and is thus the best answer.

➢ E is incorrect as the conclusion also describes that eating less can lead to calorie burn.

Question 5: D

Jason sold y Spruggles on day 1, and 2y on day 2. They cost £12 on day 1 and £9 on day 2, and he made £342 more on day 2. Thus y x 12 + 342 = 2y x 9. We want 3y (how many were sold altogether).

$18y = 12y + 342$

$6y = 342$

$3y = 171$

Question 6: C

The main point of the Clovis-First theory is that the Clovis were *the first inhabitants* of the Americas.

C is the only answer that would seriously challenge this point, as it has a specific time linked to it. The Clovis first theory suggests that they arrived at -11500 BC, and if there was a human settlement present 500 years before this time then this disproves the theory.

The other answers all link to the background information given, and could all be legitimate in a 'weaken' question, but none others seriously challenge that the Clovis were the first inhabitants.

Question 7: A

➢ Although elimination may seem quite a long process on the surface, it can be done rather quickly.

➢ Simon has 5 letters in his name so is limited to Hyde and Rush, and cannot be Rush because of the letter s. **Simon Hyde**

➢ Liam has 4 letters in his name so is limited to Doyle, Floyd and Shore, and must be Shore because of the letter l. **Liam Shore**

➢ Dylan has 5 letters in his name and must thus be Rush. **Dylan Rush**

➢ Eric must be Doyle or Floyd. It cannot be Doyle due to the letter e. **Eric Floyd.** Thus Ian's surname must be Doyle. **Ian Doyle**

Question 8: D

D is clearly the correct answer, especially given that the question tells you that it is a sarcastic comment. If you find it hard to spot sarcasm, then we can rule out B and C as the comment links to the quote about children rather than the other 2, and the sarcasm of 'no' means we should agree with whatever the quote says, which is paraphrased in D.

Question 9: A

Answer A basically paraphrases this 'evidence' and is thus the correct answer. The statement does not relate to what wealth should bring, or anything about children. D assumes a causality which is not necessarily suggested by the statement.

Question 10: D

1. Kahneman suggests that the better you are at the job, the more time you must invest in it. However, this does not necessarily imply that people who work shorter hours will give more time to their children - they may use this 'extra' time in other ways.

2. The transcript talks about not getting happier as we get richer over a certain level. However, it does not suggest that wealth under this level will not cause stress.

Question 11: B

Anecdotal evidence is evidence based on personal accounts rather than facts or research, which this story clearly is. It is not necessarily conclusive without facts to back it up, there are no statistics and it is relevant. We can argue that it is not hearsay as it is said that she is intimately involved with the family she describes.

Question 12: B

BEWARE that the symbol for Mercury and the symbol for Venus/Copper look very similar.
We can use the process of elimination here.
We can start with the first card, and at the top. There is only 1 equivalent to moon, which is silver on card 4, but the second item is different, so we can rule out cards 1 and 4 for having a pair.
The second card has 3 equivalents for the top item - cards 6, 7 and 8. The second item is only equivalent for card 7, and the last item for cards 2 and 7 are different, thus we can rule out all these cards as having a pair.
We are left with cards 3 and 5, and we can see that these are equivalent. Thus we have 1 pair in total.

Question 13: D

The main conclusion of the argument is "*In the interests of providing the most desirable outcomes, it is clear that placebos should be used as a treatment offered by the NHS.*" Thus, if treatments (such as placebos) ensure better outcomes, they should be used, which is paraphrased in D.
A doesn't necessarily support the argument, as you do not know whether the placebo will work. B, C and E are unrelated to the fundamental point of the argument.

Question 14: B

We can start by ruling out 8 and 5, as this would break the alphabetical order rule. Let's call the missing digits x and y. From the information given in the text, we know that $4 + 0 + x + y = 8 +$ number of letters in x + number of letters in y. So $x + y = 4 +$ number of letters in x and y.
Testing this rule out gives the exclusive answer of $x + y$ being 9 and 2. Here, it is important to re-read the question and make sure you give the number of letters that make up these digits, which is 7.

Question 15: B

A and C actually strengthen the argument as they back up some of the points made in the passage. D is a point against the argument, but doesn't weaken it, and is merely a statement saying the opposite of the passage. B is clearly the best answer as it directly contradicts the point that "*sport is what people do to counter the stress and pressures of work*" which is a key point in the author's argument that the growth of extreme sports is puzzling.

Question 16: C

➢ It is useful to quickly jot down the first letter of each month on a rough sheet of paper. J F M A M J J A S O N D. You can then determine which months the birthdays can occur on, along with the number of the month in the year.

➢ Jenny's and Alice's birthdays are 2 months apart - you can determine that this is only possible if Jenny's birthday is in June. It can't be in January or July because there would be no month 2 months away beginning with "A". Alice's birthday could be in August or April.

➢ Alice's and Michael's birthdays are 5 months apart - you can determine that this is only possible if Alice's birthday is in August and Michael's is in March, using the same logic as above. Thus Jenny's birthday is in June, Michael's is in March, making them 3 months apart.

Question 17: D

This argument suggests that age makes us lack sleep and age makes us have impaired memory, so the lack of sleep must cause the impaired memory. This is an error of reasoning.

1 and 3 highlight different ways that these ideas may be connected, aside from lack of sleep causing impaired memory, highlighting weaknesses in the error of reasoning. 2 is unrelated to this error of reasoning and doesn't weaken the argument.

Question 18: B

Start by writing all the square numbers between 1 and 60. The month must be 09, the day can be between 1-30, the hour can be between 1-24 and the minute can be between 1-60.

There are 4 possible days: 1, 4, 16, 25. If we start with day 01, then there are 8 times: 4:16, 4:25, 4:36, 4:49, 16:04, 16:25, 16:36 and 16:49. There will be the same number of times for days 4 and 16, as you can have 2 different hours and 4 times per hour without including the same square number, giving 24 times from these particular days. However, for day 25, there are more possible days, as you can have 3 different hours: 1, 4 and 16. These have 4 times per hour each, giving 12 times in total from this day.

24 + 12 = 36.

Question 19: B

There is no possible way that X can be green. The bottom left region must be green; it cannot be blue, yellow or red as there would be an edge with the same colour on both sides. The region to the top right of X can be red or yellow. When it is red, X can be yellow or blue. When it is yellow, X can be red or blue. Thus the region can be yellow, blue or red.

Question 20: C

If this circle was smaller than a square, and contained within a square, it could be the opposite colour to the square and no rules would be broken. If it was not contained within a square, or larger than a square, then it would have to be a different colour to the black and white to prevent any edge having the same colour on both sides.

Question 21: A

This is probably best done by trial and error. Try using 3 lines in many different combinations, and you will eventually see that 2 colours will always be sufficient, as a segment will never share an edge with more than 2 other segments, both of which can be the other colour.

Question 22: B

We can think of the top and bottom as 2 separate circles (or any shape) with 5 separate segments. If we look at the circles individually, we can see that 2 colours cannot suffice, as there would have to be 2 adjacent. 3 colours suffice. When we superimpose this on another circle, 3 colours are sufficient to never have 2 adjacent faces the same colour.

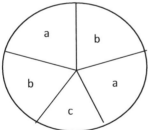

Question 23: D

Types of tile:

➢ Fully black
➢ ½ black
➢ ¼ black

➢ ¾ black
➢ Black line outlines ¼
➢ Black line outlines ½

2 X Black line outlines ½, ¼ black - this may be the one that people miss, as there are 2 forms of this shape and one cannot be rotated to form the other. If in doubt, take a rough piece of paper, draw the shape and try rotating it to see whether the tiles are   equivalent.

Question 24: B

The main conclusion of this article is that "*to reduce the harm done by alcohol, it is vital to reduce consumption.*" The author argues that one of the best ways to do this is to make alcohol more expensive. Answer B suggests that cheaper prices have led to more consumption of alcohol, backing up the point that alcohol consumption is based on price.

Answer A is irrelevant to the point about reducing consumption and answers C and D do not strengthen the argument.

Question 25: B

➢ The clocks are 41 minutes apart. The hour change must between 19 and 20. Within the same hour there would be the same digit used, or with the same first hour digit. 23 to 00 cannot be used as the 0 is used twice and 09 to 10 also uses the 0 twice.

➢ **19:ab and 20:cd:** c cannot be 0, 1 or 2 as these numbers are already used. It cannot be 4 as cd could only be 40 or 41 (41 minutes apart), and the 0 and 1 digits have been used. Therefore, c must be 3, and a is 5.

➢ **19:5b and 20:3d:** b or d cannot be 1, 2, 3, 5 or 9 as these have been used. If b is 4, then d would have to be 5 which is not possible. b can be 6 or 7 and d can be 7 or 8.

➢ **19:56-19:57 and 20:37 and 20:38.** There is no 4 used here.

Question 26: A

There is no easy way to do this- if you struggle with spatial awareness then this would be extremely difficult. Number 5 makes contact with P followed by 7 onto Q, 4 onto R and 1 onto S.

Question 27: C

We know Al is married and Charles is unmarried. Beth could be married or unmarried. C is the correct statement, as one way or the other, there was someone married looking at someone unmarried.

Question 28: D

The argument is *against* the idea that children should be exposed to harsh realities from a young age. Only D supports this, whereas the others suggest the opposite.

Question 29: E

Again this was a difficult spatial awareness question and the only quick way to solve it would be to actually make the net in the exam. The rules don't explicitly state that you may not use scissors although you should certainly approve it with your exams officer before. If spatial awareness isn't your forte- this was certainly one question to skip.

Question 30: D

1 and 3 are fundamental principles of evolution, which can be assumed when the word evolution is used. The argument says *"this characteristic must have evolved because it gave human beings a better grip underwater"* and this would not have 'evolved' if it were not advantageous. 2 is true from our previous knowledge, but is irrelevant to the argument and does not need to be assumed at any point.

Question 31: A

➢ First write down all the remaining numbers between 3 and 12, and cross each one out when used.

➢ **Bottom row:** 10+12=22, so the other 2 numbers must be add up to 7 to make 29. This means they must be 3 and 4, although we do not yet know in which order.

➢ **Top row:** 5+9=14, so the other 2 numbers must add up to 15. This can only be made from 7 and 8, although we do not yet know in which order.

➢ **Right column:** 11 must be above the 6, and we require 12 more. This means top right must be 8 and bottom right must be 4. This makes the person opposite 9 as 3.

Question 32: B

If 1% of non-cannabis users in the sample develop psychosis, and cannabis users were 41% more likely to have psychosis, 1.41% of cannabis users in the sample will have psychosis. 20% of young people report using cannabis, which is 2000 people. 2000 x 0.0141 = 2 x 14.1 = 28.2 users.

Question 33: A

Let's call the probability of psychosis for reasons other than cannabis use y. 80% of the population have probability y of getting psychosis, whereas 20% of the population have probability y plus the extra 41%.

We thus have 1.41 x 0.2y + 0.8y as the probability of getting psychosis and the percentage of getting it through cannabis use is:

$$\frac{0.41 \times 0.2y}{1.41 \times 0.2y + 0.8y} = \frac{0.082}{1.08} \approx \frac{0.08}{1.1}$$

This gives a percentage of slightly less than 8%, so A must be correct.

Question 34: B

The passage suggests that an increase in x (cannabis use) causes an increase in y (psychosis), whereas this answer provides an alternative link and suggests that y may also cause an increase in x.

A mentions age but this is not an alternative reason for the link. C suggests that the link may not be valid, which is irrelevant to what the question is asking. D is a point against an alternative link, suggesting more psychotic patients may have used cannabis than we think.

Question 35: C

Causal link is the key phrase here, and is a common theme in science and thus the BMAT. We need some evidence that A to B may be more than just a correlation.

➢ A may look tempting on the surface, but it again doesn't prove causation. There may be a separate causal factor which affects you when young or old that links to cannabis and psychosis, rather than the cannabis itself.
➢ B, if anything, is against a direct causal link between cannabis and psychosis, highlighting that other factors may be involved.
➢ C suggests a causal link, because an increase in X (cannabis strength) **caused** an increase in Y (psychosis).
➢ D is irrelevant to showing a causal link.
➢ E is irrelevant to showing a causal link.

END OF SECTION

Section 2

Question 1: H

Both the nervous system and the endocrine system are involved in homeostasis. Some of the messaging takes place using chemicals and they can receive and send messages to and from the brain.

Question 2: D

This refers to the reactivity series. A displacement reaction can take place if the element in the salt is lower down in the reactivity series than the element it is being reacted with. This only applies to 1 and 4, where Al and Zn and higher in the reactivity series than Pb and Cu respectively.

Question 3: D

Both 1 and 2 are correct in their ability to damage. However, infrared does not cause damage when penetrating matter.

Question 4: A

$$\frac{4.6 \times 10^7 + 7 \times 2 \times 10^6}{4.6 \times 10^7 - 2 \times 2 \times 10^6}$$

$$= \frac{4.6 \times 10^7 + 14 \times 10^6}{4.6 \times 10^7 - 4 \times 10^6}$$

$$= \frac{4.6 \times 10^7 + 1.4 \times 10^7}{4.6 \times 10^7 - 0.4 \times 10^7}$$

$$= \frac{6.0 \times 10^7}{4.2 \times 10^7}$$

$$= \frac{6}{4.2}$$

$$= \frac{60}{42}$$

$$= \frac{10}{7}$$

Question 5: F

Protease would break down proteins into amino acids; lipase would break down fats into fatty acids and therefore lower the pH of the solution. However, carbohydrase would function to break up the carbohydrates and would produce the non-acidic sugar products, therefore not lowering the pH.

Question 6: B

This reaction is in equilibrium, with the greater number of moles on the left hand side of the equation than the left. This means that an increase in pressure would push the equilibrium to the right, therefore producing more T product. In addition, since the forward reaction is exothermic, a lower temperature shifts the equilibrium towards the products. Catalysts have no effect on yield of product, just on reaction speed, and addition of more reactants would obviously increase the product yield.

Question 7: H

The switch being closed has turned the circuit from a series to parallel which therefore has a lower overall resistance. Since total voltage is unchanged, current must increase in accordance with V=IR. Thus P increases. With the switch open, the voltage is shared across both resistors but with it closed, the second resistor can be bypassed (short-circuited) by the new branch. This means that only the full voltage is shared by the first resistor only. Thus, Q increases and R decreases.

Question 8: F

$$4 - \frac{x^2(1-16x^2)}{(4x-1)2x^3} = 4 - \frac{(1-16x^2)}{2x(4x-1)}$$

Thus: $\frac{8x(4x-1)}{2x(4x-1)} - \frac{(1-4x)(1+4x)}{2x(4x-1)}$

$= \frac{8x(4x-1)}{2x(4x-1)} + \frac{(4x-1)(4x+1)}{2x(4x-1)}$

Thus: $\frac{8x}{2x} + \frac{(4x+1)}{2x}$

$= \frac{8x+4x}{2x} + \frac{1}{2x}$

$= 6 + \frac{1}{2x}$

Question 9: F

Sensory neurons are the longest types of neurons as they must travel all the way to the spinal cord. The relay neurons are the shortest because they are only present in the spinal cord.

Question 10: B

This involves the equation $2Na + 2H_2O \rightarrow H_2 + 2NaOH$. You can therefore work out the moles of sodium by using $Mass = moles \times M_r$:

$Moles\ of\ Sodium = \frac{1.15}{23} = 0.5\ moles$

Since the molar ratio between Sodium and hydrogen gas is 2:1, 0.25 moles of hydrogen are produced.

Therefore, Volume of Hydrogen $= 22.4 \times 0.25 = 5.6\ dm^3 = 560\ cm^3$.

Question 11: C

Remember that:

➤ Angle of incidence < Critical Angle: Light Reflected Back
➤ Angle of incidence = Critical Angle: Total Internal Reflection
➤ Angle of incidence > Critical Angle: Light leaves outside

Diagram 1 is has an angle below the critical angle. Therefore, total internal reflection does not occur and instead the light is reflected out. In diagram 2, the angle is greater than the critical angle. Therefore, total internal reflection does not occur and instead the light passes through.

Question 12: B

Label the corners of the square as A, B C and D and then see where they move in relation to the transformations performed. A reflection in the y axis therefore leads to the original orientation.

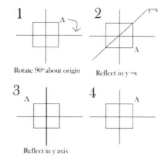

Question 13: C

The actual protein is not needed to produce it, as the intention is to allow the bacteria to produce the protein from the implanted DNA.

Question 14: A

In $MgCl_2$, the 2 valence electrons in Mg will each go to a chlorine atom, which will then mimic a filled orbital. The chloride atoms will both have the same electronic structure as argon, but when the Mg loses 2 electrons it will have the same electronic structure as neon, meaning the pair don't match.

Question 15: D

Don't get confused – this is actually easy since background radiation has been corrected for.

Source X has a half-life of 4.8 hours and thus has 5 half-lives in 24 hours.

$$Activity\ of\ X\ =\ 320\ x\ 0.5^5\ =\ 320\ x\frac{1}{32} = 10$$

Source Y has a half-life of 8 hours and thus has 3 half-lives in 24 hours.

$$Activity\ of\ Y\ =\ 480\ x\ 0.5^3 = 480\ x\frac{1}{8}\ =\ 60$$

$$Total\ Activity\ =\ 60\ +\ 10\ =\ 70$$

Question 16: D

Start off by writing the relationships mathematically: $x\ \alpha\ z^2\ and\ y\ \alpha\ \frac{1}{z^3}$

Now make the powers equivalent so we can substitute: $x^3\ \alpha\ z^6\ and\ y^2\ \alpha\frac{1}{z^6}$

$So\ y^2\ \alpha\frac{1}{x^3}\ and\ x^3\ \alpha\frac{1}{y^2}$

Question 17: A

This question requires knowledge of somatic cell nuclear transfer. 2 is incorrect, because this procedure doesn't involve sperm cells. 4 is also incorrect, because the egg with the newly transferred nucleus must begin to divide, and not differentiate (point 1). The other 3 points are valid.

Question 18: E

$NaOH + HCl \rightarrow NaCl + H_2O$

The atomic mass of $NaOH = 23\ +\ 16\ +\ 1\ =\ 40$

Theoretical maximum of NaOH in sample: $\frac{1.2}{40}\ =\ 0.03$

Moles of NaOH in sample that react is given by $n = cV$:

$\frac{50}{1000}\ x\ 0.5\ =\ 0.025$

Purity: $\frac{0.025}{0.03} = \frac{5}{6}\ =\ 83.3\%$

Question 19: D

Recall that Power = IV = I^2R. Since the resistors are in series, the overall current is given by: $\quad I = \frac{V}{R1+R2}$

Thus, Power = $(\frac{V}{R1+R2})^2 R1 = \frac{V^2R1}{(R1+R2)^2}$

Question 20: D
Smallest Cube:
We have 5 faces of the smaller shape, which is 5 x 1= **5 cm²**

Middle Cube:
Where the small cube joins the middle cube, we have a right angled triangle with lengths x, x and 1. Using Pythagoras: $1^2 = x^2 + x^2 = 2x^2$

$x = \sqrt{\frac{1}{2}} = \frac{\sqrt{2}}{2}$.

Since the triangle makes up half the side, the total length of the side = $\sqrt{2}$

There are 4 faces fully uncovered and one face partially covered by the smaller shape. Thus, the surface area = 4 x $\sqrt{2}$ x $\sqrt{2}$ = 8

Surface area of top face = $\sqrt{2}$ x $\sqrt{2}$ – 1 = 1
Surface Area of 2nd layer = 8 + 1= **9 cm²**

Largest Cube:
Using Pythagoras again: $\sqrt{2}^2 = x^2 + x^2 = 2x^2$
$2x^2 = 2$
Thus, $x = 1$. Since the triangle makes up half the side, the total length of the side = 2.
There are 5 faces fully uncovered and one face partially covered by the smaller shape. Thus, the surface area = 5 x 2 x 2 = 20
Surface area of top face = $2 x 2 - 2 = 2$
Surface Area of 3rd layer = 20 + 2= **22 cm²**

Total Surface Area: 5 + 9 + 22 = **36 cm²**

Question 21: E
The entire genome is found in every cell in the body, hence 1 and 2 are correct. Starch is broken down before it reaches the liver so 3 is incorrect.

Question 22: C
There are 2 atoms of Cr in the equation, so **d** must be 2. The equation must balance for charge, so **b** must be 8. Comparing **a** and **c** shows that these coefficients change the number of oxygens only and do not affect the number of carbons or hydrogens. Thus to have the correct number of hydrogens, **e** must be 4. **a** and **c** are 3, but this does not have to be deduced in this question.

Question 23: D
A. Rearrangement of $F = ma$
B. Rearrangement of $V = IR$
C. Rearrangement of $E_k = \frac{mv^2}{2}$
D. The relationship between wavelength and frequency is given by: $v = f\lambda$. If the y axis was wavelength, the axis should be $\frac{1}{f}$ i.e. the inverse of frequency (not a direct correlation).
E. Rearrangement of $Work\ Done = Force\ x\ Distance$

Question 24: C
We need to calculate the probability of either:
➢ 2 blue balls and 1 red ball
➢ 2 red balls and 1 blue ball
Each combination has 3 permutations (i.e. the first combination can be BBR or BRB or RBB).
Probability of 2 Blue Balls = $\frac{8}{10}$ x $\frac{7}{9}$ x $\frac{2}{8}$ x 3 = $\frac{336}{720}$
Probability of 2 Red Balls = $\frac{2}{10}$ x $\frac{1}{9}$ x $\frac{8}{8}$ x 3 = $\frac{48}{720}$
Total Probability = $\frac{48}{720} + \frac{336}{720}$
$\frac{384}{720} = \frac{8}{15}$

Question 25: C

Let's call the dominant allele A and the recessive allele a. We're looking for the proportion of Aa cats. 50% in the first cross, and 67% in the second (dead organisms don't contribute to a population).

	A	**a**
A	Aa ($\frac{1}{4}$)	aa ($\frac{1}{4}$)
A	Aa ($\frac{1}{4}$)	aa ($\frac{1}{4}$)

	A	**a**
A	AA (dead)	Aa ($\frac{1}{3}$)
a	Aa ($\frac{1}{3}$)	aa ($\frac{1}{3}$)

Question 26: B

3, then 6.

As a catalyst is not used up in the reaction, we can see that using 3 then 6 uses NO to speed up the reaction but then replenishes it at the end.

We can also see that a by-product which is not present in the net production, NO_2, is used up before the final products are formed.

Finally, the reagents used which are used up at SO_2 and ½ O_2 as seen in the reaction, and SO_3 is the only net product generated.

Question 27: E

The kinetic energy is given by $\frac{mv^2}{2}$ i.e. $\frac{4v^2}{2} = 1800$.

Thus, v = 30 ms^{-1}

Using $F = ma$, the current acceleration is $= \frac{20}{4} = 5\ ms^{-2}$

Now calculate the velocity after 2 seconds of acceleration by the same force by using: $v = u + at$: $v = 30 + 2 \times 5 = 40$

Then calculate the final kinetic energy: $\frac{4 \times 40^2}{2} = 3200$ J.

Finally, the extra kinetic energy is the difference between 3200 and 1800 = 1400 J

END OF SECTION

Section 3

"When you want to know how things really work, study them when they are coming apart." (William Gibson)

➤ This statement suggests that the function of a system is not fully represented when it is working smoothly, rather it is only when the system is put under stressful conditions that the intricacies and factors in the system can be fully appreciated. This concept can be appreciated from a tangible example of a working system, such as the internal structures in a clock. When certain cogs in the clock fall out or stop working, it is easy to identify their place in the overall system. This follows the suggestion by the phrase that a functioning system can mask certain important properties that the system possesses. A system 'coming apart' allows identification of different parts, their purposes and their importance with regards to the function of the whole system. This can be represented by the mutation of certain genes leading to defects, such as in the case of cystic fibrosis, the mutation in the chloride channel shows its vital involvement in the secretion of mucus. In addition, 'coming apart' can help identify the original function of the system, as it will presumably be unable to perform it under those conditions.

➤ This method for studying systems does however have flaws. If the entire representation of the system is based on the lack of function of the system, this means that in some cases, the original function of cannot be understood. The use of such a principle therefore relies heavily on background knowledge, as without some understanding of the system, its lack of function could mean very little. For example, a certain part of a computer can malfunction, leading to a complete breakdown; however, without understanding the basics of what each part in the computer provides to the function, there will be too many unknowns to really build up an understanding. Additionally, if more than one factor leads to the coming apart of the system, a more subtle function of a part could be masked by a less subtle function of another.

➤ This method also underestimates the complexity of the system and the number of factors that may be interacting. It also does not allow you to appreciate the system as a whole.

➤ Overall, this method could be applicable in a simple system that there is already basic knowledge in place. However, for a more complex system that includes the involvement of a number of parts, or a system that has yet to be investigated, it may not accurately represent all the interacting features.

Good surgeons should be encouraged to take on tough cases, not just safe, routine ones. Publishing an individual surgeon's mortality rates may have the opposite effect.

➤ Tough cases present a challenge to surgeons, as increased difficulty of the procedure leads to an increased risk of mortality. It is therefore suggested that due to this increased likelihood of mortality, doctors may be reluctant to take tough cases on if the records were public as their reputation would be somewhat tarnished if the operations were unsuccessful. Public exposure of surgical league tables could also prevent consent from certain patients for certain surgeons to carry out procedures. In particular, certain cases of high mortality rates in surgeons could show that the surgeon is more experienced as they have performed in situations of increased risk. The surgeon could also feel increased pressure in certain procedures, leading to potential negative outcomes. This league table could also undermine the expertise of both the GMC and the hospital, which must make regular reviews of their doctors to make sure that they are working to the best of their abilities. In addition, mortality rates give very little information about the situation in which the death took place and in a number of cases, it could have no reflection on the surgeon at all.

➤ Publishing mortality rates could also have a positive impact on surgeons. I suggested above that this could put the surgeons under increased pressure, however all doctors should feel a certain pressure as they have a responsibility to the patient. In the case of the surgeon, they hold patient's life in their hands and they should not forget that they are obligated to perform to the best of their ability. Slightly contrastingly, a league table should have no impact on the performance of a doctor, as a doctor must always act with the patient's best interest at heart, rather than worrying about their reputation. Patients also have a right to know the performance record of a doctor, as they are putting their welfare into their hands, and therefore have a right to know all the factors and risks involved in making such a decision, including their surgeon.

➤ To use such a league table for the improvement of surgical performance would be extremely beneficial; however it is unlikely that this would be the case. It is more likely that there will be a negative impact on surgical performance, or even the decision to perform the surgery in the first place. In conclusion it is probably more beneficial for both the surgeon and the patient that the league table be kept private, as it enables the doctors to act as they see fit, potentially saving more lives due to the undertaking of riskier but more beneficial operations.

"Ignorance more frequently begets confidence than does knowledge: it is those who know little, and not those who know much, who so positively assert that this or that problem will never be solved by science." (Charles Darwin)

➢ Darwin is suggesting that people who know very little about science are very confident that they know the limits of the field and therefore do not think it is able to make progress in certain areas. This can be valid idea in certain respects as little knowledge of a subject can cause a person to think they know the important aspects of a certain field. It might however actually be the case that the field has much more complexities that can offer very exciting potential developments, something which could be overlooked by an ignorant person.

➢ This concept can be represented by a self-diagnosis made by a patient online, they are only being exposed to very superficial (and potentially incorrect) areas of the field and this can lead them to incorrect conclusions.

➢ On the contrary this concept does not always apply. When looking at research environments, it is possible for a very knowledgeable researcher to have studied the subject for such an extended period of time that their sense of the bigger picture becomes clouded. This could mean that an ignorant person is then able to come and have a different take on the problem, as they have not developed the same mind-set of the knowledgeable researcher and they are less aware of possible limitations of the field, able to look 'outside the box'.

➢ This can be compared to a partnership between an engineer and a biologist in the development of medical equipment, the engineer might not know very much about biology, they are however able to give a different outlook on the problem. Similarly a scientist very knowledgeable in a field may not be able to accept that the field may be unable to progress, something that an ignorant person may be able to notice. There is also the possibility that an ignorant person may assume that science has not limits, which can be helpful and detrimental depending on the situation.

➢ Overall, this attitude does have some truth, I find however that it is cynical and a generalisation. It does not take into account that it is new and possibly initially ignorant minds that are pushing the boundaries of science, allowing it progress at a rapid pace.

In a world where we struggle to feed an ever-expanding human population, owning pets cannot be justified.

➢ This phrase suggests that the expense of owning a pet is unnecessary, rather the resources used to support a pet should be used to help support the over-expanding population. This argument is based on a number of assumptions and I will attempt to sort through them here.

➢ On one hand, it is true that the population is 'over-expanding' and we have a moral obligation to support those in need. It is also true that ownership of a pet requires a lot of resources. The pet must be fed, watered and housed appropriately. This therefore suggests that the owner must have food and money available for this particular purpose. One could argue that the money spent on the pet and the food that it consumes, could be better used elsewhere. If the resources used to support the pet were dedicated to support the needs of the expanding population, it is possible that a significant positive contribution to the cause could be made. In addition, one could argue that some expensive and exotic pets are an unnecessary luxury and too indulgent.

➢ On the other hand, the above statement assumes that the resources and food supplied to pets could be better used elsewhere. Owning a pet is an individual's decision, a choice to spend their money how they please. There is also no guarantee that even if these owners did not own a pet they would spend the equivalent amount supporting those in need, they might instead buy something equivalently recreational.

➢ An extrapolation of the above statement might lead one to say "why don't we restrict television sales and other such luxuries as the funds could be better used elsewhere?" As further support against this statement, it is important to consider the importance of pet ownership, such as the life-saving duties of guide dogs, likely to be considered a worthwhile expense. In addition, ownership of pets in some cases can reduce the wastage due to their consumption of leftover food, which would have been otherwise discarded.

➢ Additionally, the money spent on pets can be a form of support for the economy. A stronger economy allows the government to provide more assistance to those in need.

➢ In conclusion, we have a moral obligation to help those in need, however, pet ownership and lack of resources for an expanding population are not necessarily correlated, as the ownership cannot account the a massive resource deficit. Making steps to prevent food wastage or encouraging charitable donations may be a more worthwhile venture.

END OF PAPER

2014

Official Worked Solutions for the 2014 paper are freely available online at **www.uniadmissions.co.uk/bmat-past-papers**. This link will take you to a page where you can download ALL past paper as well as the 2014 solutions (you will have to provide your email for the download link).

Alternatively, you can download the papers and 2014 solutions from the link here: **https://bit.ly/2MGipmM**

2015

Section 1

Question 1: D

Plotting the information:

Stuart > Ruth > Margaret

Tim > Adrian?

We don't know where Adrian sits in relation to Margaret, but we do know that Adrian is shorter than Tim, Ruth and Stuart. So Adrian is shorter than Ruth and Stuart but not necessarily Margaret.

Question 2: E

E- is the overall conclusion from the passage as it states *" three quarters of all infections recorded last year were in people from deprived areas...and born outside"*, thus only a quarter were from more affluent populations and born in the UK.

A may be true but is a posed as a potential reason to account for the weakened immune systems that may again be the cause of the increased incidence of TB in deprived areas, so it is not a conclusion. B,C and D may again be true but are not conclusions, they are possible reasons for the overall conclusion that TB incidence has increased but in not in UK born affluent populations.

Question 3: A

The readings show the end of month readings so to find the greatest difference between September 1st and November 30th subtract the November readings from August (as August reading is end of the month, hence the value for September 1st).

The red van has the biggest difference so the answer is A.

	August	November	Difference
Red	68 240	78 853	**10 613**
Orange	64 425	73 684	9 259
Yellow	71 302	81 163	9 861
Green	64 827	75 146	10 319
Blue	73 959	83 392	9 433
Indigo	68 623	78 229	9 606
Violet	63 088	72 826	9 738

Question 4: A

A if true would best express a flaw in the overall conclusion of the passage that physical attractiveness correlates with sporting performance, as it highlights the failure of the passage to account for other potential contributing factors to sporting success.

B is not true as an objective measurement has been taken using performance in a cycling endurance race. C may be true but the passage does not claim to extend the correlation to other sports so it is not a flaw. D is never stated as no other sports are mentioned. Whilst if true, E would fail to support the findings of the research mentioned in the passage it is not a flaw in that research that has been undertaken.

Question 5: D

The top 3 scores on either attempt were no. 4 1st attempt – 7.34m, no. 13 2nd attempt 7.29m, the 3rd would be no. 4's 2nd attempt (7.26m) but they have already qualified so then next best is no. 10 1st 7.17m.

Anyone else within 50cm of third place 7.17m or 717cm qualifies, so anything above (717-50 = 667cm) 6.67m qualifies

Competitor Number	1st	2nd
1		
2	✓	
3		✓
4	Already qualified	
5		✓
6		
7	✓	
8		
9		
10	Already qualified	
11		
12		
13	Already qualified	
14		✓
15		

Which totals 5 competitors in addition to the top 3 , so 5 + 3 = 8, answer D.

Question 6: C

The passage is about the impending difficulties facing African and Asian agriculture due to future climate changes that will result in food shortages. It concludes that the development of seed banks with detailed catalogues about their trait so that farmers may be able to trial crops for the future, which is expressed in C.

A and E the basis for the problem, rather than conclusions and so are wrong. B is the hopeful outcome but not the conclusion, also the closing sentence states that seed banks are not the only answer, which B implies so it is incorrect. The main conclusion is about creating seed banks, which would then allow for D, so it is more of a secondary conclusion or suggestion rather than the main conclusion.

Question 7: C

Laying out share price for each day in a table:

Monday	Tuesday	Wednesday	Thursday	Friday
£1	£1.20	x	1.25x	£1

As Wednesday's price is unknown, use x. Thursday is 125% of x.

For Helen's shares:

Monday	Tuesday	Wednesday	Thursday	Friday
£1000	£1200	x	£1350	

This means on Tuesday the shares must have been worth £1200.

$1.25x = £1350$ so $x = £1080$ *(1350 ÷ 5 x 4 = 1080)*

Monday	Tuesday	Wednesday	Thursday	Friday
£1000	£1200	£1080	£1350	

Therefore, the price change between Tuesday and Wednesday is a decrease of 10% (1200 x 0.9 = 1080).

So for Paul: £3600 x 0.9 = £2700

Monday	Tuesday	Wednesday	Thursday	Friday
	£3600	£2700		

He therefore made a loss of £300.

Question 8: B

In a direct comparison, with non-custodial sentences the rate of reoffending was 22%, for custodial sentences it was 55%, this shows that the rate of reoffending was significantly lower for those with non-custodial sentences therefore which means that it would be a mistake, as stated in B, to give a custodial sentence when a non-custodial sentence is also appropriate as doing so would likely result in a higher chance or reoffending.

A cannot be reliably concluded as we have no specific data about serving half sentences.

That " 70% of under 18s re-offend" would suggest that there are significant problems but it does not fully support C that it is a mistake to send them to prison, we would need data comparing with those given a non-custodial sentence. C is incorrect as study 1 shows that only 72% reoffend, after 9 years. Whilst they all may do eventually we have not been given the data to show this so it cannot be reliable concluded. For E we have no data on comparing rates of reoffending for short (less than 12 months) with over 12 months so again it cannot be reliably concluded.

Question 9: D

Of the 50 000 former prisoners:

➢ In year 1 44% reoffended = 22 000
➢ In year 5 66% had reoffended which is 33,000 (a further 11,000).

Question 10: A

The phrase "instead of" implies there is the choice of whether to give a community service rather than prison sentence as they are eligible for both, which would account for the limited population where this was possible, where it reduces reoffending rates by 6%. The comparison of 55% versus 22% is simply a comparison of reoffending rates for all non-custodial versus custodial sentences but it may be that many of those given a custodial sentence were not eligible for a non-custodial so a direct comparison with the group where both were available cannot be made.

Question 11: E

E if true would strengthen the argument the most as it gives evidence compared with a control group that restorative justice is 20% more effective at reducing reoffending rates than just a community service order. This provides strong evidence that restorative evidence is effective and so would strengthen the argument that more offenders should be subjected to restorative justice to reduce reoffending rates.

A would not strengthen the argument as much as it does not provide evidence for the effects of restorative justice without a prison sentence, so there is no basis for comparison for the effects on reoffending rates.

B and C are moral and practical arguments for the argument of sending to prison rather than restorative not against, so weaken the argument. D is a financial argument rather than being based on the evidence of the report.

Question 12: D

Using the possible points of 9, 5, 3 and -2, it's possible to make:
➢ Crosswords: $9 + 5 + 5 + 3 = 22$
➢ Jigsaws: $9 + 9 + 5 - 2 = 21$
➢ Rubiks: $9 + 9 + 3 + 3 = 2$
➢ Tangrams: $9 + 9 + 9 - 2 = 25$
But it is not possible to make a score of 23 so the Solitaires score must have been wrongly calculated.

Question 13: D

The argument is essentially a discussion comparing the benefits to an individual compared with the risks to society but eventually arguing that the risk of a negative message to society is not worth the risk of allowing a previously convicted criminal to return to a high profile job, this makes the assumption that the rights of the individual are less important than the risks to society, which is stated in D.

Question 14: B

For the red paint, 20ml is left so 80 ml must be used. 20% is used as red, for 10% each purple and orange half each mix is red, so a further 5% each, so 10% total. For brown red is 1/3 of the 30%, so 10%. This means $20\% + 10\% + 10\% = 40\%$ of the mural is red from 80ml of paint, which means each 1% is 2ml.
For blue paint 10% is blue alone, for brown the blue makes up 1/3 of 30%, $= 10\%$, and for 10% each green and purple half of each mix is blue, so a further 5% each, so 10% each. This totals $10\% + 10\% + 10\% = 30\%$, and each 1% is 2ml the blue paint used is $(2ml \times 30\% =)$ 60ml. So 40 ml of the 100ml of blue paint is not used.

Question 15: A

The passage explains how human and primate brains are very similar such that we should consider research on primates to understand the effects of brain lesions may help to develop treatments, this is the conclusion, A.
B, C and D are not mentioned in the passage. E would be too strong a conclusion based on the conditional tone of the passage that research 'may' help.

Question 16: D

Including Maisy there are 16 girls, and 10 boys, 26 children total.
There are no more than two children in each family.
So the 3 older girls which have younger sisters = 6 girls.
The 2 girls with brothers (and vice versa)= 2 girls and 2 boys.
The two boys with brothers = 4 boys
Which totals 8 girls and 6 boys with brothers and sisters, i.e. 14 children, so there are 12 of the 26 children without siblings.

Question 17: A

The argument states that the development of a new blood test will help reduce the rate of recorded heart attacks in women. It states that this will be because more sensitive detection will allow for earlier treatment. However it fails to take account for the fact that with more sensitive detection, the recorded rates of heart attacks in women is likely to increase, as those not previously picked up may now be, as stated in A.

Question 18: E

For the 720 points, the span goals make up $42 \times 8 = 336$, and beat $36 \times 5 = 180$, which together total 516. This leaves $720 - 516$ points for tip goals, $= 204$ points total. This means that 102 tips were scored (as each tip scores 2)
This means that the points for span goals were roughly half, and beat and tip quarters with beat slightly smaller which is best represented in E.

Question 19: C

2 litres of sugary drink is 2000ml. 330ml contains 35g of sugar or 9 lumps. There is about 6 lots of 330ml in 2000ml. 6 x 9 lumps = 54 lumps

Question 20: C

Taking 100% of the daily-recommended sugar intake to be for example 100g. "Teenagers consume 50% more sugar on average, so 150% would be 150g. The information states that 30% is comprised of sugary drinks, which would be 30% of 150g = 45g. If this were to reduce to 1/3 as much, this would reduce by about 30g to 15g and so go from 150g to 120g, which would be the equivalent of 120% of the daily-recommended sugar, or 20% above the recommended level.

Question 21: C

If the tax resulted in a 10% reduction the 5 727 million litres of sugary drinks would reduce to about 5154 million (5727 x 0.9).

Question 22: A

If A were true it would most weaken the argument for a sales tax rather than volume tax on sugary drinks. If retailers reduced prices to remain competitive this would negate the effects of a sales tax in the attempt to reduce consumption. The increase in price with the sales tax would be counteracted by the reduction in price by the retailers to remain competitive. This would not have the same effect with a volume tax as more expensive drinks would not be as heavily taxed so there would not be the same need to reduce price by the retailers. Statements C, D and E make no comparison of the advantage of the volume versus sales tax. B only states that most food is taxed by volume but does not give any arguments for or against this rather than taxing by price.

Question 23: E

As the tourist took the shortest route, it is logical that the attraction he did not visit would be either the Tower or the Palace as these are the furthest distance from all of the others. The shortest possible route is from the hotel to the courts 60m so this will definitely be taken so it makes sense to start with this. Logically the next step is to go towards the fountain as otherwise the tourist will get to the Palace and then need to either go to the Tower, which cannot be right as one of the Tower or Palace must be missed, or back to the hotel which would add unnecessary distance.

So he goes to the fountain 80m, then the Arch 80, the Castle 90m and then Tower 110m and finally back to the hotel 110m. This totals 530m, so the Palace is the attraction that is missed as it is not possible to do the other option of missing the Tower and doing the other 5 attractions in a 530m route.

Question 24: B

The argument states that the banning of caffeine drinks may backfire on schools exam results due to the positive effects of caffeine on focus and short-term memory. However, if statements 1 and 3 were true it would weaken the argument as they suggest that caffeine that would have negative effects on sleep, which would result in lack of focus and memory. Hence the negative effects of sleep deprivation would counteract the positive effects of caffeine itself on focus.

Question 25: F

Represent the six letters as: A-8-B-X-Y-Z.

We know that:

A8B	**and**	A8
+ XYZ		+ BX
8 0 0		+ YZ
		8 0

For the three digit sums B + Z must = 10 due to the 0 digit below, the possibilities are 7 & 3 or 6 & 4 as 8 is already taken (and Y = 1 as the second column must also sum to 0 and it will have a 1 carried).

A and X must be 5 and 2 in either order as they must sum to 8 with the carried 1. They cannot be 4 and 3 as B & Z must contain exactly one of the numbers 4 & 3 therefore making the pair incompatible. So 8 + X + Z must sum to a number with 0 as the second digit (which cannot be 10 as the only way would be 8 + 1 + 1 which would use 1 twice and therefore not be allowed). Thus, it must be 20 which gives:

$$X + Z + 8 = 20$$
$$X + Z = 12$$ From earlier we know that Z the possibilities were 7, 3, 6 or 4 and for X: 5 or 2. The only combination of these that sums to 12 is 7 and 5, so we now know that X = 5 and Z = 7. Which means that B = 3 and A =2. So the full pass code is: 2-8-3-5-1-7

Question 26: C

The argument states that previous bad press surrounding saturated fats may have been based on misleading, now discredited data. The passage only discusses these particular studies but does not make any account for other studies that may have supported their results with more substantiated research, which is the assumption of the argument.

Question 27: C

As fewer than 5% owned neither device the total population that may overlap is about 95%. Plotting the potential overlap using the minimal overlap:

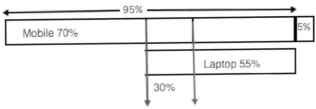

In which case the minimum overlap is 30% (70% + 50% -95%). It may be more than this if more of the mobile owners also own laptops but this is the lower bound assuming the highest potential number of children only own one device. However because the percentage that owned neither is actually **fewer** than 5%, but by how much we do not know, the potential overlap is less than 30% as the laptop owners may move further into the 5% block and so the most correct answer from the options available is 25%.

Plotting the maximum overlap:

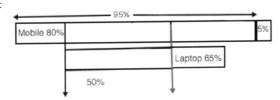

For which the overlap is 50% (80% + 65% - 95%). So, between 25 - 50%.

Question 28: A

The argument states that people are different in whether they are more alert at night or in the morning according to levels of melatonin. Consequently employers should adjust working hours to accommodate these sleeping patterns. If it were true however, that ritualistic behaviour of for example staying awake at night increases melatonin levels, this would weaken the argument, as it would suggest that the behaviour controls the melatonin levels rather than vice versa.

Question 29: A

Plotting the room:

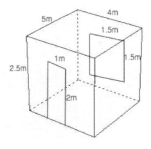

The two larger walls are 5m x 2.5m = 12.5m^2, so for both walls 25m^2
For these walls the high quality paint would need to cover 50m^2 and the low quality 75m^2. For high quality this would need 4 tins as each covers 15m^2, which would cost £60 (4 x £15). For low quality this would need 5 tins, which would cost £55 (5 x £11) so for the larger walls the low quality paint is cheaper at £55.

The two smaller walls are 4m x 2.5m = 10m^2, so for both walls 20m^2. The door is 2m^2 and the window 3m^2 so with subtracting these only 15m^2 needs to be covered. For these walls the high quality paint would need to cover 30m^2 and the low quality 45m^2. For high quality this would need 2 tins, which would cost £30 (2 x £15). For low quality this would need 3 tins which would cost £33 (3 x £11) so for the larger walls the high quality paint is cheaper at £ 30. Therefore the total cost for the paint is £55 + £30 = £85.

Question 30: C

The argument states that because working in A&E is not attractive to doctors they will choose to work in other areas and so hospitals have to pay large sums of money for temporary staff. Which is concludes then would be to pay higher wages to the A&E doctors to provide incentive to work there which would cost the same as paying the extra temporary staff and so would be of no more net cost to the health service. If it were true that many doctors work for agencies in A&E to supplement their salaries this would strengthen the argument as it implies that doctors are incentivised by money to work in A&E and so would follow that if the wages were higher for working there more doctors would be happy to do so.

Question 31: B

From the information we know that 5 must be opposite the 6. This leaves either 2 or 1 opposite the 4 and 3.

This gives the potential combinations of the opposite sides and their totals:

➤ 4:2 = 6
➤ 4:1 = 5
➤ 3:2 = 5
➤ 3:1 = 4

Which shows that it must be 4:2 and 3:1 as the other options sum to the same totals.

Which give the pairs 5:6 = 11, 4:2 = 6, 3:1 = 4.

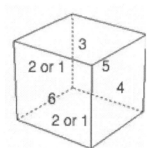 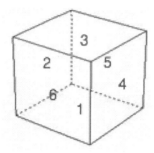

Looking at the potential answer combinations for the other die:

➤ A - 2:1 = 3, 3:6 = 9, so 5:4 = 9- not possible as two opposite sides would sum to 9.
➤ C – 4:1 = 5, 2:6 = 8, so 5:3- not possible as two opposite sides would sum to 8.
➤ D – 4:1 = 5, 3:6 = 9, so 2:5 = 7, not possible as 7 cannot be a total.
➤ E – 5:1 = 6, 2:6 = 8, so 3:4 = 7, not possible as 7 cannot be a total.
➤ F – 5:1 = 6, 3:6 = 9, so 2:4 = 6, not possible as two opposite sides would sum to 6.
➤ G – 5:1 = 6, 4:6 = 10, so 2:3 = 5 not possible as the first die has 6 as one of its totals.
➤ B – 2:1 = 3, 4:6 = 10, so 3:5 = 8- this is the answer as it has all different totals to itself and to the opposite die.

Question 32: G

Statement 1: from 2011-2013, 4265 defendants were convicted. But each year had about a 98% success rate of prosecution so about another 2% a year were tried and subsequently cleared. The additional 6% of 4265 is about 255 which added together is more than 430 so this is true.

Statement 2: in 2012 98% or 1552 were convicted, 1% = (1552 / 98= 15 remainder 82, so nearly 16) which means that the 2.1% acquitted was nearly 32, but definitely over 31, so this is true.

Statement 3: The total sentences handed out for includes all the suspended (220 + 178 + 140 = 538) plus the prison sentences (88 + 86 + 74 = 248). As 248 is less than half of 538, the prison sentences are represent less than 1/3 of the total sentences, so the suspended sentences represent over 2/3 so statement 3 is also correct.

All three statements are correct.

Question 33: B

If the north of England had not risen by 6.6% to 566, the number of people convicted would have been (94% of 566 =) 532. If this had, along with the rest of the country fallen by 11.7%, it would have decreased to (88% of 532 =) 469, which would have been 97 fewer convictions. 1371 – 97 = 1274.

Question 34: C

Statement 1: is not supported as the passage only mentions how many convictions there were in West Yorkshire for 2013, not for 2012. Even for 2013 it does not say whether this was the highest increase for any region for this year.

Statement 2: is not supported as although it states that the North of England rose by 6.6% between 2012 and 2013 this is for the North of England as a whole not just for West Yorkshire, so it is not possible to confidently say that this was definitely the increase for West Yorkshire specifically, as other areas in the North may have changed by more or less and this is an average for the area.

Statement 3 is supported as in 2013 there were 1371 convictions, 566 of which were in the North which is over $\frac{2}{5}$. So, only statement 3 is supported.

Question 35: B

The argument presented in the readers comment is that rather than the North of England being particularly bad for cruelty they simply report it more as they are concerned about animal welfare. If it were true that a higher proportion of complaints resulted in conviction in the north than other regions, B, this would weaken the argument. This is because a higher conviction to complaint ratio would imply that there was not simply a higher proportion of complaints, as the reader would argue, but genuinely a higher rate of animal cruelty in the north. All of the other options do not directly address the reader's argument that the rate or reporting is higher rather than the acts of cruelty.

END OF SECTION

Section 2

Question 1: E

The reflex arc after placing a hand on a hot object
➢ Sensory neuron transmits impulse to the CNS – A
➢ Relay neuron pass electrical impulse from sensory to motor neurons – D
➢ Motor neuron transmits electrical impulse to muscle cells – C
➢ Muscle cell contracts – B
➢ So E is not part of the reflex.

Question 2: C

Alkenes will decolorise bromine water due to the presence of the C=C bond makes them unsaturated.

	Formula	Structure
1	C_2H_4	$H_2C \equiv CH_2$
2	Polypropene is a polymer and has the repeating structure of $(C_3H_6)_n$	
3	$CH_2C(CH_3)_2$	
4	CH_3CH_2I	$H_3C - CH_2 - I$

So only 3 and 1 have C=C double bond so only they will decolourise the bromine water hence C is the answer

Question 3: B

Dark, matt surfaces are better at absorbing radiation than white surfaces so black will be the better emitter of radiation. Black surfaces are also better at emitting radiation than white surfaces.

White, shiny surfaces are better reflectors of radiation so they will be better clothes in winter as they will reflect the radiation back into the person on the inside to keep them warm, and radiate the heat away less on the outside. This is true because in winter the ambient temperature is likely to be lower than the body temperature, hence the answer is B.

Question 4: B

There are 3 black beads and the total number of beads in the bag is 8.
➢ The odds of picking a black bead the first time is $\frac{3}{8}$.
➢ Having picked this the odds of picking another black bead is $\frac{2}{7}$ (the total being one less as the previous bead has been removed).
➢ To find the probability of picking two black beads multiply these odds: $\frac{3}{8} \times \frac{2}{7} = \frac{6}{56} = \frac{3}{28}$

Question 5: A

Anaerobic respiration is: Glucose $\Rightarrow$ lactic acid (+ little energy)
So there is no formation of carbon dioxide, use of oxygen or water formed, hence the answer is A.

Question 6: A

The energy change for the forward reaction is **a** as this is the overall energy change from reactants to products, hence the reverse reaction is **–a,** hence the answer is A.

Question 7: D

As the question says, a step down transformer decreases the voltage of an alternating current (a.c.) electricity supply. Decreasing the voltage does not decrease the power, so it **stays the same**. But as the voltage goes **down**, the current goes **up**, hence the answer is D.

Question 8: E

Drawing the triangle PQR:

To work out the tangent of angle PQR we must first workout the length of the line joining P to the midpoint of QR (M).

Using Pythagoras:

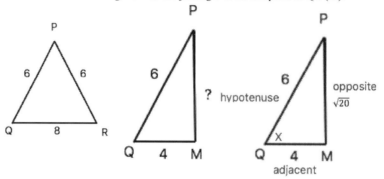

- $a^2 + b^2 = C^2$
- $QM^2 + PM^2 = PQ^2$
- $4^2 + PM^2 = 6^2$
- $PM^2 = 6^2 - 4^2$
- $PM^2 = 36 - 16$
- $PM^2 = 20$
- $PM = \sqrt{20}$

Using $Tangent = \frac{Opposite}{Adjacent}$

Hence, $\frac{\sqrt{20}}{4} = \frac{\sqrt{5 \times 4}}{4}$

Hence, $\frac{\sqrt{5}\sqrt{4}}{4} = \frac{2\sqrt{5}}{4}$

$= \frac{\sqrt{5}}{2}$

Question 9: D

The white mouse must be recessive, hence genotype cc.

The possibilities for the black mouse, 1 are CC or Cc

Punnet squares for both possibilities:

	Black	
	C	C
White c	Cc	Cc
c	Cc	Cc

Genotype 100% Cc + Phenotype 100% black

	Black	
	C	c
White c	Cc	cc
c	Cc	cc

Genotype 50% Cc, 50% cc + Phenotype 50% black, 50% white

As all of our first generation offspring are black we know that the black mouse 1, is genotype CC, and mouse 2 is genotype Cc.

A cross of these mice:

	Mouse 1	
	C	C
Mouse 2 C	CC	CC
c	Cc	Cc

Genotype 50% CC, 50% Cc + Phenotype 100% black.

Hence 50% heterozygous (CC), black only offspring and heterozygous and homozygous genotype, hence the answer is D.

Question 10: D

Rubidium is an alkali metal and in group 1, knowing the physical and chemical properties of group 1 we know that the group 1 metals become more reactive as you move down the group, therefore is it logical to assume that being far down the group, rubidium is very reactive and would need to be stored under oil, D.

A-   Is not correct as rubidium is more reactive than hydrogen and so electrolysis of rubidium chloride produces hydrogen not rubidium.

B-   Is not correct as knowing the group 1 trends that melting and boiling points decrease moving down the group and so rubidium **does not** have a higher melting or boiling point than sodium.

C-   Again, knowing the group 1 trends we know that rubidium reacts vigorously with water, increasing in reactivity down the group with rubidium being lower than sodium, so it is incorrect that rubidium reacts more slowly with water than sodium.

E-   The chemical formula for rubidium sulphate is Rb_2SO_4, not $RbSO_4$ so this is incorrect.

Question 11: A

Statement 1 is correct – that neutrons emitted in nuclear fission can cause further fission

Statement 2 states "*the half life of a radioactive substance is half the time taken for all its nuclei to decay*", which is incorrect. The half-life of a substance is actually **the time taken for the number of radioactive isotope in a sample to halve**, which is slightly different

Question 12: B

As X is a whole number greater than zero, and most of the options are true for all values of X we can take X to be 1. Hence:

$a = \frac{3}{5+1} = \frac{3}{6} = \frac{1}{2}$

$b = \frac{3+1}{5} = \frac{4}{5}$

$c = \frac{3+1}{5+1} = \frac{4}{6} = \frac{2}{3}$

$so\ a < c < b$

To check that G is not correct try with another value, $X = 2$

$a = \frac{3}{5+2} = \frac{3}{7}$

$b = \frac{3+2}{5} = \frac{5}{5} = 1$

$c = \frac{3+2}{5+2} = \frac{5}{7}$

Where still $a < c < b$, hence B is correct

Question 13: B

On a hot day, a human would sweat more and lose more water, thereby making the blood plasma more concentrated. This would mean than more water is reabsorbed into the blood and the urine becomes more concentrated so there would be **less** water in the urine on a hot day compared with a cold day. Whilst the concentration of urea in urine would be affected by the volume of water, the mass of the urea will be unaffected by external temperature so the mass of urea in urine will be the **same** on a hot day compared with a cold day.

Question 14: C

Cycloalkanes are examples of alkanes, which only have single bonds, which means they are saturated, hence C is correct.

A-   is incorrect as they have the general formula C_nH_{2n}

B-   they do not rapidly react with bromine water. Alkenes with double C=C bonds react to decolourise bromine water

D-   they burn in excess oxygen to produce carbon dioxide and water, not hydrogen, so this is incorrect

E-   incorrect as they are part of a homologous series as they have similar properties and the same general formula

F-   cycloalkanes are not giant covalent structures.

Question 15: A

The vertical force upwards can be deduced by subtracting the downwards force 20N from the lift from the wings 25N, (25N – 20N =) 5N.

Using the equation *force = mass x acceleration* and given mass = 2kg:

$5N = 2kg\ x\ a$

Thus $a = 2.5\ ms^{-2}$ upwards

The horizontal acceleration can be deduced by subtracting the air resistance (drag) force 40N from the force from the engine 50N, (50N – 40N =) 10N.

$10N = 2kg\ x\ a$

Thus $a = 5\ m/s^{-2}$ to the right

Question 16: E

➤ A : B : C

➤ 1 : $\frac{2}{3}$: $\frac{4}{5}$

➤ C = £3000, which = $\frac{4}{5}$ so divide 3000 by 4 and multiply by 5 to find:

➤ 1, = £3750, which = A.

➤ B = $\frac{2}{3}$, so $\frac{2}{3}$ of 3750 = 2500,

Total amount collected by charity = $A + B + C$
= £3750 + £3000 + £2500 = £9250

Question 17: H

Process 1 is photosynthesis, which involves neither respiratory nor digestive enzymes. Process 2 is when the plants are eaten by the animals. Process 3 is when the carbon in animals is taken in by decomposers such as microorganisms by feeding, so both 2 and 3 involve digestive enzymes. Process 4 is when the decomposers respire, releasing the carbon as carbon dioxide to the atmosphere and so uses respiratory enzymes.

Question 18: C

The equation shown is an exothermic reaction, as shown by the negative enthalpy change. This means there is a rise in temperature as the reactions transfer energy to the surroundings. Adding a catalyst would increase not decrease the rate of reaction so A is incorrect. The state if chemical 'T' being a gas means the rate of reaction would be faster because, the surface area is increased, more particles are exposed to the other reactant, there is a greater chance of particle collision and there is greater chance of the particles colliding. Hence B is incorrect.

Increasing the temperature will increase the rate of reaction because at a higher temperature the reactant particles will have more energy and move more quickly and so collide more and more collisions result in reaction, so D is incorrect. E is incorrect because the volume of gas may not change, as all the products and reactants are gases, so there may just be a change in what the gases are rather than the volume.
The activation energy is the energy required to start the reaction, the higher the activation energy the slower the rate of reaction. This is because a high activation energy will mean that only a few particles will have enough energy to collide so the reaction will be slow, hence C is correct.

Question 19: G

V and Y represent mass numbers, which is the number of protons plus neutrons. W and X are atomic numbers, which is the number of protons.
Beta emission involves changes to the nucleus whereby a neutron is converted to a proton. This means from M to N the proton number will increase by 1, so W +1 = X.
During alpha emission from N to Q the nucleus loses two protons and two neutrons, this means the mass number decreases by 4 and the atomic number decreases by 2. This means that Y will decrease by 4 when N decays to Q so Y = V-4.

Question 20: D

➤ m = mean

➤ n = number of pupils

➤ therefore the total score = mn

The expression for the number of pupils when another pupil is added is $n + 1$ and the mean is $m - 2$. The extra pupil scores n, so now the total score is $mn + n$.
Hence the expression for when the extra pupil takes the test and scores n:

$\frac{mn+n}{n+1} = m - 2$

$mn + n = (m - 2)(n + 1)$

$mn + n = mn + m - 2n - 2$

$n = m - 2n - 2$

$3n = m - 2$

$n = \frac{m - 2}{3}$

Question 21: A

The question concerns the nature of aerobic bacteria and white blood cells.

1- the structure of their DNA is a double helix so this is correct
2- they both do not possess a cell wall so this is incorrect
3- they both do not possess a nucleus so this is incorrect
4- they both do possess a cell membrane so this is correct

Hence, 1 and 3 are correct.

Question 22: C

The lower number is the mass number, which is the number of protons and electrons in an uncharged particle.

Thus, $^{35}_{17}Cl^-$, $^{40}_{18}Ar$ and $^{39}_{19}K^+$ all have 18 electrons in the arrangement 2.8.8 so they are the same.

	Number of Electrons
$^{35}_{17}Cl^-$	18
$^{35}_{17}Cl^+$	16
$^{40}_{18}Ar$	18
$^{39}_{19}K^+$	18
$^{40}_{20}Ca^+$	19
$^{41}_{19}K^-$	20

Question 23: D

The reaction time is 1.4s (0.7s x 2).

The speed of the car is 20m/s

➢ *Distance = speed x time*
➢ Distance = 20 m/s x 1.4s, = 28m travelled in the reaction time.

For the braking distance the car is decelerating at a constant speed from 20 to 0 in 3.3s so can take 10 m/s for this time as this will be the average speed.

$Distance = 10 \times 3.3 = 33m$

Adding the braking distance and reaction time distance: 28m + 33m = 61m

Question 24: D

$$= \frac{2x+3}{2x-3} + \frac{2x-3}{2x+3} - 2$$

$$= \frac{(2x+3)(2x+3)}{(2x-3)(2x+3)} + \frac{(2x-3)(2x-3)}{(2x-3)(2x+3)} - \frac{2(2x-3)(2x+3)}{(2x-3)(2x+3)}$$

$$= \frac{4x^2 + 6x + 6x + 9 + 4x^2 - 6x - 6x + 9 - 2(4x^2 + 6x - 6x - 9)}{(2x-3)(2x+3)}$$

$$= \frac{4x^2 + 6x + 6x + 9 + 4x^2 - 6x - 6x + 9 - 8x^2 - 12x + 12x + 18}{(2x-3)(2x+3)}$$

$$= \frac{9 + 9 + 18}{(2x-3)(2x+3)}$$

$$= \frac{36}{(2x-3)(2x+3)}$$

Question 25: G

Looking at the columns of chromosomes:

XAA = X:A 1:2 (or 0.5:1) so XAA is male, which makes answers A, B, C and D incorrect.

XYAA , the Y is irrelevant to the sex of the fruit fly so the ratio of X:A is again 1:2, or 0.5:1, so XYAA is also male, which makes answers, E and F incorrect.

At this point we are left with answers G and H, for which notably XXAA and XYAA are given as the same sex so to save time it would be wise to skip to the final column, XXYYAA.

XXAA has ratio of X: A, 1:1, so it is female

XXYAA has ratio of X:A , 1:1 so it is female.

XXYYAA has ratio of X:A, 2:2, 1:1 so it is also female, so H is incorrect, G is the correct row.

Question 26: B

To work out the excess of oxygen first work out the moles of CH_4 and CO_2.

$$Moles = \frac{Mass}{Molecular\ Mass}$$

Molecular Mass of:

➢ $CH_4 = 12 + (4 \times 1) = 16$
➢ $CO_2 = 12 + (16 \times 2) = 44$
➢ $O_2 = 32$

Moles of $CH_4 = \frac{1.6}{16} = 0.1$

Moles of $CO_2 = \frac{4.4}{44} = 0.1$

So we know that the number of moles of CH_4 to make CO_2 is 0.1.

The ratio of O_2 to CH_4/CO_2 is 2:1, so therefore we must have twice as much oxygen as CH_4/CO_2.

$0.1 \times 2 = 0.2$ so there are 0.2 moles of O_2.

Mass = Molecular Mass x Moles

$0.2 \times 32 = 6.4g$

So the mass of O_2 used is 6.4g which means 1.6g of the 8g is left unreacted.

Question 27: D

Considering each statement in turn:

1. *force = mass x acceleration*
        $5.0N = 4.0$ kg $\times 1.25ms^{-2}$
2. *speed = wavelength x frequency*
        5.0 ms$^{-1} = 1.25$m $\times 4.0$ Hz
3. *voltage = current x resistance*
        4.0 V $= 1.25$A $\times 5.0\ \Omega$

So 1 and 2 are true.

END OF SECTION

Section 3

"Computers are useless. They can only give you answers." (Pablo Picasso)

➢ In the statement, Picasso argues that computers are useless because they are only able to supply answers of questions posed to them. A calculator, as an example of a computer is entirely useless until one enters a sum for it to solve, but it requires the human input to perform any sort of function.

➢ As an artist Picasso must have thought computers to be against his industry of creativity and indeed at the time they would have been able to contribute very little to the arts. He may have been making a wider statement about the evolution of creativity that it will never be possible for mechanical devices to be truly creative- a thought still held by many.

➢ This argument however, whilst debatable in its truth at the time when Picasso made it, is almost impossible to defend today. Computers in their many forms are almost essential to modern day life in the western world. Most people in developed countries own mobile phones, along with the plethora of computer devices; laptops, tablets, even some cars have computers.

➢ We rely on computers for a huge number of daily functions that those who lived before them could not have anticipated. From maintaining contact with people across the world to simply arranging meetings, from organising our online banking to ordering clothes and food on websites we use computers in almost every aspect of modern life.

➢ Even in medicine, computers are often used to store patient notes and be used to keep track of lab results and as a viewing platform for medical scans. Even stripped down to the bare essentials of search engines providing answers to questions, as Picasso would argue, they are still useful to us in many areas. From searching locations, to medicine where the new recommendations for treating conditions can be found via computers.

➢ The real limits of technology are changing all the time. The evolution of technology has developed from simple calculator arithmetic functions, to now virtual realities and artificial intelligence that rivals our human abilities. Computers are on the brink of being used to drive cars more safely than humans, and be able to design and create in the way that Picasso would have never thought possible. It is possible now to interact with robots in a way that could change the way we function in the future, with computers even becoming more efficient than humans in many jobs, industrial labour for example.

➢ There are still limitations of technology, in the medical world, computers cannot replace doctors as yet. The human ability to be able to look at another human and assess how unwell they may be is a long way off being mechanised. Searching flu symptoms into an online search engine, a brain tumour appears as one of the diagnoses. It takes a person to ask the right questions, even if computers may be able to help with the answers.

➢ In conclusion, Picasso's argument may have been relevant to him and his contemporaries the context of the era and abilities of computers at the time. However, it would have been difficult for him to anticipate the exponential increase in abilities of the at the time primitive computers to now almost essential devices in every day life.

"That which can be asserted without evidence, can be dismissed without evidence."(Christopher Hitchens)

➢ Hitchens means that things that are said to be true but have no physical evidence, can be dismissed as they cannot be proven. He probably is referring to religion, the proof of the existence of which has no material evidence that can be replicated and documented in a scientific fashion. In the world of science, for example, something wouldn't be simply accepted because of folklore and people believing it to be true. The very heart of the Christian religion, the existence of God for example has not been positively proven.

➢ Arguing against this however, there are many domains where beliefs are held without material evidence. In a court of law for example a person's testimony is often held without argument. It is not often the quality of the evidence but the person who gives it that is taken into account. Why then with religion do we not believe the thousands of people, both today and historically, who we would judge to be honest caring people who deeply believe that their god exists. Especially when we will sometimes take one person at their word for deciding whether someone should be convicted or exonerated for a crime.

➢ It also brings into question what we consider to be 'evidence'. Does something have to be proven by multiple people, reportable and repeatable as with scientific research? There are many scientific principles held to be 'proven with evidence' that only have one experiment to support them. The MMR vaccine being linked to autism scandal, for example was previously supported by evidence until it was later discredited. Bloodletting was the commonplace of medicine centuries ago and was believed to have evidence supporting its efficacy, where now it seems pure madness. How many things then do we hold to be true that will later be disproven as scientific techniques evolve?

➢ Equally there are probably many things that we do not as yet have evidence for that may emerge over time. We do not yet know the full mechanism of some drugs such as paracetamol for example. Do we dismiss its worth because we cannot prove it? Hitchens' statement would argue that we should, when clearly this would be a mistake.

➢ I somewhat agree with this statement. It depends on the context to which it refers. It is potentially valid for example with new medicines. We would be wrong to accept their safety and efficacy without solid scientific evidence.

➢ However, there are also many areas of medicine that cannot be dismissed just because they don't have solid material evidence. Many personal accounts of outer body and near-death experiences where they have the option to 'move towards the light' cannot be simply dismissed because they have no physical proof. It may be that such things will never be proven, or indeed disproven in the same way that we take paracetamol to be effective simply on anecdotal evidence.

➢ Ultimately it depends on where the burden of proof lies. Is it with those who propose the hypothesis that something may be true, to then confirm their theory with evidence? Or is it for those who doubt their statement to disprove it? Hitchens would argue the former, but many issues such as widely held concepts of faith and religion are arguably the latter.

When treating an individual patient, a physician must also think of the wider society.

➢ The statement means that doctors should not only consider their patient's needs but also that of the other people within society when considering treatment. It would be ridiculous, for example, for a doctor to spend all day by a patient's bedside when that may be in their specific best interests when they have their other patients to consider as well. It also means this in wider context with respect to finite resources such as medicines, hospital resources such as radiological scans and even human tissue such as blood and organs. It is not feasible in a resource-limited system such as the NHS to prioritise the patient with no regard to the expense of other patients and wider society.

➢ It is however a doctors role to place the needs of their patients above all else. Doctors have historically prioritised patients even over the law. Helping patients addicted to drugs or alcohol for example, whilst reporting them and forcing them to change their lifestyle might be what is the best for society, doctors will often ignore these societal priorities and treat the patients regardless.

➢ Equally in a society where we have a growing population many children in foster care and waiting for adoption it is not in society's interests to treat an infertile couple with IVF where otherwise they may adopt these children and accept their infertility. But it would be against the foundations of medicine to begin to deny patients the treatment that was available and best for them to force them to accept conditions such as infertility, merely to satisfy the needs of society.

➢ There are many times where the patient's interests can conflict with those of the population. Antibiotics for example, while resistance is increasing they will become increasingly precious. It would be in society's interest to ration these for only the very unwell that won't survive without them. But this would mean that many patients suffer for longer with illnesses that could easily be treated with antibiotics that are being saved for the future society's benefit.

➢ Equally with treating people that are, or are likely to become dangerous people such as serial killers. It is clearly in society's interest to not treat these people and allow them to die, but this would be completely against the ethical principle of non-maleficence and against doctors' nature and medical training.

➢ Vaccination is another key example. Often it is actually not in the individual patient's interest to receive a vaccination, they are more likely to experience side effects from the vaccine than to acquire the infection it prevents in most cases. But it is necessary to vaccinate everyone to protect the most vulnerable members of society such as the young, elderly and immunosuppressed.

➤ In conclusion, there are times that the patient's best interests contradict that of society and doctors may have to use common sense to decide where best to spend their time and limited resources. It would for many go against the foundations of medicine to deny some patients the treatment they need to satisfy the needs of society. It may however, end up as a necessity, to ration some resources such as antibiotics for those most at need to prevent the rise of resistance making them then useless to everyone else in society.

Just because behaviour occurs amongst animals in the wild does not mean it should be allowed within domesticated populations of the same species.

➤ This statement claims that there are differences in the behaviours of domesticated animals to those living in the wild. It consequently argues that just because a certain type of behaviour exists in wild animals, that does not mean it should be automatically allowed in domesticated animals as well.

➤ It could refer to the hierarchical nature of wild animals that operate in a 'survival of the fittest' modality. It is common for animals to fight each other in the wild, for example, for simple status such as with wolves to be head of the wolf pack. This behaviour would not be acceptable in domesticated animals. It would be very difficult, for example, if every time someone attempted to walk their dog in the park it attacked other dogs to assert status, and this was allowed or even encouraged as they behaviour was fitting with its analogous species in the wild. Quite apart from the danger this might pose to the other dogs if this behaviour was encouraged, it could even lead to a change in behaviour of the fighting dog such as aggression towards humans that would make it not only impractical but dangerous to keep it as a pet.

➤ Equally with hunting where animals such as wolves and big cats would hunt their food in the wild it would be obscene to allow the domesticated versions to keep attacking other people's house bunnies or guinea pigs to satisfy their carnal nature.

➤ Arguing against the statement however, one could argue that if we are having to continually disrupt our domesticated animal's behaviour to fit with what works practically within our society, then it may be best to not keep these animals as pets. Changing the nature of these animals through behavioural modification, known as taming is arguably immoral and unfair to these animals. Training dogs to be more docile and fight their animalistic instincts on the face of it seems immoral and unfair. One could argue that just because we can force these animals to change their behaviours does not mean that we should.

➤ With breeding programmes for example, our pets have been slowly bred to be less like their historic ancestors to be more aesthetically pleasing. Whilst the taming of certain behaviours to prevent harm to other animals and even humans can be justified, it is hard to accept the alteration of animals' natural breeding behaviour for aesthetics. This has even gone to extremes where selective breeding of pugs and other dogs have been bred to their detriment as they often have breathing and other health problems.

➤ However, it is also impossible to avoid the fact that domesticated animals would ultimately not be kept as pets had their behaviour not been altered through time. So it could be argued that these changes are necessary for their existence, and so merely a form of survival adaptation. Indeed they have been so successful that they don't rival the top of the food chain humans and are even protected by them.

➤ In conclusion while modifying the behaviours of domesticated animals is ideologically difficult, it is vital to continue to allow their existence as a valued and cared for part of our society.

END OF PAPER

2016

Section 1

Question 1: D

In order to solve this question, we have to fill in the blanks in the table. Years $1 - 3$ are easy as there is only one gap to fill, years 4 and 5 are a little more difficult. From the totals we can calculate the total number of boys and girls in all 5 years.

We can calculated the total number of students in year 4: 120-24-26-40-24=16

We also know that that the probability of a boy being in year 4 is 1/12, applying this we can calculate the number of bin year 4: 72/12=6.

Since we know that there are 16 students in year 4 and 6 of them are boys, the probability of a student being a boy is 6/16 or 3/8.

Year	Boys	Girls	Total
1	18	6	24
2	16	24	40
3	8	8	16
4	6	10	16
5	24	0	24
Total	72	48	120

Question 2: B

Answer A is incorrect as the forest fires in Indonesia represent only a fraction of the net allowance of CO_2 emission and there is no information on further culprits, therefore failing as a valid the conclusion.

Answer B is correct as the text specifically mentions that only CO_2 from burnt plant material is taken up by new vegetation. Due to the fact, that there is also peat being burnt in the forest fires allows the conclusion that some of the CO_2 will not be taken up by regrowth.

Answer C is incorrect as it has no backing from the text presented in the question.

Answer D is incorrect, as it basically ties in with Answer A, they both ignore other contributing factors to CO_2 emission.

Question 3: B

This question is pretty straight forward. First, calculate the price per stay and compare.

Stay 1:

Hotel: $50 + $50 + $40 + $40 + $40 = $220

Car: 5 x $5 = $25

Total: $25 + $220 = $245

Stay 2:

Hotel: 8 x $40 = $320

Car: $25 + $5 = $30

Total: $350

Difference: $350 - $245 = $105

Question 4: B

Answer A is incorrect as this is a direct rephrasing of the second sentence, therefore not a flaw. Answer B is correct, as if the ability to synchronize to particularly slow music was an innate ability of musicians that occurs naturally, there's a chance that non-musicians would have the same trait, despite not having taken up music. Answer C is incorrect as it is not supported by the text and is also irrelevant. Answer D is incorrect as whilst it might generally speaking be true, the text passage here deals with synchronization to slow music, not with any other abilities.

Question 5: A

The first challenge in this question is to correctly identify the axes of the diagram.

Looking at the distribution of results, the Y axis must represent the written paper in steps of 10 and the X axis must represent the score of the practical paper in steps of 5. Both start at 0.

Moving from left to right, the dots therefore represent the following students:

Ina – Liz – Els – Joe – Fio – Gho – Amy – Kai – Ben – Den – Haz.

Con would fall between Fio and Gho.

Question 6: D

Answer A is incorrect as the new changes do not make it more difficult to seek justice per se, but only for claims of £200 000 or more.

Answer B is incorrect since fees associated with the court cases have been in place for a long time and the text only takes issue with the recent rise, not fees as a whole.

Answer C is incorrect for similar reasons as answer A. The statement is too general for a rather narrow answer.

Answer D is correct since it is in keeping with the text as a whole.

Answer E is incorrect as they address and issue beyond the scope of the text.

Question 7: D

To answer this, we need to use a two step approach. First, it is easiest to work backwards from the answers provided to determine the amount of taxes Paul would be paying for the respective incomes. We know that Paul must be paying $4800 ($5600 - $800). In addition to that, we also know that Paul makes more money this year than he did the year before, but pays less tax. This is only possible by crossing the 50 years of age boundary. For this reason, the correct answer is D rather than C, despite the net tax payment being the right amount in both cases.

Income	Tax if under 50	Tax if over 50
$24000	$3000	$2000
$29000	$4000	$3000
$33000	$4800	$3800
$38000	$6500	$4800
$43000	$7700	$5800

Question 8: C

For this question we have to look at figure 1 as well as figure 3. From Fig 1 we know that there were 800 units of products 1, 2 and 3 sold in April to June and none of product products 4, 5 and 6 since they have not been released yet.

Looking at Fig 3 we can determine the cost for each set of **100 units**:
Income from product 1 was 8x£1500=£12000
Income from product 2 was 8x£2000=£16000
Income from product 3 was 8x£1500=£12000
Total Income: £40000

Question 9: D

From the diagram in Figure 2 we know that in the 4 month that product 2 has been on sale by June, a total of 1700 units had been sold. We also know from the question that the 900 units sold in March represent 2/3 of the sales for the months of March to May.

If we define X as the number of units sold from March to May, we can express this as
900 = X x 2/3, solving this for X gives X = 1350.

Therefore, the number of units of product 2 sold in June must be 1700 – 1350 = 350.
We know from Figure 3 that 100 units of product 2 are worth £2000, which means that in the month of June the sale of product 2 generated £700 income (£2000 x 3.5 = £7000).

Question 10: D

From figure 1 we know that there were 600 units of product 6 sold in November and December. Using the data from figure 3 this equates to an income of 6x£4000=£24000. From Figure 2 we can see that the income produced from product 6 in December was £6000 and therefore the income from November must be £24000-£6000=£18000. This value is equivalent to 450 units of product 6, since £18000/£6000=4.5 and since prices are given for batches of 100 units, the result is 450 units sold in November.

Question 11: E

To solve this, you will have to calculate the total number of units sold in a year and divide this by the months since release. For product 1 this is 12, product 2 it is 10, product 3 it is 8, for product 4 it is 6, for product 5 this is 4 and for product 6 this is 2.

Product	Total sale since release	Average monthly sale since release
Product 1	3000	250
Product 2	3200	320
Product 3	2600	325
Product 4	1200	200
Product 5	1400	350
Product 6	600	300

Question 12: D

When Helen goes to bed it is 21mins to 2300, meaning it is 2239. When Helen wakes up it is 23mins to 0400, meaning 0337. Therefore Helen has slept from 2239 to 0337 which means she has slept 4hrs and 48 minutes.

Question 13: A

Answer A is correct since sports and entertainment are provided as specific exemptions to high IQ professions that are well paid by the text.

Answer B is incorrect since the IQ is independent of education.

Answer C is incorrect as the statement misinterprets the argument of the text that specifically states that the research did not consider parental profession when analysing IQs.

Answer D is incorrect as it argues a point beyond the scope of the text.

Question 14: D

The easiest way to approach this question is to calculate the balance per month.

Month	Balance	Month	Balance
January	$1300	July	$1100
February	$1100	August	$1300
March	$1300	September	$1200
April	$1300	October	$1500
May	$1700	November	$1400
June	$1500	December	$1400

Sam has **over $1300** in her account in May, June, October, November and December.

Question 15: B

Answer A is incorrect since the text makes no claim about unhealthy lifestyle accelerating synapse loss in old age.

Answer B is correct since the text claims that healthy lifestyles can produce additional synapses meaning that after age related synapse loss there are a higher number of synapses still available.

Answer C is incorrect as it has nothing to do with the text.

Answer D is incorrect since the link between quality of life and existing synapses is not mentioned in the text.

Answer E is incorrect since it does not provide a relevant conclusion to the topic addressed in the text.

Question 16: A

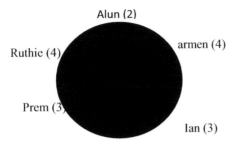

Alun (2)

Ruthie (4)

armen (4)

Prem (3)

Ian (3)

After Alun's throw, he has 0 coins, Ruthie has 7 coins and Carmen has 7 coins.
After Carmen's throw, she has 2 coins, Alun has 1 coin and Ian has 7 coins.
After Ian's throw he has 2 coins, Carmen has 4 coins and Prem has 6 coins.
After Prem's throw she has 1 coin, Ian has 3 coins and Ruthie has 9.
After Ruthie's throw he has 4 coins, Alun has 2 coins and Prem has 3 coins.

Question 17: A

Answer A is correct since only the regulation of use of these new antibiotics will maintain their effect on bacteria.

Answer B is incorrect since it neither strengthens or weakens the argument but rather provides more background information.

Answer C is incorrect since it has nothing to do with the topic discussed in the text.

Answer D is incorrect since it has little relevance for the incentive provided to pharmaceutical companies to research new antibiotics.

Question 18: D

To solve this, it is easiest to calculate the interest for the 1st year for the individual mortgages this person qualifies for.

Since he already has £25000 and wants a house worth £150000, he needs to cover approximately 83% of the house through a mortgage. This means he does not qualify for Mortgages 1 and 2.

To calculate the interest for year 1, we need to apply the respective percentages to the mortgage amount of £125000.

Mortgage No	Year 1 Interest	Arrangement Fee	Year 1 Total Cost
3	£6250	£500	£6750
4	£3750	£2000	£5750
5	£5000	£1000	£6000

Mortgage 4 is the cheapest in year 1, resulting in answer D.

Question 19: C

To calculate this, we need to calculate the growth for Texas, where the number of new wells in 2012 represents roughly 40% (40.11%) of the wells drilled in from 2005 – 2011.
40% of 2694 = 1077.6 which roughly equates to answer C.

Question 20: C

To answer this, we need to calculate the net water use per well. This is easiest using approximate numbers:
For Louisiana: 12000/2500= 4.8 million gallons per well.
For Utah: 600/1200=0.5 million gallons per well.
Difference between Louisiana and Utah: 4.8 – 0.5 = 4.3 million gallons per well. Since we used approximation we will use the next closest answer, which is C.

Question 21: E

Statement 1 is correct since the risk is specifically mentioned in the text as the last sentence in the 'Chemicals used' paragraph and according to the table Texas used 110000 million gallons of water, equating to 99.2% of the fracking fluid. Therefore, they must have used 880 million gallons of chemicals.

Statement 2 is incorrect since the total amount of pollution produced equates to 28 coal power plants per the text which means that one power plant equates to $100,551,000/28 = 3,591,107$ tonnes of carbon dioxide, not 36,000,000 tonnes carbon dioxide.

Statement 3 is correct since the text states that 26 000 million gallons are enough to supply 200,000 Colorado households for a year meaning that each household uses
26,000 million/200,000 = 130 000 gallons of water.

Question 22: H

Statement 1 does not provide an explanation for the difference in water consumption per well since the amount of water required for the process is independent from the amount of water available as mentioned by the text quoting the draught in Texas.

Statement 2 does not provide an explanation as the development in technology as it does not explain the degree of variability in water consumption, even if we assume that the fundamental understanding of technological progress is to equate in a reduction of water consumption.

Statement 3 does not provide an explanation since the per well water consumption for states varies, as demonstrated by question 20.

Question 23: D

Answer A is incorrect since it ignores the square as demonstrated by the proximity of the right and down pointing diagonal to the circle.
Answer B is incorrect since the angle of what is supposed to be the larger triangle is incorrect.
Answer C is incorrect since it misrepresents the position of the circle on the different cards.
Answer D is correct.
Answer E is incorrect since it ignores the square card.

Question 24: C

Answer A is incorrect since it is irrelevant as the text has already demonstrated the effectiveness of high impact walking irrespective of age and gender.
Answer B is incorrect since it is too specific and does not account for a general weakening of the argument.
Answer C is correct since it provides a general explanation for why the weight loss effect of playing sports is less marked.
Answer D is incorrect since it is irrelevant for the argument in the text that addresses the effect on BMI rather the motivation for performing a certain type of exercise.

Question 25: D

To answer this question we have to calculate the overall time that elapses between the start of the showings at 1015 and the end of the showings at 2245 which is 12.5hrs or 750mins.
Since the time between films is the same throughout the day, we can simply add up the duration of all the showings: 2 x (117 + 109 + 119) = 690mins
The total amount of time spent on breaks therefore is 750 – 690 = 60mins.
Since 2245 marks the end of the 6th showing 60mins has to be distributed over 5 equal rest intervals: 60/5 = 12mins.

Question 26: B

Answer A is incorrect since it has no relevance for the text that addresses the effectiveness of the mentioned self-help books not the attainability of a happy life.
Answer B is correct since this is the population that would most benefit from the purchase of a self-help book about happiness.
Answer C is incorrect since this possibility is already accounted for by the text through the use of 'are more **likely** to be anxious…'.
Answer D is incorrect since it has no relevance for the text that specifically addresses the connection between self-help books and anxiety and not the lack of a connection.

Question 27: F

There are two ways to answer this question.

Firstly, you can approach this from a theoretical perspective. In order to maximize the difference between systole and diastole, we have to have a very high systolic reading or a very low diastolic reading. Therefore, you can scan the diastolic column for the lowest readings (78 and 81) and find the combination with the highest associated systolic reading. To then ensure that you have not missed anything, check your result against the highest systolic reading.

The second option is to work backwards from the solutions looking at the differences for all the days of the respective pulse number presented in the answer.

Question 28: D

Answer A is incorrect since the text specifically mentions a degree of uncertainty in the last sentence.

Answer B is incorrect since it oversimplifies the issue with regards to the information presented in the text.

Answer C is incorrect since the text is not about cancer prevention but about what brings cancer about.

Answer D is correct since only with some influence on pollution and stress can a patient take control over them developing cancer as claimed in the last sentence.

Question 29: B

In order to find the solution for this, it is easiest to follow the instructions step by step drawing them out:

The second approach is to use the folding pattern for help. We know that the number of strands doubles on each fold, moving from 1 to 2 to 4 to 8. Since the green mark is added at the 2-strand-stage, half of the marks on the respective strands will lie between the two green marks. This means, that we will add 2 blue marks as there are 4 strings and 4 purple marks as there are 8 strands. Then, we need to add the single mark of red since it will provide the mid-point of the overall string that must necessarily lie between the two green marks.

Question 30: D

Answer A is incorrect since it is an absolute value ignoring the concept of the text itself.

Answer B is incorrect since the text does not aim to provide a time frame but merely highlight the possibility of consciousness developing in machines.

Answer C is incorrect for similar reasons as Answer A.

Answer D is correct since the connection between consciousness and brain function is unclear even when dealing with living creatures such as animals.

Question 31: D

The best way to approach this is to set up an equation.

X = points achieved by the Argents

Z = number of fesses achieved by the Sables

Y = number of fesses achieved by the Argents

X = 11Y since the Argents achieves twice as many pales as fesses and X is total number of points

X – 1 = 9Z since the Sables achieved the same number of fesses as pales

Z – 1 = Y since the Argents achieved one less fess than the Sables

Z – 1 = Y since the Argents achieve one less fess than the Sables.

Therefore Z = Y + 1.

Combining the points achieve we get 9Z + 1 = 11Y.

Insert the value of Z we have calculated above:

9Y + 10 = 11Y and solve for Y to get 5 = Y

Since Y is the number of fesses scored by the Argents, and they achieved twice as many pales as fesses, they must have scored 10 pales which is answer D.

Question 32: A

The answer from this can be found in the Table in the column giving total amount of polyunsaturated. For Olive Oil this is 10% and for Sunflower oil this is 65.7%. Since they both occur in a 50:50 mix the easiest way to find the solution is to add half of the content of each meaning 5% from the Olive Oil and 32.85% from the Sunflower Oil This leads to 37.85% which can be rounded up to 37.9%.

Question 33: A

Answer A is correct since it specifically says so in the text in the second paragraph in the last two lines.

Answer B is incorrect since the constituents of diets in hunter-gatherer societies around the world vary according to climate.

Answer C is incorrect since the third paragraph specifically mentions that the fall in heart disease is associated with the consumption of polyunsaturated fats.

Answer D is incorrect since the text specifically contradicts this by pointing out the cardio protective nature of high polyunsaturated fat oils.

Answer E is incorrect since the heart protective effect has improved with a higher ratio.

Question 34: B

According to the text the average adult ingests 143 grams of fat of which 30% are vegetable oil which equates to 43g.

Since 100g of Canola oil contain 0.6g of Erucic acid, this equates to roughly 0.26g (260mg) of Erucic acid which equates to just over 50% of the daily suggested intake.

Question 35: D

We know from the table that Sunflower oil does not contain any omega 3 so can only contribute Omega-6.

Similarly we know that the ratio of Omega-6 to Omega-3 in flaxseed oil is 0.2:1. For this is easiest to work with 100g of flaxseed oil and work out how much sunflower oil has to be added to achieve the 2:1 ratio.

Since Flaxseed oil contains 53.3g of Omega-3 we need to achieve 106.6g of Omega-6 in the mix.

12.7g of that are already contained in the flaxseed bolus, meaning we need to add the equivalent of roughly 94g of Omega-6 from sunflower oil. The amount of sunflower oil is (94/65.7)x100g = 143g.

This means the mixture of oils will weigh 243g. Looking at this we can approximate that the flaxseed oil representing slightly less than half of the mass. The closest percentage to this is 41% from answer D.

Calculation: 100/243 = 0.411 = 41%.

END OF SECTION

Section 2

Question 1: D

Answer A is incorrect since the vessels are named incorrectly.

Answer B is incorrect since the vessels are named correctly, but the content of urea is not allocated correctly.

Answer C is incorrect since the vessels are named incorrectly.

Answer D is correct since the structures are named correctly and so is the urea concentration.

Answer E is incorrect since the structures are named incorrectly.

Answer F is incorrect since the structures are named incorrectly.

Question 2: F

Statement 1 is incorrect - no element is in both period 3 and group 12.

Statement 2 is correct since the element has up to 3 electrons to donate and oxygen can take up 2 with the smallest common denominator being 6.

Statement 3 is correct since Br can take up one electron and the compound has 3 to donate.

Statement 4 is correct since elements will contain the same number of protons as they contain electrons meaning that the atomic number must be 13.

Statement 5 is incorrect since alkali metals are only in group 1.

Question 3: D

From the diagram, we know that two pieces of the material displace 100 cm^3, therefore 1 piece must displace 50 cm^3.

The weight difference from no pieces of material to one piece of material is 300g. Therefore, the correct formula is 300g/50cm^3.

Question 4: A

First, we need to establish the slope of the line that passes through the two points given in the question using the $y = mx + b$ equation:

$m = \frac{(y_2 - y_1)}{(x_2 - x_1)}$ Meaning m $= \frac{2}{3}$

To determine b we need to insert one of the two points into the equation: $y = \frac{2}{3}X + b$:

Using $\left(\frac{6}{9}\right)$ this gives $9 = \frac{2}{3} x 6 + b$; solve for b to find b = 5.

Therefore the equation is $y = \frac{2}{3}X + 5$.

Since two lines are parallel if they have the same slope, the only possible answer is A where the slope (m) is 2/3.

Question 5: B

W must be a chromosome since it is the origin of the removed DNA.

X must be a restriction enzyme since it removes cuts the DNA strand to remove the DNA fragment.

Y must be a restriction enzyme since it cuts a gap into the DNA of the other organism.

Z must be a ligase since only ligase enzymes can fuse DNA strands which will result in insertion of chromosome DNA into the other organism.

Question 6: E

Answer A is correct since removing water from a solution of calcium carbonate and water will leave solid calcium carbonate.

Answer B is correct since the two substance will have different vaporisation points due to different molecular size.

Answer C is correct since silicon dioxide is a solid that poorly dissolves in water allowing filtration as an effective means of separation.

Answer D is correct since the vaporisation point of sodium chloride and water is different.

Answer E is incorrect since ethanol and water form a solution that cannot be separated by mechanical means but only by thermic separation techniques.

Question 7: D

Statement 1 is correct since the difference in weight is due to different numbers of neutrons in the core that will not change the chemical properties of the element.

Statement 2 is correct since the atomic number is due to the number of protons.

Statement 3 is incorrect since all the isotope nuclei contain 28 protons but a varying number of neutrons.

Question 8: B

The easiest way to solve this is via an equation where we assume that N consists of a number of people all weighing exactly 75kg and that Jim, Karen and Leroy each weigh 90kg.

$$78 = \frac{[75N + (3 \times 90)]}{N + 3}$$

Solving this equation for N, leads to N = 12, which is solution B.

Question 9: F

From statement 1 we know that the enzyme must be in the stomach as the stomach is the only place in the human body with a pH of below 4.

From statement 2 we know that the enzyme must be a protease as it digests proteins into amino acids.

From statement 3 we know that the enzyme must be inside the body.

Answer A is incorrect as amylase digests sugars.

Answer B is incorrect as amylase digests sugar.

Answer C is incorrect as lipase digests fat.

Answer D is incorrect as lipase digests fat.

Answer E is incorrect since the pH in the small intestine is above 4.

Question 10: F

To solve this we need to count elements and add the respective molecular weight.

N = 14, there are 2 in the equation leading to 28.

H = 1, there are 20 in the equation leading to 20.

Fe = 56, there is 1 in the equation leading to 56.

S = 32, there are 2 in the equation leading to 64.

O = 16, there are 14 in the equation leading to 224.

In total this adds up to 392, which is answer F.

Question 11: B

Rotating the coil at a faster constant speed will accelerate the voltage change thereby increasing the amplitude and since the coil will also complete its turns faster, the frequency of the waves will increase as well. Only answer B ticks these boxes.

Question 12: D

This is a two part question. First we need to find the radius of the arterial lumen and then apply the equation defining the area of a circle to this.

Since the overall diameter of the artery is 1.6cm and the walls are 1mm thick, the inside diameter is 1.4cm, leaving the radius to be 7mm.

Since $A = \pi r^2$, $A = \pi \times 7^2 = 49\pi \ mm^2$

Question 13: B

Carbon dioxide is only produced in aerobic respiration since it requires oxygen which per definition is absent in anaerobic respiration. Glucose represents the energy substrate in either type of respiration. Lactate is only produced in anaerobic respiration.

Question 14: C

Answer A is incorrect since the cathode would give form Hydrogen as the Calcium is more reactive.

Answer B is incorrect as the anode would give oxygen since the nitrate is a complex ion.

Answer C is correct.

Answer D is incorrect since products are allocated to the wrong electrodes.

Answer E is incorrect since molten NaCl does not contain any hydrogen.

Question 15: D

Firstly, all waves named in the question travel at the speed of light which is defined as 3×10^8 m/s or 300 000 km/s.

Since the distance from satellite to transmitter and receiver is 45,000 km each, the waves must travel 90 000 km, which will take $= \frac{90,000}{300,000} = 0.3s$

Secondly, we know from the question that the waves have a frequency of 1.5×10^{10}Hz, which defines them as microwaves, which leads to answer D.

Question 16: C

To solve this question, we have to work in several steps as we need to calculate both the length of RS and the length of PQ.

To calculate RS we need to know that tan is defined as opposite/adjacent, in this case RS/6. Since the question gives us the tan as 4/3, we know that RS must be 8.

RS/6 = 4/3 = 8/6.

Appling this we can calculate the area to be $A = (5x8) + \left(\frac{6x8}{2}\right)$

$= 40 + 24 = 64cm^2$

Question 17: G

Statement 1 is correct, this is known as a gain of function mutation.

Statement 2 is correct, this is known as a loss of function mutation.

Statement 3 is correct, this is known as a loss of function mutation.

Statement 4 is correct, this is known as a gain of function mutation.

Question 18: E

We know from the question that the acid is able to donate two protons therefore we know that we must use twice the volume of NaOH, which is able accept 1 proton.

Since the concentration of NaOH is half of that of the diprotic acid, we need 4 times the amount of the solution provided in the question to neutralize the acid.

Since the volume of the acid is given as 30cm³, we need 120cm³ (4x30cm³) of NaOH.

Question 19: D

Since we don't know the resistance of the body, the best formula to use here to find the current is I = Coulombs/time in seconds.

First, we need to calculate the charge in Coulombs (1 Coulomb = 1 joule per volt):

$\frac{125J}{500V} = 0.25$ C

Therefore $I = \frac{0.25C}{0.01s} = 25A$

Question 20: C

The main point here is not to lose the overview over the different components of the equation. Then there is no real trick to this, other than going through it step by step:

$$\frac{a}{b} = \frac{c}{d} + \frac{e}{f}$$

$$\frac{a}{b} = \frac{fc + de}{df}$$

$$dfa = bfc + bde$$

$$dfa - bfc = bde$$

$$\frac{f(da - bc)}{da - bc} = \frac{bde}{da - bc}$$

$$f = \frac{bde}{da - bc}$$

Question 21: A

Statement 1 is correct as it basically aims at the doubling of the genetic material which happens through the process of mitosis.
Statement 2 is incorrect since growth happens through the production of proteins which does not occur during mitosis.
Statement 3 is incorrect since again this happens through protein production, which is halted during mitosis.
Statement 4 is correct since stem cell division follows the same principle as asexual reproduction.

Question 22: B

Due to the increased concentration of the acid, twice that of the original sample, the reaction speed will be higher, but as the net amount of the acid in the sample is the same as in the original experiment, half the mass at twice the concentration, the equilibrium will be the same. Only line B fits this bill.

Line A equates to higher concentration and higher net amount.
Line C equates to lower concentration of acid and the same net amount.
Line D equates to a lower net amount of acid.
Line E equates to a lower concentration as well as a lower net amount of acid.

Question 23: E

From the information given in the question we know that the object must weigh 1.5kg on earth (15N at g=10N/kg).
The mass of the object will remain unchanged as it is transported to the planet. The difference will be due to the different gravitational fields.
Since we know from the text that the object has a weight of 3N on the planet, it will have a kinetic energy of 30N after a fall of 10m.

Question 24: D

This question deliberately aims to confuse you with additional information. The population distribution of the different blood groups is not needed to answer this question. There are 4 different blood groups A, B, AB and 0. One individual can have one of those 4 blood types. Since the question asks about the probability of the one criminal having both A and B antigens, he/she must have blood group AB, the likelihood of which is 25% or ¼.

Question 25: H

From the diagram, we can deduce that a grey coat colour must be dominant and a white coat colour must be recessive. Therefore, any grey mouse can potentially be heterozygous. There are 12 grey mice in the diagram, leaving answer H.

Question 26: D

From the equation, we know that the volume of the final product will be equivalent to the volume contribution of X. Since we use 100cm³ of reagent X and the reaction occurs under exclusion of air and to completion, the final volume in the syringe must be 100cm³.

Question 27: C

The easiest way to answer this question is to calculate backwards to determine the wavelength of the light through air.
We know that 360nm equates to ¾ of the wavelength through air; if X = wavelength through air: 360nm = X x 0.75
X = 480nm. If Y is defined as the wavelength of light through glass, we can express this as
Y = 480nm x 2/3 which gives Y = 321.6nm.

END OF SECTION

Section 3

'You can resist an invading army; you cannot resist an idea whose time has come.' (Victor Hugo)

➢ In this question, Hugo basically evaluates the strength of physical confrontation and violence versus intellect and words when it comes to changing the status quo. This position has to be connected to be taken in connection with his time. It has to be taken into account that he lived in France just after the Revolution and during the reinstitution of the monarchy in France. You can also remove the quote completely from the historical background and look more at the conflict between forced change and passive change with the invading army representing forced change and the idea a more passive and natural form of change. The forced change does not necessarily have to come in the form of an invading army consisting of battalions of soldiers but can also represent something as simple as a prescribed mind-set or political conviction that is propagated by a political regime.

➢ Arguing against this statement, it is obvious that sufficient force seems to be able to suppress most ideas, especially if execution of opposing violence is well publicised and very large scale. One example of this would be the wide ranging suppression of individuals in dictatorships such as North Korea where dissidence is not only punished physically but also enforced by a high degree of isolation. In the end, if it is possible to enforce a perception that any form of dissent will result in personal injury and in the injury of loved ones, violence can suppress ideology, especially if punishment is highly excessive.

➢ A further point to consider when arguing against the statement in the question is the definition of "idea whose time has come". Considering the vagueness of this point, it provides a good target for arguing against the statement as it renders the statement as a whole rather moot.

➢ The idea of physical and psychological violence versus ideas is also discussed in other works of literature such as 1984, some of the means by which though control can be achieved in the book provide good examples to add to an essay.

➢ Addressing the point of the power of ideas, there are many examples in history than can be used for this. The American Revolution presents one of them, so does the resistance against Nazi occupation in France or other countries during the Second World War In all cases ideology refused to be intimidated by violence. In this context however it is essential to highlight that ideas usually achieve power by encouraging to violence thereby almost resulting in a circular argument. The probably most obvious exception to this rule is Ghandi's peaceful disobedience promoting Indian independence without use of violence.

➢ In this question you can generally speaking take multiple routes, either staying rather close to historical examples or move in a more philosophical direction, looking at the issue with a broader perspective in mind. The challenge of the latter option is that it will be easier to lose track of your arguments and more difficult to maintain relevance of the points you make with regards to the original question.

➢ This is a very interesting question as it challenges the idea of progress in general to a certain extend as it attempts to quantify the influence of eternal stressors on the intellectual development of a society.

Science is not a follower of fashion nor of other social or cultural trends.

➤ The statement attempts to explain and understand the reasoning behind scientific progress and the interaction between science and culture and other influences on social progress. The basic assumption of this sentence is that science develops independent of other social driving forces that may or may not be subject to trends and temporary interests.

➤ In order to write a good essay you have to be clear for yourself what you understand by science, fashion and social/cultural trend.

  • Science in general is defined as the endeavour of searching for new information and truths in order to widen the horizon of our knowledge and out understanding of the world around us. In addition to that, science also has a certain set of rules that define the value and the truthfulness of the information that is being found.

  • Fashion as well as cultural trends both essentially go in the same direction. They both describe temporary and variable perceptions and interests within society. What is important in this is that trends as well as fashions tend to be variable from individual to individual or from group to group. This makes it different from science since science claims to hold an overarching always valid truth.

➤ The basic idea when arguing to the contrary is to provide arguments of why science is indeed influenced by fashions and trends. There are several ways this can be done. On one hand, you can argue from a sort of social perspective putting fashion and trends into context with social interests. This is a good starting point as it makes it clear how science and the areas of research that individuals are interested in are influenced by what fascinates the masses at any given point in time. One example for this is for example the great degree of progress in military technology in the early 20th century when Europe was ravaged by war and there was a great degree of militarism throughout many levels of society.

➤ On the other hand, you can argue in a different direction illustrating the relationship between science and trends by starting at scientific discovery and arguing its influence on trends and fashion. One example for this would be the progress in nuclear technology in the 1950s when society began dreaming about nuclear powered cars etc and the strive for new technology and the idea of progress through technology was very wide spread, in particular in the US.

➤ Other examples include the great social interest in geography and the natural sciences in general during the age of discovery of the Americas. The idea of widening frontiers and pushing civilization onwards became a great social driving force contributing to large amounts of funding as well as large movements of populations to these newly discovered lands. This also directly reflects social issues in the European heartland that made a new live abroad more favourable and attractive.

➤ When phrasing your agreement or disagreement, it pays to be direct with your opinion since this will make it easier to present your point in a direct and efficient fashion. However, be sure to have argued your points appropriately before voicing your opinion.

The option of taking strike action should not be available to doctors as they have a special duty of care to their patients.

➤ In general, when arguing this point you have to be aware of the implications of this question. You have to keep in mind that the essay you write will be shown to any admitting university that may or may not use your essay during the interview. For this reason, it may be wise to be a little careful with what you write in this type of essay. This applies to essays of this scope in general.

➤ Now, when t comes to arguing this particular essay, there are several bases you have to appreciate. Firstly, there is the nature of the relationship between doctors and their patients. As doctors we have a complicated relationship with our patients in the sense that we know many intimate details about our patients but at the same time we need to keep a professional distance to make the decisions that we deem in the patient's best interest. This tight bond between patients and doctors makes for a relationship between recipient and donor of a service that is not found outside the health care setting. Secondly you have to consider the purpose of strike action.

➤ Doctors going on strike is never an easy decision, particularly due to the sometimes live and death nature of health care. With doctors not working, the care that these vulnerable patients need may not be deliverable. This can be argued to be a direct violation of the basic ethical pillar of 'do no harm', since the doctor knowingly and without direct external pressure decides to withhold treatment that may alleviate the patient's suffering. The other ethical pillar that can be argued to be violated by doctors going on strike is the idea of beneficence, since it is the doctors duty to always act in the patient's best interest, which withholding of treatment most definitely is not.

➤ The most significant justifying reason for supporting doctors strikes is the fact that poor working conditions will necessarily have a direct influence on the doctors performance which in turn will have negative and detrimental effects on patient care. Good examples of this are fatigue due to long working hours or unsafe levels of staff due to general unattractiveness of the profession or widespread symptoms of exhaustion in the work force.

➤ Another point to consider is the reason why doctor strikes are an issue in the NHS and not as much in other countries and health care systems. This will give you a good chance to demonstrate that you have an understanding of how the NHS works and how health care is provided in different countries.

> ➤ You can also consider addressing the reasoning for strikes in general and why every work force should have the right to protest unjust working conditions by going on strike. This will necessarily take you away from the question somewhat which can be a challenge but may lead to an interesting essay. In this context it is absolutely paramount to not lose track of your argumentative chain and to relate your main body back to the question in your conclusion.
> ➤ You should definitely consider to use the recent junior doctor strikes as pivot point of the essay. It provides a good pivot point to organize your essay around and it will also give you real life examples of how doctor strikes can be made as safe as possible and how potential harm to patients can be reduced as much as possible.

If we truly care about the welfare of animals, we must recognise them as fellow members of our communities with their own political rights and status.

> ➤ This can be a quiet difficult topic to argue since the question covers a wider range of different topics that need to be addressed or at least be clear in your head in order to write a good essay.
> ➤ The first point to define is the idea of animal welfare. What do we understand by animal welfare and what implications does this have on the interactions between humans and animals. At this point it would be helpful to be aware of what rules exist already and what defines current considerations of animal welfare. Having a basic knowledge of this will strengthen your essay. However, if you do not have any idea of animal welfare, you can always stay on a more general and philosophical level.
> ➤ The second point to be clear about is the meaning of political rights and status. Make sure you have a vague idea of what this would mean and also how this could be put into action, in particular since you are required to argue that we do not need to allocate animals political right and status. Points to consider in this context are also that with right always come duties and how these can be applied to animals.
> ➤ In the context of this question it can also be helpful to consider the idea of conscience and intellect in animals. Try and connect this to the idea of political rights and status and the attached duties and possibilities. Taking the right to vote for example, how are cats and dogs supposed to understand what politicians say and even if they were to understand what they say, how do we know if they have the intellectual faculties to make a decision about such matters.
> ➤ Another avenue you might want to explore is the matter of ownership. When it comes to animals, whether they are pets or animals that themselves are a resource such as in the milk and meat industry, what role does the owner of the animal play. If we make them members of our community with political rights and status, does that make them ownerless and what implications does this have? Or do they retain their ownership and if so, how does that then impact the de facto role they take in society. Since animals will always require some degree of human support either through for example access to medical care or in order to gain access to food, how does this arrangement work form a legal perspective?
> ➤ In general it could be helpful for this essay title to imagine animals as humans that are somewhat limited in their ability to communicate with the rest of society and how this would shift the role they play as members of society.
> ➤ When it comes to political institutions again, it would be useful to have a general understanding of what institutions already exist that deal with animal welfare and animal rights and the protection and care of animals. Examples of here could be NGOs such as Greenpeace or organisations such as the RSPCA.
> ➤ Finally, much like the previous topic, keep in mind that this is a difficult and somewhat charged subject and that your admitting university will receive a copy of your essay that may well be discussed, at least on a theoretical level, in your interview. Thing you should avoid are extreme claims such as animal snot needing special rights since they have no mind and are just things or property etc.

END OF PAPER

2017

Section 1

Question 1: C

Start by converting everything to ml so that you don't get confused. Hugh had 900ml of yellow paint left, meaning he used (1500 – 900 =) 600ml of it. This means that he also used 600ml of red paint to create the orange paint, as they were mixed in a 1:1 ratio. If the room was 40% orange and represents (600 + 600) = 1200ml of paint, this means that the 60% pink represents 1800ml of paint. A quarter of the pink paint was red, meaning 450ml of red paint was used to make the pink.

600ml (in the orange) + 450ml (in the pink) = 1050ml of red pain used overall

1500ml – 1050ml = 450ml of red paint left

Question 2: D

A is definitely not true; there is definitely no causal evidence from this statement alone that use of antidepressants leads to *worse* outcomes for patients (i.e. we cannot know if the disability claimants are taking antidepressants or whether they represent a different population). There is also no evidence from the information that there is short term efficacy of this medication.

B is not necessarily correct; we have no idea whether doctors are being 'increasingly encouraged' to prescribe these drugs. There may be other reasons for the increase in numbers seen over time; there may be more people suffering with their mental health in the present, there may be more people seeking help for it than previously, or more people aware of the claims there are entitled to make.

C is also not necessarily correct. Similarly to above, there may be other reasons for the mental health figures getting worse over time. including an increasing incidence of mental health issues. There is a causal link between the prescriptions and claimants in this answer, which cannot be drawn from the figures alone.

D can be drawn as a conclusion from this stem alone. The figures do indeed show that on a population level, the long term mental health in the UK is declining, and that this is not improving despite increasing use of antidepressant drugs. This answer doesn't assume any causal link between the drug prescriptions and disability claimants, instead just focussing on the numbers.

Question 3: D

A satisfies the bedrooms, garage, garden, distance to grocery store; but not distance to sports facilities. 4/5
B satisfies the garden and distance to sports facilities; but not bedrooms, garage or distance to grocery store. 2/5
C satisfies the bedrooms, garage, garden and distance to sports facilities; but not distance to grocery store. 4/5
D satisfies the bedrooms, garage, garden, distance to grocery store; but not distance to sport facilities. 4/5
E satisfies the bedrooms, distance to grocery store and sports facilities; but not garage or garden. 3/5
Out of those satisfying 4/5 wishes, D is the cheapest.

Question 4: A

The key argument against nuclear power in this passage is that it 'poses and unacceptable risk to the environment and to humanity' and that stations 'create tens of thousands of tons of lethal, high-level radioactive waste.' Thus, the phrase that would weaken this argument best would minimise these concerns.

A would weaken the argument well; it highlights that there is a way to counter-act the unacceptable risk to the environment/humanity that the author talks of.
B is incorrect; the passage suggests that we should be using less nuclear power so the author would be satisfied with this statement. It doesn't tackle the issues that are raised against nuclear power.

C gives an argument against wind power, but doesn't counter the key arguments raised against nuclear power that are highlighted above.
D is not correct; the argument actually acknowledges that nuclear power is 'less air-polluting that fossil fuels' and this is not a key reason for concern for the author.

Question 5: D

This question is obviously quite hard to explain with text alone. These questions often require trial and error and visualisation in your mind of how each shape would appear in different orientations. You are allowed to bring scissors to the exam and cut shapes out, although that may not help so much with this question compared to previous ones regarding cube folding, etc.

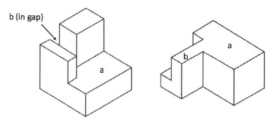

Look at the shape of 'a' on the diagram. It is nearly a rectangle with one width slightly smaller than the other. If both 'a's were appositioned directly there would be a perfect fit, whereas none of the other shapes given would be able to do this.

Also, when looking at the gap made by 'b' in the diagram, you can see it would need to be filled by a thin part of the block, which D is able to do.

Question 6: C

For any disease in medicine, there are 'risk factors' that increase your likelihood of developing a particular condition. For example, other risk factors for prostate cancer include increasing age, having a family member with the disease, or being black. However, even with no risk factors, there is a chance (albeit smaller) that one may develop the disease. Conversely, even with every risk factor for a disease, there is no guarantee that one will definitely develop the disease.

The key aspect of this argument, and simultaneously the flaw, is the suggestion that men can prevent the development of prostate cancer by minimising weight gain, as highlighted by C

A and D address other cancers which is not an issue tackled by this argument.
B doesn't address the causal link between changes to diet and exercise and prostate cancer itself.

Question 7: B

Adding up all the students at the school gets to 144, therefore there are 72 girls.
None of the girls swam and 40 girls played rounders.
28 pupils ran, of which 14 were girls.
Therefore 72 girls in total – 40 (rounders) – 14 (running) = 18 girls who played football

Question 8: C

The reconvinction frequency rate is the number of offences per 100 offenders in the cohort. Thus, over a 9 year follow-up period, there were 1057.5/100 = ~10.6 reconvictions per person in the cohort. However, the question asks specifically for the frequency of those who reoffended in the first place (74%), so we must do 10.6/0.74 = 14.3.

(Alternative method if it helps you understand it better: 42, 721 offenders in total, of which 74% were reconvicted, which is 31613 reconvicted. There were 10.6 reconvictions per offender in the total cohort, so 10.6 x 42721 = 452842 for total number of reconvinctions. If we want a figure for the number of reconvinctions per re-offender, 31613 reoffended, so we need to do 452842/31613 = 14.3)

Question 9: E

1 is true. The information explains that the figures are cumulative. 43% were reconvicted in the first year and cumulatively 55.2% by the second year, so the proportion convicted in the first year is 43/55.2 = ~0.78 = ~78%. (If you'd rather work with numbers, 43% is 18370 and 55.2% is 23582. 18370/23582 = ~78%)

2 is not true. As the figures are cumulative, 43% were reconvicted in the first year, and 61.9% by the third year, meaning only 61.9% - 43% = 18.9% were reconvicted in years 2 and 3. This is less than a third of the 74% that were reconvicted over 9 years. (Again, working with numbers, 31613 were reconvincted overall, 18370 over the first year, 26444 over three years; therefore 8074 over years 2 and 3. This is less than a third of 31613.)

3 is true. The reconviction frequency rate was 185.1 per 100 offenders at this time, meaning 1.851 per offender. There were therefore 42721 x 1.851 = 79077 reconvictions.

Question 10: C
The key information for this question is: 'offenders who received sentences of less than 12 months… committed 39% of all offences that led to a conviction in the first year of the follow-up.'

As shown above there were ~79077 reconvictions in the first year. 39% of these were by those who received sentences of less than 12 months.
79077 x 0.39 = 30840 = ~31000

Question 11: G
The wording of the question, 'might feasibly,' means that any reasonable suggestion that could account for a decline is acceptable; it doesn't need specific evidence or to be supported by the data.

1 makes sense because out of those who are going to eventually be reconvincted each year, a proportion had already done so and were in prison, so there is a smaller cohort each year.

2 is correct because we are looking at a specific cohort only (i.e. nobody can be added to the population size) and **reconvictions** rather than any new convictions. Naturally, some of the cohort will pass away.

3 could be true; if there was indeed harsher sentencing, this could feasibly deter those who would have likely reoffended with lighter sentencing.

Question 12: D
Unfortunately, there is not always a neat mathematical way to solve these types of questions; trial and error is probably best.
A cannot be true because the passage states that 'the performer who plays the role of Gracie cannot play any other characters.'
B cannot be true because both Teddy and Guard 1 are in scene 1.
C cannot be true because of the statement 'the performer who plays Rose also plays Guard 1' which would mean one performer would have to play Rose, Guard 1 and Sarah. Rose and Sarah are both in scene 10
E cannot be true because Sarah needs to be played be a female, and Graham needs to be played by a male
F cannot be true for similar reasons to C; the performer would have to play Rose, Guard 1 and Guard 2. Both guards are in scene 1.

Question 13: B
This is a classic BMAT correlation does not necessarily equal causation question. The passage even states that the drop in selenium levels *correlates* with the three extinction events.

A is nothing to do with extinction, which is the main topic of the passage.
B is correct because it acknowledges that we do not know the exact causation but that the factors could be linked.

C is incorrect because there is no mention of evidence regarding other trace elements, so we do not know how they compare in importance.
D is incorrect because it implies definite causation.

Question 14: B
The Citrons gained 57 seats and lost 29 hence now have 80 + 57 - 29 = 108. 108/240 = ~0.45
The Jonquils gained 26 seats and lost 80 hence now have 126 + 26 – 80 = 72. 72/240 = ~0.3
The Saffrons gained 51 seats and lost 25 hence now have 34 + 51 – 25 = 60. 60/240 = ~0.25
The pie chart that shows this best is B (the part which is exactly a quarter is probably most helpful).

Question 15: A
The argument presented is in the following format:
- There are two potential hypotheses for an observation, one of which is correct.
- One reason can be ruled out based on objective evidence.
- Therefore, the other reason must be correct.

A is presented in this way.
B doesn't give evidence to rule out the first potential hypotheses, only evidence to support the other reason.
C doesn't give 2 potential hypotheses for an observation.
D doesn't rule out one of the hypotheses objectively, only subjectively.

Question 16: D

All the digits add up to 38, and each digit is different. There are 8 digits in total, so each number except one from 1-9 are used. Now use trial and error to see which digit can be excluded to get to a total of 38.

A $2 + 3 + 4 + 5 + 6 + 7 + 8 + 9 = 44$
B $1 + 2 + 4 + 5 + 6 + 7 + 8 + 9 = 42$
C $1 + 2 + 3 + 4 + 6 + 7 + 8 + 9 = 40$
D $1 + 2 + 3 + 4 + 5 + 6 + 8 + 9 = 38$
E $1 + 2 + 3 + 4 + 5 + 6 + 7 + 8 = 36$

Question 17: D

The argument is very specifically suggesting that being a childhood star can cause problems in later life, so do not extrapolate this argument to include any other groups of people.

1 is not correct because the author doesn't ever argue that break-ups and mental breakdowns are exclusive and unique to childhood stars; instead they are merely suggesting that the *risk* would increase as a childhood star.

2 is incorrect because the author does not suggest any link between being adored as a child and addictions/broken relationships. The author uses the word 'adored' to emotionally charge the argument rather than as a logical explanation.

Question 18: B

Cocoa powder used = 100 + 85 = 185g. 315g remaining.
Eggs used = 2. 10 remaining.
Sugar used = 330 + 200 = 530g. 70g remaining
No lemons used. 5 remaining
Milk used = 250ml. 2250ml remaining
Butter used = 400 + 150 = 550g. 50g remaining
Flour used = 400 + 225 = 625g. 375g remaining
There is only 50g of butter left, and 8 pancakes requires 50g of butter, so this is the maximum that can be made (there are no other limiting factors at this point).

Question 19: B

The claim in the headline is that a fifth of the papers contained errors in the spreadsheets. Therefore, if they are not actually errors, the support that the data gives to the claim would be weakened - this is why B is correct.

A is not correct because the headline isn't referring to any other scientific fields.
C is not correct because even if the results were , that wouldn't discount the fact that there were errors.

D is not correct because there is no way of knowing whether this would actually reduce the number of papers containing errors.

Question 20: B

1 is incorrect, we don't have data for total number of papers published for an individual year so we cannot make this conclusion.
2 is correct; looking at the graph the figure was 50 for 2009 and just above 100 for 2011
3 is not correct; Nature published 23 with errors and BMC Bioinformatics with 21.

Question 21: D

The papers with a higher % than the average (1st graph) are Nature (23 papers), Genes Dev (55), Genome Res (68), Genome Biol (63) Nature Genet (9), Nucleic Acid Res (67) and BMC Genomics (158).
Adding these together gets to 443 papers, and out of 704 this represents 63%.

Question 22: C
'The authors found that the number of genomics papers packaged with error-ridden spreadsheets increased by 20% a year over the period, far above the 10% annual growth rate in the number of genomics papers published'

Often with these proportion questions, it is easier for the human brain to convert into figures first, which doesn't actually change the calculation. So, imagine that in 2015, there were 100 papers published and 80 were affected.

2016: 80 x 1.2 = 96 papers affected. 100 x 1.1 = 110 papers published
2017: 96 x 1.2 = 115 papers affected. 110 x 1.1 = 121 papers published
2018: 115 x 1.2 = 138 papers affected. 121 x 1.1 = 133 papers published
Therefore, every paper would be affected by 2018.

Question 23: E
Looking at the shaded area, we can work out that the puzzle pieces that would fit would require 4 bits that stick out to fill the gaps and 2 that go inwards. Additionally, in the middle to join the 2 pieces, there would need to be a bit that stuck out and a part that stuck in. Therefore, in total we should expect 5 parts that stick out and 3 parts that go inwards. This is only fulfilled by E.

The diagram above, using x and y, shows one way that the pair of pieces could fit into the jigsaw puzzle.

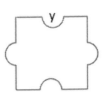 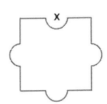 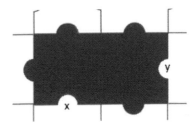

Question 24: E
The argument suggests a exclusive, causal inverse link between the number of children a woman has and their life expectancy based on a correlation. Only E acknowledges that there could be other factors that influence this correlation.

A is not correct because the argument doesn't assume or mention women's recognition of the toll of childbirth.
B is incorrect because the argument actually acknowledges that it is referring to women in rich nations.

C is incorrect because the argument is not about infant mortality.

D is not correct as the argument doesn't suggest this, only that the *risk* is increased with multiple pregnancies.

Question 25: E
The distance for the original journey between the two towns is not known, so let's call this x. We don't know the time take for the original journey, so let's call this y. Obviously, distance = speed x time.
Original journey: x = 30y
The distance is increased by 4km in the new journey, the speed is reduced by 3km/h and the journey time increases by 25%
New journey: x + 4 = 27 x 1.25y
Solve the simultaneous equations. x = 30y, and x + 4 = 33.75y so 4 = 3.75y and y = 1.067. x = 32
New journey is x + 4 = 36km

Question 26: C
The passage says that the government suggest the health checks **could** equate to prevention of over 2,000 heart attacks and strokes. However, even though the researchers identified many with risk factors, they need to follow up long-term to see whether there is actually prevention. Thus no firm conclusions can be drawn from the health checks yet, which is why C is correct.

A is not correct because we don't know if this number of heart attacks and strokes have actually been prevented.
B cannot be concluded, we don't know whether patients' health-related behaviours changed or not.
D may be subjectively true but is not specifically implied by the argument here, without the context of 'health checks'
E is incorrect, there is no suggestion to think this from the information given.

Question 27: B

The lowest common multiple of 400g, 300g and 200g is 1200g therefore Charlie would have ordered 1200g of each fish. This would make 24 plates. 3 packs of prawns are needed, and the third is free. Therefore $4.08 x 2 = $8.16. 4 packs of squid are needed, and one is free. Therefore $4.08 x 3 = $12.24. 6 packs of cockles, whelks and smoked salmon are needed and 2 are free for each. Therefore $4.08 x 3 x 4 = $48.96 → 8.16 + 12.24 + 48.96 = $69.36 in total. This is making 24 dishes, therefore 69.36/24 = $2.89 each

Question 28: E

1 is true. The second line, especially the phrase 'encourages the police officers to better regulate their own behaviour', suggests that the author thinks the 'violence' and thus level of force used can often be reduced.

2 is incorrect. The first line states that police departments across the world should use body-worn cameras, but it doesn't assume that there is any interaction between police departments in different countries.

3 is correct. The last two lines highlight this. The argument is suggesting that police-worn cameras are better because there is always a clear warning about filming from the start, which implies the assumption of those being filmed by bystanders not always being consensual.

Question 29: E

The dice, with the orientation on the left, is missing a dot from its 4, and a dot from its 6. The dice, with the orientation on the right, is missing the same dot on the 6 but we do not know which of the faces with 2 is missing a dot to make it a 3.

If we look at the orientation of both, the face to the left of the three dots on the 6 face must be the 5. This means the face to the left of the two dots on the 6 face must be 2 to make opposite faces add up to 7. Therefore, the dice without missing dots is as in the diagram above. Therefore, on the right dice, the bottom face is a 1, and the face that cannot be seen on the right is the 4 (with a missing dot). If you rotate this, you can get E but none of the others.

Question 30: C

This is yet another correlation does not equal causation question. Stronger brain connectivity is linked to 'positive' lifestyle traits; however, the causation may be in the other direction, or there may be confounding variables which influence both factors in the same way. Only C addresses this flaw.

Question 31: D

You can build a circle based on working out the preference of whoever is adjacent to someone. For example, on the left of Jess is someone who likes cricket and rugby, who is the same person who is on the right of David. You can go around the circle like this. Also, it helps to know that Jess is next to 2 people that share a common interest; so she must like 1 of cricket and rugby, and 1 of football and golf.

	George (F&S)	
David (C&H)		Eli (H&S)
Amir (C&R)		Peter (F&G)
	Jess (G&R)	

Question 32: C

This question should hopefully be straightforward, with the use of the word '*could*'. There is no evidence required but clearly both of the answers have the potential to increase the problem of London's road congestion.

Question 33: E

6 underground strikes at £10 million each = £60 million. GVA due to congestion = £5.5 billion = £5500 million. 5500/60 = 92x greater cost

Question 34: C

1 is not true. The decline of car ownership in London has occured at the same time as a reduction in London's road congestion problem, but you cannot infer a causal link. 2 is not true, there is no evidence or suggestion provided that this could be the case. 3 is true. This is heavily suggested by the third paragraph and confirmed by phrases such as 'demand for the bus service has started to decline.'

Question 35: B

To increase the GVA, there must be an increase in the production of goods and services (as per the definition).
If the widening of the congestion charge zone somehow did not increase the production of goods and services, then the proposal would not have the desired effect. With B, there may be less of the public visiting profitable retail areas (deterred by the congestion charge), meaning that the GVA may not increase – hence this is correct.

END OF SECTION

Section 2

Question 1: A

P is the gallbladder, which releases bile.

Q is the stomach, which secretes acidic HCl (thus hydrogen ions) and proteases

R is the pancreas, which secretes insulin and enzymes to digest proteins, fats and carbohydrates (protease, lipase and amylase respectively).

Question 2: E

The anode and cathode are made of copper therefore this will be the reacting species. Remember that oxidation occurs at the anode and reduction occurs at the cathode. Therefore there is the loss of electrons at the anode to generate the copper ion, and gain of electrons at the anode to convert the copper ion into the element.

(OIL RIG = Oxidation Is Loss, Reduction is Gain)

Question 3: A

1 is incorrect; microwaves have a larger (the value is **LESS** negative) wavelength than visible light. They do have a smaller **frequency**, but this is not what is stated.

2 is incorrect; all electromagnetic waves travel at the same speed through a vacuum. (This speed value is the 'c' seen in the equation $E = mc^2$).

3 is incorrect; Gamma rays have the smallest wavelength of any electromagnetic wave (it is the **MOST** negative).

4 is incorrect; X-rays are used in hospital radiography to look for broken bones.

Question 4: F

$$(\sqrt{5} - 2)(\sqrt{5} - 2) = 5 - 2\sqrt{5} - 2\sqrt{5} + 4 = 9 - 4\sqrt{5}$$

Question 5: G

This question is about the conversion of DNA into proteins, which requires both transcription and translation.

Humans have 2 alleles of every gene, one from the mother and one from the father. These can be the same, or different, and can be either recessive or dominant. The particular allele represents the exact DNA code; that is the order of each nucleic acid base (A, C, G, T) in the DNA. This is converted to mRNA by transcription, and then each triplet of bases in the mRNA code is read and converted into a specific chain of amino acids. The exact amino acid encoded, influenced by the order of bases, can influence things like the tertiary structure of the final protein and thus the shape of the active site of the enzyme. In this disease, any of these steps could go wrong (and thus be different to normal) to give a faulty enzyme which is unable to produce healthy white blood cells, therefore all the answers are correct.

Question 6: D

1 contains 16 protons (atomic number), therefore 18 neutrons (to give a mass of 34) and 18 electrons (to give a 2- charge).

2 contains 17 protons, therefore 20 neutrons and 18 electrons.

3 contains 18 protons, therefore 22 neutrons and 18 electrons.

4 contains 19 electrons, therefore 20 neutrons and 18 electrons.

5 contains 20 electrons, therefore 20 neutrons and 20 electrons.

Question 7: F

1 is correct. The rate of evaporation increases with the temperature of the liquid because molecules have more kinetic energy and move faster on average, thus more can escape.

2 is incorrect. Still air would cause the air above the puddle to become saturated. Windier conditions allow more evaporation because there is the maintenance of a gradient between high water concentration and low water concentration above the puddle.

3 is correct. The increased surface area means that more molecules can escape at the same time.

Question 8: A

We want to pick a patient suffering for a migraine twice, so we use AND and times together the 2 probabilities.

The probability of picking the first patient is 5/20; if this occurred the probability of picking the second would be 4/19.

Therefore 5/20 x 4/19 = 20/180 = 1/9

Question 9: E

There is water in all the experiments and glucose in 2, 3 and 4. Even with no/minimal gradient, random entropy means that both water and glucose will move through the partially permeable membrane, and there will be some (~equal and opposite) movement in the other direction to equilibrate.

Question 10: C

This is a classic metal + acid → salt + hydrogen reaction

$Mg + 2HCl → MgCl_2 + H_2$

As the reaction occurs, the magnesium chloride and hydrogen are formed (the latter causing the bubbling) which reduces the concentration of the reactants. Therefore, only 2 is correct. The activation energy is irrelevant (this regards the start of the reaction) and the particles do not have less energy, it is just a concentration issue.

Question 11: C

Remember, in series, the current is always the same at each resistor (hence 4 is true), whereas the total voltage and the total resistance is additive. 2 is correct because using $V = IR$, doubling the resistance at R_1 with an equal current means that the voltage must be doubled here. (In parallel, the voltage is constant and the current is additive.)

Question 12: F

Firstly, you can rule out A/B/C as PS is clearly over 1.8cm!

Thanks to QT and RS being parallel, the traingles PQT and PRS are proportional to one another. That means the ratio of PT:QT is the same as PS:RS. Let's call PT x and the proportion between the sides y.

x = 0.3y and 1.8 + x = 1.5y therefore 1.8 + 0.3y = 1.5y and 1.8 = 1.2y. Thus y is 1.5 and x is 0.45.

PT is 0.45cm therefore PS is 2.25cm.

Question 13: E

1 is true. The diploid cell contained 54 chromosomes therefore the haploid gamete cell must have contained half, which is 27.

2 is true; early embryonic cells need to be able to produce every different cell type in the body from an un-differentiated state, and they are therefore regarded as stem cells.

3 is not true. The gametes that are used at the start must have been produced by meosis to become haploid (and then become diploid again upon fusion).

Question 14: E

1 Fe (0) + $CuCl_2$ (+2; -1 -1) → $FeCl_2$ (+2; -1, -1) + Cu (0)

2 Cu_2O (+1 +1; -2) → Cu (0) + CuO (+2; -2_

3 Cl_2 (0) + H_2O (+1, +1; -2) → HCl (+1; -1) + $HClO$ (+1; +1; -2)

4 $BaCl_2$ (+2; -1, -1) + Na_2SO_4 (+1, +1; +6; -2, -2, -2, -2) → $BaSO_4$ (+2; +6; -2, -2, -2, -2) + $2NaCl$ (+1; -1)

5 Hg_2Cl_2 (+1, +1; -1, -1) → Hg (0) + $HgCl_2$ (+2; -1, -1)

As you can see, there is reduction and oxidation of Cu in 2, Cl in 3 and Hg in 5

Question 15: F

Nuclear fission comes up quite regularly in BMAT questions and you just need to remember how it works.

When an atom of U-238 is exposed to neutron radiation, its nucleus occasionally captures a neutron, making it U-239. It then needs to undergo two β decays which converts 2 neutrons into 2 protons, now giving the atomic number for plutonium.

Question 16: A

Just be careful with using standard form, and simplify your calculations as much as possible (e.g. 3.6/7 → 3.5/7 to make 0.5) and you should be fine. Firstly you can rearrange to get $M = gR^2/G$.

$R^2 = 6 \times 10^6 = 36 \times 10^{12} = 3.6 \times 10^{13}$

$gR^2 = 10 \times 3.6 \times 10^{13} = 3.6 \times 10^{14}$

$gR^2/G = 3.6 \times 10^{14} / 7 \times 10^{11} = \sim 0.5 \times 10^{25} \sim 5 \times 10^{24}$

Question 17: G

This question refers to coronary arteries on the surface of the heart.

1 is not true; glucose cannot freely diffuse from the blood stream, it requires *facilitated* diffusion with specific transporters.

2 is true, arteries carry blood away from the heart at high pressure.

3 is true, there are smooth muscle cells in the tunica media of the arterial vessels to allow vasodilation and vasoconstriction.

Question 18: C

You can count each element on both sides of the equation to see if it is balanced, which works for A-C, whereas D and E have other problems.

A: 7C 12H 5O 1Mg → 6C 12H 5O 1Mg

B: 4C 6H 5O 1Mg → 4C 7H 5O 1Mg

C: 7C 12H 7O 1Mg → 7C 12H 7O 1Mg

D: The formula for propanoic acid is not correct

E: Hopefully intuitively you can see that $Mg_3C_3O_2$ is not a compound likely to form; it would require a C_3O_2 ion to be 6-...

Question 19: E

It takes 0.01s longer for the sound wave to travel to the far wall than the near wall. The microphone is placed next to the sound source, so the sound must travel to the wall and back before it is detected.

Distance to be reflected from near wall: $2 \times 2 = 4$m

Distance to be reflected from far wall: $8 \times 8 = 16$m

Therefore it takes 0.01s to travel an extra 12m. $12/0.01 = 1200$m/s

Question 20: E

Make the denominators the same so that you can add and subtract. Times the top and bottom of the first by x-1, the second by 2x, and the third by 2(x-1).

$$\frac{1}{2x}+\frac{1}{x-1}-\frac{1}{x}=\frac{x-1}{2x(x-1)}+\frac{2x}{2x(x-1)}-\frac{2(x-1)}{2x(x-1)}=\frac{x-1+2x-2x+2}{2x(x-1)}=\frac{x+1}{2x(x-1)}$$

Question 21: D

Males and females are affected similarly, ruling out this being an X-linked inheritance pattern.

A female and male with freckles are able to produce offspring without freckles, showing that it cannot be inherited recessively. Thus it is inherited in an Autosomal Dominant fashion.

Take F to be the allele for freckles and f the allele for no freckles.

Parent 1 is Ff and 2 is ff, so the offspring have a 50% chance of being either Ff (freckles) or ff (no freckles).

Parent 5 is Ff and 6 is Ff, so the offspring have a 25% chance of being FF (freckles), 50% chance of Ff (freckles) or 25% chance of ff (no freckles).

Question 22: B

Mr of hydrated copper(II) sulphate is $64 + 32 + (16 \times 4) + 10 + (16 \times 5) = 250$

Mass/Mr $= 10/250 = 0.04$ moles of hydrated copper sulphate in 100cm^3

Therefore, in 1 dm^3 (which is 1000cm^3), there would be 0.4moles. Conc is 0.400 mol/dm^3.

Question 23: F

Newton's third law states that forces come in pairs, if object 1 exerts a force on object 2, then object 2 exerts an equal and opposite force on object 1. So if the floor is exerting force P on the table, there must be an equal and opposite force which would be the force that the table exerts on the floor.

Question 24: C

Length of the third side of the triangle using Pythagoras's theorem is $\sqrt{(9^2 - 6^2)} = \sqrt{45} = 3\sqrt{5}$

Area of the triangle is ½ x 6 x $3\sqrt{5} = 9\sqrt{5}$

Area of the circle is $\frac{1}{4}\pi6^2 = 9\pi$

Therefore, together $9\pi + 9\sqrt{5}$

Question 25: G

1 is not correct. The gene is inactive at warmer temperatures, but this is not the same as it denaturing. (Remember denaturing is generally irreversible so if this gene denatured, it would never be active again and the cat could never get darker).

2 is correct. In the colder environment, the enzyme is active which causes the coat colour to darken, which is why the extremities of the body are darker.

3 is correct. The gene confers the potential to have a darker coat, and whether the gene is active or not depends on the environment.

Question 26: B

$2Na + H_2O \rightarrow Na_2O + H_2$

Mass/Mr $= 0.23/23 = 0.1$moles of sodium, hence 0.05 mol H_2

Volume $= n \times 24 = 0.05 \times 24 = 0.12$dm^3

Question 27: C

For a straight line graph, y = mx where m is the gradient. y is the kinetic energy and x is the square of the speed which we can sub in below.

KE $= 0.5mv^2$. We have the mass to sub in as 2.5kg, so KE $= 1.25v^2$ and y=1.25x, so the gradient is 1.25.

END OF SECTION

Section 3

'He who has never learned to obey cannot be a good commander'. (Aristotle)

➢ This quote is referring to the quality of leadership. Some may believe that good leaders were 'born to lead' and have always had the relevant characteristics and strengths to become an effective, charismatic leader. However, Aristotle argues that the best of leaders are those who have learnt to lead by developing their attributes and assets based around other successful leaders.

➢ As with any profession or role in life, natural talent is not enough for success. A talented sports player will need to follow the advice and potentially strict rules of their coach if they are to thrive and potentially become a professional. A 'gifted' musician or writer will still need to work hard and learn from other successful pioneers in their profession in order to prosper. Leadership is an art form like any other and without learning from others talented in this art it would be difficult to succeed.

➢ Moreover, if one has not experienced something for themselves, they are less likely to be able to be good at it. For example, it is hard for someone who has never rode a bike to teach someone else, or someone who has not had kids to give advice about mothering. In a similar vein, until one has experienced good leadership themselves and has understood what it feels like to be led well, it would be difficult to truly know whether one's own leadership is good or not. Also, it is important to understand what it feels like from the 'other side'; that is, the empathy allowed by previously being commanded should help leadership.

➢ One could argue that there are other factors that are more important in a leader than learning how to follow, and that these can be developed independently. For example, you could elaborate on some of these points:
  o Ability to inspire confidence and provide direction
  o Planning, organising and setting targets
  o Listening, supporting and giving constructive criticism if relevant
  o Accepting responsibility for mistakes
  o Being assertive and always looking forward
  o Managing time, risk and people well

➢ You could link this quote back to medicine in the conclusion; and although they can't mark you down for an opinion, possibly err on the side of arguing that learning to follow is key to be a good leader. As a doctor, one starts with foundation training where every decision is reviewed and checked by a senior doctor, and where it is vital to follow the experience of someone who as worked for many years more. This 'leader' should hopefully also teach many attributes of the job that cannot be taught from a book at medical school. In order to thrive as a doctor, and then become a good leader in the future as a consultant, one must take on board all the advice that is given to them early on in the job.

The only moral obligation a scientist has is to reveal the truth.

➢ At its most basic level, science can be considered as the pursuit of truth. The scientific method involves systematic observation, experimentation and the formulation, testing and modification of hypotheses; all scientists should use this empirical technique if they are to be successful at their role. This statement suggests that using scientific research in order to advocate for a particular way of thinking, or other roles such as educating the public, are not a moral obligations of a scientist.

➢ It is clear cut that the role of a doctor is to look after their patients, the role of a lawyer is to represent their client in legal matters; similarly, the role of a scientist is considered to be using experimentation to acquire knowledge. One could argue that the role of scientist is solely to reveal the truth, as this is what their expertise is, and then the role of others with varying professions, such as policy makers and politicians, to decide how that truth is applied and used. Moreover, some may believe that advocacy may hurt the credibility of scientists; by having a preconception of how things should be, they may consciously or unconsciously bias their experimentation in a particular way (although strictly speaking, using strong scientific methods should minimise thus, such as the use of double blinding and randomisation in a clinical trial).

➢ There are events in history whereby successful scientists have used their power and privileged position to go beyond their pursuit of truth and harm others. A few examples:
  o Dr Reiter and Dr Wegener (both who have diseases named after them) were German physicians who committed war crimes under the Nazi regime by authorising medical experimentation on concentration camp prisoners

➢ One could argue that a scientist has the moral obligation to advocate for a particular scientific truth, especially when there could be a major danger to society. An example could be the case of global warming. Despite much rigorous scientific data showing that increasing carbon dioxide levels in the are causing atmospheric temperature increases, many politicians and global influencers deny the changes. As such, many countries and governments are not taking steps to reduce their emissions, which will make things worse and may eventually cause destruction of the planet as we know it. If scientists publically came out to promote and support changing our lifestyles to save the Earth it may significantly improve things for the generations below; that is, the potential of making the world a better place means taking action an ethical obligation.

➢ Also, one could argue that a scientist has a moral obligation to educate others and/or lead. They have likely developed a thorough, deep understanding in their area of expertise and it would benefit society for them to spread this knowledge. Alternatively, one could argue that if their research is publically funded, i.e. paid for by the public, they have an obligation to give back to the public via education.

➢ You could conclude by arguing that the baseline moral obligation of a scientist is to reveal the truth and to be a good scientist (especially if this is their sole job description), but that the best and most influential people are those who are able to successful apply (or at least think of how to apply) their knowledge, and those who are willing to educate the public.

The health care profession is wrong to treat ageing as if it were a disease.

➢ The statement is arguing that because ageing is a natural, normal and inevitable process which everyone must go through, whereas disease is an abnormal deviation from the norm which only affects a proportion of the population, we should not equate ageing as a disease. That is, something that is universal cannot be abnormal.

➢ Instead of considering ageing as a disease, it may be more useful for the health care profession to consider ageing as a risk factor for chronic diseases, and then target the specific pathology caused by that disease (which is associated but not solely caused by the ageing).

➢ Another reason to not consider ageing as a disease may be due to its certainty. Some may believe that treating ageing as a disease would lead to the misallocation of limited resources to a futile cause, which may just prolong periods of pain and/or illness.

➢ In contrast: disease is defined as a disorder in structure or function of a bodily system causing harm, and with this definition, ageing does seem to fit the bill (if a healthy adult human is considered 'normal'). For example, ageing is associated with:
  ○ Bones become more brittle, muscle mass decreases, joints degenerate and become less flexible
  ○ Wrinkling of the skin, greying of hair, hearing loss, loss of near vision
  ○ Cognitive decline, potentially leading to dementia, cardiovascular changes, increase risk of cancer
  ○ Inability to control bowel and bladder movements

➢ Moreover, if we treat ageing as a disease that can be treated and prevented, we can use scientific research to target fundamental pathways in this process, and potentially find methods to halt or slow its development. That is, treating it as a disease legitimises medical efforts to try and cure it or eliminate other conditions associated with it. Before the advent of vaccinations and antibiotics, it was rare to live over forty years old; by targeting ageing itself there is the potential to make us even more healthy.

➢ Targeting fundamental ageing mechanisms rather than 'age-related disease' may be beneficial to medicine; it may be futile to fight chronic diseases without striving to first understand their ultimate cause. Like with other diseases, modern experimental techniques have found molecular targets and deleterious changes related to ageing: particular genes involved in the process, and structures at the ends of chromosomes called telomeres which act as biological clocks for human cellular ageing. People's bodies age at different rates according to an interaction between genes and the environment.

➢ The goal of biomedical research is to enable people to be as healthy as possible, for as long as possible. With modern technology and methodology, it should be possible to better delineate the exact pathways involved in ageing. If we found more good evidence that ageing is a preventable or curable process, then it would make a lot of sense to treat it as such and fund and develop procedures to slow the process. However, in contrast, if the evidence suggested that resources were better spent directly targeting specific pathologies in age-related conditions, then it would not be worthwhile to treat ageing as a disease.

END OF PAPER

BMAT Practice Papers

Preparing for the BMAT

Before going any further, it's important that you understand the optimal way to prepare for the BMAT. Rather than jumping straight into doing mock papers, it's essential that you start by understanding the components and the theory behind the BMAT by using an BMAT textbook. Once you've finished the non-timed practice questions, you can progress to past BMAT papers. These are freely available online at **www.uniadmissions.co.uk/bmat-past-papers** and serve as excellent practice. You're strongly advised to use these in combination with the *BMAT Past Worked Solutions* Book so that you can improve your weaknesses. Finally, once you've exhausted past papers, move onto the mock papers in this book.

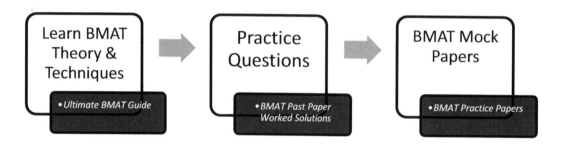

Already seen them all?

So, you've run out of past papers? Well hopefully that is where this book comes in. It contains eight unique mock papers; each compiled by expert BMAT tuors at *UniAdmissions* who scored in the top 10% nationally.

Having successfully gained a place on their course of choice, our tutors are intimately familiar with the BMAT and its associated admission procedures. So, the novel questions presented to you here are of the correct style and difficulty to continue your revision and stretch you to meet the demands of the BMAT.

General Advice

Practice

This is the best way of familiarising yourself with the style of questions and the timing for this section. Although the exam will essentially only test GCSE level knowledge, you are unlikely to be familiar with the style of questions in all sections when you first encounter them. Therefore, you want to be comfortable at using this before you sit the test.

Practising questions will put you at ease and make you more comfortable with the exam. The more comfortable you are, the less you will panic on the test day and the more likely you are to score highly. Initially, work through the questions at your own pace, and spend time carefully reading the questions and looking at any additional data. When it becomes closer to the test, **make sure you practice the questions under exam conditions**.

Repeat Questions

When checking through answers, pay particular attention to questions you have got wrong. If there is a worked answer, look through that carefully until you feel confident that you understand the reasoning, and then repeat the question without help to check that you can do it. If only the answer is given, have another look at the question and try to work out why that answer is correct. This is the best way to learn from your mistakes, and means you are less likely to make similar mistakes when it comes to the test. The same applies for questions which you were unsure of and made an educated guess which was correct, even if you got it right. When working through this book, **make sure you highlight any questions you are unsure of**, this means you know to spend more time looking over them once marked.

Manage your Time:

It is highly likely that you will be juggling your revision alongside your normal school studies. Whilst it is tempting to put your A-levels on the back burner falling behind in your school subjects is not a good idea, don't forget that to meet the conditions of your offer should you get one you will need at least one A*. So, time management is key!

Make sure you set aside a dedicated 90 minutes (and much more closer to the exam) to commit to your revision each day. The key here is not to sacrifice too many of your extracurricular activities, everybody needs some down time, but instead to be efficient. Take a look at our list of top tips for increasing revision efficiency below:

1. Create a comfortable work station
2. Declutter and stay tidy
3. Treat yourself to some nice stationery
4. See if music works for you → if not, find somewhere peaceful and quiet to work
5. Turn off your mobile or at least put it into silent mode
6. Silence social media alerts
7. Keep the TV off and out of sight
8. Stay organised with to do lists and revision timetables – more importantly, stick to them!
9. Keep to your set study times and don't bite off more than you can chew
10. Study while you're commuting
11. Adopt a positive mental attitude
12. Get into a routine
13. Consider forming a study group to focus on the harder exam concepts
14. Plan rest and reward days into your timetable – these are excellent incentive for you to stay on track with your study plans!

Keep Fit & Eat Well:

'A car won't work if you fill it with the wrong fuel' - your body is exactly the same. You cannot hope to perform unless you remain fit and well. The best way to do this is not underestimate the importance of healthy eating. Beige, starchy foods will make you sluggish; instead start the day with a hearty breakfast like porridge. Aim for the recommended 'five a day' intake of fruit/veg and stock up on the oily fish or blueberries – the so called "super foods".

When hitting the books, it's essential to keep your brain hydrated. If you get dehydrated you'll find yourself lethargic and possibly developing a headache, neither of which will do any favours for your revision. Invest in a good water bottle that you know the total volume of and keep sipping through the day. Don't forget that the amount of water you should be aiming to drink varies depending on your mass, so calculate your own personal recommended intake as follows: 30 ml per kg per day.

It is well known that exercise boosts your wellbeing and instils a sense of discipline. All of which will reflect well in your revision. It's well worth devoting half an hour a day to some exercise, get your heart rate up, break a sweat, and get those endorphins flowing.

Sleep

It's no secret that when revising you need to keep well rested. Don't be tempted to stay up late revising as sleep actually plays an important part in consolidating long term memory. Instead aim for a minimum of 7 hours good sleep each night, in a dark room without any glow from electronic appliances. Install flux (https://justgetflux.com) on your laptop to prevent your computer from disrupting your circadian rhythm. Aim to go to bed the same time each night and no hitting snooze on the alarm clock in the morning!

Revision Timetable

Still struggling to get organised? Then try filling in the example revision timetable below, remember to factor in enough time for short breaks, and stick to it! Remember to schedule in several breaks throughout the day and actually use them to do something you enjoy e.g. TV, reading, YouTube etc.

	8AM	10AM	12PM	2PM	4PM	6PM	8PM
MONDAY							
TUESDAY							
WEDNESDAY							
THURSDAY							
FRIDAY							
SATURDAY							
SUNDAY							
EXAMPLE DAY		School			Biology	Critical Thinking	Physics

Top tip! Ensure that you take a watch that can show you the time in seconds into the exam. This will allow you have a much more accurate idea of the time you're spending on a question. In general, if you've spent >150 seconds on a section 1 question or >90 seconds on a section 2 questions – move on regardless of how close you think you are to solving it.

Getting the most out of Mock Papers

Mock exams can prove invaluable if tackled correctly. Not only do they encourage you to start revision earlier, they also allow you to **practice and perfect your revision technique**. They are often the best way of improving your knowledge base or reinforcing what you have learnt. Probably the best reason for attempting mock papers is to familiarise yourself with the exam conditions of the BMAT as they are particularly tough.

Start Revision Earlier

Thirty five percent of students agree that they procrastinate to a degree that is detrimental to their exam performance. This is partly explained by the fact that they often seem a long way in the future. In the scientific literature this is well recognised, Dr. Piers Steel, an expert on the field of motivation states that *'the further away an event is, the less impact it has on your decisions'*.

Mock exams are therefore a way of giving you a target to work towards and motivate you in the run up to the real thing – every time you do one treat it as the real deal! If you do well then it's a reassuring sign; if you do poorly then it will motivate you to work harder (and earlier!).

Practice and perfect revision techniques

In case you haven't realised already, revision is a skill all to itself, and can take some time to learn. For example, the most common revision techniques including **highlighting and/or re-reading are quite ineffective** ways of committing things to memory. Unless you are thinking critically about something you are much less likely to remember it or indeed understand it.

Mock exams, therefore allow you to test your revision strategies as you go along. Try spacing out your revision sessions so you have time to forget what you have learnt in-between. This may sound counterintuitive but the second time you remember it for longer. Try teaching another student what you have learnt, this forces you to structure the information in a logical way that may aid memory. Always try to question what you have learnt and appraise its validity. Not only does this aid memory but it is also a useful skill for BMAT section 3, Oxbridge interview, and beyond.

Improve your knowledge

The act of applying what you have learnt reinforces that piece of knowledge. A question may ask you to think about a relatively basic concept in a novel way (not cited in textbooks), and so deepen your understanding. Exams rarely test word for word what is in the syllabus, so when running through mock papers try to understand how the basic facts are applied and tested in the exam. As you go through the mocks or past papers take note of your performance and see if you consistently under-perform in specific areas, thus highlighting areas for future study.

Get familiar with exam conditions

Pressure can cause all sorts of trouble for even the most brilliant students. The BMAT is a particularly time pressured exam with high stakes – your future (without exaggerating) does depend on your result to a great extent. The real key to the BMAT is overcoming this pressure and remaining calm to allow you to think efficiently.

Mock exams are therefore an excellent opportunity to devise and perfect your own exam techniques to beat the pressure and meet the demands of the exam. **Don't treat mock exams like practice questions – it's imperative you do them under time conditions.**

> ***Remember!*** It's better that you make all the mistakes you possibly can now in mock papers and then learn from them so as not to repeat them in the real exam.

Before using this Book

Do the ground work

➢ Read in detail: the background, methods, and aims of the BMAT as well logistical considerations such as how to take the BMAT in practice. A good place to start is a BMAT textbook like *The Ultimate BMAT Guide* (flick to the back to get a free copy!) which covers all the groundwork but it's also worth looking through the official BMAT site (www.admissionstesting.org/bmat).

➢ It is generally a good idea to start re-capping all your GCSE maths and science.

➢ Practice substituting formulas together to reach a more useful one expressing known variables e.g. $P = IV$ and $V = IR$ can be combined to give $P = V^2/R$ and $P = I^2R$. Remember that calculators are not permitted in the exam, so get comfortable doing more complex long addition, multiplication, division, and subtraction.

➢ Get comfortable rapidly converting between percentages, decimals, and fractions.

➢ These are all things which are easiest to do alongside your revision for exams before the summer break. Not only gaining a head start on your BMAT revision but also complimenting your year 12 studies well.

➢ Discuss scientific problems with others - propose experiments and state what you think the result would be. Be ready to defend your argument. This will rapidly build your scientific understanding for section 2 but also prepare you well for an oxbridge interview.

➢ Read through the BMAT syllabus before you start tackling whole papers. This is absolutely essential. It contains several stated formulae, constants, and facts that you are expected to apply - or may just be an answer in their own right. Familiarising yourself with the syllabus is also a quick way of teaching yourself the additional information other exam boards may learn which you do not. Sifting through the whole BMAT syllabus is a time-consuming process so we have done it for you. **Be sure to flick through the syllabus checklist** later on, which also doubles up as a great revision aid for the night before!

Ease in gently

With the ground work laid, there's still no point in adopting exam conditions straight away. Instead invest in a beginner's guide to the BMAT, which will not only describe in detail the background and theory of the exam, but take you through section by section what is expected. *The Ultimate BMAT Guide: 800 Practice Questions* is the most popular BMAT textbook – you can get a free copy by flicking to the back of this book.

When you are ready to move on to past papers, take your time and puzzle your way through all the questions. Really try to understand solutions. A past paper question won't be repeated in your real exam, so don't rote learn methods or facts. Instead, focus on applying prior knowledge to formulate your own approach.

If you're really struggling and have to take a sneak peek at the answers, then practice thinking of alternative solutions, or arguments for essays. It is unlikely that your answer will be more elegant or succinct than the model answer, but it is still a good task for encouraging creativity with your thinking. Get used to thinking outside the box!

Accelerate and Intensify

Start adopting exam conditions after you've done two past papers. Don't forget that **it's the time pressure that makes the BMAT hard** – if you had as long as you wanted to sit the exam you would probably get 100%. If you're struggling to find comprehensive answers to past papers then *BMAT Past Papers Worked Solutions* contains detailed explained answers to every BMAT past paper question and essay (flick to the back to get a free copy).

Doing all the past papers from 2009 – present is a good target for your revision. Note that the BMAT syllabus changed in 2009 so questions before this date may no longer be relevant. In any case, choose a paper and proceed with strict exam conditions. Take a short break and then mark your answers before reviewing your progress. For revision purposes, as you go along, keep track of those questions that you guess – these are equally as important to review as those you get wrong.

Once you've exhausted all the past papers, move on to tackling the unique mock papers in this book. In general, you should aim to complete one to two mock papers every night in the ten days preceding your exam.

How to use Practice papers

If you have done everything this book has described so far then you should be well equipped to meet the demands of the BMAT, and therefore **the mock papers in the rest of this book should ONLY be completed under exam conditions**.

This means:

➢ Absolute silence – no TV or music
➢ Absolute focus – no distractions such as eating your dinner
➢ Strict time constraints – no pausing half way through
➢ No checking the answers as you go
➢ Give yourself a maximum of three minutes between sections – keep the pressure up
➢ Complete the entire paper before marking
➢ Mark harshly

In practice this means setting aside two hours in an evening to find a quiet spot without interruptions and tackle the paper. Completing one mock paper every evening in the week running up to the exam would be an ideal target.

➢ Tackle the paper as you would in the exam.
➢ Return to mark your answers, but mark harshly if there's any ambiguity.
➢ Highlight any areas of concern.
➢ If warranted read up on the areas you felt you underperformed to reinforce your knowledge.
➢ If you inadvertently learnt anything new by muddling through a question, go and tell somebody about it to reinforce what you've discovered.

Finally relax… the BMAT is an exhausting exam, concentrating so hard continually for two hours will take its toll. So, being able to relax and switch off is essential to keep yourself sharp for exam day! Make sure you reward yourself after you finish marking your exam.

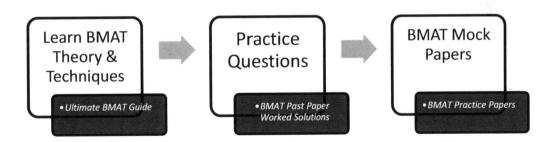

Scoring Tables

SECTION 1	1st Attempt	2nd Attempt	3rd Attempt
Mock A			
Mock B			
Mock C			
Mock D			
Mock E			
Mock F			
Mock G			
Mock H			

SECTION 2	1st Attempt	2nd Attempt	3rd Attempt
Mock A			
Mock B			
Mock C			
Mock D			
Mock E			
Mock F			
Mock G			
Mock H			

Fortunately for our mock papers our tutors have compiled model answers for you to compare your essays against! If you're repeating a mock paper, its best to attempt a different essay title to give yourself maximum experience with the various styles of BMAT essays.

SECTION 3	Essay 1	Essay 2	Essay 3	Essay 4
Mock A				
Mock B				
Mock C				
Mock D				
Mock E				
Mock F				
Mock G				
Mock H				

Mock Paper A

Section 1

Question 1:

A square sheet of paper is 20cms long. How many times must it be folded in half before it covers an area of 12.5cm^2?

A) 3   B) 4   C) 5   D) 6   E) 7

Question 2:

Mountain climbing is viewed by some as an extreme sport, while for others it is simply an exhilarating pastime that offers the ultimate challenge of strength, endurance, and sacrifice. It can be highly dangerous, even fatal, especially when the climber is out of his or her depth, or simply gets overwhelmed by weather, terrain, ice, or other dangers of the mountain. Inexperience, poor planning, and inadequate equipment can all contribute to injury or death, so knowing what to do right matters.

Despite all the negatives, when done right, mountain climbing is an exciting, exhilarating, and rewarding experience. This article is an overview beginner's guide and outlines the initial basics to learn. Each step is deserving of an article in its own right, and entire tomes have been written on climbing mountains, so you're advised to spend a good deal of your beginner's learning immersed in reading widely. This basic overview will give you an idea of what is involved in a climb.

Which statement best summarises this paragraph?
A)   Mountain climbing is an extreme sport fraught with dangers.
B)   Without extensive experience embarking on a mountain climb is fatal.
C)   A comprehensive literature search is the key to enjoying mountain climbing.
D)   Mountain climbing is difficult and is a skill that matures with age if pursued.
E)   The terrain is the biggest unknown when climbing a mountain and therefore presents the biggest danger.

Question 3:

50% of an isolated population contract a new strain of resistant Malaria. Only 20% are symptomatic of which 10% are female. What percentage of the total population do symptomatic males represent?

A)   1%   B)   9%   C)   10%   D)   80%

Question 4:

John is a UK citizen yet is looking to buy a holiday home in the South of France. He is purchasing his new home through an agency. Unlike a normal estate agent, they offer monthly discount sales of up to 30%. As a French company, the agency sells in Euros. John decides to hold off on his purchase until the sale in the interest of saving money. What is the major assumption made in doing this?

A)   The house he likes will not be bought in the meantime.
B)   The agency will not be declared bankrupt.
C)   The value of the pound will fall more than 30%.
D)   The value of the pound will fall less than 30%.
E)   The value of the euro may increase by up to 35% in the coming weeks.

Question 5:

In childcare professions, by law, there must be an adult to child ratio of no more than 1:4. Child minders are hired on a salary of £8.50 an hour. What is the maximum number of children that can be continually supervised for a period of 24 hours on a budget of £1,000?

A)   1   B)   8   C)   12   D)   16   E)   468

Question 6:

A table of admission prices for the local cinema is shown below:

	Peak	Off-peak
Adult	£11	£9.50
Child	£7	£5.50
Concession	£7	£5.50
Student	£5	£5

How much would a group of 3 adults, 5 children, a concession and 4 students save by visiting at an off-peak time rather than a peak time?

A) £11.50   B) £13.50   C) £15.50   D) £17.50   E) £18.50

Question 7:

All musicians play instruments. All oboe players are musicians. Oboes and pianos are instruments. Karen is a musician. Which statement is true?

A) Karen plays two instruments.
B) All musicians are oboe players.
C) All instruments are pianos or oboes.
D) Karen is an oboe player.
E) None of the above.

Question 8:

Flow mediated dilatation is a method used to assess vascular function within the body. It essentially adopts the use of an ultrasound scan to measure the percentage increase in the width of an artery before and after occlusion with a blood pressure cuff. Ultrasound scans are taken by one sonographer, and the average lumen diameter is then measured by an analyst. What is a potential flaw in the methodology of this technique?

A) Results will not be comparable within an individual if different arteries start at different diameters.
B) Results will not be comparable between individuals if they have different baseline arterial diameters.
C) Ultrasound is an outdated technique with no use in modern medicine.
D) This methodology is subject to human error.
E) This methodology is not repeatable.

Question 9:

If it takes 20 minutes to board an aeroplane, 15 minutes to disembark and the flight lasts two and a half hours. In the event of a delay it is not uncommon to add 20 minutes to the flight time. Megan is catching the flight in question as she needs to attend a meeting at 5pm. The location of the meeting is 15 minutes from the airport without traffic; 25 minutes with. Which of the following statements is valid considering this information?

A) If Megan wants to be on time for her meeting, given all possibilities described, the latest she can begin boarding at the departure airport is 1.30pm.
B) If Megan starts boarding at 1.40pm she will certainly be late.
C) If Megan aims to start boarding at 1.10pm she will arrive in time whether the plane is delayed or not.
D) If Megan wishes to be on time she doesn't have to worry about the plane being delayed as she can make up the time during the transport time from the arrival airport to the meeting.

Question 10:

A cask of whiskey holds a total volume of 500L. Every two and a half minutes half of the total volume is collected and discarded. How many minutes will it take for the entire cask to be emptied?

A) 80   B) 160   C) 200   D) 240   E) ∞

Question 11:

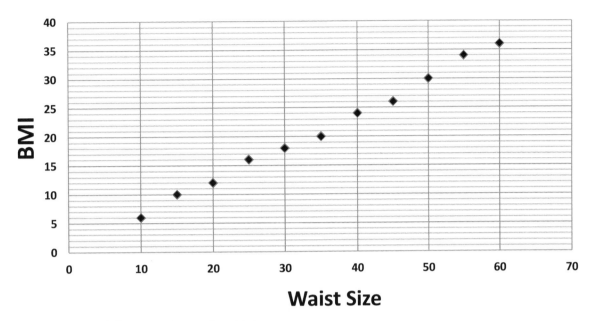

Scatter plot of Waist size vs BMI for some people

B) Having a larger BMI causes an increase in waist size.

C) Waist size is reciprocal to BMI.

D) No conclusions can be drawn from this graph.

E) None of the above are correct.

Question 12:

B is right of A. C is left of B. D is in front of C. E is in front of B. Where is D in relation to E?

A) D is behind E.   D) D is to the left of E.

B) E is behind D.   E) E is to the left of D.

C) D is to the right of E.

Question 13:

Car A has a fuel tank capacity of 30 gallons and achieves 40mpg. Car B on the other hand has a fuel tank capacity of 50 gallons but only achieves 30mpg. Both cars drive until they run out of fuel. If car A starts with a full tank of petrol and travels 200 miles further than car B, how full was car B's fuel tank?

A) 1/5   B) 1/4   C) 1/3   D) 1/2   E) 2/3

The information below relates to questions 14 and 15:

The art of change ringing adopts the use of 6 bells, numbered 1 to 6 in order of weight (1 being the lightest). Initially the bells are rung in this order: 1, 2, 3, 4, 5, 6 however the aim is to ultimately ring all the possible combinations of a 6-number sequence. The rules for doing this are very simple: each bell can only move a maximum of one place in the sequence every time it rings.

Question 14:

What is the total possible number of permutations of 6 bells?

A) 160   B) 220   C) 660   D) 720   E) 1160

Question 15:

Based on the information provided which of the following could be a possible series of bell sequences?

A) 1 2 3 4 5
    2 1 4 3 5
    2 3 1 5 4

B) 1 2 3 4 5
    2 1 4 3 5
    2 4 1 5 3

C) 1 2 3 4 5
    4 2 1 3 5
    4 2 3 1 5

D) 1 2 3 4 5
    1 4 3 2 5
    1 2 3 4 5

E) 1 2 3 4 5
    4 1 3 2 5
    5 3 1 2 4

Question 16:

The keypad to a safe comprises the digits 1 - 9. The code itself can be of indeterminate length. The code is therefore set by choosing a reference number so that when a code is entered the average of all the numbers entered must equal the chosen reference number.

Which of the following is true?

A)   If the reference number was set greater than 9, the safe would be locked forever.
B)   This safe is extremely insecure as if random digits were pressed for long enough it would average out at the correct reference number.
C)   More than one number is always required to achieve the reference number.
D)   All of the above are true.
E)   None of the above are true.

Question 17:

The use of antibiotics is one of the major paradoxes in modern medicine. Antibiotics themselves provide a selection pressure to drive the evolution of antibiotic resistant strains of bacteria. This is largely due to the rapid growth rate of bacterial colonies and asexual cell division. As such a widespread initiative is in place to limit the prescription of antibiotics.
Which of the following is a fair assumption?

A)   Antibiotic resistance is impossible to avoid as it is driven by evolution.
B)   If bacteria reproduced at a slower rate antibiotic resistance would not be such an issue.
C)   Medicine always creates more problems than it solves.
D)   In the past antibiotics were used frivolously.
E)   All of the above could be possible.

The information below relates to questions 18 – 22:

The Spaghetti Bolognese recipe below serves 10 people and each portion contains 300 kcal.

➤ 1kg mince
➤ 220g pancetta, diced
➤ 30g crushed garlic

➤ 1kg tinned tomatoes
➤ 300g diced onions
➤ 300g sliced mushrooms

➤ 200g grated cheese

Question 18:
What quantity of cheese is required to prepare a meal for 350 people?

A)   0.7kg   B)   7kg   C)   70kg   D)   700kg   E)   7000kg

Question 19:

If 12 portions represent 120% of an individual's recommended calorific intake, what is that individuals recommended calorific intake?

A)   2,600kcal   B)   2,800kcal   C)   3,000kcal   D)   3,200kcal   E)   3,400kcal

Question 20:

The recommended ratio of pasta to Bolognese is 4:1. If cooking for 30 people how much pasta should be used?

A)   30.3kg   B)   36.6kg   C)   42.9kg   D)   49.2kg   E)   55.5kg

Question 21:

What is the ratio of onions to the rest of the ingredients if garlic and pancetta are ignored?

A) 1/2.05   B) 1/3.9   C) 1/6.7   D) 1/9.3   E) 1/10

Question 22:

It takes 4 minutes to prepare the ingredients per portion, and a further 8 minutes per portion to cook. Simon has ample preparation space but is limited to cooking 8 portions at a time. What is the shortest period of time it would take him to turn all the ingredients into a meal for 25 people, assuming he didn't start cooking until all the ingredients were prepared?

A) 3 hours   B) 3 hours 40   C) 4 hours   D) 4 hours 40   E) 5 hours

Question 23:

A company sells custom design t-shirts. A breakdown of their costs is shown below:

Number of Items	Cost per Item	
	Black and white	**Colour**
0 – 99	£3.00	£5.00
100 - 499	£2.50	£4.50
500 - 999	£2.00	£4.00
1000+	£1.00	£3.00

Customers with a never before printed design must also pay a surcharge of £50 to cover the cost of building a jig. What is the total cost for an order of unique stag do t-shirts: 50 in colour, and 200 in black and white?

A) £650   B) £700   C) £750   D) £800   E) £850

Question 24:

The Scouts is a movement for young people first established by Lord Baden Powell. As the founder he was the first chief scout of the association. Since his initial appointment there have been a number of notable chief scouts including Peter Duncan and Bear Grylls. Some of the first camping trips conducted by Lord Powell's scout troop were on Brown Sea Island.

Now the Scout movement is a worldwide global phenomenon giving children from all backgrounds the opportunity not only to embark upon adventure but also to engage in the understanding and teaching of foreign culture. Traditionally religion formed the back bone of the scouting movement which was reflected in the scouts promise: "I promise to do my duty to god and to the queen".

Which of the following applies to the scout movement?

A)   Scouts work for the Queen.
B)   The scout network is aimed at adventurous individuals.
C)   Chief scout is appointed by the Queen.
D)   You have to be religious to be a scout.
E)   None of the above.

Question 25:

Three rats are placed in a maze that is in the shape of an equilateral triangle. They pick a direction at random and walk along the side of a triangle. Sophie thinks they are less likely to collide than not. Is she correct?

A)   Yes, because mice naturally keep away from each other.
B)   No, they are more likely to collide than not.
C)   No, they are equally likely to collide than not collide.
D)   Yes, because the probability they collide is 0.25.
E)   None of the above.

Question 26:

The use of human cadavers in the teaching of anatomy is hotly debated. Whilst many argue that it is an invaluable teaching resource, demonstrating far more than a text book can. Others describe how it is an outdated method which puts unfair stress on an already bereaved family. One of the biggest pros for using human tissue in anatomical teaching is the variation that it displays. Whilst textbooks demonstrate a standard model averaged over many 100s of specimens, many argue that it is the variation between cadavers that really reinforces anatomical knowledge.

The opposition argues that it is a cruel process that damages the grieving process of the effected family. For the use of the cadaver often occupies a period of up to 12 months. As such the relative in question is returned to the bereaved family for burial around the time it would be expected that they were recovering as described in the grieving model.

Does the article support or reject the use of cadavers in anatomical teaching?

A) Supports the use
B) Rejects the use
C) Impartial
D) Can't tell
E) None of the above

Question 27:

A ferry is carrying its full capacity. At the time of departure (7am) the travel time to the nearest hour is announced as 13 hours. What is the latest that the ferry could arrive at its destination?

A) 08.00   B) 20.00   C) 20.29   D) 20.30   E) 20.50

Question 28:

A game is played using a circle of 55 stepping stones. A die is rolled showing the numbers 1 - 6. The number on the die tells you how many steps you may take during your go. The only rule is that during your go you must take your steps in the routine two steps forward, 1 step back.

What is the minimum number or rolls required to win?

A) 28   B) 55   C) 110   D) 165   E) 200

Question 29:

On a race track there are 3 cars recording average lap times of 40 seconds, 60 seconds, and 70 seconds. They all started simultaneously 4 minutes ago. How much longer will the race need to continue for them to all cross the start line again at the same time?

A) 23.33 hours   B) 46.67 hours   C) 60.00 hours   D) 83.33 hours   E) 106.67 hours

Question 30:

A class of 60 2nd year medical students are conducting an experiment to measure the velocity of nerve conduction along their radial arteries. This work builds on a previous result obtained demonstrating the effects of how right-handed men have faster nerve conduction velocities than gender matched left handed individuals. 60% of the class are female of which 3% were unable to take part due to underlying heart conditions. 2 of the male members of the class were also unable to take part. On average the female cohort had faster nerve conduction velocities than men in their dominant arm.

Right handed women have the fastest nerve conduction velocities.

A) True   B) False   C) Can't tell

Question 31:

Mark is making a double tetrahedron dice by joining two square based pyramids together at their bases. Each square based pyramid is 5cm wide and 8cm tall. What area of card would have been required to produce the nets for the whole die?

A) 150cm^2 B) 180 cm^2 C) 210 cm^2 D) 240 cm^2 E) 270 cm^2

Question 32:

A serial dilution is performed by lining up 10 wells and filling each one with 9ml of distilled water. 1 ml of a concentrated solvent is then added to the first well and mixed. 1 ml of this new solution is drawn from the first well and added to the second and mixed. The process is repeated until all 10 wells have been used.

If the solvent starts off at concentration x, what will its final concentration be after 10 wells of serial dilution?

A) $x/10^9$ B) $x/10^{10}$ C) $x/10^{11}$ D) $x/10^{12}$ E) $x/10^{13}$

Question 33:

A student decides to measure the volume of all the blood in his body. He does this by injecting a known quantity of substrate into his arm, waiting a period of 20 minutes, then drawing a blood sample and measuring the concentration of the substrate in his blood. What assumption has he made here?

A)   The substrate is only soluble in blood.
B)   The substrate is not bioavailable.
C)   The substrate is not excreted.
D)   The substrate is not degraded.
E)   All of the above.

Question 34:

Jason is ordering a buffet for a party. The buffet company can provide a basic spread at £10 per head. However more luxurious items carry a surcharge. Jason is particularly interested in cup-cakes and shell fish. With these items included the buffet company provides a new quote of £10 per head. In addition to simply ordering the food Jason must also purchase cutlery and plates. Plates come in packs of 20 for £8 whilst cutlery is sold in bundles of 60 sets for £10.

With a budget of £2,300 (to the nearest 10 people) what is the maximum number of people Jason can provide food on a plate for?

A)   180   B)   190   C)   209   D)   210   E)   220

Question 35:

What were once methods of hunting have now become popular sports. Examples include archery, the javelin throw, the discus throw and even throwing a boomerang. Why such dangerous hobbies have begun to thrive is now being investigated by social scientists. One such explanation is that it is because they are dangerous we find them appealing in the first place. Others argue that it is a throwback to our ancestral heritage, where as a hunter gatherer being a proficient hunter was something to show off and flaunt. Whilst this may be the case it is well observed that many find the chase of a hunt exciting if not controversial.

Sports like archery provide excitement analogous to that of the chase during a hunter gatherer hunt.

A)   True
B)   False
C)   Can't tell

END OF SECTION

Section 2

Question 1:

A crocodile's tail weighs 30kg. Its head weighs as much as the tail and one half of the body and legs. The body and legs together weigh as much as the tail and head combined.

What is the total weight of the crocodile?

A) 220kg   B) 240kg   C) 260kg   D) 280kg   E) 300kg

Question 2:

A body is travelling at x ms^{-1} with y J of kinetic energy. After a period of retardation the kinetic energy of the body is $1/16y$. Assuming that the mass of the body has remained constant what is its new velocity?

A) $1/196x$ B) $1/16x$ C) $1/8x$ D) $1/4x$ E) $4x$

Question 3:

Which of the following cannot be classified as an organ?

1. Blood   3. Larynx   5. Prostate   7. Skin
2. Bone   4. Pituitary Gland   6. Skeletal Muscle

A) 1 and 6   B) 2 and 3   C) 5 and 7   D) 1 and 5   E) 1,4, 5 and 6

Question 4:

An increase in aerobic respiratory rate could be associated with which of the following physiological changes?

1.   A larger percentage of water vapour in expired air   4.   Perspiration
2.   Increased expired CO_2 5.   Vasodilatation
3.   Increased inspired O_2

A) 3 only   C) 1, 2 and 3 only   E) All of the above
B) 1 and 2 only   D) 2, 3 and 5

Question 5:

The nephron is to the kidney, as the _____ is to striated muscle:

A) Actin filament   D) Sarcomere
B) Artery   E) Vein
C) Myofibril

Question 6:

A diabetic patient's glucagon and insulin levels are measured over 4 hours. During this time the patient is given two large boluses of glucose. A graphical representation of this is shown below.

At which times would you expect the patients' blood glucose to be greatest?

A)   05:00 and 12:00
B)   07:00 and 14.00
C)   08:00 and 15:00
D)   10:00 and 13:00
E)   06:00, 10:00 and 16:00

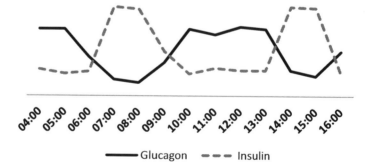

Question 7:

In addition to the A, B or O classification, blood groups can also be distinguished by the presence of Rhesus antigen (Rh). Care must be taken in blood transfusion as once blood types are mixed a Rh -ve individual will mount an immune response against Rh +ve blood. This is particular well exemplified in haemolytic disease of the newborn – where a Rh-ve mother carries a Rh+ve foetus.

Applying what is written here and your knowledge of the human immune system, explain why the mother's first child would be relatively safe and unaffected, yet further offspring would be at high risk.

A)   The first pregnancy is always such a shock to the body it compromises the immune system.

B)   Antibodies take longer than 9 months to produce and mature to an active state.

C)   First born children are immunologically privileged.

D)   There is a high risk of haemorrhage to both mother and child during birth.

E)   Plasma T cells require time to multiply to lethal levels.

Question 8:

The pH of a solution has the greatest effect on which type of interaction?

A)   Van der Waals

B)   Induced dipole

C)   Ionic bonding

D)   Metallic interaction

E)   Hydrogen bonding

Question 9:

At present a large effort is being made to produce tailored patient care. One of the ultimate goals of this is to be able to grow personal, genetically identical organs for those with end stage organ failure. This process will first require the harbouring of what cell type?

A)   Cells from the organ that is failing

B)   Haematopoietic stem cells

C)   Embryonic stem cells

D)   Adult stem cells

E)   All of the above

Question 10:

When comparing different isotopes of the same element, which of the following may change?

1.   Atomic number

2.   Mass number

3.   Number of electrons

4.   Chemical reactivity

A)   1 only

B)   1 and 2 only

C)   3 only

D)   2 and 3 only

E)   All of the above

Question 11:

From which of the following elemental groups are you most likely to find a catalyst?

A)   Alkali Metals

B)   d-block elements

C)   Alkaline Earth Metals

D)   Noble Gases

E)   Halogens

Question 12:

1.338kg of francium are mixed in a reaction vessel with an excess of distilled water. What volume will the hydrogen produced occupy at room temperature and pressure?

A) $20.4dm^3$ B) $36dm^3$ C) $40.8dm^3$ D) $60.12dm^3$ E) $72dm^3$

Question 13:

The composition of a compound is Carbon 30%, Hydrogen 40%, Fluorine 20%, and Chlorine 10%.
What is the empirical formula of this compound?

A) CH_2FCl

B) $C_3H_2F_2Cl$

C) C_3H_4FCl

D) $C_3H_4F_2Cl$

E) $C_4H_4F_2Cl$

Question 14:

What is the actual molecular formula of the compound in question 13 if the M_r is 340.5?

A) $C_3H_4F_2Cl$

B) $C_6H_8F_4Cl_2$

C) $C_9H_{12}F_6Cl_3$

D) $C_{12}H_{16}F_8Cl_4$

E) $C_{15}H_{20}F_{10}Cl_5$

Question 15:

1.2×10^{10} kg of sugar is dissolved in 4×10^{12} L of distilled water. What is the concentration?

A) 3×10^{-2} g/dL

B) 3×10^{-1} g/dL

C) 3×10^{1} g/dL

D) 3×10^{2} g/dL

E) 3×10^{3} g/dL

Question 16:

Which of the following is not essential for the progression of an exothermic chemical reaction?

A) Presence of a catalyst

B) Increase in entropy

C) Achieving activation energy

D) Attaining an electron configuration more closely resembling that of a noble gas

E) None of the above

Question 17:

What is a common use of cationic surfactants?

A) Shampoo

B) Lubricant

C) Cosmetics

D) Detergents

E) All of the above

Question 18:

Which of the following is a unit equivalent to the Volt?

A) $A.\Omega^{-1}$

B) $J.C^{-1}$

C) $W.s^{-1}$

D) $C.s$

E) $W.C.\Omega$

Question 19:

Complete the sentence below:

A voltmeter is connected in _____ and therefore has _____ resistance; whereas an ammeter is connected in _____ and has _____ resistance.

A) Parallel, zero, parallel, infinite

B) Parallel, zero, series, infinite

C) Parallel, infinite, series, zero

D) Series, zero, parallel, infinite

E) Series, infinite, parallel, zero

Question 20:

A body "A" of mass 12kg travelling at 15m/s undergoes inelastic collision with a fixed, stationary object "B" of mass 20kg over a period of 0.5 seconds. After the collision body A has a new velocity of 3m/s. What force must have been dissipated during the collision?

A) 288N

B) 298N

C) 308N

D) 318N

E) 328N

Question 21:

What process is illustrated here: $^{14}_{6}C \rightarrow\ ^{14}_{7}N + x$

A) Thermal decomposition

B) Alpha decay

C) Beta decay

D) Gamma decay

Question 22:

A radio dish is broadcasting messages into deep space on a 20 Hz radio frequency of wavelength 3km. With every hour how much further does the signal travel into deep space?

A) 200,000 km

B) 216,000 km

C) 232,000 km

D) 248,000 km

E) 264,000 km

Question 23:

A formula: $\sqrt[3]{\dfrac{z(x+y)(l+m-n)}{3}}$ is given. Would you expect this formula to calculate:

A)   A length   C)   A volume   E)   A geometric average

B)   An area   D)   A volume of rotation

Question 24:

Evaluate the following: $\dfrac{4.2 \times 10^{10} - 4.2 \times 10^{6}}{2 \times 10^{3}}$

A) 2.09979×10^{6} C) 2.09979×10^{8} E) 2.09979×10^{10}

B) 2.09979×10^{7} D) 2.09979×10^{9}

Question 25:

Calculate a – b

A)   0°
B)   5°
C)   10°
D)   15°
E)   20°

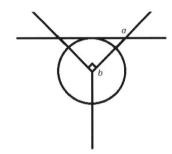

Question 26:

Jack has a bag with a complete set of snooker balls (15 red, 1 yellow, 1 green, 1 brown, 1 blue, 1 pink and 1 black ball) within it. Blindfolded Jack draws two balls from the bag.
What is the probability that he draws a blue and a black ball in any order?

A)   2/41   B)   2/210   C)   1/210   D)   1/105   E)   2/441

Question 27:

An experiment is repeated using an identical methodology and upon further review it is proven to demonstrate identical scientific practice. If the result obtained is different to the first, this would be due to:

A)   Calibration Bias   D)   Serial dilution

B)   Systematic Bias   E)   Inaccuracies in the methodology

C)   Random Chance

END OF SECTION

Section 3

1)   *Doctors should be wearing white coats as it helps produce a placebo effect making the treatment more effective.*

Explain what is meant by this statement. Argue to the contrary. To what extend do you agree with the statement? What points can you see that contradict this statement?

2)   *"Medicine is a science of uncertainty and an art of probability."*

William Osler

Explain what this statement means. Argue to the contrary. To what extent do you agree with the statement?

3)   *"The New England Journal of Medicine reports that 9 out of 10 doctors agree that 1 out of 10 doctors is an idiot."*

Jay Leno

What do you understand by this statement? Explain why the assumption above may be inaccurate and argue to the contrary.

4)   *"My father was a research scientist in tropical medicine, so I always assumed I would be a scientist, too. I felt that medicine was too vague and inexact, so I chose physics."*

Stephen Hawking

Explain what this statement means. Argue to the contrary. To what extent do you agree with the statement?

END OF PAPER

Mock Paper B

Section 1

Question 1:

"If vaccinations are now compulsory because society has decided that they should be forced, then society should pay for them." Which of the following statements would weaken the argument?

A) Many people disagree that vaccinations should be compulsory.
B) The cost of vaccinations is too high to be funded locally.
C) Vaccinations are supported by many local communities and GPs.
D) Healthcare workers do not want vaccinations.
E) None of the above

Question 2:

Josh is painting the outside walls of his house. The paint he has chosen is sold only in 10L tins. Each tin costs £4.99. Assuming a litre of paint covers an area of 5m², and the total surface area of Josh's outside walls is 1050m²; what is the total cost of the paint required if Josh wants to apply 3 coats?

A)   £104.79   B)   £209.58   C)   £314.37   D)   £419.16   E)   £523.95

Question 3:

The stars of the night sky have remained unchanged for many hundreds of years, which allows sailors to navigate using the North Star still to this day. However, this only applies within the northern hemisphere as the populations of the southern hemisphere are subject to an alternative night sky.

An asterism can be used to locate the North Star, it comes by many names including the plough, the saucepan, and the big dipper. Whilst the North Star's position remains fixed in the sky (allowing it to point north reliably always) the rest of the stars traverse around the North Star in a singular motion. In a very long time, the North Star will one day move from its location due to the movement of the Earth.

Which of the following is **NOT** an assumption made in this argument?

A) The Earth is rotating on its axis.
B) Sailors still have need to navigate using the stars.
C) An analogous southern star is used to navigate in the Southern hemisphere.
D) The plough is not the only method of locating the North Star.
E) None of the above.

Question 4:

John wishes to deposit a cheque. The bank's opening times are 9am until 5pm Monday to Friday, 10am until 4pm on Saturdays, and the bank is closed on Sundays. It takes on average 42 bank hours for the money from a cheque to become available.

If John needs the money by 8pm Tuesday, what is the latest he can cash the cheque?

A) 5pm the Saturday before
B) 5pm the Friday before
C) 1pm the Thursday before
D) 1pm the Wednesday before
E) 9am the Tuesday before

Question 5:

How many different diamonds are there in the image shown to the right?

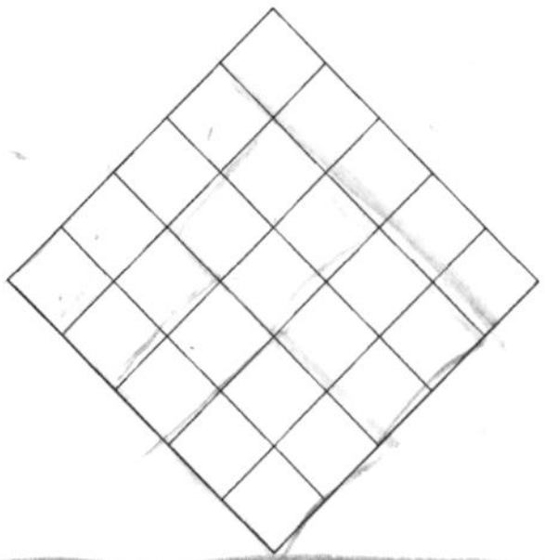

A) 25
B) 32
C) 48
D) 58
E) 63

(handwritten: 25)

Question 6:

In 4 years time I will be one third the age that my brother will be next year. In 20 years time he will be double my age. How old am I?

(handwritten: simultaneous equations / system of equations)

A) 4
B) 9
C) 15
D) 17
E) 23

(handwritten working: $3x + 11 = 2x + 20$, $x = 9$, $y = 3x + 12 - 1 \rightarrow y = 3x + 11$, $y = 2x + 40 - 20 \rightarrow y = 2x + 20$)

Question 7:

Aneurysmal disease has been proven to induce systemic inflammatory effects, reaching far beyond the site of the aneurysm. The inflammatory mediator responsible for these processes remains unknown, however the effects of systemic inflammation have been well categorised and observed experimentally in pig models.

This inflammation induces an aberration of endothelial function within the inner most layer of blood vessel walls. The endothelium not only represents the lining of blood vessels but also acts as a transducer converting the haemodynamic forces of blood into a biological response. An example of this is the NO pathway, which uses the shear stress induced by increased blood flow to drive the formation of NO. NO diffuses from the endothelium into the smooth muscle surrounding blood vessels to promote vasodilatation and therefore acts to reduce blood flow.

Failure of this process induces high risk of vascular damage and therefore cardiovascular diseases such as thrombosis and atherosclerosis.

What is a valid implication from the text above?

A) Aneurysmal disease does not affect the NO pathway.
B) Aneurysms directly increase the likelihood of cardiovascular disease.
C) Aneurysms are the opposite of transducers.
D) Observations of this kind should be made in humans to see if the results can be replicated.
E) Aneurysms induce high blood flow.

Question 8:

A traffic surveyor is stood at a T-junction between a main road and a side street. He is only interested in traffic leaving the side street. He logs the class of vehicle, the colour and the direction of travel once on the main road. During an 8-hour period he observes a total of 346 vehicles including bikes. Of which 200 were travelling west whilst the rest travelled east. The over whelming majority of vehicles seen were cars at 90% with bikes, vans and articulated lorries together comprising the remaining 10%. Red was the most common colour observed whilst green was the least. Black and white vehicles were seen in equal quantities.

Which of the following is an accurate inference based on his survey?

A) Global sales are highest for those vehicles which are coloured red.
B) Cars are the most popular vehicle on all roads.
C) Green vehicle sales are down in the area that the surveyor was based.
D) The daily average rate of traffic out of a T junction in Britain is 346 vehicles over 8 hours.
E) To the east of the junction is a dead end.

Question 9:

William, Xavier, and Yolanda race in a 100m race. All of them run at a constant speed during the race. William beats Xavier by 20m. Xavier beats Yolanda by 20m. How many metres does William beat Yolanda?

(handwritten: 100m, 80m, system of equations)

A) 30m
B) 36m
C) 40m
D) 60m
E) 64m

(handwritten working: $100 \cdot \frac{1}{80}$, $y = 8000$, $s = \frac{d}{t}$, $\frac{100y}{80}$, $t = \frac{d}{s}$, $\frac{100}{W} = \frac{80}{X}$, $\frac{100}{X} = \frac{80}{Y}$, $100X = 80W$, $X = \frac{80}{100}W$, $100y = 80x$, $X = 80/100\,y$, $\frac{100}{W} = \frac{80}{80/100\,y}$)

~ 404 ~

Question 10:

A television is delivered in a box that has volume 60% larger than that of the television. The television is 150cm x 100cm x 10m. How much surplus volume is there?

A)   0.09m^2 B)   0.9 m^2 C)   9 m^2 D)   90 m^2 E)   900 m^2

Question 11:

Matthew and David are deciding where they would like to go camping Friday to Sunday. Upon completing their research, they discover the following:

➤   Whitmore Bay charges £5.50 a night and does not require a booking. The site provides showers, washing up facilities and easy access to a beach

➤   Port Eynon charges £5 a night and a booking is compulsory. However, the site does not provide showers but does have 240V sockets free of charge

➤   Jackson Bay charges £7 a night and is billed as a luxury site with compulsory booking, private showers, toilets, mobile phone charging facilities and kitchens.

David presents the following suggestion:

As Port Eynon is the farthest distance to travel the benefit of its cheap nightly rate is negated by the cost of petrol. Instead he recommends they visit Jackson Bay as it is the shortest distance to travel and will therefore be the cheapest.

Which of the following best illustrates a flaw in this argument?

A)   Whitmore bay may be only a few miles further which means the total cost would be less than visiting Jackson Bay.
B)   With kitchen facilities available they will be tempted to buy more food increasing the cost.
C)   The campsite may be fully booked.
D)   There may be a booking fee driving the cost up above that of the other campsites.
E)   All of the above.

Question 12:

The manufacture of any new pharmaceutical is not permitted without scrupulous testing and analysis. This has led to the widespread, and controversial use of animal models in science. Whilst it is possible to test cyto-toxicity on simple cell cultures, to truly predict the effect of a drug within a physiological system it must be trialled in a whole organism. With animals cheap to maintain, readily available, rapidly reproducing and not subject to the same strict ethical laws they have become an invaluable component of modern scientific practice.

Which of the following best illustrates the main conclusion of this argument?

A)   New pharmaceuticals cannot be approved without animal experimentation.
B)   Cell culture experiments are unhelpful.
C)   Modern medicine would not have achieved its current standard without animal experimentation.
D)   Logistically animals are easier to keep than humans for mandatory experiments.
E)   All of the above.

The information below relates to questions 13 – 17:

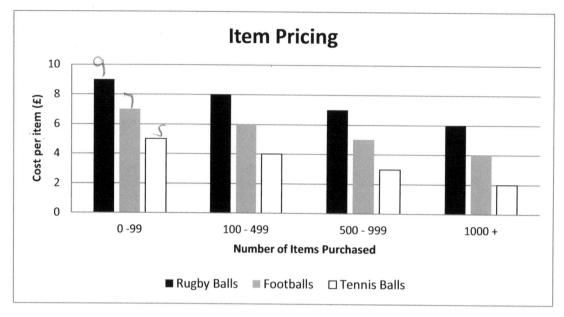

The graph above shows item pricing from a wholesaler. The wholesaler is happy to deliver for a cost of £35 to companies or £5 to individuals. Any order over the cost of £100 qualifies for free delivery. Items are defined as how they come to the wholesaler therefore 1 item = 2 rugby balls or 1 football or 5 tennis balls.

Question 13:
What is the total cost to an individual purchasing 12 rugby balls and 120 tennis balls?

A) £174   B) £179   C) £208   D) £534   E) £588

Question 14:
A private gym wishes to purchase 10 of everything, how short are they of the free delivery boundary?

A) £5.00   D) £10.01
B) £5.01   E) They are already over the minimum
C) £10.00

Question 15:
What is the most number of balls that can be bought by an individual with £1,000 pounds.

A) 200   B) 250   C) 500   D) 1,000   E) 1,250

Question 16:
The wholesaler sells all his products for a profit of 120%. If he sells £1,320 worth of goods at his prices, what did he spend on acquiring them himself?

A) £400   B) £600   C) £800   D) £1,100   E) £1,120

Question 17:
If the wholesaler pays 25% tax on the amount over £12,000 pounds; how much tax does he pay when receiving an order of 2,000 of each item?

A) £2,000   B) £3,000   C) £4,000   D) £5,000   E) £6,000

Question 18:

There are four houses on a street. Lucy, Vicky, and Shannon live in adjacent houses. Shannon has a black dog named Chrissie, Lucy has a white Persian cat and Vicky has a red parrot that shouts obscenities. The owner of a four-legged pet has a blue door. Vicky has a neighbour with a red door. Either a cat or bird owner has a white door. Lucy lives opposite a green door. Vicky and Shannon are not neighbours. What colour is Lucy's door?

A) Green

B) Red

C) White

D) Blue

E) Cannot tell

Question 19:

A train driver runs a service between Cardiff and Merthyr. On average a one-way trip takes 40 minutes to drive but he requires 5 minutes to unload passengers and a further 5 minutes to pick up new ones. As the crow flies the distance between Cardiff and Merthyr is 22 miles.

Assuming he works an 8-hour shift with two 20-minute breaks, and when he arrives to work the first train is already loaded with passengers how far does he travel?

A) 132   B) 143   C) 154   D) 176   E) 198

Question 20:

The massive volume of traffic that travels down the M4 corridor regularly leads to congestion at times of commute morning and evening. A case is being made by local councils in congestion areas to introduce relief lanes thus widening the motorway in an attempt to relieve the congestion. This would involve introducing either a new 2 or 4 lanes to the motorway on average costing 1 million pound per lane per 10 miles.

Many conservationist groups are concerned as this will involve the destruction of large areas of countryside either side of the motorway. They argue that the side of a motorway is a unique habitat with many rare species residing there.

The local councils argue that with many hundreds if not thousands of cars siding idle on the motorway pumping pollutants out into the surrounding areas, it is better for the wildlife if the congestion is eased and traffic can flow through. The councils have also remarked that if congestion is eased there would be less money needed to repair the roads from car incidents with could in theory be given to the conservationist groups as a grant.

Which of the following is assumed in this passage?

A) Wildlife living on the side of the motorway cannot be re-homed.

B) Congestion causes car incidents.

C) Relief lanes have been proven to improve traffic jams.

D) A and B.

E) B and C.

F) All of the above.

G) None of the above.

Question 21:

Apples and oranges are sold in packs of 5 for the price of £1 and £1.25 respectively. Alternatively, apples can be purchased individually for 30p and oranges can be purchased individually for 50p. Helen is making a fruit salad, she remarks that her order would have cost her an extra £6.25 if she had purchased the fruit individually.

Which of the following could have been her order?

A) 15 apples 10 oranges

B) 15 apples 15 oranges

C) 25 apples 10 oranges

D) 25 apples, 15 oranges

E) 30 apples, 30 oranges

Question 22:

Janet is conducting an experiment to assess the sensitivity of a bacterial culture to a range of antibiotics. She grows the bacteria so they cover an entire Petri dish and then pipettes a single drop of differing antibiotic at different locations. A schematic of her results is shown right where black represents growth of bacteria.

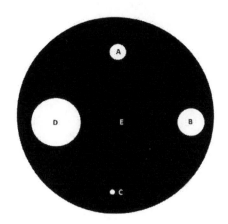

Which of the following best describes Janet's results?

A) This strain of bacteria is susceptible to all antibiotics used.
B) This strain of bacteria is susceptible to none of the antibiotics used.
C) E was the most effective antibiotic.
D) C was the most ineffective.
E) D is the most likely to be used in further testing.

Question 23:

Laura is blowing up balloons for a birthday party. The average volume of a balloon is 300cm³ and Laura's maximum forced expiratory rate in a single breathe is 4.5l/min. What is the fastest Laura could inflate 25 balloons assuming it takes her 0.5 secs to breathe in per balloon in and somebody else ties the balloons for her?

A) 112.5 seconds
B) 122.5 seconds
C) 132.5 seconds
D) 142.5 seconds
E) 152.5 seconds

Question 24:

George reasons that A is equal to B which is not equal to C. In which case C is equal to D which is equal to E. Which of the following, if true, would most *weaken* George's argument?

A) A does not equal D.
B) B is equal to E.
C) A and B are not equal.
D) C is equal to 0.
E) None of the above

Question 25:

In a single day how many times do the hour, minute and second hands of analogue clock all point to the same number?

A) 12
B) 24
C) 36
D) 48
E) 72

Question 26:

"People who practice extreme sports should have to buy private health insurance."

Which of the following statements most strongly supports this argument?

A) Exercise is healthy and private insurance offers better reward schemes.
B) Extreme spots have a higher likelihood of injury.
C) Healthcare should be free for all.
D) People that practice extreme sports are more likely to be wealthy.

Question 27:

Explorers in the US in the 18th Century had to contest with a great variety of obstacles ranging from natural to man-made. Natural obstacles included the very nature and set up of the land, presenting explorers with the sheer size of the land mass, the lack of reliable mapping as well as the lack of paths and bridges. On a human level, challenges included the threat from outlaws and other hostile groups. Due to the nature of the settling situation, availability of medical assistance was sketchy and there was a constant threat of diseases and fatal results of injuries.

Which of the following statements is correct with regards to the above text?
A) Medical supply was good in the US in the 18th Century.
B) The land was easy to navigate.
C) There were few outlaws threatening the individual.
D) Crossing rivers could be difficult.
E) All the above.

Question 28:

The statement "The human race is not dependent on electricity" assumes what?

A) We have no other energy resource.
B) Electricity is cheap.
C) Electrical appliances dominate our lives.
D) Electricity is now the accepted energy source and is therefore the only one available.
E) All of the above.

Question 29:

Wine is sold in cases of 6 bottles. A bottle of wine holds 70cl of fluid whereas a wine glass holds 175ml. Cases of wine are currently on offer for £42 a case buy one get one free. If Elin is hosting a 3-course dinner party for 27 of her friends, and she would like to provide everyone with a glass of wine per course, how much will the wine cost her?

A) £42   B) £84   C) £126   D) £168   E) £210

Question 30:

Hannah buys a television series in boxset. It contains a full 7 series with each series comprising 12 episodes. Rounded to the nearest 10 each episode lasts 40 minutes. $84 \times 40 = 3360/60 = 56$

What is the shortest amount of time it could possibly take to watch all the episodes back to back?

A) 49 hours   B) 51 hours   C) 53 hours   D) 56 hours   E) 60 hours

Question 31:

Many are familiar with the story that aided in the discovery of the "germ". Semmelweis worked in a hospital where maternal death rates during labour were astronomically high. He noticed that medical students often went straight from dissection of cadavers to the maternity wards. As an experiment Semmelweis split the student cohort in half. Half did their maternity rotation instead before dissection whereas the other half maintained their traditional routine. In the new routine, maternity ward before dissection, Semmelweis recorded an enormous reduction in maternal deaths and thus the concept of the pathogen was born.

What is best exemplified in this passage?

A) Science is a process of trial and error.
B) Great discoveries come from pattern recognition.
C) Provision of healthcare is closely associated with technological advancements.
D) Experiments always require a control.
E) All of the above.

Question 32:

Jack sits at a table opposite a stranger. The stranger says here I have 3 precious jewels: a diamond, a sapphire, and an emerald. He tells Jack that if he makes a truthful statement Jack will get one of the stones, if he lies he will get nothing.

What must Jack say to ensure he gets the sapphire?

A) Tell the stranger his name.
B) Tell the stranger he must give him the sapphire.
C) Tell the stranger he wants the emerald.
D) Tell the stranger he does not want the emerald or the diamond.
E) Tell the stranger he will not give him the emerald or the diamond.

Question 33:

Simon invests 100 pounds in a saver account that awards compound interest on a 6-monthly basis at 50%. Simon's current account awards compound interest on a yearly basis at 90%.

After 2 years will Simon's investment in the saver account yield more money than it would have in the current account?

A) Yes   B) No   C) Can't tell

Question 34:

My mobile phone has a 4-number pin code using the values $1 - 9$. To determine this, I use a standard algorithm of multiplying the first two numbers, subtracting the third and then dividing by the fourth. I change the code by changing the answer to this algorithm – I call this the key. What is the largest possible key?

A) 42   B) 55   C) 70   D) 80   E) 81

Question 35:

A group of scientists investigates the role of different nutrients after exercise. They set up two groups of averagely fit individuals consisting of the same number of both males and females aged $20 - 25$ and weighing between 70 and 85 kilos. Each group will conduct the same 1hr exercise routine of resistance training, consisting of various weighted movements. After the workout they will receive a shake with vanilla flavour that has identical consistency and colour in all cases. Group A will receive a shake containing 50 g of protein and 50g of carbohydrates. Group B will receive a shake containing 100 g of protein and 50 g of carbohydrates. All participants have their lean body mass measured before starting the experiment.
Which of the following statements is correct?

A) The experiment compares the response of men and women to endurance training.
B) The experiment is flawed as it does not take into consideration that men and women respond differently to exercise.
C) The experiment does not consider age.
D) The experiment mainly looks at the role of protein after exercise.
E) None of the above.

END OF SECTION

Section 2

Question 1:

GLUT2 is an essential, ATP independent, mediator in the liver's uptake of plasma glucose. This is an example of:

A) Active transport   D) Facilitated Diffusion
B) Diffusion   E) Osmosis
C) Exocytosis

Question 2:

The molecular weight of glucose is 180 g/mol. 5.76Kg of glucose is split evenly between two cell cultures under anaerobic conditions. One cell culture is taken from human cardiac muscle, whilst the other is a yeast culture. What will be the difference (in moles) between the amount of CO_2 produced between the two cultures?

A) 0 mol   B) 4 mol   C) 8 mol   D) 12 mol   E) 16 mol

Question 3:

Which of the following cell types will have the greatest flux along endocytotic pathways?

A) Kidney cells   D) Red blood cells
B) Liver cells   E) White blood cell
C) Nerve cells

Question 4:

Compared to the Krebs cycle, the Calvin cycle demonstrates which of the following differences?

A) CO_2 as a substrate rather than a product   D) Net loss of ATP
B) Photon dependent   E) All of the above
C) Utilisation of different electron transporters

Question 5:

Pepsin and trypsin are both digestive enzymes. Pepsin acts in the stomach whereas trypsin is secreted by the pancreas. Which graph below (trypsin in black and pepsin in grey) would most accurately demonstrate their relative activity against pH?

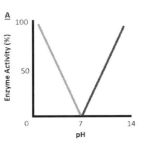

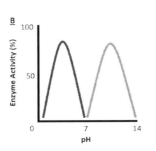

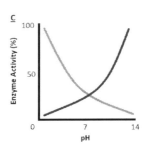

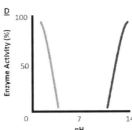

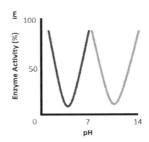

Question 6:

MRSA is an example of:

A) Natural selection
B) Genetic engineering
C) Sexual reproduction

D) Lamarckism
E) Co-dominance

Question 7:

What is the electron configuration of magnesium in $MgCl_2$?

A) 2,8
B) 2,8,2
C) 2,8,4

D) 2,8,8
E) None of the above

Question 8:

A calcium sample is run in a mass spectrometer. It is later discovered that the sample was contaminated with the most abundant isotope of chromium. A section of the trace is shown below. What was the actual abundance of the most common calcium isotope?

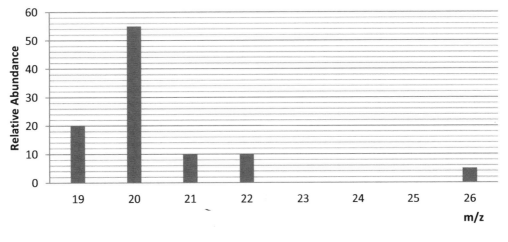

A) 1/9   B) 6/17   C) 1/2   D) 11/19   E) 17/19

Question 9:

A warehouse receives 15 tonnes of arsenic in bulk. Assuming that the sample is at least 80% pure, what is the minimum amount, in moles, of arsenic that they have obtained?

A) 1.6×10^5 B) 2×10^5 C) 1.6×10^6 D) 2×10^6 E) 1.6×10^7

Question 10:

A sample of silicon is run in a mass spectrometer. The resultant trace shows m/z peaks at 26 and 30 with relative abundance 60% and 30% respectively. What other isotope of silicon must have been in the sample to give an average atomic mass of 28?

A) 28   B) 30   C) 32   D) 34   E) 36

Question 11:

72.9g of pure magnesium ribbon is mixed in a reaction vessel with the equivalent of 54g of steam. The ensuing reaction produces $72dm^3$ of hydrogen. Which of the following statements is true?

A) This is a complete reaction
B) This is a partial reaction
C) There is an excess of steam

D) There is an excess of magnesium
E) Magnesium hydroxide is a product

Question 12:

Which species is the reducing agent in: $3Cu^{2+} + 3S^{2-} + 8H^+ + 8NO_3^- \rightarrow 3Cu^{2+} + 3SO_4^{2-} + 8NO + 4H_2O$

A) Cu^{2+} B) S^{2-} C) H^+ D) NO_3^- E) H_2O

Question 13:

Which of the following is not true of alkanes?

A) C_nH2_{n+2}
B) Saturated
C) Reactive
D) Produce only CO_2 and water when burnt in an excess of oxygen
E) None of the above

Question 14:

A rubber balloon is inflated and rubbed against a sample of animal fur for a period of 15 seconds. At the end of this process the balloon is carrying a charge of -5 coulombs. What magnitude of current must have been induced during the process of rubbing the balloon against the animal fur; and in which direction was it flowing?

A) 0.33A into the balloon
B) 0.33A into the fur
C) 0.33A in no net direction

D) 75A into the balloon
E) 75A into the fur

Question 15:

Which of the following is a unit equivalent to the Amp?

A) $V.\Omega$ B) $(W.V)/s$ C) $C.\Omega$ D) $(J.s^{-1})/V$ E) $C.s$

Question 16:

The output of a step-down transformer is measured at 24V and 10A. Given that the transformer is 80% efficient what must the initial power input have been?

A) 240W   B) 260W   C) 280W   D) 300W   E) 320W

Question 17:

An electric winch system hoists a mass of 20kg 30 metres into the air over a period of 20 seconds. What is the power output of the winch assuming the system is 100% efficient?

A) 100W   B) 200W   C) 300W   D) 400W   E) 500W

Question 18:

6×10^{10} atoms of a radioactive substance remain. The activity of the substance is quantified as 3.6×10^9. What is the decay constant of this material?

A) 0.00006   B) 0.0006   C) 0.006   D) 0.06   E) 0.6

Question 19:

An 80W filament bulb draws 0.5A of household electricity. What is the efficiency of the bulb?

A) 25%   B) 33%   C) 50%   D) 66%   E) 75%

Question 20 :

Rearrange the following equation in terms of t: $x = \frac{\sqrt{b^3 - 9st}}{13j} + \int_{-z}^{z} 9a - 7$

A) $t = \frac{(13jx - \int_{-z}^{z} 9a - 7)^2 - b^3}{9s}$

B) $t = \frac{13jx^2}{b^3 - 9s} - \int_{-z}^{z} 9a - 7$

C) $t = x - \frac{\sqrt{b^3 - 9s}}{13j} - \int_{-z}^{z} 9a - 7$

D) $t = \frac{x^2}{\frac{b^3 - 9s}{13j} + \int_{-z}^{z} 9a - 7}$

E) $t = \frac{[13j(x - \int_{-z}^{z} 9a - 7)]^2 - b^3}{-9s}$

Question 21:

An investment of £500 is made in a compound interest account. At the end of 3 years the balance reads £1687.50. What is the interest rate?

A) 20%   B) 35%   C) 50%   D) 65%   E) 80

Question 22:

What is the equation of the line of best fit for the scatter graph below?

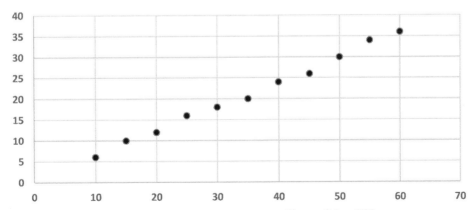

A) y = 0.2x + 0.35   D) y = 0.4x – 0.35
B) y = 0.2x – 0.35   E) y = 0.6x + 0.35
C) y = 0.4x + 0.35

Question 23 :

Simplify: $m = \sqrt{\frac{9xy^3z^5}{3x^9yz^4}} - m$ $\rightarrow$ $m = \sqrt{\frac{3y^2z}{x^6}} - m$

A) $m = \sqrt{\frac{3y^2z}{x^8}} - m$

B) $m^2 = \frac{3y^2z}{x^8} - m$

C) $2m = \sqrt{\frac{3y^2z}{x^8}}$

D) $2m^2 = 3x^{-8}y^2z$

E) $4m^2 = 3x^{-8}y^2z$

Question 24:

Which of the following is a suitable descriptive statistic for non-normally distributed data?

A) Mean

B) Normal range

C) Confidence interval

D) Interquartile range

E) Mode

Question 25:

Which best describes the purpose of statistics?

A) Evaluate acceptable scientific practice.

B) Reduce the ability of others to criticise the data

C) To quickly analyse data

D) Calculate values representative of the population from a subset sample

E) To allow for universal comparison of scientific methods

Question 26:

A rotating disc has two wells, in which bacteria are cultured. The first well is 10 cm from the centre whereas the second well is 20 cm from the centre. If the inner well completes a revolution in 1 second, how much faster is the outer well travelling?

A) 0.314m/s   B) 0.628m/s   C) 0.942m/s   D) 1.256m/s   E) 1.590m/s

Question 27:

Which is the equivalent function to: $y = 9x^{-\frac{1}{3}}$?

A) $y = \frac{1}{x}$

B) $y = \sqrt[3]{9x}$

C) $y = \frac{1}{\sqrt[3]{9x}}$

D) $y = \frac{9}{\sqrt[3]{x}}$

E) $y = \frac{3}{\sqrt[3]{x}}$

END OF SECTION

Section 3

1) *"Progress is made by trial and failure; the failures are generally a hundred times more numerous than the successes; yet they are usually left unchronicled."*

Williams Ramsey

Explain what this statement means. Argue to the contrary. To what extent do you agree with the statement?

2) *"He who studies medicine without books sails an uncharted sea, but he who studies medicine without patients does not go to sea at all."*

William Osler

Explain what this statement means. Argue to the contrary. To what extent do you agree with the statement?

3) *'"Medicine is the restoration of discordant elements; sickness is the discord of the elements infused into the living body"*

Leonardo da Vinci

Explain what this statement means. Argue to the contrary. To what extent do you think this simplification holds true within modern medicine?

4) *"Modern medicine is a negation of health. It isn't organized to serve human health, but only itself, as an institution. It makes more people sick than it heals."*

Ivan Illich

What does this statement mean? Argue to the contrary, that the primary duty of a doctor is not to prolong life. To what extent do you agree with this statement?

but to explain illness / death to patient + family members

END OF PAPER

① means modern med. is not doing its job: save lives; only makes more people sick

② · doctor does not probng life, rather trys to ←
· wrong! → modern med = save countless lives / done research to save more lives → Ebola virus Africa, etc.

③ · side effects from med. = more sick people BUT
these institutions are working towards coming up with new medications that are better / fewer side effects
Analogy → tech takes time; med takes time

~ 415 ~

Mock Paper C

Section 1

Question 1:

Adam, Beth and Charlie are going on holiday together. A single room costs £60 per night, a double room costs £105 per night and a four-person room costs £215 per night. It is possible to opt out from the cleaning service and to pay £12 less each night per room.

What is the minimum amount the three friends could pay for their holiday for a three-night stay at the hotel?

A) £122   B) £144   C) £203   D) £423   E) £432

Question 2:

I have two 96ml glasses of squash. The first is comprised of $\frac{1}{6}$ squash and $\frac{5}{6}$ water. The second is comprised of $\frac{1}{4}$ water and $\frac{3}{4}$ squash. I take 48ml from the first glass and add it to glass two. I then take 72ml from glass two and add it to glass one.

How much squash is now in each glass?

A) 16ml squash in glass one and 72ml squash in glass two.
B) 40ml squash in glass one and 32ml squash in glass two.
C) 48ml squash in glass one and 32ml squash in glass two.
D) 48ml squash in glass one and 40ml squash in glass two.
E) 80ml squash in glass one and 40ml squash in glass two.

Question 3:

It may amount to millions of pounds each year of taxpayers' money; however, it is strongly advisable for the HPV vaccination in schools to remain. The vaccine, given to teenage girls, has the potential to significantly reduce cervical cancer deaths and furthermore, the vaccines will decrease the requirement for biopsies and invasive procedures related to the follow-up tests. Extensive clinical trials and continued monitoring suggest that both Gardasil and Cervarix are safe and tolerated well by recipients. Moreover, studies demonstrate that a large majority of teenage girls and their parents are in support of the vaccine.

Which of the following is the conclusion of the above argument?

A) HPV vaccines are safe and well tolerated
B) It is strongly advisable for the HPV vaccination in schools to remain
C) The HPV vaccine amounts to millions of pounds each year of taxpayers' money
D) The vaccine has the potential to significantly reduce cervical cancer deaths
E) Vaccinations are vital to disease prevention across the population

Question 4:

Anna cycles to school, which takes 30 minutes. James takes the bus, which leaves from the same place as Anna, but 6 minutes later and gets to school at the same time as Anna. It takes the bus 12 minutes to get to the post office, which is 3km away. The speed of the bus is $\frac{5}{4}$ the speed of the bike. One day Anna leaves 4 minutes late.

How far does she get before she is overtaken by the bus?

A) 1.5km   B) 2km   C) 3km   D) 4km   E) 6km

Question 5:

The set two maths teacher is trying to work out who needs to be moved up to set one and who to award a certificate at the end of term. The students must fulfil certain criteria:

Reward	Criteria
Move to set one	Attendance over 95%
	Average test mark over 92
	Less than 5% homework handed in late
Awarded a Certificate	Absences below 4%
	Average test mark over 89
	At least 98% homework handed in on time

	Terry	Alex	Bahara	Lucy	Shiv
Attendance %	97	92	97	100	98
Average test mark %	89	93	94	95	86
Homework handed in on time %	96	92	100	96	98

Who would move up a set and who would receive a certificate?

A)   Bahara would move up a set and receive a certificate.
B)   Bahara and Lucy would move up a set and Bahara would receive a certificate.
C)   Bahara, Terry and Lucy would move up a set and Bahara and Shiv would receive a certificate.
D)   Lucy would move up a set and Bahara would receive a certificate.
E)   Lucy would move up a set and Bahara and Terry would receive a certificate.

Question 6:

18 years ago, A was 25 years younger than B is now. In 21 years time, A will be 28 years older than B was 14 years ago. How old is A now if A is $\frac{5}{6}$B?

A)   27   B)   28   C)   35   D)   42   E)   46

Question 7:

The time now is 10.45am. I am preparing a meal for 16 guests who will arrive tomorrow for afternoon tea. I want to make 3 scones for each guest, which can be baked in batches of 6. Each batch takes 35 minutes to prepare and 25 minutes to cook in the oven and I can start the next batch while the previous batch is in the oven. I also want to make 2 cupcakes for each guest, which can be baked in batches of 8. It takes 15 minutes to prepare the mixture for each batch and 20 minutes to cook them in the oven. I will also make 3 cucumber sandwiches for each guest. 6 cucumber sandwiches take 5 minutes to prepare.

What will the time be when I finish making all the food for tomorrow?

A)   4:35pm   B)   5.55pm   C)   6:00pm   D)   6:05pm   E)   7:20pm

Question 8:

Pyramid	Base edge (m)	Volume (m³)
1	3	33
2	4	64
3	2	8
4	6	120
5	2	8
6	6	120
7	4	64

What is the difference between the height of the smallest and tallest pyramids?

A) 1m   B) 5m   C) 4m   D) 6m   E) 8m

Question 9:

The wage of Employees at Star Bakery is calculated as: £210 + (Age x 1.2) – 0.8 (100 - % attendance).
Jessica is 35 and her attendance is 96%. Samira is 65 and her attendance is 89%.

What is the difference between their wages?

A) £30.40   B) £60.50   C) £248.80   D) £263.20   E) £279.20

Question 10:

It is important that research universities demonstrate convincing support of teaching. Undergraduates comprise an overwhelming proportion of all students and universities should make an effort to cater to the requirements of the majority of their student body. After all, many of these students may choose to pursue a path involving research and a strong education would provide students with skills equipped towards a career in research.

What is the conclusion of the above argument?

A)   Undergraduates comprise an overwhelming proportion of all students.
B)   A strong education would provide a strong foundation and skills equipped towards a career in research.
C)   Research universities should strongly support teaching.
D)   Institutions should provide undergraduates with a high-quality learning experience.
E)   Research has a greater impact than teaching and limited universities funds should mainly be invested in research.

Question 11:

American football has reached a level of violence that puts its players at too high a level of risk. It has been suggested that the NFL, the governing body for American football should get rid of the iconic helmets. The hard-plastic helmets all have to meet minimum impact-resistance standards intended to enhance safety, however in reality they gave players a false sense of security that only resulted in harder collisions. Some players now suffer from early onset dementia, mood swings and depression. The proposal to ban helmets for good should be supported. Moreover, it would prevent costly legal settlements involving the NFL and ex-players suffering from head trauma.

What is the conclusion of the above argument?

A)   Sports players should not be exposed to unnecessary danger.
B)   Helmets give players a false sense of security.
C)   Players can suffer from early onset dementia, mood swings and depression.
D)   The proposal to ban helmets should be supported.
E)   American football is too violent and puts its players at risk.

Question 12:

At the final stop (stop 6), 10 people get off the tube. At the previous stop (stop 5) $\frac{1}{2}$ of passengers got off. At stop 4, $\frac{3}{5}$ of passengers got off. At stop 3, $\frac{1}{3}$ of passengers got off and at stops 1 and 2, $\frac{1}{6}$ of passengers got off.

How many passengers got on at the first stop?

A) 10   B) 36   C) 90   D) 108   E) 3600

Question 13:

Everyone likes English. Some students born in spring like Maths and some like Biology. All students born in winter like Music and some like Art. Of those born in autumn, no one likes Biology, and everyone likes Art.

Which of the following is true?

A)   Some students born in spring like both Biology and Maths.
B)   Students born in spring, winter, and autumn all like Art.
C)   No one born in winter or autumn likes Biology.
D)   No one who likes Biology also likes Art.
E)   Some students born in winter like 3 subjects.

Question 14:

Until the twentieth century, the whole purpose of art was to create beautiful, flawless works. Artists attained a level of skill and craft that took decades to perfect and could not be mirrored by those who had not taken great pains to master it. The serenity and beauty produced from movements such as impressionism has however culminated in repulsive and horrific displays of rotting carcasses designed to provoke an emotional response rather than admiration. These works cannot be described as beautiful by either the public or art critics. While these works may be engaging on an intellectual or academic level, they no longer constitute art.

Which of the following is an assumption of the above argument?

A)   Beauty is a defining property of art.
B)   All modern art is ugly.
C)   Twenty first century artists do not study for decades.
D)   The impressionist movement created beautiful works of art.
E)   Some modern art provokes an emotional response.

Question 15:

The cost of sunglasses is reduced over the bank holiday weekend. On Saturday, the price of the sunglasses on Friday is reduced by 10%. On Sunday the price of the sunglasses on Saturday is reduced by 10%. On Monday, the price of the sunglasses on Sunday is reduced by a further 10%. What percentage of the price on Friday is the price of the sunglasses on Monday?

A) 55.12%   B) 59.10%   C) 63.80%   D) 70.34%   E) 72.9%

Question 16:

Putting the digit 7 on the right-hand side of a two-digit number causes the number to increase by 565. What is the value of the two-digit number?

A) 27   B) 52   C) 62   D) 66   E) 627

Question 17:

When folded, which box can be made from the net shown below?

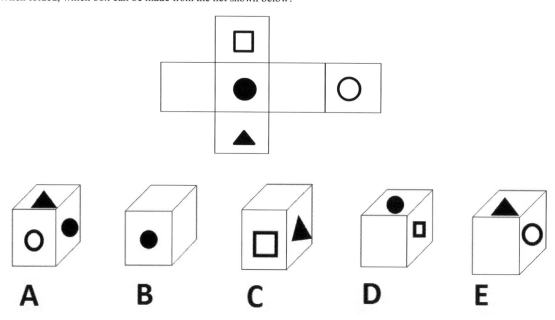

A   B   C   D   E

Question 18:

The grid is comprised of 49 squares. The shape's area is 588cm². What is its perimeter in cm?

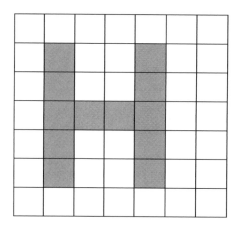

A) 26   B) 49   C) 84   D) 126   E) 182

Questions 19-21 refer to the following information:

$$BMI = weight\ (kg) \div height^2\ (m^2)$$

Men	BMR= (10 x weight in kg) + (6.25 x height in cm) – (5 x age in years) + 5
Women	BMR= (10 x weight in kg) + (6.25 x height in cm) – (5 x age in years) -161

Recommended Intake:

Amount of Exercise	Daily Kilocalories required
Little to no exercise	BMR x 1.2
Light exercise 1-3 days per week	BMR x 1.375
Moderate exercise 3-5 days per week	BMR x 1.55
Heavy exercise 6-7 days per week	BMR x 1.725
Very heavy exercise twice per day	BMR x 1.9

Question 19:

A child weighs 35kg and is 120cm tall. What is the BMI of the child to the nearest two decimal places?

A) 0.0024   B) 24.28   C) 24.31   D) 42.01   E) 42.33

Question 20:

What is the BMR of a 32-year-old woman weighing 80kg and measuring 1.7m in height?

A) 643.7 kcal   B) 1537 kcal   C) 1541.5 kcal   D) 1707.5 kcal   E) 2707.5 kcal

Question 21:

What is the recommended intake of a 45-year-old man weighing 80kg and measuring 1.7m in height who does little to no exercise each week?

A) 1642.5 kcal   B) 1771.8 kcal   C) 1851 kcal   D) 1971 kcal   E) 2712.5 kcal

Questions 22 and 23 relate to the following passage:

The achievement levels of teenagers could be higher if school started later. Teenagers are getting too little sleep because they attend schools that start at 8:30am or earlier. A low level of sleep disrupts the body's circadian rhythms and can contribute to health problems such as obesity and depression. Some doctors are now urging schools to start later in order for teenagers to get sufficient sleep; ideally 8.5 to 9.5 hours each night. During adolescence, the hormone melatonin is released comparatively later in the day and the secretion levels climb at night. Consequently, teenagers can have trouble getting to sleep earlier in the night before sufficient melatonin is present. A school in America tested this idea by starting one hour later and the percentage of GCSEs at grades A* - C increased by 16%. Schools in the UK should follow by example and shift start times later.

Question 22:

Which of the following is a flaw in the above argument?

A) Slippery slope.
B) Hasty generalisation.
C) It confuses correlation with cause.
D) Circular argument.
E) Schools should not prioritise academic achievement.

Question 23:

Which of the following, if true, would most strengthen the above argument?

A) Teenagers who are more alert will disrupt the class more.
B) Getting more sleep at night is proven to increase activity levels.
C) American schools and British schools teach the same curriculum.
D) The school in America did not alter any other aspects of the school during the trial for example curriculum, teachers, and number of students in each class.
E) Teachers get tired towards the end of the school day and are less effective.

Question 24:

The UK energy market is highly competitive. In an effort to attract more business and increase revenue, the company EnergyFirst has invested significant funds into its publicity sector. Last month, they doubled their advertising expenditures, becoming the energy company to invest the greatest proportion of investment into advertising. As a result, it is expected that EnergyFirst will expand its customer base at a rate exceeding its competitors in the ensuing months. Other energy companies are likely to follow by example.

Which of the following, if true, is most likely to weaken the above argument?

A)   Other companies invest more money into good customer service.
B)   Research into the energy industry demonstrates a low correlation between advertising investment and new customers.
C)   The UK energy market is not highly competitive.
D)   EnergyFirst currently has the smallest customer base.
E)   Visual advertising heavily influences customers.

Question 25:

The consumption of large quantities of red meat is suggested to have negative health ramifications. Carnitine is a compound present in red meat and a link has been discovered between carnitine and the development of atherosclerosis, involving the hardening, and narrowing of arteries. Intestinal bacteria convert carnitine to trimethylamine-N-oxide, which has properties that are damaging to the heart. Moreover, red meat consumption has been associated with a reduced life expectancy. It may be that charring meat generates toxins that elevate the chance of developing stomach cancer. If people want to be healthy, a vegetarian diet is preferable to a diet including meat. Vegetarians often have lower cholesterol and blood pressure and a reduced risk of heart disease.

Which of the following is an assumption of the above argument?

A)   Diet is essential to health and we should all want to be healthy.
B)   Vegetarians do the same amount of exercise as meat eaters.
C)   Meat has no health benefits.
D)   People who eat red meat die earlier.
E)   Red meat is the best source of iron.

Question 26:

Auckland is 11 hours ahead of London. Calgary is 7 hours behind London. Boston is 5 hours behind London. The flight from Auckland to London is 22 hours, but the plane must stop for 2 hours in Hong Kong. The flight from London to Calgary is 8 hours 30 minutes. The flight from Calgary to Boston is 6 hours 30 minutes. Sam leaves Auckland at 10am for London. On arrival to London, he waits 3 hours then gets the plane to Calgary. Once in Calgary, he waits 1.5 hours and gets the plane to Boston. What time is it when Sam arrives in Boston?

A)   13:30pm   B)   22.30pm   C)   01:00am   D)   01:30am   E)   03:30am

Question 27:

Light A flashes every 18 seconds, light B flashes every 33 seconds and light C flashes every 27 seconds. The three lights all flashed at the same time 5 minutes ago.

How long will it be until they next all flash simultaneously?

A)   33 seconds   B)   294 seconds   C)   300 seconds   D)   333 seconds   E)   594 seconds

Question 28:

There are 30 students in a class. They must all play at least 1 instrument, but no more than 3 instruments. 70% play the piano, 40% play the violin, 20% play the guitar and 10% play the saxophone. Which of the following statements must be true?

1.   3 or more students play piano and violin.
2.   12 students or less play the piano and the violin.
3.   9 or more students do not play piano or violin.

A)   1 only   B)   1 and 2   C)   2 only   D)   2 and 3   E)   3 only

Question 29:

Neil, Simon and Lucy are playing a game to see who can role the highest number with two dice. They start with £50 each. The losers must halve their money and give it to the winner of each game. If it is a draw, the two winners share the loser's money. If all three tie, then they keep their money. Neil wins game 1, Simon and Lucy win game 2 and 5 and Lucy wins games 3 and 4.

How much money does Lucy gain?

A) £15.63   B) £75.00   C) £75.16   D) £78.13   E) £128.10

Question 30:

Drivers in the age group 17-19 comprise 1.5% of all drivers; however, 12% of all collisions involve young drivers in this age category. The RAC Foundation wants a graduated licensing system with a 1-year probationary period with restrictions on what new drivers can do on roads. Additionally, driving instructors need to emphasise the dangers of driving too fast and driving tests should be designed to make new drivers more focused on noticing potential hazards. These changes are essential and could stop 4,500 injuries on an annual basis.

What is the assumption of the above argument?

A) Young drivers are more likely to have more passengers than other age groups.
B) Young drivers spend more hours driving than older drivers.
C) Young drivers are responsible for the collisions.
D) The cars that young people drive are unsafe.
E) Most young drivers involved in accidents are male.

Question 31:

The mean weight of 6 apples is 180g. The lightest apple weighs 167g. What is the highest possible weight of the heaviest apple?

A) 193g   B) 225g   C) 235g   D) 245g   E) 255g

Question 32:

"Sugar should be taxed like alcohol and cigarettes."

Which of the following arguments most supports this claim?

A) Sugar can cause diabetes.
B) Sugar has high addictive potential and is associated with various health concerns.
C) High sugar diets increase obesity.
D) People that eat a lot of sugar are more likely to start abusing alcohol.
E) None of the above.

Question 33:

The farmer had 184 sheep to sell. He intended to sell each for £112. However, he sold less than one quarter on day 1. As a result, he reduced the price by $\frac{1}{8}$. On the second day he sold twice as many and made £3528 more than on day 1, leaving him with less than $\frac{1}{3}$ of original. How many sheep did he sell altogether?

A) 42   B) 84   C) 98   D) 112   E) 1

Question 34:

There is no empirical evidence that human activities directly result in global warming and this is used as a reason against decreasing carbon emissions. However, many scientists believe that human activity is highly likely to cause global warming since higher levels of greenhouse gases cause the atmosphere to thicken, retaining heat. It therefore seems sensible that we should not wait for proof considering the catastrophic effects of climate change, regardless of subsequent findings. Similarly, if a tree branch had a significant chance of falling on you, it would be sensible to move away immediately.

What is the main conclusion?

A)   Many scientists believe that human activity is highly likely to cause global warming.

B)   We should not wait for proof of climate change.

C)   If a tree branch had a significant chance of falling on you, it would be sensible to move away immediately.

D)   The effects of climate change are catastrophic.

E)   There is no empirical evidence that human activities directly result in global warming, so we should not reduce carbon emissions.

Question 35:

The average of 8 numbers is y. If 13 and 31 are added, the mean of the 10 numbers is also y. What is y?

A)   11   B)   22   C)   25   D)   27   E)   44

END OF SECTION

Section 2

Question 1:

Which of the following are correct regarding polymers?

1.   Sucrose is formed by the condensation of hundreds of monosaccharides.
2.   Lactose is found in milk and is formed by condensation of two glucose molecules.
3.   Glucose has two isomers.
4.   Glycogen, starch and cellulose are all polysaccharides formed by condensation of multiple glucose molecules.
5.   People with lactose intolerance lack lactase and can experience diarrhoea after drinking milk.

A)   1 only   D)   3, 4 and 5
B)   1, 2 and 3   E)   4 and 5 only
C)   1 and 3 only

Question 2:

Which of the following statements regarding enzymes are correct?

1.   Enzymes are denatured at high temperatures or extreme pH values.
2.   Amylase is produced in the salivary glands only and converts starch to sugars.
3.   Lipases catalyse the breakdown of oils and fats into glycerol and fatty acids. This takes place in the small intestine.
4.   Bile is stored in the pancreas and travels down the bile duct to neutralise stomach acid.
5.   Isomerase can be used to convert glucose into fructose for use in slimming products.

A)   1 and 3 only   D)   2 and 4 only
B)   1, 3 and 4 only   E)   3 and 5 only
C)   1, 3 and 5

Question 3:
Which of the following describes the role of the colon?

A) Food is combined with bile and digestive enzymes.
B) Storage of faeces.
C) Reabsorption of water.
D) Faeces leave the alimentary canal.
E) Any digested food is absorbed into the lymph and blood.

Question 4:
Which of the following are true?

1. A nerve impulse is transmitted along the nerve axon as an electrical impulse and across the synapse by diffusion of chemical neurotransmitters.
2. Drugs that block synaptic transmission can cause complete paralysis.
3. The fatty sheath around the axon slows the speed at which nerve impulses are transmitted.
4. The peripheral nervous system includes the brain and spinal cord.
5. A reflex arc bypasses the brain and enables a fast, autonomic response.

A) 1 and 2   D) 2, 4 and 5
B) 1, 2 and 3   E) 3, 4 and 5
C) 1, 2 and 5

Question 5:
Which of the following statements are true regarding the transition elements?
1. Iron (II) compounds are light green.
2. Transition elements are neither malleable nor ductile.
3. Transition metal carbonates may undergo thermal decomposition.
4. Transition metal hydroxides are soluble in water.
5. When Cu^{2+} ions are mixed with sodium hydroxide solution, a blue precipitate is formed.

A) 1 and 2   B) 1 and 3   C) 1, 3 and 5   D) 3 and 5   E) 5 only

Question 6:
What is the value of C when the equation is balanced?

$$\underline{5}\,PhCH_3 + \underline{A}\,KMnO_4 + \underline{9}\,H_2SO_4 = \underline{5}\,PhCOOH + \underline{B}\,K_2SO_4 + \underline{C}\,MnSO_4 + \underline{14}\,H_2O$$

A) 3   B) 4   C) 5   D) 7   E) 9

Question 7:
Tongue-rolling is controlled by the dominant allele T, while non-rolling is controlled by the recessive allele, t.
Red-green colour blindness is controlled by a sex-linked gene on the X chromosome. Normal colour vision is controlled by dominant allele B, while red-green colour blindness is controlled by the recessive allele, b.
The mother of a family is colour blind and heterozygous for tongue-rolling, while the father has normal colour vision and is a non-roller.

Which of the following statements are correct?

1. More males than females in a population are red-green colour blind.
2. 50% of children will be non-rollers.
3. All the male children will be colour-blind.

A) 1 and 2 only   D) 2 and 3 only
B) 1, 2 and 3   E) 3 only
C) 2 only

Question 8:

Make y the subject of the formula: $\frac{y+x}{x} = \frac{x}{a} + \frac{a}{x}$

A) $y = \frac{x^2}{a} + a$

B) $y = \frac{x^2 + a^2 - ax}{a}$

C) $y = \frac{-ax}{x^2 + a^2}$

D) $y = \frac{x^2}{ax} + a - x$

E) $y = a^2 - ax$

Question 9:

What is the mass in grams of calcium chloride, $CaCl_2$, in 25cm³ of a solution with a concentration of 0.1 mol.l⁻¹? (Ar of Ca is 40 and Ar of Cl is 35)

A) 0.28g   B) 0.46g   C) 0.48g   D) 0.72g   E) 1.28g

Question 10:

Consider the equations: A: y = 3x and B: $y = \frac{6}{x} - 7$. At what values of x do the two equations intersect?

A) x=2 and x=9

B) x=3 and x=6

C) x=6 and x=27

D) x=6

E) x=18

Question 11:

Which of the following statements regarding the circulatory system are correct?

1.   The pulmonary artery carries oxygenated blood from the right ventricle to the lungs.
2.   The aorta has a high content of elastic tissue and carries oxygenated blood from the left ventricle around the body.
3.   The mitral valve is between the pulmonary vein and the left atrium.
4.   The vena cava carries deoxygenated blood from the body to the right atrium.

A) 1 and 3   B) 1 and 2   C) 2 only   D) 2 and 4   E) 3 only

Question 12:

A compound with a molar mass of 120 g.mol⁻¹ contains 12g of carbon, 2g of hydrogen and 16g oxygen. What is the molecular formula of the compound?

A) CH_2O B) $C_2H_4O_2$ C) C_4H_2O D) $C_4H_8O_4$ E) $C_8H_{16}O_8$

Question 13:

Rupert plays one game of tennis and one game of squash.

The probability that he will win the tennis game is $\frac{3}{4}$

The probability that he will win the squash game is $\frac{1}{3}$

What is the probability that he will win one game only?

A) $\frac{3}{12}$ B) $\frac{7}{12}$ C) $\frac{4}{5}$ D) $\frac{13}{12}$ E) $\frac{7}{6}$

Question 14:

What is the median of the following numbers: $\frac{7}{36}$; $0.\dot{3}$; $\frac{11}{18}$; 0.25; 0.75; $\frac{62}{72}$; $\frac{7}{7}$

A) $\frac{7}{36}$

B) $0.\dot{3}$

C) $\frac{11}{18}$

D) $\frac{62}{72}$

E) 0.75

Question 15:

16.4g of nitrobenzene is produced from 13g of benzene in excess nitric acid: $C_6H_6 + HNO_3 -> C_6H_5NO_2 + H_2O$

What is the percentage yield of nitrobenzene ($C_6H_5NO_2$)?

A) 65%   B) 67%   C) 72%   D) 78%   E) 80%

Question 16:

Which of the following points regarding electromagnetic waves are correct?

1. Radiowaves have the longest wavelength and the lowest frequency.
2. Infrared has a shorter wavelength than visible light and is used in optical fibre communication, and heater and night vision equipment.
3. All of the waves from gamma to radio waves travel at the speed of light (about 300,000,000 m/s).
4. Infrared radiation is used to sterilise food and to kill cancer cells.
5. Darker skins absorb more UV light, so less ultraviolet radiation reaches the deeper tissues.

A) 1 and 2   B) 1 and 3   C) 1, 3 and 5   D) 2 and 3   E) 2 and 4

Question 17:

Two carriages of a train collide and then start moving together in the same direction. Carriage 1 has mass 12,000 kg and moves at $5ms^{-1}$ before the collision. Carriage 2 has mass 8,000 kg and is stationary before the collision.
What is the velocity of the two carriages after the collision?

A) 2 ms^{-1} B) 3 ms^{-1} C) 4 ms^{-1} D) 4.5 ms^{-1} E) 5 ms^{-1}

Question 18:

Which of the following statements are true?

1. Control rods are used to absorb electrons in a nuclear reactor to control the chain reaction.
2. Nuclear fusion is commonly used as an energy source.
3. An alpha particle is comprised of two protons and two neutrons and is the same as a helium nucleus.
4. When $^{14}_{6}C$ undergoes beta decay, an electron and $^{14}_{7}N$ are produced.
5. Beta particles are less ionising than gamma rays and more ionising than alpha particles.

A) 1 and 2   D) 3, 4 and 5
B) 1 and 3   E) None of the statements are true
C) 3 and 4

Question 19:

Simplify fully: $\frac{(3x^{\frac{1}{2}})^3}{3x^2}$

A. $\frac{3x}{\sqrt{x}}$ B. $\frac{9}{x}$ C. $3x^{\frac{1}{2}}$ D. $3x\sqrt{x}$ E. $\frac{9}{\sqrt{x}}$

Question 20:

Which of the following are true?

1. Lightning, as well as nitrogen-fixing bacteria, converts nitrogen gas to nitrate compounds.
2. Decomposers return nitrogen to the soil as ammonia.
3. The shells of marine animals contain calcium carbonate, which is derived from dietary carbon.
4. Nitrogen is used to make the amino acids found in proteins.

A) 1 only   D) 2, 3 and 4
B) 1 and 2   E) They are all true
C) 2 and 3

Question 21:

Write $\frac{\sqrt{20}-2}{\sqrt{5}+3}$ in the form: $p\sqrt{5}+q$

A) $2\sqrt{5}-4$ B) $3\sqrt{5}-4$ C) $3\sqrt{5}-5$ D) $4\sqrt{5}-6$ E) $5\sqrt{5}+4$

Question 22:

Which of the following statements are false?

1.   Simple molecules do not conduct electricity because there are no free electrons and there is no overall charge.
2.   The carbon and silicon atoms in silica are arranged in a giant lattice structure and it has a very high melting point.
3.   Ionic compounds do not conduct electricity when dissolved in water or when melted because the ions are too far apart.
4.   Alloys are harder than pure metals.

A) 1 and 2   B) 1, 2 and 4   C) 1, 2, 3 and 4   D) 2 and 4   E) 3 only

Question 23:

The graph below shows a circle with radius 5 and centre (0,0).

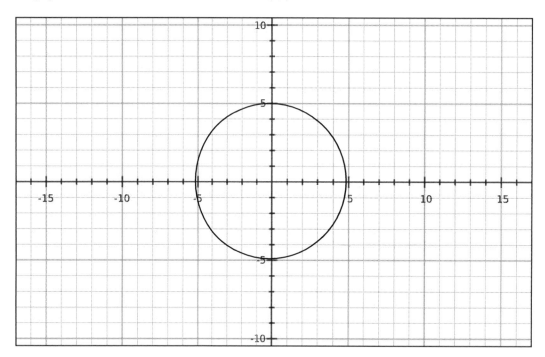

What are the values of x when the line y=3x-5 meets the circle?

A)   x=0 or x=3   D)   x=1.5 or x=-3
B)   x=0 or x=3.5   E)   x=1.5 or x=-2
C)   x=1 or x=3.5

Question 24:

Which if the following statements regarding heat transfer are correct?

1.   Heat energy is transferred from hotter to colder places by conduction because particles in liquid and gases move more quickly when heated.
2.   Liquid and gas particles in hot areas are less dense than in cold areas.
3.   Radiation does not need particles to travel.
4.   Dull surfaces are good at absorbing and poor at reflecting infrared radiation, whereas shiny surfaces are poor at absorbing, but good at reflecting infrared radiation.

   A) 1 and 2   A. 1, 2 and 4   B. 2 and 3   C. 2 and 4   D. 4 only

Question 25:

The following points refer to the halogens:

1. Iodine is a grey solid and can be used to sterilise wounds. It forms a purple vapour when warmed.
2. The melting and boiling points increase as you go up the group.
3. Fluorine is very dangerous and reacts instantly with iron wool, whereas iodine must be strongly heated as well as the iron wool for a reaction to occur and the reaction is slow.
4. When bromine is added to sodium chloride, the bromine displaces chlorine from sodium chloride.
5. The hydrogen atom and chlorine atom in hydrogen chloride are joined by a covalent bond.

Which of the above statements are false?

A) 1, 3 and 5   C) 2 and 4   E) 3, 4 and 5

B) 1, 2 and 3   D) 3 only

Question 26:

Consider the triangle right where BE=4cm, EC=2cm and AC=9cm.

What is the length of side DE?

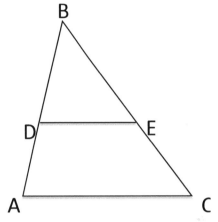

A) 4cm   C) 6cm   E) 8cm

B) 5.5cm   D) 7.5cm

Question 27:

A ball is projected vertically upwards with an initial speed of 40 ms^{-1}. What is the maximum height reached? (Take gravity to be 10 ms^{-2} and assume negligible air resistance).

A) 25m   B) 45m   C) 60m   D) 75m   E) 80m

END OF SECTION

Section 3

1) 'The NHS should not treat obese patients'

Explain what this statement means. Argue to the contrary, that we **should** treat obese patients. To what extent do you agree with this statement?

2) 'We should all become vegetarian'

Explain what this statement means. Argue to the contrary, that we **should not** all become vegetarian. To what extent do you agree with this statement?

3) 'Certain vaccines should be mandatory'

Explain what this statement means. Argue to the contrary, that vaccines **should not** be mandatory. To what extent do you agree with this statement?

4) 'Compassion is the most important quality of a healthcare professional'

Explain what this statement means. Argue to the contrary, that there are more important qualities than compassion for health professionals. To what extent do you agree with this statement?

END OF PAPER

Mock Paper D

Section 1

Question 1:

"Competitors need to be able to run 200 metres in under 25 seconds to qualify for a tournament. James, Steven and Joe are attempting to qualify. Steven and Joe run faster than James. James' best time over 200 metres is 26.2 seconds." Which response is definitely true?

A)   Only Joe qualifies.

B)   James does not qualify.

C)   Joe and Steven both qualify.

D)   Joe qualifies.

E) No one qualifies.

Question 2:

You spend £5.60 in total on a sandwich, a packet of crisps and a watermelon. The watermelon cost twice as much as the sandwich, and the sandwich cost twice the price of the crisps.
How much did the watermelon cost?

A)   £1.20   B)   £2.60   C)   £2.80   D)   £3.20   E)   £3.60

Question 3:

Jane, Chloe and Sam are all going by train to a football match. Chloe gets the 2:15pm train. Sam's journey takes twice as long Jane's. Sam catches the 3:00pm train. Jane leaves 20 minutes after Chloe and arrives at 3:25pm.
When will Sam arrive?

A)   3:50pm   B)   4:10pm   C)   4:15pm   D)   4:30pm   E)   4:40pm

Question 4:

Michael has eleven sweets. He gives three sweets to Hannah. Hannah now has twice the number of sweets Michael has remaining. How many sweets did Hannah have before the transaction?

A)   11   B)   12   C)   13   D)   14   E)   15

Question 5:

Alex gets a pay rise of 5% plus an extra £6 per week. The flat rate of income tax on his salary is decreased from 14% to 12% at the same time. Alex's old weekly take-home pay after tax is £250 per week.
What will his new weekly take-home pay be, to the nearest whole pound?

A) £260   B) £267   C) £273   D) £279   E) £285

Question 6:

You have four boxes, each containing two cubes. Box A contains two white cubes, Box B contains two black cubes, and Boxes C and D both contain one white cube and one black cube. You pick a box at random and take out one cube. It is a white cube. You then draw another cube from the same box.
What is the probability that this cube is not white?

A)   ½   B)   ⅓   C)   ⅔   D)   ¼   E)   ¾

Question 7:

Anderson & Co. hire out heavy plant machinery at a cost of £500 per day. There is a surcharge for heavy usage, at a rate of £10 per minute of usage over 80 minutes. Concordia & Co. charge £600 per day for similar machinery, plus £5 for every minute of usage. For what duration of usage are the costs the same for both companies?

A)   100 minutes

B)   130 minutes

C)   140 minutes

D)   170 minutes

E)   180 minutes

Question 8:

Simon is discussing with Seth whether or not a candidate is suitable for a job. When pressed for a weakness at interview, the candidate told Simon that he is a slow eater. Simon argues that this will reduce the candidate's productivity, since he will be inclined to take longer lunch breaks.

Which statement **best** supports Simon's argument?

A) Slow eaters will take longer to eat lunch
B) Longer lunch breaks are a distraction
C) Eating more slowly will reduce the time available to work
D) Eating slowly is a weakness
E) Eating slowly will lead to less time to work efficiently

Question 9:

Three pieces of music are on repeat in different rooms of a house. One piece of music is three minutes long, one is four minutes long and the final one is 100 seconds long. All pieces of music start playing at exactly the same time. How long is it until they are next starting together again?

A) 12 minutes   C) 20 minutes   E) 300 minutes
B) 15 minutes   D) 60 minutes

Question 10:

A car leaves Salisbury at 8:22am and travels 180 miles to Lincoln, arriving at 12:07pm. Near Warwick, the driver stopped for a 14 minute break.

What was its average speed, whilst travelling, in kilometres per hour? It should be assumed that the conversion from miles to kilometres is 1:1.6.

A) 51kph   B) 67kph   C) 77kph   D) 82kph   E) 86kph

Questions 11 and 12 refer to the following data:

Five respondents were asked to estimate the value of three bottles of wine, in pounds sterling.

Respondent	Wine 1	Wine 2	Wine 3
1	13	16	25
2	17	16	23
3	11	17	21
4	13	15	14
5	15	19	29
Actual retail value	8	25	23

Question 11:

What is the mean error made when guessing the value of wine 1?

A) £4.80   B) £5.60   C) £5.80   D) £6.20   E) £6.40

Question 12:

Which respondent guessed most accurately?

A) Respondent 1   C) Respondent 3   E) Respondent 5
B) Respondent 2   D) Respondent 4

Question 13:

"Recently in Kansas, a number of farm animals have been found killed in the fields. The nature of the injuries is mysterious, but consistent with tales of alien activity. Local people talk of a number of UFO sightings, and claim extra terrestrial responsibility. Official investigations into these claims have dismissed them, offering rational explanations for the reported phenomena. However, these official investigations have failed to deal with the point that, even if the UFO sightings can be explained in rational terms, the injuries on the carcasses of the farm animals cannot be. Extra terrestrial beings must therefore be responsible for these attacks."

Which of the following best expresses the main conclusion of this argument?

A)   Sightings of UFOs cannot be explained by rational means
B)   Recent attacks must have been carried out by extraterrestrial beings
C)   The injuries on the carcasses are not due to normal predators
D)   UFO sightings are common in Kansas
E)   Official investigations were a cover-up

Question 14:

"To make a cake you must prepare the ingredients and then bake it in the oven. You purchase the required ingredients from the shop, however your oven is broken. Therefore you cannot make a cake."

Which of the following arguments has the same structure?

A)   To get a good job, you must have a strong CV then impress the recruiter at interview. Your CV was not as good as other applicants; therefore you didn't get the job.
B)   To get to Paris, you must either fly or take the Eurostar. There are flight delays due to dense fog, therefore you must take the Eurostar.
C)   To borrow a library book, you must go to the library and show your library card. At the library, you realise you have forgotten your library card. Therefore you cannot borrow a book.
D)   To clean a bedroom window, you need a ladder and a hosepipe. Since you don't have the right equipment, you cannot clean the window.
E)   Bears eat both fruit and fish. The river is frozen, so the bear cannot eat fish.

Question 15:

"Making model ships requires patience, skill and experience. Patience and skill without experience is common – but often such people give up prematurely, since skill without experience is insufficient to make model ships, and patience can quickly be exhausted."

Which of the following summarises the main argument?

A)   Most people lack the skill needed to make model ships
B)   Making model ships requires experience
C)   The most important thing is to get experience
D)   Most people make model ships for a short time but give up due to a lack of skill
E)   Successful model ship makers need to have several positive traits

Question 16:

"Joseph has a bag of building blocks of various shapes and colours. Some of the cubic ones are black. Some of the black ones are pyramid shaped. All blue ones are cylindrical. There is a green one of each shape. There are some pink shapes."

Which of the following is definitely **NOT** true?

A)   Joseph has pink cylindrical blocks
B)   Joseph doesn't have pink cylindrical blocks
C)   Joseph has blue cubic blocks

D)   Joseph has a green pyramid
E)   Joseph doesn't have a black sphere

Question 17:

Sam notes that the time on a normal analogue clock is 1540hrs.
What is the smaller angle between the hands on the clock?

A)   110°   B)   120°   C)   130°   D)   140°   E)   150°

Question 18:
A fair 6-faced die has 2 sides painted red. The die is rolled 3 times.
What is the probability that at least one red side has been rolled?

A) $^8/_{27}$ B) $^{19}/_{27}$ C) $^{21}/_{27}$ D) $^{24}/_{27}$ E) 1

Question 19:
"In a particular furniture warehouse, all chairs have four legs. No tables have five legs, nor do any have three. Beds have no less than four legs, but one bed has eight as they must have a multiple of four legs. Sofas have four or six legs. Wardrobes have an even number of legs, and sideboards have and odd number. No other furniture has legs. Brian picks a piece of furniture out, and it has six legs."
What can be deduced about this piece of furniture?

A) It is a table
B) It could be either a wardrobe or a sideboard
C) It must be either a table or a sofa
D) It must be either a table, a sofa or a wardrobe
E) It could be either a bed, a table or a sofa

Question 20:
Two friends live 42 miles away from each other. They walk at 3mph towards each other. One of them has a pet falcon which starts to fly at 18mph as soon as the friends set off. The falcon flies back and forth between the two friends until the friends meet. How many miles does the falcon travel in total?

A) 63   B) 84   C) 114   D) 126   E) 252

Question 21:
"Antibiotic resistance is on the increase. As a result, many antibiotics in our vast armoury are becoming ineffective against common infections. Probably the most significant contributor to this is the use of antibiotics in farming, as this exposes bacteria to antibiotics for no good reason, giving the opportunity for resistance to develop. If this worrying trend continues, we might, in 30 years time, be back in the Victorian situation, where people die from skin or chest infections we consider mild today."
Which of the following best represents the overall conclusion of the passage?

A) Antibiotic resistance is a serious issue
B) Antibiotics use in farming is essential
C) The use of antibiotics in farming could cause us serious harm
D) Victorians used to die from diseases we can treat today
E) Antibiotics can treat skin infections

Question 22:
A complete set of maths equipment includes a pen, a pencil, a geometry set and a pad of paper. Pens cost £1.50, pencils cost 50p, paper pads cost £1 and geometry sets cost £3. Sam, Dave and George each want complete sets, but Mr Browett persuades them to share some items. Sam and Dave agree to share a paper pad and a geometry set. George must have his own pen, but agrees that he and Sam can share a pencil.

What is the total amount spent?
A) £12.00   B) £13.50   C) £16.50   D) £17.50   E) £18.00

Question 23:

The figure below shows 12 individual planks arranged such that 5 squares are made with them.
To make 7 squares in total, which two planks need to be moved?

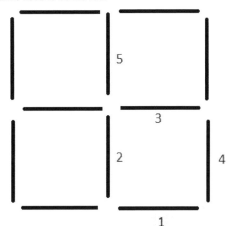

A) 1 and 2   B) 1 and 3   C) 1 and 4   D) 3 and 5   E) 4 and 5

Question 24:

A cube has six sides of different colours. The red side is opposite to black. The blue side is adjacent to white. The purple side is adjacent to blue. The final side is yellow.
Which colour is opposite purple?

A) Red   B) Black   C) Blue   D) White   E) Yellow

Question 25:

UK chocolate businesses purchase 36,000,000kg of cocoa beans each year. Each gram costs the UK 0.3p, from which the supplier takes 20% commission. Of what is left, the local government takes 60% and the distribution company gets 30%.
How much are the cocoa farmers left with per year?

A) £3.68m   B) £6.82m   C) £8.64m   D) £10.8m   E) £11.4m

Question 26:

"Some people with stomach pains and diarrhoea have Giardiasis."
Which of the following statements is supported?
A) Some people have stomach pains, but do not have Giardiasis.
B) Some people with stomach pains and diarrhoea do not have Giardiasis.
C) Kate has Giardiasis. Therefore she has stomach pains.
D) Giardiasis is defined as stomach pains and diarrhoea together.
E) None of the above

Question 27:

Catherine has 6 pairs of red socks, 6 pairs of blue socks and 6 pairs of grey socks in her drawer. Unfortunately, they are not paired together. The light in her room is broken so she cannot see what colour the socks are. She decides to keep taking socks from the drawer until she has a matching pair.

What is the minimum number of socks she needs to take from the drawer to guarantee at least one matching pair can be made?

A) 2   B) 3   C) 4   D) 5   E) 6

Question 28:

Luca and Giovanni are waiters. One month, Luca worked 100 hours at normal pay and 20 hours at overtime pay. Giovanni worked 80 hours at normal pay and 60 hours at overtime pay. Neither received any tips. Luca earned €2000; Giovanni earned €2700.

What is the overtime rate of pay?

A) €10 per hour   C) €20 per hour   E) €30 per hour
B) €15 per hour   D) €25 per hour

Question 29:

"Train A leaves Bristol at 1000hrs and travels at 90mph. Train B leaves Newcastle at 1030hrs and travels at 70mph. The distance between the two cities is 405 miles. Due to a mistake, both trains are travelling on the same track."

Calculate the distance from Bristol, to the nearest whole mile at which the trains will collide.

A) 158 miles   B) 203 miles   C) 228 miles   D) 248 miles   E) 263 miles

Question 30:

"100 pieces of rabbit food will feed one pregnant rabbit and two normal rabbits for a day. 175 pieces of food will feed two pregnant and three normal rabbits for a day. There is no excess food."

Which statement is **FALSE**?
A) A normal rabbit can be fed for longer than a day with 30 pieces of food.
B) 70 pieces of food are sufficient to feed a pregnant rabbit for a day.
C) A pregnant rabbit needs twice as many pieces per day as a normal rabbit.
D) Two pregnant and four normal rabbits will need 200 pieces of food for a day.
E) Three pregnant and ten normal rabbits will need 450 pieces of food for a day.

Question 31:

Michael bought a painting at an auction for £60. After 6 months, he realised the value of the painting had increased, so he sold it for £90. Realising a mistake, he wanted to buy the painting back, which he was able to do for £110. A year later, he then re-sold the painting for £130.

What is Michael's total profit on the painting?

A) £20   B) £30   C) £40   D) £50   E) £60

Question 32:

"Insect pests such as aphids and weevils can be a problem for farmers, as the feed on crops, causing destruction. Thus many farmers spray their crops with pesticides to kill these insects, increasing their crop yield. However, there are also predatory insects such as wasps and beetles that naturally prey on these pests – which are also killed by pesticides. Therefore it would be better to let these natural predators control the pests, rather than by spraying needless chemicals."

Which of the following best describes the flaw in this logic?

A) Many pesticides are expensive, so should not be used unless necessary
B) It fails to consider other problems the pesticides may cause
C) It does not explain why weevils are a problem
D) It fails to assess the effectiveness of natural predators compared to pesticides
E) It does not consider the benefits of using fewer pesticides

Question 33:

A parliament contains 400 members. Last election, there was a majority of 43% of the popular vote to the liberal party. However, as a first-past-the-post system of voting was in effect, they gained 298 seats in parliament.

How many excess members did they have, relative to a straight proportional representation system?

A) 72   B) 98   C) 112   D) 126   E) 148

Question 34:
A cube is painted such that no two faces that touch may be the same colour.

What is the minimum number of colours required for this?

A) 2   B) 3   C) 4   D) 5   E) 6

Question 35:
4 people need to cross a river, one of whom is on a horse. . They make a stable raft, but find it can only take the weight of either two people or the rider alone. The raft must have someone in it to cross the river in order to propel and steer it.

What is the minimum number of journeys the raft must make across the river to get all 4 people to the other side?

A) 3   B) 5   C) 7   D) 9   E) 11

END OF SECTION

Section 2

Question 1:
Which of the following is **NOT** present in the Bowman's capsule?

A) Urea   C) Sodium   E) Haemoglobin
B) Glucose   D) Water

Question 2:
The primary ions responsible for an action potential on a muscle cell membrane are Sodium and Potassium. Sodium concentration is higher than that of potassium outside the cell. Potassium concentration is higher than that of sodium inside the cell. Depolarisation occurs when the membrane potential increases (become more positive).
Which of the following **must** be true when a muscle cell membrane depolarises?

A) More potassium moves into the muscle cell than sodium.
B) More sodium moves into the muscle cell than potassium.
C) There is no net flow of sodium or potassium ions.
D) The membrane potential becomes more negative
E) None of the above

Question 3:
Calculate the radius of a sphere which has a surface area three times as great as its volume.

A) 0.5   C) 1.5   E) 2.5
B) 1   D) 2   F) More information is needed

Question 4:
A mechanical winch lifts up a bag of grain in a mill from the floor into a hopper.
Assuming that the machine is 100% efficient and lifts the bag vertically only, which of the following statements are **TRUE**?
1.   This increases gravitational potential energy
2.   The gravitational potential energy is independent of the mass of the grain
3.   The work done is the difference between the gravitational potential energy at the hopper and when the grain is on the floor
4.   The work done is the difference between the kinetic energy of the grain in the hopper and on the floor

A) 1 only   C) 1 and 4   E) 1, 2 and 4
B) 1 and 3   D) 1, 2 and 3   F) None of the above

Question 5:

A barometer records atmospheric pressure as 10^5 Pa. Recalling that the diameter of the Earth is 1.2×10^7 m, **estimate** the mass of the atmosphere. [Assume g = 10 ms^{-2}, the earth is spherical and that π=3]

A) 4.5×10^8 kg

B) 4.5×10^{10} kg

C) 4.5×10^{12} kg

D) 4.5×10^{13} kg

E) 4.5×10^{18} kg

F) More information is required

Question 6:

Which of the following in NOT a polymer?

A) Polythene

B) Glycogen

C) Collagen

D) Starch

E) DNA

F) Triglyceride

Question 7:

SIADH is a metabolic disorder caused by an excess of Anti-Diuretic Hormone (ADH) release by the posterior pituitary gland.

Which row best describes the urine produced by a patient with SIADH?

	Volume	Salt Concentration	Glucose
A)	High	Low	Low
B)	High	High	Low
C)	High	High	High
D)	Low	Low	Low
E)	Low	High	Low
F)	Low	High	High

Question 8:

A 6kg missile is fired and decelerates at 6ms^{-2}.

What is the difference in resistive force compared to a 2kg missile fired and decelerating at 8ms^{-2}?

1	Sodium	4	Zinc
2	Potassium	5	Copper
3	Aluminium	6	Magnesium

A) 8N

B) 12N

C) 16

D) 20

Question 9:

Place the following substances in order from most to least reactive:

A) 1 » 2 » 6 » 3 » 4 » 5

B) 1 » 2 » 6 » 3 » 5 » 4

C) 2 » 1 » 6 » 3 » 4 » 5

D) 2 » 1 » 6 » 3 » 5 » 4

E) 2 » 6 » 1 » 3 » 4 » 5

Question 10:

The normal cardiac cycle has two phases, systole and diastole.

During diastole, which of the following is **FALSE**?

A) The aortic valve is closed

B) The ventricles are relaxing

C) There is blood in the ventricles

D) The pressure in the aorta increases

E) There is blood in the ventricles

Question 11:

The figure below shows a schematic of a wiring system. All the bulbs have equal resistance. The power supply is 24V.

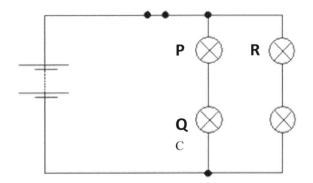

If headlight Q is replaced by a new one with twice the resistance, with the switch closed, which of these combinations of voltage drop across the four bulbs is possible?

	P	Q	R	S
A)	8V	16V	12V	12V
B)	8V	16V	16V	8V
C)	8V	16V	8V	16V
D)	12V	24V	24V	24V
E)	12V	12V	12V	12V
F)	16V	8V	12V	12V
G)	16V	8V	8V	16V
H)	24V	24V	24V	24V
I)	4V	8V	6V	6V
J)	8V	4V	6V	6V

Question 12:

A cup has 144ml of pure deionised water. How many electrons are in the cup due to the water? [Avogadro Constant = 6 x 10^{23}]

A) 8.64 x 10^{24} C) 1.2 x 10^{24} E) 4.8 x 10^{25}

B) 8.64 x 10^{25} D) 4.8 x 10^{24}

Question 13:

Below is a graph showing the concentration of product over time as substrate concentration is increased.   Some enzyme inhibitors are introduced.

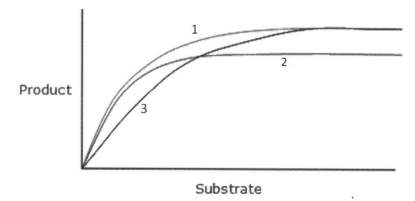

Which, if any, line represents the effect of competitive inhibition?

A) Line 1   C) Line 3
B) Line 2   D) None of these lines

Question 14:

Which of the following is **NOT** present in the plasma membrane?

A) Extrinsic proteins   C) Phospholipids   E) Nucleic Acids
B) Intrinsic proteins   D) Glycoproteins   F) They are all present

Question 15:

There are 1000 international airports in the world. If 4 flights take off every hour from each airport, estimate the annual number of commercial flights worldwide, to the nearest 1 million.

A) 20 million   C) 37 million   E) 42 million
B) 35 million   D) 40 million   F) 44 million

Question 16:

Steve's sports car requires 2.28kg of octane to travel to Pete's house 10 miles away. Calculate the mass of CO_2 produced during the journey.

A) 0.88 kg   B) 1.66 kg   C) 2.64 kg   D) 3.52 kg   E) 5.28 kg   F) 7.04 kg

Question 17:
Given:
F + G + H = 1
F + G − H = 2
F − G − H = 3

Calculate the value of FGH.

A) -2   B) -0.5   C) 0   D) 0.5   E) 2

Question 18:

A pulmonary embolism occurs when a main artery supplying the lungs becomes blocked by a clot that has travelled from somewhere else in the body.

Which option best describes the path of a blood clot that originated in the leg and has caused a pulmonary embolism?

A) Inferior Vena cava	F) Left ventricle	
B) Superior Vena cava	G) Pulmonary artery	
C) Right atrium	H) Pulmonary vein	
D) Right ventricle	I) Aorta	
E) Left atrium	J) Coronary artery	

A) C, D, H, G	C) I, E, F, G	E) A, C, D, J, G
B) B, C, D, H, G	D) A, C, D, G	F) A, C, D, J, E, F, G

Question 19:

The concentration of chloride in the blood is 100mM. The concentration of thyroxine is 1×10^{-10}kM. Calculate the ratio of thyroxine to chloride ions in the blood.

A) Chloride is 100,000,000 times more concentrated than thyroxine
B) Chloride is 1,000,000 times more concentrated than thyroxine
C) Chloride is 1000 times more concentrated than thyroxine
D) Concentrations of chloride and thyroxine are equal
E) Thyroxine is 1000 times more concentrated than chloride
F) Thyroxine is 1,000,000 times more concentrated than chloride

Question 20:

Put the following types of electromagnetic waves in ascending order of wavelength:

	Shortest - Longest			
A)	Visible Light	Ultraviolet	Infrared	X Ray
B)	Visible Light	Infrared	Ultraviolet	X Ray
C)	Infrared	Visible Light	Ultraviolet	X Ray
D)	Infrared	Visible Light	X Ray	Ultraviolet
E)	X Ray	Ultraviolet	Visible Light	Infrared
F)	X Ray	Ultraviolet	Infrared	Visible Light
G)	Ultraviolet	X Ray	Visible Light	Infrared

Question 21:

How many seconds are there in 66 weeks? [n! = 1 x 2 x 3 x... x n].

A) 7!	B) 8!	C) 9!	D) 10!	E) 11!	F) 12!

Question 22:

Which of the following is **NOT** a hormone?

A) Insulin	D) Cortisol	G) None of the above
B) Glycogen	E) Thyroxine	
C) Noradrenaline	F) Progesterone	

Question 23:

In a lights display, a 100W water fountain shoots 1L of water vertically upward every second.

What is the maximum height attained by the jet of water, as measured from where it first leaves the fountain? Assume that there is no air resistance, that the fountain is 100% efficient and g=10 ms^{-2}

A) 2m
B) 5m
C) 10m
D) 20m
E) The initial speed of the jet is required to calculate the maximum height

Question 24:

Which of the following statements regarding neural reflexes is **FALSE**?

A) Reflexes are usually faster than voluntary decisions
B) Reflex actions are faster than endocrine responses
C) The heat-withdrawal reflex is an example of a spinal reflex
D) Reflexes are completely unaffected by the brain
E) Reflexes are present in simple animals
F) Reflexes have both a sensory and motor component

Question 25:

Study the diagram, comprising regular pentagons.
What is the product of **a** and **b**?

A) 580°
B) 1,111°
C) 3,888°
D) 7,420°
E) 9,255°
F) 15552°

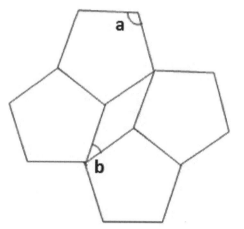

Question 26:

The table below shows the results of a study investigating antibiotic resistance in staphylococcus populations.

Antibiotic	Number of Bacteria tested	Number of Resistant Bacteria
Benzyl-penicillin	10^{11}	98
Chloramphenicol	10^9	1200
Metronidazole	10^8	256
Erythtomycin	10^5	2

A single staphylococcus bacterium is chosen at random from a similar population. Resistance to any one antibiotic is independent of resistance to others.

Calculate the probability that the bacterium selected will be resistant to all four drugs.

A) 1 in 10^{12} B) 1 in 10^6 C) 1 in 10^{20} D) 1 in 10^{25} E) 1 in 10^{30} F) 1 in 10^{35}

Question 27:

Which of the following units is **NOT** a measure of power?

A) W
B) Js^{-1}
C) Nms^{-1}
D) VA
E) V$^2\Omega^{-1}$
F) None of the above

END OF SECTION

Section 3

1)   *'The concept of medical euthanasia is dangerous and should never be permitted within the UK'*

Explain the reasoning behind this statement. Suggest an argument against this statement. To what extent, should legislation regarding the prohibition of medical euthanasia in the UK be changed?

2)   *'The obstruction of stem-cell research is directly responsible for death arising from stem-cell treatable diseases.'*

Explain what this argument means. Argue the contrary. To what extend do you agree with the statement?

3)   *'Imagination is more important than knowledge'*

Albert Einstein

Explain how this statement could be interpreted in a medical setting. Argue to the contrary that knowledge is more important than imagination in medicine. To what extent do you agree with the statement?

4)   *"The most important quality of a good doctor is a thorough understanding of science"*

Explain what this statement means. Argue in favour of this statement. To what extent do you agree with it?

END OF PAPER

Mock Paper E

Section 1

Question 1:

"Peter books a return flight to Dubai for £725. The flight is refundable, but there is a fee of £45 payable for cancelling. Peter notices as time passes, the remaining tickets on the same plane are becoming cheaper. He decides to cancel his flight, booking a new one for £530 through the same provider. Once again he sees prices have fallen, so he cancels this flight but can only buy a new one for £495."

What is his overall saving, relative to the original price paid?

A) £110   B) £140   C) £150   D) £195   E) £230

Question 2:

"You have three bags, each containing four balls numbered with single digit numbers. Bag A contains even numbers only, Bag B contains odd numbers only, and Bag C contains the numbers 2, 5, 6 and 8. You take a ball from Bag B and put it into Bag C; then you then take a ball from Bag C and put it into Bag A. You draw a ball at random from Bag A."

What is the probability that this ball is an odd number?

A) $^1/_{25}$ B) $^2/_{25}$ C) $^3/_{25}$ D) $^4/_{25}$ E) $^1/_5$

Question 3:

The price of bread rises by 40% due to a poor grain harvest. This is later reduced by 20% due to a government farming subsidy. Dave buys three loaves of bread and gets a fourth free because of a discount in the shop.

How much did he pay per loaf of bread? Express your answer as a percentage of the original price.

A) 66%   B) 84%   C) 92%   D) 98%   E) 110%

Question 4:

Sam notes that the time on a normal analogue clock is 2120hrs. What is the smaller angle between the hands on the clock?

A) 130°   B) 140°   C) 150°   D) 160°   E) 170°

Question 5:

Sam needs to measure out exactly 4 litres of water into a tank. He has two pieces of equipment – a bucket that holds 5 litres and a one that holds 3 litres, with no intermediate markings.

Is it possible to measure out 4 litres? If so, how much water is needed in total in order to measure the 4 litres?

A) 4 litres
B) 7 litres
C) 8 litres
D) 10 litres
E) Not possible with this equipment

Question 6:

"A librarian is sorting books into their correct locations. All history books belong to the right of all science books. Science books are divided into five locations: engineering, biology, chemistry, physics and mathematics (in an uninterrupted order from right to left). Art books are located to the right of mathematics between engineering and sport, and sport books between art and history. Literature books are to the right of art books."

What can be certainly said about the location of literature books?

A) They are located between art and history books
B) They are located to the left of history books
C) They are located between mathematics and art
D) They are located to the right of engineering
E) They are not located to the left of sport

Question 7:

"Many people choose not to buy brand new cars, as buying brand new has significant disadvantages. Most importantly, a car's value drops substantially the moment it is first driven on the road. Even though a car is virtually unchanged by these first few miles, the potential resale value is significantly reduced. Therefore it is better to buy second hand cars, as their value does not drop so much immediately after purchase."

Which of the following best represents the main conclusion of this passage?

A) There are many equal reasons to avoid buying brand new cars
B) Cars that have driven lots of miles should be avoided
C) The rapid loss of value of new cars makes buying second-hand a wise choice
D) Second hand cars are at least as good as new ones
E) New cars should not be driven to ensure they keep their resale value

Question 8:

James is a wine dealer specialising in French wine. From his original stock of 2,000 bottles in one cellar, he sells 10% to one customer and 20% of the remaining wine to another customer. He makes £11,200 profit from the two transactions combined. What is the average profit per bottle?

A) £18   B) £20   C) £22   D) £24   E) £26

Question 9:

"Many good quality pieces of old furniture are considered 'timeless' – they are used and enjoyed by many people today, and this is expected to continue for many generations to come. However, most of this furniture dates back to previous eras, and modern furniture does not fall under the 'timeless' category of being enjoyed for many years to come."

Which of the following is the main flaw in the argument?
A) There may be many factors which make furniture good
B) There used to be more furniture makers than today
C) No evidence is given to tell us old furniture is better than new
D) Old furniture is desirable for other reasons than its quality
E) We cannot yet tell whether new furniture will become 'timeless'

Question 10:

"Red wine is thought to be much healthier than beer because it contains many antioxidants, which have been shown to be beneficial to health. Many red wines are produced in Southern France and Italy, therefore it is no surprise that residents there have a greater life expectancy than in the UK and Germany, which are predominantly beer producing countries."

Which of the following is an assumption of the above argument?

A) Italian people drink red wine   D) Beer is not produced in Italy
B) Antioxidants are beneficial for health   E) Italian life expectancy is greater than in the UK
C) British people prefer beer to red wine

Question 11:

Hannah, Jane and Tom are travelling to London to see a musical. Hannah catches the train at 1430. Jane leaves at the same time as Hannah, but catches a bus which takes 40% longer then Hannah's train. Tom also takes a train, and the journey time is 10 minutes less then Hannah's journey, but he leaves 45 minutes after Jane leaves. He arrives in London at 1620.

At what time will Jane arrive in London?

A) 1545   B) 1600   C) 1615   D) 1700   E) 1715

Question 12:

At a show, there are two different ticket prices for different seats. The cost is £10 for a standard seat, and £16 for a premium view seat. The total revenue from a show is £6,600, and the total attendance was 600.

How many premium view seats were purchased?

A) 60   B) 100   C) 140   D) 180   E) 240

Question 13:

The moon orbits the Earth once every 28 days. Between 20th January and 23rd May inclusive, how many degrees has the Moon turned through? This is not a leap year.

A)　1540°　　　　B)　1560°　　　　C)　1580°　　　　D)　1600°　　　　E)　1620°

Question 14:

Drama academies are special schools students can go to in order to learn performing arts. These schools are only available to the most skilled young performers, and aim to give students the best training in the arts, whilst still covering mainstream academic subjects. However, many parents are reluctant for their children to attend such academies, as they feel the academic teaching will be worse than at a standard school.

Which of the following, if true, would most weaken the above argument?

A)　Most top actors attended a drama academy as children
B)　There is as much time dedicated to academic work in drama academies as there is in normal schools
C)　The academic work comprises a greater proportion of the study time than drama related activities
D)　Most children are keen to attend a drama academy if given the opportunity
E)　80% of students at drama academies attain higher than average GCSE scores

Question 15:

Anil and Suresh both leave point A at the same time. Anil travels 5km East then 10km North. Anil then travels a further 1km North before heading 3km West. Suresh travels East for 2km less than Anil's total journey distance. He then heads 13km North, before pausing and travelling back 2km South. How far, as the crow flies, are the two men now apart?

A)　11km　　　　B)　12km　　　　C)　13km　　　　D)　15km　　　　E)　17km

Question 16:

Building foundations are covered by 14cm of concrete. A builder thinks this is too thick, and grinds down the concrete by an amount three times the thickness of the concrete which he eventually leaves.

What is the remaining thickness of concrete?

A)　1.5cm　　　　B)　2.0cm　　　　C)　2.5cm　　　　D)　3.0cm　　　　E)　3.5cm

Question 17:

Chris leaves his house to go and visit Laura, who lives 3 miles away. He leaves at 1730 and walks at 4mph towards Laura's house, stopping for one 5-minute to chat to a friend. Meanwhile Sarah also wants to visit Laura. She sets off from her house 6 miles away at 1810, driving in her car and averaging a speed of 24mph.

Who reaches the house first and with how long do they wait for the other person?
A)　Chris, and waits 5 mins for Sarah
B)　Chris, and waits 10 mins for Sarah
C)　Sarah, and waits 5 mins for Chris
D)　Sarah, and waits 10 mins for Chris
E)　They both arrive at the same time

Question 18:

"Illegal film and music downloads have increased greatly in recent years. This causes significant harm to the relevant industries. Many people justify this to themselves by telling themselves they are only diverting money away from wealthy and successful singers and actors, who do not need any more money anyway. But in reality, illegal downloads are deeply harming the music industry, making many studio workers redundant and making it difficult for less famous performers to make a living."

Which of the following best summarises the conclusion of this argument?
A)　Unemployment is a problem in the music industry
B)　Taking profits away from successful musicians does more harm than good
C)　Studio workers are most affected by illegal downloads
D)　Illegal downloads cause more harm than people often think
E)　Buying music legally helps keep the music industry productive

Question 19:

"40,000 litres of water will extinguish two typical house fires. 70,000 litres of water will extinguish two house fires and three garden fires. There is no surplus water"

Which statement is **NOT** true?

A)   A garden fire can be extinguished with 12,000 litres, with water to spare.
B)   20,000 litres is sufficient to extinguish a normal house fire.
C)   A garden fire requires only half as much water to extinguish as a house fire.
D)   Two house and four garden fires will need 80,000 litres to extinguish.
E)   Three house and ten garden fires will need 140,000 litres to extinguish.

Question 20:

A car travels at 20ms^{-1} for 30 seconds. It then accelerates at a constant rate of 2ms^{-2} for 5 seconds, then proceeds at the new speed for 20 seconds before braking with constant deceleration of 3ms^{-2} to a stop. What distance is covered in total?

A)   1325m   B)   1350m   C)   1375m   D)   1425m   E)   1475m

Question 21:

"Plans are in place to install antennas underground, so that users of underground trains will be able to pick up mobile reception. There are, as usual, winners and losers from this policy. Supporters of the policy argue that it will lead to an increase in workforce productivity and increase convenience in day-to-day life. Critics respond by saying that it will lead to an annoying environment whilst travelling, it will facilitate the ease of conducting a terrorist threat and it will decrease levels of sociability. The latter camp seems to have the greatest support and so a re-consideration of the policy is urged."

Which of the following **best** summarises the conclusion of this passage?

A)   The disadvantages of installing underground antennas outweigh the benefits
B)   The cost of the scheme is likely to be prohibitive
C)   The policy must be dropped, since a majority does not want it
D)   More people don't want this scheme than do want it
E)   A detailed consultation process should take place

Question 22:

"Ecosystems in the oceans are changing. Recently, restrictions on fishing have been imposed to tackle the decline in fish populations. As a result, farm fishing and the price of fish have increased, whilst the seas recover. It is hoped that these changes will lead to a brighter future for all."

Which of the following are **TWO** assumptions of this argument?

A)   People will still buy farmed fish at a higher price
B)   The population of wild fish can recover
C)   Fishermen will benefit from working on this scheme
D)   Ecosystems have been altered as a result of climate change
E)   Heavy sea fishing is to blame for the changes in the ecosystem

Question 23:

Brian is tossing a coin. He tosses the coin 5 times. What is the probability of tossing exactly 2 heads?

A) $^{1}/_{16}$ B) $^{5}/_{32}$ C) $^{4}/_{16}$ D) $^{5}/_{16}$ E) $^{7}/_{16}$

Question 24:

The amount of a cleaning powder to be added to a bucket of water is determined by the volume of water, such that exactly 40g is added to each litre. A bucket contains 5 litres of water, and is required to have cleaning powder added. However, the markings on the bucket are only accurate to the nearest 2%. Calculate the difference between the maximum and minimum amounts of cleaning powder which might be required to be added to make up the solution correctly.

A)   4g   B)   6g   C)   8g   D)   12g   E)   20g

Question 25:

International telephone calls are charged at a rate per minute. For a call between two European countries, the rate is 22p per minute off-peak and 32p per minute at peak hours, rounded up to the nearest whole minute. In addition, there is a connection fee of 18p for every call.

What is the cost of an off-peak call from France to Germany, lasting 1.4 hours?

A) £18.48   B) £18.66   C) £26.88   D) £27.06   E) £30.98

Question 26:

"UV radiation is harmful to the skin, and can lead to the development of skin cancers. Despite this, many people sunbathe and use tanning salons, exposing themselves to dangerous radiation. If people took more sensible decisions about their health, many serious diseases, such as skin cancers, could be avoided."

What is the main conclusion of this passage?

A) UV radiation is harmful to the skin
B) Many people like to get tanned, despite the risks
C) People do not always consider the health risks of choices they make
D) Skin cancer is a serious disease
E) Sunbathing is risky, and people should avoid it

Question 27:

Jim washes windows for pocket money. Washing a window takes two minutes. Between one house and the next, it takes Jim 15 minutes to pack up, walk to the next house and get ready to start washing again. Each resident pays Jim £3 per house, regardless of how many windows the house has. In one day, Jim washes 8 houses, with an average of 11 windows per house.

What is his equivalent hourly pay rate?

A) £4.38   B) £4.86   C) £5.12   D) £5.62   E) £6.12

Question 28:

"Bottled water is becomingly increasingly popular, but it is hard to see why. Bottled water costs many hundreds of times more than a virtually identical product from the tap, and bears a significant environmental cost of transportation. Those who argue in favour of bottled water may point out that the flavour is slightly better – but would you pay 300 times the price for a car with just a few added features?"

Which of the following, if true, would most weaken the above argument?

A) Bottled water has many health benefits in addition to tasting nicer
B) Bottled water does not taste any different to tap water
C) The cost of transportation is only a fraction of the costs associated with bottling and selling water
D) Some people do buy very expensive cars
E) Buying bottled water supports a big industry, providing many jobs to people

Question 29:

"There are no marathon runners that aren't skinny, nor no cyclists that aren't marathon runners."
Which of the following **must** be true?

A) Cyclists do not run marathons   D) Marathon runners must all be cyclists
B) Cyclists are all skinny   E) All of the above
C) Any skinny person is also a cyclist

Question 30:

"Langham is East of Hadleigh but West of Frampton. Oakton is midway between Langham and Stour. Frampton is West of Stour. Manley is not East of Langham."
Which of the following **cannot** be concluded?

A)   Oakton is East of Langham and Hadleigh.
B)   Frampton is West of Stour and East of Manley.
C)   Stour is East of Hadleigh and Langham.
D)   Oakton is East of Langham and West of Frampton
E)   Manley is West of Oakton and West of Frampton.

Question 31:

A pot of paint gives sufficient paint to cover 12m² of wall area. The inner surface of a planetarium must be painted. The planetarium consists of a hemispheric dome of internal diameter 14 metres. How many pots of paint are required to give the dome two full coats of paint? [Assume π=3]

A)   25   B)   36   C)   49   D)   64   E)   98

Question 32:

A planetarium has just been painted as in **31**, above. Assuming each pot of paint is 2 litres, and that the solid component of the paint is 40%, calculate the percentage decrease in the volume of the planetarium, due to the painting.

A)   0.0029%   B)   0.0057%   C)   0.029%   D)   0.057%   E)   2.86%

Question 33:

A sweet wrapping machine takes 400ms to wrap a sweet. How many sweets can it wrap in 2 hours?

A)   3,000   B)   7,000   C)   9,000   D)   14,000   E)   18,000

Question 34:

A train travels from Crabtree to Eppingsworth. There are four stations in between at which the train stops. The time taken to travel between each of these stations decreases by one-fifth for each leg of the journey. Travelling from Station 3 to Station 4 takes 16 minutes.
What is the total journey time?

A)   80 minutes   C)   105 minutes   E)   183 minute
B)   89 minutes   D)   125 minutes

Question 35:

Scott believes the number of Ford Escorts on the roads decreases by 25% every year. In 2005, there were 36,000 Escorts remaining. How many does Scott expect will remain in 2015?

A)   227   B)   2,027   C)   4,500   D)   7,700   E)   8,600

END OF SECTION

Section 2

Question 1:
The buoyancy force of an object is the produce of its volume, density and the gravitational constant, g. A boat weighing 600 kg with a density of $1000 kgm^{-3}$ and hull volume of 950 litres is placed in a lake. What is the minimum mass that, if added to the boat, will cause it to sink? Use $g = 10ms^{-1}$.

A) 3.55 kg
B) 35 kg
C) 350 kg

D) 355 kg
E) 3,550 kg
F) None, the boat has already sunk

Question 2:
Which of the following below is **NOT** an example of an oxidation reaction?

A) $Li^+ + H_2O \rightarrow Li^+ + OH^- + \frac{1}{2}H_2$
B) $N_2 \rightarrow 2N^+ + 2e^-$
C) $2CH_4 + 2O_2 \rightarrow 2CH_2O + 2H_2O$

D) $2N_2 + O_2 \rightarrow 2N_2O$
E) $I_2 + 2e^- \rightarrow 2I^-$
F) All of the above are oxidation reactions

Question 3:
Which of the following statements, regarding normal human digestion, is **FALSE**?

A) Amylase is an enzyme which breaks down starch
B) Amylase is produced by the pancreas
C) Bile is stored in the gallbladder

D) The small intestine is the longest part of the gut
E) Insulin is released in response to feeding
F) None of the above

Question 4:
Mr Khan fires a bullet at a speed of 310 ms^{-1} from a height of 1.93m parallel to the floor. Mr Weeks drops an identical bullet from the same height.

What is the time difference between the bullets first making contact with the floor?[Assume that there is negligible air resistance; $g= 10$ ms^{-2}]

A) 0 s
B) 0.2 s
C) 1.93 s

D) 2.1 s
E) More information is needed

Question 5:
A 1.4kg fish swims through water at a constant speed of 2ms^{-1}. Resistive forces against the fish are 2N. Assuming $g = 10$ms^{-2}, how much work does the fish do in one hour?

A) 7,200 J
B) 10,080 J
C) 14,400 J

D) 19,880 J
E) 22,500 J
F) More information is needed

Question 6:
Jane is one mile into a marathon. Which of the following statements is **NOT** true, relative to before she started?

A) Blood flow to the skin is increased
B) Blood flow to the muscles is increased
C) Blood flow to the gut is decreased

D) Blood flow to the kidneys is decreased
E) Cardiac Output Increases
F) None of the above

Question 7:
Balance the following chemical equation. What is the value of **x**?

$$w\ HIO_3 + 4FeI_2 + x\ HCl \rightarrow y\ FeCl_3 + z\ ICl + 15H_2O$$

A) 4   B) 5   C) 9   D) 15   E) 22   F) 25

Question 8:
A newly discovered species of beetle is found to have 29.6% Adenine (A) bases in its genome. What is the percentage of Cytosine (C) bases in the beetle's DNA?

A) 20.4%

B) 29.6%

C) 40.8%

D) 59.2%

E) 70.6%

F) More information is required

Question 9:
Study the following diagram of the human heart. What is true about structure **A**?

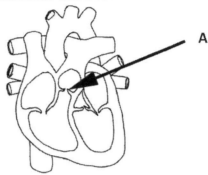

A) It is closed during systole

B) It prevents blood flowing into the left ventricle during systole

C) It prevents blood flowing into the right ventricle during systole

D) It prevents blood flowing into the left ventricle during diastole

E) It opens due to left ventricular pressure being greater than aortic pressure.

F) It is open when the right ventricle is emptying

Question 10:
Carbon monoxide binds irreversibly to the oxygen binding site of haemoglobin. Which of the following statements is true regarding carbon monoxide poisoning?

A) Carbon monoxide poisoning has no serious consequences

B) Haemoglobin is heavier, as both oxygen and carbon monoxide bind to it

C) Affected individuals have a raised heart rate

D) The CO_2 carrying capacity of the blood is decreased

E) The O_2 carrying capacity of the blood is unchanged as it dissolves in the plasma instead

Question 11:
A crane is 40 m tall. The lifting arm is 5m long and the counterbalance arm is 2m long. The beam joining the two weighs 350kg, and is of uniform thickness. The lifting arm lifts a 2000 kg mass. What counterbalance mass is required to balance exactly around the centre point? Use $g = 10$ ms^{-2}.

A) 4,220 kg

B) 4,820 kg

C) 5,013 kg

D) 5,263 kg

E) 10,525 kg

Question 12:
For Christmas, Mr James decorates his house with 20 strings of 150 bulbs each. Each 150-bulb string of lights is rated at 50 Watts. Mr James turns the lights on at 8pm and off at 6am each night. The lights are used for 20 days in total.

If 100 kJ of energy costs 2p, how much is the total cost Mr James has to pay?

A) £2160.00

B) £144.00

C) £14.40

D) £0.72

E) £0.24

Question 13:
Calculate the perimeter of a regular polygon each interior angle is 150° and each side is 15 cm.

A) 75 cm

B) 150 cm

C) 180 cm

D) 225 cm

E) 1,500 cm

F) More information is needed.

Question 14:

The diagram shown below depicts an electrical circuit with multiple resistors, each with equal resistance, Z. The total resistance between A and B is 22 MΩ. Calculate the value of Z.

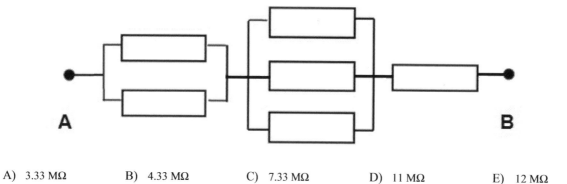

A) 3.33 MΩ B) 4.33 MΩ C) 7.33 MΩ D) 11 MΩ E)   12 MΩ

Question 15:

A cylindrical candle of diameter 4cm burns steadily at a rate of 1cm per hour. Assuming the candle is composed entirely of paraffin wax ($C_{24}H_{52}$) of density 900 kgm^{-3} and undergoes complete combustion, how much energy is transferred in 30 minutes? You may assume the molar combustion energy is 11,000 kJmol^{-1}, and that $\pi=3$.

A) 140,000J   C) 185,000J   E) 215,000J
B) 175,000J   D) 200,500J   F) 348,000J

Question 16:

A different candle to that in question **15** is used to heat a bucket of water. The candle burns for 45 minutes, releasing 250KJ of energy as heat. It is used to heat a 2 litre bucket of water at 25°C.

Assuming the bucket is completely insulated, what is the water temperature after 45 minutes? (For reference: One calorie heats one cm^3 of water by one degree Celsius, 1kCal = 4,200J).

A) 35°C   C) 55°C   E) 75°C
B) 45°C   D) 65°C   F) 85°C

Question 17:

A person responds to the starting gun of a race and begins to run. Place the following order of events in the most likely chronological sequence. Which option is a correct sequence?

1	Blood CO_2 increases	5	Impulses travel along relay neurones
2	The eardrum vibrates to the sound	6	Quadriceps muscles contract
3	Impulses travel along motor neurones	7	Glycogen is converted into glucose
4	Impulses travel along sensory neurones	8	Creatine phosphate rapidly re-phosphorylates ADP

A) $2 \rightarrow 5 \rightarrow 4 \rightarrow 3 \rightarrow 6 \rightarrow 7$ D) $2 \rightarrow 4 \rightarrow 3 \rightarrow 1 \rightarrow 6 \rightarrow 7$
B) $2 \rightarrow 4 \rightarrow 3 \rightarrow 8 \rightarrow 6 \rightarrow 1$ E) $2 \rightarrow 4 \rightarrow 3 \rightarrow 6 \rightarrow 8 \rightarrow 7$
C) $2 \rightarrow 3 \rightarrow 4 \rightarrow 6 \rightarrow 7 \rightarrow 1$

Question 18:

On analysis, an organic substance is found to contain 41.4% Carbon, 55.2% Oxygen and 3.45% Hydrogen by mass. Which of the following could be the chemical formula of this substance?

A) $C_3O_3H_6$ C) $C_4O_2H_4$ E) $C_4O_2H_8$
B) $C_3O_3H_{12}$ D) $C_4O_4H_4$ F) More information needed

Question 19:

Simplify and solve: $(e - a)(e + b)(e - c)(e + d)...(e - z)$?

A) 0

B) e^{26}

C) $e^{26}(a-b+c-d...+z)$

D) $e^{26}(a+b-c+d...-z)$

E) $e^{26}(abcd...z)$

F) None of the above.

Question 20:

Which of the following best describes the events that occur during expiration?

A) The ribs move up and in; the diaphragm moves down.

B) The ribs move down and in; the diaphragm moves up.

C) The ribs move up and in; the diaphragm moves up.

D) The ribs move down and out; the diaphragm moves down.

E) The ribs move up and out; the diaphragm moves down.

F) The ribs move up and out; the diaphragm moves up.

Question 21:

Simplify fully: $1 + \left(3\sqrt{2} - 1\right)^2 + \left(3 + \sqrt{2}\right)^2$

A) $30 + 6\sqrt{2} - 2\sqrt{18}$

B) $30 + 6\sqrt{2} + 2\sqrt{18}$

C) $3[2(\sqrt{2} - 1) + 2]$

D) 24

E) 29

F) 31

Question 22:

200 cm^3 of a 1.8 moldm^{-3} solution of sodium nitrate ($NaNO_3$) is used in a chemical reaction. How many moles of sodium nitrate is this?

A) 0.09 mol

B) 0.36 mol

C) 9.00 mol

D) 36.0 mol

E) 360 mol

Question 23:

A tourist at Victoria Falls accidentally drops her 400g camera. It falls 125 metres into the water below. Assuming resistive forces to be zero and $g = 10\text{ms}^{-1}$, what is the momentum of the camera the instant before it strikes the water? [Momentum = mass x velocity]

A) 4 kgms^{-1}

B) 13 kgms^{-1}

C) 16 kgms^{-1}

D) 20 kgms^{-1}

E) 50 kgms^{-1}

F) $20,000 \text{ kgms}^{-1}$

Question 24:

Antibiotics can have serious side effects such as liver failure and renal failure. Therefore, scientists are always trying to develop antibiotics to minimise these effects by targeting specific cellular components. Which of these cellular components offers the best way to treat infections and minimise side effects?

A) Mitochondrion

B) Cell membrane

C) Nucleic acid

D) Cytoskeleton

E) Flagellum

Question 25:

A is a group 3 element and B is a group 6 element. Which row best describes what happens to A when it reacts with B?

	Electrons are	Size of Atom
A)	Gained	Increases
B)	Gained	Decreases
C)	Gained	Unchanged
D)	Lost	Increases
E)	Lost	Decreases
F)	Lost	Unchanged

Question 26:

Each vertex of a square lies directly on the edge of a circle with a radius of 1cm. Calculate the area of the circle that is not occupied by the square. Use $\pi = 3$.

A) 0.25cm^2

B) 0.5cm^2

C) 0.75cm^2

D) 1.0cm^2

E) 1.25cm^2

F) 1.5cm^2

Question 27:

A funicular railway like the one illustrated lifts a full carriage weighing 3600kg up an incline. The distance travelled is 200m, and the vertical ascent, **v**, is 80m. Ten passengers weighing an average of 72kg disembark, then the carriage descends. As a result of efficient design, the energy from the descent is stored to drive the next ascent.

 Assuming the same load of 10 passengers then enters the car, how powerful an engine is required to move the carriage at 4ms⁻¹?

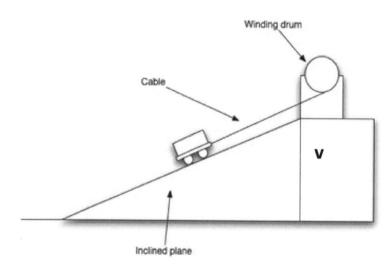

A)   9.2 kW

B)   11.5 kW

C)   28.8 kW

D)   46.1 kW

E)   57.6 kW

END OF SECTION

Section 3

1) *'Doctors know best and should decide which treatment a patient receives'*

Explain what this statement alludes to. Argue to the contrary that patients know best and should be able to choose their management plan. To what extent do you agree with this statement?

2) *'World peace will be achieved in the future'*

Explain what this statement means. Argue to the contrary, that world peace will never be achieved. To what extent, if any, do you agree with the statement?

3) *'Medicine is a science; not an art'*

Explain what this statement means. Argue to the contrary that medicine is in fact an art using examples to illustrate your answer. To what extent, if any, is medicine a science?

4) *'People should live healthier lives to reduce the financial burden of healthcare to the taxpayer.'*

Explain what this statement means. Argue to the contrary. To what extent do you agree with the statement?

END OF PAPER

Mock Paper F

Section 1

Question 1:

Every year, there are tens of thousands of motor crashes, causing a serious number of fatalities. Indeed, this represents the leading cause of death in the UK that is not a disease. In spite of this horrendous statistic, there are still thousands of uninsured drivers. The government is under moral obligation to clamp down on uninsured drivers, to reduce the incidence of such crashes. That they have not acted is arguably the most outrageous failing of the present government.

Which of the following is the best statement of a **flaw** in this passage?

A)　It has made unsupported claims that the government's failure to act is morally outrageous.

B)　It has not provided any evidence to support its claims that motor crashes are the leading cause of death in the UK outside of diseases.

C)　Even if motor crashes were prevented, it would not save lives of people who die from other causes.

D)　It has implied that lack of insurance is related to the incidence of motor crashes.

E)　It has fabricated an obligation on the government's part to intervene and reduce the numbers of uninsured drivers.

Question 2:

Several years ago the Brazilian government held a referendum of the populace, to decide whether they should enact a law banning the ownership of guns. The Brazilian people voted strongly against this proposal. When asked why this had happened, one commentator said he believed the reason was that 90% of criminals who use guns to commit crimes buy their weapons on the black market, illegally. Thus, if Brazil were to ban the legal sale of guns, this would remove the ability of law-abiding citizens to purchase protection, whilst doing little to remove weapons from the hands of criminals.

Some commentators have pointed to this statistic, and claimed that the UK should also legalise guns, to allow citizens to protect themselves. However, in the UK the black market for weapons is not as widespread as in Brazil. Most people in the UK have little reason to fear gun attacks, and legalising the sale of guns would simply make it much easier for criminals to acquire weapons.

Which of the following best expresses the main conclusion of this passage?

A)　The UK should not follow Brazil's lead on gun legislation.

B)　Efforts to reduce gun ownership should focus on the black market.

C)　Violent crime is a more pressing concern in Brazil than the UK.

D)　Legalising the sale of guns in the UK would result in widespread ownership.

E)　Criminals will always find a way to obtain firearms.

Question 3:

Hannah is buying tiles for her new bathroom. She wants to use the same tiles on the floor and all 4 walls and for all the walls to be completely tiled apart from the door. The bathroom is 2.4 metres high, 2 metres wide and 2 metres long, and the door is 2 metres high, 80cm wide and at the end of one of the 4 identical walls. The tiles she wants to use are 40cm x 40cm. How many of these tiles does she need to tile the whole bathroom?

A)　110　　　　　　B)　120　　　　　　C)　135　　　　　　D)　145　　　　　　E)　150

Question 4:

Jane and Trevor are both travelling south, from York to London. Jane is driving, whilst Trevor is travelling by train. The speed limit on the roads between York and London is 70mph, whilst the train travels at 90mph. Thus, we should expect that Trevor will arrive first.

Which of the following would weaken this passage's conclusion?

A) The train takes a direct route, whilst the road from York to London goes through several major cities and zig-zags somewhat on its way down the country.
B) Trevor left before Jane.
C) Jane is a conscientious driver, who never exceeds the speed limit.
D) Trevor's train makes a lot of stops on the way, and spends several minutes at each stop waiting for new passengers to board.
E) Meanwhile, Raheem is making the same journey by plane, and will arrive before either Trevor or Jane.

Question 5:

A recipe for 20 cupcakes needs 200g of butter, 200g of sugar, 200g of flour and 4 eggs. Jeremy has two 250g packs of butter, a bag of 600g of sugar, a kilogram bag of flour and a pack of 12 eggs.

How many cupcakes can he make and how many eggs does he have left over?

A) 50, 2   B) 50, 3   C) 60, 0   D) 60, 2   E) 60, 3

Question 6:

ABC taxis charges a rate of 15p per minute, plus £4. XYZ taxis charges a rate of £4 plus 30p per mile. I live 6 miles from the station.

What would the taxi's average speed have to be on my journey home from the station for the two taxi firms to charge exactly the same fare?

A) 25   B) 30   C) 45   D) 55   E) 60

Question 7:

King Arthur has been issued a challenge by Mordac, his nephew who rules the adjacent Kingdom. Mordac has challenged King Arthur to select a knight to complete a series of challenging obstacles, battling a number of dark creatures along the way, in a test known as the Adzol. The King's squire reports that there are tales told by the elders of the court meaning that only a knight with tremendous courage will succeed in Adzol, and all others will fail. He therefore suggests that Arthur should select Lancelot, the most courageous of all Arthur's Knights. The squire argues that due to what the Elders have said, Lancelot will succeed in the task, but all others will fail.

Which of the following is **NOT** an assumption in the squire's reasoning?

A) Lancelot has sufficient courage to succeed in the Adzol.
B) No other knights in Arthur's command also have tremendous courage, so will all fail Adzol.
C) Great courage is required to be successful in the Adzol.
D) The tales told by the elders of the court are correct.
E) None of the above – they are all assumptions.

Question 8:

A historian is examining a recently excavated hall beneath a medieval castle. She finds that there are a series of arch-shaped gaps along one length of the wall, surrounded by a different pattern of bricks to that seen elsewhere in the walls. These are found to represent where windows where once located, looking out onto one side of the castle. However, the site is now underground. Underground halls in castles never contain windows, so the historian reasons that this hall must once have been located above the ground. Therefore, the ground level must have changed since this castle was built.

Which of the following represents the main conclusion of this passage?

A) Windows are never found in underground halls.
B) Arch-shaped gaps always indicate that windows were once present.
C) It is unexpected for windows to be found in halls in castles.
D) The hall was once located above ground.
E) The ground level must have changed since this hall was built.

Question 9:

Adam's grandmother has sent him to the shop to buy bread rolls. Usually, bread rolls are 30p for a pack of 6 and so his grandmother has given him the exact amount to buy a certain number of bread rolls. However, today there is a special offer whereby if you buy 3 or more packs of rolls, the price per roll is reduced by 1p. He can now buy 1 more pack than before and get no change.
How many bread rolls was he originally supposed to buying?

A) 4   B) 5   C) 6   D) 24   E) 30

Question 10:

The England men's cricket team have recently been knocked out of the world cup after a very poor performance that saw them eliminated at the group stage, managing only 1 win and losing against teams well below them in the rankings. The board of English cricket is sitting down to discuss why the team's performance was so poor, and what can be done to ensure that future world cups have a more positive outcome. The chairman of the board says that the current crop of players is not good enough, and that the team's performance should improve soon, as more able players come through the ranks in the county teams, so no action is needed.

However, the sporting director takes a different view, saying that England have not gone further than the group stage of any cricket world cup for the last 25 years, during which time numerous players have come and gone from the team. The sporting director argues that this long period of poor performance indicates that there is a problem with English cricket, meaning that not enough talented players are being produced in the country. He argues that therefore, steps should be taken to reform English cricket to actively foster the development of more talented players.

Which of the following, if true, would most strengthen the sporting director's argument?

A) The English cricket team is regarded as one of the best in the world, with some of the most talented players.
B) England have been steadily falling lower in the world cricket rankings for the last 25 years, due to poor performances across the board in various cricket competitions.
C) A skilled batsman, who was ranked as the 4th best player in the world, has recently retired from the England team. Now, there are no English cricket players in the top 10 of the world cricket player rankings, which is the first time this has happened in over 70 years.
D) Despite not performing well in world cups, England have performed well in other cricket competitions over the last 20 years.
E) Cricket was invented in England, so everybody expects that England should have a lot of good players in their team.

Question 11:

Karl is making cupcakes for a wedding. It takes him 25 minutes to prepare each batch of cakes. Only 12 can go in the oven at a time and each batch takes 20 minutes in the oven.

What is the latest time Karl can start if he needs to make 100 cupcakes by 4pm?

A) 11:55am   B) 12:20pm   C) 12:40pm   D) 13:20pm   E) 14:00pm

Question 12:

	Boys Absenteeism	Girls Absenteeism	Pupils on Roll	Average
Hazelwood Grammar	7%	Boys' School	300	7%
Heather Park Academy	5%	6%	1000	5.60%
Holland Wood Comprehensive	5%	6%	500	5.60%
Hurlington Academy	Girls' School		200	
Average		7%		

Some of the information is missing from the table above. What is the rate of girls' absenteeism at Hurlington Academy?

A)   6.5%   B)   7%   C)   9%   D)   11.5%   E)   13%

Question 13:

Up until the 20th century, all watches were made by hand, by watchmakers. Watchmaking is considered one of the most difficult and delicate of manufacturing skills, requiring immense patience, meticulous attention to detail and an extremely steady hand. However, due to the advent of more accurate technology, most watches are now produced by machines, and only a minority are made by hand, for specialist collectors. Thus, some watchmakers now work for the watch industry, and only perform *repairs* on watches that are initially produced by machines.

Which of the following *cannot* be reliably concluded from this passage?

A)   Most watches are now produced by machines, not by hand.
B)   Watchmaking is considered one of the most difficult of manufacturing skills
C)   Most watchmakers now work for the watch industry, performing repairs on watches rather than producing new ones.
D)   The advent of more accurate technology caused the situation today, where most watches are made by machines.
E)   Some watches are now made by hand for specialist collectors.

Question 14:

Many vegetarians claim that they do not eat meat, poultry or fish because it is unethical to kill a sentient being. Most agree that this argument is logical. However, some Pescatarians have also used this argument, that they do not eat meat because they do not believe in killing sentient beings, but they are happy to eat fish. This argument is clearly illogical. There is powerful evidence that fish fulfil just as much of the criteria for being sentient as do most commonly eaten animals, such as chicken or pigs, but that all these animals lack certain criteria for being "sentient" that humans possess. Thus, pescatarians should either accept the killing of beings less sentient than humans, and thus be happy to eat meat and poultry, or they should not accept the killing of any partially sentient beings, and thus not be happy to eat fish.

Which of the following best illustrates the main **conclusion** of this passage?

A)   The argument that it is unethical to eat meat due to not wishing to kill sentient beings but eating fish is acceptable is illogical.
B)   Pescatarians cannot use logic.
C)   Fish are just as sentient as chicken and pigs, and all these beings are less sentient than humans.
D)   It is not unethical to eat meat, poultry or fish.
E)   It is unethical to eat all forms of meat, including fish and poultry.

Question 15:

Recent research into cultural attitudes in British has revealed a striking hypocrisy. When asked whether foreign people travelling to British on holiday should learn some English, 60% of respondents answered yes. However, when asked if they would attempt to learn some of the language before travelling to a country which did not speak English, only 15% of the respondents answered yes. This is a shocking double-standard on the part of the British public, and is symptomatic of a deeper underlying issue that British people feel themselves superior to other cultures.

Which of the following can be reliably concluded from this passage?

A) 60% of people in Britain think that foreign people travelling to Britain for a holiday should learn English, but would not learn the language themselves when going on holiday to a country which did not speak English.

B) The British public do not feel that it is important to learn some of the language before travelling to a country which does not speak English.

C) There are numerous issues of racism amongst the British public, stemming from the fact they feel themselves superior to other cultures.

D) Less than 10% of the British public would attempt to learn some of the language before travelling to a country which did not speak English.

E) Some in Britain think that foreign people travelling to Britain for a holiday should learn English, but would not learn the language themselves when going on holiday to a country which did not speak English.

Question 16:

Harriet is a headmistress and she is making 400 information packs for the sixth form open evening. Each information pack needs to have 2 double sided sheets of A4 of general information about the school. She also needs to produce 50 A5 single sided sheets about each of the 30 A Level courses on offer. Single sided A5 costs £0.01 per sheet. Double sided costs twice as much as single sided. A4 printing costs 1.5 times as much as A5.

How much does she spend altogether on the printing?

A) £27   B) £31   C) £35   D) £39   E) £43

Question 17:

Kirkleatham Town football club are currently leading the league. One week they play a crucial match against Redcar Rovers, who are second placed. The points tally of the teams in the table means that if Kirkleatham Town win this game, they will win the league. Before the game, the manager of Kirkleatham Town says that Redcar Rovers are a tough opponent, and that if his team do not play with desire and commitment, they will not win the game. After the game, the manager is asked for comment on the game, and says he was pleased that his team played with so much desire, and showed high levels of commitment. Therefore, Kirkleatham will win the league.

Which of the following best illustrates a flaw in this passage?

A) It has assumed that Kirkleatham will not win the game if they do not play with desire and commitment.

B) It has assumed that if Kirkleatham play with desire and commitment, they will win the game.

C) It has assumed that Kirkleatham played with desire and commitment.

D) It has assumed that Redcar Rovers are a tough opponent, and that Kirkleatham will not be able to easily win the game.

E) It has assumed that if Kirkleatham win the match against Redcar Rovers, they will win the league.

Question 18:

Two councillors are considering planning proposals for a new housing estate, to be built on the edge of Bluedown Village. Councillor Johnson argues for a proposal to be built upon brownfield land, land which has previously been built on, rather than greenbelt land, which has not previously been built on. He argues that this will both lower the cost of building the estate, as the land would already have some underlying infrastructure and would not need as much preparation, and will ensure a minimal impact on wildlife around the area.

Which of the following would most weaken the councillor's argument?

A) Brownfield land is often not as appealing as greenbelt land visually, and it is likely that houses built on brownfield land will not sell for as high a price as houses built on greenbelt land.

B) An area of brownfield land on the edge of the village, originally built as an outdoor leisure complex, has since become run down, and ironically is now a haven for various types of rare newts, lizards and birds.

C) Much of the brownfield land around the edge of the village has undergone substantial underground development, with a good system of electricity cables, gas pipes and plumbing in place.

D) The village is surrounded by several greenbelt areas designated as areas of outstanding natural beauty, supporting an abundance of wildlife.

E) The village mayor, who has ultimate control over the planning proposal, agrees with councillor Johnson's argument. Thus, it is likely his recommendations will be followed

Question 19:

	Pool A	Pool B	Pool C	Pool D
1st	France	Argentina	England	South Africa
2nd	Holland	Mexico	Nigeria	Brazil
3rd	United States	Denmark	Germany	Japan
4th	India	Korea	Ghana	Algeria
5th	Australia	Switzerland	Portugal	Serbia
6th	Greece	New Zealand	Honduras	Uruguay
7th	Chile	Slovakia	Cameroon	Paraguay

The table above shows the final standings in the pool stages of a football competition. The top 2 teams from each pool progress into the quarterfinals. The fixtures for the quarterfinals are determined as follows:

QF1: Winners Pool A vs. Runner up Pool B   **QF3:** Winners Pool C vs. Runner up Pool D
QF2: Winners Pool B vs. Runner up Pool C   **QF4:** Winners Pool D vs. Runner up Pool A

The winners of QF1 then play the winners of QF3 in one semifinal, and the winners of QF2 and winners of QF4 play each other in the other semifinal. The winners of the semi-finals progress to the final.
Which of these teams could England play in the final?

A) Nigeria   B) France   C) Mexico   D) Denmark   E) Brazil

Question 20:

The pie chart shows the voting intentions of some constituents interviewed by a polling group, prior to an upcoming election.

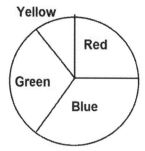

How many times more people said their intention was to vote for the red party than the yellow party?

A) 2   B) 3   C) 4   D) 5   E) 6

Question 21:

A pizza takeaway is having a sale. If you spend £30 or more at full price, you can get 40% off.
Prices are as follows:
- Basic cheese and tomato pizza: £8 small, £10 large
- All other toppings are £1 each
- Sides are: Garlic bread £3, Potato wedges £2.50, Chips £1.50 and Dips £1 each

Ellie and Mike want to order a large pizza with mushrooms and ham, garlic bread, 2 portions of chips and a dip.

Which of these additional items can they order to minimise the amount they have to pay?

A) Small pizza with pineapple and onion   D) 4 portions of potato wedges
B) Large pizza with mushroom   E) Garlic bread
C) Barbecue dip

Question 22:

	Goals Scored	Goals Conceded
City	10	4
United	8	5
Rovers	1	10

The table above shows the goal scoring record of teams in a football tournament. Each team plays the other teams twice, once at home and once away. Here are the results of the first 4 matches:
➢ United 2 – 2 City   ➢ City 2 – 1 Rovers
➢ Rovers 0 – 3 City   ➢ Rovers 0 – 3 United

What were the results of the final two fixtures?

A) United 2 – 0 Rovers, City 0 – 0 United
B) United 1 – 0 Rovers, City 1 – 1 United
C) United 0 – 0 Rovers, City 2 – 1 United
D) United 1 – 0 Rovers, City 2 – 2 United
E) United 2 – 0 Rovers, City 3 – 1 United

Question 23:

The M1 Abrams tank is widely regarded as the most fearsome tank in the world. Highly advanced depleted uranium composite armour makes it difficult to damage from range, whilst a good top speed in excess of 50kmph and a large fuel capacity make it difficult to catch and contain the tank in an operational context. Whilst the tank does have weak spots that can be exploited at close range, a formidable 122m smoothbore gun as the main armament makes this an incredibly dangerous tactic for opposing tanks.

Country X is developing a new main battle tank to boost the prowess of their armoured formations, and have released a statement describing how they will implement next-generation armour into this new tank, to boost its defensive capacity. The government of country X believe this will allow their new tank to compete with the best tanks in the world. However, this view is mistaken. The M1 Abrams clearly demonstrates that a *combination* of different factors, including protection, manoeuvrability and firepower, are responsible for its status as the world's most formidable tank. Simply increasing the defensive capabilities of a tank is not sufficient. Thus, Country X's government is clearly incorrect in this matter.

Which of the following best illustrates the main conclusion of this passage?

A) Increasing the defensive capacity of a tank is not sufficient to make it equal to the best tanks in the world.
B) Multiple factors are required to make a tank equal to the best tanks in the world.
C) The new tank will not be as good as the M1 Abrams, as its defensive capacity will not be as good.
D) The view of Country X's government, that increasing the defensive capacity of a tank will make it equal to the best in the world, is clearly incorrect.
E) No tank is able to compete with the M1 Abrams, which will always be the world's most formidable tank.

Question 24:

The table below shows the balances of my bank accounts in pounds. Interest is paid at the end of the calendar year. My salary, which is the same every month, is paid into my current account on the 2nd of each month. All the money I have is in one or other of my bank accounts.

	Current Account	Savings	ISA
1st March	1300	5203	2941
1st April	3249	2948	2941
1st May	4398	9384	0
1st June	3948	8292	0

In which month did I spend the most money?

A) February
B) March
C) April
D) May
E) 2 or more months are the same

Question 25:

On Monday, my son developed a disease; no one else in the house has the disease. The doctor gave me some medicine and told me that everyone in the house who does not have the disease should also take half the dose. We need to take the medicine for 10 days, and the dosage is based on weight.

Weight	Dosage
Under 30kg	0.1ml per kg, 3 times a day
30kg – 60kg	0.2ml per kg, 4 times a day
60kg +	0.1ml per kg, 6 times a day

My son is 40 kg. I also have a daughter who is 20 kg. I am 75 kg and my husband is 80 kg. How many 200 ml bottles of medicine will we need for the whole 10 days?

A) 4   B) 5   C) 6   D) 7   E) 8

Question 26:

At Tina's nursery school, they have red, yellow or blue plastic cutlery. They have just enough forks and just enough knives for the 21 children there. There are the same number of forks as knives of each colour. Twice as many pieces of cutlery are yellow as blue. Half as many pieces of cutlery are red as blue. Tina takes a fork and a knife at random. What is the probability that she will get her favourite combination, a red fork and a yellow knife?

A)   4/49   B)   1/9   C)   36/49   D)   3/9   E)   3/49

Question 27:

The UK's taxation and public spending is horrendously flawed, with various immoral features. One example of such a flaw is the subsidy of public transport with money raised via taxation. According to recent research, public transport is only used by 65% of the population, and since there is no economic benefit stemming from a good public transport system, the other 35% of the population gets no benefit from public transport, but are still required to pay towards it via taxation. The system is in urgent need of reform, such that taxation is only used to support services and systems which are of benefit to everyone.

Which of the following is the best application of the principle used in this passage?

A)   Only 48% of the population have ever visited an art gallery, so public funds should not be used to subsidise art galleries, as not all the population use it.
B)   Primary and Secondary education provides an economic benefit to the whole country, so public funds should be used to support schools.
C)   Although many people never use a hospital, we should still use public funds to provide them, because many people cannot afford private healthcare, and thus we need a publically available health service for those people.
D)   There is no evidence that the fire service provides any benefit to the majority of the public, who will never experience a house fire in their lifetime. Thus, the fire service should not be publically funded via taxation.
E)   The Police service is a vital service for the country so should be publically funded regardless of how few people benefit from its presence.

Question 28:

SpicNSpan Inc is a cleaning company offering a range of cleaning services across the UK. The board has recently acquired a new chairman, who has called a meeting of the board to assess how the company can move forwards, expanding its services and increasing its market share. One of the things the new chairman is looking at is the types of services the company provides. He argues that their "All inclusive" service, where customers pay a fixed amount to clean a house throughout as a one-off event, are more popular than their "Hourly" services, where customers pay for a cleaner to carry out a certain number of hours each week. The new chairman argues that they should therefore focus on the "All inclusive services", rather than the "Hourly" services, in order to increase profits.

Which of the following best illustrates a flaw in the Chairman's argument?

A)   The company offers other services which may bring in even more profit than All Inclusive Services
B)   The fact that All inclusive services are more popular than Hourly services does not mean that they are more profitable. Hourly services may be more profitable.
C)   He has assumed that hourly services are more popular than All inclusive services
D)   He has assumed that all inclusive services are more popular than hourly services
E)   The rest of the board may have other strategies to increase profits, which are better than the new Chairman's.

Question 29:

The effects of fossil fuels such as Oil, Coal and Natural gas on the environment are plain and clear for everybody to see. The long-term use of such non-renewable fuels to produce power has led to devastating climate change, and will continue to cause damage as long as it continues. With this in mind, the European Commission has devised a set of targets to promote energy production by different types of fuels. However, there is a glaring problem with these targets. Shockingly, the Commission has targeted a "150% increase in the amount of energy produced by Nuclear Power by 2025". This is an outrageous misjudgement, because Nuclear Power is a non-renewable fuel, just like Oil, Coal and Natural gas. If we wish to protect the environment and halt climate change, we need to switch to *renewable* fuels, which are proven not to cause damage to the environment, NOT non-renewables such as Nuclear.

Which of the following best illustrates a flaw in this passage?

A)   It has assumed that all non-renewable power sources cause environmental damage.
B)   It has assumed that renewable energy sources do not cause environmental damage.
C)   It has assumed that the targets will be met, when in fact there is no guarantee that this will happen.
D)   It has neglected to consider other problems with the targets set by the Commission.
E)   It has assumed that the climate change caused by burning of oil, coal and natural gas cannot be offset or prevented by other strategies.

Question 30:

Despite the overwhelming evidence which certifies that vaccines are a miracle of modern medicine, and are responsible for saving a great number of lives, there remains a stubborn section of society that refuses to take vaccinations against important diseases, insisting that they are unsafe and ineffective. This group maintain this view in spite of extremely strong evidence that vaccines are safe, and against advice given by doctors. This group is particularly strong in the USA, where they pose a very real concern. Over the last 5 years, the proportion of the population that is unvaccinated has been rising by 1 each year, such that now a staggering 6% of Americans have not received any vaccinations.

Experts have advised that due to the way diseases are spread, if less than 90% of the population at any given time is unvaccinated, then it is almost certain that we will see an outbreak of Measles, a highly contagious and damaging disease. Thus, we expect that there will likely be an outbreak of measles in the next 5 years in the USA, and we should take steps to prepare for this.

Which of the following, if true, would most *strengthen* this argument?

A)   New and powerful evidence of the safety of vaccinations is due to be released to the public next year.
B)   Measles is a highly damaging disease, which frequently causes death or severe permanent injury in those affected.
C)   Throughout the last half-century, the number of people who are not vaccinated has risen and fallen continuously. Usually, the increases in non-vaccinated individuals occur over a 6-year period, after which time vaccination becomes more popular, and this number falls.
D)   The number of doctors advising against vaccination has been rising for the last 10 years, and shows no signs of decreasing.
E)   The rise in unvaccinated individuals has been increasing steadily for 5 years. The only time such a rate of increase has occurred in history was during the 1950s/1960s. In this case, a similar rate of increase in non-vaccinated individuals was maintained for a staggering 13 years.

Question 31:

It is well established that modern humans evolved in Africa, around 2 million years ago, and that the first humans were mainly hunter-gatherers, living off hunted meat and plant foods collected from their environment. However, this poses an interesting question. Humans are relatively weak, small, feeble creatures, and around 2 million years ago most wildlife in Africa consisted of large, powerful creatures. Thus, it is unclear how humans were able to hunt successfully, and obtain meat for food. One theory is that humans are well-built for long-distance running, largely thanks to our ability to control our temperature via sweating.

This theory reasons that humans were able to pursue animals such as antelope, which run when challenged, and were able to keep on running until the antelope collapsed through heat exhaustion. Meanwhile, the humans were kept cool via sweating, and were able to then go in and butcher the defenceless antelope.

Recent evidence has emerged supporting this theory, showing that human feet are well-developed for long-distance running, with fleshy areas in the correct orientation to absorb the impact without causing joint damage, and a heart well evolved to keep pumping at a moderately fast pace for long periods. With the emergence of this powerful new evidence, we should accept this theory, known as "the persistence running theory" as true.

Which of the following identifies a flaw in this argument?

A) The emergence of evidence in support of the persistence running theory does not mean that this theory is true.
B) There is little evidence that the human body is well setup for long-distance running.
C) It has neglected to consider other theories for how humans obtained meat during their early evolution.
D) There are numerous issues with the theory of persistence running, but many of these have been resolved thanks to the new evidence that has emerged.
E) It has not considered evidence that humans evolved in Europe, where there are smaller animals which humans may have easily been able to tackle.

Question 32:

		PREDICTED					
		A	**B**	**C**	**D**	**E**	**U**
ACTUAL	**A**	7	4	2	1	0	0
	B	3	8	2	2	1	0
	C	2	4	5	7	3	1
	D	2	2	2	6	5	0
	E	1	2	2	1	7	2
	U	1	1	0	3	5	6

The table above shows the actual and predicted AS grades for 100 AS mathematics students at Greentown Sixth Form. Each student is only predicted one grade. What percentage of students had their grades correctly predicted?

A) 14%   B) 16%   C) 39%   D) 61%   E) 78%

Question 33:

In one year, Mike lowers his workers' wages by x%. The next year, he lowers their wages by x%. The year after this, he raises the wages by x%. In the final year, he raises their wages by x%. In all these stages, x is a constant positive number.

Compared to the workers' original wages before any raising or lowering, what are their new wages?

A) The same as the original wages
B) Lower than the original wages
C) Higher than the original wages
D) Can't tell from the provided information even if we know what x was
E) Can't tell from the provided information but would be able to tell if we knew what x was.

Question 34:

The medical scientific establishment has a long established system for naming body parts and medical phenomena. This system is based upon ease of understanding, such that a body part, or a process of the body, is named based on its clinical relevance. This means that features are named in a way which will help doctors understand and explain to patients what the body part is, or what is wrong with it in the case of a disease. However, this poses significant problems for scientific medical research. Often, the most important features of a body part from a scientific point of view are not the most clinically important features, leading to confusion within the scientific literature, as medical researchers misunderstand the purpose of a discussion, due to confusing nomenclature. Whilst it is important for doctors to be able to explain things clearly to patients, it is relatively easy for this to happen in spite of confusing nomenclature, whereas confusing names cause serious problems in the scientific world. Thus, the naming system for medical features should be edited, to reflect the scientifically important features of body parts, rather than the clinically important ones.

Which of the following best illustrates the main conclusion of this passage?

A) The naming system based on clinically important features causes problems in scientific literature.
B) Changing the naming system would allow faster progress to be made in scientific medical research.
C) The naming system should be changed to reflect the features of body parts which are most important scientifically.
D) The current naming system is sufficient and should not be changed to help lazy scientists who cannot be bothered to do fact-checking.
E) It is more important to have good doctor-patient relations than good progress in scientific research.

Question 35:

Applicants for language teacher training have to specify which languages they studied as part of their degree. 180 people applied for teacher training. Of these, 128 did french as part of their degree. Half as many as did French did Spanish. Three quarters as many as did Spanish did not do either French or Spanish.

How many must have done both French and Spanish?

A) 12   B) 24   C) 36   D) 48   E) 60

END OF SECTION

Section 2

Question 1:
Why do cells undergo mitosis?

1. Asexual Reproduction
2. Sexual Reproduction
3. Growth of the human embryo
4. Replacement of dead cells

A) 1 only
B) 2 only
C) 3 only
D) 4 only
E) 2 and 3
F) 1, 2, and 3
G) 1, 3, and 4
H) 2, 3, and 4

Question 2:
A ball of radius 2 m and density 3 kg/m^3 is released from the top of a frictionless ramp of height 20m and rolls down. What is its speed at the bottom? Take $\pi = 3$ and $g = 10$m^{-2}.

A) 1 ms^{-1}
B) 4 ms^{-1}
C) 7 ms^{-1}
D) 9 ms^{-1}
E) 14 ms^{-1}
F) 20 ms^{-1}

Question 3:

In a healthy person, which one of the following has the highest blood pressure?

A) The vena cava
B) The systemic capillaries
C) The pulmonary artery
D) The pulmonary vein
E) The aorta
F) The coronary artery

Question 4:

Which of the following statements is true regarding waves?

A) Waves can transfer mass in the direction of propagation.
B) All waves have the same energy.
C) All light waves have the same energy.
D) Waves can interfere with each other.
E) None of the above.

The following information applies to questions 5 - 6:

Professor Huang accidentally touches a hot pan and her hand moves away in a reflex action. The diagram below shows a schematic of the reflex arc involved.

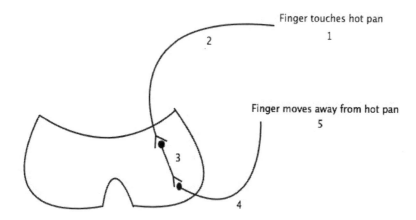

Question 5:

Which option correctly identifies the labels in the pathway?

	Muscle	Sensory Neurone	Receptor	Motor Neurone
A)	1	2	3	4
B)	2	3	1	5
C)	5	2	1	4
D)	1	4	5	2
E)	3	4	5	2
F)	4	2	1	3

Question 6:

Which one of the following statements is correct?

1. Information passes between 1 and 2 chemically.
2. Information passes between 2 and 3 electrically.
3. Information passes between 3 and 4 chemically.

A) 1 only
B) 2 only
C) 3 only
D) 1 and 2
E) 2 and 3
F) 1 and 3
G) All of the above
H) None of the above

Question 7:

Which of the following correctly describes the product of the reaction between hydrochloric acid and but-2-ene?

A) CH_3-CH_2-$C(Cl)H$-CH_3

B) CH_3-$C(Cl)$-CH_2-CH_3

C) $C(Cl)H_2$-CH_2-CH_2-CH_3

D) CH_3-CH_2-CH_2-$C(Cl)H_2$

E) None of the above.

Question 8:

Rearrange $\frac{(7x+10)}{(9x+5)} = 3z^2 + 2$, to make x the subject.

G. $x = \frac{15 z^2}{7 - 9(3z^2+2)}$

H. $x = \frac{15 z^2}{7 + 9(3z^2+2)}$

I. $x = -\frac{15 z^2}{7 - 9(3z^2+2)}$

J. $x = -\frac{15 z^2}{7 + 9(3z^2+2)}$

K. $x = -\frac{15 z^2}{7 + 3(3z^2+2)}$

L. $x = \frac{15 z^2}{7 + 3(3z^2+2)}$

Question 9:

The electrolysis of brine can be represented by the following equation: $2\ NaCl + 2\ X = 2\ Y + Z + Cl_2$

What are the correct formulae for X, Y and Z?

	X	Y	Z
A)	H_2O	H_2	O_2
B)	H_2O	NaOH	O_2
C)	H_2O	NaOH	H_2
D)	H_2	H_2O	O_2
E)	H_2	NaOH	O_2
F)	H_2	NaOH	H_2
G)	NaOH	H_2O	H_2
H)	NaOH	H_2O	O_2

Question 10:

Element $^{188}_{90}X$ decays into two equal daughter nuclei after a single alpha decay and the release of gamma radiation. What is the daughter element?

A) $^{91}_{45}D$

B) $^{92}_{44}D$

C) $^{184}_{88}D$

D) $^{186}_{90}D$

E) $^{186}_{45}D$

Question 11:

An unknown element has two isotopes: ^{76}X and ^{78}X. $A_r = 76.5$. Which of the statements below are true of X?

1. ^{76}X is three times as abundant as ^{78}X.
2. ^{78}X is three times as abundant as ^{76}X.
3. ^{76}X is more stable than ^{78}X.

A) 1 only

B) 2 only

C) 3 only

D) 1 and 3

E) 2 and 3

F) None of the above.

Question 12:

For the following reaction, which of the statements below is true?

$$6CO_{2\ (g)} + 6H_2O \rightarrow C_6H_{12}O_6 + 6O_{2\ (g)}$$

A) Increasing the concentration of the products will increase the reaction rate.

B) Whether this reaction will proceed at room temperature is independent of the entropy.

C) The reaction rate can be monitored by measuring the volume of gas released.

D) This reaction represents aerobic respiration.

E) This reaction represents anaerobic respiration.

Question 13:
Which of the following are true about the formation of polymers?

1.   They are formed from saturated molecules.
2.   Water is released when polymers form.
3.   Polymers only form linear molecules.

A)   Only 1   D)   1 and 2   G)   All of the above.
B)   Only 2   E)   1 and 3   H)   None of the above.
C)   Only 3   F)   2 and 3

Question 14:
The diagram below shows a series of identical sports fields:

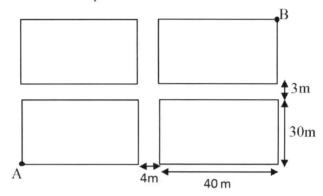

Calculate the shortest distance between points A and B.

A)   100 m   C)   146 m   E)   154 m
B)   105 m   D)   148 m   F)   None of the above.

Question 15:
Calculate $\dfrac{1.25 \times 10^{10} + 1.25 \times 10^{9}}{2.5 \times 10^{8}}$

    A)   0   D)   110   G) 5.5×10^{8}
    B)   1   E) 1.25×10^{8}
    C)   55   F) 5.5×10^{7}

The following information applies to questions 16 - 17:

Duchenne muscular dystrophy (DMD) is inherited in an X-linked recessive pattern [transmitted on the X chromosome and requires the absence of normal X chromosomes to result in disease]. A man with DMD has two boys with a woman carrier.

Question 16:
What is the probability that both boys have DMD?

A)   100%   B)   75%   C)   50%   D)   25%   E)   12.5%   F)   0%

Question 17:
If the same couple had two more children, what is the probability that they are both girls with DMD?

A)   100%   B)   75%   C)   50%   D)   25%   E)   12.5%   F)   0%

Question 18:

Which row of the table is correct regarding the cell shown below?

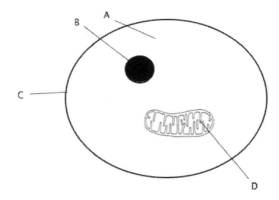

	Most Chemical Reactions occur here	Involved in Energy Release	Cell Type
A)	A	B	Animal
B)	A	B	Bacterial
C)	A	D	Animal
D)	B	D	Bacterial
E)	B	B	Animal
F)	B	A	Bacterial
G)	D	D	Animal
H)	D	B	Bacterial

Question 19:

Solve $y = 2x - 1$ and $y = x^2 - 1$ for x and y.

A) (0, -1) and (2, 3)   C) (1, 4) and (3, 2)   E) (3, -1) and (3, 1)

B) (1, -1) and (2, 2)   D) (2, -3) and (4, 5)   F) (4, -2) and (-2, 4)

Question 20:

Tim stands at the waterfront and holds a 30 cm ruler horizontally at eye level one metre in front of him. It lines up so it appears to be exactly the same length as a cruise ship 1 km out to sea. How long is the cruise ship?

A) 299.7 m   B) 300.0 m   C) 333.3 m   D) 29,970 m   E) 30,000 m

Question 21:

Which of the following Energy-Temperature graphs best represents the melting of ice to water?

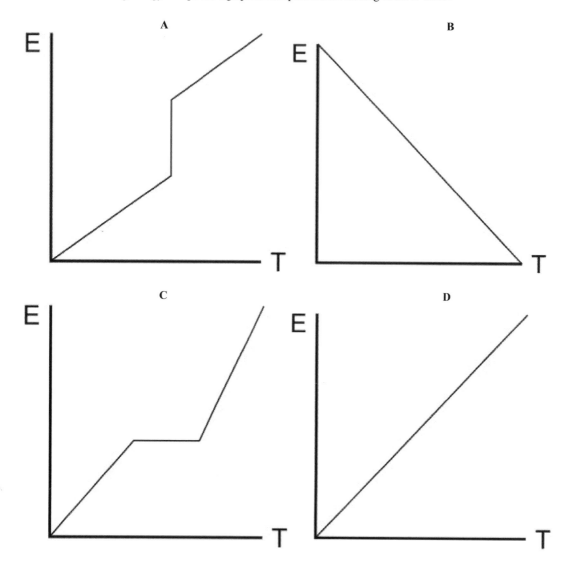

Question 22:

Which of the following statements about white blood cells is correct?

1.   They act by engulfing pathogens such as bacteria.
2.   They are able to kill pathogens.
3.   They transport carbon dioxide away from dying cells.

A)	Only 1	C)	Only 3	E)	2 and 3	G)	All
B)	Only 2	D)	1 and 2	F)	1 and 3	H)	None

Question 23:

Which of the following statements is true regarding the Doppler Effect?

A)   The Doppler Effect applies only to sounds.
B)   The Doppler Effect makes ambulances appear to have a higher frequency when driving towards you.
C)   The Doppler Effect makes ambulances sound higher-pitched when driving away from you.
D)   The Doppler Effect means you never hear the real siren sound as an ambulance drives past.

Question 24:

A 1.2 V battery is rated at 2500 mA hours and is used to power a 30 W light. How many batteries will it take to power the light for 1 hour?

A)   1   B)   6   C)   10   D)   60   E)   100

Question 25:

When electricity flows through a metal, which of the following are true?

1.   Ions move through the metal to create a current.
2.   The lattice in the metal is broken.
3.   Only electrons which were already free of their atoms will flow.

A)   1 only   C)   3 only   E)   1 and 3
B)   2 only   D)   1 and 2   F)   2 and 3

Question 26:

A man cycles along a road at the rate shown in the graph below.

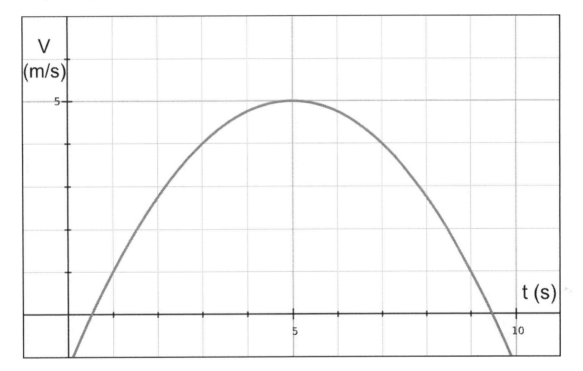

Calculate his displacement at t = 10 seconds.

A)   5 m   C)   25 m   E)   35 m
B)   10 m   D)   30 m   F)   40m

Question 27:

Bob is twice as old as Kerry, and Kerry is three times as old as Bob's son. Their ages combined make 50 years. How old was Bob when his son was born?

A)   15   B)   20   C)   25   D)   30   E)   35

END OF SECTION

Section 3

1)   *'A doctor should never disclose medical information about his patients'*

What does this statement mean? Argue to the contrary using examples to strengthen your response. To what extent do you agree with this statement?

2)   *'Science is nothing more than just a thought process'*

Explain what this statement means. Argue to the contrary, that science is much more than just a thought process. To what extent, if any, do you agree with the statement?

3)   *'With an ageing population, it's necessary to increase the individual's contribution to the healthcare system in order to maintain standards.'*

Explain what this statement means. Argue to the contrary. To what extent do you agree with the statement?

4)   *'Assisted suicide allows those suffering from incurable diseases to die with dignity and without unnecessary pain.'*

Explain what this statement means. Argue the contrary. To what extent do you agree with the statement?

END OF PAPER

Mock Paper G

Section 1

Question 1:

Irish Folk Band, the Willow, have recently signed a contract with a new manager, and are organising a new musical tour. They and their manager are discussing which country would be best to organise their tour in. The lead singer of the willow would like to organise a tour in Germany, which has a rich history of folk music. However, the new manager finds that ticket sales for folk music concerts in Germany have been steadily declining for several years, whilst France has recently seen a significant increase in ticket sales for folk music concerts. The manager says that this means the group's ticket sales would be higher if they organise a tour in France, than if they organise one in Germany.

Which of the following is an assumption that the manager has made?

A) The band should prioritise profits and organise a tour in the most profitable country possible.
B) The band should not embark upon a new tour and should instead focus on record sales.
C) The decrease of ticket sales in Germany and the increase in France means that the band will sell fewer tickets in Germany than in France.
D) There will not be other countries which are even more profitable than France to organise the tour in.
E) Folk music is popular in France.

Question 2:

Wendy is sending 50 invitations to her housewarming party by first class post. Every envelope contains an invitation weighing 70g, and some who are going to family and friends who live further away also contain a sheet of directions, which weigh 25g. The table below gives the prices of sending letters of certain weights by first or second class post.

If the total cost of sending the invitations is £33, how many of the invitations contain the extra information?

	First Class	Second Class
Less than 50g	£0.50	£0.30
Less than 75g	£0.60	£0.40
Less than 100g	£0.70	£0.50
Less than 125g	£0.80	£0.60
Less than 150g	£0.90	£0.70

A) 15   B) 20   C) 25   D) 30   E) 35

Question 3:

Grace and Rose have both been attending an afterschool gymnastics class, which finishes at 5pm. After the class has finished, Grace and Rose cool down and change out of their gym clothes before heading home. Both girls depart at 5:15pm. Grace and Rose both live a 1.5 mile walk away from the local gymnasium. Therefore, they will definitely arrive home at the same time.

Which of the following is **NOT** an assumption made in this argument?

A) Both girls will walk at the same speed.
B) Both girls departed at the same time.
C) The gymnastics class is being held at the local gymnasium.
D) Grace will not get lost on the way home.
E) Both girls are walking home.

Question 4:

John is a train enthusiast, who has been studying the directions in which trains travel after departing from various London Stations. He finds that Trains departing from King's Cross station in London head North on the East Coast Mainline, and travel to Edinburgh. Trains departing from Waterloo Station head West on the Southwest Mainline and travel to Plymouth. Trains departing from Victoria Station head South and travel to Kent.

John surmises that presently, in order to travel on a train from London to Edinburgh, he must get on at King's Cross Station.

Which of the following is an assumption that John has made?

A) The East Coast mainline has the fastest trains.
B) It would not be quicker to take a train from Waterloo to Southampton Airport, then travel to Edinburgh on an Aeroplane.
C) Rail lines will not be built that will allow trains to travel from Waterloo Station or Victoria Station to Edinburgh.
D) King's Cross trains do not have any other destinations other than Edinburgh.
E) There are no other train stations in London from which trains may travel to Edinburgh.

Question 5:

Summer and Shaniqua are playing a game of "noughts and crosses". Each player is assigned either "noughts" (O) or "crosses" (X) and they take it in turns to choose an empty box of the 3x3 grid to put their symbol in. The winner is the first person to get a line of 3 of their symbol in any direction in the grid (vertically, horizontally or diagonally). Summer starts the game. The current position is shown below:

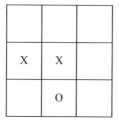

Assuming Shaniqua now plays her symbol in the square which will stop Summer being able to win the game straight away, Summer should play in either of which 2 boxes to ensure she is able to win the game on the next turn no matter what Shaniqua does?

A) 1 and 3   B) 1 and 5   C) 1 and 6   D) 2 and 4   E) 3 and 5

Question 6:

Tanks and armoured vehicles were a hugely influential factor in all battles in World War 2. German tanks were highly superior to the tanks used by France, and this was an essential reason why Germany was able to defeat France in 1940. However, Germany was later defeated in World War 2 by the Soviet Union. Germany lost a number of key battles such as the Battle of Stalingrad and the Battle of Kursk. These victories were essential for the eventual victory of the Soviet Union over Germany. Therefore, the Soviet Union's tanks in the battles of Stalingrad and Kursk must have been superior to those of Germany.

Which of the following is an assumption made in this argument?

A) Tanks were hugely influential in the Battle of Stalingrad.
B) The Battles of Stalingrad and Kursk were essential for the Soviet Union's victory over Germany.
C) The reasons why the Soviet Union defeated Germany in battle were the same as the reasons why Germany defeated France in battle.
D) German tanks being superior to those used by France was an essential reason why Germany was able to defeat France.
E) If the Soviet Union's tanks were superior to Germany's tanks, the Soviet Union's armoured vehicles must also have been superior to Germany's armoured vehicles.

Question 7:

In the Battle of Waterloo, in 1815, French Emperor Napoleon Bonaparte's army was defeated by a British army commanded by British General Arthur Wellesley, Duke of Wellington. Essential to The British army's victory was the arrival of a group of Prussian reinforcements led by Field Marshal Von Blucher, which joined up with The British army and allowed them to overwhelm Bonaparte's left flank. Bonaparte had been aware of the threat posed by Von Blucher's Prussians, and had detached a force of French soldiers several days earlier under the command of Field Marshal Grouchy, with orders to engage the Prussians led by Von Blucher, and prevent them joining up with The British Army.

However, whilst dining at a local inn, Grouchy mistook the sounds of gunfire for thunder, and believed that the battle had been cancelled. He therefore disobeyed his orders and did not engage the Prussians commanded by Von Blucher. Therefore, if Field Marshal Grouchy had not made this mistake and had engaged the Prussian force as commanded, The British would not have won the Battle of waterloo.

Which is the best statement of a flaw in this argument?

A) It implies Field Marshal Grouchy was an incompetent commander, when in fact he was a highly respected general of the day.
B) It assumes that had Grouchy engaged the Prussian force, he would have been able to successfully prevent them joining up with the British army.
C) It assumes that the British army would not have been victorious without the arrival of the Prussian reinforcements.
D) It ignores the other mistakes made by Napoleon which contributed to the British army being victorious in the Battle of Waterloo.
E) It implies that thunder and gunshot sounds are frequently mistaken by generals.

Question 8:

A cruise ship is sailing from Southampton to Barcelona, making several stops along the way at Calais and Bordeux, in France, Bilbao in Spain, and Porto in Portugal. At each stop, the ship must wait in a queue to be assigned a Dock at which it can pull in, refuel and resupply. The busier the port, the longer the ship will have to queue to be assigned a Dock. The Captain of the ship is planning the journey, and knows he must work out which ports will have the longest queues.

The Captain made the same journey last year, and found out that Bilbao was the busiest port in Europe during the course of the journey. He also knows that Bordeux is the busiest port in France, and that Porto is the busiest port in Portugal. Whilst he is planning the journey, he discovers that Calais is busier than Porto. The Captain concludes that he must plan for Bilbao to have the longest queue in the journey, Bordeux to have the second longest queue, Calais to have the third longest queue, and Porto to have the fourth longest queue.

Which of the following best illustrates a flaw in the Captain's Reasoning?

A) Porto is less busy than Calais, but may be busier than Bordeux.
B) The rankings may have changed and Bilbao may no longer be the busiest port in Europe.
C) Just because a port is busier does not necessarily mean it will have the longest queues.
D) The ship may not have time to make all the stops.
E) The captain has forgotten to consider how many passengers will embark and disembark at each stop.

Question 9:

A packaging company wishes to make cardboard boxes by taking a flat 1.2 m by 1.2 m square piece of cardboard, cutting square sections out of each corner as shown by the picture below and folding up the sections remaining on each side to make a box. The company experiments with different size boxes by cutting differently sized squares from the corners each time. It makes a box with 10 cm by 10 cm squares cut out of each corner, a box with 20 cm by 20 cm squares cut out of each corner and so on up to one with 50 cm by 50 cm squares cut out of each corner.

Which side length cut out would result in a box with the largest volume?

A) 10 cm   B) 20 cm   C) 30 cm   D) 40 cm   E) 50 cm

Question 10:

The aeroplane was a marvel of modern engineering when it was first developed in the early 20th Century, and was testament to human ingenuity. Throughout the 20th Century, the aeroplane allowed humans to travel more freely and widely than ever before, and allowed people to see and appreciate the stunning natural beauty the world has to offer. However, Aeroplanes also produce lots of pollution, such as Carbon Dioxide and Sulphur Oxide. High levels of Carbon Dioxide in the atmosphere are currently causing global warming, which is destroying or damaging many natural environments throughout the world.

Therefore it is clear that the aeroplane, which once offered such opportunity to appreciate the world's natural beauty, has been largely responsible for damage to various natural environments throughout the world. We must now seek to curb air traffic in order to save the world's remaining natural environments.

Which of the following is the best statement of a flaw in this argument?

A) It assumes that aeroplanes are a major reason for the high levels of Carbon Dioxide in the atmosphere which are currently causing global warming.
B) It assumes that aeroplanes offer greater opportunity to appreciate the world's natural environments.
C) It assumes that high levels of Carbon Dioxide are responsible for global warming.
D) It does not consider the effects of Sulphur Dioxide pollution released by aeroplanes.
E) It implies that we should take action to prevent damage to the world's natural environments.

Question 12:

Professors from the department of Pathology at Oxford University are conducting research into possible new treatments for malaria, which is caused by a microbe known as Plasmodium. Research from Sierra Leone, a third world country with a high rate of malaria, has found that liver cells in malaria patients are reactive to the antibody Tarpulin. Plasmodium is known to infect liver cells, and thus liver cells would react to Tarpulin if Plasmodium itself was reactive to Tarpulin. Thus, the professors at Oxford begin to research how Tarpulin can be used to target Plasmodium and treat malaria.

However, this research will not be successful, because liver cells would also react to Tarpulin if the wrong solution is used whilst conducting the experiments. Since malaria is not prevalent in Oxford, the professors must rely on the data from Sierra Leone. If the experiments in Sierra Leone used the wrong solutions, then the liver cells would react to Tarpulin even if Plasmodium does not react to Tarpulin.

Which of the following best illustrates a flaw in this argument?

A) From the fact that Plasmodium infects liver cells, it cannot be inferred that infected liver cells would react to Tarpulin if Plasmodium does.
B) From the fact that the research was carried out in Sierra Leone, it cannot be inferred that the wrong solutions were used.
C) From the fact that the wrong solutions are used, it cannot be inferred that the liver cells would react to Tarpulin.
D) From the fact that Plasmodium is reactive to Tarpulin it cannot be assumed that Tarpulin can be used to combat Plasmodium.
E) From the fact that Liver cells react to Tarpulin, it cannot be inferred that Plasmodium is reactive to Tarpulin.

Question 12:

Ancient Egypt was one of the world's most powerful nations for several thousand years, and wondrous structures such as the Sphinxes and the Great Pyramids serve as a permanent reminder of its stature. Many other powerful nations throughout the ages have also built magnificent structures, such as the Colosseum built by the Romans, the Hanging Gardens of Babylon built by the Persians and the Great Wall of China built by the Chinese. As well as building magnificent structures, Rome, Persia and China had one other thing in common, namely a very strong military. Thus, history clearly shows us that in ancient times, for a nation to be a powerful nation, it must have had a very strong military. In addition to building great structures such as the pyramids, Ancient Egypt must have also possessed a very strong military.

Which of the following best illustrates the main conclusion of this argument?

A) In order to be a powerful nation, a nation must build magnificent structures.
B) In Ancient times a very strong military was required to be a powerful nation.
C) Ancient Egypt built magnificent structures; therefore it must have been a powerful nation.
D) Rome, Persia and China were all powerful nations.
E) Ancient Egypt was a powerful nation; therefore it must have had a very strong military.

Question 13:

Global warming is widely presented in modern society as a cause for significant concern. One particular area often thought to be at risk is the Ice caps of the North and South Poles, which are often presented to be at risk of melting due to increased temperature. Environmentalist groups often campaign for energy consumption to be reduced, thus reducing CO_2 emissions, the leading cause of global warming. However, recent research shows that the North and South Poles are actually becoming cooler, not warmer, thanks to mysterious and unexplained weather patterns. Clearly, high energy consumption is not contributing to damage to the Polar Ice caps.

Which of the following statements can be reliably inferred from this argument?

A)   There is no point in reducing energy consumption for environmental reasons.
B)   Reducing energy consumption will not reduce CO_2 emissions.
C)   We should trust the recent research stating that the North and South poles are becoming cooler.
D)   Reducing energy consumption will not contribute to saving the polar ice caps.
E)   We should not be concerned about damage to the Polar Ice caps.

Question 14:

In 1957 the drug Thalidomide was released, and used to relieve nausea and morning sickness during pregnancy. The pharmaceutical company which released Thalidomide had carried out extensive testing of the drug, and had carried out more tests than was required for new drugs in the 1950s. No adverse affects were reported, and the drug was thought to be safe and effective. However, after it was released, Thalidomide was found to be responsible for severe deformities in thousands of babies whose mothers had taken the drug whilst pregnant with them. When further research was carried out, it was found that the molecules in Thalidomide could adopt 2 molecular structures, known as isomers. One of these isomers was perfectly safe, but the other caused significant biological problems in pregnant women and had been responsible for the deformities in the babies. The company producing Thalidomide had not been aware of this 2nd isomer when developing the drug.

Which of the following is a conclusion that can be drawn from this passage?

A)   The company producing Thalidomide had acted irresponsibly by not carrying out the required level of testing for the drug.
B)   No isomers of Thalidomide are safe.
C)   The drug testing requirements in 1950s were not sufficient to identify all possible isomers of a given drug.
D)   Thalidomide was not effective at relieving nausea and morning sickness.
E)   The dangerous isomer of Thalidomide was not effective at relieving nausea and morning sickness.

Question 15:

A teacher is trying to arrange the 5 students in her class into a seating plan. Her classroom contains 2 tables, arranged one behind the other, which each sit 3 people. Ashley must sit on the front row on the left hand side nearest the board because she has poor eyesight. Bella and Caitlin must not be sat in the same row as each other because they talk and disrupt the class. Danielle needs to be sat next to an empty seat as she sometimes has help from a teaching assistant. Emily should be sat on the end of a row because she has poor mobility and it is hard for her to get into a middle seat.

Who is sitting in the front right seat?

A) Empty   B) Bella   C) Caitlin   D) Danielle   E) Emily

Question 16:

The release of CO_2 from consumption of fossil fuels is the main reason behind global warming, which is causing significant damage to many natural environments throughout the world. One significant source of CO_2 emissions is cars, which release CO_2 as they use up petrol. In order to tackle this problem, many car companies have begun to design cars with engines that do not use as much petrol. However, engines which use less petrol are not as powerful, and less powerful cars are not attractive to the public. If a car company produces cars which are not attractive to the public, they will not be profitable.

Which of the following best illustrates the main conclusion of this argument?

A)   Car companies which produce cars that use less petrol will not be profitable.
B)   The public prefer more powerful cars.
C)   Car companies should prioritise profits over helping the environment.
D)   Car companies should seek to produce engines that use less petrol but are still just as powerful.
E)   The public are not interested in helping the environment.

Question 17:

Penicillin is one of the major success stories of modern medicine. Since its discovery in 1928, it has grown to become a crucial foundation of medicine, saving countless lives and introducing the age of antibiotics. Alexander Fleming is today given most of the credit for introducing and developing antibiotics, but in fact Fleming played a relatively minor role. Fleming initially discovered Penicillin, but was unable to demonstrate its clinical effectiveness, or discern ways of reliably and consistently producing it. 2 other scientists called Howard Florey and Ernst Chain were actually responsible for developing Penicillin to the point where it could be reliably produced and used in medicine, to treat infections in patients. Clearly, the credit for the wonders worked by Penicillin should not go to Fleming, but to Florey and Chain.

Which of the following best illustrates the main conclusion of this argument?

A) Fleming was unable to develop penicillin to the point of being a viable medical treatment.
B) The credit for Penicillin's effects on medicine should go to Ernst Chain and Howard Florey, not to Alexander Fleming.
C) Without Chain and Florey, Penicillin would not have been developed into a viable treatment.
D) Alexander Fleming only played a small role in the process of Penicillin becoming a feature of modern medicine.
E) Alexander Fleming is not given enough credit for his role in the development of penicillin.

Question 18:

I write my 4 digit pin number down in a coded format, by multiplying the first and second number together, dividing by the third number than subtracting the fourth number. If my code is 3, which of these could my pin number be?

A) 3461   B) 9864   C) 5423   D) 7848   E) 6849

Question 19:

Worcestershire Aquatic Centre is a business seeking to recruit a new dolphin trainer. They interview several candidates, and find that there are 2 candidates which are clearly more suitable than the others. They give both of these candidates a 2nd interview, with further questions about their experience and qualifications.

They discern that Candidate 1 has a proven capability to perform well to crowds, which is likely to bring in more profit to the Aquatic Centre as more people will come and watch a more entertaining dolphin show. However, unlike Candidate 1, Candidate 2 has experience at handling dolphins, and a proven ability to maximise their welfare standards. The manager of the aquatic centre tells the recruiting officer to prioritise profits, and therefore to hire Candidate 1.

Which of the following statements, if true, would most *weaken* the manager's argument?

A) Market research conducted by an external organisation showed that 60% of members of the public would be more likely to attend a dolphin show presented by a charismatic host.
B) Candidate 1's performance experience was not in the aquatic industry.
C) Other aquatic centres with poor welfare standards have been subject to negative media attention and subsequent boycotts.
D) A local charity-run aquatic centre have decided to prioritise donkey welfare and their manager recommends such a strategy.
E) A well-respected business analyst predicts that profit will rise under Candidate 2.

Question 20:

Rental yield for buy to let properties is calculated by dividing the potential rent per year paid for a house by the amount it cost to buy the house and get it in a rentable condition. Tina is considering 5 houses as possible buy to let investments. House A is in good condition and could be rented as it is for £700 a month, and costs £168,000 to buy. House B is also in good condition but is a student house so Tina would need to buy furniture for it. The house would cost £190,000 to buy and £10,000 to furnish, but could be rented for 40 weeks of the year to 4 students at a rent of £125 a week each. House C needs a lot of work doing. It costs £100,000 but would need £44,000 of renovations, and would rent for £600 a month. House D costs £200,000 and would need £40,000 of renovations, and would rent out for £2000 a month. House E costs £80,000 and would need £20,000 of renovations, and could be rented out for £200 a week.

Which house has the highest rental yield?

A) A   B) B   C) C   D) D   E) E

Question 21:
There has recently been a new election in the UK, and the new government is pondering what policy to adopt on the railway system in the UK. The Chancellor argues that the best policy is to have an entirely privatised railway system, which will encourage different train companies to be competitive, and try and attract customers by providing the best service at the lowest price, thus driving down costs and increasing quality for customers. However, the Transport Minister argues that this is a short-sighted policy. She argues that privatised companies will only run services on the most profitable lines, where there are lots of passengers.
Under this system, train companies may not choose to run many services to rural areas. This will lead to rural communities being cut off, with a consequent lack of opportunities for people in these communities. She argues that public funding should be put towards rail services in order to ensure that people in rural communities are adequately served by rail services.

Which of the following, if true, would most strengthen the Transport Minister's argument?

A) The Transport Minister has ultimate power over railway policy, and she can overrule the Chancellor if she sees fit.
B) Many train services to rural communities currently have low passenger numbers, and are unlikely to be profitable.
C) French rail services receive high level of public funding, and users of these services enjoy good quality and low prices.
D) American railway services are privatised with no public funding, and yet rural communities in America are well served by railway services.
E) The Prime Minister agrees with the Transport Minister's line of argument. He sympathises with rural communities and does not believe in a privatised rail system.

Question 22:
Niall can choose whether to join the gym as member or pay per session. Gym membership costs £30 per month but attending classes or gym sessions is free. Pay as you go gym sessions cost £4 and attending classes are £2 each. Niall works out that it will cost him £2 more if he pays per session than it will to buy membership. Which of these is a possible combination of Niall's gym sessions and classes for one month?

A) 5 gym sessions, 4 classes
B) 4 gym sessions, 4 classes
C) 5 gym sessions, 6 classes

D) 4 gym sessions, 6 classes
E) 5 gym sessions, 8 classes

Question 23:
If the mean of 5 numbers is 8, the median is 6 and the mode is 4, what must the two largest numbers in the set of numbers add up to?

A) 13   B) 16   C) 22   D) 26   E) 28

Question 24:
The North York Moors is one of several National Parks in England. The Management team has been awarded a grant from the National Lottery looking for a way to attract more visitors to the Moors. Sam suggests that they invest in enhancing the natural landscapes present in the Moors, thus creating more beauty, and making people more inspired to visit. However, Lucy disagrees, and feels that they should invest in more visitor centres and information points. Lucy's argument that whilst this will be more costly in terms of staffing these centres, the increase in visitor numbers will bring in more income for the Moors, and will counteract this extra cost.

Which of the following, if true, would most *weaken* Lucy's argument?

A) Information Centres in other national parks do not generally generate as much revenue as they cost to staff.
B) National Lottery grants have a history of being badly spent my National Parks such as the North York Moors.
C) There are large numbers of people who are interested in volunteering to help the North York Moors and would be happy to staff visitor centres.
D) Another National Park, the Yorkshire Dales, has recently opened up 5 new visitor centres and seen their profits increase significantly.
E) The North York Moors is currently struggling to attract visitors.

Question 25:

How many different squares (of either 1, 2, 3 or 4 grid squares in side length) can be made using the grid below?

A) 16   C) 25   E) 30

B) 20   D) 26

Question 26:

When I board my train at York at 15:30, the announcer tells me it is 120 minutes to London Kings Cross. Assuming the announcement is accurate to the nearest 10 minutes and that the train is on time, what is the earliest time I might arrive at my destination which is 10 minutes walk from Kings Cross?

A) 17:20   B) 17:25   C) 17:30   D) 17:35   E) 17:40

Question 27:

My sister does 2 loads of washing per week plus an extra one for everyone who is living in the house that week. When her son is away at university, she buys a new carton of washing powder every 6 weeks, but when her son is home she has to buy a new one every 5 weeks. How many people are living in the house when her son is home?

A) 2   B) 3   C) 4   D) 5   E) 6

Question 28:

A train shuttle service runs between the city centre and the airport between 5:30am and 11:30pm on weekdays and 6:00am and midnight on weekends. There are two trains used to operate the service and each journey from the airport to the city centre or vice versa takes 24 minutes. It takes 4 minutes for the train to unload and reload at the airport and 2 minutes for the train to unload and reload at the city centre.

What is the maximum number of single journeys that can be made by the shuttle service in one day?

A) 36   B) 48   C) 60   D) 72   E) 96

Question 29:

Northern Line trains arrive into Waterloo station every 6 minutes, Jubilee Line trains every 2.5 minutes and Bakerloo Line trains every 4 minutes.

If trains from all 3 lines arrived into the station 4 minutes ago, how long will it be before they do so again?

A) 20 minutes   C) 30 minutes   E) 60 minutes

B) 26 minutes   D) 56 minutes

Question 30:

Sam is deciding whether to make her wedding invitations herself or get them professionally made at a cost of £1 each. She decides to work out how much it will cost to make them herself. Each invitation uses 1 sheet of cream card, 4 sheets of red paper and 1 metre of gold ribbon. She will also uses a gold sticker on each invitation and stamp them with a stamper she will buy. The stamper needs a pad of ink which will last for 70 invitations. The table of stationery costs is shown below:

Product	Price
Red paper (pack of 100)	£2
15m roll of gold ribbon	£3
Pack of 30 gold stickers	£1
Stamper	£8
Ink pad	£4
Cream card (pack of 20)	£2

She wants to send 90 invitations and wants to have enough supplies for 4 spares only. How much will she save by making the invitations herself?

A) £15   B) £19   C) £29   D) £31   E) £33

Question 31:

Half of the boys in Mrs Nelson's class have brown eyes and two thirds of the class have brown hair. At least as many boys in the class as girls have brown hair. There are at least as many boys as girls in the class. There are 36 children in the class in total.

What's the minimum number of boys that have both brown hair and brown eyes?

A) 2   B) 3   C) 4   D) 5   E) 6

Question 32:

Mandy is making orange squash for her daughter's birthday party. She wants to have a 300ml glass of squash for each of the 8 children attending and a 400ml glass of squash each for her and for 2 parents who are helping out. She has 600ml of the concentrated squash. What ratio of water:concentrated squash should she use in the dilution to ensure she has the right amount to go around?

A) 7:1   B) 6:1   C) 5:1   D) 4:1   E) 3:1

Question 33:

A group of students is picking A Level options from the table below. In each "block", they choose 1 option. They cannot choose the same option in more than one block, and they have to take a subject in every block. How many possible combinations of options are there?

Block A	Block B	Block C	Block D
English Literature	History	Biology	Chemistry
German	Physics	English Language	Economics
Mathematics	Psychology	Geography	Philosophy
Psychology	Spanish	Mathematics	Sociology

A) 16   B) 48   C) 64   D) 224   E) 2

Question 34:

Hannah opens a savings account for her 14 year old son who will go to university in 4 years time. She works out that he will need £19000 to live on at current prices so deposits this amount in the savings account when she opens it. The cost of living rises with inflation by 3% a year, and Hannah's money earns 4% a year on balances under £20000 or 5% a year on balances over £20000. The table below shows how much an increase of various percentages equates to for various amounts. The interest is calculated and paid into the account at the end of each year.

How much more money will Hannah's son have than he needs to live on to the nearest £100 (if he assumes the cost of living does not rise further and no further interest is earned after the start of his course in 4 years time)?

	3%	4%	5%
£19,000.00	£19,570.00	£19,760.00	£19,950.00
£19,570.00	£20,157.10	£20,352.80	£20,548.50
£19,760.00	£20,352.80	£20,550.40	£20,748.00
£19,950.00	£20,548.50	£20,748.00	£20,947.50
£20,157.10	£20,761.81	£20,963.38	£21,164.96
£20,352.80	£20,963.38	£21,166.91	£21,370.44
£20,548.50	£21,164.96	£21,370.44	£21,575.93
£20,550.40	£21,166.91	£21,372.42	£21,577.92
£20,748.00	£21,370.44	£21,577.92	£21,785.40
£20,761.81	£21,384.67	£21,592.29	£21,799.90
£21,372.42	£22,013.59	£22,227.31	£22,441.04
£21,577.92	£22,225.26	£22,441.04	£22,656.82

A) 800   B) 1000   C) 1300   D) 1400   E) 1500

Question 35:

Amaia needs to arrive at her university interview by 11am. She has to travel from her house, a 10 minute walk from Southtown station, to Northtown University. Using the timetables below, what is the latest time she should she leave her house to arrive on time?

Train Timetable

Location	Time of Arrival		
Southtown	0913	0943	1013
Westtown	0924	0954	1024
Northtown West	0950	1020	1050
Northtown Central	0958	1028	1058
Easttown	1009	1039	1109

Bus Timetable

Location	Time of Arrival			
Northtown West station	0959	1009	1019	1029
Northtown Shopping Centre	1009	1019	1029	1039
Northtown Central station	1020	1030	1040	1050
Northtown Football Club	1023	1033	1043	1053
Northtown University	1033	1043	1053	1103

A) 0903   B) 0913   C) 0923   D) 0933   E) 0943

END OF SECTION

Section 2

Question 1:

Which of the following statements are true?

1. Natural selection always favours organisms that are faster or stronger.
2. Genetic variation leads to different adaptations to the environment.
3. Variation is purely due to genetics.

A) Only 1   D) 1 and 2   G) All of the above.
B) Only 2   E) 2 and 3   H) None of the above.
C) Only 3   F) 1 and 3

Question 2:

Which of the following statements are true about the electrolysis of brine?

1. It describes the reduction of 2 chloride ions to Cl_2.
2. The amount of NaOH produced increases in proportion with the amount of NaCl present in solution, provided there is enough H_2 present to dissolve the NaCl.
3. The redox reaction of the electrolysis of brine results in the production of dissolved NaOH, which is a strong acid.

A) Only 1   C) Only 3   E) 1 and 3   G) All of the above.
B) Only 2   D) 1 and 2   F) 2 and 3   H) None of the above.

The following information applies to questions 3 – 4:

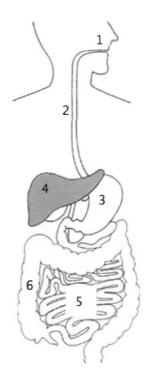

Question 3:

Which of the following numbers indicate where amylase functions?

A) 1 only   C) 1 and 3   E) 2 and 4   G) 5 and 6
B) only   D) 1 and 5   F) 3 and 4

Question 4:

In which of the following does the majority of chemical digestion occur?

A) 1   C) 3   E) 5   G) None of the
B) 2   D) 4   F) 6   above.

Question 5:

Which of the following correctly describes the product of the reaction between propene and hydrofluoric acid (HF)?

A) $C(F)H_3$-CH_2-CH_3 D) CH_3-$C(F)H_2$-CH_3
B) CH_3-$C(F)H$-CH_3 E) None of the above.
C) CH_3-$C(F)H_2$-CH_2

Question 6:

Which of the following statements is **FALSE**?

A) A nuclear power plant may have an accident if free neutrons in a fuel rod aren't captured.
B) Humans cannot currently harness the energy from nuclear fusion.
C) Uncontrolled nuclear fission leads to a large explosion.
D) Mass is conserved during nuclear explosions caused by nuclear bombs.
E) Nuclear fusion produces much more energy than nuclear fission.

Question 7:

Which of the following are true about the reaction between alkenes and hydrogen halides?

1. The product formed is fully saturated.
2. The hydrogen halide binds at the alkene's saturated double bond.
3. The hydrogen halide forms ionic bonds with the alkene.

A) Only 1
B) Only 2
C) Only 3

D) 1 and 2
E) 2 and 3
F) 1 and 3

G) All of the above.
H) None of the above.

Question 8:

Rearrange the following to make m the subject.

$$T = 4\pi \sqrt{\frac{(M + 3m)l}{3(M + 2m)g}}$$

A) $m = \frac{16\pi^2 lM - 3gMT^2}{48\pi^2 l - 6gT^2}$

B) $m = \frac{16\pi^2 lM - 3gMT^2}{6gT^2 - 48\pi^2 l}$

C) $m = \frac{3gMT^2 - 16\pi^2 lM}{6gT^2 - 48\pi^2 l}$

D) $m = \frac{4\pi^2 lM - 3gMT^2}{6gT^2 - 16\pi^2 l}$

E) $m = \left(\frac{16\pi^2 lM - 3gMT^2}{6gT^2 - 48\pi^2 l}\right)^2$

Question 9:

Which of the following correctly describes the product of the polymerisation of chloroethene molecules?

The following information applies to questions 10 – 11:

The diagram below shows the genetic inheritance of colour-blindness, which is inherited in a sex-linked recessive manner [transmitted on the X chromosome and requires the absence of normal X chromosomes to result in disease]. X^B is the normal allele and X^b is the colour-blind allele.

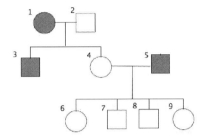

Question 10:

What is the genotype of the individual marked 4?

A) $X^B X^b$ B) $X^B X^B$ C) $X^b X^b$ D) $X^B Y$ E) $X^b Y$

Question 11:

If 8 were to reproduce with a heterozygote female, what is the probability of producing a colour-blind boy?

A)   100%   B)   75%   C)   50%   D)   25%   E)   12.5%   F)   0%

Question 12:

The mean of a set of 11 numbers is 6. Two numbers are removed and the mean is now 5. Which of the following is not a possible combination of removed numbers?

A.   1 and 20   C.   10 and 11   E.   19 and 2
B.   6 and 9   D.   15 and 6

Question 13:

For the following reaction, which of the statements below are true?

$$N_{2(g)} + 3\,H_{2(g)} \rightleftharpoons 2\,NH_{3(g)}$$

1.   Increasing pressure will cause the equilibrium to shift to the right.
2.   Increasing pressure will form more ammonia gas.
3.   Increasing the concentration of N_2 will create more ammonia.

A)   1 only   C)   3 only   E)   2 and 3
B)   2 only   D)   1 and 2   F)   All of the above.
A.   None of the above.

Question 14:

Find the values of angles b and c.

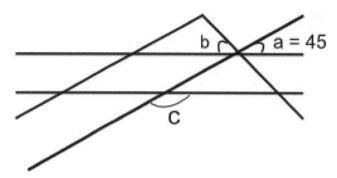

A) 45° and 135°

B) 45° and 130°

C) 50° and 135°

D) 55° and 130°

E) More information needed.

Question 15

When sodium and chlorine react to form salt, which of the following best represents the bonding and electron configurations of the products and reactants?

	Sodium (s)		Chlorine (g)		Salt (s)	
	Intra-element bond	Element electron configuration	Intra-element bond	Element electron configuration	Compound bond	Compound electron configuration
A)	Ionic	2, 8, 1	Covalent	2, 8, 8, 1	Ionic	2, 8, 1 : 2, 8, 8, 1
B)	Metallic	2, 7	Covalent	2, 8, 1	Ionic	2, 8 : 2, 8
C)	Covalent	2, 8, 2	Ionic	2, 8, 8	Covalent	2, 8 : 2, 8, 8
D)	Ionic	2, 7	Ionic	2, 8, 8, 7	Covalent	2, 7 : 2, 8, 8, 7
E)	Metallic	2, 8, 1	Covalent	2, 8, 7	Ionic	2, 8 : 2, 8, 8

Question 16:

Evaluate: $\dfrac{3.4 \times 10^{11} + 3.4 \times 10^{10}}{6.8 \times 10^{12}}$

A) $x\ 10^{-12}$

B) 5.5×10^{-2}

C) 5.5×10^{1}

D) 5.5×10^{2}

E) 5.5×10^{10}

F) 5.5×10^{12}

The following information applies to questions 17 – 18:

In pea plants, colour and stem length are inherited in an autosomal manner. The allele for yellow colour, Y, is dominant to the allele for green colour, y. Furthermore, the allele for tall stem length, T, is dominant to short stem length, t.

When a pea plant of unknown genotype is crossed with a green short-stemmed pea plant, the progeny are 25% yellow + tall-stemmed plants, 25% yellow + short-stemmed plants, 25% green + tall-stemmed plants and 25% green + short-stemmed plants.

Question 17

What is the genotype of the unknown pea plant?

A)   Yytt

B)   YyTt

C)   YyTT

D)   yyTt

E)   yyTT

F)   yytt

Question 18:

Taking both colour and height into account, how many different combinations of genotypes and phenotypes are possible?

A)   6 genotypes and 3 phenotypes

B)   8 genotypes and 3 phenotypes

C)   8 genotypes and 4 phenotypes

D)   9 genotypes and 4 phenotypes

E)   9 genotypes and 3 phenotypes

F)   10 genotypes and 3 phenotypes

Question 19:

Which of the following statements is true regarding electrolysis?

A)   Using an AC-current is most effective.

B)   Using a DC-current is most effective.

C)   An AC-current causes cations to gather at the cathode.

D)   A DC-current would plate the anode in copper from a copper sulphate solution.

E)   No current is used in electrolysis.

Question 20:

Evaluate the following expression:

$$\left(\left(\frac{6}{8}\times\frac{7}{3}\right) \div \left(\frac{7}{5}\times\frac{2}{6}\right)\right) \text{ x } 0.40 \text{ x } 15\% \text{ x } 5\% \text{ x } \pi \text{ x } \left(\sqrt{e^2}\right) \text{ x } 0.20 \text{ x } (e\pi)^{-1}$$

A) $\frac{4}{55}$

B) $\frac{8}{770}$

C) $\frac{9}{4,000}$

D) $\frac{8}{54,321}$

E) $\frac{9}{67,800}$

Question 21:

Which will have a greater current, a circuit with two identical resistors in series or one with the same two resistors in parallel?

A)   Series will have greater current than parallel.

B)   Parallel will have greater current than series.

C)   Same current in both.

D)   It depends on the battery.

Question 22:

A 2000 kg car is driving down the road at 36 km per hour. A deer runs out into the road 105 m in front of the car. It takes the driver 0.5 seconds to react to the deer and hit the brakes. The car stops just in time. What is average braking force exerted?

A)   20 N

B)   100 N

C)   200 N

D)   1,000 N

E)   2,000 N

Question 23:

What is the **MOST** important reason for each cell in the human body to have an adequate blood supply?

A)   To allow protein synthesis.

B)   To receive essential minerals and vitamins for life.

C)   To kill invading bacteria.

D)   To allow aerobic respiration to take place.

E)   To maintain an optimum cellular temperature.

F)   To maintain an optimum cellular pH.

Question 24:

A man drives along a road as shown in the figure below.

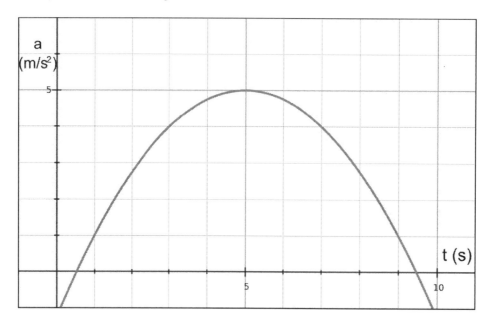

Which of the following statements is true?

A) He drives a total of 30 m.

B) He has an average velocity of 30 m/s.

C) He has a final velocity of 30 m/s.

D) He has an average acceleration of 30 m/s^2.

E) His velocity decreases between 5 and 9 seconds.

Question 25:

A circle has a radius of 3 metres. A line passes through the circle's centre and intersects with a tangent 4 metres from its tangent point. How far is this point of intersection from the centre of the circle?

A) 1 metres   B) 3 metres   C) 5 metres   D) 7 metres   E) 9 metres

Question 26:

Which of the following statements is **FALSE**?

A) Energy cannot be created or destroyed.

B) Energy can be turned into matter.

C) Efficiency is the ratio of useful energy to wasted energy.

D) Energy can be dispersed through a vacuum.

E) There are always losses when energy is transformed from one type to another.

Question 27

Which of the following statements is **FALSE**?

A) A beam of light exits a pane of glass at a different angle than it entered.

B) A beam of light reflects at an angle dependent on the angle of incidence.

C) Light travels a shorter distance to reach the bottom of a pool filled with water than a pool without water.

D) Any neutrally charged atom has the potential to emit light.

E) Photons are particles without a mass.

END OF SECTION

Section 3

1)   *'Doctors will eventually become obsolete as a result of advancing medical technologies.'*

Explain what this statement means. Argue to the contrary. To what extent do you agree with the statement?

2)   *"Science is a procedure for testing and rejecting hypotheses, not a compendium of certain knowledge."*

Stephen Jay Gould

What do you understand from the statement above? Explain why it might be argued that science does rely on a compendium of certain knowledge? To what extent is science defined by the challenging of preconceived hypotheses?

3)   *'Animal euthanasia should be made illegal'*

Explain what this statement means. Argue to the contrary that animal euthanasia should remain legal. To what extent do you agree with the statement?

4)   *'The primary duty of a doctor is to prolong life as much as possible'*

What does this statement mean? Argue to the contrary, that the primary duty of a doctor is not to prolong life. To what extent do you agree with this statement?

END OF PAPER

Mock Paper H

Section 1

Question 1:

A chemical change may add something to a substance, or subtract something from it, or it may both subtract and add, making a new substance with entirely different properties. Sulphur and carbon are two stable solids. The chemical union of the two forms a volatile liquid. A substance may be at one time a solid, at another a liquid, at another a gas, and yet not undergo any chemical change, because in each case the chemical composition is identical.

Which of the following statements cannot be reliably concluded from the above passage?

A)   The chemical composition of a compound may influence its physical nature.
B)   Substances can exist as solid, liquid or gas, without their chemical composition changing.
C)   Chemicals can be combined to create a new substance with similar or very different properties.
D)   Combining two substances in one state can lead to the production of a compound in a completely different state.
E)   The transition from solid to liquid is not a chemical one.

Question 2:

In the sequence B Y F U I R K P ? ? Which two letters come next?

A) U N   B) M N   C) L O   D) H O   E) N M

Question 3:

An insect differs from a horse, for example, as much as a modern printing press differs from the press Franklin used. Both machines are made of iron, steel, wood, etc., and both print; but the plan of their structure differs throughout, and some parts are wanting in the simpler press, which are present and absolutely essential in the other. So with the two sorts of animals; they are built up originally out of protoplasm, or the original jelly-like germinal matter, which fills the cells composing their tissues, and nearly the same chemical elements occur in both, but the mode in which these are combined, the arrangement of their products: the muscular, nervous and skin tissues, differ in the two animals.

Which of the following statements can be reliably concluded from the above passage?

A) The printing press has adapted from the press Franklin used, due to the designers observing differences in nature.
B) Horses and insects differ as they are made up of completely different chemical elements.
C) The muscular, nervous and skin tissues are what define an organism.
D) Chemical elements make up protoplasm, which is the building block for all major organisms.
E) It is the manner in which chemicals are arranged that determine an organism as a final product.

Question 4:

What day comes two days after the day, which comes four days after the day, which comes immediately after the day, which comes two days before Monday?

A)   Monday   B)   Tuesday   C)   Thursday   D)   Saturday   E)   Sunday

Question 5:

Cellulose is distinguished by its inherent constructive functions, and these functions take effect in the plastic or colloidal condition of the substance. These properties are equally conspicuous in the synthetical derivatives of the compound.

Which of the following statements would weaken the above passage?

A) Cellulose has a constructive role in nature.
B) Synthetic cellulose is made from natural cellulose.
C) Synthetic and natural cellulose are structurally very similar.
D) Synthetic cellulose only actually shares some of its properties with natural cellulose.
E) Synthetics cellulose is more useful in industry than natural cellulose.

Question 6:

If John gives Michael £20, the ratio of their money is 2:1. If Michael gives John £5, the ratio of John's money to Michael's is 5:1. How much money do they have combined?

A) £180   B) £120   C) £90   D) £210   E) £150

Question 7:

From the primitive pine-torch to the paraffin candle, how wide an interval! Between them how vast a contrast! The means adopted by man to illuminate his home at night, stamp at once his position in the scale of civilisation. The fluid bitumen of the far East, blazing in rude vessels of baked earth; the Etruscan lamp, exquisite in form, yet ill adapted to its office; the whale, seal, or bear fat, filling the hut of the Esquimaux or Lap with odour rather than light; the huge wax candle on the glittering altar, the range of gas lamps in our streets, all have their stories to tell.

Which of the following statements best summarises the above passage?

A) Burning animal fat was the original way to produce fire.
B) The use of fire has spread to all corners of the Earth.
C) Using fire for light is what defines us as being human.
D) Each light source over the globe is able to tell its own tale.
E) The development and evolution of the use of fire helps to define mankind as a civilisation.

Question 8:

972 patients ordered food for lunch. They could choose roast chicken, mac and cheese, vegetable chilli or cottage pie. Half chose the roast chicken, 1/3 chose the mac and cheese and 1/12 chose the cottage pie.

How many opted for the vegetarian option?

A) 81   B) 92   C) 68   D) 95   E) 102

Question 9:

It was a little late to search for the philosophers' stone in 1669, yet it was in such a search that phosphorus was discovered. Wilhelm Homberg (1652-1715) described it in the following manner: "a man little known, of low birth, with a bizarre and mysterious nature in all he did, found this luminous matter while searching for something else."

What can be reliably concluded about the above passage?

A) Phosphorous was easy to identify as a result of its luminous nature.
B) Phosphorous was found as a result of this man's low social status.
C) Phosphorous was identified by accident, in the search for the philosophers' stone.
D) Wilhelm Homberg discovered phosphorous.
E) Phosphorous was discovered in the 18th century.

Question 10:

How many minutes past noon is it, if 3 times this many minutes before 3pm is 28 minutes later than this many minutes past noon?

A) 54   B) 32   C) 45   D) 38   E) 18

Question 11:

Everyone is familiar with the main facts of such a life-story as that of a moth or butterfly. The form of the adult insect is dominated by the wings—two pairs of scaly wings, carried respectively on the middle and hindmost of the three segments that make up the *thorax* or central region of the insect's body. Each of these three segments carries a pair of legs.

Which of the following statements can be concluded from the above statement?

A) The wings of the insects alternate patterns when the insect flies.
B) The wings that attach to the segments of the insect's body are the most prominent feature of the butterfly or moth.
C) Wings attach to each of the three segments of the thorax.
D) Moths and butterflies are very similar in that each segment of their thorax carries a pair of legs.
E) Scaly wings protect these creatures from predators.

Question 12:

John and Mary are selling cakes at a cake sale. John has 8 cupcakes and 56 brownies, where as Mary has 12 cupcakes and 24 brownies. What is the difference between the percentages of brownies in the two stalls?

A) 6%   B) 115/3%   C) 19.25%   D) 125/6%   E) 22.2%

Question 13:

In 2007 AD, Halley's Comet and Comet Encke were observed in the same calendar year. Halley's Comet is observed on average once every 73 years; Comet Encke is observed on average once every 104 years. Based on this, estimate the calendar year in which both Halley's Comet and Comet Encke are next observed in the same year.

A) 9559 AD   B) 2114 AD   C) 5643 AD   D) 3562 AD   E) 1757 AD

Question 14:

The supreme court of Judicature at Athens punished a boy for putting out the eyes of a poor bird; and parents and masters should never overlook an instance of cruelty to anything that has life, however minute, and seemingly contemptible the object may be.

Which of the following statements best summarises the above passage?
A) The boy was prosecuted because the bird is a large enough organism.
B) Putting out the eyes of an organism is the most unacceptable form of animal cruelty.
C) The more important to mankind the animal, the worse the animal cruelty crime is.
D) Any cruelty to any creature is an action that should not be tolerated.
E) It is only acceptable to harm an animal so long as it benefits a human.

Question 15:

In a school there are 40 more girls than there are boys. The boys make up a percentage of 40% of the school. What is the number of students in the school?

A) 150   B) 200   C) 300   D) 500   E) 720

Question 16:

5 cars are travelling down a road in a line. The red car is following the blue car; the yellow car is in front of the green car. The purple car is between the green car and the blue car. What colour is the car second in line?

A) Red   B) Blue   C) Yellow   D) Green   E) Purple

Question 17:

To get to school, Joanne takes the school bus every morning. If she misses this, then she can take the public bus to school. The school bus arrives at 08:15, which if she misses will come again at 08:37. The public bus comes every 17 minutes, starting at 06:56. The school bus takes 24 minutes to get to her school; the public bus takes 18 minutes. If she arrives at the bus stop at 08:25, which bus must she catch to get to school first?

A) The 08:37 school bus   B) The 08:26 public bus   C) The 08: 38 public bus   D) The 08: 31 public bus

Question 18:

Puddle ducks are typically birds of fresh, shallow marshes and rivers rather than of large lakes and bays. They are good divers, but usually feed by dabbling or tipping rather than submerging. The speculum, or coloured wing patch, is generally iridescent and bright, and often a tell-tale field mark. Any duck feeding in croplands will likely be a puddle duck, for most of this group are sure-footed and can walk and run well on land. Their diet is mostly vegetable, and grain-fed mallards or pintails or acorn-fattened wood ducks are highly regarded as food.

Which of the following statements summarises the above passage best?

A) Other ducks are often eaten by puddle ducks in both large lakes and shallower waters.
B) Puddle ducks feed mainly without diving to gain vegetarian food sources.
C) Puddle ducks are the most common duck seen in croplands because they are vegetarian.
D) Other ducks are prone to predate on puddle ducks.
E) Puddle ducks live in large lakes as they can access vegetable food sources easily.

Question 19:

When the earth had to be prepared for the habitation of man, a veil, as it were, of intermediate being was spread between him and its darkness, in which were joined, in a subdued measure, the stability and the insensibility of the Earth, and the passion and perishing of mankind.

Which of the following statements best summarises the above statement?

A) The veil discussed is what links the good and evil of the human race.
B) Without this veil, mankind would not exist.
C) The Earth has more good than evil.
D) The veil keeps the human race alive.
E) Mankind would be better off without such a veil.

Question 20:

What is the value of ? in the following sequence:

3 1 6 8
8 4 5 0
4 2 7 8
9 2 3 ?

A) 5   B) 4   C) 8   D) 2   E) 7

Question 21:

Metformin has been thought to inhibit the process of fat cell growth. This is because *in vitro* metformin causes fat cells to stop growing. However, when a metformin inhibitor is used alongside metformin, the fat cells still don't grow. Thus we can conclude that metformin does not inhibit fat cell growth.

Which of the following statements highlights the flaw in the argument?

A) Metformin doesn't inhibit fat cell growth.
B) The mechanism by which metformin inhibits fat cell growth is poorly understood.
C) We are not aware of how this inhibitor acts to inhibit the actions of metformin.
D) Fat cell growth has not been quantified here.
E) Metformin does not inhibit fat cell growth *in vivo*.

Question 22:

All dancers are strong. Some dancers are pretty. Alexandra is strong, and Katie is pretty.

Choose a correct statement.

A) Alexandra is a dancer
B) Katie is not a dancer
C) A dancer can be strong and pretty
D) A dancer can be strong and ugly

Question 23:

During the last fifteen years the subject of bacteriology has developed with a marvellous rapidity. At the beginning of the ninth decade of the century bacteria were scarcely heard of outside of scientific circles, and very little was known about them even among scientists. Today they are almost household words, and everyone who reads is beginning to recognise that they have important relations to everyday life.

Which of the following statements would best support the above passage?
A) Bacteriology has improved due to the advancements in our ability to see and study such organisms.
B) Bacteria are too small to see in everyday life.
C) The development of antibiotics has helped us to understand bacteria better.
D) Every household understands the problems with bacterial infections.
E) Bacteria were much scarcer in the ninth decade than they are today.

Question 24:

Read the following statements. Which of the options is correct regarding whether the conclusions drawn from the statements are true or not?

Statements:
➢ No man is a lion.
➢ Joseph is a man.

Conclusions:
➢ **I:** Joseph is not a lion.
➢ **II:** All men are not Joseph.

A) Conclusion I is TRUE and Conclusion II is TRUE
B) Conclusion I is TRUE and FALSE
C) Conclusion I is TRUE and Conclusion II is CAN'T TELL
D) Conclusion I is FALSE and Conclusion II is CAN'T TELL
E) Conclusion I is FALSE and Conclusion II is TRUE

Question 25:

We may define a food as any substance, which will repair the functional waste of the body, increase its growth, or maintain the heat, muscular, and nervous energy. In its most comprehensive sense, the oxygen of the air is a food; as although it is admitted by the lungs, it passes into the blood, and there re-acts upon the other food, which has passed through the stomach. It is usual, however, to restrict the term food to such nutriment as enters the body by the intestinal canal. Water is often spoken of as being distinct from food, but for this there is no sufficient reason.

Which of the following statements highlights the weakness in the passage?
A) Oxygen also is absorbed in the digestive tract.
B) Water is only made up of two elements, which is why it is not classified as food.
C) Water is needed for bodily functions and therefore must be food.
D) It is not explained why water is not classified as a food.
E) Any substance that is involved in a physiological process must have originated from food.

Question 26:

Billy is James's father. 4 years ago, Billy's age was 4 times that of James. After 6 years, the ages of the Billy and James are in the ratio of 5:2. How old is Billy?

A) 12   B) 13   C) 14   D) 15   E) 16

Question 27:

The word soap appears to have been originally applied to the product obtained by treating tallow with ashes. In its strictly chemical sense it refers to combinations of fatty acids with metallic bases, a definition which includes not only sodium stearate, oleate and palmitate, which form the bulk of the soaps of commerce, but also the linoleates of lead, manganese, etc., used as driers, and various pharmaceutical preparations, *e.g.*, mercury oleate, zinc oleate and lead plaster, together with a number of other metallic salts of fatty acids.

What can be reliably concluded about the above passage?

A) All metallic salts of fatty acids are classified as soaps.
B) Soaps are only used in industry for commercial use, driers and as pharmaceutical preparations.
C) Treating tallow with acids forms a soap as it results in fatty acids combining with a metallic acid.
D) All soaps are fatty acids combined with metallic bases.
E) All metals form soaps used in industry.

Question 28:

The average marks scored by 12 students is 73. If the scores of Bea, Bay and Boe are included, the average becomes 73.6. If Bea scored 68 marks and Boe scored 6 more than Bay, what was Bay's score?

A) 75   B) 76   C) 77   D) 78   E) 79

Question 29:

A building company employs 90 men to work for 8 hours per day to complete some building work. The company wants to finish work in 200 days but after 120 days, the work is only a third complete. If the men start working 12-hour days, how many more men are required to complete the work on time?

A) 170   B) 180   C) 190   D) 200   E) 210

Question 30:

The organs that form the digestive tract are the mouth, pharynx, oesophagus, stomach, intestines and the annexed glands, viz.: the salivary, liver, and pancreas. The development of these organs differs in the different species of animals. For example, solipeds possess a small, simple stomach and capacious, complicated intestines. Just the opposite is true of ruminants. The different species of ruminants possess a large, complicated stomach, and comparatively simple intestines. In swine we meet with a more highly developed stomach than that of solipeds and a more simple intestinal tract. Of all domestic animals the most simple digestive tract occurs in the dog.

What can be reliably concluded from the above passage?

A) Dogs have the simplest digestive tracts of domesticated animals due to their size.
B) Solipeds and ruminants differ only in their digestive tracts.
C) The more complex the digestive tract, the more complex the organism.
D) Mammals have varying digestive tracts that are adapted for their environments.
E) The mammalian digestive tract is vital for the survival of the animal.

Question 31:

If every alternative letter starting from A of the English alphabet is written in lower case, and the rest are all written in upper case, how would the day "Wednesday" be written?

A)   wEdNEsdAy
B)   weDNesDay
C)   WEdnESdAY
D)   weDneSDaY
E)   WedNesdAY

Question 32:

A man covers a distance in 1hr 24min by covering 2/3 of the distance at 4 km/h and the rest at 5km/h. What distance does he cover?

A) 5km   B) 6km   C) 7km   D) 8km

Question 33:

In regions of a comparatively low altitude many birds, as is well known, fly to the far North to find the proper climatic conditions in which to rear their broods and spend their summer vacation, some of them going to the subarctic provinces and others beyond. How different among the sublime heights of the Rockies! Here they are required to make a journey of only a few miles, say from five to one hundred or slightly more, according to the locality selected, up the defiles and canons or over the ridges, to find the conditions as to temperature, food, nesting sites, etc., that are precisely to their taste.

Which of the following statements can be reliably concluded from the above passage?

A) A journey of 100 miles is too far for these birds to travel for food.
B) Rearing their young is the most important part of these birds' lives.
C) These birds fly north as all their needs are in a localised area.
D) Nesting in the Rockies keeps the birds away from predators.
E) The birds struggle to survive in the harsh cold temperatures further north.

Question 34:

Three ladies X, Y and Z marry three men A, B and C. X is married to A, Y is not married to an engineer, Z is not married to a doctor, C is not a doctor and A is a lawyer. Then which of the following statements is correct?

A)   Y is married to C who is an engineer
B)   Z is married to C who is a doctor
C)   X is married to a doctor
D)   None of these

Question 35:

A woman works for a company for 60 days where she is paid £100 per day, but she has to pay £20 for each day of her absence. If she gets £5040 at the end, she was absent for work for how many days?

A)   4   B)   6   C)   8   D)   10   E)   12

END OF SECTION

Section 2

Question 1:
Hydrogen Bicarbonate (HCO_3^-) acts as a buffer in the blood i.e. to keep the PH close to 7.
Which statement is true regarding bicarbonate?

A) It is alkaline.
B) It is an acidic molecule.
C) If the pH of the blood drops below 7, bicarbonate will release the H^+ ion to stabilise the pH.
D) It is only released when the pH drops below 7.
E) It is bound to protein in the blood.

Question 2:
Which of the statements regarding this series circuit is true?

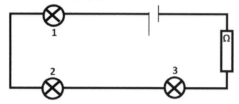

A) Current is different at different points in the circuit.
B) Potential difference is shared between the three lightbulbs.
C) Resistance is constant throughout the circuit.
D) The current is higher in bulb 1 than in bulbs 2 and 3.

Question 3:
The below statements are about breathing. Which of them are correct?
1. The diaphragm plays no part in breathing.
2. The intercostal muscles relax during exhalation to allow the ribcage to move inwards and downwards.
3. The total pressure inside the chest decreases relative to the pressure outside the body during inhaling to draw air inside the lungs.

A) 1 only   C) 3 only   E) 1 and 3
B) 2 only   D) 2 and 3   F) None

Question 4:
Bill wants to lay down laminate flooring in his living room, which has an in-built circular fish tank that he will have to lay the flooring around. He has decided to buy planks that he can cut to fit the dimensions of his room. He must, however, buy whole planks and cut them down himself. The room's dimensions are given below, as are those of one plank.

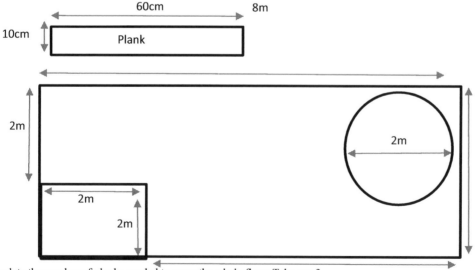

Calculate the number of planks needed to cover the whole floor. Take $\pi = 3$.

A) 30   B) 417   C) 600   D) 589   E) 43

Question 5:

In pregnancy the foetus is supplied with blood from the mother via the umbilical cord. This cord is comprised of one vein and two arteries. The table below shows which vessel carries which type of blood in which direction.

	Vessel	Direction	Blood
1.	Vein	Mother to foetus	Oxygenated
2.	Artery	Foetus to Mother	Deoxygenated
3.	Artery	Foetus to Mother	Oxygenated
4.	Vein	Mother to Foetus	Deoxygenated

Which options are correct?

A) 1 only   C) 3 only   E) 1 and 2   G) 4 and 1

B) 2 only   D) 4 only   F) 2 and 3   H) 3 and 1

Question 6:

Solve $y = x^2 - 3x + 4$ and $y - x = 1$ as (x,y).

A) (-1, 2) and (3,4)   C) (7,-2) and (6,5)   E) (1,-1) and (-7,-1)

B) (1,2) and (3,4)   D) (2,-3) and (4,-1)

Question 7:

A ball of mass 5kg is at rest at the top of a 5m slope. Calculate the velocity of the ball as it travels down the slope. Take $g = 10kgm^{-1}$ and assume there is no resistance.

A) 10   B) 45   C) 100   D) 5   E) 6

Question 8:

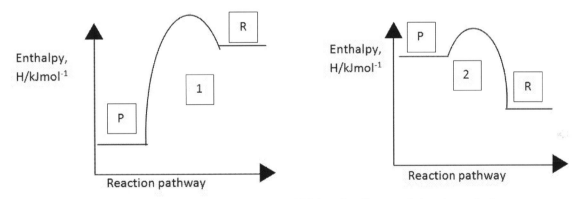

The two graphs shown above are Enthalpy profile diagrams. Which best describes an endothermic reaction?

	Graph	ΔH	Heat energy	Stability of reactants
A)	1	Negative	Absorbed from surroundings	P is more stable than R
B)	2	Negative	Released to surroundings	R is more stable than P
C)	1	Positive	Absorbed from surroundings	P is more stable than R
D)	2	Positive	Absorbed from surroundings	R is more stable than P

Question 9:

What is the function of the kidneys?

1. Ultrafiltration
2. Kill bacteria in the blood
3. Reabsorption
4. Release of waste
5. Store water
6. Produce hormones
7. Blood glucose regulation

A) 1 only   C) 3 only   E) 5 only   G) 3 and 5   I) 4, 5 and 6

B) 2 only   D) 4 only   F) 6 and 7   H) 1, 3 and 4

Question 10:

Mike and Vanessa are two healthy adults. They have two children. Their first child, Rory, was born with Haemophilia B, an X linked recessive disorder that causes problems with blood clotting. They have just had another baby, a girl and want to get her tested for the condition. What is the likelihood of the baby girl having the condition?

A) 0%   B) 25%   C) 50%   D) 75%   E) 34%

Question 11:

Pyrite, also known as Fool's Gold, is an ore of Iron containing sulphur in the form of iron (II) disulphide, FeS_2.
By mass 75% of this ore is FeS_2.

Calculate the maximum mass of iron that can be extracted from 480kg of ore.
[A_r: Fe = 55; S = 32]

A) 167.7kg   B) 200kg   C) 360.5kg   D) 118kg   E) 120.2kg

Question 12:

$X_{(s)} + FeSO_{4(aq)} \rightarrow XSO_{4(aq)} + Fe_{(S)}$
Which metal can be correctly be substituted in X's place?

A) Tin (Sn)   C) Lead (Pb)   E) Copper (Cu)
B) Zinc (Zn)   D) Silver (Ag)

Question 13:

Which of the following statements is true regarding Red Shift?

A) The further a distant galaxy or celestial object is, the further down the red end of the light spectrum it's light will be.
B) The closer a galaxy gets, the longer it's wavelengths get, thus moving down the red end of the spectrum.
C) Red shift means that we never see the real light from distant galaxies.
D) We can never tell how far away galaxies are using red shift.

Question 14:

Which is true regarding X-rays?

A) X-rays do not pass through denser materials like bone and that's why they show up as white on the X-ray film.
B) X-rays pass through bone but not skin and soft tissue, and that's why bones show up white on the X-ray film.
C) X-rays don't ionise cells and thus are safe.
D) Gamma rays are safer than X-rays.

Question 15:

Rearrange $\frac{(16x+11)}{(4x+5)} = 4y^2 + 2$ to make x the subject

A) $x = \frac{20y^2-1}{[16-4\,(4y^2+2)]}$ C) $= \frac{6y^2-1}{[16-4\,(4y^2+2)]}$ E) $= \frac{7y^2-1}{[6-14\,(6+7)]}$

B) $= \frac{20y^2-8}{[16-6\,(4y^2+2)]}$ D) $= \frac{21y^2-1}{[16-4\,(2y^2+2)]}$

Question 16:

The element shown below is Germanium. It has an ionic charge of 4+. How many electrons does one atom of Germanium have?

$$73$$
$$Ge$$
$$32c$$

A) 32   B) 73   C) 36   D) 41   E) 4

Question 17:

Bacteria invade the body and produce toxins that kill cells.

What are some of the first line defences the body has to prevent bacteria entering?

1. Mucus lining the airways
2. Heat produced by the body
3. Skin

4. Antibodies produced by the immune system
5. Toxins produced by the body
6. Hydrochloric acid in the stomach

A) 1 only

B) 2 only

C) 3 only

D) 1, 3, 4 and 6

E) 4, 5 and 6

F) 1, 3 and 6

G) 2 and 4

Question 18:

If $(3p + 5)^2 = 24p + 49$, calculate p.

A) -5 or -9   B) -3 or -6   C) -4 or 6   D) -6 or 4   E) 4 or -2

Question 19:

Reaction rates are explained by the Collision Theory. This theory states that particles have to collide (hard enough) in order to react.

Which of the statements below are true?

1. An increase in the temperature of the system can cause more collisions with greater force, therefore causing more reactions.
2. Smaller particles are don't collide or react as well.
3. Increasing the pressure of the system does not cause more collisions.
4. The collision theory only applies to gasses.
5. Increasing the surface area of the reactant increases the chances of collision.

A) 1 only

B) 2 only

C) 3 only

D) 4 only

E) 2 and 3

F) 3, 4 and 5

G) 1 and 5

Question 20:

When someone is lost in the mountains, the rescue team often wraps an aluminium covered plastic sheet around them in order to keep them warm.

Which of the statements are true regarding the effects on heat loss may have on their body?

1. There is less heat loss through conduction.
2. Air is trapped closer to the body and this means that there is less heat loss due to convection.
3. Aluminium absorbs more sunlight and thus this keeps the person warm as more heat is absorbed.

A) 1 only

B) 2 only

C) 3 only

D) 1 and 2

E) 1, 2, 3

F) 2 and 3

G) None

Question 21:

Which of the following is true with regards to osmosis?

A) It does not require a concentration gradient

B) It can apply to any substance, not just water

C) It is the movement of water across a partially permeable membrane

D) It is an active process

E) Transporters move water molecules across the membrane of cells

Question 22

For the following reaction, which of the statements is true?

$$CH_{4(g)} + 2O_{2(g)} \rightarrow 2H_2O_{(aq)} + CO_{2(g)}$$

A) This is an example of complete combustion.

B) By increasing the concentration of CO_2 you can increase the rate of combustion

C) The reaction is anaerobic

D) Combustion of a gas always produces a liquid like water

E) If you remove some of the oxygen you get more product.

Question 23:

Which of the following is a unit of resistance?

A) $V.A^{-1}$ B) $C.A$ C) $C.\Omega$ D) $V.\Omega^{-1}$ E) $W.V^{-1}$ F) J

Question 24:

To screw a piece of wood into a plank of wood, Bob uses a 20cm spanner. The moment of the force used to twist the screw into the plank is 40Nm. How much force does Bob need to exert on the screw?

A) 2   B) 0.2   C) 80   D) 200   E) 0.5   F) 820

Question 25:

The carbon cycle is the cycle regarding the intake and release of carbon by organisms. Which of these statements are true?

A) Plants take carbon via photosynthesis and taking nutrients from the soil, which have come from decayed organisms.
B) Animals give off carbon via respiration, waste, eating and death.
C) The CO_2 in the air comes from burning of plant/animal products and respiration from living organisms only.
D) Trees do not store any carbon as they give it all off as carbon dioxide.

Question 26:

Refraction occurs when a wave passes from a material of low density to a material of high density or vice versa.

Which of these statements regarding refraction is true?
A) If a wave hits a different medium at an angle, the wave does not change direction.
B) If a wave hits a boundary face on, it slows down but carries on in the same direction. Thus it has a shorter wavelength but the same frequency.
C) Waves can be refracted even if they hit the boundary head on.
D) Light is the only type of wave that can be refracted.
E) Glass to air slows down the wave.

Question 27:

Enzymes are thought to work by two mechanisms – lock and key or the induced fit theory. The Lock and Key theory states that the active site of an enzyme is already perfectly shaped for the substrate, whereas the induced fit theory states that the enzyme's active site moulds itself around the substrate's shape. Which of these statements is true?

A) Enzymes are substrate specific.
B) The induced fit theory allows multiple, different types of substrates to be acted on by one enzyme.
C) The induced fit theory allows multiple, different types of enzymes to work on the same substrate.
D) The lock and key theory does not allow space for catatonic reactions (breaking the substrate up.

END OF SECTION

Section 3

1) *"Time and time again, throughout the history of medical practice, what was once considered as "scientific" eventually becomes regarded as "bad practice"."* - **David Stewart**

What does this statement mean? Give some examples of times when scientific practice has become bad practice and describe how this has had an impact on medicine.

2) *"Formerly, when religion was strong and science was weak, men mistook magic for medicine; now, when science is strong and religion is weak, men mistake medicine for magic."* – **Thomas Szasz**

What does this statement mean? Do you think it is correct in assuming all men mistake medicine for magic?

3) *'Approximately 26.9% of the adult population in the UK is obese. We should be offering bariatric surgery to every obese person that walks through the doors.'*

Explain what this statement means. Argue to the contrary. To what extent do you agree with the statement?

4) *'Placebos may solve the problem of patients demanding medication they do not need.'*

Explain what this statement means. Argue the contrary. To what extent do you agree with the statement?

END OF PAPER

PRACTICE PAPER

ANSWERS

Answer Key

Paper A		Paper B		Paper C		Paper D	
Section 1	Section 2	Section 1	Section 2	Section 1	Section 2	Section 1	Section 2
1 C	1 B	1 A	1 D	1 D	1 D	1 B	1 E
2 D	2 D	2 C	2 E	2 D	2 C	2 D	2 B
3 B	3 A	3 C	3 E	3 B	3 C	3 E	3 B
4 D	4 E	4 D	4 E	4 B	4 C	4 C	4 B
5 D	5 D	5 D	5 D	5 B	5 C	5 C	5 E
6 B	6 B	6 B	6 A	6 C	6 E	6 C	6 F
7 A	7 D	7 D	7 A	7 D	7 B	7 E	7 E
8 D	8 E	8 C	8 D	8 D	8 B	8 E	8 D
9 C	9 D	9 B	9 A	9 A	9 A	9 D	9 C
10 E	10 D	10 A	10 D	10 C	10 C	10 D	10 D
11 E	11 B	11 E	11 A	11 D	11 D	11 C	11 A
12 D	12 E	12 D	12 B	12 D	12 D	12 C	12 E
13 E	13 D	13 A	13 C	13 E	13 B	13 B	13 B
14 D	14 C	14 E	14 A	14 A	14 C	14 C	14 E
15 E	15 B	15 E	15 D	15 E	15 E	15 E	15 B
16 A	16 A	16 B	16 D	16 C	16 C	16 C	16 F
17 D	17 E	17 A	17 C	17 E	17 B	17 C	17 D
18 B	18 B	18 B	18 D	18 E	18 C	18 B	18 D
19 C	19 C	19 E	19 D	19 E	19 E	19 D	19 B
20 B	20 A	20 E	20 E	20 C	20 E	20 D	20 E
21 D	21 C	21 D	21 C	21 D	21 A	21 C	21 E
22 E	22 B	22 E	22 E	22 B	22 E	22 B	22 B
23 D	23 C	23 A	23 E	23 D	23 A	23 C	23 C
24 E	24 B	24 B	24 D	24 B	24 D	24 D	24 D
25 B	25 A	25 B	25 D	25 A	25 C	25 C	25 C
26 C	26 B	26 B	26 B	26 A	26 C	26 E	26 D
27 C	27 C	27 D	27 C	27 B	27 E	27 C	27 F
28 B		28 C		28 B		28 D	
29 B		29 B		29 D		29 D	
30 C		30 A		30 C		30 E	
31 C		31 E		31 D		31 D	
32 B		32 C		32 B		32 D	
33 E		33 A		33 E		33 D	
34 D		34 D		34 B		34 B	
35 A		35 D		35 B		35 C	

	Paper E		Paper F		Paper G		Paper H	
	Section 1	Section 2	Section 1	Section 2	Section 1	Section 2	Section 1	Section 2
1	B	D	D	G	C	B	A	A
2	B	E	A	D	D	H	C	B
3	B	F	C	E	B	D	E	D
4	D	A	D	D	E	E	D	B
5	B	C	A	C	C	B	D	E
6	D	D	B	C	C	D	E	B
7	C	F	E	A	B	A	E	A
8	B	A	E	A	B	B	A	C
9	E	F	D	C	B	A	C	H
10	A	C	B	B	A	A	D	A
11	C	D	A	A	B	D	B	A
12	B	B	D	C	E	B	D	B
13	C	C	C	H	D	F	A	A
14	E	E	A	B	C	E	D	A
15	D	B	E	C	E	E	B	A
16	E	C	D	D	A	B	D	A
17	A	E	B	E	B	B	C	F
18	D	F	B	C	E	D	B	D
19	E	A	A	A	C	B	A	G
20	E	B	B	B	E	C	E	E
21	D	F	B	A	B	B	C	C
22	B&E	B	E	D	C	D	C	A
23	D	D	D	B	D	D	A	A
24	C	E	D	C	A	C	C	D
25	B	E	B	C	E	C	D	B
26	C	D	A	D	C	E	C	B
27	C	B	D	C	D	A	D	A
28	A		B		B		C	
29	A		A		D		B	
30	D		E		D		D	
31	A		A		C		B	
32	A		C		D		B	
33	E		B		C		C	
34	C		C		D		C	
35	B		E		D		C	

Raw to Scaled Scores

Section 1								Section 2					
1	1	11	2.8	21	5.4	31	8.3	1	1	11	3.5	21	6.6
2	1	12	3.0	22	5.7	32	8.5	2	1	12	3.7	22	6.9
3	1	13	3.2	23	6.0	33	8.7	3	1.3	13	4	23	7.3
4	1	14	3.5	24	6.3	34	9	4	1.6	14	4.3	24	7.6
5	1.2	15	3.7	25	6.6	35	9	5	1.9	15	4.6	25	8
6	1.5	16	4.0	26	6.9			6	2.2	16	5	26	8.5
7	1.8	17	4.2	27	7.1			7	2.5	17	5.3	27	9
8	2.0	18	4.5	28	7.4			8	2.8	18	5.6		
9	2.3	19	4.8	29	7.7			9	3.0	19	5.9		
10	2.5	20	5.1	30	8.0			10	3.2	20	6.2		

Mock Paper A Answers

Section 1

Question 1: C

The simplest solution is to calculate the total area at the start as 20 x 20 = 400cm². Then recognise that with every fold the area will be reduced by half therefore the area will decrease as follows: 400, 200, 100, 50, 25, 12.5 – requiring a total of 5 folds.

Question 2: D

This is the only correct as it is the only statement that doesn't categorically state a fact that was discussed in conditional tense in the paragraph.

Question 3: B

Off the 50% carrying the parasite 20% are symptomatic. Therefore 0.5 x 0.2 = 10% of the total population are infected and symptomatic. Of which 0.1 x 0.9 = 9% are male.

Question 4: D

The most important part of the question to note is the figure of 30% reduction during sale time. Although A and B are possible the question asks specifically with regard to cost. Therefore, it is only worth waiting for the sale period if the sterling to euro exchange rate does not depreciate more than the magnitude of the sale. As such solution D is the only correct answer as it describes anticipating a loss in sterling value less than 30% against the euro.

Question 5: D

Begin by calculating the number of childminders that can be hired for a 24-hour period as 24 x 8.5 = 204. Therefore, a total of 4 childminders can be hired continually for 24 hours with £184 left over – as the question states the hire has to be for a whole 24-hour period and therefore the remainder £184 cannot be used. As such D is the correct answer of 4 x 4 = 16.

Question 6: B

The simplest way to approach this question is to recognise that there is a difference of £1.50 between peak and off-peak prices for all individuals except students. The total savings can therefore be calculated as (3 + 5 + 1) x 1.5 = 9 x 1.5 = 13.5.

Question 7: A

Karen is a musician, so she must play an instrument, but we do not know how many instruments she plays. Although all oboe players are musicians, it does not mean all musician play the oboe. Similarly, oboes and pianos are instruments, but they are not the only instruments. So, statements b and c are incorrect. Karen is a musician but that merely means that she plays an instrument, we do not know if it is the oboe. So, statement d is incorrect.

Question 8: D

Answers A and B are simply incorrect as the measurement taken is a percentage increase (/decrease) which will normalise baseline diameters therefore allowing for comparison over multiple time points. You should be aware from your studies that ultrasound is an invaluable technique in distinguishing between adjacent tissue types. Any methodology is repeatable if it is correctly chronicled and followed therefore leaving the correct answer of D.

Question 9: C

If both the flight and travel from the airport are delayed this will be the longest the journey could possible take – producing a total journey time of 20 + 15 + 150 + 20 + 25 = 230 minutes or 3 hours 50 minutes. Therefore given all possible eventualities, to arrive at 5pm, boarding should begin at 13.10pm. Answer D is incorrect as a delayed plan would add 20 minutes to the journey whilst the transport to the meeting at the other end takes a minimum of 15 minutes – even if Megan could teleport instantaneously from the airport to the meeting she would be 5 minutes later than if there wasn't a plane delay.

Question 10: E

This is almost a trick question and simply an application of exponential decay. Recall that an exponential decay is asymptotic to 0 as no matter how small the volume within the cask becomes, only half of it is ever removed. It could be argued that this process cannot continue once a single molecule of whiskey is left – and when splitting that single molecule in half it is no longer whiskey. However, the question does not ask "how long till all the whiskey is gone" but rather "how many minutes will it take for the entire cask to be emptied" and therefore the process can continue infinitely – even if the only thing left in the cask is a collection of quarks … or half that.

Question 11: E

With these questions it is important to only consider information displayed in the graph and not involve any assumptions provided by your prior knowledge. Therefore, this question is questioning your ability to consider correlation as opposed to causation. The graph simply shows that waist size and BMI are positively correlated with one another and that is it. The nature of a scatter plot does not allow you to deduce which of the variables (if any) drives the result observed in the dependent variable. However, the fact that they are correlated is an important result and therefore D is also incorrect.

Question 12: D

A sky view of the arrangement leads to:

C	A	B					A	C	B
D		E		*or*				D	E

In both, D is to the left of E, thus is the only correct answer.

Question 13: E

The question can be expressed as $(40 \times 30) - x(50 \times 30) = 200 = 1{,}200 - x1{,}500$. Therefore $x = 2/3$.

Question 14: D

The information given is very much a red herring in this question. This can be solved using your own application of sequence theory. If a sequence is 6 numbers long, the 6 numbers can be re-arranged into a total of 6! possible sequences = $1 \times 2 \times 3 \times 4 \times 5 \times 6 = 720$

Question 15: E

Although this is an extremely abstract question, all the information needed to answer it is provided. The key rule to have isolated from the information is that with each progressive sequence, a bell can only move one position at most. Therefore, looking through each in turn – for example we can exclude A as the 3 goes from position 4 to position 2. This leaves the only correct answer of B. Recall that the information states bells can only move a maximum of one place – they can of course move 0 places.

Question 16: A

As the largest digit on the number pad is 9, even if 9 was pressed for an infinitely long time the entered code would still average out at no larger than 9. Therefore, it would be impossible to achieve a reference number larger than 9. Indeed, this is an extremely insecure safe but not for the reason described in B (for if the same incorrect number was pressed indefinitely it would never average out as the correct one) but rather because the safe could in theory be opened with a single digit.

Question 17: D

A is incorrect as it ignores the section of the text that states the evolution of resistant strains is driven by the presence of antibiotics themselves. The text states that the rate of bacterial reproduction is a large contributing factor and therefore not wholly responsible – hence B is incorrect. Since this is just one example (and only the information in the text should be considered for these questions) for C to make such a general statement is complete unjustified.

Question 18: B

The fastest way to solve this question is to calculate the quantity of cheese per portion as $200/10 = 20$. Which for 350 people would require $350 \times 20 = 7000$g or 7kg.

Question 19: C

Calculate the calorific content of 12 portions as $12 \times 300 = 3{,}600$kcal. As this represents 120%, evaluate what the initial amount would be as $(3{,}600/120) \times 100 = 3{,}000$kcal.

Question 20: B

Begin by calculating the initial weight of all the ingredients in the Bolognese sauce which comes to a total of 3.05kg. Therefore when cooking for 10 people $3.05 \times 4 = 12.2$kg of pasta should be used. Which in turn means for 30 people $3 \times 12.2 = 36.6$kg should be used.

Question 21: D

Calculate the new weight of ingredients in the Bolognese sauce excluding garlic and pancetta which produces a total of 2.8kg. Note that onions represent 0.3kg per 10 people and as such the ratio can be represented as 0.3/2.8 or alternatively dividing top and bottom by 0.3 → 1/9.3

Question 22: E
Begin with calculating total preparation time as 25 x 4 = 100 mins. The fact that Simon can only cook 8 portions at a time is somewhat a red herring as it doesn't impact the calculation. Total cooking time can be calculated as a further 25 x 8 = 200 mins. Producing a total time of 300mins or 5 hours.

Question 23: D
Due to the quantities colour t-shirts are priced at £5 and black and white at £2.50. Therefore, the order will incur a total cost of 50 + (50 x 5) + (200 x 2.5) = £800.

Question 24: E
Answers A and B directly conflict with information presented in the text whilst C and D may well be true but there is insufficient information in the text to address these points. This is an important reminder that although you may well have been a scout and be able to comment on options C and D, you may only consider these arguments in terms of the text provided.

Question 25: B
There are two directions: clockwise and anticlockwise and rats will only collide if they pick opposing directions. Rat A - clockwise, Rat B - clockwise, Rat C - clockwise = 0.5 x 0.5 x 0.5 = 0.125.
Rat A - anticlockwise, Rat B - anticlockwise, Rat C - anticlockwise = 0.5 x 0.5 x 0.5 = 0.125.
0.125 + 0.125 = 0.25. So, the probability they do not collide is 0.25.

Question 26: C
The article seen here is a particularly good effort at a discursive text as it is completely impartial. Note that the article simply states the facts from either side in equal measure. Nowhere does the author present their opinion on the matter nor do they insinuate their beliefs in anyway.

Question 27: C
The journey time is rounded to the nearest hour (13). Therefore, the longest it could possibly be is 13 hours 29 minutes or it would be round up to 8 hours. Therefore, the latest the ferry will arrive, assuming the travel time estimate is accurate, is given as 20.29.

Question 28: B
The correct answer is 28. Assume, although very unlikely, that you roll a 2 every single go – you will never need to take a step back, only your two forward. Therefore, rolling a lower number in this case is beneficial.

Question 29: B
The simplest way to calculate this is to find the lowest common multiple of the given laptops which is 40 x 60 x 70 = 168,000 seconds = 2,800 minutes = 46.67 hours – although they will not all be on the same lap number at this time.

Question 30: C

A large amount of subtly different data is described here. Of note is the first experiment which describes how nerve conduction is faster in right handed men than it is in left handed men. This result is not transferable to women until it is proven! The experiment currently being conducted only considers dominant hand in men vs. women. That could be either hand or not necessarily the females' right hand. For example, all of the females in this experiment could have been left handed and there is no information in the text to say otherwise, therefore we cannot tell.

Question 31: C
Recognise that two square based pyramids will comprise 8 triangles of base width 5 and height 8; plus, two 5 x 5 squares. Thus, giving at total area of 8(8/2 x 5) + 2(5 x 5) = 210 cm^2

Question 32: B
In order to approach this question first realise that in the first well 1ml of solvent is being combined with 9ml of distilled water producing 1eq of solute in 10ml – hence the first well produces a dilution by a factor of 10. With each progressive dilution the concentration is reduced by a further factor of 10 – hence by well 10 the concentration is at $x/10^{10}$

Question 33: E

The compartments of the human body are occupied by numerous fluids, as the student is only interested in measuring the volume of blood, it is essential he chooses a solute that will only dissolve in blood. So as his known quantity of solute remains no, it must be neither removed nor added during its time in the body. Hence all of the written assumptions must be made and many more.

Question 34: D

The fastest way to approach this question is by calculating the total price per head for the cheapest option as $10 + (8/20) + (10/60) = 10.567$. To make the maths simpler this can be rounded safely to £11 a head at this stage. $2,300/11$ is approximately equal to 209 which when rounded to the nearest 10 is 210 people.

Question 35: A

Whilst this passage is attempting to weigh up two sides of an argument, it has a clear one-sided approach focussing heavily on the excitement of dangerous sports. It even states that hunting is recognised as exciting by some. Since the previous sentence discussed the link between archery and hunting, the statement is a fair extrapolation to make.

END OF SECTION

Section 2

Question 1: B

Let tail = T, body and legs = B and head = H.
As described in the question $H = T + 0.5B$ and $B = T + H$.
We have already been told that $T = 30Kg$.
Therefore, substitute the second equation into the first as $H = 30 + 0.5(30 + H)$.
Re-arranging reveals that $-0.5H = 45Kg$ and therefore the weight of the head is 90Kg, the body and legs 120Kg and as we were told the tail weighs 30Kg.
Thus, giving a total weight of 240Kg

Question 2: D

Recall that kinetic energy can be calculated as $E = 0.5mv^2$. Therefore, if mass remains constant it is the v^2 term that must be reduced to a sixteenth. In other words, $v^2 = 1/16$ and therefore the correct velocity is $1/4x$.

Question 3: A

An organ is defined as comprising multiple tissue types. As blood and skeletal muscle are themselves tissues they cannot be classified as organs.

Question 4: E

This question is best considered in terms of the aerobic respiration equation. With that in mind it becomes apparent that increased forward drive through the reaction will produce large amounts of water and CO_2 whilst demanding an increased supply of O_2. Further from this equation we realise that aerobic respiration produces large amounts of heat, and as such it is expected – in the interest of thermoregulation – that the body will both perspire and vasodilate in attempt to increase heat loss. Therefore, E is the correct answer.

Question 5: D

Recall that the nephron is the smallest functional unit of the kidney. The question therefore is asking you what is the smallest basic functional unit of striated muscle? To which the answer is the sarcomere. Note that a myofibril is a collection of many sarcomeres and is therefore not the correct answer.

Question 6: B

Insulin is a polypeptide hormone released by the pancreas in response to elevated plasma glucose levels. Therefore, it can be expected that plasma glucose concentration will be proportional to the concentration of insulin in the blood. Furthermore, recall that glucagon also released by the pancreas mobilises glucose stores. Therefore, the greatest concentration of plasma glucose would be expected at the time when glucagon is highest during a period of elevated insulin.

Question 7: D

Answers a and c are both nonsense and can be eliminated straight away. You will know from your study of the immune system that it is plasma B cells that produce antibodies and that plasma T cells do not exist. Also recall that an immune response can be mounted as quickly as within a fortnight which leaves the only correct answer d. The passage states that only once blood types are mixed is the immune response initiated, therefore answer d provides an explanation as to how this happens but also why the first-born child is unaffected.

Question 8: E

Recall that pH is a logarithmic scale of proton concentration and therefore will have the largest effect on hydrogen bonding.

Question 9: D

An organ consists of many cell types which once differentiated are committed to that single cell line. Therefore, a totipotent stem cell is required to produce the multiple cell types required. In order to ensure that the organ is an exact genetic match, stem cells from the individual in question must be used. Unless that individual is an embryo, adult stem cells must be used.

Question 10: D

Isotopes of an element all contain the same number of protons but a different number of neutrons. As atomic number refers solely to the number of protons it will not change. However as mass number is the sum of atomic number and neutron number – it would be expected to change. If an isotope contains one extra proton, then assuming that the charge of that isotope is 0, then it must also contain one extra electron. Chemical properties are the same for all isotopes. Therefore, the correct answer is D.

Question 11: B

The transition metals are the most abundant catalysts – presumably due to their ability to achieve a variable number of stable states. Therefore, the correct answer is the d-block elements.

Question 12: E

Begin by writing down the balanced equation that describes the reaction of francium with water: $2Fr + 2H_2O \rightarrow 2FrOH + H_2$. Next calculate the moles of francium entering the reaction as $1338/223 = 6$. We therefore know from the stoichiometry of the equation that this reaction will produce 3 moles of hydrogen. Recall that 1 mole of gas at room temperature and pressure occupies $24dm^3$. Therefore, the hydrogen produced in this reaction will occupy $3 \times 24 = 72dm^3$.

Question 13: D

The simplest way to approach this type of question is to assume that there are 10 atoms within the compound. In this case that produces the following result: $C_3H_4F_2Cl$. Next look to see if any of the subscript numbers are divisible by a common factor. Also, if there are any decimals, multiply up by a common factor until only integers are present. In this case the correct answer is achieved straight away.

Question 14: C

This question requires you to have a correct answer from the previous question, although these questions are unfair in the fact that this current question cannot be answered without success in the first part – there are always one or two of these per paper. Simply calculate the Mr of your empirical formula: 113.5. And then divide 340.5 by this: $340.5/113.5 = 3$. Therefore, multiply your empirical formula up by a factor of 3.

Question 15: B

The calculation in this question is simple: concentration = mass/volume, what this question is really testing is the manipulation of unorthodox units. Begin by noting the use of g/dL in the final answers and therefore begin by converting the quantities in the question into these units. 1.2×10^{10} kg = 1.2×10^{13} grams and with 10 decilitres in a litre, 4×10^{12} L = 4×10^{13} dL. $\frac{(1.2 \times 10^{13})}{(4 \times 10^{13})} = 3 \times 10^{-1}$ g/dL.

Question 16: A

A catalyst is not essential for the progression of a chemical reaction, it only acts to lower the activation energy and therefore increase the likelihood and rate of reaction.

Question 17: E

Cationic surfactants represent a class of molecule that demonstrates both hydrophilic and hydrophobic domains. This allows it to act as an emulsifying agent which is particularly useful in the disruption of grease or lipid deposits. Therefore, cationic surfactants have applications in all of the products listed.

Question 18: B

Recall that V = E/Q; therefore, when substituting SI units into these equation it is discovered that $V = J/C = JC^{-1}$.

Question 19: C

Recall that voltmeters are always connected in parallel – and so that they don't draw any current from the circuit have an infinite resistance. Ammeters on the other hand are connected in series and therefore must not perturb the flow of given, meaning they have zero resistance.

Question 20: A

Much of the information in this question is not needed and is simply put there to distract you. This question can be most quickly solved using the equation F=ma or force = mass x acceleration. As object A is the only things moving in this scenario it is the only source of energy to be considered. Its mass will be the same before and after the collision and so we need only calculate the magnitude of retardation. Given as $(15 - 3)/0.5 = 24 \text{ms}^{-2}$. Therefore, when plugging into the first equation we realise that F = 12 x 24 = 288N of force dissipated. Alternatively, this question could be solved by calculating the rate of change of momentum.

Question 21: C

Note the atomic masses and numbers in the equation. Whilst the atomic mass has remained constant the atomic number has increased by one and hence the element has changed. The only explanation for this is that a neutron has turned into a proton (and an electron which is represented by x). Therefore, the correct answer is C – beta radioactive decay.

Question 22: B

Begin by calculating the velocity of the wave as speed = wavelength x frequency = 3 x 20 = 60km/s. Which in a time period of one hour (3600s) would equate to a total distance of 60 x 3600 = 216,000km.

Question 23: C

The numerator of the fraction consists of 3 distinct terms or 3 distinct dimensions. As all other functions within the equation are constants one would consider this the volume of a complex 3D shape.

Question 24: B

Expand the larger scientific number so that it reads 10 to the power 6 like so: $4.2 \times 10^{10} = 42000 \times 10^6$. Now that the powers are the same across the numerators, a simple subtraction can be performed $(42000 - 4.2) \times 10^6 = 41995.8 \times 10^6$ which can be simplified to 4.19958×10^{10}. Next consider the division which can be competed in a two-step process, first divide the numerator by 2 like so $(4.19958/2) \times 10^{10} = 2.09979 \times 10^{10}$ and then subtract the powers like so $2.09979 \times 10^{(10-3)} = 2.09979 \times 10^7$.

Question 25: A

Note the triangle formed by the right-angle lines and the tangent. Recall that as this is a right-angle triangle then the other two angles must be 45º. As angles along a straight line add up to 180º a must equal 180 – 45 = 135º. Angles around the origin must add up to 360º and therefore b = (360 – 90)/2 = 135º. Therefore, the correct answer is A.

Question 26: B

The probability of drawing a blue ball (1/21) and then a black ball (1/20) is 1/21 x 1/20 = 1/420. However, note that it is also possible that these balls could also be drawn out in the opposite order. Therefore, the probability must be multiplied by two like so 1/420 x 2 = 2/420 = 1/210.

Question 27: C

The question states that the repeat experiment is identical to the first in all aspects apart from the result. Therefore, although a number of the options may be true like calibration bias, it would have been applied to both experiments and therefore should not affect the result. As such the difference in results is simply due to random chance.

END OF SECTION

Section 3

Doctors should be wearing white coats as it helps produce a placebo effect making the treatment more effective.

➤ This statement addresses the role of the patient's personal experience in his/her cure or treatment of their disease. It is an interesting topic since the role of psychological factors in the treatment of disease is largely unexplored. There is a growing body of evidence that supports the effectiveness of placebo treatments for some diseases when it comes to managing patient symptoms, but there is very little that addresses the role of attire and visual appearance of doctors.

➤ It is also important to immediately question the truth of the statement. There is some evidence to suggest that there is such a thing as a "white coat effect" that influences patient's behaviour and the perception of their problems when they are faced with a doctor. Questioning the statement is very important as it demonstrates that you reflect on the issue.

➤ When answering this question, there are several factors to consider. There is the role that clothing plays in the definition of professions. How does the attire of an individual influence the way he or she is perceived by those receiving his/her service? Some examples here are police officers or judges where the uniforms are heavily tied to the public perception of their profession. Police officers are a particularly good non-medical example since there are uniformed and non-uniformed officers that play different roles playing on the different public perception of uniform and civilian clothing and the fact that without the uniform the police officer is not recognisable. Then the question arises if this should apply for doctors too. Does it make a difference if doctors have clothing that visually separate them from other people in the hospital and does this have an influence on the patient's experience of treatment. Other points to consider when addressing the role of attire is the depiction of doctors in the public sphere. This includes TV shows, books, news etc.

➤ Arguing against the statement is more difficult than it seems simply because you should make sure that you provide a diverse answer that addresses several aspects. On one hand, there is the connection of a specific attire to a specific professional role as described above. On the other hand, there is the question whether attire is relevant to influence the patient's experience to improve health outcome. The whole point of a placebo effect in this context is that it improves the outcome.

➤ Another point to consider when arguing against this statement is the power distribution that comes with the uniforms and whether that is something that is beneficial for the patient-doctor relationship.

➤ Arguing to the contrary, you can look at situations where a professionalization of the doctor-patient relationship can be beneficial. Now, to be clear, the relationship between doctor and patients should always be professional, but in the context of this question, you can use the role of attire and its role in establishing this professionalization. Examples for this include conversations about life-style changes and the role the patient can play in improving his/her own health, especially if this involves giving something up. In this case, attire can give the doctor legitimacy and a certain degree of authority.

"Medicine is a science of uncertainty and an art of probability."

➤ This statement basically addresses the fact that there is no such thing as certainty in medicine. People are different and individual and so is their experience of disease. For this reason, the statement argues that all a doctor can do in terms of approaching a sense of certainty, the doctor should weigh up different probabilities and possibilities of disease. The argument also suggests, that weighing up the different options of diagnoses is an art, rather than acquired knowledge. This suggests some degree of natural talent. It also provides a degree of contrast between the aspect of science that provides the theoretical basis for pretty much every decision we make in medicine and the art of the application of knowledge. It also acknowledges that science always contains a degree of uncertainty, even when individuals believe in the absolute truth of their theory/knowledge.

➤ Arguing to the contrary basically aims at increasing the perceived role of science and certainty versus that of art and uncertainty or flexibility. The main problem with this statement is the general perception associated with the words "science" and art. They naturally lie on different ends of a spectrum with science being associated with facts and certainty and art being associated with softer skills and an absence of certainty.

➤ If you choose to go into a more example oriented direction, there are several points you can raise to write a good and strong essay. One example is the treatment of infectious diseases with antibiotics, especially in severe cases. Often you will find that the disease is treated with a broad-spectrum antibiotic that is likely to target the causative agent based on local experiences and local occurrence of diseases. This is then later adjusted if necessary once a precise identification of the causative agent was possible. Other examples include the stratification of disease causes. One example here is smoking and lung cancer. Whilst it is generally accepted that smoking increases the risk of lung cancer, there are still non-smokers that get lung cancer and life-time smokers that do not. This pattern can be applied to a variety of parameters to result in similar results.

➤ In general, you can keep this essay very philosophical and abstract, or you can aim more at direct examples to illustrate your points. Both options have strengths and weaknesses. A theoretical essay will stay more with the overall style of the statement, whilst a more example oriented essay will be easier to write and to keep track of. However, it will also be more difficult to find appropriate examples.

➤ In the conclusion, when you give your opinion, it pays to be very direct on one hand, but also to be very specific. Depending on which route you took for your main body arguments, this may be easier or more difficult. You can also pick up the idea of medicine being art again as this is an interesting point and ties in with the idea that medicine cannot be exclusively learned from books but should also contain a component of patient interaction.

"The New England Journal of Medicine reports that 9 out of 10 doctors agree that 1 out of 10 doctors is an idiot."

➤ This statement addresses how scientific research will never find complete acceptance in the field. It also suggests that no matter what is being published by even the most highly acclaimed scientific journals, there is always a risk of error. In the end, it illustrates that medical research usually is a game of probabilities as there is never complete certainty when it comes to the pattern of diseases or the optimal treatment of disease.

➤ On a face value level, this question is simple. It is very vague in its assertion, not defining which one of the 10 doctors think that the other is an idiot. Do they all think the same person is an idiot or do they think different persons are idiots? This an important issue to raise when answering this question as it presents a fundamental flaw of the question, especially if one is to apply it to general medical research and research practice. Even in the sometimes uncertain realm of medical research, parameters such as populations of subjects are always clearly defined, which is what gives any form of research value. If this was not the case, research would be completely arbitrary. Coming back to the question then, if all 9 doctors believe that that the one specific one of them is an idiot and they have no prior contact and no connection to each other, then chances are that this single doctor is actually an idiot. If, however, there is no pattern whatsoever to the claim that one doctor is an idiot, then that weakens the claim. Especially if the whole concept is then widened to a population level.

➤ Arguing to the contrary has different obvious points. You can either stay very close to the actual wording of the question which will lead you down a similar road as I have illustrated above, or you can use the question as a parable for the way we conduct scientific research. This will require you to have a good understanding of scientific method.

➢ If you decide to go down the scientific method pathway, you will have several things to point of. Firstly, is the definition of populations, as this is completely ignored in the question. In any form of research defining the pool of data you draw from is essential as only this will deliver accurate and usable information. It is all about reducing vagueness as much as possible. Secondly you need to define the research criteria. What exactly is meant by idiot in this case for example? Only if you formulate a clear-cut goal can you then acquire the data needed to come to a meaningful result. The term 'asking the right questions' comes to mind. Thirdly you need to ensure repeatability. For this you need to define your populations very broadly and in appropriate sizes. You must make sure that there is as little connection between the subjects as possible as this will reduce and bias from personal relationships.

➢ These two options should help you write a strong essay, especially since they can be combined in essentially any way you choose.

"My father was a research scientist in tropical medicine, so I always assumed I would be a scientist, too. I felt that medicine was too vague and inexact, so I chose physics."

➢ There are several components to this question that you need to be aware of if you want to write a good essay. On one hand is the person Stephen Hawking himself. Being a world renowned theoretical physicist, gives the whole quote an almost comical note. This is something you should be aware of as you will always have to point out problems with the questions. Moving beyond this, there are several other points you should be aware of. One is the vagueness of the statement. This obviously is due to the fact that it is taken from what probably was a whole speech, rather than this single passage. Again, something you should point out. Then there is the subject matter of tropical medicine. Tropical medicine is in part still a very new field and a field which much room for exploration simply because there is such a wealth of different life forms in tropical areas that can cause diseases, some of which may never have been observed before. This necessarily adds to the perceived uncertainty. In addition, keep in mind that Stephen Hawking is now 75 years old, which places his father's professional career to the first half of the 20th century, a lot has happened in medicine since then. Secondly you should address the subject matter of physics. Whilst some fields in physics have very little uncertainty and vagueness, a lot of areas are very precise such as gravity or mechanics. So, it is important to make that distinction as physics is such a broad topic.

➢ When it comes to arguing to the contrary, there are several perspectives you can take. On one hand, you can argue that tropical medicine is more accurate than Hawking gives it credit for. An easy way to do this is to enlarge it to general medicine as the clear majority of general medical principles will still apply in tropical medicine, what will change will be different pathogens and the environmental factors influencing pathology and healing processes. If we accept that in general medicine is a fairly exact science, we can use that to support the same claims about tropical medicine.

➢ Another point of attack would be to point out the vagueness of some areas of physics. Easy targets here are String theory and relativity theory. Neither of those can be supported by non-mathematical evidence at this point and even the potential discovery of the Higgs boson in the Cern super collider does not provide enough answers to these questions yet. There are many other examples in the field of physics that are vague, that's why they have given rise to completely separate job description: the theoretical physicist.

➢ Finally, you can also consider Hawking's personal history with medicine. His suffering from ALS for decades and being bound to a wheelchair after far outliving any suggested life-expectancy it is understandable that he considers medicine as somewhat vague. This could well have influenced him in this statement.

END OF PAPER

Mock Paper B Answers

Section 1

Question 1: A

If society disagree that vaccinations should be compulsory, then they will not fund them. So, statement A is correct. It attacks the conclusion. Statement b - society does not necessarily mean local so this does not address the argument. Statement c strengthens, not weakens, the argument for vaccinations. Statement d – the wants of healthcare workers do not affect whether vaccinations are necessary.

Question 2: C

Start by calculating the area of wall that may be painted per tin of paint as $10 \times 5 = 50m^2$. Therefore, to paint the whole area $1050/50 = 21$ tins of paint are required per coat. As such to complete 3 coats it will cost Josh $3 \times 21 \times 4.99 = 314.37$.

Question 3: C

A is a correct assumption as procession is a function of rotational motion. B is a necessary assumption or rather inference of the first sentence. The second sentence only says that an asterism can be used, not that it is the only possible method. Nothing is mentioned of navigating the Southern Hemisphere and therefore C is not a valid assumption.

Question 4: D

Recognise that "bank hours" refers only to hours that the bank is open – which Mon to Fri is 8 hours whereas it is only 6 hours on a Saturday. Although John needs the money by 8pm the bank closes at 5 and that 3 hours difference cannot be used. Hence working backwards John will need 8 hours on the Tuesday, 8 hours on the Monday, Sunday is closed, 6 hours on the Saturday, 8 hours on Friday and 8 hours on Thursday and 4 hours on the Wednesday. With a closing time of 5pm, the latest John can cash the cheque on Wednesday is 1pm.

Question 5: D

First thing to recognise here of course is that individual diamonds can be combined to form larger diamonds with the 5×5 diamond the biggest of them all. To avoid counting them all and risking losing count, instead deduced the number of triangles per corner and per side; then multiply up by 4.

Question 6: B

Let my current age = m and my brother's current age = g. The first section of this question can therefore be expressed as $m + 4 = 1/3(g + 1)$ whereas the second half can be represented as $2(m + 20) = g + 20$. Therefore, this problem can be solved as simultaneous equations. Rearranged the second equation reads $m = 1/2g - 10$; when substituted into the first equation we form $1/2g – 10 + 4 = 1/3(g + 1)$. Expand and simplify to $1/2g – 6 = 1/3g + 1/3 \to 1/6g = 6\frac{1}{3}$ which therefore means my brother's current age $= 6\frac{1}{3} / (1/6) = 114/3 = 38$. Which means that my current age $= 1/2(38) – 10 = 9$.

Question 7: D

A is categorically wrong as the first two paragraphs discuss how aneurysms produce inflammation which in turn blunts endothelial NO action. B is incorrect as it states aneurysms directly promote CVD, this is not a direct process. It is the blunted NO which directly produces the CVD. C can be ignored as nowhere are aneurysms categorised like this. E is incorrect as the text states that aneurysms reduce NO which will reduce vasodilatation, thus increasing basal vasoconstriction and thus reducing blood flow. Leaving the correct answer of D which is of course true as observations are not transferable between species until tested scientifically.

Question 8: C

Any statement which refers to national or global figures is instantly incorrect as the text does not mention any statistical analysis has taken place. In order to produce national statistics from a small sample size such as this requires statistical analysis. Whilst E could possibly be true it cannot be stated as there are so many possibilities – perhaps the time of the survey was during rush hour in which case the majority of the traffic would have been travelling in the same direction anyway to reach an industrialised area.

Question 9: B

The runners aren't apart at a constant distance; they get further apart as they run. Xavier and Yolanda are less than 20m apart at the time William finishes. Each runner beats the next runner by the same distance, so they must have the same difference between speeds. When William finishes at 100m and Xavier is at 80m. When Xavier crosses the finish line then Yolanda is at 80m. We need to know where Yolanda is when Xavier is at 80m. William's speed = distance/time = 100/T. Xavier's speed = 80/T. So, Xavier has 80% of William's speed. This makes Yolanda's speed 80% of Xavier's and 64% (80% x 80% = 64%) of William's. So, when William is at 64m when William finishes. 100m - 64m = 36m, thus William beats Yolanda by 36m.

Question 10: A

This question can be solved quickly if you first realise that there is no need to calculate both volumes and subtract the larger from the smaller, instead only convert the television dimensions into metres and then calculate 60% of that.

Question 11: E

From the information provided all the flaws listed are valid since David's main point is that he has chosen the cheapest. A could be true as there is an additional cost of £3 for staying at Whitmore, therefore if the vehicle they are using achieves sufficient miles per gallon then travelling the extra few miles could cost less than £3 in terms of petrol. B again is possible which would argue against it being cheap, as would D. And if C is true then David's argument is flawed altogether.

Question 12: D

C is irrelevant as nowhere does the passage mention standards of modern medical practice. A may be incorrect as nowhere does the article explicitly say that animal testing is the only accepted method of drug approval. B categorically conflicts with the first sentence of the second paragraph.

Question 13: A

Begin by converting all the quantities into terms of items as that is the terminology used on the graph axis. Therefore 12 rugby balls = 6 items and 120 tennis balls = 24 items. Reading from the graph reveals their respective prices as £9 and £5. Therefore, the total cost of products in the order is (6 x 9) + (24 x 5) = 174. Since this is significantly more than £100 the delivery charge is waived.

Question 14: E

Calculate the cost of 10 of everything as (2 x 5) x (10 x 7) x (5 x 9) = £125. Recall that delivery charge is waived at £100 and this therefore a trick question and no delivery charge is applied anyway.

Question 15: E

Tennis balls are sold in the largest pack and so they must be considered. Begin by dividing 1000/5 using the value from the first column = 200. As this is above the range 0 -99 look up the item value in the 100 -499 range where a £1 discount is applied per item. Therefore, in actually fact 1000/4 = 250 items can be purchased which equates to a total of 250 x 5 = 1250 balls.

Question 16: B

Recognise that 120% profit is equivalent to 220% of the original price. In which case the initial purchase price = (1,320/220) x 100 = £600.

Question 17: A

Note that here the question uses the term item and therefore simply read the costs directly off the graph giving a total order cost of (2 x 2000) + (4 x 2000) + (6 x 2000) = 24,000. Recall though that he only pays tax on the amount over £12,000 which in this case is £12,000. Therefore, he pays 12,000/4 = £3,000 tax.

Question 18: B

Lucy must live between Vicky and Shannon. Lucy is Vicky's neighbour, so Shannon cannot have a red door. Vicky lives next to someone with a red door, so Lucy must have the red door. This leaves Shannon with the blue door and Lucy with the white. The green door is across the road and so does not belong to any of them.

Question 19: E

First calculate an average complete one-way journey time as 40 + 5 + 5 = 50 minutes. Deducting his breaks, he works a total of 7 hours 20 or 440 minutes. Since the first train is already loaded his first run will only take 45 minutes leaving 395 minutes to complete his working day. 395/50 = 7 remainder 45. Note that 45 minutes is not enough to fully unload the train, but it is enough to load the train and drive the distance. Therefore, the driver will complete a total of 9 journeys equalling a distance of 198 miles.

Question 20: E

A is not actually a valid assumption as we do not know what proposal conservationists might be bringing to the local councils, they have only expressed their concern. They may well be bringing a proposal to ask for funding to rehome all the species in the affected environment. B is essential to the final paragraph whilst C must be assumed otherwise the councils would not be presenting these proposals at all.

Question 21: D

As there is not really information in the question to calculate the answer quickly. Instead consider each answer in term and calculate the differences to find the correct price difference in the question:

A) $(3x \ 1) + (2 \times 1.25) - (15 \times 0.3) + (10 \times 0.5)$ etc…

Question 22: E

Based on the information in the question options A - D are simply wrong. A is incorrect as antibiotic E has not affected growth at all. B is incorrect as the other antibiotics have significantly affected growth. E was the least effective antibiotic. C was not the most ineffective as it did disrupt growth slightly whereas E had no effect at all. D will now be taken and the experiment repeated with D at numerous concentrations to find the optimum dose.

Question 23: A

$1L = 1000$ cubic centimetres and therefore the total volume of air Laura needs to produce is $25 \times 0.3 = 7.5L$. With a total of 25 balloons she will take $25 \times 0.5 = 12.5$ seconds breathing in and a further $7.5/4.5 = 1\frac{2}{3}$ minutes inflating the balloons. This yields a total time of 1 minute $40 + 12.5$ seconds = 1 minute 52.5 seconds or 112.5 seconds.

Question 24: B

Quickly represent the question schematically as $(A = B) \neq (C = D = E)$. We can now observe that A in fact supports George's argument, C also supports George's argument and D may well be true, but it would have no effect in disrupting the argument, simply only imply D and E are both also equal to 0. However, as E is equal to C it should therefore not equal B.

Question 25: B

This question is much less complicated than it sounds. Begin by just considering a single hour. Throughout the hour of 1 the hour hand will be pointing at 1. Only during the 5th minute of that hour will the minute hand point to the 1 whereas every 5th second of the minute the second-hand points to the 1. All these events will only coincide once. As there are 24 hours in a day 00:00 through to 23:59 this event will happen 24 times.

Question 26: B

A) Potentially correct, but extreme sports also carry higher risks of injury.
B) True.
C) True, but irrelevant for the question.
D) Potentially correct, but irrelevant to the question.

Question 27: D

A) False – we are not told about the healthcare directly but are told that injury and disease posed a threat.
B) False – the terrain was difficult, and mapping was poor.
C) False – outlaws were a significant threat.
D) True – as the text states, there was a marked lack of bridges.

Question 28: C

Whilst B may be true it is not a reason for dependence, only a supporting factor. Dependence implies that we have no choice but to use electricity. Hence A is wrong as gas is readily available; hence D is wrong for the same reason. This leaves the correct answer C which is the only statement which truly describes our absolute necessity for electricity – since electrical appliance by definition only function with electricity.

Question 29: B

First note that 27 guests plus Elin herself means that 28 people will be eating the 3 courses which will require a total of $28 \times 3 = 84$ glasses of wine. This is a total volume of $84 \times 175 = 14.7L = 21$ bottles. As wine is only sold in cases of 6, Elin will have to buy 24 bottles so as not to run out. Recall the buy one get one free offer so she only pays for 2 cases.

Question 30: A

Recognise that when rounded to the nearest 10 the shortest an episode could last is 35 minutes. Hence a total of $7 \times 12 = 84$ episodes would take a total of $84 \times 35 = 2,940$ minutes = 49 hours.

Question 31: E

Points A and B are the best exemplified through this passage. Often great discoveries come from accidental observations and then exact processes are refined through many experiments in a trial and error fashion until the correct methodology is achieved. The passage demonstrates how as our understanding of the world around us advance so too does our ability to provide healthcare. D can be observed in the passage as the 50/50 split.

Question 32: C

Whilst A and D are true they do not force the stranger to give him the sapphire – remember Jack can be given any stone for a truthful statement. B and C are both lies and will earn Jack nothing. Instead if Jack states E then the stranger has no choice to hand over the sapphire else it would be a false statement.

Question 33: A

Despite the enormous interest rate in Simon's current account it is only awarded twice, whereas in the saver account it is awarded 4 times. Hence earnings from the saver account = 100 x 1.5^4 = £506 whereas earnings from the current account would have stood at £361.

Question 34: D

The largest possible key can be obtained were the first two numbers are at a maximum because they are multiplied together → 9 x 9 = 81. Subtract the smallest number to yield 81 – 1 = 80 and again divide by the smallest number which is 1 hence 80 is the largest possible key.

Question 35: D

A) Incorrect. The text clearly states that the exercise routine is resistance training based.

B) False. Both groups contain equal numbers of men and women per the text.

C) False. Both groups are age matched in the range of 20 to 25 years.

D) Correct. As the only difference between the two shakes is the protein content.

END OF SECTION

Section 2

Question 1: D

As the question states that GLUT2 is ATP independent then answer A) active transport is instantly incorrect as it is ATP dependent. Osmosis is applicable only to water molecules and is therefore incorrect. Exocytosis refers to the movement of molecules out of a cell and is therefore incorrect. Simple diffusion is incorrect as the question states that GLUT2 is essential for the process. This leaves the correct answer of facilitated diffusion.

Question 2: E

In order to answer this question you must recall that anaerobic respiration in humans produces only lactate and energy, whilst in yeast the anaerobic respiratory process yields a molecule of ethanol and CO_2 per glucose molecule. Therefore, there will be 0 mol of CO_2 produced in the human cell culture and you need only work out the moles of CO_2 produced by the yeast cell culture. There is a total of 5.76/0.18 = 32 mol of glucose, of which half is supplied to the yeast cell culture. With a stoichiometric ratio of 1:1 in the anaerobic respiration equation a total of 16 mol of CO_2 will be produced.

Question 3: E

Firstly, recall that endocytosis is a process of molecular transport into cells that result in vesicular formation. This question requires you to realise the special case of this which is phagocytosis – conducted by white blood cells in the ingestion of pathogens.

Question 4: E

All of the above statements are true of the Calvin cycle with regards to the Krebs cycle. As the main driver of photosynthesis, we know that the Calvin cycle requires both CO_2 and light in order to conduct ATP dependent reactions. As opposed to the Krebs cycle in man however, the Calvin cycle adopts the use of NADPH as the intermediate in electron transport.

Question 5: D

Option D is one of only 2 graphs that demonstrate a quadratic relationship with the peak enzyme activity correctly placed – pepsin from the stomach close to pH 1, and trypsin secreted by the pancreas and therefore alkaline around pH 13. The curves traced in option c however are far too broad over the pH range to represent enzyme activity. As the pH scale is logarithmic, even a change of 1 or 0.5 can be devastating to enzyme activity.

Question 6: A

This question was taken directly from the BMAT syllabus where many examples are listed for different principles. Reading the BMAT syllabus and highlighting these is a very good idea as well as learning the definitions listed.

Question 7: A

Initially the electron configuration of Mg is 2,8,2. In binding to two chlorine atoms it is effectively ionised to Mg^{2+} and it loses two electrons to leave a complete outer shell and thus the correct answer is 2,8.

Question 8: D

The first thing to note in this trace is that the m/z axis has been cut short. From looking up the mass of calcium in the periodic table one would expect to see the x axis centred around 40. However here the trace is only displaying those isotopes with valence 2 ($z = 2$) hence the values are half the size. Therefore (from the periodic table) when dividing the most abundant isotope of chromium by two, 52/2 = 26, we confirm that the outlier bar on the right is indeed the contaminant. Therefore, to calculate the actual abundance of Mr 40 calcium ignore the chromium like so: 55/95 = 11/19.

Question 9: A

Begin by converting the total weight of arsenic into grams like so $15 \times 10^6 = 1.5 \times 10^7$. Then divide by the Mr of arsenic which is 75 (2sf) giving 2×10^5. Don't forget that the sample is at worst 80% pure. Therefore, there will be a minimum of $(2 \times 10^5) \times 0.8 = 1.4 \times 10^5$ moles of pure arsenic.

Question 10: D

Recall that average atomic mass is calculated as the sum of (isotope mass x relative abundance). Therefore $28 = (26 \times 0.6) + (30 \times 0.3) + 0.1x$. Rearranging this equation reveals that $0.1x = 3.4$ and that the mystery isotope therefore has an atomic mass of 34.

Question 11: A

First recall that when a group 2 metal is reacted with steam a metal oxide is formed and therefore the following chemical equation can be drawn: $Mg + H_2O_{(g)} \rightarrow MgO + H_2$. Note the stoichiometric ratio which is simply 1. Next calculate that there is 72/24 = 3 mol of hydrogen produced. Therefore, assuming that there is 3 mol of all other reactants and the reaction is complete one would expect $3 \times 24.3 = 72.9$g of magnesium and $3 \times 18 = 54$g of steam. This is indeed the case and therefore the reaction is complete.

Question 12: B

The reducing agent is the species which is itself reduced in this instance from looking at the oxidation states we can see that that species is S^{2-}. As after the reaction has taken place it has an oxidation state of +6 which would require a loss of negative charge i.e. electrons.

Question 13: C

The highly stable bonds between carbon atoms, and between carbon and hydrogen atoms renders alkanes relatively unreactive. This is important to note as it highlights the major difference between alkanes and alkenes.

Question 14: A

Recall that current = charge/time. The question provides both charge and time in the correct units and so the calculation is relatively simple with no unit conversions required. Therefore current = 5/15 = 1/3 = 0.33A. As the question states that the balloon has a negative charge it has therefore gained electrons. Given that a current is defined as a net movement of electrons, in this situation the current must be flowing into the balloon.

Question 15: D

Given that Power = IV it can be deduced that I = P/V. Recall that power given in Watts is a measure of the energy transferred per second and therefore has the alternative units Js^{-1}. When substituting these units into the power equation re-arranged for Amps it is revealed that $I = (Js^{-1})/V = A$.

Question 16: D

For a transformer that is 100% efficient power in must equal power out, recalling that P=IV. Therefore, the transformer has a power output of 24 x 10 = 240W which is 80% of the initial input. As such the initial power input was (240/80) x 100 = 300W.

Question 17: C

Begin by calculating the energy required to hoist the mass, this is calculated using the potential energy equation: mgh. Energy = mass x g x height = 20 x 10 x 30 = 6000N. The power output of the motor is calculated as the joules dissipated per second = 6000/20 = 300W

Question 18: D

In order to solve this problem recall that activity = decay constant x number of remaining atoms. Therefore, the decay constant can be calculated simply as 0.36/6 = 0.06.

Question 19: D

Recall that household electricity is available in the UK at 240V. Begin by calculating the wattage that the bulb is receiving as 0.5 x 240 = 120W. Given that the energy rating of the bulb is 80W, we can assume that this bulb is only 80/120 = 66% efficient.

Question 20: E

Begin by subtracting the integral from both sides producing $x - \int_{-z}^{z} 9a - 7 = \frac{\sqrt{b^3 - 9st}}{13j}$. Next multiply both sides by 13j and square, rendering $[13j(x - \int_{-z}^{z} 9a - 7)]^2 = b^3 - 9st$. Finally subtract b^3 from both sides and divide by -9s leaving the correct answer: $\frac{[13j(x - \int_{-z}^{z} 9a - 7)]^2 - b^3}{-9s} = t$.

Question 21: C

The formula for calculating compound interest can be given as investment x (interest rate[years]) or in short hand for this situation: $1687.5 = 500x^3$. Therefore, in order to calculate the interest rate the above formula must be rearranged to $\sqrt[3]{1687.5/500} = 1.5$ revealing an interest rate of 50%.

Question 22: E

Begin by drawing your line of best fit, remembering not to force it through the origin. Begin fitting the general equation y = mx + c to your line. Calculate the gradient as $\Delta y/\Delta x$ and read the y intercept off your annotated graph.

Question 23: E

In order to start rearranging the fraction begin by adding m to both sides and squaring to yield $4m^2 = \frac{9xy^3z^5}{3x^9yz^4}$. Now it is clear to see that this can be most simply displayed in terms of powers. Therefore, E is the correct answer.

Question 24: D

Non-normally distributed data doesn't demonstrate a 50-50 split of data points either side of the mean. Therefore, standard data analysis techniques like normal range are inappropriate (as the formula for normal range is mean$\pm$ 1.96SD). Instead the interquartile range is used.

Question 25: D

Random chance is a large issue particularly in medicine. Clinical trials are inherently flawed as they only consider a very small percentage of the population which is far outweighed by the genetic variation demonstrated within the human genome. Therefore, statistics must be used to transform sample data into data representative of the entire population.

Question 26: B

Begin by calculating the speed of the innermost well as the circumference of travel over time = 20 x 3.14 = 62.8m/s. Calculate the outermost well speed in the same manner = 40 x 3.14 = 125.6m/s. 125.6 – 62.8 = 62.8m/s faster.

Question 27: C

The question states that the repeat experiment is identical to the first in all aspects apart from the result. Therefore, although a number of the options may be true like calibration bias, it would have been applied to both experiments and therefore should not affect the result. As such the difference in results is simply due to random chance.

END OF SECTION

Section 3

"Progress is made by trial and failure; the failures are generally a hundred times more numerous than the successes; yet they are usually left unchronicled."

➢ This statement aims at several aspects of science. On one hand, it aims at scientific method. It demonstrates that science itself is based on trial and error and that to come to the right answer we should test theories repeatedly, adjusting them all the time to become more precise and more in keeping with our results. In the end, it is very rare that a theory survives unchanged. It also stresses that the progress of science is slow and laborious as it requires a constant string of trial an error experiments before providing any results. The second component the statement addresses is the way the scientific progress is seen in the public and even amongst scientists. The common perception is that only success counts and if a theory cannot be proven it is a failure. This of course is a problem since every failure provides a new angle to start from on the hunt for success. Failure becomes necessary for success to be possible.

➢ Since this statement basically has two components, when arguing to the contrary, you will have to demonstrate either that failures stand in a different relationship to success or that the reporting of failures is equal to that of successes. Either is going to be difficult as the stamen itself forms a self-fulfilling prophecy. You cannot disprove it, provided it has some truth to it, since you will not find any evidence for it. So, you will have to focus on a more theoretical level to fund support to argue against the quote.

➢ You can argue that failures, being part of research arte always reflected to some extend in the presentation of data in research papers. They will also appear in the analysis components of any piece of research as failures are essential for the progress of research as it narrows the field of possible answers.

➢ Another perspective you can approach this topic from is to separate the failure and the success. Failure of one theory, even if it had been thought to be correct at some point will lead to evolution of a different idea that builds on the conceptual failures of the previous idea. Thereby, one idea facilitates the other and the failing of one concept will directly result in a new concept that then in turn will either remain a success or become a failure at some point down the line.

➢ You can also consider the role failure plays in our society. It is generally seen as a bad thing and as something to be avoided. This of course does also apply to the scientific community. But at the same time, failure can also provide a new stepping stone for future success, provided lessons are learned from the cause of the failure that can then be applied for future projects.

"He who studies medicine without books sails an uncharted sea, but he who studies medicine without patients does not go to sea at all."

➤ This statement aims directly at the connection between science and soft skills when it comes to the practice of medicine. It claims that medicine is more than just a science that can be learned by theories alone but has a large human component that gives the scientific aspect of medicine meaning. Without the application of the theoretical knowledge, the subject studied has no value at all, at least it cannot be called medicine.

➤ When addressing this statement, there are several tension points you should consider. Firstly, there is the uncharted sea. This symbol has several aspects. On one hand, it is threatening and dangerous as the sailor cannot know where difficult streams and lurking rocks are located. On the other hand, it also has a component of excitement and adventure, just think of the old explores Cook and Columbus etc. Secondly the symbol of not sailing at all. In this again, you can use the sailor metaphor to become fully clear on what he means. Imagine a sailor that has excellent navigational skills but lacks the courage to apply them and so wanders to the harbour every day to stare across the sea. All his skill is wasted as he never sets foot on the waves.

➤ In order to argue against this stamen, you should focus on the first part, the sailing an uncharted sea. This is because the second part holds a deep truth that is difficult to disprove. Even when it comes to medical research, you will have to interact with patients that you draw data from for your research. However, it is simple to argue why books play a vital role in medicine. In this case books are synonymous to all forms of theoretical learning.

➤ The main focus here should be that of safety of the patients. Without books it is only a matter of time until the doctor makes a mistake, crashing his proverbial ship on a proverbial sandbank. Since the focus of medical treatment is the improvement of the patient's condition or quality of life, uncertainty and adventurism have no role in this. At this point it is essential to keep the text on a general level, as medical progress does also come from ignoring the common wisdom. Remember the theory of 4 humours from the Middle Ages, if it hadn't been for somebody breaking with this common wisdom and basis of teaching, modern medicine would never have been born...

➤ If you write an essay about this topic, make sure that you have a very clear position that will give you a good basis to argue from. Also make sure you have fully understood the statement. Due to the use of metaphors this can be tricky, but on the other hand, if you have understood the question, you can use similar metaphors and they will tie in nicely with the question. This will then give your whole essay a smoother appearance and make it better to read.

'"Medicine is the restoration of discordant elements; sickness is the discord of the elements infused into the living body"

➤ To understand this statement, you have to understand Leonardo da Vinci. Being an Renaissance artist and scientist, artist and architect, he had a very varied background but also lived in the late 15th and early 16th century which will obviously have influenced his perception of medicine. The idea of discordant elements that are infused in the body and that have to be rebalanced is clear evidence of that since it basically rephrases the theory of the 4 humours that was the basis of pretty much all school medicine pretty much up until the 19th century, when pathogens were discovered and described in their properties to cause disease. In general, however, the statement is to be understood in a sense that disease represents a damaging influence on the body, who's default is health, and the job of medicine is to rectify this influence to restore health.

➤ When arguing against this stamen, there are several possible angles of attack. On one hand, there is the historical aspect of the 4 humours mentioned above. This is pretty straight forwards since you can easily demonstrate why da Vinci would be influenced by this theory and how this theory was inherently false. On the other hand, you can attack the idea of corruption through diseases causing influence by a more general discussion of disease patterns. Whilst it is true that infectious diseases are cause by the insertion of pathogens into the healthy organism, there are a vast amount of diseases that are not. One good example are genetic diseases. Especially those that are inherited in a recessive pattern meaning that parent generations must be carriers and therefore be 'corrupted' as well without displaying the actual disease. If you want to go down the mutation route as well, you can point out that not all change causes negative outcomes since mutations form the basis of evolution and thereby the basis of how we as a species came to be.

➢ The second point of attack to argue against da Vinci here is the role of medicine. Whilst it is generally true that the aim of medicine is to cure the patient, sometimes this is either not possible due to a lack of ability of the medical profession, i.e. we just don't have a cure, or it is not desirable since the risk to the patient if undergoing treatment outweighs the risk of the disease or the benefit of treatment. Good examples here are chemotherapy in the frail and elderly. Another example is that of mutations as the motor of evolution as mentioned above.

➢ To the last part of the question, this statement still holds some truth to modern medicine, consider for example cancer that can be caused by poisonous external influences such as smoking or radiation, in this case the treatment will involve on one hand the removal of the negative stimulus, if possible, and on the other hand the treatment of the negative impact this stimulus has left.

"Modern medicine is a negation of health. It isn't organized to serve human health, but only itself, as an institution. It makes more people sick than it heals."

➢ With this statement, some background knowledge can be very helpful. Ivan Illich was a Croatian-Austrian Priest and philosopher that lived during the 20th century. He is generally known for is critic of the institutions of Western culture such as schools or in this case modern medicine. Looking at the statement itself, it is clear, that the basic message is that medicine has no interest in curing humans but rather prolongs their suffering to sell them as much treatment as possible to fund its own interests.

➢ In order to argue against this stamen, it can be helpful to detect components of truth in it that can then be refuted. One point that can be raised in connection to this statement is that of medicalisation. By labelling everything that does not conform 100% with the ideal of health in medical terms produces a population of sick people that then require treatment.

➢ Another point where the statement holds true is in different health care systems such as the one in the US where maintaining a sick status provides continued income to the doctor and the medical professionals involved in treatment. This is less of an issue in publicly funded environment such as the NHS where there is a stricter regulation of resources and therefore less option for artificial prolongation of treatment requirements.

➢ Arguing against the statement is fairly easy, especially when arguing from the perspective of the NHS. In the NHS, healthcare is provided free of charge for residents and there are no direct barriers in place to block access to health care. This in itself proves Illich wrong since it would serve the institution to make health care a luxury item that comes with the associated price tag.

➢ Arguing that the primary duty of the doctor is not to prolong life is more difficult, since two of the ethical pillars of the medical profession call for doctors not to do harm and to act in the patients' best interest, both of which aim at the prolongation of life in the majority of cases. There are some exceptions to the prolonging life idea, and it is probably safest to approach this part of the question from that angle as it will ensure that you stay on the right track and don't end up in a direction you didn't want to go.

➢ Limitations to the idea of prolonging life are pretty much all the cases falling under palliative care where the idea is to remove suffering and providing symptomatic relief rather than curing the dieses causing the symptoms. Common examples here are cancer in the elderly that are not fit enough to undergo chemotherapy or surgery. Other examples are incurable diseases such as inoperable brain tumours etc.

➢ In this question again, it helps very much to have a clear idea of what you think about the issue. It will make it easier to structure your answer appropriately and it will ensure that you don't navigate yourself into uncertain waters which is fairly easy with this topic, especially when the idea of prolonging life or not is being introduced.

END OF PAPER

Mock Paper C Answers

Section 1

Question 1: D

There are three different options for staying at the hotel. They could either pay for three single rooms for £180, one single and one double room for £165, or one four-person room for £215.

Subtracting the cleaning cost for one night would leave:

£180-(3x£12) = £144
£165-(2x£12) = £141
£215-£12 = £203

The cheapest option is one single and one double room and they want to stay three nights, giving £141x3 = £423.

Question 2: D

Glass one starts with 16ml squash and 80ml water. Glass two starts with 72ml squash and 24ml water. 48ml is half of 96ml so 8ml squash and 40ml water is transferred to glass two. Glass two now contains (8+72 = 80ml squash) and (24+40 = 64ml water). Glass two now has a total of 144ml and half of this is transferred to glass one. Glass one now has (40+8 = 48ml squash) and (32+40 = 72ml water). Therefore, glass one has 48ml squash and glass two has 40ml squash.

Question 3: B

B is the main conclusion of the argument. Options A and D both contribute reasons to support the main conclusion of the argument that the HPV vaccination should remain in schools. C is a counter argument, which is a reason given in opposition to the main conclusion. Option E represents a general principle behind the main argument.

Question 4: B

The speed of the bus can be calculated using the relationship: Speed $=\frac{distance}{time}$

$\frac{3\,km}{0.2\,h} = 15$ kmh^{-1}

The bike speed is therefore ($\frac{4}{5}$ x 15 = 12 kmh^{-1}). Considering that the bus leaves 2 minutes after the bike, it is now possible to write an expression, where d is the distance travelled when the bus overtakes the bike:

$$\frac{d\,km}{12\,km/h} = \frac{1}{30}\,h + \frac{d\,km}{15\,km/h}$$

This expression can be solved by multiplying each term by (12 kmh^{-1} x 15 kmh^{-1}):

15d km = 6 km +12d km
3d km = 6 km
d = 2 km

Therefore, the bus overtakes the bike after travelling 2 km.

Question 5: B

Firstly, determine who will move up to set one. Terry, Bahara, Lucy and Shiv all have attendance over 95%. Alex, Bahara and Lucy all have an average test mark over 92. Terry, Bahara, Lucy and Shiv all have less than 5% homework handed in late. Therefore, Bahara and Lucy will both move up a set. Secondly, determine who will receive a certificate. Terry, Bahara, Lucy and Shiv have absences below 4%. Alex, Bahara and Lucy have an average test mark over 89. Bahara and Shiv have at least 98% homework handed in on time. Therefore, only Bahara will receive a certificate.

Question 6: C

Firstly, construct two algebraic equations: A-18=B-25 and A$=\frac{5}{6}$B

Solve these two equations as simultaneous equations by substituting $\frac{5}{6}$B for A in equation 1:

$\frac{5}{6}$B-18=B-25

$7=\frac{1}{6}$B

B=42

Put B=42 back into equation 2: A= 42 x $\frac{5}{6}$

A=35

Question 7: D

I need to make 48 scones, which makes up 8 batches.
8 batches would take: 35+ 7(25+10) +25 = 305 minutes

I need to make 32 cupcakes, which makes up 4 batches.
4 batches would take: 15+ (4x20) =95 minutes

I need to make 48 cucumber sandwiches
This would take (8x5) = 40 minutes

Adding 305, 95 and 40 minutes is 440 minutes in total. 440 minutes is equivalent to 7 hours and 20 minutes. Adding 7 hours and 20 minutes to 10:45am leads to 6:05pm so I will be finished at 6:05pm.

Question 8: D

The volume of a pyramid is given by the equation:

$v = \dfrac{a^2 h}{3}$ where v=volume, a=base and h=height

Rearrange to work out the height for each pyramid: $h = \dfrac{3v}{a^2}$

Pyramid	Base edge (m)	Volume (m³)	Calculation:	Height (m)
1	3	33	$\dfrac{3 \times 33}{9}$	11
2	4	64	$\dfrac{3 \times 64}{16}$	12
3	2	8	$\dfrac{3 \times 8}{4}$	6
4	6	120	$\dfrac{3 \times 120}{36}$	10
5	2	8	$\dfrac{3 \times 8}{4}$	6
6	6	120	$\dfrac{3 \times 120}{36}$	10
7	4	64	$\dfrac{3 \times 64}{16}$	12

The tallest pyramid is 12m and the smallest is 6m. Subtracting the height of the tallest pyramid from the height of the smallest pyramid leaves 6m.

Question 9: A

Work out the two wages by substituting the information provided into the formula:
Jessica's wage is: 210 + 42 - 3.2 = 248.8
Samira's wage is: 210 + 78 - 8.8 = 279.2

Subtracting 248.8 from 279.2 leave 30.4 so the difference between their wages is £30.4.

Question 10: C

The main conclusion is C. A and B both represent reasons to support the main conclusion of the argument. Option D represents an assumption that is not stated in the argument but is required to support the main conclusion that research universities should strongly support teaching. Option E is a counter argument that provides a reason to oppose the main argument.

Question 11: D

D is the main conclusion of the argument. A is a general principle of the argument, but the argument is more specific to the use of helmets rather than the wider concept of danger in sport and the responsibilities of the governing bodies to sports players. Options B and C are reasons to support the main conclusion. Option E is an intermediate conclusion, which acts as support for the next stage of the argument and as a reason to support the main conclusion.

Question 12: D

There are 10 passengers on the tube at the final stop. At stop 5 there were twice the number of passengers on the tube so 20 passengers were at stop 5. At stop 4, there were $\frac{5}{2}$ times the number of passengers at stop 5 so 50 passengers were present at stop 4. At stop 3, there were $\frac{3}{2}$ times the number of passengers at stop 4 so 75 passengers were on the tube. At stop 2, there were $\frac{6}{5}$ times the number of passengers at stop 3 so 90 passengers were present at stop 2. Similarly, at stop 1, there were $\frac{6}{5}$ times the number of passengers at stop 2 so at the first stop 108 passengers got on the tube.

Question 13: E

➢ Some students born in winter like English, Art and Music
➢ There is not enough information to tell whether some students born in spring like both Biology and Maths.
➢ We don't know what the students born in spring think about Art.
➢ We don't know what the students born in winter think about Biology.
➢ There is not enough information to know whether this is true or not.

Subject	TIME OF BIRTH		
	SPRING	AUTUMN	WINTER
English	Everyone likes	Everyone likes	Everyone likes
Biology	Some like	No one likes	
Art		Everyone likes	Some like
Music			Everyone likes
Maths	Some like		

Question 14: A

The main conclusion is option A that some works of modern art no longer constitute art. B is not an assumption made by the author as the main conclusion does not rely on *all* modern art being ugly to be valid. C is not an assumption because the argument does not rely on artists studying for decades to produce pieces of work that constitute art. This point is simply used to support the main argument. Options D and E are stated in the argument so are not assumptions. A is an assumption because it is required to be true to support the main conclusion but is not explicitly stated in the argument.

Question 15: E

Reducing the price of the sunglasses by 10% is equivalent to multiplying the price by 0.9. the price of the sunglasses is successively reduced by 10% three times and so the price on Monday is 0.9^3 the price of the sunglasses on Friday. 0.9^3 is equal to 0.729 and so the price of the sunglasses on Monday is 72.9% of the price of the sunglasses on Friday.

Question 16: C

It is easier to write out this calculation in the following format:

a b 7 –
  a b
_____

5 6 5

From the above subtraction it is clear that b must be equal to 2 because 7 minus 2 is equal to 5, which is the unit term of the answer. It is now possible to rewrite the calculation with 2 substituted for b:

a 2 7 –
  a 2
_____

5 6 5

From the above calculation it is possible to gauge certain facts. A must be greater than 5 because 1 is carried over to the second term:

ₐ ¹2 7 –
  a 2
_____

5 6 5

It is now clear than a must be equal to 6 because 12 minus 6 is equal to 6, which is the tens value of the answer.

Question 17: E

Look at the flat cube net and note the shapes that are adjacent to each other. Sides that are joining on the net will be beside each other on the formed cube. Work through to deduce option E can be formed from the cube net shown.

Question 18: E

The H shape is comprised of 12 squares. The shape's area of 588 can be divided by 12 to give 49, which is the area of each individual square. The square root of 49 is 7 and so the side length of each individual square is 7cm. The perimeter of the shape is comprised of 26 sides and the length of each side is 7 so the perimeter of the shape is 182cm.

Question 19: E

The information provided about the child needs to be inserted into the BMI formula: $BMI = 35 \div 1.2^2$

1.2 squared is equal to 1.44 and it may be easier to work out 3500 divided by 144. The answer needs to be worked out to 3 decimal places for an answer required to 2 decimal places. The answer to 3 decimal places is 24.305 and so the BMI to 2 decimal places is 24.31.

Question 20: C

It is important that the information is inserted into the formula given for calculating the BMR of a woman rather than a man:

BMR= (10 x weight in kg) + (6.25 x height in cm) – (5 x age in years) -161

BMR = (10 x 80) + (6.25 x 170) – (5 x 32) – 161

BMR= 800 + 1062.5 -160 -161

The BMR of the woman in the question is therefore 1541.5 kcal

Question 21: D

This time, the information needs to be inserted into the formula for calculating the BMI of a man:

BMR= (10 x weight in kg) + (6.25 x height in cm) – (5 x age in years) + 5

BMR= (10 x 80) + (6.25 x 170) – (5 x 45) +5

BMR= 800 + 1062.5 -225 +5

The BMR of the man in the question is therefore 1642.5 kcal. The man does little to no exercise each week. It is therefore required to multiply 1642.5 by 1.2, which gives a daily recommended intake of 1971 kcal.

Question 22: B

Slippery Slope describes a series of loosely connected and increasingly worse events that lead to an extreme conclusion. A is not a flaw because the author does not predict a series of undesirable outcomes. C is not a flaw. It is unlikely that correlation has been confused with cause if the American school did not change other aspects of the school day although this is not explicitly stated in the argument. D is not a flaw. A circular argument assumes what it attempts to prove and this is not the case in this argument. E is a counter argument rather than a flaw. B is the flaw in the argument. Just because moving start times later worked in one school in America does not mean that it will work in all other cases.

Question 23: D

Options A and E, if true, would weaken the argument. If the class is more disrupted this will be detrimental to learning, as will less effective teaching. B does not strengthen the main conclusion, which is based on improvement in academic achievement levels rather than activity levels. C does not strengthen the argument as the school curriculum makes no difference to the argument about the science behind teenage brains. If D is true then it suggests that the improvement in grades is a direct effect of the later school starts rather than a mere correlation.

Question 24: B

The main conclusion is that EnergyFirst is expected to expand its customer base at a rate exceeding its competitors in the ensuing months. A does not directly contradict the main argument. It demonstrates a flaw in the argument in that it ignores the fact that other companies may be stronger in other areas and attract customers by other means. However, it does not serve to weaken the main argument. C does not contradict the main conclusion; EnergyFirst could still expand its customer base at the fastest rate even if there is not much competition between energy companies. D would not weaken the argument as it refers to the rate of new customer intake rather than the number of new customers attracted. E, if true, would strengthen the argument because it suggests that visual advertising would attract new customers. B would weaken the main argument because if it were true then investing the most money in advertising would not serve to attract the most customers.

Question 25: A

Option B points out a flaw in the argument, which attributes the healthier circulatory system of vegetarians to diet, but ignores other potential contributory factors to a healthy circulatory system such as exercise. C is not an assumption: the health benefits of a vegetarian and omnivorous diet are not discussed; rather the argument is centred on the negative health ramifications. D is stated in the argument so is not an assumption and option E is a counter argument, not an assumption. Option A is required to support the main conclusion but isn't stated so is an assumption made in the argument.

Question 26: A

First, calculate the number of hours spent flying and waiting. It takes 24 hours in total from Auckland to London, 11.5 hours from London to Calgary and 8 hours from Calgary to Boston. In total this amounts to 43.5 hours of flying and waiting. Boston is 16 hours behind Auckland and so when Sam arrives in Boston it will be 27.5 hours ahead of 10am. The time in Boston will therefore be 13:30 pm.

Question 27: B

This question requires you to find the lowest common multiple. This is the product of the highest power in each prime factor category.

$18 = 3^2 \times 2$

$33 = 3 \times 11$

$27 = 3^3$

Therefore, 3^3, 11 and 2 need to be multiplied together which equals 594 seconds between simultaneous flashes. 5 minutes or 300 seconds needs to be subtracted from 594 in order to find the length of time until the next flash. The time that they will next flash simultaneously is 294 seconds.

Question 28: B

Firstly, calculate the number of students who play each instrument. 21 students play piano, 12 play violin and 3 play saxophone. Point 1 is true because the sum of 21 piano students and 12 violin students is 33, which is 3 more than the total number of students in the class. Therefore, at least 3 students must play both piano and violin. Point 2 is true because only 12 students actually play the violin so there cannot be more than 12 students playing both piano and violin. Point 3 does not have to be true because some of the 9 students that do not play piano may play the violin.

Question 29: D

One way of answering this question is to set out the result after each game:

	Neil	Simon	Lucy
Start	50	50	50
Game 1	100	25	25
Game 2	50	50	50
Game 3	25	25	100
Game 4	12.5	12.5	125
Game 5	6.25	15.625	128.125

After game 5 Lucy has £128.13, however the question is asking how much money Lucy gains. The difference between 128.125 and 50 is 78.13, so Lucy gains £78.13.

Question 30: C

Option A may explain why young drivers are involved in more accidents but does not need to be true for the main conclusion to hold. B would weaken the argument if true as drivers that spend more time driving will have a greater chance of being involved in accidents regardless of age. D is not an assumption, but if true may weaken the argument as it attributes the accidents to unsafe cars rather than unsafe driving. E is irrelevant to the main conclusion: it does not matter whether the young drivers are male or female; arguably steps should still be taken to reduce the number of accidents. Option C represents an assumption that is not stated in the argument but is required to support the main conclusion.

Question 31: D

The total weight of all of the apples is 6 multiplied by 180g, which equals 1080g. The highest value the heaviest apple could take would occur if all of the other 5 apples weighed the same as the lightest apple. 5 multiplied by 167g, the weight of the lightest apple, is 835g. The difference between the weight of all of the apples (1080g) and 835g gives the highest possible weight of the heaviest apple, which is 245g.

Question 32: B

A) True, but not far-reaching enough.

B) Correct answer. Sugar does indeed have an addictive potential as it causes the release of endorphins and the health concerns are well known. This characteristic makes it like alcohol and smoking, and potentially suitable for similar policies.

C) True, but similar to option A) and thus too limited.

D) Potentially true, but also too limited.

Question 33: E

If we make x the number of sheep sold on day 1, it is possible to write an expression for the profit made on both days:

$2(\frac{7}{8} \times 112)x = 112x + 3528$

$(98 \times 2x) = 3528 + (112x)$

$84x = 3528$

$x = 42$

The number of sheep sold on day 1 was therefore 42. Since twice the number of sheep were sold on day 2 as day 1, then the total number of sheep sold across the two days is equal to 42 multiplied by 3, which is 126 sheep.

Question 34: B

The main conclusion is that we should not wait for proof of climate change

A and D are reasons to support the main conclusion

C is an analogy

E is a counter argument

Question 35: B

The mean is the sum of all of the numbers divided by the number of terms. From the information, we know that the sum of the first 8 numbers divided by 8 is equal to 44 plus the sum of the first 8 numbers all divided by 10. An expression for this can be written like this:

$$\frac{\text{sum of 8 numbers}}{8} = y = \frac{\text{sum of 8 numbers} + 44}{10}$$

Two equations can be derived from the above expression:

10y = sum of 8 numbers +44

8y = sum of 8 numbers

If we subtract the second equation from the first, we are left with: 2y=44 → y=22

The value of y and the average of both sets of numbers is therefore 22.

END OF SECTION

Section 2

Question 1: D

Statement 1 is false. Sucrose is a disaccharide formed by the condensation of two monosaccharides (glucose and fructose).

Statement 2 is false. Lactose is a disaccharide formed by condensation of a glucose molecule with a galactose molecule.

Statement 3 is true. Glucose has two isomers: alpha-glucose and beta-glucose.

Statement 4 is true.

Statement 5 is true.

Question 2: C

Statement 1 is true. High temperatures and pH extremes cause a permanent alteration to the highly specific shape of the active site so that the substrate can no longer bind, and the enzyme no longer works.

Statement 2 is false. Amylase is produced in the salivary glands, pancreas, and small intestine.

Statement 3 is true.

Statement 4 is false. Bile is stored in the gall bladder, but it does travel down the bile duct to neutralise hydrochloric acid found in the stomach.

Statement 5 is true. Fructose is sweeter than glucose so smaller amounts can be used in food used in the slimming industry.

Question 3: C

The combining of food with bile and digestive enzymes occurs in the duodenum of the small intestine. In the ileum of the small intestine, the digested food is absorbed into the blood and lymph. The digested food then progresses into the large intestine. In the colon, water is reabsorbed. Faeces are then stored in the rectum and leave the alimentary canal via the anus.

Question 4: C

Statement 1 is true.

Statement 2 is true. For example, the drug curare, a South American plant toxin which is used in arrow poison, stops the nerve impulse from crossing the synapse and causes paralysis and can stop breathing.

Statement 3 is false. The sheath provides insulation for the nerve axon and increases the speed of impulse transmission via saltatory conduction.

Statement 4 is false. The peripheral nervous system includes motor and sensory neurons carrying impulses between receptors, effectors, and the central nervous system. The CNS consists of the spinal cord and the brain.

Statement 5 is true. A reflex arc travels from sensory neuron to relay neuron to motor neuron and is an innate mechanism designed to keep the animal safe. For example, it allows a person to quickly draw their hand away from a flame.

Question 5: C

Statement 1 is true.

Statement 2 is false. The transition metals are both malleable and ductile, they conduct heat and electricity and they form positive ions when reacted with non-metals.

Statement 3 is true. Thermal decomposition is a reaction whereby a substance breaks down into two or more other substances due to heat. When a transition metal carbonate is heated, metal oxide and carbon dioxide are produced. The carbon dioxide can be collected and will turn limewater cloudy.

An example of this reaction is: $CuCO_3 \rightarrow CuO + CO_2$

Statement 4 is false. Transition metal hydroxides are insoluble in water.

Statement 5 is true.

Question 6: E

There are 9 Sulphur atoms on the left so there must be 9 on the right. Therefore, the values of B and C must add to make 9. This can be written as an equation: B+C=9

It is now useful to try to balance the Oxygen atoms: 4A+36 = 10+4B+4C+14

Simplify to give: 12 = 4B+4C-4A

Equation 1 can now be substituted into equation 2 to give: 12 = (4x9)-4A

24 = 4A

A = 6

There are 6 Potassium atoms on the left. This means that there must also be 6 potassium atoms on the right, so B must by 3. As shown in equation 1, B and C add to make 9 so C must be 6.

_5_ PhCH₃ + _6_ KMnO₄ + _9_ H₂SO₄ = _5_ PhCOOH + _3_ K₂SO₄ + _6_ MnSO₄ + _14_ H₂O

Question 7: B

Statement 1 is true. Males have one X chromosome so if the allele is present they will be affected. Females have two X chromosomes so both need to be affected to be red-green colour blind as the condition is recessive

Statement 2 is true because according to the Punnett square below half of the children will have the homozygous recessive tt genotype and so will be non-rollers.

	T	t
t	Tt	tt
t	Tt	tt

Statement 3 is true because all of the male children will inherit an X chromosome from the mother which will carry the colour-blind allele.

Question 8: B

Start by multiplying each term by ax to give: $a(y+x)=x^2+a^2$

Expand the brackets: $ay+ax=x^2+a^2$

Subtract ax from both sides: $ay=x^2+a^2-ax$

Lastly, divide the both sides by a to get: $y = \frac{x^2+a^2-ax}{a}$

Question 9: A

This question requires the use of the equation: $C = \frac{n}{v}$ where C= concentration, n= moles and v=volume

Convert 25cm^3 into litres to get 0.025 litres and plug the values for concentration and volume into the equation to get the number of moles: $0.1 = \frac{n}{0.025}$ so n=0.0025

This question also requires the use of the equation: $n = \frac{m}{Mr}$ where m=mass, n=moles and Mr= molecular mass

The molecular mass is the sum of one calcium and two chlorine atoms which is equal to 111gmol^{-1}.

Inserting the molecular mass and number of moles into the above equation can be used to calculate the mass of calcium chloride: $m = 0.0025 \; x \; 111 = 0.28g$

Question 10: C

Solve as simultaneous equations

Start by substituting $x = \frac{y}{3}$ into equation B.

This gives $y = \frac{18}{y} - 7$

Multiply every term by y to give:

$0 = y^2 + 7y - 18$

Factorise this quadratic to give:

$0 = (y+9)(y-2)$

Where the graphs meet, y is equal to 2 and 9. Then y=3x so the graphs meet when x = 6 and x = 27

Question 11: D

Statement 1 is false because the pulmonary artery carries deoxygenated blood from the right ventricle to the lungs.
Statement 2 is true. This property of the aorta allows it to carry blood at high pressure and is why it pulsates.
Statement 3 is false because the mitral valve, otherwise known as the bicuspid valve, is between the left atrium and left ventricle.
Statement 4 is true.

Question 12: D

The Ar of Carbon is 12, Hydrogen is 1 and Oxygen is 16. Therefore, 12g of carbon is 1 mole of carbon; 2g of H is 2 moles of hydrogen and 16g of O is 1 mole of oxygen. The empirical formula is therefore CH_2O. The molecular weight is 30 g.mol^{-1}, which goes into 120 g.mol^{-1} exactly 4 times. The empirical formula must therefore be multiplied by 4 to obtain the molecular formula so the molecular formula is $C_4H_8O_4$.

Question 13: B

To win one game, Rupert must win one squash game and one tennis game. In order to calculate the probability one winning one game, it is necessary to add the probability of winning one tennis game and losing one squash game to the probability of losing one tennis game and winning one squash game. The following calculation must be performed: $(\frac{3}{4} \; x \; \frac{2}{3}) + (\frac{1}{4} \; x \; \frac{1}{3}) = \frac{7}{12}$

Question 14: C

The numbers can all be written as a fraction over 36:

➢ $0.\dot{3}$ is the same as $\frac{12}{36}$

➢ $\frac{11}{18}$ is the same as $\frac{22}{36}$

➢ 0.25 is the same as $\frac{9}{36}$

➢ 0.75 is the same as $\frac{27}{36}$

➢ $\frac{62}{72}$ is the same as $\frac{31}{36}$

➢ $\frac{7}{7}$ is the same as $\frac{36}{36}$

Ordering them from lowest to highest gives: $\frac{7}{36}$; 0.25; $0.\dot{3}$; $\frac{11}{18}$; 0.75; $\frac{62}{72}$; $\frac{7}{7}$

Therefore, the median value is $\frac{11}{18}$

Question 15: E

This question requires use of the equation: Percentage yield $= \frac{actual \; yeild \; (g)}{predicted \; yield \; (g)}$ x 100.

If all of the benzene was converted to product (100 percent yield) then 20.5g of nitrobenzene would be produced: 13g C_6H_6 x $\frac{1 \; mol \; C6H6}{78g \; C6H6}$ x $\frac{123g \; C6H5NO2}{1 \; mol \; C6H6}$ = 20.5g $C_6H_5NO_2$.

However, only 16.4g are actually produced. Using the equation, we can now calculate the percentage yield:

$\frac{16.4g}{20.5g}$ x 100 = 80% yield.

Question 16: C

Statement 1 is true.

Statement 2 is false because infrared has a longer wavelength than visible light.

Statement 3 is true.

Statement 4 is false because gamma radiation and not infrared radiation is used to sterilise food and to kill cancer cells.

Statement 5 is true because darker skins contain a higher amount of melanin pigment, which absorbs UV light.

Question 17: B

This question requires the use of the equation:

p=mv where p=momentum, m=mass and v=velocity.

The total momentum before the collision is equal to the sum of the momentum of carriage 1 (12000 x 5) and carriage 2 (8000 x 0), which is 60,000 kg ms^{-1}. Momentum is conserved before and after the collision so the total momentum after the event also equal 60,000 kg ms^{-1}. The carriages now move together so the combined mass is 20,000kg. Using the equation again, the total momentum (60,000 kg ms^{-1}) divided by the total mass (20,000 kg) gives the velocity of the train carriages after the crash, which is equal to 3 ms^{-1}.

Question 18: C

Statement 1 is false. In a nuclear reactor, every uranium nuclei split to release energy and three neutrons. An explosion could occur if all the neutrons are absorbed by further uranium nuclei as the reaction would escalate out of control. Control rods that are made of boron absorb some of the neutrons and control the chain reaction.

Statement 2 is false. Nuclear fusion occurs when a deuterium and tritium nucleus are forced together. The nuclei both carry a positive charge and consequently, very high temperatures and pressures are required to overcome the electrostatic repulsion. These temperatures and pressures are expensive and hard to repeat and so fusion is not currently suitable as a source of energy.

Statement 4 is true. During beta decay, a neutron transforms into a proton and an electron. The proton remains in the nucleus, whereas the electron is emitted and is referred to as a beta particle. The carbon-14 nucleus now has one more proton and one less neutron, so the atomic number increases by 1 and the atomic mass number remains the same.

Statement 5 is false. Beta particles are more ionising than gamma rays and less ionising than alpha particles.

Question 19: E

Firstly, deal with the term in the brackets: 3^3=27

$$(x^{½})^3 = x^{1.5}$$

$$(3x^{½})^3 = 27x^{1.5}$$

Next, divide by $3x^2$: $\frac{27}{3} = 9$

$$\frac{x^{1.5}}{x^2} = x^{-0.5} = \frac{1}{\sqrt{x}}$$

Answer= $\frac{9}{\sqrt{x}}$

Question 20: E

Statement 1 is true.

Statement 2 is true. Decomposers in the soil break down urea and the bodies of dead organisms and this results in the production of ammonia in the soil.

Statement 3 is true.

Statement 4 is true.

Question 21: A

Write $\frac{\sqrt{20}-2}{\sqrt{5}+3}$ in the form $p\sqrt{5} + q$

Firstly, multiply the term by $\frac{\sqrt{5}-3}{\sqrt{5}-3}$ (ie 1) and write $\sqrt{20}$ as $2\sqrt{5}$

This gives: $\frac{10-6\sqrt{5}-2\sqrt{5}+6}{5-9}$

This simplifies to: $\frac{16-8\sqrt{5}}{-4}$

This simplifies to: $2\sqrt{5} - 4$

Therefore p= 2 and q= -4

Question 22: E

The question is asking for which of the statements are *false*.

Statement 1 is true.

Statement 2 is true.

Statement 3 is false. Ionic compounds do conduct electricity when dissolved in water or when melted because the ions can move and carry current. On the other hand, solid ionic compounds do not conduct electricity.

Statement 4 is true. Alloys contain different sized atoms, making it harder for the layers of atoms to slide over each other.

Question 23: A

The equation for a circle, with centre at the origin and radius r is $x^2 + y^2 = r^2$

The equation of this circle is therefore $x^2 + y^2 = 25$

Solve the problem using simultaneous equations or by drawing the line onto the graph.

$x^2 + (3x-5)^2 = 25$

This simplifies to $10x^2 - 30x = 0$

$10x(x - 3) = 0$

So x=3 or x=0 where the two graphs intersect

Question 24: D

Statement 1 is false. Heat energy is transferred from hotter to colder places by convection.

Statement 2 is true.

Statement 3 is true. Radiation can travel through a vacuum like space.

Statement 4 is false. Shiny surfaces are poor at reflecting and absorbing infrared radiation and dull surfaces are good at absorbing and reflecting infrared radiation.

Question 25: C

Statement 1 is true.

Statement 2 is false. The melting and boiling points increase as you go down the group.

Statement 3 is true.

Statement 4 is false. Chloride is more reactive than bromine, so no displacement reaction occurs.

Statement 5 is true.

Question 26: C

ABC and DBE are similar triangles because all of the angles are equal.

Therefore:

$\frac{BE}{BC} = \frac{DE}{AC}$

This is the case because the side lengths of the small and large triangles are in proportion to each other. Substitute the side lengths into the expression:

$\frac{4}{6} = \frac{DE}{9}$

DE=6cm

Question 27: E

This question requires the use of the equation:

$v^2 = u^2 + 2ah$ where v=final velocity, u=initial velocity, a=acceleration and h=height

From the information provided in the question, we know that v=0ms^{-1}, u=40ms^{-1} and a=-10ms^{-2}. Inserting these values into the equation gives:

$0 = 1600 + 2(-10h)$

The maximum height reached is therefore 80m.

END OF SECTION

Section 3

'The NHS should not treat obese patients'

Explain what this statement means. Argue to the contrary. To what extent do you agree with the statement?

The statement argues that free health care should not be given to patients with a BMI of 30 or more. This essay will consider both perspectives before arriving at a conclusion.

There are several arguments to support the treatment of obesity by the NHS. The first is that it is in accordance with the definition of a disease; namely that is reduces life expectancy, negatively impacts normal body function and can be induced by genetic factors. Obesity often has a genetic basis, for example the melanocortin-4 receptor polymorphism and leptin receptor deficiency, which shift the homeostatic balance towards weight gain and are associated with hyperphagia and obesity. Obesity can also be a major feature of certain syndromes such as Prader-Willi syndrome, Bardet-Biedl syndrome and Cohen syndrome. If obesity is either classified as a disease or is an unavoidable ramification of certain syndromes then surely it should be treated by the NHS just the same as any other disease.

Moreover, if the NHS refuses to treat obese patients, it will become difficult to decide where to draw the line. Should smokers or people who drink alcohol also be denied free health treatment and how many cigarettes or units per week should qualify? Should all obese people be denied free health treatment or just in cases where it is not an unavoidable secondary result of certain syndromes? Obesity is often a consequence of mental illnesses such as depression and it may be hard to differentiate cause from effect.

On the other hand, obesity in certain cases could be considered as a self-induced condition rather than an actual illness. Individuals arguably exercise a degree of free will and are responsible for the amount of calories that they consume and the amount of exercise that they do. There is an argument that obesity is driven by structural changes in the environment and is a mass phenomenon influenced by advertising and propaganda. Perhaps societal changes in the outlook towards healthy living are required to address the obesity problem.

Many NHS organisations already ration surgery for overweight patients and will not for example pay for joint or hip replacements for patients with a BMI of over 30. Surely NHS funds and taxpayers money is better spent on people who make an effort to maintain a good level of health, for instance patients who are subject to largely unpreventable and serious diseases such as certain cancers.

I would suggest that each case should be considered on an individual basis and that obesity treatments should be included on the NHS where they may act to significantly improve the patient's life in the longer term.

'We should all become vegetarian'

Explain what this statement means. Argue to the contrary, that we should not all become vegetarian. To what extent do you agree with this statement?

This statement is saying that everyone should stop eating meat. This essay will consider both perspectives before arriving at a conclusion.

Some animals are raised in poor living conditions. Circumstances can be cramped and due to growth rate maximisation, animals can develop serious joint problems. Pig tails are cut, chickens have their toenails and beaks clipped and cows are dehorned without painkillers. The slaughter process can also be stressful and inhumane. Halal meat is not stunned before the jugular vein is slit and death is not instantaneous. It could be considered unethical to kill animals for food in this way when vegetarian options are available. Moreover, if farmers grew crops in place of livestock, this would generate more food and potentially alleviate world hunger.

Farming meat also has environmental implications. The overgrazing of livestock entails significant deforestation, which destroys natural habitats and endangers wild species. Enteric fermentation generates huge greenhouse gas emissions and ammonia and hydrogen sulphide leach poisonous nitrate into the water.

A vegetarian diet also has notable health benefits. Diets high in animal protein can cause excretion of calcium, oxalate and uric acid, which contribute to the development of kidney stones and gallstones. Vegetarians absorb more calcium: meat has a high renal acid content which the body neutralises with calcium leached from bones, which can weaken them. A diet rich in legumes, nuts and soy proteins can improve glycaemic control in diabetics. Moreover, growing crops instead of farming livestock can reduce antibiotic use and minimise the development of resistance.

However, there are advantages to eating meat. Meat contains healthy saturated fats that enrich the function of the immune and nervous system. Meat is the best source of vitamin B12 required for nervous and digestive system function and is a better source of iron than vegetables (the body absorbs 15-35% of heme iron found in meat compared to only 2-20% of the non-heme iron found in vegetable sources). Most plants do not contain sufficient levels of essential amino acids.

Moreover, a vegetarian diet can actually have negative environmental consequences. For example, some herbicides utilised on genetically modified crops are toxic to wild plants and animals are often killed during harvest. Eating meat could be considered as natural rather than cruel or unethical. Moreover, the problem of world hunger could partly be attributed to economics and distribution as opposed to insufficient amounts of food.

I would argue that it is ethically acceptable to eat meat so long as it is raised in a satisfactory way. It provides important nutrients, especially for growing children. However, it would be better if we reduced the amount of meat we eat in order to reduce the environmental impact of enteric fermentation and deforestation.

'Certain vaccines should be mandatory'

Explain what this statement means. Argue to the contrary. To what extent do you agree with the statement?

Vaccines are antigenic substances derived from the infectious microorganism itself that provide immunity against a disease. The statement argues that some vaccines should be compulsory. This essay will consider both perspectives before arriving at a conclusion.

Vaccines can protect the individuals that receive them against terrible debilitating diseases. Moreover, vaccines can also protect others in the population. If a certain proportion of the population are protected, herd immunity can be achieved. This means that people who cannot be vaccinated, for instance if they are immunocompromised or undergoing chemotherapy, will not contract the disease. Vaccines can also protect later generations. For instance, mothers vaccinated against rubella reduce the chance of their unborn children acquiring birth defects such as loss of vision, heart defects, cataracts and mental disabilities. Some vaccines have completely eradicated diseases for example the last case of Smallpox occurred in Somalia in 1977. Rinderpest, a disease of cattle, has also been eradicated and the instance of Polio has been substantially reduced.

Although many vaccines are available on the NHS and are funded by the taxpayer, they ultimately cost less to administer than the expense involved in time off work to care for a sick child, long term disability care and medical costs.

Nonetheless, vaccines sometime have serious and occasionally fatal consequences. About one in a million children are at risk of anaphylactic shock. The rotavirus vaccination can result in a type of bowel blockage known as intussusception; and the DPT and MMR vaccines have been associated with seizures, coma and permanent brain damage. Some physicians have raised concerns over the ingredients used in vaccinations. For example, thimerosal has been linked to autism, aluminium taken in excess can cause neurological harm and formaldehyde is a carcinogen that can result in coma, convulsions and death.

It could also be argued that the decision to be vaccinated should constitute a personal medical decision and individuals should be allowed to exert freedom of choice. There are also religious objections to vaccinations. For example the Amish object to vaccines and mandatory vaccinations. The Catholic Church is also opposed to the ingredients of certain vaccinations. For example, the MMR vaccine is cultivated in cells derived from two foetuses aborted in the 1960s.

However, the chance of serious side effects is incredibly small and furthermore the ingredients in vaccines are safe in the tiny amounts used: the exposure of children to aluminium is higher in breast milk than it is in vaccines. The FDA (food and drug administration) requires vaccines to be tested for up to 10 years before they are licensed and even after licensing, they continue to be monitored. In my view, the wider benefits of vaccines outweigh the minimal risk of poor side effects. In addition, it could be argued that personal decisions should be restricted when they affect the health of others. Therefore, I am supportive of certain vaccinations being mandatory.

'Compassion is the most important quality of a healthcare professional'

Explain what this statement means. Argue to the contrary. To what extent do you agree with the statement?

The statement argues that in careers involved in caring for the sick; kindness and empathy for the patients is the most important professional attribute. I will provide reasons for why this might be the case, whilst also discussing the importance of a sound scientific knowledge. I will then decide which quality I believe to be the most important.

A lack of compassion in care homes and hospitals could be held partly responsible for inexcusable cases of patient neglect. For example, care home members have been mocked and tortured and in hospitals, patients have been left surrounded in their own urine and forced to drink water from flower vases. Arguably this neglect has arisen from a lack of care and compassion from the healthcare professionals. However, at the same time it must partly be attributed to understaffing, lack of resources and training. Moreover, it is difficult to assess someone's level of compassion and it is uncertain whether this is something that can actually be taught.

A greater level of compassion would lead to better diagnoses. A large aspect of healthcare involves listening and communicating to patients. If a doctor has more empathy, patients are more likely to trust their doctor and disclose more personal information. An empathetic manner has also been shown to reduce patient anxiety and lead to faster patient recovery.

On the other hand, too much empathy could actually hamper healthcare professionals. Doctors and other health workers often require a degree of objectivity in order to make optimal decisions that may go against the patient's wishes. A level of detachment would also help professionals to remain calm in stressful clinical situations. Clearly, it is desirable for doctors and other healthcare professionals to have a detailed and comprehensive medical knowledge contributing to faster diagnoses, more skilled treatments and faster recoveries.

I would argue that scientific knowledge is the most important quality of a doctor especially because it is a necessity in order to practice medicine. However, compassion is also a highly important quality in a healthcare worker and is what separates an adequate doctor or nurse from an exceptional one.

END OF PAPER

Mock Paper D Answers

Section 1

Question 1: B
James runs 26.2 seconds, which is outside the qualifying time, therefore he does not qualify.

Question 2: D
Using s as the sandwich price, c for the crisps and w for the watermelon, the equation to solve is £5.60 $= s + c + w$.
Substituting in the information that $w = 2s$ and $s = 2c$:
£5.60 $= s + 2s + s/2$ or £5.60 $= 3.5 s$
$s = $ £1.60
$Hence, w = 2 \, x \, £160 = £320$

Question 3: E
Jane leaves at 2:35pm and arrives at 3:25pm, taking 50 minutes. Sam's journey takes twice as long, so leaving at 3:00pm it takes 100 minutes, giving an arrival time of 4:40pm.

Question 4: C
After the transaction, Michael has eight sweets. Therefore Hannah has 16 sweets after the transaction and hence 13 sweets before.

Question 5: C
Find original pay: £250/0.86 = 290 basic original pay. Add the rise: (290 x 1.05) + 6 = £311 new basic pay. Subtract the income tax at 12% = 311 x 0.88 = £273 new pay rate

Question 6: C
Given the first cube is a white cube, you are drawing from one of three boxes, boxes A, C or D. Boxes C and D will have just had their only white cube removed, whereas box A will have one white cube remaining. Therefore the probability of drawing a second white cube is $^1/_3$, thus the probability of non-white (i.e. black) is $^2/_3$.

Question 7: E
This is a simultaneous equations question.
500 + 10(x – 80) = 600 + 5x; true when x ≥ 80.
500 + 10x – 800 = 600 + 5x
» 5x = 900
» x = 180, therefore after 180 minutes

Question 8: E
The keyword here is **efficiency**. Simon's argument is that a slow eater will be less productive. Whilst eating slowly might be a weakness (D) and lunch breaks might be considered a distraction (B), they do not directly support Simon's argument. Although eating slow may lead to longer lunches (A) and reduce the time available to work (C), this doesn't necessarily mean the individual will be less productive – the lunch break might make them more efficient than other individuals. In order for Simon to assume slow eaters will be less productive, it must follow that slower eaters will have less time to work **efficiently** (E).

Question 9: D
This is a LCM question. We need to find the lowest common multiple of the song lengths. The LCM of 100, 180 and 240 is 3,600 seconds – equal to 60 minutes. For ease of arithmetic, you may choose to work reduce all numbers by a factor of 10.

Question 10: D
The journey is 3 hours and 45 mins, minus a 14 minute break gives 3hrs 31 mins travel time, or 211 minutes. Therefore the average speed is 51mph, or 82kmh by using the stated conversion factor.

Question 11: C
The mean guess is £13.80, which is £5.80 too high.

Question 12: C
The overall error for respondent 3 is £13, which is the least.

Question 13; B

The passage suggests that the attacks were carried out by extra terrestrial beings. Though the supposed UFO sightings have rational explanations, the writer feels this is insufficient to dismiss his idea.

Question 14: C

The initial argument suggests that two things must be present for an action to happen. If only one is absent, the action cannot happen. Argument C has the same form, the others do not.

Question 15: E

Building model ships requires several positive traits. The passage does not tell us which is the most important or most commonly lacked skill, only that more than one skill is required for success.

Question 16: C

Joseph does not have blue cubic blocks, since all his blue blocks are cylindrical.

Question 17: C

Each hour is a 1/12 of a complete turn of the clock face, equalling 30°. In an hour, the hour hand rotates 30°, so 10° every 20 minutes. The distance between the numbers 4 and 8 on a clock face is 4 x 30 = 120. There is still 1/3 of the distance between 3 and 4 to go, so you need to add 10° to get the total angle.

Question 18: B

The chance of red is 2/6 = 1/3. To get no reds at all, it must be non-red for each of three independent rolls. The probability of this is $(2/3)^3$ = 8/27. Therefore the probability of at least one red is 1 – 8/27 = 19/27

Question 19: D

These three furniture items are compatible with having 6 legs. All the other statements are false.

Question 20: D

Work this out by time. The friends are closing on each other at a total of 6mph overall, therefore the 42 miles take 7 hours. In seven hours, the falcon, flying at 18mph covers 18 x 7 = 126 miles.

Question 21: C

The passage tells us that antibiotic resistance could lead to people dying from Victorian diseases, and that liberal use of antibiotics in farming is the "most significant" contributor to this. Therefore it would be true to say that this use of antibiotics could cause serious harm.

Question 22: B

Calculate the overall cost of three stationery sets, then subtract any items not bought. For each item shared between two people, there is one of that item not required. The overall cost is £6.00 per person, £18.00 overall. Subtract one geometry set (£3), one paper pad (£1) and one pencil (50p) to give £13.50 overall cost.

Question 23: C

Moving planks 1 and 4 to form a cross inside one of the other squares will solve the problem. Two squares are broken (the bottom right hand corner and the overall large square) but four new small ones are created, bringing the total up to seven.

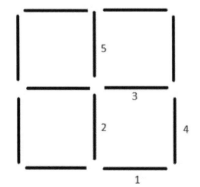

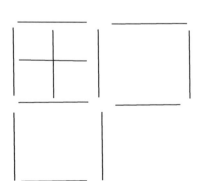

Question 24: D

The purple square is opposite white, since both are adjacent to blue on opposite sides. White and purple cannot be adjacent to each other since the position of the opposite black and red sides makes that impossible.

Question 25: C
We take the overall price to the UK and subtract money which does not go to the farmers. 36,000,000kg at 300p/kg gives £108m. Subtract commission 108 x 0.8, then take 10% of the remaining proceeds as the farmers' share, giving £8.64m

Question 26: E
None of the responses can be reliably deduced from the statement regarding the Giardiasis. It cannot be A, because the statement relates to both stomach pain and diarrhoea. It cannot be B as the statement only refers to individuals with Giardisis – we do not know if people with these symptoms do not have Giardiasis. It cannot be C and D due to the natures of the statement itself.

Question 27: C
Catherine must choose four socks. If choosing three or fewer, it is possible that they could each be of different colour. When choosing four, it is certain that at least two socks will make a matching pair, but possible that there will be two pairs.

Question 28: D
This is another simultaneous equations question. Solve to find x, the normal rate of pay.
$100x + 20y = 2000$ » $60y = 6000 – 300x$ (substitute this into the equation representing Giovanni's pay)
$80x + 60y = 2700$
» $80x + (6000 – 300x) = 2700$
» $220x = 3300$
$x = 15$
To get the value for overtime rate of pay, substitute x back into one of the two equations and solve for y
$100(15) + 20y = 2000$
$Y = 25$
So the overtime rate of pay is 25 euros per hour.

Question 29: E
The easiest way to do this is via simultaneous equations to solve for the time when both trains collide (x). Let A be the distance travelled by the Bristol train and B the distance travelled by the Newcastle train. Using Distance = speed x time. :
$A = 90x + 45$ (to account for the extra ½ hour which this train has been travelling relative to B, at 90miles an hour) and B = $70x$. The collision will occur when the total distance travelled by both trains is = 405 i.e. $A + B = 405$.
Therefore, $90x + 45 + 70x = 405$. $X = 2.25$ hours. Thus, collision happens at 12:45.
Substitute $x = 2.25$ into the first equation to give the distance from Bristol: $A = 90 \times 2.25 + 45 = 247.5$

Question 30: E
Form simultaneous equations to do this quickly. Let x be pregnant rabbits and y be non-pregnant rabbits. $100 = x + 2y$ and $175 = 2x + 3y$. Substituting for x gives $175 = 2(100-2y) + 3y$. Solve for $y = 25$ and therefore $x = 50$ when substituted back into $100 = x + 2y$. Only E doesn't concord with these values.

Question 31: D
Michael pays £60 and £110 = £170 for the painting. He sells it for £90 and £130 = £220. Thus, he makes a profit of £220 - £170 = £50.

Question 32: D
The principle problem is that it does not compare the relative effectiveness of pesticides and natural predators. It might be that pesticides are far more effective at controlling pests, despite the unnecessary excess killing.

Question 33: D
Proportionately, there would be 172 members. Therefore there is an excess of 298 – 172 = 126 members.

Question 34: B
If each pair of opposite faces is painted one colour, this requirement can be satisfied with a minimum of three colours.

Question 35: C
Two people cross dropping one (first crossing) on the other side. One returns (second) and then takes another person over (third). When one returns this time (fourth) he leaves two on the correct bank and returns for the rider alone on the wrong side. The rider comes back alone leaving one person on the wrong side (five) and one of the people (six) return to collect the final person. These two return together completing the 7th crossing.

END OF SECTION

Section 2

Question 1: E
Haemoglobin is contained within red blood cells and is not free in the blood. Additionally, as a protein it is too large to normally pass through the glomerular filtration barrier. All the other substances are freely filtered.

Question 2: B
In order for the membrane potential to become more positive, there must be a net movement of positive ions into the muscle cell (so it becomes more positive compared to its resting state). Since there is a greater concentration of sodium ions outside, more sodium than potassium must move inwards.

Question 3: B
Equate the volume with the surface area in the proportion instructed by the question. $3(^4/_3\pi r^3) = 4\pi r^2$, simplifies to $r = 1$.

Question 4: B
Gravitational potential energy increases as the grain is lifted further from floor; this is equal to the work done against gravity to attain the higher position. The potential energy equal to $mg\Delta h$, so it is dependent upon the mass of the grain that is lifted.

Question 5: E
This is a tricky question that requires a conceptual leap. Only the top candidates will get this correct.

Surface Area of Earth $= 4\pi r^2$
$= 4 \times 3 \times (0.6 \times 10^7)^2$
$= 12 \times (6 \times 10^6)^2$
$= 12 \times 36 \times 10^{12}$
$= 3.6 \times 10^{14}$

Since $= \frac{Force}{Area}$, $Atmospheric\ Pressure = \frac{Force\ exerted\ by atmosphere}{Surface\ Area\ of\ Earth}$

Therefore: $Force = 10^5 \times 3.6 \times 10^{14}$
$= 3.6 \times 10^{19}\ N$

The force exerted by the atmosphere is equal to its weight therefore:
$Force = Weight = mass \times g$

Hence, $Atmospheric\ Mass = \frac{3.6 \times 10^{19}}{10} = 3.6 \times 10^{18}\ Kg$

Question 6: F
A polymer consists of repeating monomeric subunits. Polythene consists of multiple ethenes; glycogen of glucose; collagen of amino acids, starch of glucose; DNA of nucleotide bases, but triglycerides are not composed of monomeric subunits.

Question 7: E
Increased ADH causes more water reabsorption. This concentrates the sodium in the urine by reducing urine volume. In the healthy kidney, all glucose is reabsorbed and none is excreted into the urine.

Question 8: D
F = ma; therefore the difference in force is equal to $m_1a_1 - m_2a_2$. This equals $(6 \times 6) - (2 \times 8) = 20N$

Question 9: C
It's important to know your reactivity series as its easy marks. Remember that potassium is more reactive than sodium, as it has a greater number of electron shells, with the outermost single electron being more loosely attracted to the nucleus because of this, and hence more likely to be lost. Following this pattern, sodium is the next most reactive and copper the least.

Question 10: D
Diastole is the relaxation phase of the cardiac cycle. In diastole the pressure in the aorta decreases as the contractile force from the ventricles is reduced. All of the other statements are true; the aortic valve closes after ventricular systole. All four chambers of the heart have blood in them throughout the cardiac cycle.

Question 11: A
Because the two sides of the circuit are in parallel, both sets of lights experience a 24v voltage drop across them. In lights R and S this is shared equally between them, but in lights P and Q, the new light with twice the resistance takes twice the voltage in accordance with Ohm's Law (V= IR).

Question 12: E
144ml of water is 144g, which is the equivalent of 8 moles. 8 times Avogadro's constant gives the number of molecules present, which is 4.8×10^{24}. There are 10 protons and 10 electrons in each water molecule, hence there are 4.8×10^{25} electrons.

Question 13: B
Competitive inhibition occurs when the inhibitor prevents a reaction by binding to the enzyme active site. Hence, a higher concentration of the substrate can result in the same overall rate of reaction. i.e. the substrate outcompetes the competitor.

Non competitive inhibition is where the inhibitor binds to the enzyme (not at the active site) and prevents the reaction from taking place. Increasing the substrate concentration therefore does not increase the reaction rate i.e. the substrate cannot outcompete the competitor as the enzymes are disabled and the competitor is not binding to the active site.

In this graph, line 1 shows the normal reaction without inhibition, line 2 shows competitive inhibitor and line 3 shows non-competitive inhibition.

Question 14: E
Nucleic acids are only found in the nucleus (DNA & RNA) and cytoplasm (RNA). They are not a component of the plasma membrane, whereas the other molecules are.

Question 15: B
$Number\ of\ annual\ flights\ =\ Flights\ per\ hour\ x\ Number\ of\ hours\ in\ one\ year\ x\ Number\ of\ airports$
$=\ 4\ x\ (24\ x\ 365)\ x\ 1000$
$=\ 96\ x\ 365\ x\ (1000)$
$\approx\ 100\ x\ 365\ x\ 10\ x\ 100$
$=\ 365\ x\ 10^5 = 36.5\ Million$

However, this is an overestimate since we have multiplied by 100 instead of 96. Hence, the actual answer will be slightly lower. 35 Million is the only other viable option available.
365x24=8760 is the number of hours in a year, then 8760 x number of flights per hour (4) = 35040 flights per year per airport. Multiply by the number of airports – 42 million to the nearest million.

Question 16: F
Write the equation to calculate molar ratios:
$C_8H_{18} + 12.5\ O_2 \rightarrow 8CO_2 + 9H_2O$
Travelling 10 miles uses: 228 x 10 = 2,280g of Octane.
M_r of Octane = 12 x 8 + 18 x 1 = 114
Number of moles of octane used = 2,280/114 = 20 moles.
Thus, 160 moles of CO_2 must be produced.
M_r of CO_2 = 12 + 16 x 2= 44
Mass of CO_2 produced = 44 x 160
= 7,040 g = 7.04 kg

Question 17: D
Add the first and last equations together to give: 2F = 4, thus F = 2.
Then add the second and third equations to give 2F – 2H= 5. Thus, H = -0.5
Finally, substitute back in to the first equation to give 2 + G – 0.5 = 1. Thus, G = -0.5
Therefore, FGH = 2 x -0.5 x -0.5 = 0.5.

Question 18: D
The main artery to the lungs is the pulmonary artery, which gets blocked. The clot must therefore travel through the inferior vena cava and right side of the heart. It does not enter the superior vena cava or left (systemic) circulation.

Question 19: B
Note that the units are the same (M = moldm^{-3}), only the orders of magnitude are different. Convert the orders of magnitude to discover a 10^6 difference with more chloride than thyroxine

Question 20: E
This is a simple recall question. X rays have the shortest wavelength whilst microwaves have the longest wavelengths with visible light being somewhere in the middle. It is well worth your time remembering the basic positions of the components of the electromagnetic spectrum as it frequently gets tested in the BMAT.

Question 21: E

The way to solve this is to break the calculation down into parts, almost working backwards. The number of seconds in 66 weeks is given by:

$= 60 \times 60 \times 24 \times 7 \times 66$:

$= (10 \times 6) \times (12 \times 5) \times (4 \times 6) \times 7 \times (11 \times 6)$

$= 1 \times 4 \times 5 \times 6 \times 6 \times 6 \times 7 \times 10 \times 11 \times 12$

$= 1 \times 4 \times 5 \times 6 \times (6) \times 7 \times 10 \times 11 \times (12 \times 6)$

$= 1 \times 4 \times 5 \times 6 \times (3 \times 2) \times 7 \times 10 \times 11 \times (72)$

$= 1 \times 2 \times 3 \times 4 \times 5 \times 6 \times 7 \times 10 \times 11 \times (9 \times 8)$

$= 1 \times 2 \times 3 \times 4 \times 5 \times 6 \times 7 \times 8 \times 9 \times 10 \times 11$

Question 22: B

Glycogen is not a hormone, it is a polysaccharide storage product primarily found in muscle and the liver.

Question 23: C

The conceptual leap required for this question is that since the system is 100% efficient, energy won't be created or destroyed but merely transferred.

Thus, *Energy input into water = Gravitational potential energy at top of stream*

$100\ J\ per\ Second = mg\Delta h$

$h = \frac{100}{mg}$

$h = \frac{100}{1 \times 10} = 10\ m$

Question 24: answer missing

Reflexes can be influenced by the brain e.g. if you willingly pick up a hot plate, you will be able to withstand much greater heat than if you touch it by accident and discover it is hot. Reflex actions are fast as they usually bypass the brain. Since they are mediated by nerves, they are much faster than endocrine responses. Most animals show basic reflexes like the heat-withdrawal reflex which requires both sensory and motor components.

Question 25: C

Remember the interior angles of a pentagon add up to 540° (three internal triangles), so each interior angle is 540/5 = 108°. Therefore angle **a** is 108°. Recalling that angles within a quadrilateral sum to 360°, we can calculate **b**. The larger angle in the central quadrilateral is 360° – 2 x 108° (angles at a point) = 144°. Therefore the remaining angle, **b** = (360 – 2(144)]/2 = 36°. The product of 36 and 108 is 3,888°.

Question 26: D

The key here is to note that the answers are several orders of magnitude apart so you can round the numbers to make your calculations easier:

Probability of bacteria being resistant to every antibiotic =

$P\ (Res\ to\ Antibiotic\ 1) \times P\ (Res\ to\ Antibiotic\ 2) \times P\ (Res\ to\ Antibiotic\ 3) \times P\ (Res\ to\ Antibiotic\ 4)$

$= \frac{100}{10^{11}} \times \frac{1000}{10^{9}} \times \frac{100}{10^{8}} \times \frac{1}{10^{5}}$

$= \frac{10^{8}}{10^{33}} = \frac{1}{10^{25}}$

Question 27: F

All the above units are measures of power, the amount of work done per unit time.

END OF SECTION

Section 3

'The concept of medical euthanasia is dangerous and should never be permitted within the UK'

Explain the reasoning behind this statement. Suggest an argument against this statement. To what extent, should legislation regarding the prohibition of medical euthanasia in the UK be changed?

Euthanasia, from the Greek for '*mercy-killing*', involves the active painless taking of patients' lives for those facing incurable physical, mental and social torment by disease. The practice is legal in certain parts of the world, including Switzerland, although the above statement argues that changes to the current law in the UK would have negative impact on the foundation of medical practice and welfare of doctors, patients and their families.

This is because the action of actively ending the life of patients may result in guilt-filled psychological effects on doctors. This is especially so as the profession in most cases aims to extend quantity of life.

In addition, the line between suffering patients who would and would not 'deserve' euthanasia is ambiguous. Some patients who have diseases that have a detrimental impact on their lives may opt for euthanasia, when actually there may be therapy or support to aid them in leading fulfilling lives. It would be also difficult to assess whether patients have the competence to judge their willingness for euthanasia, for example underlying depression that may be influencing their choice.

However, doctors should always have the quality of life of a patient at the forefront of treatment. In many debilitating disorders where patients require constant attention, physical support and lack fulfilling stimulation, the quality of life of patients is abysmal, for example in locked-in syndrome. For those who want to end their life, the relief of this suffering by the medical profession could be considered a caring act.

Moreover, euthanasia may relieve the burden of constant care faced by family members. Currently many patients make the journey to countries where medical euthanasia is legal, but as a result, family members face prosecution in good-willingly helping the patients to do so. Legalisation would relive the strain on family members of not only losing their patient relative but also of this criminal prospect.

However, in conclusion, the legislation should not be changed. Ultimately, it is too difficult to assess cases in which medical euthanasia would be acceptable. In addition, its legalisation may permit family members burdened with caring for a patient to pressurise patients into euthanasia. As a result, the effect of this concept revolving on 'mercy' towards a patient may end up being one of detriment.

'The obstruction of stem-cell research is directly responsible for death arising from stem-cell treatable diseases.'

Explain what this argument means. Argue the contrary. To what extend do you agree with the statement?

Stem-cell research offers the opportunity for a vast increase in our understanding of disease and generally of how the human body works. Being able to conduct research into the different areas that are involved in the growth of humans from sperm and egg and the formation of complex organisms will contribute greatly to our understanding not only of genetic and birth defects and their origin, but also to our understanding of disease. On top of that stem-cell research might theoretically offer the key to the cure of a variety of diseases such as spinal cord damage, organ failure or genetic defects. Being able to grow organs for transplantation for example provides a huge opportunity for the whole of mankind. Not having to wait for a needed organ will make the patient's life better and more bearable and also bring about a cure faster. Whilst the current status quo does allow some research into stem-cells, there are heavy restrictions when it comes to the use of human stem cells, not only because of the question at which point a baby becomes a human being.

Opposing the idea of deregulation of stem-cell research stands the main concern that all life is valuable and that includes the potential life represented by a foetus, the ultimate source of stem-cells used in research. The extraction of stem-cells for research is essentially nothing else as the destruction of potential human life, which is not acceptable. This concern is based on the idea that with the fusion of sperm and egg a new human being is being created that needs to be asked for consent in the participation of any research. As this is not possible for a foetus, some think that stem-cell research is unethical and in breach of the main principles underlying medical research and medical care in general. Another idea is that of 'the ends don't justify the means'. Whilst it is a reasonable prospect to expect that progress in stem-cell research will lead to the discovery of new treatment forms for a variety of severely impacting diseases, this does not justify the use of what essentially are non-consented human beings for research.

Stem-cell research most definitely holds a great deal of potential when it comes to the increase in medical knowledge. Being able to research into the area could potentially provide us with a greater understanding of a variety of biological processes as well with a cure to many forms of disease. None the less it is questionable if this alone justifies ignoring the fact that a foetus can potentially be killed by the extraction of stem cells which will directly cause the destruction of potentially healthy life. Ultimately there are arguments for and against stem-cell research. The majority hinge on the understanding of life as an untouchable unit. If we assume that a foetus only becomes a living human being after a certain period of time, this justifies research on the non-human state prior to that date. This is the commonly used legal interpretation of the issue facing researchers today.

'Imagination is more important than knowledge'

Albert Einstein

Explain how this statement could be interpreted in a medical setting. Argue to the contrary that knowledge is more important than imagination in medicine. To what extent do you agree with the statement?

Einstein alludes to creativity and imagination, rather than pure knowledge, being the driving force for scientific discovery. Medically, this could analogized as it being more important for doctors to be able to assimilate different thoughts and concepts together to come up with a differential diagnosis/ holistic management plan than it is to simply memorize all the underlying textbook theory.

On the contrary, unlike with many scientific discoveries, you cannot come up with a reliable and safe differential or management plan simply based on a hunch without a comprehensive underlying knowledge base.

If something goes wrong in scientific discovery, you can simply try again. With patients, this is not an option.

Guidelines and treatment recommendations e.g. from NICE/ specific medical bodies have largely removed the need for creativity in the management of patients. On the contrary, the gold standard is the practice of evidence-based medicine based on sound underlying knowledge.

It could be said that a strong knowledge base would help you diagnose/manage the most common conditions, while creativity and imagination help with the rarer conditions. Seeing as common is common, on average knowledge would be more important than imagination in treating most patients.

On balance, I believe that a strong knowledge base would make a good doctor, while a strong knowledge base together with imagination and creativity would make a great doctor. Knowledge can be seen as the core foundation needed by all doctors to practice safely and reliably. However the human body does not behave like a textbook and most conditions can present atypically. In this case, being able to integrate both these skills and use your underlying knowledge of pathophysiology to imagine how a condition may manifest atypically could be the key between spotting and missing a life-threatening diagnosis. E.g. An atypical presentation of Tuberculosis.

"The most important quality of a good doctor is a thorough understanding of science".

Explain what this statement means. Argue in favour of this statement. To what extent do you agree with it?

You should start the essay by showing your understanding of the question and pointing the reader in the direction you wish to take for the essay. Clarify any assumptions you will make.

The question talks about the most important quality of a good doctor – this implies that there are other important qualities of a good doctor, and it might help to identify what some of these are. It is not sufficient to simply argue that a thorough understanding of science is important – you must explain why it is more important than other important qualities.

Other qualities might include good communication skills, the ability to understand people's emotions, practical clinical skills, an understanding of how the healthcare system works and an understanding of your own limitations.

Address the reasons why a thorough understanding of science is important, then explain why these are more significant than the reasons supporting other important traits. Consider that the practice of medicine is based upon the scientific method, and all treatments to be funded by a modern healthcare system must be supported by evidence. Doctors need a thorough understanding of science in order to be able to work by these principles.

Note that science informs the correct treatment. Other skills might improve the benefit to patients, but without an appropriate treatment, informed by science, there will be no benefit.

A lack of knowledge or understanding of the science underlying medicine might lead to dangerous mistakes being made. This would go against the principle "first, do no harm" as first outlined in the Hippocratic Oath. In addition, patient safety is a fundamental and overarching principle of a safe and effective NHS.

State your position: to what extent do you agree with the statement and why. You might want to mitigate your support by reference to the importance of personal skills: better communication makes it more likely to find out what is wrong to inform the correct treatment, it makes it more likely the patient will understand and follow your suggested treatment regime, it might improve the placebo component of treatment (which is significant) and it might increase the patient's trust in you, leading to a better long-term relationship and thus benefiting their health.

Likewise with reference to practical skills, such as surgery, you might suggest that even if the right treatment is selected, without the right practical skills the treatment will not be beneficial to the patient and indeed could cause significant harm.

END OF PAPER

Mock Paper E Answers

Section 1

Question 1: B
The total saving on the final booking relative to the first is £230, but the cost of two cancellations must be deducted (£90) giving a total saving of £140.

Question 2: B
There are originally no odd numbered balls in Bag A. But as a result of the transfer, there could be an odd ball in Bag A. Therefore the probability of drawing an odd ball is found by multiplying the probability of selecting the new ball ($^1/_5$) from Bag A by the probability that that ball is odd ($^2/_5$ – given by adding the one odd ball in the bag C originally to the odd ball introduced from Bag B) giving an probability of $^2/_{25}$ that the selected ball from Bag A is odd.

Question 3: B
Assume the price of bread is 100p. 100 x 1.4 x 0.8 = 112p after the subsidy. The cost of three loaves is therefore 336p (divided four ways this equals 84p per loaf). Hence, as a percentage of the original price, this is 84%.

Question 4: D
At 2120hrs, the minute hand is pointing to 4 and the hour hand is pointing one third of the way past 9 towards 10. 360°/12 = 30° – this is the number of degrees per hour division. Between the two hands then, there are 5 hour divisions plus an extra $^1/_3$. Therefore the angle is (30x5)+(30/3) = 160°

Question 5: B
There is a 3l and 5l bucket – therefore 4 litres can be measured from the difference between the buckets as follows. Fill the 5l bucket, decant 3l into the smaller bucket and then you are left with 2l in the large bucket. Pour this into the tank. Repeat the process again, decanting the remaining 2l into the tank once again to make 4l in total. The first time, 5 litres was required. The second time, the 3 litres from the second bucket could be tipped back into the 5l bucket, and then filled up with fresh water to measure the final 2 litres in. Therefore 4 + 3 = 7 litres of water is sufficient to fill the tank with 4l.

Question 6: D
To answer this question, make a timeline showing the locations of the different genres of books. Place each book on the timeline as appropriate, making sure to indicate where more than one location is a possibility. From that, you will see that literature books are located to the right of engineering. This is true since they are to the right of art (which we know is right of mathematics (and therefore engineering, since the run between the sciences is uninterrupted)). The other statements, whilst potentially true, cannot be deduced for certain.

Question 7: C
The passage tells us that brand new cars lose value quickly, despite the car being virtually unchanged. Therefore in the absence of any contradictory information, it is reasonable to conclude that buying second hand cars is a wise choice.

Question 8: B
First, calculate how many bottles are sold. 2000 – (2000x0.9x0.8) = 560 bottles. Then divide the total profit by the number of units to give the profit per unit, which comes to 11200/560 = £20 per bottle.

Question 9: E
The definition of timelessness requires something to be tested by time. Something that modern furniture cannot fulfil. Therefore statement E expresses a significant flaw in the reasoning. The other statements do not refer to the 'timelessness' aspect of furniture, therefore they are not directly relevant to the argument.

Question 10: A
The passage talks about the benefits of drinking red wine, not about living near to vineyards. The passage does not state that Italians drink more wine than Germans, therefore the assumption that they do is central to the argument.

Question 11: C
Tom arrives at 1620, and leaves 45 mins after Jane leaves. Therefore he also leaves 45 mins after Hannah leaves, since Jane and Hannah leave together. Since his journey is 10 mins faster than Hannah's, he arrives only 35 minutes after Hannah arrives (which happens to be 1620). Therefore Hannah arrives 35 minutes earlier than this, at 1545. Since she left at 1430, her journey took 75 minutes. Jane's journey took 40% longer (1.4 x 75 = 105 minutes). Therefore leaving at the same time as Hannah, 1430, Jane arrived 105 minutes later at 1615.

Question 12: B

This is a simultaneous equations question. Let **x** be the number of standard tickets sold, and **y** be the number of premium tickets sold.

Therefore: $x + y = 600$; $10x + 16y = 6,600$

$x = 600 - y$ » substitute: $10(600 - y) + 16y = 6600$

$6y = 600$

$y = 100$, therefore 100 premium tickets were sold.

Question 13: C

Between 20th January and 23rd May, there are 123 days. In 123 days, the moon makes $123/28 = 4.39$ orbits. This is equal to $4.39 \times 360° = 1580°$

Question 14: E

You are looking for a strong opposition to the proposition that students at drama academies are not taught well academically. The strongest opposition would be evidence that such students perform academically well in some objective measure. Evidence of significantly above average GCSE results provides this.

Question 15: D

You should definitely draw this one out on paper. Trace out the paths and you find that both people have a net displacement of 11km to the North. Therefore since Anil is only net 2km East, and Suresh is 17km East of the starting point, there is a 15km separation between them

Question 16: E

If three times the final amount of concrete is ground off by the builder, three quarters of the original thickness is removed, hence one quarter remains. $14/4 = 3.5$cm

Question 17: A

Walking at 4mph, 3 miles takes ¾ hour = 45 mins. Adding the 5 minute stop, Chris will arrive at 1820, since he set off at 1730. At 24mph, 6 miles takes ¼ hour, 15 mins. Therefore setting off at 1810, Sarah will arrive at Laura's at 1825. Therefore Chris arrives 5 minutes earlier than Sarah.

Question 18: D

The passage tells us that illegal downloads are causing harm to the music industry. Whilst it gives an example, this does not mean the stated example is the principal issue. The conclusion that best fits the passage as a whole is to say illegal downloading is more harmful than many people think, given their willingness to undertake it.

Question 19: E

First, calculate the amount of water needed for each type of fire. Use algebra:

Use x as the amount of water used to extinguish a house fire. $40,000L = 2x$, so $x = 20,000L$. Then, take y as the amount of water needed to extinguish a garden fire, so $70,000L = 2x + 3y$. $30,000L = 3y$, $y = 10,000L$.

Knowing this, A is correct, B is correct, C is correct and D is correct. Only E is false.

Three house and ten garden fires require 160,000 litres to extinguish, not 140,000.

Question 20: E

To answer this question, we need to use SUVAT equations, specifically:

$s = ½ (u + v) t$ and $s = vt + (1/2a \times t^2)$

We can calculate the distance the car initially moves before accelerating by 20ms⁻¹ x 30s = 600m. Using the $s = vt + (1/2 a \times t^2)$, substitute in the values we know to find the distance travelled during the acceleration.

$s = 20 \times 5 + (1/2 \times 2 \times 25) = 125$m

We can calculate our new speed after the moment of acceleration by:

$20 + (5 \times 2) = 30$ms⁻¹, and the distance travelled at this new velocity by 20 x 30 = 600m

The distance travelled during the deceleration is found by:

$s = ½ (30 + 0) t$, as the car is coming to the stop so the final velocity is 0.

We find t by doing 30ms⁻¹ / 3ms⁻² = 10s

So, substituting this in give the distance = 150m. Add all these distances together to give the final distance covered = 1475m.

Question 21: D

The passage only talks about people's opinions on the scheme, and not about any action which could potentially be taken. Therefore the best summary is to say that more people oppose the scheme than support it.

Question 22: B + E

The question asks for two responses, therefore you must mark two and get them both correct for one mark. The suggestion is made that reducing wild fishing will improve fish populations. This assertion carries two major assumptions – that the fishing originally caused the decline, and that the decline is reversible, and can therefore recover if the threat is removed. Select these two responses for a mark.

Question 23: D

To calculate this, you need to work out how many possible combinations there are, and how many of them contain exactly two heads. Since there are 2 possibilities and 5 trials, the number of potential outcomes is $2^5 = 32$. For two heads, any combination of two coins can show heads – and since there are 5 coins tossed, there are 10 possible combinations of exactly two heads. Therefore the probability is $^{10}/_{32}$, which is equivalent to $^5/_{16}$.

Question 24: C

For a 5 litre bucket with a 2% margin for error, the maximum possible volume is 1.02 x 5 = 5.10l, and the minimum is 0.98 x 5 = 4.90l. Therefore there is a 200ml difference between the maximum and minimum volume possible. Therefore the range of cleaning powder required is 0.2 x 40 = 8g.

Question 25: B

To calculate the cost of the call, you need to first work out its duration in minutes and multiply by the off-peak rate per minute. Then you add on the connection fee of 18p. A call of 1.4 hours = 1 hour 24 minutes = 84 mins. (84 x 22 =1848 = £18.48). £18.48 + 18p = £18.66.

Question 26: C

The passage tells us about the risks of sunbathing, and that many people do not see the danger in it. The final sentence shows us that the conclusion is a wide-ranging one, not a specific observation about UV radiation. Therefore Answer C is correct, it best sums up the passage as a whole.

Question 27: C

First calculate the total pay, then divide this by the number of hours Jim works for an hourly rate. Total pay = 3 x 8 = £24 covers the total pay for the 8 houses. Total time taken to wash all the windows = (11 windows x 8 x 2mins per window) + (15mins to pack up and walk to the next house x 7) = 281 minutes = 4.68 hours. [the total time is equal to the number of windows in total multiplied by the time taken to clean each window, plus the time travelling between the houses, which is 15 multiplied by the 7 journeys required].

Once you have worked out the number of hours Jim worked, you can find the hourly rate by £24/4.68hours = £5.12per hour.

Question 28: A

The passage argues that bottled water is pointless, as is almost identical to tap water. If bottled water had an additional benefit, such as being good for health, it might be that it makes sense to drink bottled water.

Question 29: A

Be careful – this sentence contains a triple negative. If the sentence read "...nor any cyclists that are marathon runners", it would be clear that no cyclists run marathons too. Changing the sentence to "...nor NO cyclists that AREN'T marathon runners" introduces a double negative, hence the meaning is not changed. Therefore it still means no cyclists run marathons, hence **A** is true.

Question 30: D

Draw this one out on a line. You will see that whilst we know Oakton is East of Langham, we cannot conclude its whereabouts in relation to Frampton – it could be either East or West. Therefore **D** cannot be said with certainty.

Question 31: A

Firstly calculate the surface area of the dome, then divide by the surface area covered by one pot to calculate the number of pots needed. A dome is half a sphere, so the area is given by $\frac{4\pi r^2}{2} = 12 \times \frac{49}{2} = 294$. Since one pot covers 12m^2, 24.5 pots are required to cover the whole dome, therefore 25 must be purchased.

Question 32: A

25 Pots x 2 Litres per pot = 50 litres of paint.

This gives a solid volume of 50 x 0.4 = 20 when the paint dries = 0.02m^3.

The volume of the hemisphere is: $\frac{\frac{4}{3}\pi r^3}{2} = \frac{4r^3}{2}$

$= 2 \times 7^3 = 686$

Hence the overall percentage decrease is: $\frac{0.02}{686}x100$

$= \frac{2}{686} = \frac{1}{343}$ 0.0029%

Question 33: E

400ms = 400 x 10^{-3}seconds, or 0.4s. Since there are 60 x 60 x 2 = 7200 seconds in 2 hours, and it can wrap 7200/0.4 = 18,000 sweets in 2 hours.

Question 34: C

If there are four stations between Crabtree and Eppingsworth, there are five independent journeys between them. Calculate the times between each of the stations, then sum them for the total journey time. The journey between stations 3 and 4 is leg 4 of the journey, and it takes 16 minutes. Therefore leg 5, the final component takes 16x0.8 = 12.8 minutes. Leg 3 takes 16/0.8 minutes = 20 minutes, leg 2 takes 20/0.8 = 25 minutes. Leg 1 takes 25/0.8 = 31.25 minutes. Summing these together gives: 31.25 + 25 + 20 + 16 + 12.8 = 105.05, about 105 minutes, since we do not measure rail journeys in fractions of minutes.

Question 35: B

This is a deceptively easy question – a poor candidate would get bogged down in a long and arduous calculation. A good student will spot that the answer options are relatively far apart enough to allow you to **estimate rather than calculate**.

There are 36,000 Ford Escorts in 2005, and you are asked for the number remaining 10 years on. Therefore 36,000 needs to be reduced by 25%, then this answer needs to be reduced by 25% and so on, such that 10 reductions of 25% have taken place. We can calculate this compound decrease in the number of escorts on the road by: 36,000 x (0.75)^{10}

The easier way to do this is just to roughly remove a quarter ten times as shown in the table below:

Year	Value (x10^3)	Year	Value (x10^3)
0	36	6	6
1	27	7	4.5
2	21	8	3.3
3	16	9	2.5
4	12	10	1.9
5	9		

Thus, the final answer must be close to 1,900 – option B is the only viable answer.

END OF SECTION

Section 2

Question 1: D

Firstly, convert Litres → m^3: 950 Litres = 0.95 m^3

Buoyancy Force = Volume x Density x g.

= 0.95 x 1000 x 10 = 9,500 N

Weight of the boat = mg= 600 x 10 = 6,000 N

Since buoyancy force > Weight, the boat will float.

The difference between Buoyancy Force + weight = 9500 – 6000 = 3,500N

Hence adding mass of 350kg (=3,500N as *g* is 10) will balance both forces.

Adding further mass will cause the boat to sink. Hence, the answer is 355kg (350kg won't cause sinking – merely balance the force).

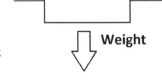

Question 2: E

Recall that reduction is the gain of electrons whilst oxidation is a loss. Also remember that oxidation the gain of oxygen, while reduction is loss. Only Iodine is gaining electrons and so shows reduction.

Question 3: F

None of the above, they are all true facts about digestion.

Question 4: A

Remember that you can separate the vertical and horizontal components of both bullets. Both bullets actually have zero vertical velocity at t=0. Thus, only gravity affects them- and it does so equally. Therefore, rather counter-intuitively, they hit the floor at the same time.

Question 5: C

You don't need to know the mass of the fish for this one, since there is no acceleration or deceleration taking place. The resistive forces are equivalent to the force of thrust of the fish. Recall that work done = force x distance. Travelling at 2ms^{-1}, the fish travels 60 seconds x 60 minutes x 2 ms^{-1} = 7200 m in one hour. Therefore the work done against resistive forces is f x d = 2N x 7200 = 14,400J

Question 6: D

Blood flow to the kidneys is constant - not exercise dependent. Overall cardiac output increases since heart rate and stroke volume increase (because there is greater oxygen demand from exercising muscle). There is more blood flow to the muscles to fuel them and to the skin to help lose excess heat. Blood flow to the gut decreases to increase availability to muscles. Blood flow to vital organs such as the kidney and brain remains constant.

Question 7: F

To balance the equation, start working from what you're given – the oxygen. Since you know there are 15 oxygen atoms on the right, there must be the same on the left. Therefore **w** = 5. You also know that there are 30 Hydrogen atoms on the right hand side, and so you can work out x. 30-5 leaves 25 atoms unaccounted for, so x=25.

Question 8: A

Since A-T and C-G are the DNA base pairings, 29.6% Adenine implies 29.6% Thymine as well. Therefore the remaining 100 – 59.2 = 40.8% is shared between Guanine and Cytosine equally, so there is 20.4% cytosine.

Question 9: F

Structure A is the right semi-lunar valve, the pulmonary valve. It opens in systole to allow flow of blood from the right ventricle into the pulmonary artery and to the lungs. It closes in diastole to ensure the right ventricle fills only from the right atrium, maintaining a one-way flow of blood. Therefore F is true, it opens when the right atrium is emptying. None of the other statements are true.

Question 10: C

Since CO binds to the oxygen binding site of haemoglobin, it reduces oxygen binding and therefore oxygen carrying capacity of blood. Hence, the blood becomes less oxygenated. Since more blood needs to flow to deliver the same amount of oxygen, this must be accomplished by an increased in heart rate. Haemoglobin does not become heavier as the CO binds **instead** of oxygen rather than in **addition** to. Carbon Dioxide is carried in plasma so is unaffected by carbon monoxide poisoning which affects haemoglobin.

Question 11: D

A Moment of force = Force x Perpendicular distance to pivot

If the lifting arm is a uniform 5m long, the weight exerts $2000 \ x \ 10 \ x \ 5 \ = \ 100{,}000 \ Nm$ of torque. In addition,

there is a $250 \ x \ 10 \ x \ 2.5 \ = \ 6{,}250 \ Nm$ contribution from the weight of the beam ($\frac{5}{7}$ the mass, acting through the centre of

mass of the beam).

On the other side, the remaining $\frac{2}{7}$ of the beam makes a $100 \ x \ 10 \ x \ 1 \ = \ 1{,}000 \ Nm$ contribution.

Therefore, the counterbalance must make a $(100{,}000 + 6{,}250) - 1{,}000 = 105{,}250 \ Nm$ contribution. As the

counterbalance arm is 2 m long, this requires a weight of $\frac{105{,}250}{2} \ = \ 52{,}625 \ N$ weight, or a mass of 5,263 kg.

The crane's height is a distracter and not needed for this question

Question 12: B

Work out the total energy transferred - 20 x 50W =1,000W of overall power by the 20 strings of lights when on. As W = Js^{-1},
can use the time the lights are on to find the energy used over this time period. 8pm – 6am is 10 hours, so in seconds is 10x60x60
= 36,000s. When multiplying this by the power of all sets of lights, gives the energy used as:

1000 W x 36,000 s = 36,000,000 J of energy, or 36,000kJ. Multiply this by 20 to account for the lights being on for 20 days =
gives 720,000 kJ

As 100 kJ of energy costs 2p, need to do 720,000/100 = 7,200. Multiply this by 2p = 14,400p. Convert to pounds by dividing
by 100 = £144.

Question 13: C

The formula for the sum of internal angles in a regular polygon is given by: $180(n - 2)$, where n is the number of sides of the
polygon.

Thus: $180(n - 2) \ = \ 150 \ x \ n$

$180n - 360 \ = \ 150n$

$3n \ = \ 36$

$n = \ 12$

Each side is 15cm so the perimeter is 12 x 15cm = 180cm.

Question 14: E

For Resistors in parallel, $\frac{1}{R_T} = \frac{R_1 \ x \ R_2 ...}{R_1 + R_2 ...}$

For the first segment: $\frac{1}{R} = \frac{1}{Z} + \frac{1}{Z} = \frac{2}{Z}$

For the third segment: R = Z

For the second segment: $\frac{1}{R} = \frac{1}{Z} + \frac{1}{Z} + \frac{1}{Z} = \frac{3}{Z}$

Thus the total resistance is: $Z + \frac{Z}{2} + \frac{Z}{3} \ = \ 22.$

$\frac{6Z + 3Z + 2Z}{6} \ = \ 22$

$11Z \ = \ 22 \ x \ 6$

$Z \ = \frac{132}{11} \ = \ 12M\Omega$

Question 15: B

The volume of candle burned in 0.5 hour = $0.5 \ x \ (\pi \ x \ 2^2) \ = \ 6cm^{-3}$

$6cm^{-3} = 6 \ x \ 10^{-3} \ m^3$

Since $Density = \frac{mass}{volume}$, in this case $900 \ kgm^{-3} = \frac{mass}{6 \ x \ 10^{-3} \ m^3}$

Thus, Mass burned = $900 \ x \ 6 \ x \ 10^{-3} \ = 5400 \ x \ 10^{-3} kg = 5.4 \ g$

The Mr of $C_{24}H_{52}$ = $12 \ x \ 24 + 52 \ x \ 1 \ = \ 340.$

Thus the number of moles burned $= \frac{5.4}{340} \ = \ 0.016 \ moles.$

Total Energy transferred = $0.016 \ x \ 11{,}000$

$= \ 16 \ x \ 10^{-3} \ x \ 11 \ x \ 10^3 = 11 \ x \ 16$

$= \ 176 \ kJ \ = \ 175{,}000 \ J$

Question 16: C

$250\ kJ\ =\ 25\ x\ 10^4\ J$

$25\ x\ 10^4\ J\ =\ \frac{25\ x10^4}{4.2\ x\ 10^3}\ kCal$

$=\ \frac{250}{4.2}\ \approx\ 60\ kCal$

$2\ Litres\ =\ 2000\ cm^3$

Thus, each cm^3 of water is heated by $\frac{60\ kCal}{2000}\ =\ -30°C$

$Final\ Temperature\ =\ initial\ temperature\ +\ change\ in\ temperature$

$=\ 25\ +\ 30\ =\ 55°C$

Question 17: E

E is the correct sequence. Remember sensory neurone take sensory information to the brain, and motor neurones take information away.

Question 18: F

The information given can only be used to work out the empirical formula. You would need to know the molar mass in order to calculate the chemical formula.

Question 19: A

Don't be afraid of how difficult this initially looks. If you follow the pattern, you get (e-e) which = 0. Anything multiplied by 0 gives zero.

Question 20: B

Intra-thoracic volume must decrease during expiration. Thus, the intercostal muscles relax causing the ribs must move down and in. The diaphragm moves up as well.

Question 21: F

$1 + (3\sqrt{2} – 1)^2 + (3 + \sqrt{2})^2$

$= 1 + (18 – 2(3\sqrt{2}) +1) + (9 + 2(3\sqrt{2}) + 2)$

$= 31 – 6\sqrt{2} + 6\sqrt{2} = \underline{31}$

Question 22: B

The trick in this question is to conserve your units to prevent silly mistakes from creeping in. $200\ cm^{-3}\ =\ 0.2\ dm^{-3}$

$Number\ of\ moles\ =\ concentration\ x\ volume$ so: $0.2\ x\ 1.8\ =\ 0.36\ mol$

Question 23: D

Using $s = ut + 0.5at^2$.

$125 = 0.5x10x\ t^2$

$125 = 5\ t^2$

$t = 5$ seconds

Therefore the final speed is 50ms^{-1}, and the momentum is about 50ms^{-1} x 0.4g = 20kgms^{-1}.

Question 24: E

The most effective method in minimising side effects would be to only target bacteria. Only bacteria have a flagellum.

Question 25: E

Group 6 elements are non-metals whilst group 3 elements are metals. Thus, the group 3 element must lose electrons when it reacts with the group 6 element. The donation of electrons from its outer shell will decrease atomic size.

Question 26: D

It is extremely helpful to draw diagrams to simplify this.

Shaded area = area of circle – area of square

The area of the circle is $\pi r^2 = 3\ x\ 1^2 = 3cm^2$.

We don't know the side length of the square, but we do know the length of the diagonal is 2 cm, splitting the shape into two triangles.

The hypotenuse is therefore = radius x 2 = 2

Using Pythagoras' theorem, $2^2 = x^2 + x^2$ (where x = length and width of square)

Hence $2x^2 = 4$

$x^2 = 2$= the area of the square

Therefore, the shaded area = 3 – 2 = 1cm^2

Question 27: B

The energy stored on descent is equal to the change in gravitational potential energy. The same energy is required to increase the height again, with an excess of the energy needed to lift the passengers. The energy needed to lift the passengers is therefore $= mg\Delta h = 72 \times 10 \times 10 \times 80 = 576kJ$. If the carriage moves at $4ms^{-1}$ for 200m, it takes 50 seconds to ascend. Therefore the rate of energy transfer is $576/50 = 11.52kJs^{-2} = 11.52kW$.

END OF SECTION

Section 3

'Doctors know best and should decide which treatment a patient receives'

Explain what this statement alludes to. Argue to the contrary that patients know best and should be able to choose their management plan. To what extent do you agree with this statement?

The statement describes paternalistic medicine in which it is assumed that since a doctor has been through years of training and experience, they are in the best place to consider the patients best interests and choose the most appropriate management plan. Patient autonomy is one of the core tenants of medicine. Since it is the patient's body, they should be fully informed about the range of treatments available, and their respective side effects, before deciding which treatment to receive.

There is already a power imbalance between doctors and patients, with patients often being in the most vulnerable stage of their lives. There is potential for this imbalance to be abused by some doctors – patients will often blindly follow whatever a doctor tells them simply because they are in the position of power. Paternalistic medicine further distorts this power balance towards doctors while patient autonomy helps the patient reclaim some power.

Even if all doctors are assumed to be benevolent, they are simply acting in the best interests of their patients, to the best of their knowledge. With a paternalistic approach, management is purely based on medical facts and a patient's lifestyle factors are not taken into account. This could have a large impact on the efficacy of management. E.g. Negative side effects leading to poor compliance.

Not only would the involvement of patients in treatment choices lead to better compliance (and therefore efficacy), involving patients in the decision making process may demystify a very complex and foreign topic and help ease any worries or concerns they have, leading to a more holistic, beneficial form of management.

I agree with the statement to the extent that doctors have been trained in medicine and are vastly experienced. However, I believe this experience would be best utilized by presenting the patient with the full range of treatment options in a manner they can understand, together with respective success rates/ side effects so that patients can make an informed decision as to which treatment they would like to receive. With this approach, patient autonomy is maintained, the patient would be able to consider which management plan is most compatible with their lifestyle and at the same time, the experience and knowledge of doctors would be provided at all times to help both parties to come to an informed decision together about the best overall management plan.

'World peace will be achieved in the future'

Explain what this statement means. Argue to the contrary, that world peace will never be achieved. To what extent, if any, do you agree with the statement?

This statement is ambiguous to define. It could mean no wars between nations, militaries and organisations or it could mean a world without murder or violence between humans. I would take world peace to mean an end to armed conflicts between nations and organisations but not necessarily a cessation of violence between individuals as this would be labelled as crime (as the scale of these acts cannot be considered a world problem or event).

Divisive issues will always arise between nations. There have been countless organisations (including the U.N., the league of nations and even the catholic church) that have attempted to bring nations to the negotiating table to work around these issues and to not rely on military force. However, despite their increasing power and influence over time, all have failed to prevent war.

Countries and organisation have spent vast sums of money purchasing arms and cumulative destructive ability has never been higher. Countries and organisations will always do this in order to ensure that their neighbours/rivals are not more powerful than them. To remove the enormous military capability that countries possess individual countries would need to be persuaded to disarm giving their rivals an advantage and removing the mutually assured destruction that underlies most of the stability in the modern day. Even if worldwide disarmament were achieved it would be difficult to maintain the world in this unstable state as rival countries or organisations could restart even minor arms races.

Man is intrinsically violent. Evolution has crafted man to identify any threats, whether from other species or other humans and to be able to react to these with force in order to survive. On the other hand there has been a steady drop in violence over the last few centuries which shows no sign of reversing in the future. This is true for both wars as well as crimes, it is easy to mislead by the raw data which is confounded by the huge increase in population over the last few hundred years but it is undoubtedly safer to be alive now than at any other time in history.
The world is becoming increasingly interconnected both in terms of economics and populations so wars have more impact now than ever. This provides a large disincentive for conflicts to arise that will only increase as human society integrates further.
As education increases the use of violence will no longer be accepted as a method conflict resolution.

Gains have been made in reducing conflict due to the interconnectivity of world trade and indeed its differing societies. Furthermore there are many powerful organisations that are designed to prevent conflict from taking place.
On the other hand these organisations have all failed to prevent wars. Furthermore disarming nations could actually lead to a decrease in stability as in provides an incentive for any nation to resist de-armament as doing so gives them a huge advantage over their rivals. Thus in conclusion I would argue that world peace is possible but unlikely.

'Medicine is a science; not an art' **Explain what this statement means. Argue to the contrary that medicine is in fact an art using examples to illustrate your answer. To what extent, if any, is medicine a science?**

The statement suggests that medicine like science, is objective and quantifiable; a patient's symptoms and observations can be 'measured' to hypothesise a diagnosis which can then be treated empirically, in essence, evidence based medicine, rather than like an art, which is more subjective.

However, many would argue medicine is still an art. Patients don't behave like textbook examples, with the multiplicity of pathophysiological mechanisms meaning there are several different ways to arrive at the same disease, and the same disease manifesting in several different ways. This would be too complex to analyse using a simple measurement based approach and so a certain amount of intuition and creativity is needed when approaching patients.

Furthermore, pathophysiology is only half of medicine, with the actual interaction with the patient being as, if not more important. While there may be algorithms for communications skills, there is still no real science behind people skills and good bedside manner. To see how medicine could be considered an art, you simply have to watch a skilled physician communicate effortlessly with a patient.

Art can be used to describe any skill, which takes time and patience to master – many clinical skills easily fit into these criteria. The most extreme example that could be considered is a plastic surgeon, someone who is to all extents and purposes an artist with the human body as their canvass.

Despite this, I do still believe medicine is to a large extent still a science, and increasingly so. In every patient interaction, a clinician uses their underlying understanding of the first principles of science to come up with a hypothesis of a patients illness, which is then tested with further investigations before being managed empirically with evidence based medicine; the entire encounter is an illustration of the scientific process. Furthermore, management guidelines from NICE and other organisations are increasingly removing the intuitive aspect of management previously seen in medicine, and increasingly so as numerous studies continue to prove both efficacy and patient safety is improved with the use of guideline based medicine over intuitive medicine.

'People should live healthier lives to reduce the financial burden of healthcare to the taxpayer.' **Explain what this statement means. Argue to the contrary. To what extent do you agree with the statement?**

The statement basically argues that by providing a financial encouragement to the individual to live healthy, the overall situation of public health will change. This is supposed to have a positive effect on the overall balance sheet of health care as financial encouragement is still cheaper than the treatment of disease. The thought is therefore that it is justified to pay money to reduce the occurrence of disease overall, at least in areas where disease can be prevented or reduced in incidence by mere healthy living. As statistically speaking this represents an increasingly great amount of overall burden on medical facilities, the argument does seem to have some validity.

There are some problems that arise from phrases like the one above. This lends itself well for counter arguments. One argument is the question, if it is acceptable to essentially influence the individual's free will and right to make their own decisions by providing them with an active encouragement to alter their behaviour. In a way this can be compared to training animals in the form that they receive a treat every time they do what their owner sees as positive and desirable. Another issue arises from the fact that only some diseases can be prevented by healthier life-style choices and that there is by no means a guarantee that healthy living will keep you free of disease.

Even if we ignore the counter arguments mentioned above, there are some issues with ideas such as these, the most obvious being the question of regulation. How can we monitor whether an individual is actually making healthy life-style choices and therefore is deserving of the promised bonus? Monitoring through medical tests will produce both, administrative as well as purely operative costs which will have to be taken into consideration when arguing that a bonus system will save money. Detailed monitoring of the individual's behaviour on the other hand is extremely difficult and most importantly raises the critical question of the individual's right of privacy from the government. Concerning the impact on the provision of healthcare, a financial bonus system might therefore reduce the overall cost of life-style associated diseases, but will cause a myriad of other costs and necessary medical processes that will make its effectiveness questionable.

All in all a financial bonus system seems like it might have a positive effect on the overall expenditure on healthcare, but upon closer inspection it becomes clear that the issue is a very complex one, this in turn makes its usefulness as a tool for the conservation of resources all the more questionable. Whilst it is a good idea to encourage the individual to life healthy and to do their best to avoid diseases that can be avoided, a system relying on financial encouragement might not be the right path to success.

END OF PAPER

Mock Paper F Answers

Section 1

Question 1: D

Answer C) is completely irrelevant, so is not a flaw. Answer B) is not a flaw because when assessing an argument, anything that is stated (i.e. not concluded from other reasons in the passage) is accepted as true. We do not require evidence or sources for any statistics presented. Answers A) and E) are both claiming that something is immoral, which is thus expressing an opinion on the part of the arguer. This is not a flaw, the arguer is at liberty to claim something is immoral, and to claim that the government is morally obliged to act, and that it has not done so. Also, E) claims that *arguably* this is the most outrageous flaw of the government, clearly expressing an opinion, which is thus not required to be supported. However, answer D) identifies a valid flaw. The argument rests on us accepting that if there were less uninsured drivers, there would be less crashes. This is not necessarily correct, so D) is a flaw in the passage.

Question 2: A

The sentence 'Thus, the situation in Brazil is not applicable to the UK, and legalising gun ownership in the UK would be a bad move' gives the main conclusion of the argument and this is summarised in Answer A). Answer B) is partially supported by the passage, but the main conclusion concerns the situation in the UK and the passage states that there is little black market in the UK. Answer C) is incorrect as the passage only talks about gun ownership, not violent crime more generally. Answer D) is not fully supported by the passage, which states only that legalising guns would result in it being *easier* for criminals to acquire guns, not that there would be a large increase in their number. Answer E) is not the main conclusion as it focuses on an aspect of the evidence from Brazil, rather than the main conclusion which focuses on gun legislation in the UK.

Question 3:C

For each of the walls where there is no door, the wall is 6 tiles high and 5 tiles wide, which is 30 tiles. The wall where the door is requires a row of 2 tiles above the door, then there is a width of wall of 120cm which requires completely tiling, which is 6 tiles high and 3 tiles wide, hence this wall requires a total of 20 tiles. Hence a total of 110 tiles are required for the walls. The floor is 2 metres by 2 metres, so 5 tiles by 5 tiles, hence 25 tiles are required for the floor. Hence the answer is 135.

Question 4: D

Answer E) is irrelevant to which of Trevor and Jane will arrive first, so does not weaken the conclusion. Answers A), B) and C) all strengthen the answer, giving further reasons why we might expect Trevor to arrive first. Answer D), however, would slow Trevor down, meaning that it was more likely that Jane would arrive first. Thus, Answer D) weakens the passage's conclusion, and hence Answer D) is the answer.

Question 5: A

He has enough butter to make 2.5 times as many cupcakes as the recipe, which is 50
He has enough sugar to make 3 times as many cupcakes as the recipe, which is 60
He has enough flour to make 5 times as many cupcakes as the recipe, which is 100
He has enough eggs to make 34 times as many cupcakes as the recipe, which is 60
The lowest of these is 50, so he makes 50 cupcakes. He needs 2.5 x 4 eggs to do this, which is 10 eggs. Therefore he has 2 eggs left over.

Question 6: B

Let the number of minutes the journey takes be t. Therefore, ABC charge 400+15t pence for the journey. We can calculate that XYZ taxis charge 400+(30x6) pence, = 580 pence. Therefore, for both journeys to cost the same, 580=400+15t. 180=15t, therefore t=12. Therefore the 6 mile journey needs to take 12 minutes. 6 miles in 12 minutes is 30 miles per hour, so the answer is B.

Question 7: E

We can see that all of answers A) through D) are essential for the conclusion to be valid from the squire's reasoning. Lancelot must have great courage, this must be a requirement for the Adzol, and no other knights must have sufficient courage, in order for us to be certain that Lancelot will succeed but all of Arthur's other knights will fail. Thus A) and B) can be clearly identified as assumptions. C) and D) require a bit more thought, but we can see that nothing in the passage explicitly states the Elders' tales are correct. If the elders are not correct, then great courage may not be required to be successful in the Adzol. Thus, both C) and D) are also assumptions. Hence, the answer is E).

Question 8: E

B) is incorrect, as the passage does not say that arch-shaped gaps *always* indicate where windows once stood, simply that *these arches* do. C) is also incorrect, as the passage simply states that windows are not found in *underground halls*. A) is a reason in the passage, and is not a conclusion. D) and E) could both be described as conclusions from this passage, but we see that if we accept D) as true (along with the fact that the hall is now underground), we have good reason to believe that E) is true, whereas E) being true does not necessarily mean that D) is true. Thus, E) is the *main* conclusion, whilst D) is an *intermediate conclusion*, which supports the main conclusion.

Question 9: D

Usually bread rolls cost 30p for a pack, but if the cost per bread roll is reduced by 1p then they will cost 24p. Hence we need to find z, where $24(z+1)=30z$, where z is the original number of packs that could have been afforded. $24z+24=30z$, hence $24=6z$, so $z=4$. Hence he was originally supposed to be buying 4 packets of bread rolls, which is 6 x 4 = 24 rolls.

Question 10: B

Answer E) is an irrelevant statement that says nothing about whether England *do* have good players. Answers A) and D) actually weaken the sporting director's arguments, suggesting that England may have a good team, and it may just be poor performances in world cups, and not a lack of talented players. This leaves B) and C). C) may appear to strengthen the sporting director's argument, but on closer inspection we see that in fact it says that for the last 70 years, England have had at least 1 player in the top 10 in the world. This does *not* strengthen the argument that England have been lacking talent for the last 25 years, and may actually reinforce the chairman's argument that it is simply the *current* crop of players that are not good enough. Answer B), however, does strengthen the argument, suggesting that England's performances have been poor over the last 20 years, thus strengthening the argument that there may be a lack of talented players that has been ongoing for a couple of decades, as claimed by the sporting director.

Question 11: A

He can prepare each batch of cakes while the previous one is in the oven but it takes longer so we have to allow 25 minutes for each batch, plus 20 minutes for the last batch to cook while no further batch is being prepared. There are 12 in each batch, so for 100 cupcakes there needs to be 9 batches. Hence the total time needed is 25 minutes x 9, + 20 minutes. This is 245 minutes, or 4 hours 5 minutes. Hence to be ready by 4pm he needs to start at 11:55am, so the answer is A.

Question 12: D

We can first work out the rate of girls' absenteeism. First we need to work out how many of the pupils at Heather Park Academy and Holland Wood Comprehensive are girls. Let g be the number of girls in Heather Park Academy. Then $0.06(g)+0.05(1000-g)=(1000)(0.056)$. Then $0.06g-0.05g=56-50$. Then $0.01g=6$, so g = 600. Hence 600 pupils at Heather Park Academy are girls. The proportions at Holland Wood Comprehensive are the same but there are half as many pupils, so 900 pupils at the two schools combined are girls.

The average absenteeism of girls is 7%. We know that 900 of the 1100 girls have an average absenteeism rate of 6%. Let the average absenteeism rate of girls at Hurlington Academy be r. Then $900 \times 0.06 +200r = 0.07 \times 1100$. Hence $54+200r=77$. $77-54 = 200r$. $23/200 = r$. $r=0.115$. Hence, the rate of absenteeism amongst girls at Hurlington Academy is 11.5%

Question 13: C

A), B) D) and E) are all directly stated in the passage, so can all be reliably concluded. Perhaps the trickiest of these to see is answer D), which is true because the passage says "*due to*" the advent of more accurate technology, thus clearly identifying that this had *caused* the switch to the situation of most watches being made by machine. C), however, is *not* necessarily true. The passage states that most *watches* are produced by machines, but only states that *some* watchmakers now only perform repairs. This does not necessarily mean that most watchmakers do not produce watches. It could be that only a handful are required in the entirety of the watch industry for repairs, and that the numbers still producing watches exceeds those in the repair business. Thus, C) cannot be reliably concluded from the passage.

Question 14: A

B) is not a valid conclusion from the passage, because the fact that someone uses an illogical argument (as some pescatarians are claimed to in this passage) does not mean that they cannot use logic. D) and E) are not conclusions from this passage because the passage is not saying anything about the ethicality of eating meat, but simply commenting that one argument used against doing so is not logical. Answers C) and A) are both valid conclusions from the passage, but we see that if we accept C) as being true, it gives us good cause to believe that A) is true, but this does not apply the other way round. Thus, C) is an intermediation conclusion, whilst A) is the main conclusion.

Question 15: E

The research conducted does not ask about whether it is *important* to learn some of the language before travelling abroad, simply whether participants *would*, so B) cannot be concluded. D) is incorrect because the passage states *15%* would, which is clearly not less than 10%. The passage states that this is symptomatic of a deeper underlying issue, but does not say that many issues of racism stem from this, so C) cannot be concluded. Now, the passage states that 60% of people feel foreign people should learn English before travelling to Britain, and 15% of people would attempt to learn the language before travelling to a country which did not speak English. However, this 15% could be some of the same people as the 60%, in which case A) would be incorrect. Thus, A) cannot be reliably concluded. However, there must be at least 45% of people who feel that foreign people should learn English, but would not learn a foreign language themselves, so E) *can* be reliably concluded.

Question 16: D

She needs to print 400 x 2 = 800 double sided A4 sheets, which will cost 0.01 x 2 x 1.5 = £0.03 each. Hence the total cost of this is 800 x 0.03 = £24. She also needs to print 1500 single sided A5 sheets, costing £0.01 each, giving a total of 1500 x 0.01 = £15. Hence the total cost is £39.

Question 17: B

The passage has stated that if Kirkleatham win the game they will win the league, so E) is not an assumption. Meanwhile, the manager has stated A), C) and D), and the passage has not claimed anything about whether Kirkleatham can easily win the game, so A) and D) are not assumptions. However, B) does identify an assumption in the passage. The fact that Kirkleatham will not win the game without playing with desire and commitment does *not* mean that they will win the game if they do play with desire and commitment. And we can see that for the argument's conclusion (that Kirkleatham *will* definitely win the league) to be valid from its reasoning, this is required to be true. Thus, B) identifies an assumption in the passage.

Question 18: B

Answers A) and E) are not relevant, because neither affect the strength of the councillor's argument from a critical thinking point of view. The councillor's argument says nothing about house prices, simply the cost of building the estate and the effects on wildlife, so A) is not relevant. E) is not relevant because additional support, or likelihood that it will be heeded, does nothing to affect the strength of a given argument. C) and D) actually strengthen the councillor's argument, suggesting that brownfield land does have good infrastructure (C)) and that the greenbelt areas do have a lot of wildlife (D)). B) does weaken the councillor's argument, as it suggests that building on brownfield land may also have adverse impacts on wildlife.

Question 19: A

England won Pool C so they will be in Quarterfinal 3, where they will play Brazil. If they win, they will play the winner of Quarterfinal 1. Hence they can only meet teams from Quarterfinals 2 or 4 in the final. These teams are Argentina, Nigeria, South Africa or Holland. Hence the only one of these 5 teams they can play in the final is Nigeria.

Question 20: B

We can tell the amounts for the green party and the blue party are both 1/3 of the total, and that the amount for the red party is 1/4 of the total. Hence 1/12 is left, so the amount for the yellow party must be 1/12. Hence the red party have 3 times the intended vote of the yellow party.

Question 21: B

A large pizza wish mushrooms and ham is £12, garlic bread is £3, chips are £1.50 x 2 = £3, a dip is £1, hence the current total is £19. The cheapest way to order this is to get the price up to exactly £30 as this will reduce the price to £18. This takes £11. Only one of these options costs £11, which is a large pizza with mushroom. Hence the answer is B.

Question 22: E

In Rovers' first 3 games, they have scored 1 goal and had 8 goals scored against them. In total they scored 1 goal and had 10 goals scored against them, so they must have lost their last game against United 2-0.
In City's first 3 games, they scored 7 goals and had 3 goals scored against them. In total they scored 10 goals and had 4 goals scored against them. Hence they must have won their game against United 3-1. Hence the answer is E.

Question 23: D

The question simply describes how a combination of factors are responsible for the M1 being the world's most formidable tank, so the view of country X is incorrect. It does *not* claim that it is impossible for a tank to be as good as the M1 Abrams, so E) is not a valid conclusion. Equally, it does not say the new tank's armour will not be as good as the Abrams (in fact it is implied that it may well be as good), so C) is also incorrect. A) B) and D) are all valid conclusions from this passage, but we can see that A) and B) contribute towards supporting the conclusion in D). Thus, D) is the main conclusion of this passage, whereas A) and B) are *intermediate* conclusions given to support this main conclusion.

Question 24: D

We can simply add up the amounts in the bank accounts and find the difference between each month – it doesn't matter that the salary is paid in as it is the same every month. Doing this, we find out the biggest difference is between 1st May and 1st June, hence the answer is D, May.

Question 25 : B

Firstly we can work out the full dose for the son. He needs to take 0.2ml per kg of weight for each dose, and he is 40kg, so this is 8ml. He takes 40 doses altogether, so in total he needs 320ml of medicine.

Then we work out the doses for everyone else, add them together and half them. The daughter's full dose would be 2ml, 30 times, which is 60ml altogether. My dose would be 7.5ml, 60 times, which is 450ml. My husband's dose would be 8ml, 60 times, which is 480ml. Altogether, this is 990ml. However, only half of these dosages is needed, which is 495ml. Hence the total needed is 320 + 495, which is 815ml. Hence 5 200ml bottles of medicine are needed for the full course.

Question 26: A

There are 21 forks and 21 knives. If half as many are red as blue, and half as many are blue as yellow, they are in the ratio red:blue:yellow 1:2:4. Hence of the 21, 3 are red, 6 are blue and 12 are yellow. Hence the probability of getting a yellow knife is 12/21 = 4/7. The probability of getting a red fork is 3/21 = 1/7. Hence the probability of getting both is (1/7) x (4/7) = 4/49.

Question 27: D

The Principle used in the passage is that public funds raised through taxation (which is compulsory) should not be used for any services unless they benefit everyone, such that nobody is forced to pay for services that do not benefit them. Answer D) is the best application of this principle, as it directly follows it. Answer B) mentions public funds being used to support a service that benefits the whole country, but this does not necessarily mean that they *shouldn't* be used to support services that don't benefit everyone, so answer B) is not as directly an application of the principle as answer D). Answer E) is not the same principle because this is talking about funds being used for services that benefit the *country*, rather than everyone in it. Meanwhile, Answer A) is talking about how many people *use* a certain service, rather than how many people *benefit* from it, so this is not the same principle. Answer C) is completely different talking about funds being used because some cannot afford private health service, regardless of how many people are benefitting from the public health service.

Question 28: B

The chairman has *stated* that All inclusive services are more popular than Hourly services. He has not deduced this from any evidence, and thus he has assumed nothing about their popularity. Thus, C) and D) are incorrect. The chairman's argument is simply that focusing on All inclusive services will bring in more profit than Hourly services, as he says they should focus on All inclusive *rather* than Hourly services. Thus, any reference to other services or other profit-raising strategies are irrelevant, so A) and E) are irrelevant. However, B) correctly identifies the chairman's flaw. Just because All inclusive is more *popular* than Hourly services does not mean they are more *profitable*, and if they are not then the chairman's conclusion is no longer valid. Thus, B) correctly identifies the flaw in his argument.

Question 29: A

B) is not an assumption because the passage *states* that renewable sources do not cause damage, so we accept this as true. E) is not a flaw, because again the passage has stated that the use of these fuels to produce power will continue to cause climate change *as long as it continues*, thus we must accept as true that it cannot be halted or prevented whilst these fuels are used. C) and D) are irrelevant to the argument's conclusion that if we wish to stop damage to the environment, we need to switch to renewable fuels, and thus they are not flaws. However, at no point is it stated that *all* non-renewable fuel sources cause environmental damage, it is only stated the non-renewables *such as* oil, coal and natural gas do. Thus, we have no guarantee that Nuclear fuel will cause environmental damage, and if it doesn't, the passage's conclusion no longer stands. Thus, A) is a valid assumption in the passage.

Question 30: E

Answers A) and D) do *not* strengthen or weaken the argument because the question states that this increase in non-vaccinated individuals has occurred despite powerful evidence of vaccine safety, and in spite of advice from doctors. This suggests that people who do not vaccinate pay little attention to evidence of advice from doctors, so we should not expect these factors to have much of an effect. B) is completely irrelevant to whether the rate of increase will continue. C) actually weakens the argument, suggesting that such increases are common, and normally stop after 6 years. If the current increase was to follow suit, it would stop next year and vaccination rates would not fall below 90%. E), however, implies that this kind of increase has happened only once before, and in this case it continued for 13 years. If the current increase was to follow *this* pattern, it would continue for another 8 years, where vaccination rates would be below 90%. Thus, E) strengthens the argument's conclusion that we should expect an outbreak of measles.

Question 31: A

Answers B) and E) are both in contradiction with stated points in the question, which states that there is now powerful evidence human bodies *are* set up for long-distance running, and that it is well established that humans evolved in Africa. Answer C) is irrelevant, because the presence of other theories does *not* necessarily affect whether we should believe this theory based on the new evidence. Answer D) actually strengthens the argument, suggesting that the new evidence does provide powerful reasons to believe this theory. Answer A) however is a valid flaw, that evidence supporting a theory does not necessarily *prove* that it is true. Thus, the answer is A).

Question 32: C

The percentage of students who had their grades predicted correctly is the same as the number who had their grades predicted correctly as there are 100. Hence we simply need to add up the numbers on the diagonal of the table, where actual grade is the same as predicted. This adds up to 39, hence the answer is C.

Question 33: B

The 2 wage reductions mean that when the wage increases happen, the raises will be x% of a smaller number than the decreases were. Thus, the wage will not rise as high as the original level.

If you are struggling to visualise this, the easiest way to do it is to substitute a number for x. Let us do the calculation treating x as 10. The first wage drop is by 10%. Thus, the wage is now 90% of the original wage.

The second wage drop is also by 10%, but at this point, the wage is only 90% of the original wage. Thus, the drop will be by 10% *of 90%* of the original wage, resulting in a new wage of 81% of the original wage (10% of 90% is 9%)

Then, we have the first increase, which will be 10% of this new wage (81% of the original wage). Thus, after the first increase, the new wage will be 81% +(10% of 81%) of the original wage. Thus, it will be 81% + 8.1%, which is 89.1% of the original wage.

Now we have the second increase. Another 10% is added, this time of 89.1% of the original wage. We now have an increase of 8.91% (10% of 89.1). Thus, after the final raise, the wage will be 89.1% + 8.91%, which is 98.01% of the original wage. Thus, the new wage is lower than the original wage.

Question 34: C

The passage says nothing about whether it is more important to have good doctor-patient relations than scientific progress, so answer E) is not a valid conclusion. Answer D) is a direct argument against the question, which claims the naming system *should* be changed, so answer D) is not a conclusion. The passage states that the confusing system causes problems in scientific literature, but this does *not* necessarily mean that changing it would allow faster progress in scientific research, so answer B) is not a valid conclusion. Answers A) and C) are both valid points from the passage, but we can see that Answer A) is a *reason* stated in the passage, which helps support the statement in C). Thus, Answer C) is the main conclusion.

Question 35: E

180 people applied in total. 128 did French, 64 did Spanish, 48 did neither French nor Spanish. We hence know that exactly 132 people did either French, Spanish or both. Since 128 people did French, only 4 people can have done Spanish without doing French. Hence 60 people must have done both French and Spanish.

END OF SECTION

Section 2

Question 1: G

The replacement of dying, damaged, and lost cells, the growth of the embryonic cell to a multi-cellular organism, and asexual reproduction are the three main reasons why cells divide through mitosis.

Question 2; F

This question will discriminate between students who spot short-cuts built into questions to save valuable time and those that simply dive straight in without appraising the question.

The key here is that due to the conservation of energy, all the gravitational potential energy, mgh, at the top of the ramp will be converted to kinetic energy, $\frac{1}{2}mv^2$, at the bottom.

Thus, we can calculate the final velocity using the following: $mgh = \frac{1}{2}mv^2$

Note that the mass cancels so there is no need to use the density and volume information in order to calculate mass. Hence we get: $2gh = v^2$

$V^2 = 2 \times 10 \times 20 = 400$

Therefore, $v = 20$ ms^{-1}

Question 3: E

Blood pressure in the aorta is the highest of any vessel in the body, as blood has just been ejected from the left ventricle to go to the body. The pressure in the left ventricle (and hence the Aorta) is higher than that in the right ventricle (and hence the Pulmonary Artery) because the pressure must be sufficient to pump to the entire body, rather than just the lungs.

Question 4: D

Waves do not transfer mass, but their net neutral motions can interfere with each other to cause standing waves or other interference patterns. The energy of a wave depends on frequency, so waves have many different energies. Gamma rays have the highest energy for light, while visible light is lower in energy.

Question 5: C

A sensory receptor (1) senses the heat of the pan. This information is passed down the sensory neurone (2) through a relay neurone to the motor neurone (4), which then causes the muscle (5) to contract, pulling the finger away.

Question 6: C

The receptor is directly coupled to the sensory neurone, so the communication here is electrical. All information between neurones passes via synapses, which use neurotransmitters to convey the information chemically. This occurs between the sensory neurone and the relay neurone, and between the relay neurone and the motor neurone. Therefore, the answer is C).

Question 7: A

This is an example of an addition reaction: the chloride and hydrogen atoms are added at the unsaturated bond of the but-2-ene, which is between the 2nd and the 3rd C-atom. If you're unsure about this type of question draw it out and the answer will be obvious.

Question 8: A

Multiply by the denominator to give: $(7x + 10) = (3z^2 + 2)(9x + 5)$

Partially expand brackets on right side: $(7x + 10) = 9x(3z^2 + 2) + 5(3z^2 + 2)$

Take x terms across to left side: $7x - 9x(3z^2 + 2) = 5(3z^2 + 2) - 10$

Take x outside the brackets: $x[7 - 9(3z^2 + 2)] = 5(3z^2 + 2) - 10$

Thus: $x = \frac{5(3z^2 + 2) - 10}{7 - 9(3z^2 + 2)}$

Simplify to give: $x = \frac{(15z^2)}{[7 - 9(3z^2 + 2)]}$

Question 9: C

The electrolysis reaction for brine is: $2\,NaCl + 2\,H_2O = 2\,NaOH + H_2 + Cl_2$

Thus, keeping in mind the stoichiometry of the given equation, the solution must be C.

Question 10: B

An alpha particle is a helium nucleus consisting of 2 protons and 2 neutrons. An alpha decay therefore reduces the atomic (proton) number by 2 and the mass number by 4. After a single alpha decay, the resulting proton number is 88 and the resulting mass number is 184. As this then splits in to two, the resulting element has a proton number of 44 and a mass number of 92. Gamma radiation does not alter the subatomic particle make-up of an atom.

Question 11: A

If the two isotopes were in equal abundance, the A_r would be 77, half-way between the two isotope masses (the average). The A_r is 76.5 (a weighted average), one quarter of the way between the isotopes, so there must be three times as much of the lighter isotope to move the A_r closer to its mass of 76 (0.75x76 + 0.25x78 = 76.5).

Though there is more of ^{76}X than ^{78}X, this does not necessarily imply that ^{78}X is lost through decay, as opposed to naturally less abundant from the beginning, so there is no way to know the relative stability of the isotopes.

Question 12: C

Increasing the concentration of the reactants (not products) would affect reaction rate, which can be monitored by measuring the gas volume released (proportional to molar concentration). This is the reaction for photosynthesis, which does not occur spontaneously and is endothermic.

Question 13: H

Most polymers are made up of alkenes, which are unsaturated molecules. Polymerisation does not release water, as it is an addition reaction. Depending on the monomer molecule, polymers can take a variety of shapes.

Question 14: B

The shortest distance between points A and B is a direct line. Using Pythagoras:

The diagonal of a sports field = $\sqrt{40^2 + 30^2} = \sqrt{1,600 + 900} = \sqrt{2,500} = 50$.

The diagonal between the sports fields = $\sqrt{4^2 + 3^2} = \sqrt{16 + 9} = \sqrt{25} = 5$.

Thus, the shortest distance between A and B = $50 + 5 + 50 = 105\ m$.

Question 15: C

Let $y = 1.25 \times 10^8$; this is not necessary, but helpful, as the question can then be expressed as: $\frac{100y + 10y}{2y} = \frac{110y}{2y} = 55$

Question 16: D

Taking the diseased allele to be X^D and X as the normal allele, we can model the scenario in the Punnett square below:

		Carrier Mother	
		X^D	X
Diseased	X^D	$X^D X^D$	$X^D X$
Father	Y	$X^D Y$	XY

Boys are XY and girls are XX. 50% of the boys produced would have DMD. So the probability that both boys would have the disease is 0.5 x 0.5 = 0.25

Question 17: E

We can see from the Punnett square that the probability of having a girl with DMD is 25% ($X^D X^D$). The probability that both are girls with DMD is 0.25 x 0.25 = 0.125.

Question 18: C

Chemical reactions take place in the cytoplasm, and the mitochondrion is the site for aerobic respiration releasing energy. The lack of a cell wall means that this is an animal cell.

Question 19: A

Equate y to give:

$2x - 1 = x^2 - 1$

$\rightarrow x^2 - 2x = 0$

$\rightarrow x(x - 2) = 0$

Thus, x = 2 and x = 0

There is no need to substitute back to get the y values as only option A satisfies the x values.

Question 20: B

The ruler and the cruise ship look to be the same size because their edges are in line with Tim's line of sight. His eyes form the apex of two similar triangles. All the sides of two similar triangles are in the same ratio since the angles are the same, therefore:

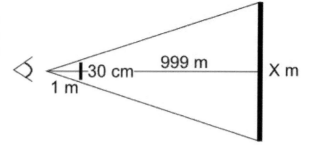

$$\frac{0.3m}{X\ m} = \frac{1\ m}{1\ m + 999\ m}$$

Thus, $X\ m = 1000\ m \times \frac{0.3\ m}{1\ m}$

$1000 \times 0.3 = 300\ m$

Question 21: A

As energy is added to ice, the molecules increase their vibrations and the temperature increases. As the ice begins to melt, all energy goes into breaking the bonds to form water, and none goes to increasing the temperature. Once all bonds are broken, the energy again goes to increasing the temperature of the water.

Question 22: D

White blood cells can engulf/phagocytose pathogens in order to kill them. CO_2 is transported in the plasma, not in blood cells.

Question 23: B

The Doppler Effect applies to all waves including members of the electromagnetic spectrum. A wave emitted from a moving object, like the sound from a siren, will be compressed as it moves toward you, causing sounds or light of higher frequency (pitch, energy). When an ambulance drives toward you, the siren will become higher in pitch, as it drives past it will move neither towards or away from you, so there will be no Doppler Effect, then the Doppler Effect will cause a lower pitch as it drives away and the waves are stretched to longer wavelengths.

Question 24: C

Power= Current x Voltage

Thus, one battery generates: 1.2 V x 2500 mAh

= 1.2 V x 2.5 Ah = 3 Watt hours

The light uses: 30 W x 1 h = 30 Wh

Therefore, it will take 30 Wh / 3 Wh = 10 batteries to power the light for one hour.

Question 25: C

Electric conduction is a consequence of metallic bonding: metal atoms lose their valence electrons to obtain their optimum energy state, with the cations forming a lattice held together by electromagnetic attraction to the cloud of free electrons. These free electrons can then conduct electricity, as they are not bound to any particular atom. This does not require the movement of ions or the breaking of the cation lattice.

Question 26: D

The distance travelled is the area under the curve (v x t = d at every v and t, sum for each t to find the total d, which also is the area; think of the case for constant v if confused).

Each square corresponds to 1 m (1 m/s x 1 s = 1 m), so counting squares gives an approximate distance of 30 m travelled: 31 m in a positive direction and 1 m in the direction (negative velocity).

Question 27: C

Let Bob = B, Kerry = K and Son = S.

$B = 2K, K = 3S$ and $B + K + S = 50$

$50 = 2K + K + \frac{K}{3} = \frac{6K}{3} + \frac{3K}{3} + \frac{K}{3}$

$50 = \frac{10K}{3}$

Hence: $10K = 150$

$K = 15$

$B = 2 \times 15 = 30$

$S = \frac{15}{3} = 5$

So: Bob's age when his son was born = 30 – 5= 25.

Section 3

'A doctor should never disclose medical information about his patients'. **What does this statement mean? Argue to the contrary using examples to strengthen your response. To what extent do you agree with this statement?**

This statement addresses the issue of patient confidentiality, saying that there are no circumstance in which a doctor should reveal information about a patient's health to a third-party (be it friends, family, or stranger) without the patients informed knowledge and consent. For the purpose of this essay, I will assume that this excludes other medical staff who may also be involved in the patients care and argue that there are several situations in which it is valid to disclose patient information.

An important distinction has to be made between instances in which a doctor is legally required to disclose medical information and instances when a doctor may choose to disclose medical information about a patient.

There are several instances in which it is legally required for doctors to disclose patient information, regardless of patient consent. For example, there is a list of notifiable diseases including meningitis and tuberculosis that have to be reported in the interest of infection control and patient safety. Other legal requirements include information about road traffic accidents, potential terrorist activities, and coroner requests or in the court of law.

There are also other instances in which a doctor may be legally permitted to disclose information but need not necessarily do so. These tend to have their own host of ethical challenges and ultimately reside in the discretion of the doctor. For example, if a doctor suspects that a patient may have mental health issues or has mentioned something suggesting they may harm themselves or others, a doctor is legally permitted to bring it up with the relevant authorities, if they feel it is relevant. Another example may be cases in which a doctor suspects potential child abuse, and the doctor may choose to bring this up further with the authorities without the parents consent.

To a large extent, I do agree with the statement as patient confidentiality is one of the core tenants of medicine; patients place a huge amount of trust in doctors during one of the most vulnerable points of their life and this responsibility needs to be respected. However, as highlighted above, there are several cases in which a doctor is legally required/ permitted to disclose information about patients, and these tend to be in cases where disclosure results in greater good overall (to either the patient or the public) than patient confidentiality.

With an ageing population, it's necessary to increase the individual's contribution to the healthcare system in order to maintain standards.'. **Explain what this statement means. Argue to the contrary. To what extent do you agree with the statement?**

Due to generally decreasing numbers of births and a constantly increasing longevity, our population grows increasingly old. As people live longer and there are less people born, the percentage of elderly individuals in the population grows in relation to the percentage of younger individuals that work and finance health care through their taxes. With increasing age, medical needs of the individual increase as well. Statistically speaking, the biggest need for medical attention and therefore the biggest consumption of healthcare assets happens in the last third of the individual's life. In order to maintain the current standard of support of the increasing amounts of elderly, it becomes necessary to spend more money on healthcare which in turn makes higher contributions of the individual necessary. It is important to recognise that when the NHS was established in 1948 the structure of society as a whole was a completely different one compared to what it is today.

There are several issues that arise from this statement that lend themselves well for counterarguments. One of the issues is the general structure of our healthcare system. This does not aim at the idea of free health care for all, but more at the idea of individual processes being unnecessarily complicated and inefficient consuming disproportional amounts of resources. Streamlining the structure of the NHS as well as the procedures in place can provide means of reducing overall expenditure. One other valid counterargument lies in the principle idea of the NHS to put an end to health care inequalities based on the individuals income. By increasing the individual contribution without adjustment of wages etc, a disproportionally heavy burden is placed on the less well of in comparison to the rich which essentially re-instates healthcare inequality on the basis of income.

There is this general idea of a responsibility between generations. Our parents and grandparents worked their whole life building the society we live in today. In a way we harvest the fruits of their labour today and for this reason it is our responsibility to care for the generations that came before us. It is therefore our duty to care for them to the best of our abilities. Issues such as the ones our healthcare system is facing today, make it increasingly difficult to fulfil this responsibility, especially with chronic morbidity becoming increasingly common, even in younger generations with diseases such as diabetes and metabolic syndrome affecting increasingly individuals in the second third of their life. Due to this increase in overall workload for the healthcare system, it seems unavoidable that the individuals contribution must be increased in order to maintain the same standard of care delivered today, unless the disease patterns of the population change.

'Science is nothing more than just a thought process'

Explain what this statement means. Argue to the contrary, that science is much more than just a thought process. To what extent, if any, do you agree with the statement?

Science is obviously driven by the interpretation of information and the subsequent generation of hypothesis, this is what is referred to by the thought process in the title. *A precise definition of the meaning of the question is essential in directing subsequent discussion.*

However these observations are built on experimentation and observation that do not always require a specific thought process behind them. *Signpost the topics you are going to raise later.*

In addition to this we must consider that if something is discovered unintentionally is it still a scientific discovery. Is it important that the discovery advances Science or that it is made in a scientific manner.

Some aspects of science can be performed completely in the mind, mathematics and theoretical physics rely on insight generated by thinking and applying core axioms on which they are founded. *What is there in support of the statement, don't back away from offering strong arguments as they will be of value when coming to a conclusion.*

The scientific method is the generation and testing of hypothesis with the intention of proving them to be false, this is the basis of all science and it is very clearly a thought process, all of the experimentation done is directed by this.

There is a need for investigation in some areas where we have little knowledge and what would ideally be a spearhead of inquiry is often a broad net of hope. A good example of this is Genome Wide Association Studies (GWAS), these look for associations between gene variants and disease yet despite this there is no thought process behind this deeper than the need to generate information and provide new target genes for research.

In the same manner high throughput screening of chemical compounds to assess their function as potential pharmaceutical agents has no specified aim on undertaking. Any success that comes from such a process cannot be entirely attributed to the thought process, there is a large degree of serendipity.

Science in its purest form is a thought process but in reality it cannot be. There is a requirement for information to be generated so insight can be found. In many cases we must proceed without an expectation of the insight we will gain because without committing to an investigation we would not be able to have insight. This thought process is different from that stated in the title as it does not require a precise interpretation only a general line of inquiry.

In conclusion science is a thought process imposed on experimentation and observation, without the other each has no individual utility but information must be generated before it can be interpreted.

'Assisted suicide allows those suffering from incurable diseases to die with dignity and without unnecessary pain.'

Explain what this statement means. Argue the contrary. To what extent do you agree with the statement?

The idea of euthanasia as a tool for pain control and palliative care is nothing new. Over the past years it has been raised over and over again in order to legalize the practice as is the case in some other countries such as the Switzerland. The argument behind the idea is that every individual has the right to decide what is to be done with their body and their life independent of what others might think provided that he or she is in control of their mental powers and are free of undue external influence. The idea of euthanasia as a possible form of treatment revolves around the concept that a medical professional has to accept the individuals decision for whatever form of treatment, even if this decision is considered unwise or might even lead to the individuals death. An additional component arises from the idea of dignity in general and in particular of dying in dignity. Everybody has a deep desire to be comfortable at the time of their death and the prospect of unbearable pain or complete loss of self is a wide-spread fear.

Some of the strongest arguments opposing euthanasia are philosophical ones. It evolves around the idea of how far human beings have the right to take another humans life. In addition to that euthanasia goes against everything that the medical profession stands for. The cure of disease and the saving of life and good health. Death will never cure a disease nor will it ever maintain life or good health. Other opposing arguments include the idea of actual consent to commit suicide. What happens with patients that are not able to communicate with the doctors or to what extend is it a possibility that relatives are 'talked into' feeling as a burden and wanting to end their lives for that reason. A final concern arises from the idea that allowing euthanasia for terminally ill patients or as a form of pain control will open the door to other arguments arguing for life in some cases not being worth living.

The idea of euthanasia may have a place in the theoretical realms of arguing medical care and the treatment of terminally ill patients. In real practice however it has little place. Whilst it can be argued that in some cases an individual might be happier dead rather than being alive, the dangers and concerns arising from the concept of euthanasia seem to outweigh these concerns and ultimately euthanasia goes against the very core ideas that form the basis of the medical profession.

END OF PAPER

Mock Paper G Answers

Section 1

Question 1: C

Answer B) is completely irrelevant to what the manager is saying, so is incorrect. A) and E) are also incorrect as the manager is simply talking about ticket sales. He has not mentioned anything about the relevant popularity of folk music, or how much the band should value profit. D) is incorrect as the manager is simply saying that the band will have higher ticket sales in France than in Germany, so other countries are not relevant.

C) is correct as Germany could still have higher ticket sales for folk music than France despite the recent changes in ticket sales.

Question 2: D

Let the number of invitations with the extra information in be m. Invitations with extra information in cost £0.70 to send and invitations without cost £0.60. Therefore the total cost of posting is £0.70m + £0.60(50-m) and this is equal to £33. 33=0.70m-0.60m+30. 3=0.1m therefore m=30. So the number of invitations with extra information in is 30. Therefore the answer is D.

Question 3: B

Only B) is not an assumption, as it is stated in the question that both Grace and Rose departed at 5:15. The other answers are all assumptions. At no point has it been stated that both the girls are walking, or that they will walk at the same speed. If either of these points are incorrect, we cannot definitely state that they will arrive home at the same time. Therefore A) and E) are assumptions. Also, it has not been stated that the gymnastics class is being held at the local gymnasium. If this is not the case, then we cannot know how far Grace and Rose have to walk, and therefore cannot state that they will arrive home at the same time. Therefore, C) is an assumption. Equally, if Grace gets lost, she may arrive home after Rose, so D) is an assumption.

Question 4: E

Answer A) is completely irrelevant to John's conclusions, as the speed of travel has no effect on the train's destination. D) is also irrelevant as other destinations from King's Cross station also bear no effect on John's conclusion. Meanwhile, B) is incorrect as John's conclusions refer to travelling to Edinburgh by train, so the possibility of travelling by aeroplane has no effect. C) is not an assumption because John's conclusion is in the present tense, referring to journeys made at the moment, so future developments have no effect.

E) is an assumption John has made. Only two other stations in London have been mentioned. At no point has it been mentioned that there are no other stations in London that John could travel from.

Question 5: C

The question says that Shaniqua plays in the square which will stop Summer being able to win straight away, so Shaniqua must play in 4. Summer then needs to play in a square where there will be 2 different options to make a line on the turn afterwards, so that Shaniqua cannot block both of them. If Summer plays in 1, she can make a line by playing in either 5 or 6 the next turn, so Shaniqua cannot stop her winning. If Summer plays in 2, she cannot make a line on the next turn at all. If Summer plays in 3, she can only make a line by playing in 6 the next turn and so Shaniqua can stop her. If Summer plays in 5, she can only make a line by playing in 5 the next turn and so Shaniqua can stop her. If Summer plays in 6, she can make a line by playing in either 1 or 3 the next turn, so Shaniqua cannot stop her winning. Therefore she either needs to play in 1 or 6 to be able to be certain of winning the next time.

Question 6: C

B) and D) are both stated in the question. A) is also stated as the question states that Tanks were a hugely influential factor in ALL battles in World War 2.

E) is not stated but is not an assumption as it is not required to be true for the argument's conclusion to be valid.

C) However, is required to be true for the conclusion to be valid and yet is never stated in the question, so it is an assumption.

Question 7: B

D) is irrelevant to the argument's conclusion, whilst A) and E) are also irrelevant as the argument does not directly imply either of these things (and even if it did they are irrelevant to the argument's conclusions so are not flaws).

C) is incorrect because the argument states that the Prussian arrival was essential to the British victory, so C) is not an assumption.

B), however, is never stated in the question, but is needed to be true for the argument's conclusion to be valid.

Question 8: B

D) and E) are both entirely irrelevant to waiting times, so are not flaws.

C) is not correct, as the question states that busier ports have longer queuing times. A) is also incorrect as the question states that Bordeux is the busiest port in France, so Calais is definitely less busy than Bordeux. Therefore, Porto cannot be busier than Bordeux but less busy than Calais.

B) is a flaw, as the fact that Bilbao was busiest last year does not necessarily mean it will be busy this year.

Question 9: B

The volume of the box with 10cm squares cut out is $10*100*100 = 100000cm^3$

The volume of the box with 20cm squares cut out is $20*80*80 = 128000cm^3$

The volume of the box with 30cm squares cut out is $30*60*60 = 108000cm^3$

The volume of the box with 40cm squares cut out is $40*40*40 = 64000cm^3$

The volume of the box with 50cm squares cut out is $50*20*20 = 20000cm^3$

Therefore the biggest box is the one with the 20cm squares cut out, so the answer is B.

Question 10: A

At no point is A) stated, but if aeroplanes are not a major source of carbon dioxide then it does not follow that they are largely responsible for the damage caused by global warming. Therefore A) is a valid assumption.

B) and C) are both stated in the question, whilst D) is irrelevant to the conclusion. E), meanwhile, is stated, as the question states that *we must now seek to curb air traffic in order to save the world's remaining natural environments*.

Question 11: B

A) and C) can be inferred, as the question states that these things would happen. Meanwhile, D) and E) actually serve to reinforce the argument's conclusion that the research into a new cure will not be successful. Therefore, they are not flaws in the argument's reasoning.

The point raised by B) does weaken the argument, and is a valid flaw in the argument's reasoning.

Question 12: E

A), C) and D) are all irrelevant to the argument's main conclusion, namely that Egypt was a powerful nation and must therefore have had a very strong military.

B) is a conclusion from the argument, but goes on to support E). If a nation required a very strong military to be a powerful nation, then it follows that if Egypt was a powerful nation it must have had a very strong military. Therefore, B) is an intermediate conclusion within the argument. E) is the *main* conclusion of the argument.

Question 13: D

At no point does the argument state or imply that we should not be concerned about damage to the polar ice caps, or that reducing energy consumption will not reduce CO2 emissions. Therefore, B) and E) are incorrect.

C) could be described as an assumption made in the argument, and is therefore not a conclusion.

A) goes beyond what the argument says. The argument does not say there are no environmental benefits to reducing energy consumption; it merely says it will not help the Polar Ice Caps. Therefore A) is incorrect and C) is a valid conclusion from the argument.

Question 14: C

A), B) and D) are all in direct contradiction to statements made in the passage, so cannot be conclusions. E), meanwhile, does not contradict the argument, but at no point does the argument say that the dangerous isomer was not effective at relieving nausea, so E) is not a conclusion.

However, the fact that the company followed the required level of testing and still did not detect the dangerous isomer does suggest that the required level of testing was not sufficient to identify isomers, so C) is correct.

Question 15: E

Ashley has to be sat in the front left seat so there are only two seats left in the front row. Bella and Caitlin have to be sat in different rows, so one of them must be sat in the front row and one in the back row. Now there is only one seat left in the front row, so there is not room for Danielle and her teaching assistant to both sit there. Therefore Danielle and the teaching assistant must take the two remaining seats in the back row. Therefore Emily must sit on the front row as there are no seats remaining in the back row. Emily cannot sit in the middle seat due to her mobility issues, so she must sit in the front right seat.

Question 16: A

At no point is it stated or implied that car companies should prioritise profits over the environment, so C) is incorrect. Neither is it stated that the public do not care about helping the environment, so E) is incorrect.

B) is a reason given in the argument, whilst D) is impossible if we accept the argument's reasons as true, so neither of these are conclusions.

Question 17: B

E) is contradictory to the main conclusion of the argument.

A), C) and D) are all reasons which go on to support the main conclusion of the argument, which is given in B). If we accept A), C) and D) as true, then it follows readily that the statement given in B) is true. Therefore, B) is the main conclusion.

Question 18: E

We can work out the code for each number and see which one equals 3.

The code for A is (3x4) = 12, divided by 6 = 2, minus 1 = 1

The code for B is (9x8) = 72, divided by 6 = 12, minus 4 = 8

The code for C is (5x4) = 20, divided by 2 = 10, minus 3 = 7

The code for D is (7x8) = 56, divided by 4 = 14, minus 8 = 6

The code for E is (6x8) = 48, divided by 4 = 12, minus 9 = 3

Therefore the pin number with the code 3 is E, 6839.

Question 19: C

The manager's conclusion (that the centre should hire Candidate 1 in order to maximise profits) relies on the assumption that performance experience, rather than welfare experience, will maximise profits. A valid flaw will mean that this assumption is not valid.

A) supports this assumption and so is not a flaw. B) is irrelevant as this assumption does not rest on the performance experience being with dolphins specifically. D) is irrelevant as it concerns a charity's outlook and is thus not relevant to a profit-making business. If E) were indeed a correct prediction then it could still be that profit would rise by *more* with Candidate 1, so the manager may still be correct.

C) is a flaw as it expresses a way in which profit may be higher if the business prioritises welfare standards over performance standards, as a boycott of the business could potentially greatly reduce profits.

Question 20: E

We can calculate all the rental yields as follows:

House A: (700x12)/168000 = 0.05

House B: (40x125x4)/200000 = 20000/200000 = 0.10

House C: (600*12)/144000 = 7200/144000 = 0.05

House D: (2000*12)/240000 = 24000/240000 = 0.10

House E: (200*52)/100000 = 10400/100000. We can see by observation that this is > 0.1 as 10000/100000 would equal 0.1, therefore there is no need to work this out to be able to say that this is the house with the highest yield.

Question 21: B

B) is an underlying assumption in the Transport Minister's argument. If rural areas have plenty of passengers, her assertion that rail companies will not run many services to these areas does not follow from her reasoning. Therefore, if B) is true, it strengthens the transport minister's argument.

Meanwhile, D) would actually weaken the transport minister's argument, suggesting that privatisation would not lead to less service for rural areas.

C) is irrelevant as the transport minister is arguing about how rural communities will be cut off by a privatised system. She is not referring to the quality or price of rail services under a publically subsidised system.

A) and E) are completely irrelevant points, which have no effect at all on the strength of the Transport Minister's argument.

Question 22: C

If it will cost Niall £2 more to pay per session than to buy membership, the combination of classes and gym sessions he is going to attend must cost £32. The only one of these combinations which costs £32 is C (5 x gym sessions at £4, 6 x classes at £2). Hence the answer is C.

Question 23: D

If the median is 6, the 3rd number when the numbers are written in order is 6. If the mode is 4, there must be at least two 4s (and can only be two 4s, because there are only 2 numbers less than 6 due to what we know about the median). Therefore, the smallest three numbers in the set are 4, 4, 6. For the mean to be 8, the numbers must add up to 5 times 8 = 40. Therefore the largest two numbers must add up to 40-(4+4+6)=26.

Question 24: A

B) and E) are irrelevant points which do not affect the strength of Lucy's argument.

C) and D) would both serve to strengthen Lucy's argument. C) suggests that running costs will be low, whilst D) suggests that visitor centres are profitable. Both of these, if true, serve to suggest that opening up visitor centres will be profitable for the park, therefore supporting Lucy's argument.

A), however, would weaken Lucy's argument by suggesting that visitor centres will not be profitable.

Question 25: E

There are 16 squares of dimension 1. There are 9 squares of dimension 2 (one in each corner, one halfway across each side and one right in the middle). There are 4 squares of dimension 3 (one in each corner). There is 1 square of dimension 4. Therefore, the total number of squares is $16 + 9 + 4 + 1 = 30$.

Question 26: D

If the announcement is accurate to the nearest 10 minutes, this means that the soonest the train will arrive in London is 115 minutes after the announcement, which is 17:25. The final destination is 10 minutes from King's Cross, so the earliest time I might arrive there is 17:35.

Question 27: C

When the sister's son is home, she has to buy a carton of washing powder 1.2 times as often, so she must be doing 1.2 times as many loads of washing. If we let x be the number of people living at home when the son is home, the number of loads of washing when he is home is 2+x, whereas the number when is not at home is 2+x-1=1+x. Therefore 2+x must equal 1.2(1+x). Rearranging this we get 0.8=0.2x, so x is 0.8 divided by 0.2, which is 4. So 4 people are living in the house when the son is home.

Question 28: D

The total time each train can run is 18 hours a day. Each journey takes the train 30 minutes (24+4+2). So each train can make 36 journeys a day. Therefore the total journeys made by the shuttle service per day will be 2x36 (because there are 2 trains) so the answer is 72.

Question 29: D

The lowest common multiple of 6, 4 and 2.5 is 60. Hence trains from all 3 lines will arrive at the same time every 60 minutes. If the last time they did was 4 minutes ago, it will hence be 60-4=56 minutes until they do so again. Therefore the answer is D.

Question 30: D

If Sam buys the invitations, she will spend £90 (90 x £1 each).

If Sam makes the invitations, she will need enough supplies for 94 invitations, which will be:

- 4 packs of red paper at £2 each = £8
- 7 rolls of ribbon at £3 each = £21
- 4 packs of gold stickers at £1 each = £4
- 1 stamper = £8
- 2 ink pads at £4 each = £8
- 5 packs of cream card at £2 each = £10

Adding these up, we get that the total spent on making the invitations herself is £59. Therefore she saves £90-£59=£31 by making them rather than buying them. Therefore the answer is D.

Question 31: B

If there are at least as many boys as girls in the class of 36, then there are at least 18 boys and at least 9 boys have brown eyes. If two thirds of the class have brown hair and at least as many boys as girls have brown hair, at least two thirds of the boys in the class have brown hair, so at least 12 boys have brown hair. There are only 18 boys in the class, so of the 9 boys who have brown eyes and 12 who have brown hair, at least 3 of these must be the same boys. So at least 3 boys in the class have both brown hair and brown eyes.

Question 32: C

The total amount of dilute squash needed is (300ml x 8) + (400ml x 3) = 3600ml. Mandy has 600ml of concentrated squash so she needs 3000ml of water to make up the right amount. There should be 3000ml:600ml of water:concentrated squash so the ratio needed is 5:1

Question 33: D

If we assumed that none of the options in any block were the same, there would be 4x4x4x4 = 256 different sets of options. However, Psychology and Mathematics are in 2 different blocks so there will be some options that cannot be taken together. There are 16 sets of 4 options that involve Mathematics twice (Mathematics in Blocks A and C and then any combination of options in the other blocks) and 16 that involve Psychology twice (Psychology in Blocks A and B and any combination of options in the other blocks), so we need to take off 32 from the total options we calculated. Hence the total number of sets of possible options is 256-32=224.

Question 34: C

We need to work out both what the cost of living will be in 4 years time and how much savings there will be in 4 years time.

The cost of living rises by 3% a year, so using the table we can see that at the end of Year 1 the cost of living is £19,570.00. Then at the end of Year 2, the cost of living is £20,157.10. At the end of Year 3, the cost of living is £20,761.81. At the end of Year 4, the cost of living is £21,384.67.

The savings rise by 4% each year until they exceed £20,000, so after Year 1 they are £19,760.00. After Year 2, they are £20,550.40. This means that for the last 2 years, they will yield 5%. So after Year 3, they will be £21,577.92, and after Year 4, they will be £22,656.82.

We can then work out the difference between these two amounts. We only need an approximation, so we can do £22,656 - £21,384 = £1272, which to the nearest £100 is £1300. Hence the answer is C.

Question 35: D

Amaia needs to arrive by 1100 so the latest bus she can catch is the one which arrives at Northtown University at 1053. This arrives at Northtown West station at 1019, or Northtown Central station at 1040. The latest train she can catch to make this bus is the one that gets to Northtown Central for 1028, and leaves Southtown at 0943. Amaia lives a 10 minute walk from Southtown station so she will need to leave her house at 0933 to get to the interview on time. Therefore the answer is D.

END OF SECTION

Section 2

Question 1: B

Natural selection favours those who are best suited for survival – this can mean faster and stronger organisms, but not always. For example, snails are pervasive, despite being weak and slow. Variation can arise due to both genetic and environmental components.

Question 2: H

Chloride is oxidised during this process to form Cl_2. Although the first part of 2) is correct, H_2O is required to dissolve the NaCl (not H_2 which is a product of the reaction). NaOH is a strong base.

Question 3: D

The enzyme amylase catalyses the breakdown of starch into sugars in the mouth (1) and the small intestine (5).

Question 4: E

Whilst there is some enzymatic digestion in 1 and 3, the vast majority occurs in the small intestine (5). The liver facilitates digestion via the production of bile, and the large intestine is primarily responsible for the absorption of water.

Question 5: B

This is an example of an addition reaction, the fluorine and hydrogen atoms are added at the unsaturated bond. If you're unsure about this type of question draw it out and the answer will be obvious.

Question 6: D

The energy in a nuclear bomb comes from $E = mc^2$. When two nuclei fuse, the combined mass is slightly smaller than the two individual nuclei, and the mass lost is converted to energy according to Einstein's equation. Fusion releases much more energy than fission, as in the sun, and humans cannot harness this energy yet. Uncontrolled fission causes the explosion in an atom bomb and is created by a neutron-induced chain reaction. In power plants these neutrons are tightly controlled, so as not to overload the reactors and cause an explosion.

Question 7: A
The hydrogen halide binds to the alkene's unsaturated double bond. This results in a fully saturated product that consists purely of covalent bonds.

Question 8: B

$$\left(\frac{T}{4\pi}\right)^2 = \frac{l(M + 3m)}{3g(M + 2m)}$$

$$\frac{T^2}{16\pi^2} \times \frac{3g}{l} = \frac{M + 3m}{M + 2m}$$

$$3gT^2(M + 2m) = 16l\pi^2(M + 3m)$$

$$3gT^2M + 6gT^2m = 16l\pi^2M + 48l\pi^2m$$

$$6gT^2m - 48l\pi^2m = 16l\pi^2M - 3gT^2M$$

$$m(6gT^2 - 48l\pi^2) = 16l\pi^2M - 3gT^2M$$

$$m = \frac{16l\pi^2M - 3gT^2M}{6gT^2 - 48l\pi^2}$$

Question 9: A
The polymerisation reaction opens the double bond between the two C atoms to allow the formation of a long chain of monomers.

Question 10: A
Replotting the genetic diagram with genotype information produces the diagram:

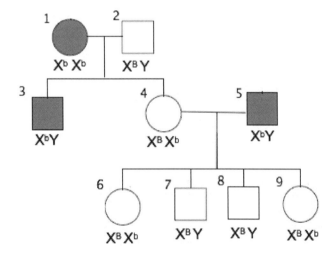

If squares were female, all of 5's circular male offspring would be affected. Circles must be females, so 1 must be homozygous recessive.

Question 11: D
The genotype of a heterozygote female is be $X^B X^b$, and the genotype of 8 is $X^B Y$. Plotting the information in a Punnett square:

		Female Heterozygote	
		X^B	X^b
Individual 8	X^B	$X^B X^B$	$X^B X^b$
(Unaffected Male)	Y	$X^B Y$	$X^b Y$

The progeny produced are 25% $X^B X^B$ (homozygous normal female), 25% $X^B X^b$ (heterozygous carrier female), 25% $X^B Y$ (normal male) and 25% $X^b Y$ (affected male). So the chance of producing a colour blind boy is 25%.

Question 12: B

The mean is the sum of all the numbers in the set divided by the number of members in the set. The sum of all the numbers in the original set must be: 11 numbers x mean of 6 = 66. The sum of all the numbers once two are removed must then be: 9 numbers x mean of 5 = 45. Thus any two numbers which sum to 66 – 45 = 21 could have been removed from the set.

Question 13: F

All of the above are true. Every mole of gas occupies the same volume. The left side therefore occupies 4 volumes, and the right side occupies 2 volumes. Increasing pressure will favour the lower volume side, and the equilibrium will shift right to produce ammonia and decrease the overall volume that the products and reactants occupy. If more N_2 gas is added, equilibrium will shift to react away this gas and lower the concentration again, with the result that more ammonia will be formed.

Question 14: E

From the rules of angles made by intersections with parallel lines, all of the angles marked with the same letter are equal. There is no way to find if d = 90°, only that b + d = c = 180° – a = 135°, so b is unknown.

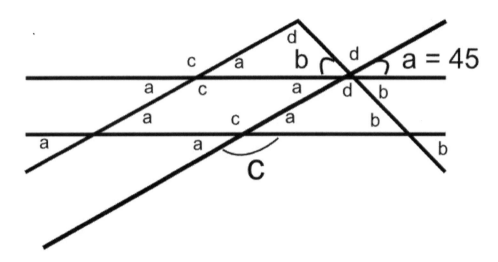

Question 15: E

Sodium is element 11 on the periodic table, a group 1 element, so has electron configuration: 2, 8, 1. It forms a metallic bond with other sodium atoms. Chlorine is element 17 in group 7, so has 17 electrons and 7 valence electrons, giving configuration: 2, 8, 7. Chlorine forms the covalent gas Cl_2, sharing one electron for a full valence shell.

Salt (NaCl) is an ionic compound, where sodium gives its single valence electron to chlorine so both atoms have full outer electron shells (8 electrons, so 2, 8:2, 8, 8).

Question 16: B

Let $y = 3.4 \times 10^{10}$; this is not necessary, but helpful, as the question can then be expressed as:
$$\frac{10y + y}{200y} = \frac{11y}{200y} = \frac{11}{200} = \frac{5.5}{100}$$
$$= 5.5 \times 10^{-2}$$

Question 17: B

As the known parent has both recessive genotypes, it can only have the gametes, y and t. The next generation has a phenotypic ratio of 1:1:1:1. As both recessive and dominant traits are present in the progeny, the unknown parent's genotype must contain both the recessive and dominant alleles. Hence the unknown parent's genotype must be YyTt as this would produce the gamete combinations of YT, Yt, yT and yt, which when combined with the known yt gametes would result in YyTt, Yytt, yyTt and yytt in equal ratios.

Question 18: D

The possible genotypes are: YYTT (yellow, tall), YyTT (yellow, tall), yyTT (green, tall), YYTt (yellow, tall), YYtt (yellow, short), YyTt (yellow, tall) Yytt (yellow, short), yyTt (green, tall), yytt (green, short). Thus, 9 different genotypes and 4 different phenotypes are possible.

Question 19: B

During electrolysis a current is used to draw charged ions to electrodes. The anode is positively charged and draws anions like sulphate, and the cathode is negatively charged and attracts positively charged cations like copper. For electrolysis to work well, the electrodes need to keep their positive or negative charge. If an alternating AC-current was used, the anode and cathode would repeatedly switch places, and the ions would make no net movement toward either electrode.

Question 20: C

Transform all numbers into fractions then follow the order of operations to simplify. Move the surds next to each other and evaluate systematically:

$$= \left(\left(\frac{6}{8} x \frac{7}{3}\right) \div \left(\frac{7}{5} x \frac{2}{6}\right)\right) x \frac{4}{10} x \frac{15}{100} x \frac{5}{100} x \frac{5}{25} x \pi x \left(\sqrt{e^2}\right) x e\pi^{-1}$$

$$= \left(\frac{42}{24} \div \frac{14}{30}\right) x \frac{4 \, x \, 3 \, x \, 25}{10 \, x \, 20 \, x \, 100 \, x \, 25} x \pi x \pi^{-1} x e^{-1} x e$$

$$= \left(\frac{21}{12} \div \frac{7}{15}\right) x \frac{12}{200 \, x \, 100} x \frac{\pi}{\pi} x \frac{e}{e}$$

$$= \left(\frac{21}{12} x \frac{15}{7}\right) x \frac{3}{50 \, x \, 100}$$

$$= \frac{45}{12} x \frac{3}{5000}$$

$$= \frac{9}{4} x \frac{1}{1000}$$

$$= \frac{9}{4000}$$

Question 21: B

R of series circuit= R + R = 2R

$$R \text{ parallel} = \frac{1}{\frac{1}{R}+\frac{1}{R}} = \frac{1}{\frac{2}{R}} = \frac{R}{2}$$

Thus, the parallel circuit has a smaller resistance than the series circuit.

Since $I = \frac{V}{R}$, the parallel circuit will have a greater current than the series.

Question 22: D

Firstly, convert 36km/h to m/s to conserve units:

$$\frac{36,000 \, m}{3600 \, seconds} = \frac{360 \, m}{36s} = 10\frac{m}{s}$$

Before the driver can react the car travels at 36 km/hour for 0.5 seconds. Thus, it covers a distance of 0.5 x 10 = 5 metres. There are 100 m left to the deer and the car must slow from 10 m/s to 0 m/s.

Using: $v^2 = u^2 + 2as$ gives: $0 = 10^2 + 2 \, x \, a \, x \, 100$

Thus, $-200a = 100$

Thus, $-200a = 100$

Thus, $a = -0.5 \, ms^{-2}$

Finally, using $F = ma$: 2000 x 0.5 = 1,000 N

Question 23: D

Whilst getting vitamins, killing bacteria, protein synthesis, and maintaining cellular pH and temperature are all important processes that require a blood supply, the MOST important reason for having a blood supply is the delivery of oxygen and removal of CO_2. This allows aerobic respiration to take place, which produces energy for all of the cell's metabolic processes. None of these processes can be sustained for a meaningful period of time without the energy made available from respiration.

Question 24: C

Although the magnitude of acceleration decreases after 5 seconds he is still increasing his velocity. In this case, the velocity is given by the area under the curve. Summing the velocity gained over each second gives the final velocity, with squares here corresponding to 1 m/s² x 1 s = 1 m/s. He only ever loses velocity between 0-0.5 s and 9.5-10 s.

Question 25: C

The radius and tangent to a circle always form a right angle, so using Pythagoras:

$3^2 + 4^2 = X^2$
$X = 5$ m

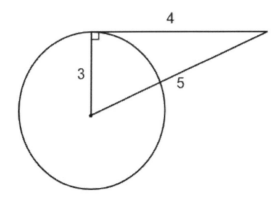

Question 26: E

Statements A, B and C are correct. Although energy is usually wasted when transformed, e.g. in power plants and engines, there are times when energy can be transformed without any losses to extraneous energy forms, e.g. a ball free-falling through a vacuum loses potential energy and gains kinetic energy without losses to other energy forms. Energy can be dispersed through vacuums e.g. solar heat energy through space.

Question 27: A

A beam of light is refracted toward the normal to the glass-air interface when it enters the glass, but refracted away by the same amount when it exists, for overall no net change. The angle of reflection of a beam is equal to (and thus dependent on) the angle of incidence. Beams of light entering a denser medium are refracted toward the normal to the interface, so light entering a pool of water would descend more steeply. Excited electrons in any atom can emit a massless photon to reduce their energy.

END OF SECTION

Section 3

'Doctors will eventually become obsolete as a result of advancing medical technologies.'

Explain what this statement means. Argue to the contrary. To what extent do you agree with the statement?

Technological advances in medicine are becoming increasingly important for modern diagnostics. Examples include X-ray technology and genetics, which both play an increasingly important role in medical diagnostics. Different forms of X-Ray imaging play a vital role in all medical investigations, whilst genetics are becoming increasingly important in finding tailored cures for complex diseases such as HIV, bacterial antibiotic resistance or even for cancer. The central idea underlying the statement is that knowledge and the ability to detect information on different patient parameters is all that is needed to cure him or her.

There is no doubt that increasing technological abilities play a central role in the treatment and diagnosis of disease. And in a sense, it is correct that the ability to accumulate a wide range of data will aid in the diagnosis of disease and therefore contributes to the treatment. But curing a disease goes beyond the mere diagnosis. A diagnosis will only allow the medical professional to decide on the path to take for treatment of the individual disease, but it will not necessarily facilitate a cure. Using HIV as an example, the ability to genetically detect implantation of the retrovirus in the cells DNA aids in making a diagnosis and the existence of anti-retroviral medication will allow for the suppression of the disease and allow the patient to lead an almost normal life, but even with all this knowledge and all these capabilities, owed to technological progress, a cure is not possible. But even in diseases that can theoretically be cured, the mere ability to cure it is worth little without the treating medical professional making the right decisions concerning investigations and treatments. Ultimately, the component of human error will never be outweighed by any technological ability and vice versa, the complexity of the human mind will never be replaced by machines.

The statement has some merit as technology is a central component of modern diagnostics and therefore is an important tool in pointing the treating medical professional into the right direction. It will however never replace experience and intuition from the side of the treating doctor as there will always be some degree of biological variance between patients that makes sole reliance on technological data difficult and potentially inaccurate. In essence, whilst technological means of diagnosis, and through this technological advance, play an important role in modern medicine and the cure of patients, it is unlikely for technological achievements to ever.

'Animal euthanasia should be made illegal'. **Explain what this statement means. Argue to the contrary that animal euthanasia should remain legal. To what extent do you agree with the statement?**

The statement argues that animal euthanasia, the act of humanely ending an animal's life or withholding life saving treatment, should be illegalised, usually on the grounds that it is cruel, unethical or inhumane. This essay will argue to the contrary that there are many instances in which animal euthanasia may be the best available option.

Firstly, there are many cases in which an animal may have a terminal illness such as cancer or rabies that would no longer respond to treatment. In these cases, it may be more humane to end the animals life in as swift and pain free a way as possible than to let it suffer helplessly for longer. Similarly, the animal may have a non-terminal illness, but one that may significantly affect its quality of life, or perhaps the owner could not afford the treatment. An animal's health and quality of life may significantly deteriorate in old age, and again, in these cases, euthanasia may be seen as the most humane option.

In some cases, the animal may suffer from behavioural problems that may mean that they are no longer suitable as a pet. While these problems may usually be corrected with the right care, in many instances, they fail to improve. A change in owner circumstances may mean that animals may have lack of a suitable home or caretaker. In both these cases, some people may believe that euthanasia is the kindest way to deal with them.

A large proportion of medical research is based on animal models. Euthanasia may be seen as the most humane way of killing animals in order to provide specimens for dissection, or to end animal suffering after experiments, with the balance being that the knowledge gained may result in better treatments overall for humans or animals.

In some instances such as with the foot and mouth outbreak in cows or tuberculosis in badgers, culling by euthanasia is crucial in spreading the spread of disease. Furthermore, population control of particular species e.g. Deers by euthanasia is crucial in maintaining particular ecosystems.

While I do believe it is morally wrong to end an animal's life unnecessarily, I largely disagree with the statement, I believe there are many instances, such as those described above, in which it is in the best interests of animals to be euthanized, and in these cases, it should remain legal.

"Science is a procedure for testing and rejecting hypotheses, not a compendium of certain knowledge."

Stephen Jay Gould

What do you understand from the statement above? Explain why it might be argued that science does rely on a compendium of certain knowledge? To what extent is science defined by the challenging of preconceived hypotheses?

"A compendium of certain knowledge" refers to science as a self-contained, consistent body of information. Gould argues that rather than being static, science is the constant challenging of current ideas about the processes that underlie the universe. It is this idea of scientific controversy and continuous adaptation to new experimental findings that enables science to be driven forward, via the rejection of previous hypotheses and formulation of new ones. However, many fundamental scientific ideas have been supported on so many occasions that they could be considered close to empirical truths. Since Darwin's formulation of the theory of natural selection, many natural observations have verified this theory, for example the observed selection of camouflaged peppered moths over white moths in the smog-filled Victorian Britain. Further understanding of inheritance, as being mediated through chromosomes has since led to the acceptance of natural selection within the scientific community.

Moreover, some scientific work is focused on determining the details of an accepted body of knowledge. Over twenty years of sequencing was undertaken by the Human Genomes Project to derive the full human genetic code. Rather than challenging any hypothesis, science in this case was working from the understanding that the human genome was the means of inheritance through its sequence of bases. However, even though some scientific fundamentals are accepted, this does not mean they will eventually be challenged. The notion of classical mechanical physics was accepted as the ultimate explanation for all universal processes, for hundreds of years. It was only at the beginning of the 20th century where particles were shown to be able to display wave-like properties, that the concept of quantum physics was born. It is important to note that 'fundamental' scientific theories are *theories* – even scientific acceptance of natural selection has been updated to take into account new discoveries such as that of the inheritance of chemical alterations to DNA, epigenetics.

Newton summarises the work of science accurately: "If I have seen further it is by standing on the shoulders of giants". Scientific work often starts from the basis of a current body of knowledge, but ultimately it is the process of challenging this body that defines it – science is an ever-evolving, unconstrained field. However, as organisms limited to our five senses, perhaps science could be considered a compendium of knowledge, limited by our sense-based perception of the universe.

'The primary duty of a doctor is to prolong life as much as possible'

What does this statement mean? Argue to the contrary, that the primary duty of a doctor is not to prolong life. To what extent do you agree with this statement?

This statement interprets medical beneficence as treating a patient so as to maximise their lifespan, irrespective of quality of life or the patient interests. This essay will argue that while prolonging life is a crucial part of a doctor's duty, it is not always the best, or most appropriate target of treatment.

Firstly, this statement assumes that all patients want to live longer. While it is a inherent quality of nature to want to survive, many patients, particularly elderly patients may feel they have already lived a long and fulfilling life and may not want to (in their eyes) artificially prolong their life further. They may rather wish to die naturally in their home surrounded by friends and family than in a medical setting. The statement does not take into account any religious beliefs a patient may hold. For example, choosing to prolong the life of an unconscious patient with a blood transfusion may not be in the patients best interest if they were a Jehovah's witness, a sect of Christianity in which all blood products are banned as this may both compromise their beliefs and lead to them being banished from their community.

The statement also fails to address the issue of quality of life, instead simply choosing to focus on quantity. Is it worth having an extra five years of life if you were to be in excruciating pain or in a permanent vegetative state for the rest of it? The use of QALYs (quality adjusted life years) helps address this aspect, and for many patient, one extra year of disease free health may be more valuable than five extra years suffering. In underfunded and stretched NHS where resources are finite, resource allocation also has to be considered when thinking about the duties of a doctor. On balance, some may say it is more just to spend £10,000 on providing life saving vaccination for hundreds of children than it may be to extend a 90 year olds life by a year.

While I do believe that prolonging life is, and always has been, one of the core aims of medicine, I do not agree that it is the primary duty of a doctor. I believe quality of life is equally as, if not more, important and should be taken into account. Furthermore, patient autonomy is one of the core principles of medicine and the patient's wishes should always be taken into account; if a patient applied for a DNR form previous to deteriorating, it would not be appropriate of a doctor to try prolong life as much as possible.

END OF PAPER

Mock Paper H Answers

Section 1

Question 1: A

In this question we are looking at what cannot be reliably concluded from the passage. B and E conclude the state of a substance is not dependent on its chemical properties. C and D discuss how combining two substances can produce a new substance with very different physical properties. The passage refers to how the chemistry of a compound does not necessarily affect the physical properties of that compound. Thus, the answer must be A, which claims the chemical composition of a compound influences its physical nature.

Question 2: C

In this sequence, each alternate letter goes forward starting with B or backwards starting with Y. They start by jumping 4 letters, then 3, then 2 and finally the letters we are trying to find will have jumped by just 1. Thus, the letter after K is L, and the letter before P is O, so the answer is C.

Question 3: E

In this question we are looking at what can be reliably concluded from the passage. The passage is referring to products being made from similar parts; it is the way in which these parts are arranged that actually determines the final product. Thus, B cannot be right. D is also not correct, as there is no mention of protoplasm being the building block for life. E is the correct answer therefore.

Question 4: D

With these questions it is easiest to start at the end of the question and work backwards. The day two days before Monday is Saturday. The day immediately after that is Sunday. The day that comes four days after Sunday is Thursday, and two days after that is Saturday. Thus, the answer is D.

Question 5: D

In this question we are looking at what could weaken the passage above. The passage is discussing synthetic and natural cellulose and how their functions depend on whether the cellulose is plastic or colloidal. However, it states that the properties of natural and synthetic cellulose are equally similar. Therefore, any statement claiming some of the properties between the two forms of cellulose are different would weaken the passage, thus the answer is D.

Question 6: E

In order to work out this question, we need to make some simultaneous equations to relate John and Michael's money. If the amount of money John has at the start is J, and the amount that Michael has is M, we get the following equations:
$J - 20 = 2(M + 20)$ and $J + 5 = 5(M - 5)$, which is simplified to:
$J = 2M + 60$ and $J = 5M - 30$.
Substituting in, to work out M gives:
$2M + 60 = 5M - 30$, thus $3M = 90$ and $M = 30$.
Substituting in $M = 30$ to one of the equations gives:
$J = 60 + 60 = 120$.
Thus, $J + M = 150$, so the answer is E.

Question 7: E

In this question we want a summary of the passage. This passage refers to the use of fire in civilisations to create light. Through the passage it talks about the evolution of the use of fire, finishing with a reference to gas lamps in the street. Thus a good conclusion will refer to how the use of fire has changed over time, but also how lighting one's home is a key factor of civilisation. The answer must therefore be E, which discusses the evolution of fire use and also its importance in civilisation.

Question 8: A

$972/2 = 486$, thus 486 patients did not have chicken.
$972/3 = 324$, thus 162 patients did not have the chicken or the mac and cheese.
$972/12 = 81$, thus $972 - 486 - 324 - 81 = 81$, which is the number of patients that had the vegetarian option. Therefore the answer is A.

Question 9: C

In this question we are looking at what can be reliably concluded from the passage. The passage does not tell us exactly how phosphorous was discovered, but we know that it was not Wilhelm Homberg who discovered it, thus A, B and D cannot be correct. 1669 is not in the 18th century thus E is also false. The passage describes how the element phosphorous was discovered by accident, by a man of low social status. Therefore, C is the only correct answer.

Question 10: D

For this question refer to the times in minutes, rather than hours, so 3pm is 180 minutes. x is the number of minutes past noon that we are trying to find. Therefore x + 28 will give the same amount of minutes past noon as 180-3x.

$x + 28 = 180 - 3x$

$4x = 152$

$x = 38$, thus the answer is D.

Question 11: B

In this question we are trying to find a suitable conclusion to the passage. A and E are completely irrelevant to the passage. C is incorrect as wings only attach to the posterior two segments of the insect's body. While D is correct, the legs are not referenced as being the most important part of the insect's body. Thus the answer must be B, which states the wings are the most dominant part of the body.

Question 12: D

For John: 56/64 x 100 = 87.5% or 7/8

For Mary: 24/36 x 100 = 66.7% or 2/3

Therefore we need to work out 7/8 – 2/3

21/24 – 16/24 = 5/24. Multiply by 100 to get the actual percentage:

500/24 = 125/6, thus the answer is D.

Question 13: A

To calculate this one needs to find the lowest common multiple of both 73 and 104, and then add that value to 2007. The lowest common multiple of 73 and 104 is 7592, which when added to 2007 gives 9559AD.

Question 14: D

In this question we are trying to find a suitable summary of the passage. There is no mention how important an animal is to mankind determining whether cruelty is acceptable, thus C and E are wrong. The passage states that the nature of the cruelty and the type of organism involved is irrelevant and should be punished regardless, thus the answer cannot be A or B, and must be D.

Question 15: B

If the number of girls is 40 more than the number of boys, and the boys make up 40% of the total number of students, then the discrepancy of 40 between boys and girls must represent 20%. Therefore, 1%=2 students and therefore the total number of students is 200.

Question 16: D

With this question it is easiest to start by putting the purple car between the green and the blue car. Since the red car is behind the blue car and the yellow car is in front of the green car, we know that the order of the cars must be:

Yellow, green, purple, blue and then red at the back.

Thus the second car in line is the green car and the answer is D.

Question 17: C

Based on the information, the school bus will get her to school at 09:01. The public bus arrives at 08:21, which she will miss, and the next bus will arrive at 08:38, which will take 18 minutes to arrive, meaning she will at school at 08:56, so the public bus at 08:38 will get her to school first.

Question 18: B

In this question we want a summary of the passage. The passage talks mostly about the feeding habits of the puddle duck, thus a summary discussing the predation of puddle ducks is irrelevant, meaning A and D are not correct. E is wrong as puddle ducks mainly live in shallow waters, and this is not because of their eating habits so C is also wrong. Thus, B is the correct answer as the ducks feed on mainly vegetarian food sources, and although they can dive for food, this is not their main route of feeding.

Question 19: A

In this question we want a summary of the passage. The veil discussed is clearly involved in a connection between the good and evil of the earth and therefore of mankind. Thus for a good summarising sentence, we want this connection to be discussed. Therefore, the answer must be A, which states that the veil links the good and evil of the human race, as is discussed in the passage.

Question 20: E

There is a specific sequence linking these numbers. Multiplying the first and third numbers of each row gives a number that makes up the second and fourth numbers of the same row.

9 x 3 = 27, thus the missing number is 7 and the answer is E.

Question 21: C

In this question we are looking to find the flaw in the argument. Answers D and E are irrelevant to the question. While B is correct it does not explain why the metformin inhibitor would have not had any effect on metformin's inhibition of fat cell growth. The key problem here is we are not given any information about the metformin inhibitor mentioned, and thus are not able to judge how it would affect metformin's fat cell growth inhibition, thus the answer is C.

Question 22:

We do not know whether Alexandra and Katie are dancers, so **A** and **B** are wrong. We do not know whether any dancers are ugly, so **D** is wrong.

Question 23: A

In this question, we are looking for a statement, which would support the passage. The reason that bacteriology has improved so much over recent centuries is due to our ability to study bacteria. Thus, any statement claiming our ability to study such organisms will support the passage. Therefore the answer must be either A or C. Techniques have improved out understanding of bacteria better than using antibiotics has, thus the answer must be A.

Question 24: C

Joseph is a man and no man is a lion i.e. all men are NOT lions hence Joseph is not a lion.

The statement is '1 to many' connection i.e. it puts Joseph in the bigger group of men BUT does not state that all men are or are not Joseph hence can't tell.

Question 25: D

In this question we are asked choose a sentence that most weakens the above argument. The main problem with this passage is that it struggles to give a reason for why water is not classified as a food. This is because the passage states that food is any nutriment that enters the body through the intestinal canal. Thus, water should be classified as food, but the passage claims that water is not food and does not give a reason for this. The answer is therefore D.

Question 26: C

Son age now= S

Father age now= F

$F- 4 = 4(S-4) \rightarrow F = 4S -12$

$F+6/S+6 =5/2 \rightarrow 2F +12 = 5S+ 30$

$8S -12 = 5S +30$

$S=14$

$F=44$

Question 27: D

In this question we are asked to find a conclusion for the passage. Not necessarily all metals and all metal salts form salts thus A and E are not correct. Treating tallow with acids forms a soap containing a metallic base not a metallic acid so E is false. Soaps are not just used for commercial use, driers and pharmaceutical preparations, which rules out B, therefore D is the correct answer as the basic definition of a soap in the passage is a fatty acid combined with a base.

Question 28: C

Total of 12 = 73*12=876

$E + A + O = (73.6*15) - 876 = 228$

$68 +A + (A+6) =228$

$2A = 154 \rightarrow A=77$

Question 29: B

90 x 8 x 120 = 86,400. This value is the man hours required to get a third of the job done, therefore 172,800 man hours are required to finish the job. If they have 80 days left and 12 hours per day then:
172,800/80 = 2,160 and 2,160/12 = 180, thus the answer is B.

Question 30: D

In this question we are asked to find a conclusion for the passage. The passage discusses how digestive tracts vary between mammals and describes how they differ between specific animals. This is presumably a result of the animals experiencing different environments and having different diets. Thus D is the correct answer.

Question 31: B

Letters a, c, e, g… will be lower case and B, D, F, H… will be upper case. Therefore, Wednesday will be written as weDNesDay, so the answer is B.

Question 32: B

Distance = 2/3S
distance=1-2/3S=1/3S
21/15 hr=2/3 S/4 + 1/3s /5
84=14/3S * 3
S= 6km

Question 33: C

In this question we are asked to find a conclusion for the passage. The passage discusses how some birds fly north to breed as there is the conditions are better suited to them and they only have to travel a few miles to gather food. Thus it is fair to conclude that the reason for these birds migration is because the food and conditions are more suited to them, therefore the answer is C.

Question 34: D

X is married to A who is a lawyer, so B and C are either a doctor or a lawyer respectively. Y is not married to an engineer, and C is not a doctor so Y must be married to B who is a doctor, and Z must be married to C who is an engineer, so the correct answer is D, as none of these options are available.

Question 35: C

If x is the number of days she works for, then:
100x – 20(60-x) = 5040
120x = 6240. x = 52, therefore she was absent for 8 days, so the answer is C.

END OF SECTION

Section 2

Question 1: A

HCO_3^- is an alkaline substance and a vital component of the physiological buffering system. If the pH of the blood drops below 7, the bicarbonate molecule will accept a H^+ whereas if the pH increase, it will release H^+, thus HCO_3^- is an alkali.

Question 2: B

The current in a series circuit is always the same at any point in the circuit according to Kirchoff's first law which states that *at any node or junction in a circuit the sum of the current flowing into that node is equal to the current leaving that same node.* Thus current is always conserved. Since a series circuit does not have any nodes or junctions, we can assume the current is constant throughout. The potential difference is shared between all the components of the circuit ($V_{total}= V_1 + V_2 + V_3...$). This is because the total work done on the charge by the battery must equal the total work done by the charge on the components. Resistance in a series circuit is the sum of all the individual resistances (R = R1 + R2 + R3…). The resistance of two or more resistors is bigger than the resistance of just one of the resistors on its own because the battery has to push charge through all of them.

Question 3: D

The diaphragm is crucial to breathing as during inhalation it contracts and expands the chest space, along with the intercostal muscles which draw the ribs upwards and outwards, effectively lowering the pressure within the thoracic cavity and drawing air into the lungs. During exhalation all the muscles relax which lets the ribs drop downwards and inwards and the diaphragm balloons upwards into the chest space. This increases the pressure within the thoracic cavity which forces air out of the lungs.

Question 4: B

There are several steps to working out this problem. The first is to work out the area of the entire floor, minus the fish tank and the cut out corner. We can see that the length of the room is 8m and the width of the room is 4m (the sides of the cut out square are 2m). Thus the area of the entire room is **32m²**.

The cut out corner is a square with the dimension 2 x 2m. Thus the area of the cut out corner is **4m²**.

The fish tank is a circle, and thus its area can be worked out using πr^2. Π is taken to be 3 and thus $3 \times 1^2 = $ **3m²**.

Therefore the floor area, Bill needs to cover is $32 - (4 + 3) = $ **25m²**.

We then need to work out the area of one plank. The dimensions of this are in cm and so we need to convert to m. 1m is 100cm and so we can say that the length of the plank is 0.6m and the width is 0.1m. Thus the area is $0.6 \times 0.1 = $ **0.06m²**.

To work out the number of planks, required, we need to divide the area of the floor space by the area of the plank. A quick way of doing this would be rounding the area of the room down to 24 and multiplying the area of the plank by 100 so it becomes 6. 24/6 = 4, then because we multiplied the area of the plank by 100, we then multiply the answer by 100 which gives us **400 planks.** The closest answer to our solution is 417, which is listed as B.

Question 5: E

Some students may think that the arteries carry oxygenated blood from the mother to the foetus and that the vein carries the deoxygenated blood from the foetus to the mother, but it is important to remember that arteries always carry blood to the heart (in this case the mother's) and veins always travel away from the heart. A prime example of this is the pulmonary system, as like the foetal-mother system, the pulmonary arteries carry deoxygenated blood to the lungs away from the heart and the pulmonary veins carry oxygenated blood back to the heart.

Question 6: B

Solve $y = x^2 - 3x + 4$ and $y - x = 1$ as (x,y).

Substitute the quadratic expression into the other non-quadratic. You'll get another equation.

$x + 1 = x^2 - 3x + 4$

Rearrange to get a quadratic equation and solve.

$x^2 - 4x + 3 = 0$

$(x - 1)(x - 3) = 0$

Therefore x = 1 or x = 3

Substitute your x values into the equation, $y - x = 1$ and solve to work out y values.

y = 2 or y = 4

Therefore the coordinates are (1, 2) and (3, 4)

Question 7: A

The gravitational potential energy of the ball at the top of the slope is *mgh*. The kinetic energy of the ball as it travels down the slope is *0.5mv²*. The gravitational potential energy = kinetic energy, therefore: mgh = 0.5mv²

The mass values on either side cancel out to leave: gh = 0.5v²

Thus we can substitute values into the equation:

$$10 \times 5 = 0.5 \times v^2$$
$$50 = 0.5 \times v^2$$
$$50/0.5 = v^2$$
$$\sqrt{100} = v$$
$$10 = v$$

Question 8: C

ΔH is positive because the enthalpy of the products is higher than the enthalpy of the reactants. This also means that the reactants are less stable than the products and because it is ENDOthermic, energy is absorbed from the surroundings.

Question 9: H

The kidneys are involved in ultrafiltration as they filter all of the blood in the body of toxins/waste products from metabolic reactions. The waste is released as urine via the bladder. Some of the water is filtered out then reabsorbed by the kidney, especially when the body is dehydrated. Although glucose is reabsorbed by the kidney, it does not play a part in glucose regulation as that is mainly done by the pancreas by secretion of insulin and glucagon. These hormones are two of many found in the body, none of which are produced by the kidneys. There are some that are produced by the adrenal cortices that sit atop the kidneys, but these are a separate anatomical structure from the kidney.

Question 10: A

Haemophilia B is an X-linked recessive disorder which means you need two copies of the faulty genes in girls to present the phenotype associated with the disease and only one copy in males as they have XY chromosomes and are thus missing the extra X chromosome which may have carried the healthy, dominant gene. As Mike, the father of the baby girl, is not affected, we can assume that the mother carries one copy of the faulty gene herself. Thus, although the baby girl will not be affected by the condition, she may be a carrier of the gene and so, can pass it on to future generations.

Question 11: A

There are several methods to work this out, one of which is shown below.

Mass of FeS_2 in the ore = 480 x 0.75 = 360kg. 1 mole of FeS_2 = 55 + 32 + 32 = 119g → this

can be rounded to 120g for ease of calculation. Number of moles of FeS_2 in the ore $= \frac{360 \times 10^3}{120}$

= 3 x 10³ mol. Mass of Fe = (3 x 10³) x 55 = 165kg. 167.7kg is closest to this value.

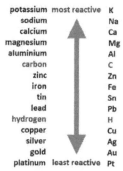

Question 12: B

Here, it is important to remember the reactivity series.

This is important as it tells you which elements are able to displace other elements in redox reactions. In this example, Zinc is the only element above Iron in the series and thus, is the only element that would be able to displace Iron.

Question 13: A

As galaxies and celestial objects move away from Earth, the wavelength of the light they emit, gets longer as it travels towards us. Thus there is a noticeable shift towards the red end of the spectrum, when we measure those waves. Scientists are able to measure the real light coming from galaxies far away using telescopes that pick up and record this light. Using red shift we can tell which galaxies are further away and which ones are closer. There is another phenomena called blue shift, which is the opposite of red shift in that, we can tell which galaxies are moving closer to us as the wavelengths of those galaxies become shorter and therefore shift to the blue end of the spectrum.

Question 14: A

X-rays are able to pass through soft, less dense material, like skin, soft tissue and air to stain the x-ray film black. They can't pass through denser material like bone and thus the x-ray film stays white. X-rays are harmful with prolonged exposure as they ionise cells and cause DNA damage that can result in conditions like cancer. Radiologists or technicians working with x-rays wear lead aprons to protect them from excess radiation. Gamma rays are different to x-rays with shorter wavelengths that are able to pass through dense material and because of this, they are considered more dangerous than x-rays.

Question 15: A

$\frac{(16x+11)}{(4x+5)} = 4y^2 + 2$

$16x + 11 = (4y^2 + 2)(4x + 5)$

$16x + 11 = 4x(4y^2 + 2) + 5(4y^2 + 2)$

$16x - 4x(4y^2 + 2) = 5(4y^2 + 2) - 11$

$x(16 - 4(4y^2 + 2)) = 20y^2 - 1$

$X = \frac{20y^2 - 1}{[16 - 4(4y^2 + 2)]}$

Question 16: A

In the diagram shown, the number at the top (73) denotes the mass number of an atom of Germanium. This is the number of protons and neutrons in the nucleus. The number at the bottom (32) is the proton number, i.e. the number of protons in the nucleus. Protons have a positive charge, neutrons have a neutral charge and electrons have a negative charge. As a stable element, Germanium must have a charge of 0 and thus the electrons and protons have to cancel out. Therefore, Germanium has 32 electrons.

Question 17: F

The first line defence of the body from invading pathogens is the skin. This is a tough keratinized layer, which is not easily broken down by bacteria. There is also flora on the skin (bacteria that live on the skin) that prevents any harmful bacteria from colonising. The next line of defence is the mucus lining the airways. It traps dirt and pathogens, to be either expelled from the body or swallowed into the gut. The next layer of defence mentioned in the answers, is hydrochloric acid found in the stomach. This has a pH of 2 and so effectively kills any pathogens that enter the body through the food. Other defences not listed include, tears (they contain lysozymes that break down the bacteria) and acidic substances by the sebaceous glands of the skin.

Some students may get confused by the antibodies. Although it is true that antibodies provide a line of defence, they are a secondary line of defence after the pathogen has got past the initial defences.

Question 18: D
The first step is to multiply out $(3p + 5)^2$
$(3p + 5)(3p + 5) = 24p + 49 \rightarrow 9p^2 + 30p + 25 = 24p + 49 \rightarrow 9p^2 + 6p - 24 = 0$. Then put the quadratic into brackets: $(3p + 6)(3p - 4) = 0$. Therefore, p must equal -6 or +4.

Question 19: G
The collision theory applies to all materials in any state that are reacting with each other. Increasing the temperature of a system increases the kinetic energy of the particles, making them move faster and this increases the force of collisions as well as the chance of two or more particles colliding. Increasing the smaller area of the substance reacting (making the particles smaller e.g. solids become fine powders or liquid become gases) increases the chances of more collisions happening. Another way to increase collisions is by increasing the pressure of a system as the molecules are in a more enclosed space and thus have less room to move without colliding with each other.

Question 20: E
Aluminium is not a good conductor of heat and thus heat is preserved. It is also shiny and thus is a poor emitter and absorber of heat. This means it won't let heat out, nor will it cause the person to overheat.

Question 21: C
Osmosis is the movement of water particles across a partially permeable membrane from an area of low concentration to an area of high concentration (of solute). It is not an active process as water can easily diffuse through bilipid layer membranes and thus does not require a specific passage.

Question 22: A
This is complete combustion as all of the methane is used to make water and carbon dioxide. It is an aerobic reaction as oxygen is present and needed to cause the combustion of the fuel. By increasing the carbon dioxide in the system you would either slow down or not affect the rate of combustion, but definitely would not speed it up. This also applies to removing oxygen from the system.

Question 23: A
Recall that $R = V/I$ which is Ohm = Volt/Ampere which is expressed as $V.A^{-1}$.

Question 24: D
The equation for work done is: $W (Nm) = F (N) \times d (m)$. Therefore, substituting the information available in the question, you can rearrange the equation and work out the answer: $40 = F \times 0.2 \rightarrow 40/0.2 = F \rightarrow 200 = F$

Question 25: B
Plants also give off carbon via respiration and death. Although some of the carbon is given off, trees and plants do store carbon in their cells and thus they are known as carbon stores.

Question 26: B
If a wave hits a boundary at an angle, it will refract because, the part of the wave that hit the boundary first slows down before the rest of it. This causes it to change direction and speed.
If a wave hits a boundary face on, it does not change direction, but slows down.
Waves slow down if they hit a boundary from low density to high density.

Question 27: A
Enzymes are always substrate specific as the active site is made up of a specific set of amino acids that determine which reaction the enzyme catalyses.

END OF SECTION

Section 3

1) **"Time and time again, throughout the history of medical practice, what was once considered as "scientific" eventually becomes regarded as "bad practice"."** – David Stewart

What does this statement mean? Give some examples of times when scientific practice has become bad practice and describe how this has had an impact on medicine.

This statement seems to suggest that as medicine and science progress, there is a constant change in the way healthcare professionals treat and manage various conditions. For example, until recently, the use of Sodium Valproate as a drug to treat epilepsy had been highly effective but it was found to cause congenital abnormalities in the growing foetus and thus it is now considered 'bad practice' to prescribe Sodium Valproate during pregnancy.

Change is crucial if we want to improve and advance the way we treat and practice in medicine. Science is constantly discovering new things and disproving previously accepted theories. With this information being easily accessible over the internet, it is key that we as healthcare providers ensure we don't lag behind and provide the newest, most effective treatments available for our patients. Not only does this increase patient satisfaction, it also increases patient safety as practices that have previously been safe in one patient population may not be safe in another.

A good example of this, along the same vein as Sodium Valproate, is Thalidomide. Used as a drug to treat conditions such as Leprosy and TB, it was found to be a highly effective antiemetic and was prescribed for women as a treatment for morning sickness. Unfortunately, the drug was not tested thoroughly enough to determine its effects on the growing foetus and many babies were born with congenital abnormalities, namely missing limbs. The outcome of this disaster was a worldwide effort to bring about stricter drug licensing and testing regulations. It is now bad practice to prescribe drugs without knowing the full extent of its adverse effects. This has brought about a practice in which everyone has benefited, both patients and healthcare providers.

Another example of scientific processes becoming bad practice is the Andrew Wakefield MMR scandal. Wakefield conducted a study on the correlation between autism and the MMR vaccine in young children. His study seemed to prove that there was a strong correlation which led to many parents refusing to vaccinate their children. This caused an epidemic of MMR across the country. After researchers reviewed his study and found that his study was in fact inaccurate and the results were forged, MMR vaccine uptake rates started to increase and the incidence of MMR decreased drastically. This has changed both scientific and medical practice as any research that is available is peer reviewed and carried out to the highest standard to avoid such instances like this.

In conclusion, change is needed, and it only serves to improve the service and the way we practice medicine today. Without 'bad practice' you do not have 'good practice' and as science and the world advances, it is important to remember that patient safety is always the most important part of medicine.

2) **"Formerly, when religion was strong and science weak, men mistook magic for medicine; now, when science is strong and religion weak, men mistake medicine for magic." – Thomas Szasz**

What does this statement mean? Do you think it is correct in assuming that all men mistake medicine for magic?

This statement suggests that the understanding of how medicine works is based upon the strength of science and what it can provide for the medical community. It portrays the idea that in the past, when religion was seen as the 'law,' people saw magic as miraculous cures for diseases and illnesses they didn't understand. Now that science has advanced and is seen in parallel with religion, or even as a competitor, people see medicine as something that provides the cures or the treatments to make them better. Because they don't fully understand the processes that occur in the body when medicine or treatment is prescribed, they perceive it as 'magic.'

In the past, we did not have access to knowledge about DNA or bacteria. Antibiotics were unheard of and any medical treatment given in the past was mainly trial and error. This resulted in patient deaths and a general sense of hopelessness – if a person fell ill, they were more likely to die. When someone happened to cure a person by chance, or they got better, it was seen as magic, forces of unknown nature and power being used to treat the individual. As science progressed and the world came to know more about the world around us and how we could utilise it to better ourselves, medicine also advanced. The medical world is still evolving, helped along by science as theories are disproved and new hypotheses are created. This leads to a level of awe as medicine starts to treat the untreatable or cure the incurable.

This is both beneficial and a hindrance to medicine. One benefit of medicine as magic is it demands a certain level of respect. Patients are willing to listen to healthcare providers and the advice we give as they know that medicine will help them. On the other hand, patients sometimes believe that they are entitled to this treatment and when it doesn't work, or cannot be prescribed, they get angry. One example of this is antibiotics. Many patients don't understand that antibiotics do not treat viral infections and demand the drugs in order to cure a cold or a sore throat. Sometimes despite an explanation of why antibiotics can't be prescribed, the patient still demands the antibiotic and the doctor has no other choice but to give in. This unnecessary overuse of antibiotics leads to antibiotic resistance which causes a grievance to the rest of the population.

In conclusion, we must be careful to educate our patients about aspects of medicine they may not understand. We must ensure that they do not hold medicine on too high a pedestal as that leads to decreased satisfaction and a sense of being let down. It is important that patients understand that medicine cannot cure everything, that medicine is the result of years of research and discovery, and that it takes time to evolve.

3) *"Approximately 26.9% of the adult population in the UK is obese. Shouldn't we be offering bariatric surgery to every obese person that walks through the doors?" Explain what this statement means. Argue to the contrary. To what extent do you agree with the statement?*

Obesity is a major cause of heart disease, stroke, osteoarthritis, diabetes and much more. These conditions can lead to death, which is why it is important for people to lose weight. Although losing weight is important, surgery is invasive and carries its own risks, like infection and bleeding. It is better to try to lose weight through diet and exercise first before trying invasive treatments. Diet and exercise are usually sufficient enough to cause a reduction in weight, and there are many ways a clinician can encourage his/her patients. Suggesting that the patient take it step by step (e.g. walking for ten minutes one day then walking fifteen minutes the next). This can make the task seem less daunting and the patient will be more likely to try and start losing weight this way. It also means that they ease themselves into it, and let their body adjust to the change. Supplementary interventions can include counselling which will probe into the patient's reasons for weight loss and encourage them.

If exercise and diet change are not working then certain drugs can be brought into play, which along with the diet and exercise help the patient reduce their BMI. However, medications have their own side effects, which can be unpleasant for the patient. These can deter the patient from taking them in which case, the clinician may need to look for alternatives or prescribe medication that counteracts the side effects.

Surgery is the last resort intervention as it is invasive and has the worst complications. Bariatric surgery is useful in that it is an easy way to help patients lose weight quickly along with exercise and diet changes. It can also make the patient more incentivised to carry out the weight loss plan as they may feel that the journey is shorter and will happen quicker.

This is counter-balanced by the fact that patients often have to lose weight before they go into surgery to minimise the risk of complications that occur afterwards. This means that although, surgery is easier and quicker than just exercise, a certain amount of weight loss will be required as a pre-requisite to the surgery and so patients will find that they have to exercise anyway.

In conclusion, surgery should not be offered to everyone who walks through the doors as not only is it expensive, it also has a lot of requirements and complications that can be very harmful to some. It is always better to encourage the patient to try non-invasive methods first then, if absolutely necessary use the invasive treatments.

4) *'Placebos may solve the problem of patients demanding medication they do not need.' Explain what this statement means. Argue the contrary. To what extent do you agree with the statement?*

Placebos are used in research as a control against drugs that are being studied. They can either be sugar pills or a drug that has already been licensed and is known to treat the disease being studied by the researchers.

Patients often come into a GP consultation or to the hospital and expect a drug to be given to them, even when they don't need it. Millions of pounds are spent giving antibiotics to patients who have a viral infection and this costs the NHS greatly. The reason these are given is because although clinicians try to explain why antibiotics are not needed for viral infection, patients refuse to listen and demand the drugs anyway. In this case a placebo would reduce the overall harm caused by giving antibiotics out wrongly. Simple sugar pills would not harm the patient and would reduce the amount of abuse clinicians get for refusing to prescribe medication.

Placebos have also been shown to have a psychological effect on patients in that, patients feel better after having taken the placebo even if it was just a sugar pill. This could have a significant effect on care as placebos could help patients get better by themselves. Taking a pill makes the patient believe they are being treated and so in effect, they will themselves better. This would save the NHS a lot of money in the long run as less drugs and time would be used treating patients for whom the illness is easy to cure without medication.

The downside to placebo treatment is that the doctor-patient relationship may be harmed if the patient realises they are being given placebos. A certain level of trust is built up between the doctor and the patient, and regaining it is hard. Trust is paramount in the successful treatment of a patient. Another downside is that it is unethical to provide a placebo to a patient and make them believe they are actually being treated. It is dishonest and goes against the values of being a good doctor.

Overall, placebos should not be used as an alternative treatment as the pros do not justify the ethical dilemmas faced by a clinician. Trust is important in the relationship a doctor holds with a patient and that should take first priority. I think doctors should try and explain to their patients the best they can, that sometimes medication is not necessary to treat a certain disease/condition.

END OF PAPER

Final Advice

Arrive well rested, well fed and well hydrated

The BMAT is an intensive test, so make sure you're ready for it. Unlike the UKCAT, you'll have to sit this at a fixed time (normally at 9AM). Thus, ensure you get a good night's sleep before the exam (there is little point cramming) and don't miss breakfast. If you're taking water into the exam then make sure you've been to the toilet before so you don't have to leave during the exam. Make sure you're well rested and fed in order to be at your best!

Move on

If you're struggling, move on. Every question has equal weighting and there is no negative marking. In the time it takes to answer on hard question, you could gain three times the marks by answering the easier ones. Be smart to score points- especially in section two where some questions are far easier than others.

Make Notes on your Essay

Some universities may ask you questions on your BMAT essay at the interview. Sometimes you may have the interview as late as March which means that you **MUST** make short notes on the essay title and your main arguments after the essay. This is especially important if you're applying to UCL and Cambridge where the essay is discussed more frequently.

Afterword

Remember that the route to a high score is your approach and practice. Don't fall into the trap that "*you can't prepare for the BMAT*"– this could not be further from the truth. With knowledge of the test, some useful time-saving techniques and plenty of practice you can dramatically boost your score.

Work hard, never give up and do yourself justice.

Good Luck!

Acknowledgements

I would like to thank Rohan and the UniAdmissions Tutors for all their hard work and advice in compiling this book, and both my parents and Meg for their continued unwavering support.

Matthew

About UniAdmissions

UniAdmissions is an educational consultancy that specialises in supporting **applications to Medical School and to Oxbridge**.

Every year, we work with hundreds of applicants and schools across the UK. From free resources to our *Ultimate Guide Books* and from intensive courses to bespoke individual tuition – with a team of **300 Expert Tutors** and a proven track record, it's easy to see why UniAdmissions is the **UK's number one admissions company**.

To find out more about our support like intensive **BMAT courses** and **BMAT tuition** check out www.uniadmissions.co.uk/bmat

Thanks for purchasing this Ultimate Guide Book. Readers like you have the power to make or break a book – hopefully you found this one useful and informative. If you have time, *UniAdmissions* would love to hear about your experiences with this book.

As thanks for your time we'll send you another ebook from our Ultimate Guide series absolutely FREE!

How to Redeem Your Free Ebook in 3 Easy Steps

1) Either scan the QR code or find the book you have on your Amazon purchase history or your email receipt to help find the book on Amazon.

2) On the product page at the Customer Reviews area, click on 'Write a customer review'

Write your review and post it! Copy the review page or take a screen shot of the review you have left.

3) Head over to www.uniadmissions.co.uk/free-book and select your chosen free ebook! You can choose from:
  ➢ The Ultimate UKCAT Guide – 1250 Practice Questions
  ➢ The Ultimate BMAT Guide – 800 Practice Questions
  ➢ The Ultimate Oxbridge Interview Guide
  ➢ The Ultimate Medical School Interview Guide
  ➢ The Ultimate Medical Personal Statement Guide
  ➢ The Ultimate Medical School Application Guide
  ➢ BMAT Past Paper Solutions
  ➢ BMAT Practice Papers
  ➢ UKCAT Practice Papers

Your ebook will then be emailed to you – it's as simple as that!

Alternatively, you can buy all the above titles at **www.uniadmissions.co.uk/our-books**